# Neurotransmitters in Plants

# Neurotransmitters in Plants

## Perspectives and Applications

Edited by
Akula Ramakrishna
Victoria V. Roshchina

CRC Press
Taylor & Francis Group
Boca Raton London New York

CRC Press is an imprint of the
Taylor & Francis Group, an **informa** business

CRC Press
Taylor & Francis Group
6000 Broken Sound Parkway NW, Suite 300
Boca Raton, FL 33487-2742



Printed on acid-free paper

International Standard Book Number-13: 978-1-1385-6077-2 (Hardback)



**Library of Congress Cataloging-in-Publication Data**

Names: Ramakrishna, Akula, editor. | Roshchina, V. V. (Viktoriëiìa
Vladimirovna), editor.
Title: Neurotransmitters in plants : perspectives and applications / editors:
Akula Ramakrishna and Victoria V. Roshchina.
Description: Boca Raton, FL : CRC Press, Taylor & Francis Group, 2018. |
Includes index.
Identifiers: LCCN 2018011298 | ISBN 9781138560772 (hardback : alk. paper)
Subjects: LCSH: Plant physiology. | Neurotransmitters.
Classification: LCC QK746 .N48 2018 | DDC 571.2--dc23
LC record available at https://lccn.loc.gov/2018011298

**Visit the Taylor & Francis Web site at**
**http://www.taylorandfrancis.com**

**and the CRC Press Web site at**
**http://www.crcpress.com**

MIX
Paper from responsible sources
FSC® C013056

Printed and bound in Great Britain by
TJ International Ltd, Padstow, Cornwall

# Contents

## SECTION III Cellular Location of Neurotransmitters: Their Reception and Signaling



## SECTION IV Role of Neurotransmitters in Relationships of Organisms in Biocenosis

## SECTION V Methodical and Practical Aspects of Neurotransmitters' Enriched Plants for Agronomy, Food and Medicine

# Preface

Neurotransmitters (neuromediators)—named so by neurophysiologists—had been first discovered in fungi and later found to function in mammals as signaling substances and regulators. Today, the compounds are found in plants and microorganisms, where they play multifunctional roles and are called biomediators. In our time, we constantly see universal chemical bases of life in the living world—from proteins, nucleic acids, and lipids to small molecules known traditionally as neurotransmitters. After the first monographs devoted to the presence and function of the compounds in plants, published in 1991 and 2001, many new works in the field appeared. In this connection, a necessity arose to combine the new information in a collective issue, showing both achievements and perspectives for investigations and human practices. This book will address the work at the margins of current research in plant biology and provide powerful tools for both analyzing and manipulating organisms. Efforts by scientists from many countries have been concentrated on the occurrences of neurotransmitters in plants, their functions in plant cells, and practical application of the knowledge about acetylcholine, catecholamines, serotonin, melatonin, histamine, gamma-aminobutyric acid, and glutamine for ecology, agriculture, medicine, and food industry.

This current book may be used by universities and academies, libraries, and research organizations and scientists who work in this field, including wider ring of biologists—biochemists, physiologists, botanists, and those working in the appropriate industry. The text appears for use in special courses for students, both graduate and post-graduate.

# Acknowledgments

Our special thanks and heartiest gratitude to all the contributors to this volume, which provides an ample treatise on the topic. We understand that the contributors are active scientists and have agreed to share their findings and perspectives with the readers, who will benefit from learning about the implications and applications of the subject matter as an emerging area of science.

We would like to thank Professor Fluck especially for extending his support for the book creation.

We are grateful to Alice Oven (Senior Editor), Jennifer Blaise (Senior Editorial Assistant), Marsha Hecht (Project Editor), John Shannon (Senior Editorial Production Manager) and the team from Taylor & Francis for their help in bringing out this volume in an attractive manner.

Both of us are thankful to our families, who have encouraged us to take up this task and permitted us to take the time off from relatives.

Dr. Ramakrishna is thankful to Monsanto Crop Breeding Center, Bengaluru, India, for its support and also for the support of the Central Food Technological Research Institute (CFTRI) in Mysore, India. Dr. Ramakrishna sincerely thanks his mentor Dr. G.A. Ravishankar and co-guide Dr. P. Giridhar for their guidance and is thankful to the staff and students of the Plant Cell Biotechnology Department at CFTRI.

**Akula Ramakrishna**

**Victoria V. Roshchina**

# Introduction

## INTRODUCTION (FROM DROPS TO OCEAN)

One drop is only a beginning,
Drops usually form an ocean.
One fact is like a streaming,
For science origin it will be done.

Today, neurotransmitters acetylcholine, dopamine, noradrenaline (norepinephrine), adrenaline (epinephrine), serotonin, and histamine have been found not only in mammalians but in all living organisms—from microorganisms to plants and animals—and play various roles in biocenosis. One needs to remember that, first of all, the compounds were found in fungi by pharmacologists Barger, Dale and Ewins in 1910–1914 (Barger and Dale 1910; Ewins 1914). First reports related to plants dealt with noradrenaline found by Buelow and Gisvold (1944). Then there was the discovery of acetylcholine (Emmelin and Feldberg 1947), histamine (Werle and Raubb 1948), catecholamines, and serotonin (Waalkes et al. 1958; Udenfriend et al. 1959). Only beginning from 1970 did publications related to attempts to study physiological roles of acetylcholine appear under the aegis of Professor Mordecai J. Jaffe. This book opens with reminiscences about this outstanding scientist by Professor Richard A. Fluck, a pioneer of the field, who, in collaboration with Professor Jaffe, showed the existence of enzyme cholinesterase in plant cells that confirm a reality of the acetylcholine occurrence exactly in plant cells. Thus, small drops started to form an ocean problem of the role of neurotransmitters in plants.

From 1970 to current time, the study of the neurotransmitter compounds was followed by many pioneers; some of them, such as Professors Gupta (India) and Saxena (Canada) are active participants in the research. Fundamental valuable contributions to this line of study were done in 1970–1990 by researchers from laboratories of Kostir et al., Kutacek and Vackova (Chechoslovakia), Smallman (Great Britain), Hartman, Grosse (Germany), Regula (former Yugoslavia, Croacia), Kopsewicz, Tretyn and Bednarska (Poland), Kendrick (the Netherlands), Momonoki and Momonoki (Japan), and Miura and Madhavan (USA). Special information and opinions that dealt with plants have been summed in monographs and reviews (Fluck and Jaffe 1974, 1976; Hartman and Gupta 1989; Roshchina 1991a,b, 2001; Tretyn and Kendrick 1991; Szopa et al. 2001; Wessler et al. 2001; Baluška et al. 2005, 2006a,b; Brenner et al. 2006; Murch 2006; Kulma and Szopa 2007; Ramakrishna et al. 2011; Erland et al. 2016; Roshchina 2016a,b). Based on monographs (Roshchina 1991a, 2001), there are attempts in Russian to prepare text manuals for lecturers who teach special courses in universities (Yurin 2004; Roshchina 2010).

Today, there is new epoch in the study of a role of neurotransmitters in plants. Via the compounds, the contacts (microorganism–microorganism, microorganism–plant, microorganism–animal, plant–animal, plant–plant, and animal–animal) in nature may occur. There is the universality in response of any cells to the compounds that play important roles in the signal system regulating various processes, especially as cellular growth and development. In any case, the compounds found in every organism can be called biomediators or biotransmitters (Roshchina 1989), rather than neurotransmitters/neuromediators. But the term *neurotransmitters* is widespread today, and we also use it for a better understanding of the reader audience (Roshchina 2010, 2016a).

Traditional directions in the field of the study in plant neurotransmitter compounds linked with regulation of differentiation and development. Special attention was paid to serotonin and melatonin as both regulators of the processes and human health. A successful contribution to the understanding of the structure and function of enzymes involved in neurotransmitters' systems was made by molecular biology. Recently, scientists were able to clone and express the acetylcholinesterase gene. New lines were seen from chapters devoted to the interactions between organisms in biocenosis via the chemical signalizers or signals such as neurotransmitters released by every living individual from microorganisms to plants and animals. Chemical signaling by neurotransmitters became the part of the allelopathic relations between organisms and between cells of one and same-plant organism. Problems arose here from the analysis of the contact between sea or river organisms, as well as the relationships between pollen and pistil, plant and microorganism, or a plant and insect. Special mechanisms of signaling will be also considered here. Earlier publications concerning neurotransmitters in plants included acetylcholine and biogenic amines, while today gamma-aminobutyric acid and glutamate become the objects of the attention of plant physiologists. Moreover, there is the bridge between human health and content of neurotransmitters and antineurotransmitters' compounds in food and medical species. Ecological significance of the actual knowledge was recently confirmed by studies in ophthalmology where histamine, earlier found in pollen of wind-pollinated plant species, may induce the increase in human allergic reactions in spring because UV-light induces the accumulation of these neurotransmitters in the microspores. Hazard and profit of neurotransmitters, as well as care of patients, perform actuality of the collaboration and cooperation between scientists from different regions of biology.

The title of the book includes the term *neuromediators*, although the finding of the compounds in every living organism leads someone to change this word in order to show universal character of the substances, not only to animals. In some cases, for example, this is exchanged for the word *biomediators* or *biotransmitters*. But for a wider audience, we have restored the traditional nomination *neurotransmitters*.

In this book, a reader may find new information about achievements and new trends in the understanding of the neurotransmitters' functions within plant organisms and possible mechanisms of external action of the compounds on the plants, animals, and microbes. Along with this, there are chapters related to other weakly investigated problems that address new approaches to future research. Particular attention is paid to medical, agricultural, and environmental practical solutions that can be used in the application of the knowledge about neurotransmitters for human benefit.

This book may serve as a collective monograph, which can be used by universities and academies, libraries, and by research organizations and scientists who work in this field. Moreover, the texts of some chapters could be applied to special courses for students, both graduate and post-graduate. We hope that the book will also attract the attention of physicians, pharmaceutical researchers, and toxicologists, as well as scientists working for the food and pharmaceutical industry.

## RECOMMENDED REFERENCES

## FIRST IN PLANT BIOLOGY

Barger, G. and Dale, H.H. (1910). A third active principle in ergot extracts. *Proceedings of Chemical Society* 26: 128–129.

Ewins, A.J. (1914). Acetylcholine, a new active principle of ergot. *Biochemical Journal* 8: 44–49.

### ACETYLCHOLINE

Emmelin, N. and Feldberg, W. (1947). The mechanism of the sting of the common nettle (*Urtica urens*). *Journal of Physiology* 106: 440–455.

Emmelin, N. and Feldberg, W. (1949). Distribution of acetylcholine and histamine in nettle plants. *New Phytologist* 48: 143–148.

## Catecholamines

Buelow, W. and Gisvold, O. (1944). 3,4-Dihydroxyphenylethylamine from *Hermidium alipes*. *Journal of the American Pharmaceutical Association* 33: 270–274.
Udenfriend, S., Lovenberg, W., and Sjoerdsma, A. (1959). Physiologically active amines in common fruits and vegetables. *Archives in Biochemistry and Biophysics* 85: 487–490.

## Serotonin

Waalkes, T.P., Sjoerdama, A., Greveling, C.R., Weissbach, H., and Udenfriend, S. (1958). Serotonin, norepinephrine, and related compounds in bananas. *Science* 127: 648–650.
West, G. (1959a). Indole derivatives in tomatoes. *Journal of Pharmacy and Pharmacology* 11: 275–277.
West, G. (1959b). Tryptamines in tomatoes. *Journal of Pharmacy and Pharmacology* 11: 319–320.

## Histamine

Werle, E. and Raub, A. (1948). Über Vorkommen, Bildung und Abbau biogener Amine bei Pflanzen unter besonderer Beruck-sichtigung des Histamins. *Biochemische Zeischrift* 318: 538–553.
Werle, E. and Zabel, A. (1948). Über die Verbreitung der Histaminase im Pflanzenreich. *Biochemische Zeitschrift* 318: 554–559.
Appel, W. and Werle, E. (1959). Nachweis von Histamin, N-Dimethylhistamin, N-Acetylhistamin und Acetylcholin in *Spinacia oleracea*. *Arzneimittelforschung* 9: 22–26.

## Enzymes

Barlow, R.B. and Dixon, R.O.D. (1973). Cholineacetyltransferase in the nettle *Urtica dioica* L. *Biochemical Journal (London)* 132: 15–18.
Fluck, R.A. and Jaffe, M.J. (1974b). The distribution of cholinesterases in plant species. *Phytochemistry* 13: 2475–2480.
Fluck, R.A. and Jaffe, M.J. (1974c). Cholinesterases from plant tissues VI. Distribution and subcellular localization in *Phaseolus aureus* Roxb. *Plant Physiology* 53: 752–758.
Fluck, R.A. and Jaffe, M.J. (1974d). Cholinesterases from plant tissues V. Cholinesterase is not pectin esterase. *Plant Physiology* 54: 797–798.
Fluck, R.A. and Jaffe, M.J. (1975). Cholinesterases from plant tissues VI. Preliminary characterization on enzymes from *Solanum melongena* L. *and Zea mays* L. *Biochimica et Biophysica Acta* 410: 130–134.

## Main Reviews and Methodological Papers

Fluck, R.A. and Jaffe, M.J. (1974a). The acetylcholine system in plants. In: E. Smith (Ed.), *Current Advances in Plant Sciences*, Vol. 5. Oxford, UK: Science Engineering, Medical and Data Ltd, pp. 1–22.
Fluck, R.A. and Jaffe, M.J. (1976). The acetylcholine system in plants. In: H. Smith (Ed.), *Commentaries in Plant Science*. Oxford, UK: Pergamon Press, pp. 119–136.
Roshchina, V.V. and Mukhin, E.N. (1986). Acetylcholine, its role in plant life. *Uspekhi Sovremennoi Biologii (Advances in Modern Biology: Reviews, Russia)* 101: 265–274.
Hartmann, E. and Gupta, R. (1989). Acetylcholine as a signaling system in plants. In: W.F. Boss and D.I. Morve (Eds.), *Second Messengers in Plant Growth and Development*. New York: Allan Liss, pp. 257–287.
Roshchina, V.V. (1989). Biomediators in chloroplasts of higher plants. I. The interaction with photosynthetic membranes. *Photosynthetica* 23: 197–206.
Roshchina, V.V. (1991a). Neurotransmitters catecholamines and serotonin in plants. *Uspekhi Sovremennoi Biologii (Advances in Modern Biology, Russia)* 111: 622–636.
Tretyn, A. and Kendrick, R.E. (1991). Acetylcholine in plants: Presence, metabolism and mechanism of action. *Botanical Review* 57: 33–73.
Kuklin, A. and Conger, B. (1995). Enhancement of somatic embryogenesis in orchard grass leaf cultures by epinephrine. *Plant Cell Reports* 14: 641–644.

Szopa, J., Wilczynski, G., Fiehn, O., Wenczel, A., and Willmitzer, L. (2001). Identification and quantification of catecholamines in potato plants (*Solanum tuberosum*) by GC-MS.

Wessler, I., Kilbinger, H., Bittinger, F., and Kirkpatrick, C.J. (2001). The non-neuronal cholinergic system: the biological role of non-neuronal acetylcholine in plants and humans. *Japanese Journal of Pharmacology*. 85: 2–10.

Baluska, F., Volkmann, D., and Menzel, D. (2005). Plant synapses: Actin-based domains for cell-to-cell communication. *Trends in Plant Science* 10: 106–111.

Brenner, E.D., Stahlberg, R., Mancuso, S., Vivanco, J.M., Baluska, F., and van Volkenburgh, E. (2006). Plant neurobiology: an integrated view of plant signaling. *Trends in Plant Science*. 11: 413–419.

Murch, S.J. (2006). Neurotransmitters, neuroregulators and neurotoxins in plants. In: F. Baluska, S. Mancuso, and D. Volkmann (Eds.), *Communication in Plants: Neuronal Aspects of Plant Life*. Berlin, Germany: Springer, pp. 137–151.

Baluska, F., Hlavacka, A., Mancuso, S., and Barlow, P.W. (2006a). Neurobiological view of plants and their body plan. In: F. Baluska, S. Mancuso, and D. Volkmann (Eds.), *Communication in Plants: Neuronal Aspects of Plant Life*. Berlin, Germany: Springer, pp. 19–35.

Baluska, F., Mancuso, S., and Volkmann, D. (Eds.). (2006b). *Communication in Plants: Neuronal Aspects of Plant Life*. Berlin, Germany: Springer.

Kulma, A. and Szopa, J. (2007). Catecholamines are active compounds in plant. *Plant Science* 172: 433–440.

Roshchina, V.V. (2010). Evolutionary considerations of neurotransmitters in microbial, plant and animal cells. In: M. Lyte and P.P.E. Freestone (Eds.), *Microbial Endocrinology: Interkingdom Signaling in Infectious Disease and Health*. New York: Springer, pp. 17–52.

Ramakrishna, A., Giridhar, P., and Ravishankar, G.A. (2011). Phytoserotonin. *Plant Signaling and Behavior* 6(6): 800–809.

Erland, L.A.E., Turi, C.E., and Saxena, P.K. (2016). Serotonin: An ancient molecule and an important regulator of plant processes. *Biotechnology Advances* 34(8): 1347–1361.

Amaroli, A. (2017). Neurotransmitters or biomediators? The cholinergic system in protozoa. *EC Microbiology* 7(2): 40–41.

Erland, L.A.E. and Saxena, P.K. (2017). Beyond a neurotransmitter: The role of serotonin in plants. *Neurotransmitter* 4: e1538. doi:10.14800/nt.1538.

Briguglio, M., Dell'Osso, B., Panzica, G., Margaroli, A., Banfi, G., Zanaboni Dina, C., Galentino, R., and Porta, M. (2018). Dietary neurotransmitters: A narrative review on current knowledge. *Nutrients* 10: 591. doi:10.3390/nu1005059.

## MONOGRAPHS

Roshchina, V.V. (1991b). *Acetylcholine and Biogenic Amines in Plants*. Pushchino, Russia: Biological Center AN SSSR, 192 p.

Roshchina, V.V. (2001). *Neurotransmitters in Plant Life*. Enfield, UK: Science Publisher, 283 p.

Roshchina, V.V. (2014). *Model Systems to Study Excretory Function of Higher Plants*. Dordrecht, the Netherlands: Springer.

Ravishankar, G.A. and Akula Ramakrishna, A. (Eds.). (2016). *Serotonin and Melatonin: Their Functional Role in Plants, Food, Phytomedicine, and Human Health*. Boca Raton, FL: CRC Press.

## EVOLUTION OF NEUROTRANSMITTERS AND THEIR FUNCTIONS

Buznikov., G.A. (1987). *Neurotransmitters in Embryogenesis*. Moscow, Russia: Nauka, 232 p.

Buznikov, G.A. (1990). *Neurotransmitters in Embryogenesis*. Chur, Switzerland: Harwood, 526 p.

Roshchina, V.V. (2016). New tendency and perspectives in evolutionary considerations of neurotransmitters in microbial, plant, and animal cells. In: M. Lyte (Ed.), *Microbial Endocrinology: Interkingdom Signaling in Infectious Disease and Health*, Advances in Experimental Medicine and Biology, Vol. 874. New York: Springer.

Liebeskind, B.J., Hofmann, H.A., Hillis, D.M., and Zakon, H.H. (2017). Evolution of animal neural systems. *Annual Review of Ecology, Evolution, and Systematics* 48: 377–398.

## MEDICINAL ASPECT OF PLANT NEUROTRANSMITTERS

Heidinder, A., Rabensteiner, D.F., Rabensteiner, J., Kieslinger, P., Horwath-Winter, J., Stabentheiner, E., Riedl, R., Wedrich, A., and O. Scmut. (2017). Decreased viability and proliferation of Chang conjunctival epithelial cells after contact with ultraviolet-irradiated pollen. *Cutaneous and Ocular Toxicology* 26. doi:10.1080/15569527.2017.1414226.

## Lecturers' Text-Manuals for Special Courses in Russian

Yurin, V.M. (2004). *Biomediatori v rastenizyakh (Biomediators in plants).* Minsk, Belarus: Belorussian State University. (in Russian).

Roshchina, V.V. (2010). Neurotransmitteri - biomediatori i regulatori rastenii. (Neurotransmitters: Plant Biomediators and Regulators—A Tutorial Text Manual) Moscow: Informica (Electron issue.) http://window.edu.ru/window_catalog/files/r68504/neirotransmitters.pdf; http://window.edu.ru/catalog/pdf2txt/504/68504/42266.

# Editors

**Akula Ramakrishna** is currently a scientist at Monsanto Breeding Station, Bangalore, India. Dr. Ramakrishna holds a master's degree from Sri Krishna Devaraya University, Anantapur, India. He started his research career in 2005 at the Department of Plant Cell Biotechnology, CFTRI, Mysuru, India, in the research group of Dr. G. A. Ravishankar. He is a recipient of Senior Research Fellow (CSIR, New Delhi). He obtained his PhD (in biochemistry) from the University of Mysore, Mysuru, India, in the development of high-frequency somatic embryogenesis and regulation of secondary metabolites in *Coffea canephora*. He worked extensively on the role of serotonin and melatonin calcium-mediated signaling in plants. He has made significant contributions to metabolic engineering of secondary metabolites from plants and abiotic stress in plants. He has worked in the area of tissue culture, *in vitro* production, and regulation of plant secondary metabolites from food value plants that include natural pigments such as caffeine, steviosides, anthocyanins, and carotenoids. He is the author of 3 books, 12 peer-reviewed publications, 2 reviews, and 8 chapters in books. His books include *Serotonin and Melatonin: Their Functional Role in Plants; Food, Phytomedicine, and Human Health: Metabolic Adaptations in Plants During Abiotic Stress; and Neurotransmitters in Plants: Perspectives and Applications*. He is a member of the Society for Biotechnologists (India). He is a recipient of the Fellow of Society for Applied Biotechnology, India (2012); Global Vegetable Research Excellence award (2017); three global technology recognition awards; Rapid Recognition Award; Test Master, Asia Vegetable R&D quarterly recognitions; and special recognition from the Monsanto company. He attended the Fifth International Symposium on Plant Neurobiology held during May 25–29, 2009, in Florence, Italy. He also attended the Technical Community of Monsanto (TCM) meeting held during June 7–8, 2016, in St. Louis, MO.

**Victoria V. Roshchina**, plant biochemist and physiologist, Dr. Sciences, Professor, PhD, senior scientist of the Russian Academy of Sciences Institute of Cell Biophysics in the Biological Center of Pushchino, Russia. Her main interests deal with the action mechanisms of biologically active compounds (neurotransmitters and antineurotransmitter substances) and the spectral methods of their analysis in plant cells. She is the author of 250 publications including 8 monographs. Among the publications is the first monograph devoted to neurotransmitters in plants published in Russian (1991, USSR Biological Center, Pushchino) and in English (2001, Science Publishers, Enfield, Plymouth).

# Contributors

**Marino B. Arnao**
Department of Plant Biology (Plant Physiology)
Faculty of Biology
University of Murcia
Murcia, Spain

**Kiran Bamel**
Department of Botany
University of Delhi
and
Department of Botany
Shivaji College
University of Delhi
Delhi, India

**Atanu Bhattacharjee**
Department of Pharmacy
Assam Down Town University
Guwahati, India

**Paramita Bhattacharjee**
Reader Department of Food Technology and Biochemical Engineering
Jadavpur University
Kolkata, India

**Soumi Chakraborty**
Department of Food Technology and Biochemical Engineering
Jadavpur University
Kolkata, India

**Kamil Ekici**
Department of Food Hygiene and Technology
Van, Turkey
Veterinary College
University of Yŭzŭncŭ Yıl

**Lauren A.E. Erland**
Department of Plant Agriculture
Gosling Research Institute for Plant Preservation
University of Guelph
Guelph, Ontario, Canada

**Richard A. Fluck**
Emeritus Professor of Biology
Franklin & Marshall College Lancaster
Lancaster, Pennsylvania

**V.M. Grischenko**
Institute of Biological Instrumentation
Russian Academy of Sciences
Pushchino, Russia

**A. Arturo Guevara-García**
Instituto de Biotecnologнa
Universidad Nacional Autynoma de Mйxico
Cuernavaca, México

**Rajendra Gupta**
Department of Botany
University of Delhi
Delhi, India

**Josefa Hernández-Ruiz**
Department of Plant Biology (Plant Physiology)
Faculty of Biology
University of Murcia
Murcia, Spain

**K.C. Jisha**
Plant Physiology and Biochemistry Division
Department of Botany
University of Calicut
Kerala, India

**Anatolii A. Kataev**
Institute of Cell Biophysics
Russian Academy of Sciences
Pushchino, Russia

**Vilma Kisnieriene**
Institute of Biosciences
Life Sciences Center
Vilnius University
Vilnius, Lithuania

**Dimitrii A. Konovalov**
Pyatigorsk Medical and Pharmaceutical Institute
Branch of Volgograd State Medical University of the Ministry of Health of Russia
Pyatigorsk, Russia

**Bogdan A. Kurchii**
IEC
Irpin, Ukraine

**Indre Lapeikaite**
Institute of Biosciences
Life Sciences Center
Vilnius University
Vilnius, Lithuania

**Jesús Salvador López-Bucio**
Instituto de Investigaciones Químico-Biológicas
Universidad Michoacana de San Nicolás de Hidalgo
Morelia, México

**José López-Bucio**
Instituto de Investigaciones Químico-Biológicas
Universidad Michoacana de San Nicolás de Hidalgo
Morelia, México

**Yoshie S. Momonoki**
Department of Molecular Microbiology
Faculty of Life Sciences
Tokyo University of Agriculture
Tokyo, Japan

**Soumya Mukherjee**
Department of Botany
Jangipur College
University of Kalyani
West Bengal, India

**Timothy A. Nelson**
Department of Biology
Seattle Pacific University
Seattle, Washington

**Alexander V. Oleskin**
General Ecology Department
Biology Faculty
Moscow State University
Moscow, Russia

**Abdullah Khalid Omer**
Sulaimani Veterinary Directorate
Veterinary Quarantine
Bashmakh International Border
Halabja, Iraq

**Ramón Pelagio-Flores**
Instituto de Investigaciones Químico-Biológicas
Universidad Michoacana de San Nicolás de Hidalgo
Morelia, México

**Vilmantas Pupkis**
Institute of Biosciences
Life Sciences Center
Vilnius University
Vilnius, Lithuania

**Jos T. Puthur**
Plant Physiology and Biochemistry Division
Department of Botany
University of Calicut
Kerala, India

**Akula Ramakrishna**
Monsanto Crop Breeding Centre
Chikkaballapur, India

**Homero Reyes de la Cruz**
Instituto de Investigaciones Quнmico-Biolуgicas
Universidad Michoacana de San Nicolбs de Hidalgo
Ciudad Universitaria
Morelia, México

**Richard L. Ridgway**
Department of Biology
Seattle Pacific University
Seattle, Washington

**Victoria V. Roshchina**
Laboratory of Microspectral Analysis of Cells and Cellular Systems
Russian Academy of Sciences Institute of Cell Biophysics
Pushchino, Russia

**Praveen K. Saxena**
Department of Plant Agriculture
Gosling Research Institute for Plant Preservation
University of Guelph
Guelph, Ontario, Canada

**A.M. Shackira**
Plant Physiology and Biochemistry Division
Department of Botany
University of Calicut
Kerala, India

**Rashmi Sharma**
Acharya Narendra Dev College
University of Delhi
New Delhi, India

**Boris A. Shenderov**
Gabrichevsky Research Institute of Epidemiology and Microbiology
Moscow, Russia

**R.Sh. Shtanchaev**
Institute Russia
Institute of Theoretical and Experimental Biophysics
Russian Academy of Sciences
Pushchino, Russia

**Kathryn L. Van Alstyne**
Shannon Point Marine Center
Anacortes, Washington

**Kosuke Yamamoto**
Department of Molecular Microbiology
Faculty of Life Sciences
Tokyo University of Agriculture
Tokyo, Japan

**Olga M. Zherelova**
Institute Russia
Institute of Theoretical and Experimental Biophysics
Russian Academy of Sciences
Pushchino, Russia

# *Section I*

## *At the Beginning of Systematic Studies of Neurotransmitters' Systems in Plants and Modern Tendency*

# 1 Cholinesterase Activity in Plants

## *Reminiscences of Working with Mordecai J. Jaffe*

*Richard A. Fluck*

**CONTENTS**



## 1.1 INTRODUCTION

In 1970, Mark (Mordecai) Jaffe reported (1) the occurrence of acetylcholine (ACh) in mung bean (*Phaseolus aureus*) seedlings, and (2) data supporting the conclusion that ACh mediates changes in ion flux across membranes and regulates phytochrome-mediated phenomena in *P. aureus* (Jaffe 1970a, 1970b). At that time, I was a graduate student at the University of California, Berkeley, studying acetylcholinesterase (AChE) activity in differentiating muscle cells *in vitro* (Fluck and Strohman 1973).

In the course of my research, I had read about the occurrence of cholinesterase (ChE) activity in non-nervous tissue, for example, presumptive liver endoderm and mesenchyme, and its positive correlation with morphogenetic movements of cells in the formation of the liver (Drews and Kussäther 1970; for a review of cholinesterase in embryonic development, see Drews 1975). Other groups had reported that ACh is present during cleavage in sea urchin embryos (*Strongylocentrotus drobachiensis*) (Buznikov et al. 1968) and may regulate cell movement during gastrulation in sea urchin embryos (Gustafson and Toneby 1970). A colleague and postdoctoral fellow at Berkeley (Lars-G. Lundin, who is now at Uppsala University) and I were intrigued by the possibility that neurotransmitters such as ACh might function in early embryos by altering transmembrane ion fluxes and gene activity.

Mark's report (Jaffe 1970a) prompted me to write to him in October 1970 to tell him about my interest in the occurrence of ACh in non-nervous tissues and its function there. When he responded, he told me that an undergraduate student in his laboratory (Philip Russ) was beginning to look for ChE activity in *P. aureus*. He also invited me to come to his laboratory to do research, and I promptly applied for a National Science Foundation (NSF) postdoctoral fellowship to support a collaboration with him.

In August 1971, the American Society of Plant Physiologists held its annual meeting at the Asilomar Conference Grounds in Pacific Grove, California, just down the coast from Berkeley. By then, I was writing my dissertation and had been awarded an NSF postdoctoral fellowship to support my research with Mark, so I drove to Pacific Grove to meet him to discuss our upcoming collaboration. Mark generously offered to find a house that I could rent in Athens for my family.

## 1.2 1971–1974: A PRODUCTIVE COLLABORATION

I began working in Mark's laboratory in November 1971, just in time to overlap for about two weeks with postdoctoral fellow Dr. Joseph Riov (he is now at The Hebrew University of Jerusalem). Dr. Riov had purified and characterized ChE from *P. aureus* roots and reported that the ChE is not identical to the animal AChE but is inhibited by the most common inhibitors of AChE (Riov and Jaffe 1973). The plant ChE also is inhibited by various growth retardants, suggesting that ACh might play a role in plant growth (Riov and Jaffe 1973). During our two weeks together in Mark's laboratory, Dr. Riov taught me how to extract and purify ChE from *P. aureus* roots.

In our first paper (Fluck and Jaffe 1974a), Mark and I reported that ChE activity is present not only in *P. aureus* roots but also leaves and stems. We also found ChE activity in root callus tissue, suspension cultures of root cells, and root nodules—all derived from *Glycine max* (soybean). We also reported evidence suggesting that ChE is found primarily in the cell wall. In subsequent papers, we reported the occurrence of ChE activity in plants from 5 of the 24 families we examined—Characeae, Cruciferae, Graminae, Legumonisae, and Solanaceae (Fluck and Jaffe 1974b)—and that ChE is not pectin esterase (Fluck and Jaffe 1974c).

Mark and I also characterized the ChE activity extracted from *Solanum melongena* (eggplant) and *Zea mays* (maize), reporting that the activity from both species is inhibited by several anticholinesterases, including the plant growth retardant AMO 1618. The most potent inhibitor of the enzymes from both species was neostigmine. The subcellular localization of the enzymes in these two species seemed to differ from that of the enzyme in *P. aureus*: Whereas 95% of ChE activity in *P. aureus* is associated with the cell wall, more than half of the activity in *S. melongena* and *Z. mays* was in other subcellular fractions. Lastly, because the activity of the enzyme from *S. melongena* but not that from *Z. mays* was inhibited at superoptimal concentrations of substrate, we concluded that the enzyme from *S. melongena* but not that from *Z. mays* was a ChE (Fluck and Jaffe 1975).

More recently, Professor Yoshie Momonoki (Tokyo University of Agriculture) and his colleagues (including Professor Kosuke Yamamoto, a contributor to this volume) have purified and cloned the gene for AChE from *Z. mays*. Based on biochemical characterization of the enzyme and in silico screening, Sagane et al. (2005) proposed that the enzyme from *Z. mays* and its homologs in other species comprise "a novel family of the enzymes that is specifically distributed in the plant kingdom."

My collaboration with Mark culminated in "The Acetylcholine System in Plants," a paper in which we reviewed evidence for (1) the occurrence of ACh, ChE activity and choline acetyltransferase activity in plants and (2) the effects of ACh and other cholinergic molecules on the physiology, growth, and development of plants. However, we found no direct evidence in the literature for the occurrence of an ACh receptor in plants (Fluck and Jaffe 1974d). To my knowledge, no such evidence has yet been published.

## 1.3 INDOLE-3-ACETYLCHOLINE AND OTHER CONJUGATES OF CHOLINE

In summer 1974, at the annual meeting of the American Society of Plant Physiologists at Cornell University, I met with Dr. Carl Pike from Franklin & Marshall College (F&M) to discuss his research on the effects of ACh in plants (Kirshner et al. 1975). In the course of our conversation, Carl told me about a job opening at F&M and encouraged me to apply for it. Within three months, I had accepted a position at the college, and my family and I had moved to Lancaster, Pennsylvania.

At F&M, I returned to the study of ACh and AChE in early animal embryos, specifically a small teleost fish, the medaka (*Oryzias latipes*). My collaborators and I reported the occurrence of ACh and AChE activity in these embryos (Fluck 1978; Fluck and Shih 1981; Fluck 1982), as well as the effects of cholinergic molecules on and movement and aggregation of dissociated blastomeres from *O. latipes* embryos (Fluck et al. 1980).

However, I continued to think about ACh and AChE in plants. In 1986, I asked Professor Phyllis Leber, an organic chemist at F&M, whether she would be interested in collaborating with me to test the following hypothesis: Indole-3-acetylcholine (IAC), that is the choline conjugate of indole-3-acetic acid (IAA), can be hydrolyzed by ChE in pea (*Pisum sativum*) stem segments, releasing IAA and thereby stimulating elongation of the segments.

Thus began a long, fruitful collaboration that included many undergraduate students who were interested in both organic chemistry and plant physiology. The collaboration resulted in papers in which we reported (1) the synthesis and bioassay of IAC (Ballal et al. 1993); (2) a versatile and efficient method for preparing IAC and the choline conjugates of other auxins, for example, 1-naphthaleneacetic (Bozsó et al. 1995); and (3) the synthesis of the choline and thiocholine conjugates of several auxins, the effects of the choline conjugates on pea-stem elongation, and the rates of hydrolysis of the thiocholine conjugates by pea ChE (Fluck et al. 2000). The results of the last study were consistent with a model in which ChE releases active auxins from these conjugates.

## 1.4 MARK JAFFE, JULY 7, 1933–OCTOBER 14, 2007

For this section of the paper, I am grateful to Amy Jaffe (Mark's widow) for providing details about Mark and his journey as a scientist. Mark and Amy met and married at Cornell University and returned there after Mark retired in 1998 (Figure 1.1). Together, they raised three children: Jennifer, Benjamin, and Samuel.

Mark was a city boy, born and raised in Manhattan, New York City, so it may seem ironic that he, who had never set foot on a farm until his second semester of graduate school, would get a PhD from Cornell University's College of Agriculture and go on to be a pioneer in the study of the role of ACh in plants and coin the term *thigmomorphogenesis* to name the effects of touch on plant growth and development (Jaffe 1973).

Amy wrote, "Having grown up in Manhattan, Mark may have had an advantage over more countrified folk when he walked in fields or woods. He noticed things that the rest of us take for granted,

**FIGURE 1.1** Mark Jaffe beholds a waterfall in Taughannock Falls State Park, which is about three miles northwest of Ithaca, New York, on the shore of Cayuga Lake. (Courtesy of Amy Jaffe.)

because we had known these plants since we were children and took their characteristics as part of our definition of that plant. So we never asked, for example, why dandelions distribute their seeds from a puffball formation." Mark did.

When Mark began to do research on ACh in plants, scientists knew much about the function of ACh in animals (and organs, tissues, and cells from animals), particularly at the neuromuscular junction. To learn more about animal physiology, Mark attended medical conferences and did research at the Colorado State University School of Veterinary Medicine and Biomedical Sciences, where he studied ACh in the eyes of dark-adapted chickens.

Amy said this about how Mark approached research and teaching: "When Mark began studying rapid movements in plants, he read about the origins of the universe so he could understand concepts of time and thus know whether the coiling of pea tendrils could rightly be called a *rapid* movement. When he began to teach a course in stress physiology (an offshoot of his interest in thigmomorphogenesis), he taught himself physics, because he understood that a knowledge of biophysics is basic to understanding stress in organisms. A student who was taking a physics course at the same time told Mark that his physics professor was teaching physics by way of mathematical formulas, whereas Mark's lectures, which emphasized phenomena, made it much easier to understand physics."

Years later, Professor Randy Wayne (Cornell University) would say, "[Mark] found more joy in the search for the answer than in knowing the answer ... Mark loved every aspect of the search—the initial crazy observation, designing equipment, doing the experiment itself, and interpreting the results to discover the underlying unity of nature. More scientists today should be like him." (Anonymous, 2007). Dr. Riov wrote this about Mark: "I have often said that I was specifically impressed by Mark's imagination, which I believe is an important trait for any scientist. Undoubtedly, this trait led Mark to study unique topics in plant biology."

At Ohio University, Mark did research on thigmomorphogenesis and phytochrome alongside the work he did on ACh and ChE in plants. He continued his research on thigmomorphogenesis after moving to Wake Forest University in 1980, where he accepted the Charles H. Babcock Professor of Botany endowed chair. After retiring from Wake Forest, he returned to his beloved Cornell University, where he lunched regularly with Randy Wayne and Professor A. Carl Leopold to discuss their common interests, which included contractile roots (Jaffe and Leopold 2007) and how paramecia sense the edge of a water droplet (Anonymous 2007).

Professor Leopold, in the obituary he wrote for the newsletter of the American Society of Plant Biologists (Leopold 2007), said this about Mark: "It was a characteristic of his research career that he would pursue highly original research on entirely new areas of study. Examples include his seminal thesis work on tendril coiling, and then his surprising finding of acetylcholine in plants, and then his origination of the new field of thigmomorphogenesis as a basic phenomenon in plant growth and differentiation. He made numerous contributions to areas of growth regulation, phytochrome actions, tissue differentiation, and signal transduction. ... Mark found joy in the search for answers to questions about underlying units of nature—a true intellectual researcher."

Mark would have relished the opportunity to write a paper for this volume. I am grateful to have had the opportunity to work with him and to honor him here.

## ACKNOWLEDGEMENTS

I thank Amy Jaffe, Joseph Riov, and Randy Wayne for contributing their memories of Mark and for helping to improve an earlier version of the paper; Philip Russ for clarifying details about his research in Dr. Jaffe's laboratory; Carl Pike and Phyllis Leber for helping to clarify certain historical details; Diane McCauley for sending me copies of selected abstracts from annual meetings of the American Society of Plant Physiologists; and Mark Olson for identifying the waterfall in Figure 1.1.

## LITERATURE CITED


Anonymous. 2007. Mordecai J. "Mark" Jaffe. *Ithaca J.*, October 17. http://www.legacy.com/obituaries/theithacajournal/obituary.aspx?pid=96305979 (accessed March 4, 2017).

Ballal, S., R. Ellias, R. A. Fluck et al. 1993. The synthesis and bioassay of indole-3-acetylcholine. *Plant Physiol. Biochem.* 31:249–255.

Bozsó, B. A., R. A. Fluck, R. A. Jameton, P. A. Leber, and J. G. Varnes. 1995. A versatile and efficient methodology for the preparation of choline ester auxin conjugates. *Phytochemistry* 40:1027–1031.

Buznikov, G. A., I. V. Chudakova, L. V. Berdysheva, and N. M. Vyazmina. 1968. The role of neurohumors in early embryogenesis II. Acetylcholine and catecholamine content in developing embryos of sea urchin. *J. Embryol. Exp. Morphol.* 20:119–128.

Drews, U. 1975. Cholinesterase in embryonic development. *Prog. Histochem. Cytochem.* 7:1–52.

Drews, U. and E. Kussäther. 1970. Histochemisch nachweisbare Cholinesterase-Aktivität bei der Entwicklung der Leberanlage des Hühnerembryos. *Z. Anat. Entwicklungsgesch.* 130:330–344.

Fluck, R. A. 1978. Acetylcholine and acetylcholinesterase activity in early embryos of the medaka *Oryzias latipes*, a teleost. *Dev. Growth Differ.* 20:17–25.

Fluck, R. A. 1982. Localization of acetylcholinesterase activity in young embryos of the medaka *Oryzias latipes*, a teleost. *Comp. Biochem. Physiol. C* 72:59–64.

Fluck, R. A. and M. J. Jaffe. 1974a. Cholinesterases from plant tissues. III. Distribution and subcellular localization in *Phaseolus aureus* Roxb. *Plant Physiol.* 53:752–758.

Fluck, R. A. and M. J. Jaffe. 1974b. The distribution of cholinesterases in plant species. *Phytochemistry* 13:2475–2480.

Fluck, R. A. and M. J. Jaffe. 1974c. Cholinesterases from plant tissue: V. Cholinesterase is not pectin esterase. *Plant Physiol.* 54:797–798.

Fluck, R. A. and M. J. Jaffe. 1974d. The acetylcholine system in plants. *Commentaries Plant Sci.* 11:119–136.

Fluck, R. A. and M. J. Jaffe. 1975. Cholinesterases from plant tissues VI. Preliminary characterization of enzymes from *Solanum melongena* L. and *Zea mays* L. *Biochim. Biophys. Acta* 410:130–134.

Fluck, R. A., P. A. Leber, J. D. Lieser et al. 2000. Choline conjugates of auxins. I. Direct evidence for the hydrolysis of choline-auxin conjugates by pea cholinesterase. *Plant Physiol. Biochem.* 38:301–308.

Fluck, R. A. and T. M. Shih. 1981. Acetylcholine in embryos of *Oryzias latipes*, a teleost: Gas chromatographic-mass spectrometric assay. *Comp. Biochem. Physiol. C* 70:129–130.

Fluck, R. A. and R. C. Strohman. 1973. Acetylcholinesterase activity in developing skeletal muscle in vitro. *Dev. Biol.* 33:417–428.

Fluck, R. A., A. J. Wynshaw-Boris, and L. M. Schneider. 1980. Cholinergic molecules modify the in vitro behavior of cells from early embryos of the medaka *Oryzias latipes*, a teleost fish. *Comp. Biochem. Physiol. C* 67:29–34.

Gustafson, T. and M. Toneby. 1970. On the role of serotonin and acetylcholine in sea urchin morphogenesis. *Exp. Cell Res.* 62:102–117.

Jaffe, M. J. 1970a. Acetylcholine: A new plant hormone regulating phytochrome mediated responses in mung bean roots. *Plant Physiol.* 46(Supplement):2.

Jaffe, M. J. 1970b. Evidence for the regulation of phytochrome-mediated processes in bean roots by the neurohumor, acetylcholine. *Plant Physiol.* 46:768–777.

Jaffe, M. J. 1973. Thigmomorphogenesis: The response of plant growth and development to mechanical stimulation. *Planta* 114:143–157.

Jaffe, M. J. and A. C. Leopold. 2007. Light activation of contractile roots of Easter lily. *J. Am. Soc. Hortic. Sci.* 132:575–582.

Kirshner, R. L., J. M. White, and C. S. Pike. 1975. Control of bean bud ATP levels by regulatory molecules and phytochrome. *Physiol. Plant.* 34:373–377.

Leopold, C. 2007. Mark Jaffe: Obituary. *Newslett. Am. Soc. Plant Biologists* 34(6):30.

Riov, J. and M. J. Jaffe. 1973. Cholinesterases from plant tissues: I. Purification and characterization of a cholinesterase from mung bean roots. *Plant Physiol.* 51:520–528.

Sagane, Y., T. Nakagawa, K. Yamamoto, S. Michikawa, S. Oguri, and Y. S. Momonoki. 2005. Molecular characterization of maize acetylcholinesterase. A novel enzyme family in the plant kingdom. *Plant Physiol.* 138:1359–1371.

# 2 Plant Acetylcholinesterase Plays an Important Role in the Response to Environmental Stimuli

*Kosuke Yamamoto and Yoshie S. Momonoki*

## CONTENTS



## 2.1 ROLE OF ACETYLCHOLINE–MEDIATED SYSTEM AT THE NEUROMUSCULAR JUNCTION IN ANIMALS

The acetylcholine (ACh)—ACh-receptor (AChR)—acetylcholinesterase (AChE) system, that is *the ACh-mediated system*, is a well-known signal transmission system in animals (Soreq and Seidman, 2001; Hill, 2003; Lai and Ip, 2003; Ferraro et al., 2011).

At the presynaptic terminus, the membrane depolarization due to arrival of an action potential leads to the opening of the voltage-dependent $Ca^{2+}$ channels. The resulting influx of $Ca^{2+}$ ions initiates the fusion of synaptic vesicles, which contain ACh, to the presynaptic membrane, leading to the release of ACh to the synaptic cleft. Then, ACh diffuses and binds to acetylcholine receptors (AChRs) localized on the postsynaptic membrane on the muscle fiber. This binding leads to openings of their intrinsic ion channels permeable to both $Na^+$ and $K^+$. The resulting cation entry depolarizes the muscle membrane locally; if that reaches a critical threshold, it opens voltage-gated $Na^+$ channels, generating an action potential that is propagated through the muscle fiber causing $Ca^{2+}$ release from the sarcoplasmic reticulum into the cytosol and muscle contraction. AChE, anchored by Collagen-Q in the basal lamina, quickly inactivates ACh released from the presynaptic membrane so that the ACh concentration in the synaptic cleft decreases rapidly and neurotransmission stops (Figure 2.1; Soreq and Seidman, 2001; Hill, 2003; Lai and Ip, 2003; Ferraro et al., 2011).

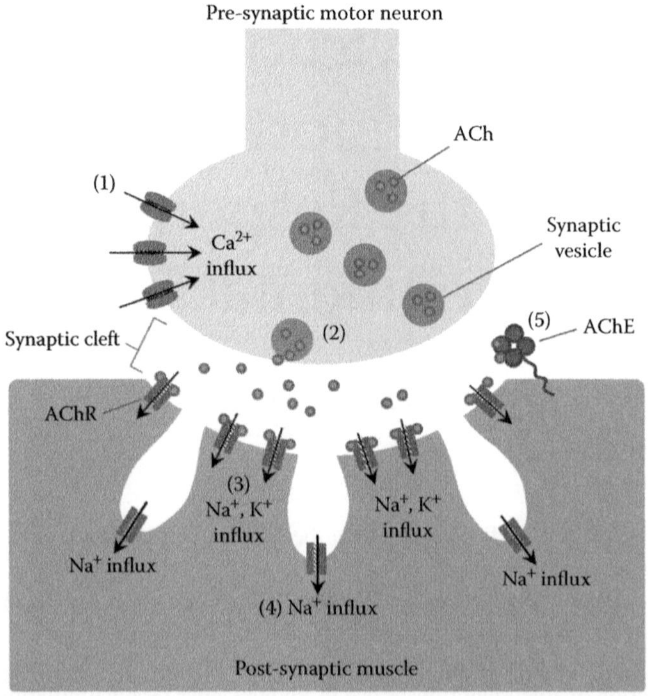


**FIGURE 2.1** Diagrammatic representation of the acetylcholine-mediated system at the neuromuscular junction in animals. (1) Action potential arriving at the presynaptic terminus allows opening of voltage gated $Ca^{2+}$ channels. (2) Influx of $Ca^{2+}$ into the presynaptic neuron triggers release of acetylcholine (ACh) to the synaptic cleft. (3) Interaction of ACh with ACh receptors (AChRs) allows influx of both $Na^+$ and $K^+$ into the postsynaptic muscle and depolarizes the postsynaptic membrane locally. (4) Voltage gated $Na^+$ channels open, triggering muscle action potential. (5) Acetylcholinesterase (AChE) quickly breaks ACh into choline and acetate so that the ACh concentration in the synaptic cleft decreases rapidly and neurotransmission stops.

## 2.2 PURIFICATION AND MOLECULAR PROPERTIES OF ACETYLCHOLINESTERASE IN PLANTS

ACh-hydrolyzing activity in plants was first discovered in extracts of Characeae algae *Nitella* (Dettbarn, 1962). In the same period, its activities were also found in mycelium of fungi *Physarium polycephalum* (Nakajima and Hatano, 1962) and yeast *Saccharomyces cerevisiae* (Jacobsohn and Azevedo, 1962). Then, a series of numerous studies of Jaffe and Fluck with co-workers in 1970–1975 reported that ~118 terrestrial species and 10 marine algae were identified as having ACh-hydrolyzing activity (Roshchina, 1991, 2001). According to Roshchina (2001), ACh-hydrolyzing activities were widely expressed in the plant kingdom: 2 species of 2 families from Pteridophytes, 10 species of a family from marine algae, 10 species of 7 families from gymnosperms, and 172 species of 78 families from angiosperms.

The values of the enzyme activity (the substrate hydrolysis rate) in higher plants were an average of 1–900 mmol/h/g fresh weight, depending on the plant species (Roshchina, 2016). Bryophytes (mosses, liverworts, and hornworts) also possessed the ACh-hydrolyzing activity of up to 0.360 mmol/h/g fresh weight (Gupta et al., 2001). Pollen also contains the ACh-hydrolyzing activity (Bednarska and Tretyn, 1989; Bednarska, 1992; Rejon et al., 2012; Roshchina, 2016). AChE was first partially purified and characterized from mung bean (Riov and Jaffe, 1973). To date, AChE has been extracted and partially purified from many plant species: *Phaseolus aureus*, *Phaseolus vulgaris*, *Pisum sativum*, *Cicer arietinum*, *Robinia pseudoacacia*, *Zea mays*, and *Urtica dioica*

(Roshchina, 2001; Malik, 2014). However, its nucleotide and amino acid sequences have been unknown until recently.

In our recent reports, AChEs were purified and identified from maize (*Z. mays* L.) seedlings as a representative of monocotyledons and from siratro (*Macroptilium atropurpureum* Urb.) seedlings as a representative of dicotyledons (Sagane et al., 2005; Yamamoto et al., 2008). These reports provided to our knowledge the first direct evidence of the AChE molecule in plants. AChEs were purified from crude extracts of etiolated maize and siratro seedlings using a series of chromatography steps as follows: a Sephadex G-200 gel filtration column, a Poros HQ/20 ion exchange column, a Poros HS/20 ion exchange column, and then a Hiload Superdex 200 pg 16/60 column. After this series of purification procedures, yields from 2 to 0.3 kg of maize and siratro seedlings were 0.8 and 0.258 mg of protein having a total activity of 0.07 and 0.2 units, respectively (Tables 2.1 and 2.2).

Next, to further purify and then analyze molecular properties of maize and siratro AChEs, both maize and siratro AChE proteins were subjected to PAGE analyses. In case of maize AChE (Sagane et al., 2005), the AChE purified by a series of chromatography steps showed two predominant bands having different Coomassie Brilliant Blue (CBB) staining intensities (designated as the U and L forms, Figure 2.2a) on native PAGE. Both the U and L bands exhibited AChE activity, as indicated by staining using acetylthiocholine (ASCh) as a substrate (Figure 2.2a). The native PAGE gel slices corresponding to the U and L bands were subjected to SDS-PAGE under reducing and non-reducing conditions. Both U and L bands migrated as a diffuse band corresponding to a molecular mass of between 42- and 44-kDa on SDS-PAGE under the reducing condition (Figure 2.2b). On the other hand, each band displayed two bands on SDS-PAGE under the non-reducing condition, one sharp band with an apparent molecular mass of 88 kDa and another diffuse band with a molecular mass

**TABLE 2.1**
**Summary of Maize AChE Purification**

| Purification Steps | Activity (Units[a]) | Total Protein (mg) | Specific Activity (Units/mg) | Purification-Fold | Yield (%) |
|---|---|---|---|---|---|
| Sephadex G-200 | 5.45 | 15.59000 | 0.4 | 1.0 | 100.0 |
| Poros HQ | 3.33 | 5.37000 | 0.6 | 1.8 | 61.09 |
| Dialysis | 3.12 | 3.71000 | 0.8 | 2.4 | 57.27 |
| Poros HS | 0.80 | 0.06530 | 12.3 | 35.0 | 14.67 |
| Superdex 200 | 0.07 | 0.00079 | 82.5 | 236.0 | 1.20 |

*Source:* Sagane, Y. et al., *Plant Physiol.*, 138, 1359–1371, 2005.
[a] One unit of activity was defined as the hydrolysis of 1 μmol of ASCh per min at 30°C.

**TABLE 2.2**
**Summary of Siratro AChE Purification**

| Purification Steps | Activity (Units[a]) | Total Protein (mg) | Specific Activity (Units/mg) | Purification-Fold | Yield (%) |
|---|---|---|---|---|---|
| Sephadex G-200 | 0.7 | 104.0 | 0.007 | 1.0 | 100.0 |
| Dialysis | 0.6 | 71.8 | 0.009 | 1.3 | 87.0 |
| Poros HS | 0.4 | 0.8 | 0.458 | 64.5 | 50.8 |
| Superdex 200 | 0.2 | 0.3 | 0.957 | 134.8 | 33.4 |

*Source:* Yamamoto, K. et al., *Planta*, 227, 809–822, 2008.
[a] One unit of activity was defined as the hydrolysis of 1 μmol of ASCh per min at 30°C.

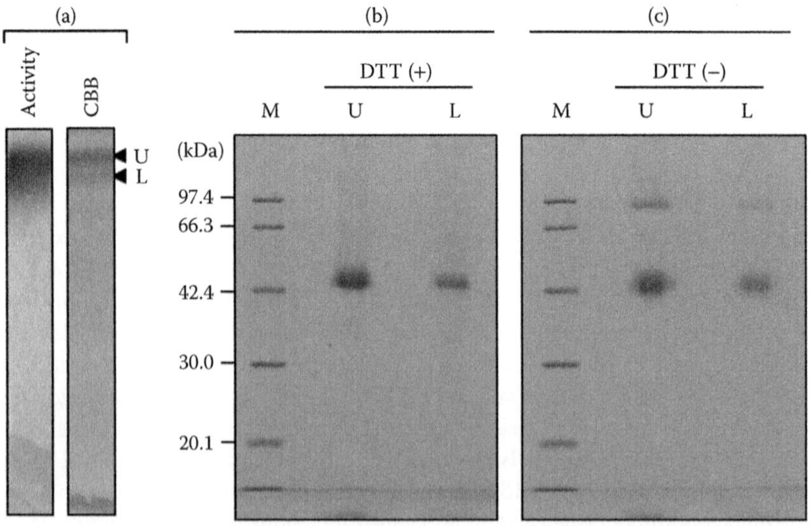


**FIGURE 2.2** Native and SDS-PAGE analysis of the purified maize AChE. (a) Native PAGE of the partially purified maize AChE. Separated proteins were stained with Coomassie Brilliant Blue (CBB) and AChE activity stain (Activity). (b) SDS-PAGE (12.5%) of the extractions (U and L, which correspond to bands in [a]) from the native PAGE gels in the presence of dithiothreitol (DTT). The lane indicated by M is molecular mass standards. (c) SDS-PAGE (12.5%) of the extractions from the native PAGE gels in the absence of DTT. (From Sagane, Y. et al., *Plant Physiol.*, 138, 1359–1371, 2005.)

of approximately 42- to 44-kDa (Figure 2.2c). These results indicate that both the U and L forms of AChE exist as a mixture of 88 kDa dimers consisting of disulfide-linked 42- to 44-kDa polypeptides and a 42- to 44-kDa monomer. Based on a gel filtration analysis, the purified maize AChE was estimated to be approximately 67 kDa and exists as a single molecular species without any differently sized forms. Therefore, the U and L forms of AChE were composed of a mixture of different dimeric forms: a disulfide-linked homodimer and a noncovalently linked homodimer, each of which consists of 42- to 44-kDa polypeptides. The N-terminal amino acid sequences for the 42- to 44-kDa U- and L-form proteins were identical, $NH_2$-AGAGGDCHFPAVFNFGDSNS. This result implied that the U and L forms of AChE are the same gene products with different posttranslational modifications, which may consist of the deletion of several residues from the C termini and/or different degrees of the glycosylation.

In case of siratro AChE (Yamamoto et al., 2008), the AChE purified by a series of chromatography steps gave two diffuse bands and one sharp band on native PAGE (Figure 2.3a). Of these, only one diffuse band demonstrated AChE activity, as indicated by staining using ASCh as a substrate (Figure 2.3a). To further purify this protein, the native PAGE gel slice containing AChE activity was collected and subjected to modified SDS-PAGE under low SDS concentration (0.01%) in the gel and in the electrophoresis buffer. The AChE migrated as several bands on the gel, but only one band exhibited AChE activity, as indicated by activity staining (Figure 2.3b). Therefore, the contaminants in the preparation could be removed by this modified SDS-PAGE procedure. The highly purified AChE in the gel slice after the purification step with modified SDS-PAGE was subjected to normal SDS-PAGE under reducing and non-reducing conditions. The purified siratro AChE migrated as two bands corresponding to molecular masses of 41- and 42-kDa on SDS-PAGE under the reducing condition (Figure 2.3c). Meanwhile, purified siratro AChE ran as one sharp band with an apparent molecular mass of 125 kDa on SDS-PAGE under the non-reducing condition (Figure 2.3d). The native molecular mass of siratro AChE was estimated to be approximately 121 kDa based on a gel filtration analysis. These results indicate that the siratro AChE exists as a 125 kDa trimer consisting of disulfide-linked 41- or 42-kDa monomers.

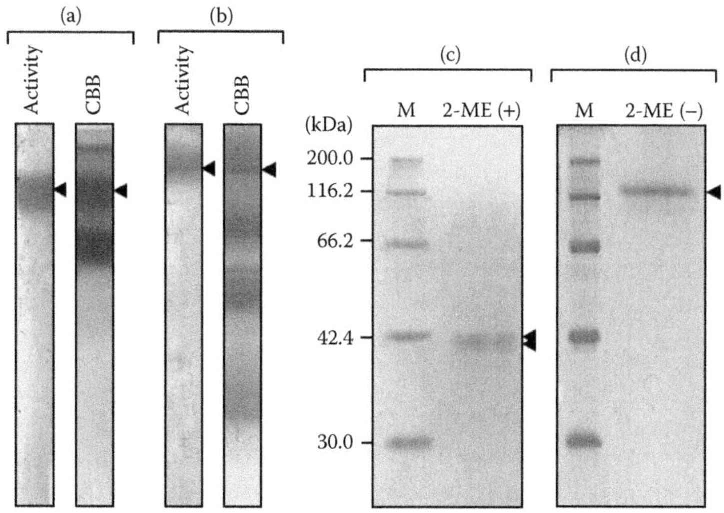


**FIGURE 2.3** Native and SDS-PAGE analysis of the purified siratro AChE. (a) Native-PAGE profile of active fractions containing AChE activity after chromatography. Separated proteins were visualized by Coomassie Brilliant Blue (CBB) staining (right panel) and AChE activity staining (left panel). (b) Modified SDS-PAGE (12.5%) profile of the purified AChE after native-PAGE in the absence of 2-mercaptoethanol (2-ME). The modified SDS-PAGE was performed using low SDS concentration (0.01%) in the electrophoresis buffer and gel. Separated proteins were stained with CBB (right panel) and AChE activity stain (left panel). (c) SDS-PAGE (12.5%) of the extracted AChE active band on modified SDS-PAGE gels in the presence of 2-ME. The lane indicated by M is molecular mass standards. (d) SDS-PAGE (12.5%) of the extracted AChE active band on the modified SDS-PAGE gels in the absence of 2-ME. (From Yamamoto, K. et al., *Planta*, 227, 809–822, 2008.)

In animals, the active AChE has been found in monomeric, disulfide-linked dimeric, and tetrameric forms depending on the animal species. Sipunculida (Talesa et al., 1993) and Insecta (Gnagey et al., 1987; Toutant, 1989) produce only the disulfide-linked dimeric form. On the other hand, in Cercopithecus monkey brain, multiple populations of AChE having different subunit structures including monomeric, dimeric, and tetrametric were found, although not all of the subunits were linked with a disulfide bond (Liao et al., 1993). The different quaternary structures noted between maize and siratro AChEs imply that AChEs in different plant species also would have various subunit arrangements.

## 2.3 IDENTIFICATION OF NUCLEOTIDE SEQUENCE ENCODING THE PLANT ACETYLCHOLINESTERASE cDNA

Amino acid composition of cholinesterase purified 200 times from pea chloroplasts (Roshchina, 1988) has been analyzed (Roshchina, 2001). When this result was compared with AChE isolated from electric ray *Torpedo californica* (Otto, 1985), there was a similarity in the number of amino acids per mole of protein (Roshchina, 2001). Madhavan et al. (1995) reported that a probe prepared from *Torpedo AChE* cDNA hybridized to *Arabidopsis thaliana* DNA sequence under permissive hybridization conditions. However, no significantly homologous genes to animal AChEs can be found in plant genome databases.

In our efforts, *AChE* cDNAs were cloned from maize (*Z. mays* L.), rice (*Oryza sativa* L.), siratro (*M. atropurpureum* Urb.) and *Salicornia europaea* plants. In maize plants, the 42- to 44-kDa polypeptide of the U form AChE, which was separated by SDS-PAGE under the reduced condition, was digested with *Staphylococcus aureus* V8 protease and generated proteolytic fragments were applied for N-terminal amino acid sequence analyses. Of the several derivative fragments, two probable internal sequences were yielded, $NH_2$-YFSQALYTFDIGQNDITSSY and $NH_2$-VEAIIPDLME. A tBLASTn search against the maize database using these amino acid sequences identified a partial-length maize mRNA nucleotide sequence (accession number, AY107095) as a candidate. A partial open reading frame, which covers all short amino acid sequences determined for the purified AChE protein, was identified in the mRNA sequence. However, the sequence does not contain termination codon, although it possesses a presumed initiating ATG codon. To determine the complete nucleotide sequence corresponding to AY107095, the full-length cDNA for maize *AChE* gene was obtained using 5′- and 3′-rapid amplification of cDNA ends (RACE) technique. The determined nucleotide sequence consists of 1471 bp. The sequence contains a putative entire open reading frame, which consists of 1182 bp and encodes 394 amino acid residues (accession number, AB093208; Sagane et al., 2005).

In siratro plants, the 41- and 42-kDa polypeptides corresponding to siratro AChE, which were separated by SDS-PAGE under the reduced condition, were digested with *S. aureus* V8 protease, generating proteolytic fragments that were then subjected to N-terminal amino acid sequence analysis. This yielded three probable internal sequences, $NH_2$-TIFQYGISPFSLDIQIVQFN, $NH_2$-RSRLPVPEEFAKALYTFDIG and $NH_2$-MATEFNKQMKD. Subsequently, the partial genomic DNA sequence of siratro *AChE* was determined using sense and antisense degenerate primers designed based on three probable internal protein sequences. A 643 bp product was amplified and sequenced. Based on its partial DNA sequence for siratro *AChE*, the full-length cDNA for the AChE was cloned by 5′- and 3′-RACE technique. The determined nucleotide sequence consists of 1441 bp. The sequence contains a putative entire open reading frame containing 1146 bp, and it encodes 382 amino acid residues (accession number, AB294246; Yamamoto et al., 2008).

In Salicornia plants, the partial cDNA sequence of *AChE* gene was amplified by reverse transcription polymerase chain reaction (RT-PCR) using degenerate primers designed based on maize and siratro *AChE* genes. PCR product of 418 bp for Salicornia *AChE* gene was yielded and sequenced. Based on its partial DNA sequence, the full-length cDNA for the AChE was cloned by 5′- and 3′-RACE technique. The full-length cDNA of Salicornia *AChE* was 1536 nucleotides, encoding a 387-residue protein (accession number, AB489863; Yamamoto et al., 2009).

In rice plants, the full-length cDNA sequence of rice AChE was found by a BLASTn search using full-length cDNA sequence of maize AChE as query. The full-length rice *AChE* cDNA was 1757-nucleotides long and encoded 391 amino acids (accession number, AK073754; Yamamoto et al., 2015).

Figure 2.4 shows alignment of the amino acid sequences of rice, maize, siratro, and Salicornia AChEs. On the basis of the protein family database (Pfam), four plant AChEs contain a conserved putative lipase GDSL family domain, but not the alpha/beta-hydrolase fold superfamily that is phylogenetically related to cholinesterases. Each of the five motif blocks is characteristic of the GDSL lipase family in bacteria and plants (Brick et al., 1995; Upton and Buckley, 1995; Akoh et al., 2004). The catalytic Ser/Asp (Glu)/His triads of the enzymes belonging to lipase GDSL family were characterized in some of these proteins, and the sequences surrounding the consensus residues are well conserved (Upton and Buckley, 1995). Therefore, the putative catalytic triad for plant AChEs, Ser/Asp (Glu)/His triad, was deduced by superimposing the sequences on those of the lipase GDSL family enzymes (Figures 2.4 and 2.5). The animal AChE also has a catalytic Ser/Asp(Glu)/His triad (Soreq and Seidman, 2001). On the other hand, based on Pfam protein family analysis, the animal AChE possesses a distinct carboxylesterase consensus domain, but not the lipase GDSL consensus sequence. Thus, plant AChEs could not belong to the well-known AChE family, which is distributed throughout the animal kingdom. Sánchez et al. (2012) reported an interesting finding that the PA4921 gene product in *Pseudomonas aeruginosa* PAO1 has ACh-hydrolyzing activity and belongs to the GDSL lipase family (SGNH hydrolase family) similar to plant AChEs. Therefore, the AChE belonging to GDSL lipase family would distribute not only in plants, but also in gram-negative bacterium.

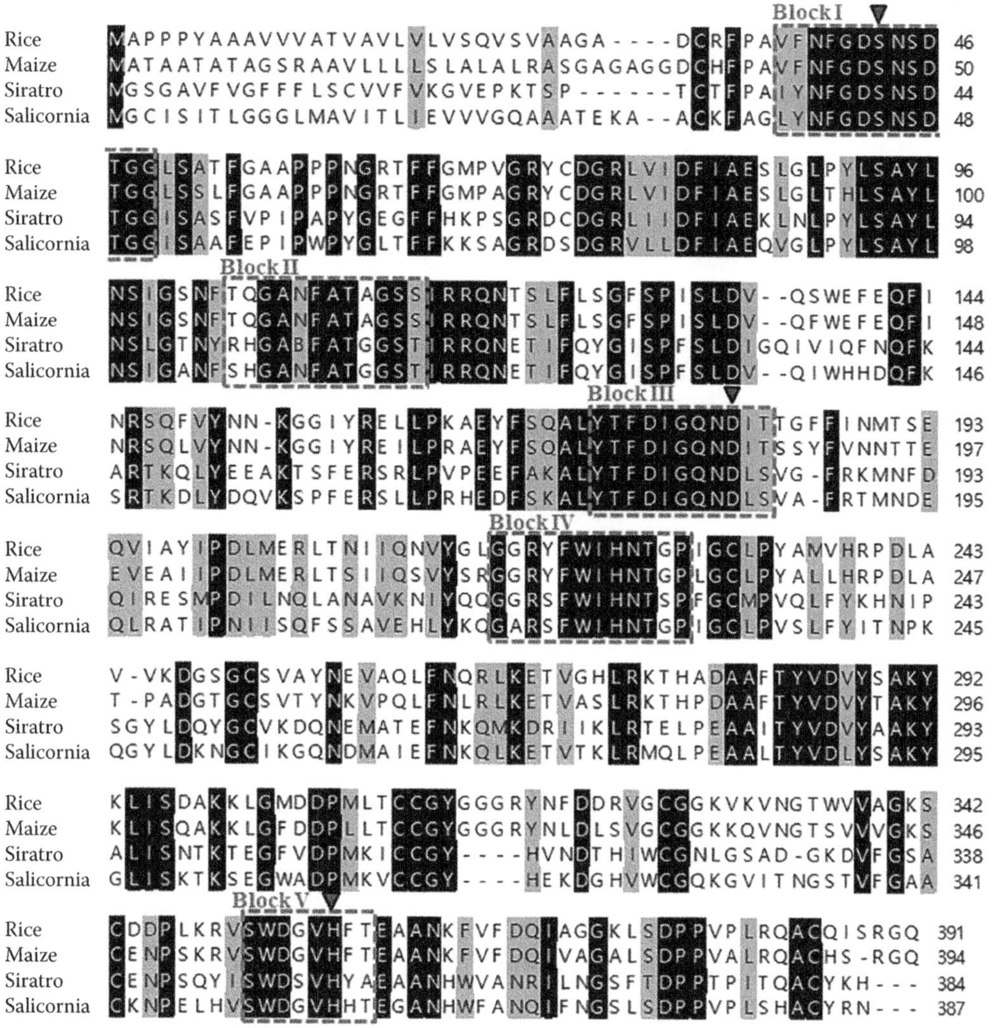


**FIGURE 2.4** Alignment of the amino acid sequences of rice, maize, siratro, and Salicornia AChEs. Conserved amino acid residues are highlighted in black. Non-identical, but functionally conserved amino acids are highlighted in gray. The blocks of sequences conserved in the GDSL family enzymes and the putative catalytic triad Ser/Asp(Glu)/His residues in the GDSL family enzymes, which were presumed previously (reference in Brick et al. 1995; Upton and Buckley 1995; Akoh et al. 2004), are indicated by boxes (dashed lines, Blocks I–V) and arrowheads, respectively. The GenBank accession numbers used in the analysis are as follows: rice AChE, AK073754; maize AChE, AB093208; siratro AChE, AB294246; Salicornia AChE, AB489863. (From Yamamoto, K. et al., *Biochem. Biophys. Res. Commun.*, 465, 488–493, 2015.)

Muralidharan et al. (2013) recently reported that the protein encoded by the *A. thaliana* homolog of the *Z. mays AChE* gene, At3g26430, exhibited lipase activity with preference to long chain substrates but did not hydrolyze choline esters. The At3g26430 protein belongs to the SGNH clan of serine hydrolases, and more specifically to the GDSL lipase family. Unfortunately, they didn't mention the other *A. thaliana* homologs of the *Z. mays AChE* gene. When the maize AChE protein sequence was used as a query sequence in TAIR BLASTP searching against *Arabidopsis* TAIR10 protein database, At1g67830 and AT5g14450 also shared 58% and 52% identities with the maize AChE,

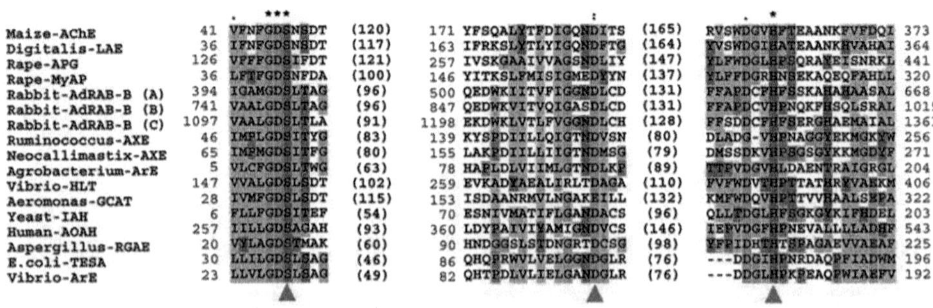

**FIGURE 2.5** Alignment of the amino acid sequences of maize AChE and selected members of the GDSL lipase family. The alignment was generated by a ClustalX analysis. Dashes indicate gaps introduced to improve the alignment. Numbers in brackets indicate intervening amino acid numbers, and the numbers without brackets indicate the positions in the sequences. The putative catalytic triad Ser/Asp (Glu)/His residues in the GDSL family enzymes, which were presumed previously (reference in Brick et al. 1995; Upton and Buckley 1995; Akoh et al. 2004), are indicated by arrowheads. The abbreviations and accession numbers of the cDNA nucleotide sequences used for this analysis are follows: Maize-AChE, *Z. mays* AChE (AB093208); Digitalis-LAE, *Digitalis lanata* lanatoside 15′-O-acetylesterase (AJ011567); Rape-APG, *Brassica napus* anter-specific Pro-rich protein APG (P40603); Rape-MyAP, *Brassica napus* myrosinase-associated protein (U39289); Rabbit-AdRAB-B, *Oryctolagus cuniculus* phospholipase (Z12841); Ruminococcus-AXE, *Ruminococcus flavefaciens* acetyl xylan esterase (AJ238716); Neocallimastix-AXE, *Neocallimastix patriciarum* acetylxylan esterase bnaIII (U66253); Agrobacterium-ArE, *Agrobacterium radiobacter* arylesterase (AF044683); Vibrio-HLT, *Vibrio parahaemolyticus* thermolabile hemolysin (M36437); Aeromonas-GCAT, *Aeromonas hydrophila* glycerophospholipid-cholesterol acyltransferase (X07279); Yeast-IAH, *Saccharomyces cerevisiae* isoamyl acetate hydrolytic enzyme (X92662); Human-AOAH, *Homo sapiens* acyloxyacyl hydrolase (BC025698); Aspergillus-RGAE, *Aspergillus aculeatus* rhamnogalacturon acetylesterase (X89714); E. coli-TESA, *Escherichia coli* acyl-coA thioesterase I (L06182); Vibrio-ArE, *Vibrio mimicus* arylesterase (X71116). (From Yamamoto, K. et al., *Planta*, 227, 809–822, 2008.)

respectively, as similar as At3g26430. We should examine whether the other *A. thaliana* homologs of the *Z. mays AChE* gene containing At1g67830 and AT5g14450 possess ACh-hydrolyzing activities. They also concluded that as compared to previously characterized plant cholinesterase activities (Riov and Jaffe, 1973; Muralidharan et al., 2005), the enzyme encoded by LOC606473 (also called NP_001105800), dubbed *ache* by Sagane et al. (2005), can be characterized as a serine hydrolase with marginal reactivity toward choline-esters. Because it is yet to be demonstrated that overexpression of this maize gene would result in ectopic accumulation of a choline-ester hydrolyzing enzyme, alternative explanations to results of Sagane et al. (2005) cannot be ruled out (Muralidharan et al., 2013). Even though plant AChEs are distinct from the animal AChE family, altered expression of the *AChE* gene in transgenic plants resulted in enhancement or suppression of ACh-hydrolyzing activity. Transgenic rice plants overexpressing rice AChE (*pUbi*::rice *AChE*) had 144.7–0.2-fold higher AChE activities than wild-type plants (Figure 2.6a; Yamamoto et al., 2015). On the other hand, transgenic rice plants silencing *AChE* gene (rice *AChE-$K_D$*) showed 85%–90% decrease in levels of AChE activities, compared with wild-type plants (Figure 2.6b; Yamamoto et al., 2016). Further, transgenic tobacco plants carrying maize AChE (*35S*::maize *AChE*) had 5.6–16.0-fold higher AChE activities than wild-type plants (Figure 2.6c; Yamamoto et al., 2011). Furthermore, transgenic tobacco plants overexpressing Salicornia AChE (*35S*::Salicornia *AChE*) showed 18.3–19.4-fold higher AChE activities compared with wild-type plants (Figure 2.6d; Yamamoto et al., 2009). These results strongly suggest that putative maize, rice, and Salicornia *AChE* genes produce the protein that possesses ACh-hydrolyzing activity.

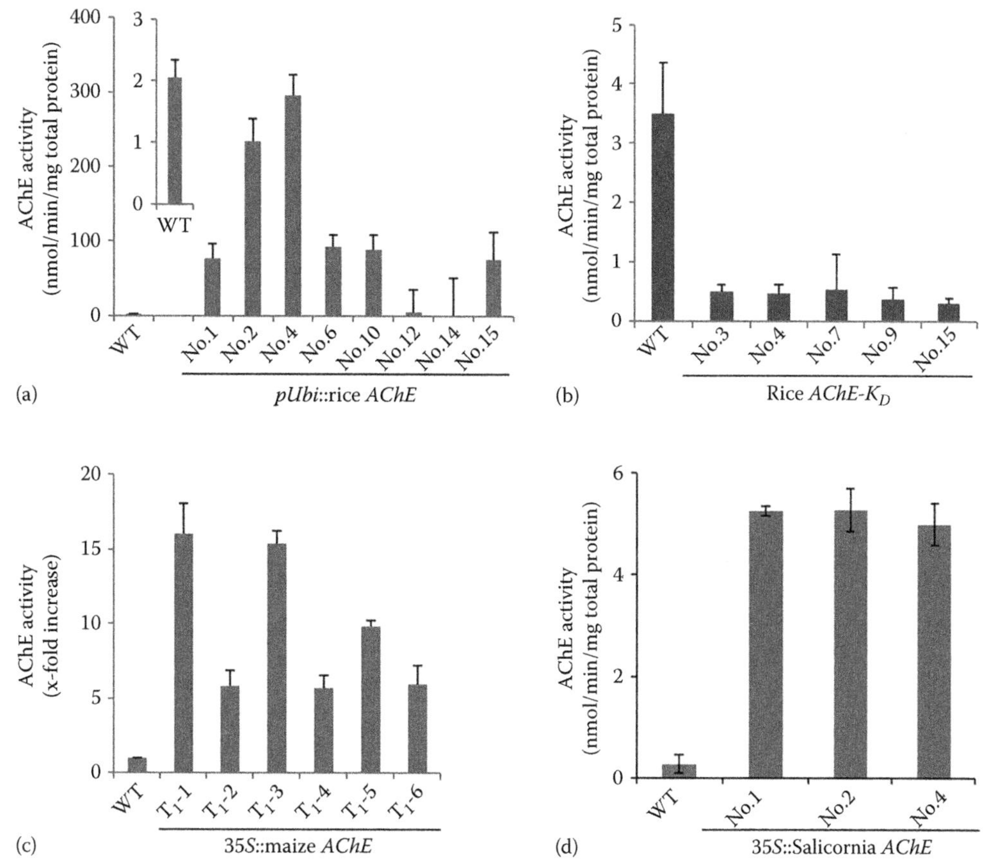


**FIGURE 2.6** AChE activity in transgenic plants. AChE activity measured by the DTNB method with acetylthiocholine (ASCh) as substrate according to our previous reports. (From Sagane, Y. et al., *Plant Physiol.*, 138, 1359–1371, 2005; Yamamoto, K. et al., *Planta*, 227, 809–822, 2008.) WT indicates AChE activity in wild-type plants. Each textured bar represents the average of three replicates; thin vertical bars indicate standard errors. (a and b) show AChE activities in transgenic rice plants overexpressing and silencing rice AChE, respectively. (From Yamamoto, K. et al., *Biochem. Biophys. Res. Commun.*, 465, 488–493, 2015; Yamamoto, K. et al., *Plant Signal. Behav.*, 11, e1163464, 2016.) (c) shows AChE activity in transgenic tobacco plants overexpressing maize AChE. (From Yamamoto, K. et al., *J. Plant Physiol.*, 168, 1987–1992, 2011.) The activity of each transgenic plant was shown as relative activity compared with that of wild-type plants. (d) shows AChE activity in transgenic tobacco plants overexpressing Salicornia AChE. (From Yamamoto, K. et al., *Plant Signal Behav.*, 4, 361–366, 2009.)

The phylogenetic tree based on the amino acid sequences of siratro AChE illustrates the clear divergence of this enzyme family into monocots, rosid- and asterid-dicots (Figure 2.7; Yamamoto et al., 2008). It is thought that the monocots/dicots divergence occurred approximately 200 million years ago (Wolfe et al., 1989), while the asterids/rosids divergence in the dicot class occurred approximately 100–150 million years ago (Yang et al., 1999), based on chloroplast and mitochondrial DNA sequences, respectively. The finding that the monocotyledonous AChE (maize) and dicotyledonous AChE (siratro) share a similar primary structure indicates that the latest time the plant kingdom could have acquired the common ancestor of the plant *AChE* gene predates the monocots/dicots phylum divergence event.

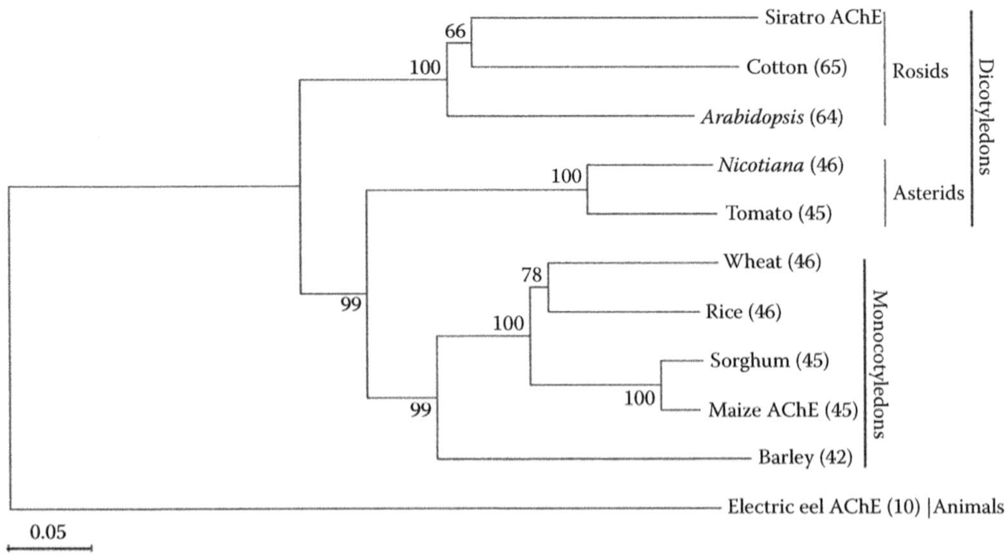


**FIGURE 2.7** Phylogenetic tree derived from amino acid sequences of the siratro *AChE* gene. The AChE homologues were identified by a tBLASTn analysis against TIGR gene indices. The following scale indicates the number of amino acid substitutions. The phylogenetic tree was constructed by the neighbor-joining method. *Numbers at branches* indicate bootstrap values. The amino acid sequence identities between the siratro AChE and each gene product is indicated as percentages in parentheses. The tentative consensus (*TC*) numbers and GenBank accession numbers used in the analysis are as follows: siratro AChE, AB294246; cotton, CGI TC68238; *Arabidopsis*, AtGI TC312464; *Nicotiana*, NbGI TC9050; tomato, LeGI TC175087; wheat, TaGI TC254536; rice, OsGI TC331846; sorghum, SbGI TC95805; maize AChE, AB093208; barley, HvGI TC146608; electric eel AChE, AF030422. (From Yamamoto, K. et al., *Planta*, 227, 809–822, 2008.)

## 2.4 COMPARISON OF ENZYMATIC PROPERTIES OF PURIFIED PLANT AND ANIMAL ACETYLCHOLINESTERASES

Cholinesterases (ChEs) are a family of enzymes that fall broadly into two types: AChE (E.C.3.1.1.7) and butyrylcholinesterase (BChE) (E.C.3.1.1.8). They are distinguished primarily by their substrate specificity: animal AChE hydrolyzes ACh faster than cholinesters having bulkier acyl chains; thus, it is much less active against the synthetic substrate, butyrylcholine (BCh). In contrast, animal BChE displays similar activity toward the two substrates; thus, animal BChE is less substrate-specific (Chatonnet and Lockridge, 1989).

The purified siratro, maize, and electric eel AChEs displayed hydrolytic activity against acetylthiocholine (ASCh), propionylthiocholine (PpSCh), ACh, and propionylcholine (PpCh) (Figure 2.8). The activities of siratro and maize AChEs increased with increasing ASCh, PpSCh, ACh, and PpCh concentrations. Both plant and animal AChEs exhibited extremely low hydrolytic activity against butyrylthiocholine (BSCh) and BCh (Figure 2.8). The maize and siratro enzymes were characterized as an AChE (E.C.3.1.1.7). The inhibition of animal AChE by excess substrate is also one of the important features that distinguish it from animal BChE, which is not inhibited by excess substrate concentration (Tougu, 2001). Substrate inhibition for both plant and animal AChEs was examined at different concentrations of ASCh and PpSCh (10 μM–1 M) (Figure 2.9). The electric eel AChE exhibited substrate inhibition at concentrations higher than 10 mM. In contrast, the siratro and maize AChEs were not inhibited by excess substrate concentrations. The siratro and maize AChEs showed both enzymatic properties of AChE in animals, which is much less active against BSCh (or BCh), and BChE in animals, which is not inhibited by excess substrate concentration. In addition, there was no significant difference between the enzymatic properties of siratro and maize AChEs.

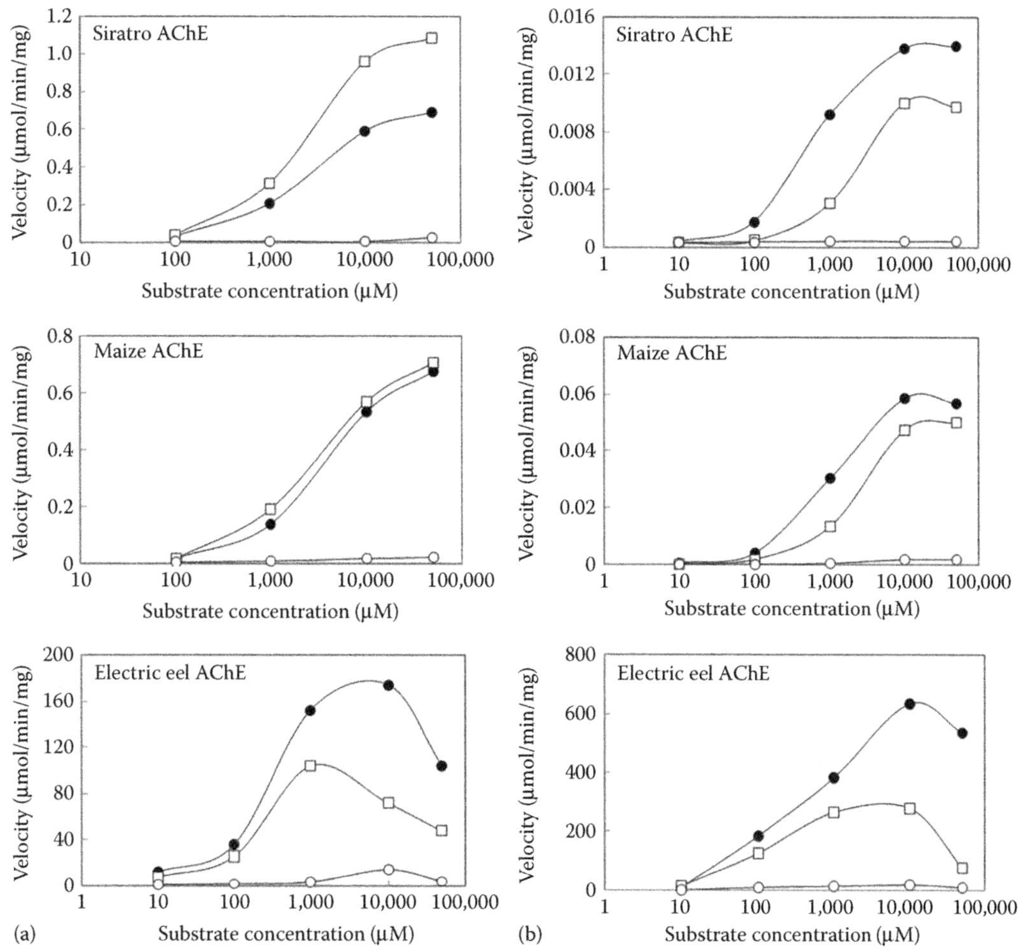


**FIGURE 2.8** Comparison of substrate selectivity of siratro, maize, and electric eel AChEs against thiocholinesters and cholinesters. (a) ASCh (closed circles), PpSCh (open squares) and BSCh (open circles) were used as thiocholinester substrates. The AChE activities were measured by the DTNB method as described in previous our reports. (From Sagane, Y. et al., *Plant Physiol.*, 138, 1359–1371, 2005; Yamamoto, K. et al., *Planta*, 227, 809–822, 2008.) The horizontal (logarithmic) axis indicates substrate concentrations. The vertical axis indicates velocity. (b) ACh (closed circles), PpCh (open squares), and BCh (open circles) were used as cholinester substrates. The AChE activities were measured by an Amplex Red Acetylcholine/Acetylcholinesterase Assay Kit as described in a previous our report. (From Yamamoto, K. et al., *Planta*, 227, 809–822, 2008.) The horizontal (logarithmic) axis indicates substrate concentrations. The vertical axis indicates velocity.

These findings correspond to AChEs in *P. sativum* (Muralidharan et al., 2005), *C. arietinum* (Gupta and Maheshwari, 1980) and *Avena sativa* (Roshchina, 2001). In most plants, however, AChE activities were inhibited by excess substrate concentration (Roshchina, 2001). Since the *intermediate* enzymes, which have enzymatic properties of both animal AChE and animal BChE, having been found in invertebrates, the existence of *intermediate* enzymes may reflect the phylogenetic evolution of cholinesterases (Roshchina, 2001). Thus, this evidence indicates the possible existence of species-specific differences in enzymatic properties of plant AChEs.

Table 2.3 shows kinetic parameters of siratro, maize, and electric eel AChEs against thiocholinesters and cholinesters. The siratro and maize AChEs exhibited slightly higher $K_m$ values against ASCh compared with the electric eel AChE and other plant AChEs such as *P. sativum* (Muralidharan et al., 2005) and *A. sativa* (Roshchina, 2001). The $K_m$ values against ACh in two plant

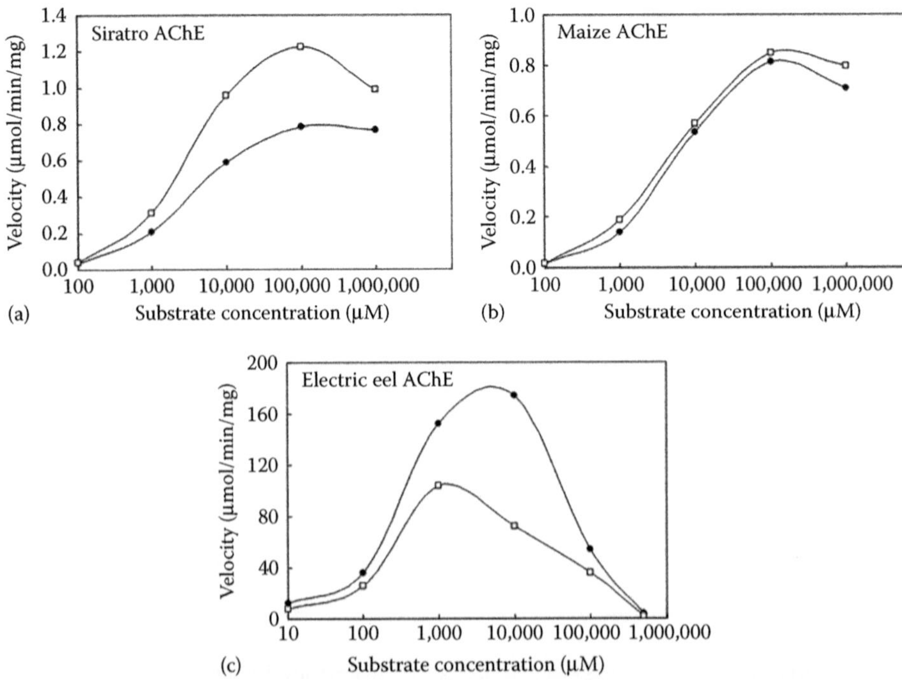


**FIGURE 2.9** Substrate concentration/velocity curves of (a) siratro, (b) maize, and (c) electric eel AChEs. The effect of substrate concentration was investigated using ASCh (closed circles), PpSCh (open squares). The horizontal (logarithmic) axis indicates substrate concentrations. The vertical axis indicates velocity. (From Yamamoto, K. et al., *Planta*, 227, 809–822, 2008.)

**TABLE 2.3**
**Kinetic Parameters of Siratro, Maize, and Electric eel AChEs against Thiocholinesters and Cholinesters**

| | Substrates | $K_m$ (mM) | $V_{max}$ (μmol $min^{-1}$ $mg^{-1}$) | $k_{cat}$ ($s^{-1}$) | $k_{cat}/K_m$ ($mM^{-1}$ $s^{-1}$) |
|---|---|---|---|---|---|
| | | | **Thiocholinesters** | | |
| Siratro | ASCh | 1.9 | 0.4 | 0.2 | 0.1 |
| | PpSCh | 2.0 | 0.7 | 0.5 | 0.2 |
| Maize | ASCh | 4.7 | 1.7 | 1.2 | 0.3 |
| | PpSCh | 3.1 | 1.7 | 1.2 | 0.4 |
| Electric eel | ASCh | 0.2 | 312.5 | 364.6 | 2278.6 |
| | PpSCh | 1.0 | 319.0 | 372.2 | 372.2 |
| | | | **Cholinesters** | | |
| Siratro | Ach | 0.08 | 0.04 | 0.03 | 0.3 |
| | PpCh | 0.02 | 0.004 | 0.003 | 0.1 |
| Maize | Ach | 0.15 | 0.12 | 0.09 | 0.6 |
| | PpCh | 0.03 | 0.01 | 0.007 | 0.2 |
| Electric eel | Ach | 0.09 | 555.6 | 648.2 | 7202.2 |
| | PpCh | 0.05 | 163.9 | 191.3 | 3825.0 |

*Source:* Yamamoto, K. et al., *Planta*, 227, 809–822, 2008.

AChEs were of the same order as that of electric eel AChE. Thus, the affinity of the active centers of two plant AChEs for ACh is comparable to electric eel AChE. However, siratro and maize AChEs exhibited low $k_{cat}$ values (substrate turnover numbers), compared with electric eel AChE. This finding implies that both plant AChEs may hydrolyze not only cholinesters but also other esters compound within a plant cell.

Neostigmine bromide is known as a specific inhibitor of the animal AChE, as it binds to the catalytic site of the enzyme (Standaert, 1990), and it also inhibits plant AChE (Momonoki, 1992; Ballal et al., 1993). Neostigmine bromide inhibited hydrolysis of ASCh with apparent IC 50 values of $6 \times 10^{-4}$ M for siratro AChE and $3 \times 10^{-5}$ M for maize AChE, respectively (Yamamoto et al., 2008). The siratro AChE exhibited somewhat elevated IC 50 value, compared with that for maize AChE. However, IC 50 values of siratro and maize AChEs to neostigmine bromide were approximately 3 and 4 orders higher than that of electric eel AChE (IC 50 = $2 \times 10^{-8}$ M), respectively. The sensitivities of siratro and maize AChEs to neostigmine bromide were approximately 1–3 orders lower than those of other plant AChEs, such as *P. sativum* (Muralidharan et al., 2005), *P. aureus* (Riov and Jaffe, 1973) and *A. sativa* (Roshchina, 2001). In comparison with electric eel AChE, both siratro and maize AChEs exhibited a low sensitivity to neostigmine bromide. Likewise, the sensitivities of most plant AChEs to neostigmine bromide were approximately 1–3 orders lower than that of mammalian erythrocytes ($1.6 \times 10^{-8}$–$5 \times 10^{-7}$ M). Meanwhile, the enzyme derived from latex produced by laticifers in *Synadenium grantii* exhibited an extremely high sensitivity to neostigmine bromide, similar to animal tissues (Roshchina, 2001). This difference in the sensitivities to neostigmine bromide among plant AChEs may reflect the phylogenetic evolution of AChEs.

## 2.5 LOCALIZATION OF ACETYLCHOLINESTERASE IN PLANTS

As the first demonstration on plants by electron microscopic histochemistry, cholinesterase was detected in plasmalemma, cell wall, and partly the cytoplasm of root cells of *P. aureus* (Fluck and Jaffe, 1974). As summarized in the monograph of Roshchina (2001), the enzyme has been found in the cell wall, plasmalemma, nucleus, chloroplasts, and the exine of pollen. Most of these data show the localization of ACh-hydrolyzing activity in plants. Our recent experiments have directly demonstrated the localization of AChE protein using the anti-maize AChE polyclonal antibody raised in rabbit and the GFP-tagged maize AChE protein.

The subcellular localization of AChE was observed by immunofluorescence in paraffin-embedded leaf and stem tissues of transgenic rice plants carrying maize AChE. The maize AChE protein was detected at extracellular spaces in the leaf and stem of the plants (Figure 2.10; Yamamoto and Momonoki, 2008). In an alternative way to confirm the earlier result, the maize AChE fused with GFP at its C-terminus (maize AChE–GFP) was overexpressed in rice plants. While the control protein, GFP–tagged GUS (GUS–GFP), was mainly localized in the cytoplasm of transgenic rice leaves, maize AChE–GFP was detected in the leaf extracellular spaces (Figure 2.11; Yamamoto et al., 2011). These results suggest that maize AChE functions in the extracellular spaces of transgenic rice plants, similar to some isoforms of animal AChE (Soreq and Seidman, 2001; Rotundo, 2003). Most of the AChE activity in the root was associated with cell wall materials (Fluck and Jaffe, 1974). The computer-assisted cellular sorting prediction program TargetP presumed that our purified maize AChE (Sagane et al., 2005) is targeted to the secretory pathway via the endoplasmic reticulum. Furthermore, the SOSUI program (http://sosui.proteome.bio.tuat.ac.jp/sosuiframe0.html), which discriminates between membrane and soluble proteins, showed that the maize AChE does not contain any likely transmembrane helical regions, which are features of proteins that associate with the lipid bilayers of the cell membrane. These findings suggested that the maize AChE would be localized at the cell wall.

The tissue localization of AChE in coleoptile nodes and mesocotyls of maize seedlings was analyzed using anti-maize AChE antibody. AChE localized in vascular bundles including endodermis and epidermis in coleoptile nodes and mesocotyls (Figure 2.12; Yamamoto and Momonoki, 2012). Additionally, AChE mainly localized at extracellular space in these

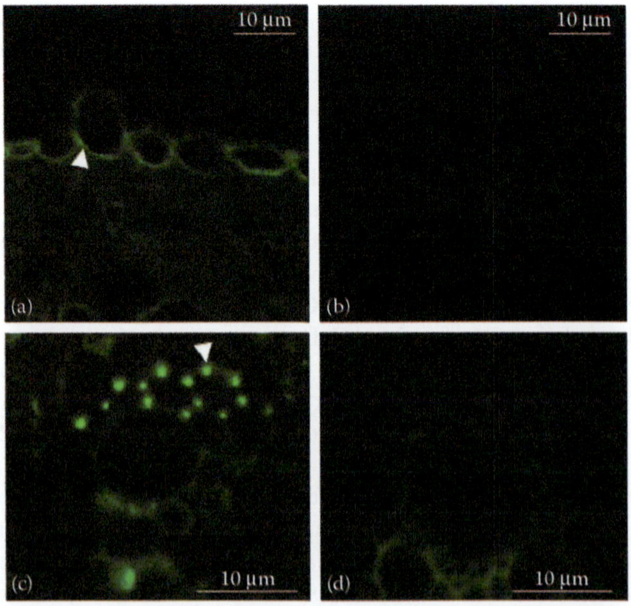


**FIGURE 2.10** Subcellular localization of maize AChE in leaf and stem of transgenic rice. (a) Leaf cross-section of transgenic rice; (b) leaf cross-section of control; (c) stem cross-section of transgenic rice; and (d) stem cross-section of control. Each section was probed with maize AChE antibody and then visualized with Alexa Fluor 488-conjugated secondary antibody. Control indicates rice plants transfected with p2K-1$^+$ vector only. Arrowheads indicate localization of maize AChE. (From Yamamoto, K. and Momonoki, Y. S., *Plant Signal. Behav.*, 3, 576–567, 2008.)

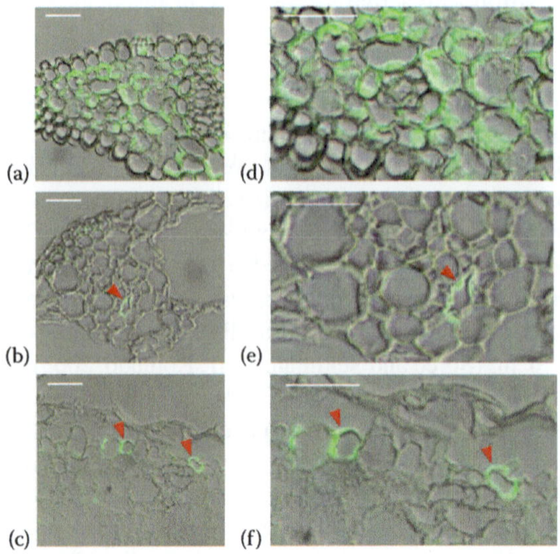


**FIGURE 2.11** Subcellular localization of GFP-tagged maize AChE in leaves of transgenic rice plants. (a) GUS–GFP detected with anti-GFP (control), (b) maize AChE–GFP detected with anti-GFP, and (c) maize AChE–GFP detected with anti-maize AChE. (d, e, and f) show magnified images in (a, b, and c), respectively. Each frame represents merged fluorescent and bright field images. Each arrowhead shows localization of maize AChE. Scale bar shows 10 μm. (From Yamamoto, K. et al., *J. Plant Physiol.*, 168, 1987–1992, 2011.)

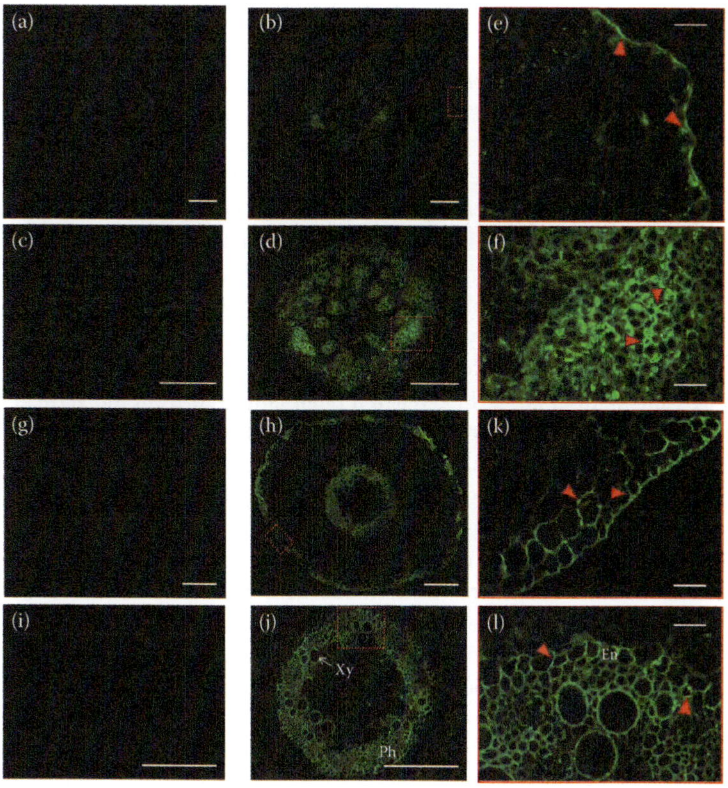


**FIGURE 2.12** Tissue-localization of maize AChE in the coleoptile node and the mesocotyl. Immunofluorescence images of cross sections (6 μm) of the paraffin-embedded coleoptile node (a–f) and mesocotyl (g–l); a, c, g, and i, secondary antibody alone (negative control); b, d, e, f, h, j, k, and l, anti-maize AChE staining; e, f, k, and l, magnified images of the regions highlighted by dashed line boxes in b, d, h, and j, respectively. Arrowheads indicate localization of maize AChE. En, endodermis; Ph, phloem; Xy, xylem. Bars = 100 μm for a, b, c, d, g, h, i, and j; 10 μm for e, f, k, and l. (From Yamamoto, K. and Momonoki, Y. S., *Plant Signal. Behav.*, 7, 1–5, 2012.)

tissues (see the arrows in Figure 2.12). Interestingly, AChE was also observed in epidermis of coleoptile nodes and mesocotyls. Momonoki et al. (1996b) reported that AChE activity was localized in the epidermis of Salicornia roots. Similar evidence has been obtained in the case of human AChE (Schallreuter et al., 2004). As additional supportive evidence, the rice AChE expressed in the epidermis of leaves of transgenic rice plants in the promoter-reporter gene studies (Yamamoto et al., unpublished results). Taken together, the plant AChE would function at the cell wall in the epidermis and around vascular bundles including the endodermis in coleoptile nodes and mesocotyls.

## 2.6 POSSIBLE FUNCTION OF ACETYLCHOLINESTERASE IN PLANTS UNDER GRAVISTIMULATION

Both ACh and ACh-hydrolyzing activity have been widely recognized in plants (Roshchina, 2001; Horiuchi et al., 2003), as has choline acetyltransferase, which is involved in ACh synthesis by many plant species (Tretyn and Kendrick, 1990; Kawashima et al., 2007). Murata et al. (2015) recently reported a strong evidence that ACh exists in *A. thaliana* tissues based on the electrospray ionization-orbitrap Fourier transform mass spectrometry (ESI-orbitrap FT-MS) analysis. The level of ACh is highest in seed, followed by root and cotyledon. Moreover, exogenously applied ACh inhibited the elongation of *Arabidopsis* root hairs. They concluded that ACh exists primarily in seed and root in *Arabidopsis* seedlings and plays a pivotal role during the initial stages of seedling development by controlling root hair elongation in *Arabidopsis* (Murata et al., 2015). Kisnierienė et al. (2009) also reported that the membrane permeability of *Nitellopsis obtusa* cells was significantly increased by high concentrations of ACh. Biochemical and pharmacological analyses have revealed *ACh-binding sites* in plant subcellular components such as plasma membranes and chloroplasts (Roshchina, 2001) and tonoplasts (Gong and Bisson, 2002). In addition, 3-2′-aminobenzhydryloxy tropane, a high affinity ligand of the muscarinic AChR, was observed to bind plasma membranes of *Vicia faba* L. and *P. sativum* L. guard cells (Meng et al., 2001). Thus, it is conceivable that in plants the ACh-mediated system, composed of ACh, AChR, and AChE, plays a significant role in signal transduction, as it does in animals.

For the last 20 years, Momonoki and co-workers have focused on the ACh-mediated system as a potential gating regulator that causes asymmetric distribution of hormones and substrates in response to gravity, heat, and changes in ACh content, AChE activity, and $Ca^{2+}$ concentration (Momonoki, 1992; Momonoki, 1997; Yamamoto et al., 2011). When a plant is moved from the vertical orientation to a horizontal position, the stem shows curvature as it returns to a vertical position by response acting against gravistimulus. Interestingly, maize seedlings that were treated first with neostigmine bromide (AChE inhibitor) did not respond to the gravistimulus (Momonoki, 1997). Furthermore, after the stimulus, the AChE activity was distributed asymmetrically at the interface between the stele and the cortex of the gravistimulated maize seedlings (Momonoki, 1997). The gravitropic bending occurs because of an asymmetric distribution of the growth hormone auxin (indole-3-acetic acid, IAA) (Ross and Wolbang, 2008). Momonoki et al. (1998, 2000) previously observed that gravistimulation leads to an asymmetric distribution of both IAA-inositol synthase and the AChE activities in maize seedlings and in rice seedlings. Its asymmetric distribution of both enzyme activities in maize and rice shoots was also inhibited by the treatment with neostigmine bromide. The role of AChE in animal nervous systems has been clearly demonstrated by the effects of its inhibitors. In addition, neostigmine bromide is an AChE inhibitor having no inhibitory effect on IAA metabolism in plants (Ballal et al., 1993). These findings suggest that AChE would play an important role in the gravitropic response of plants.

To directly confirm previous reports, the functionality of plant AChE under gravitropic stimulus was recently proved by overexpressing and silencing *AChE* gene in rice plants (Yamamoto et al., 2015, 2016). First, the rice *AChE* gene (accession no. AK073754) was fused to the maize *ubiquitin1* promoter (*pUbi*) and transformed into rice; eight independent hygromycin-resistant plants were selected by the assay of AChE activity and two T2 lines, *pUbi::AChE* No.4 with the highest expression level of rice AChE and *pUbi::AChE* No.10 with low expression level of rice AChE were chosen and studied further. As shown in Figure 2.6a, ACh-hydrolyzing activities in *pUbi::AChE* No.4 and No.10 carrying rice AChE were 144.7-fold and 37.4-fold higher, respectively, than that in wild-type plants. Furthermore, transgenic rice plants silencing rice *AChE* gene were generated by RNAi techniques (Yamamoto et al., 2016). RNA silencing construct carrying the inverted repeat of rice *AChE* fragments was made using the pANDA vector (Miki and Shimamoto, 2004; Miki et al., 2005). The RNAi plasmid was constructed and designated $AChE\text{-}K_D$, and transformed into rice; five independent transgenic lines were selected by AChE activity assay (Figure 2.6b). Two T2 lines ($AChE\text{-}K_D$ No.3 and No.15) were chosen and studied further. $AChE\text{-}K_D$ No.3 and No.15 silencing *AChE* gene showed an 85.5% and 90% decrease in levels of AChE activities respectively, compared with wild-type plants (Figure 2.6b). RNAi expression also caused reduced levels of its gene expression compared with wild-type plants (Yamamoto et al., 2016).

Next, an impact on phenotypic characteristics of transgenic rice plants under gravistimulation was assessed. The gravitropic response, relative to wild-type, was enhanced when AChE was overexpressed (Figure 2.13). Gravitropism was measured in 4-day-old dark-grown seedlings because endogenous rice AChE mainly expresses during 2–10 days after sowing (Yamamoto et al., 2015). After 180 minutes of gravistimulation, wild-type seedlings bent upward at approximately 52.4°, whereas overexpression of AChE caused curvature up to approximately 94.7° in *pUbi::AChE* No.4 and approximately 93.2° in *pUbi::AChE* No.10; this indicates a strong gravitropic response. On the other hand, the suppression of AChE caused greatly reduced gravitropic responses; the curvature of

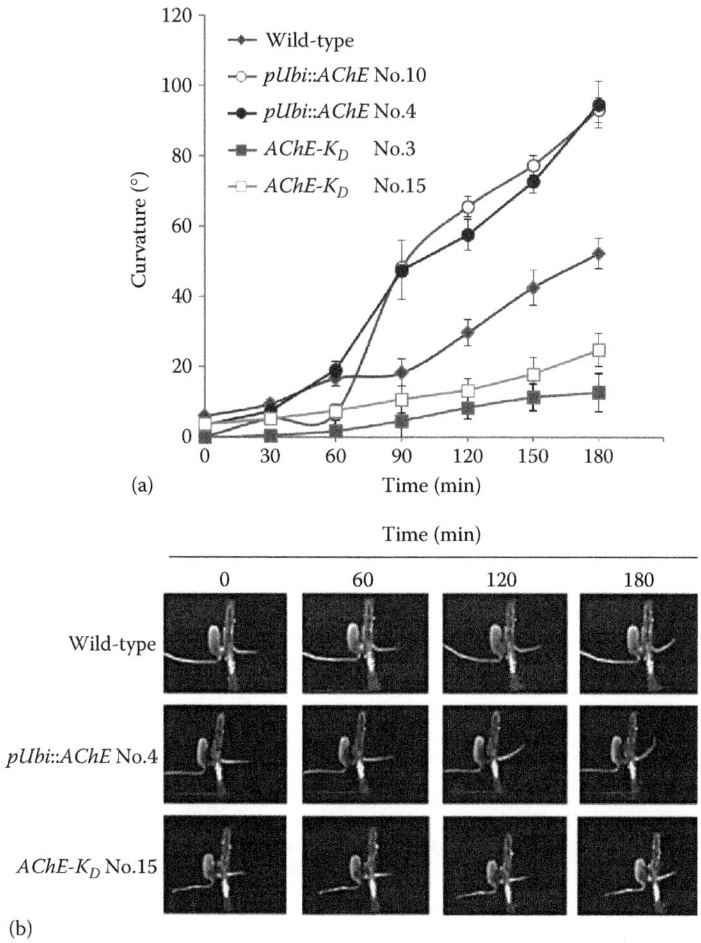


**FIGURE 2.13** Gravitropic responses of transgenic rice plants after stimulation or inhibition of AChE expression. Overexpression lines: *pUbi::AChE* No.4 and No.10. RNAi-silenced lines: *AChE-*$K_D$ No.3 and No.15. WT: wild-type. (a) Quantification of gravitropic responses in wild-type and transgenic rice seedlings (T2 generation) as described in previous our reports. (From Yamamoto, K. et al., *Biochem. Biophys. Res. Commun.*, 465, 488–493, 2015; Yamamoto, K. et al., *Plant Signal. Behav.*, 11, e1163464, 2016.) Each data point represents measurement of 10 plantlets. Thin vertical bars represent standard errors. (b) Typical phenotypes of wild-type and transgenic rice plantlets (T2 generation) under gravistimulation. Digital images were taken at designated time points to the upper of each panel.

gene-silencing plantlets was only approximately 12.8° in *AChE-*$K_D$ No.3 and approximately 24.9° in *AChE-*$K_D$ No.15 after 180 minutes of gravistimulation (Figure 2.13). Lawson and colleagues (1978) reported that higher concentration of ACh had an inhibitory effect on elongation. In addition, Di Sansebastiano et al. (2014) recently reported that an ACh concentration beyond 50 mM appeared to negatively interfere with the auxin-driven elongation process. Murata et al. (2015) had also produced similar evidence suggesting that exogenously applied ACh can inhibit seminal root elongation and root hair formation. It is conceivable, therefore, that an ACh concentration in AChE-silencing plants might be higher than that in wild-type plants, and its higher level of ACh might negatively regulate shoot gravitropism in AChE-silencing plants.

Taken together, all these results strongly suggest that the rice AChE plays an important role in the gravitropic response and is a positive regulator of shoot gravitropism in the seedlings.

## 2.7 POSSIBLE FUNCTION OF ACETYLCHOLINESTERASE IN PLANTS UNDER HEAT STRESS

Momonoki and Momonoki (1993) have reported that native tropical zone plants showed high ACh-hydrolyzing activity in their nodes, stems, pulvini, and roots after heat treatment. In siratro plants, a forage crop in the tropical and subtropical zones, its ACh-hydrolyzing activity and ACh content respond significantly to heat stress (Momonoki and Momonoki, 1992). Under heat stress, ACh-hydrolyzing activity was increased about four and twofold in the primary and secondary pulvini of siratro leaves, respectively. The highest ACh-hydrolyzing activity in both of the pulvini was three minutes after heat stress, at which time the ACh content was significantly increased in primary pulvini and leaf drooping was the deepest. In addition, Momonoki et al. (1996a) have investigated the effect of heat stress on tissue distributions of both AChE activity and $Ca^{2+}$ in the coleoptile node of dark-grown maize seedlings. At the animal neuromuscular junction, an electrical impulse causes an influx of $Ca^{2+}$ from outside the membrane via voltage-gated channels, and then the $Ca^{2+}$ acts as a trigger for ACh release into the synaptic cleft (Figure 2.1). The fluorescent-labeled $Ca^{2+}$ in the coleoptile node of maize seedlings was localized in cortex cells around the vascular system, epidermis, and adhering peripheral cortical cells. Following heat stress, $Ca^{2+}$ was found more in the cortical cells and whole endodermal cells between the cortex and stele. The AChE activity in the coleoptile node of heat-stressed maize seedlings was also enhanced in endodermal cells near the vascular system. The appearance of AChE and $Ca^{2+}$ in endodermal cells after heat stress seems to be related to function of ACh-mediated system controlling ion channels (Momonoki et al., 1996a).

In our recent report (Yamamoto et al., 2011), AChE activity was detected in coleoptiles, coleoptile nodes, mesocotyls, seeds, and roots of five-day-old etiolated maize seedlings (Figure 2.14a). AChE activities in coleoptile nodes and seeds were 1.5–4.5-fold higher than those in other organs

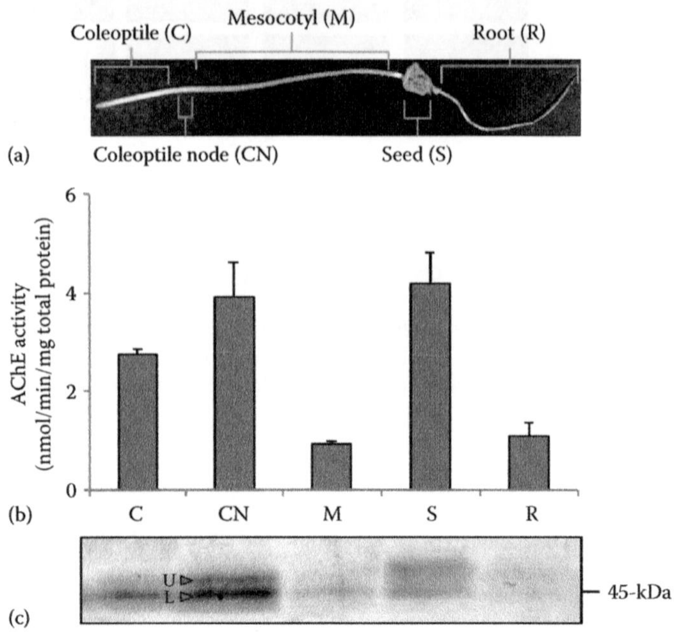


**FIGURE 2.14** AChE activity and protein levels in maize seedlings. (a) Organs (indicated by brackets) of five-day-old etiolated maize seedlings sampled for AChE. (b) AChE activity (determined by the DTNB method) in various organs of maize seedlings. Each textured bar represents the average of four replicates. Thin bars represent standard errors. (c) Gel blot of proteins (70 μg total protein/lane) in different organs of maize seedlings. Arrowheads indicate AChE, detected using anti-maize AChE antibody: U, upper-form; L, lower-form. (From Yamamoto, K. et al., *J. Plant Physiol.*, 168, 1987–1992, 2011.)

(Figure 2.14b). Gel blot analysis showed AChE protein mainly in coleoptile nodes and seeds, consistent with the distribution of AChE activity (Figure 2.14c). Both the highest activity and protein level were observed in the nodes. In addition, AChE proteins separated into two predominant bands, designated the U and L forms (Figure 2.14c). The effect of heat treatment on AChE activity was measured in coleoptile nodes of maize seedlings, since they have the highest activity and protein level. Heat treatment (45°C) caused a statistically significant increase of AChE activity (about 15%) after 10, 30, and 60 min exposures, but after 180 min, the activity returned to the pre-stress level (Figure 2.15a). Since heat treatment did not change the total protein content of coleoptile nodes (Figure 2.15b), the increase in AChE activity caused by heat might be the result of either enhanced transcript levels or posttranslational modifications. Cytochemical staining also showed that *in situ* AChE activity in coleoptile nodes of etiolated maize seedlings increased after 10, 30, and 60 min of heat stress (Figure 2.16a and b). The *in situ* AChE activity was highest after 30 min of heat stress and was concentrated near vascular bundles, the shoot apical meristem (SAM), and the base of the leaf primordium (Figure 2.16b). Horiuchi et al. (2003) reported that the upper portion of the

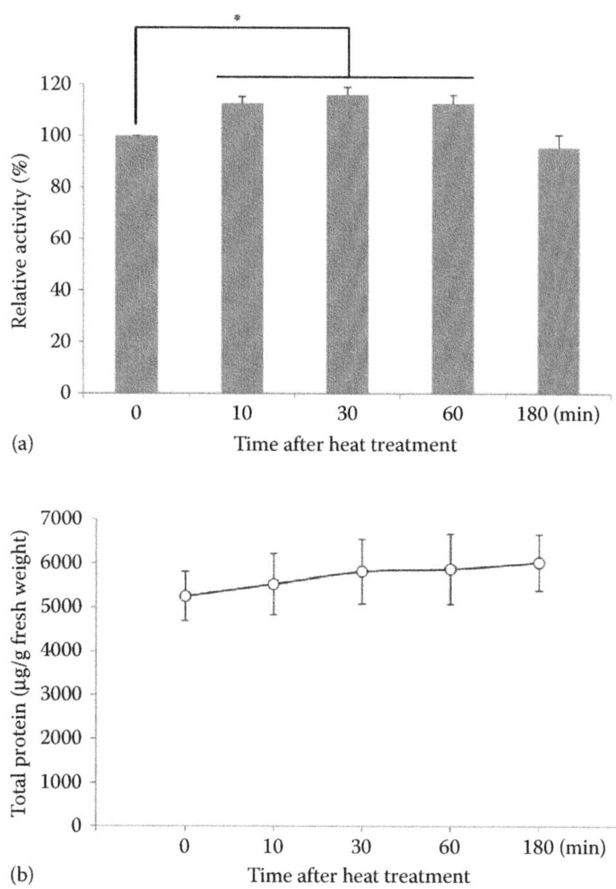


**FIGURE 2.15** Enhancement of *in vitro* AChE activity in maize coleoptile nodes by heat treatment. (a) AChE activity (determined by the DTNB method) was expressed as relative activity/mg total protein relative to the control (before heat treatment at 0 min), which was taken as 100%. Each textured bar represents the average of three replicates. Thin bars represent standard errors. The results were analyzed by Student's *t*-test; * indicates *p* values <0.01, considered significantly different from the 0 time control. (b) Total protein content (determined by the Bradford method) of coleoptile nodes over the time course of heat treatment. (From Yamamoto, K. et al., *J. Plant Physiol.*, 168, 1987–1992, 2011.)

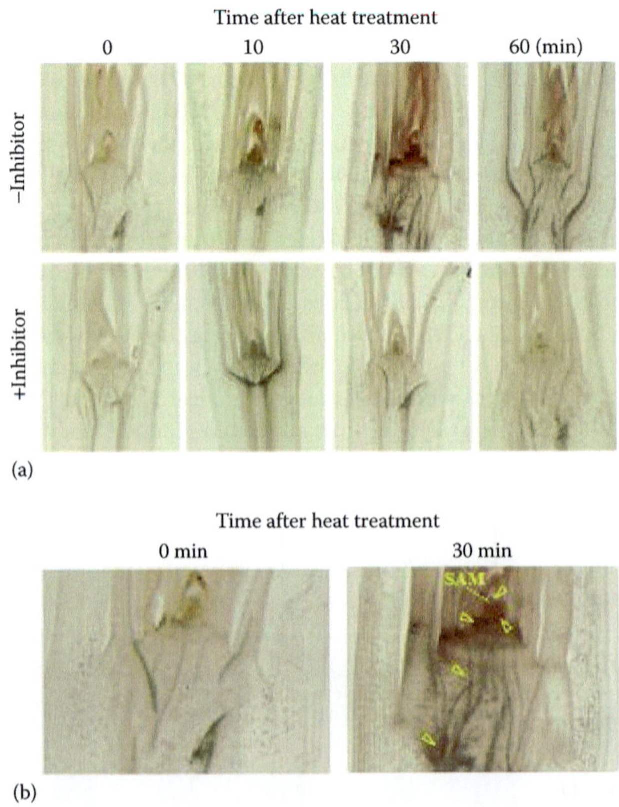


**FIGURE 2.16** Enhancement of *in situ* AChE activity in coleoptile nodes by heat treatment. (a) *in situ* AChE activity in untreated and heat-treated sections. The brown color in the micrographs indicates enzymatic activity of AChE, which detected according to the method of Karnovsky and Roots (1964). For the control (+ inhibitor) 100 μM neostigmine bromide was added prior to and during incubation with substrate. (b) Magnified images of untreated and heat-treated sections. Arrowheads indicate locations of enhanced AChE activity. SAM, shoot apical meristem. (From Yamamoto, K. et al., *J. Plant Physiol.*, 168, 1987–1992, 2011.)

bamboo (*Phyllostachys bambusoides* S. et Z.) shoot had 16- and 84-fold higher ACh content than guinea-pig ileum longitudinal muscle and rat brainstem, respectively. The very high ACh content in the upper portions of bamboo species, which grow very rapidly, is consistent with the capacity for ACh synthesis in those regions (Horiuchi et al., 2003). Further, Wiśniewska and Tretyn (2003) suggested that type I phytochrome is involved in regulation of AChE activity and that the type II photoreceptor influences the rate of *de novo* AChE synthesis, based on a genetic study of tomato mutants. Furthermore, Bamel et al. (2007) reported that ACh induced rooting and promoted secondary root formation in leaf explants of tomato. These reports indicate that ACh and AChE might be involved in differentiation, morphogenesis, and/or photomorphogenesis in plants. Thus, enhanced maize AChE activity around SAMs and leaf primordia during heat stress might relate to differentiation and/or morphogenesis.

To determine whether the effect of heat on maize AChE activity in coleoptile nodes was due to *de novo* protein synthesis, AChE mRNA and protein levels were analyzed by semi-quantitative RT-PCR and protein gel blots. Results (Figure 2.17a and b) showed that transcript and protein levels did not change compared with controls at all times tested, implying that increased AChE activity is not associated with AChE protein expression. However, the U form of AChE (Figure 2.17b) gradually disappeared with the passage of time. In addition, total protein content of maize seedlings did not change after heat treatment (Figure 2.15b). These results suggest that the U form would undergo

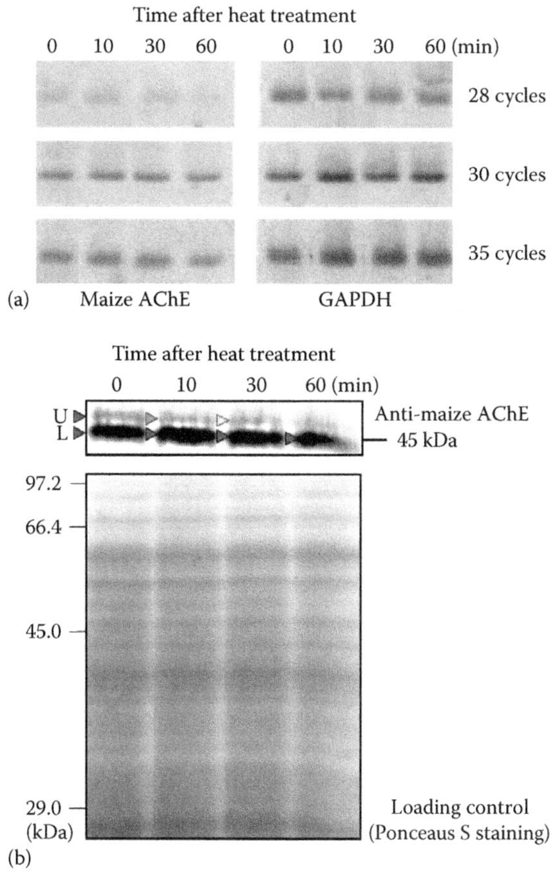


**FIGURE 2.17** Transcript and protein levels of maize AChE in coleoptile nodes after heat treatment. (a) *AChE* mRNA expression analyzed by semi-quantitative RT-PCR. Total RNA was extracted from coleoptile nodes at 0, 10, 30, and 60 min after the start of heat treatment. GAPDH amplification shows equal loading of RNA and RT-PCR efficiency. (b) Gel blot of total protein (70 μg/lane) extracted from coleoptile nodes at 0, 10, 30, and 60 min after the start of heat treatment. Upper panel probed with anti-maize AChE. Lower panel shows Ponceau S staining to confirm equal loading of samples. Arrowheads indicate U, upper- and L, lower-forms of maize AChE. (From Yamamoto, K. et al., *J. Plant Physiol.*, 168, 1987–1992, 2011.)

posttranslational modifications, which may consist of amino acid deletions from the C terminus and/or different degrees of glycosylation. These possible modifications might lead to conversion from the U- to the L-form, thereby enhancing AChE activity (Yamamoto et al., 2011).

To examine whether increased AChE activity after heat stress has a functional role in heat tolerance, maize *AChE* cDNA was overexpressed in tobacco plants. The maize *AChE* gene was introduced behind the cauliflower mosaic virus 35S promoter in the pBI121 vector and transformed into tobacco plants; six independent kanamycin-resistant transgenic lines ($T_1$ generation) were chosen based on AChE activity. The transgenic tobacco plants carrying maize AChE had 5.6–16.0-fold higher AChE activities than wild-type plants (Figure 2.6c). Three $T_1$ lines ($T_1$-1, -3, and -5) with high AChE activities were chosen for a heat tolerance assay. Phenotypically, the transgenic plants did not differ from wild-type plants under normal growth conditions at four weeks after seeding (Figure 2.18a). All wild-type plants were dead, whereas most (61%–80%) transgenic plants (included small plants tending to die) survived, within 7 days after heat (48°C for 45 min) treatment (Figure 2.18b and c). Therefore, overexpression of maize AChE enhances heat tolerance in transgenic tobacco plants, which strongly suggests that maize AChE plays a positive, important role in

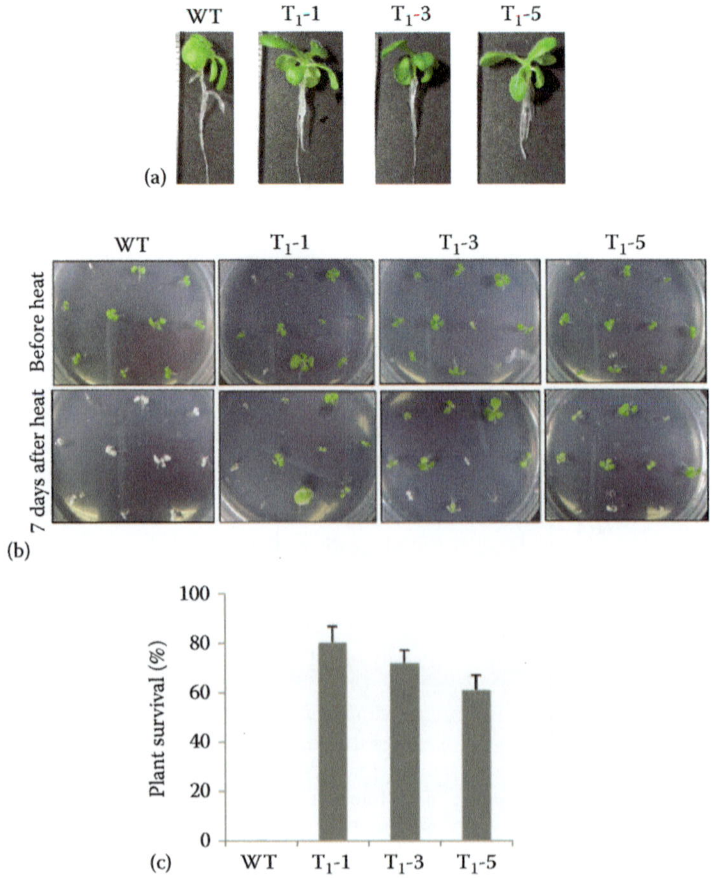


**FIGURE 2.18** Heat tolerance of transgenic tobacco plants expressing the maize *AChE* gene. (a) Vegetative growth of wild-type (WT) and transgenic plants $T_1$-1, -3 and -5 on half strength MS medium for 4 weeks. (b) Phenotypes of wild-type (WT) and transgenic tobacco plants $T_1$-1, -3, and -5 after heat treatment. Plants were grown on half strength MS medium (25°C, 16 h light/day) for 21 days, exposed to heat (48°C, 45 min), then returned to the original growth conditions for 7 days. (c) Survival of transgenic tobacco plants 7 days after the heat treatment. All wild-type (WT) plants were dead. Each textured bar represents the average of three replicates (10 plantlets per replicate). Thin bars represent standard errors. (From Yamamoto, K. et al., *J. Plant Physiol.*, 168, 1987–1992, 2011.)

maize heat tolerance. Since endogenous *AChE* mRNA and protein levels do not change in maize seedlings after heat treatment, enhancement of AChE activity mediated by posttranslational modifications would be a positive regulator of plant heat tolerance (Yamamoto et al., 2011).

Taken together, all these studies suggest that AChE in native tropical plants would be involved in acclimatization to higher temperatures and, therefore, that plant AChE would be useful in engineering plants with enhanced heat tolerance.

## 2.8 SUMMARY

The plant *AChE* genes were identified from maize, rice, siratro, and Salicornia plants. The plant *AChE* gene products have ACh-hydrolyzing activities, although they belong to the GDSL lipase family, but not the alpha/beta-hydrolase fold superfamily that is phylogenetically related to cholinesterases. In the gravistimulated plant, AChE plays would be a positive regulator of shoot gravitropism and would be useful to study plant growth and development under altered gravity conditions and might lead to better understanding of plant growth in space. In the heat-stressed plant, the plant AChE would be a positive regulator of plant heat tolerance and would be useful in engineering plants with enhanced heat tolerance. Further research will be required to clarify the precise signaling pathways involved in AChE in plants. In the future, we will attempt to identify the genes related to ACh-mediated system such as a putative ACh receptor gene in plants.

## REFERENCES


Akoh, C. C., G. C. Lee, Y. C. Liaw, T. H. Huang, and J. F. Shaw. 2004. GDSL family of serine esterases/lipases. *Prog. Lipid Res.* 43:534–552.

Ballal, S., R. Ellias, R. Fluck, R. Jameton, P. Leber, R. Lirio, and D. Salama. 1993. The synthesis and bioassay of indole-3-acetylcholine. *Plant Physiol. Biochem.* 31:249–255.

Bamel, K., S. C. Gupta, and R. Gupta. 2007. Acetylcholine causes rooting in leaf explants of *in vitro* raised tomato (*Lycopersicon esculentum* Miller) seedlings. *Life Sci.* 80:2393–2396.

Bednarska, E. 1992. The localization of nonspecific esterase and cholinesterase activity in germinating pollen and in pollen tube of *Vicia faba* L. The effect of actinomycin D and cycloheximide. *Biol. Plant* 34:229–240.

Bednarska, E. and A. Tretyn. 1989. Ultrastructural localization of acetylcholinesterase activity in the stigma of *Pharbitis nil* L. *Cell Biol. Int. Rep.* 13:275–281.

Brick, D. J., M. J. Brumlik, J. T. Buckley et al. 1995. A new family of lipolytic plant enzymes with members in rice, *Arabidopsis* and maize. *FEBS Lett.* 377:475–480.

Chatonnet, A. and O. Lockridge. 1989. Comparison of butyrylcholinesterase and acetylcholinesterase. *Biochem. J.* 260:625–634.

Dettbarn, W. D. 1962. Acetylcholinesterase activity in *Nitella*. *Nature* 194: 1175–1176.

Di Sansebastiano G. P., S. Fornaciari, F. Barozzi, G. Piro, and L. Arru. 2014. New insights on plant cell elongation: a role for acetylcholine. *Int. J. Mol. Sci.* 15:4565–4582.

Ferraro, E., F. Molinari, and L. Berghella. 2011. Molecular control of neuromuscular junction development. *J. Cachexia Sarcopenia Muscle* 3:13–23.

Fluck, R. A. and M. J. Jaffe. 1974. Cholinesterases from plant tissues VI. Distribution and subcellular localization in *Phaseolus aureus* Roxb. *Plant Physiol.* 54:797–798.

Gnagey, A. L., M. Forte, and T. L. Rosenberry. 1987. Isolation and characterization of acetylcholinesterase from *Drosophila*. *J. Biol. Chem.* 262:13290–13298.

Gong, X. Q. and M. A. Bisson. 2002. Acetylcholine-activated $Cl^-$ channel in the *Chara* tonoplast. *J. Membr. Biol.* 188:107–113.

Gupta, A., S. S. Thakur, P. L. Uniyal, and R. Gupta. 2001. A survey of Bryophytes for presence of cholinesterase activity. *Am. J. Bot.* 88:2133–2135.

Gupta, R. and S. C. Maheshwari. 1980. Preliminary characterization of a cholinesterase from roots of Bengal gram-*Cicer arietinum* L. *Plant Cell Physiol.* 21:1675–1679.

Hill, M. 2003. The neuromuscular junction disorders. *J. Neurol. Neurosurg Psychiatry* 74:ii32–ii37.

Horiuchi, Y., R. Kimura, N. Kato et al. 2003. Evolutional study on acetylcholine expression. *Life Sci.* 72:1745–1756.

Jacobsohn, K. and M. D. Azevedo. 1962. A propos de la cholinesterase vegetale. *Comptes Rendus de la Societe de Biologie* 156:202–204.

Karnovsky, M. J. and L. A. Roots. 1964. "Direct coloring" thiocholine method for cholinesterase. *J. Histochem. Cytochem.* 12:219–221.

Kawashima, K., H. Misawa, Y. Moriwaki et al. 2007. Ubiquitous expression of acetylcholine and its biological functions in life forms without nervous systems. *Life Sci.* 80:2206–2209.

Kisnierienė, V., V. Sakalauskas, A. Pleskačiauskas, V. Yurin, and O. Rukšėnas. 2009. The combined effect of $Cd^{2+}$ and ACh on action potentials of *Nitellopsis obtusa* cells. *Cen. Eur. J. Biol.* 4:343–350.

Lai, K. O. and N. Y. Ip. 2003. Central synapse and neuromuscular junction: Same players, different roles. *Trends Genet.* 19:395–402.

Lawson V. R., R. M. Brady, A. Campbell, B. G. Knox, G. D. Knox, and R. L. Walls. 1978. Interaction of acetylcholine chloride with IAA, GA3 and red light in the growth of excised apical coleoptile segments. *Bull. Torrey Bot. Club* 105:187–191.

Liao, J., B. Nørgaard-Pedersen, and U. Brodbeck. 1993. Subunit association and glycosylation of acetylcholinesterase from monkey brain. *J. Neurochem.* 61:1127–1134.

Madhavan, S., G. Sarath, and P. L. Herman. 1995. Acetylcholinesterase activity in plants. In *Enzymes of the Cholinesterase Family*, (Eds.) Quinn, D. M., Balasubramanian, A. S., and Doctor, B. P., pp. 291–292. New York: Plenum Press.

Malik, S. 2014. Cholinesterases in animals and *Arabidopsis thaliana*. *Int. J. Innovative Res. Develop.* 3:280–281.

Meng, F., X. Liu, S. Zhang, and C. Lou. 2001. Localization of muscarinic acetylcholine receptor in plant guard cells. *Chin. Sci. Bull.* 46:586–587.

Miki, D., R. Itoh, and K. Shimamoto. 2005. RNA silencing of single and multiple members in a gene family of rice. *Plant Physiol.* 138:1903–1913.

Miki, D. and K. Shimamoto. 2004. Simple RNAi vectors for stable and transient suppression of gene function in rice. *Plant Cell Physiol.* 45:490–495.

Momonoki, Y. S. 1992. Occurrence of acetylcholine-hydrolyzing activity at the stele-cortex interface. *Plant Physiol.* 99:130–133.

Momonoki, Y. S. 1997. Asymmetric distribution of acetylcholinesterase in gravistimulated maize seedlings. *Plant Physiol.* 114:47–53.

Momonoki, Y. S., C. Hineno, and K. Noguchi. 1998. Acetylcholine as a signaling system to environmental stimuli in plants. III. Asymmetric solute distribution controlled by ACh in gravistimulated maize seedlings. *Plant Prod. Sci.* 1:83–88.

Momonoki, Y. S., N. Kawai, I. Takamure, and S. Kowalczyk. 2000. Gravitropic response of acetylcholinesterase and IAA-inositol synthase in lazy rice. *Plant Prod. Sci.* 3:17–23.

Momonoki, Y. S. and T. Momonoki. 1992. The influence of heat stress on acetylcholine content and its hydrolyzing activity in *Macroptilium atropurpureum* cv. Siratro. *Jpn. J. Crop Sci.* 6:112–118.

Momonoki, Y. S. and T. Momonoki. 1993. Changes in acetylcholine-hydrolyzing activity in heat-stressed plant cultivars. *Jpn. J. Crop Sci.* 62:438–446.

Momonoki, Y. S., T. Momonoki, and J. H. Whallon. 1996a. Acetylcholine as a signaling system to environmental stimuli in plants. I. Contribution of $Ca^{2+}$ in heat-stressed *Zea mays* seedlings. *Jpn. J. Crop Sci.* 65:260–268.

Momonoki, Y. S., S. Oguri, S. Kato, and H. Kamimura. 1996b. Studies on the mechanism of salt tolerance in *Salicornia europaea* L. III. Salt accumulation and ACh function. *Jpn. J. Crop Sci.* 65:693–699.

Muralidharan, M., K. Buss, K. E. Larrimore, N. A. Segerson, L. Kannan, and T. S. Mor. 2013. The *Arabidopsis thaliana* ortholog of a purported maize cholinesterase gene encodes a GDSL-lipase. *Plant Mol. Biol.* 81:565–576.

Muralidharan, M., H. Soreq, and T. S. Mor. 2005. Characterizing pea acetylcholinesterase. *Chemico-Biol. Int.* 157–158:406–407.

Murata, J., T. Watanabe, K. Sugahara, T. Yamagaki, and T. Takahashi. 2015. High-resolution mass spectrometry for detecting Acetylcholine in *Arabidopsis. Plant Signal. Behav.* 10:e1074367.

Nakajima, H. and S. Hatano. 1962. Acetylcholinesterase in the plasmodium of the myxomycete, *Physarium polycephalum. J. Cell. Comp. Physiol.* 59:259–264.

Otto, P. 1985. Membrane acetylcholinesterase: Purification, molecular properties, and interactions with amphiphilic environments. *Biochimica et Biophysica Acta* 822:375–392.

Rejσn, J. D., A. Zgnieszka, M. I. Rodriguez-Garcia, and A. J. Castro. 2012. Profiling and functional classification of esterases in olive (*Olea europaea*) pollen during germination. *Ann. Bot.* 110:1035–1045.

Riov, J. and M. J. Jaffe. 1973. A cholinesterase from bean roots and its inhibition by growth retardants. *Experentia* 29:264–265.

Roshchina, V. V. 1988. Characterization of pea chloroplast cholinesterase: Effect of inhibitors of animal enzymes. *Photosynthetica* 22:20–26.

Roshchina, V. V. 1991. *Biomediators in Plants. Acetylcholine and Biogenic Amines.* Pushchino, Russia: Biological Center of USSR Academy of Sciences.

Roshchina, V. V. 2001. *Neurotransmitters in Plant Life.* Plymouth, UK: Science Publishers.

Roshchina, V. V. 2016. New trends and perspectives in the evolution of neurotransmitters in microbial, plant, and animal cells. In *Microbial Endocrinology: Interkingdom Signaling in Infectious Disease and Health, Advances in Experimental Medicine and Biology*, (Ed.) Lyte, M., pp. 25–77, 874. Cham, Switzerland: Springer International Publishing AG.

Ross, J. J. and C. M. Wolbang. 2008. Auxin, gibberellins and the gravitropic response of grass leaf sheath pulvini. *Plant Signal. Behav.* 3:74–75.

Rotundo, R. L. 2003. Expression and localization of acetylcholinesterase at the neuromuscular junction. *J. Neurocytol.* 32:743–766.

Sagane, Y., T. Nakagawa, K. Yamamoto, S. Michikawa, S. Oguri, and Y. S. Momonoki. 2005. Molecular characterization of maize acetylcholinesterase. A novel enzyme family in the plant kingdom. *Plant Physiol.* 138:1359–1371.
Sánchez, D. G., L. H. Otero, C. M. Hernández, A. L. Serra, S. Encarnación, C. E. Domenech, and A. T. Lisa. 2012. A *Pseudomonas aeruginosa* PAO1 acetylcholinesterase is encoded by the PA4921 gene and belongs to the SGNH hydrolase family. *Microbiol. Res.* 167:317–325.
Schallreuter, K. U., S. M. A. Elwary, N. C. J. Gibbons, H. Rokos, and J. M. Wood. 2004. Activation/deactivation of acetylcholinesterase by $H_2O_2$: More evidence for oxidative stress in vitiligo. *Biochem. Biophys. Res. Commun.* 315:502–508.
Soreq, H. and S. Seidman. 2001. Acetylcholinesterase—New roles for an old actor. *Nat. Rev. Neurosci.* 2:294–302.
Standaert, F. G. 1990. Neuromuscular physiology. In *Anesthesia*, (Ed.) Miller, R. D., pp. 650–684. New York: Churchill Livingstone.
Talesa, V., G. B. Principato, E. Giovannini, M. V. Di Giovanni, and G. Rosi. 1993. Dimeric forms of cholinesterase in *Sipunculus nudus. Eur. J. Biochem.* 215:267–275.
Tougu, V. 2001. Acetylcholinesterase: Mechanism of catalysis and inhibition. *Curr. Med. Chem.* 1:155–170.
Toutant, J. P. 1989. Insect acetylcholinesterase: Catalytic properties, tissue distribution and molecular forms. *Prog. Neurobiol.* 32:423–446.
Tretyn, A. and R. E. Kendrick. 1990. Induction of leaf unrolling by phytochrome and acetylcholine in etiolated wheat seedlings. *Photochem. Photobiol.* 52:123–129.
Upton, C. and J. T. Buckley. 1995. A new family of lipolytic enzymes? *Trends Biochem. Sci.* 20:178–179.
Wiśniewska, J. and A. Tretyn. 2003. Acetylcholinesterase activity in *Lycopersicon esculentum* and its phytochrome mutants. *Plant Physiol. Biochem.* 41:711–717.
Wolfe K. H., M. Gouy, Y. W. Yang, P. M. Sharp, and W. H. Li. 1989. Date of the monocot-dicot divergence estimated from chloroplast DNA sequence data. *Proc. Natl. Acad. Sci. USA* 86:6201–6205.
Yamamoto, K. and Y. S. Momonoki. 2008. Subcellular localization of overexpressed maize *AChE* gene in rice plant. *Plant Signal. Behav.* 3:576–567.
Yamamoto, K. and Y. S. Momonoki. 2012. Tissue localization of maize acetylcholinesterase associated with heat tolerance in plants. *Plant Signal. Behav.* 7:1–5.
Yamamoto, K., S. Oguri, S. Chiba, and Y. S. Momonoki. 2009. Molecular cloning of acetylcholinesterase gene from *Salicornia europaea* L. *Plant Signal Behav.* 4:361–366.
Yamamoto, K., S. Oguri, and Y. S. Momonoki. 2008. Characterization of trimeric acetylcholinesterase from a legume plant, *Macroptilium atropurpureum* Urb. *Planta* 227:809–822.
Yamamoto, K., H. Sakamoto, and Y. S. Momonoki. 2011. Maize acetylcholinesterase is a positive regulator of heat tolerance in plants. *J. Plant Physiol.* 168:1987–1992.
Yamamoto, K., H. Sakamoto, and Y. S. Momonoki. 2016. Altered expression of acetylcholinesterase gene in rice results in enhancement or suppression of shoot gravitropism. *Plant Signal. Behav.* 11:e1163464.
Yamamoto, K., S. Shida, Y. Honda, M. Shono, H. Miyake, S. Oguri, H. Sakamoto, and Y. S. Momonoki. 2015. Overexpression of acetylcholinesterase gene in rice results in enhancement of shoot gravitropism. *Biochem. Biophys. Res. Commun.* 465:488–493.
Yang Y. W., K. N. Lai, P. Y. Tai, and W. H. Li. 1999. Rates of nucleotide substitution in angiosperm mitochondrial DNA sequences and dates of divergence between *Brassica* and other angiosperm lineages. *J. Mol. Evol.* 48:597–604.

# 3 Neurotransmitters in Marine and Freshwater Algae

*Kathryn L. Van Alstyne, Richard L. Ridgway, and Timothy A. Nelson*

## CONTENTS



## 3.1 INTRODUCTION

Algae are a phylogenetically diverse assemblage of organisms ranging from single-celled plankton to giant kelps. The term algae itself is neither unequivocally defined nor taxonomically useful. Algae are sometimes characterized as eukaryotic, aquatic, photosynthetic, and lacking a sterile jacket of cells around their reproductive structures. However, some authors consider prokaryotic cyanobacteria to be algae (Lee 1999). Other algae have lost the ability to conduct photosynthesis, as also seen in parasitic embryophytes (i.e., land plants), and live as parasites, saprophytes, or predators (e.g., dinoflagellates). A number are mixotrophic, capable of both feeding and photosynthesizing. Terrestrial algae are common in humid microclimates or live symbiotically within or on terrestrial organisms (e.g., lichens). Some Charophytes, traditionally included as green algae (errantly in Division Chlorophyta historically), protect their reproductive cells within sterile jackets (Graham et al. 2008).

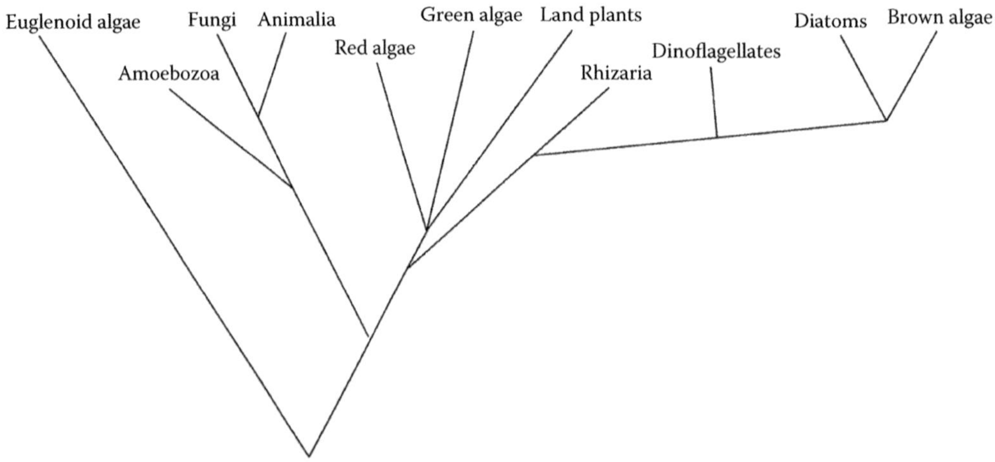


**FIGURE 3.1** Major extant eukaryotic lineages, with well-known algal taxa emphasized. Only the unikont branch (animals, fungi, and amoebozoans) lack algal representatives. Excavates, represented by the euglenoid algae (but including non-photosynthetic flagellates as well), are the earliest branching eukaryotes. The archaeplastid lineage (including red algae, green algae, and land plants) comprise all organisms that gained plastids via primary endosymbiosis of a cyanobacterium, with other groups secondarily taking up a red or green alga. The chromalveolate lineage (leading to brown algae) includes water molds, Apicomplexa, brownish or yellowish algae (including many groups not shown) and is sister to the rhizarians (Foraminifera and Chorarachniophyta, the latter an algal group represented by photosynthetic amoebae). Branch lengths are not drawn to scale of genetic differences. (Based on del Campo et al. 2014.)

Similarly vexing, algae are not monophyletic, and the term has no taxonomic utility. Most obviously, some authors and most phycology textbooks consider cyanobacteria to be algal and thus clearly different from eukaryotic algae (e.g., see Lee 1999). Even among eukaryotic algae, secondary endosymbiotic events lead to diverse nuclear and mitochondrial genomes associated with the ability to conduct photosynthesis. Representatives are found in the Excavata (e.g., euglenoids), Alveolata (e.g., dinoflagellates), Stramenopila (e.g., kelps and diatoms), and Archaeplastida (i.e., organisms that obtained a plastid via a primary endosymbiosis, including red and green algae as well as land plants), as well as several groups that remain challenging to place in deep eukaryotic phylogenies (Adl et al. 2012) (Figure 3.1). Williams (2014) notes that the Excavata have less genetic similarity to land plants than animals do. The remaining eukaryotic algae (and land plants) are as genetically distinct from each other as animals are from fungi or true amoebae.

The closest algal relatives of land plants, the green algal phyla Charophyta and Chlorophyta, include several species producing neurotransmitters or receptors for known neurotransmitters (Table 3.1). The Charophyta were recently removed from a single Phylum Chlorophyta, commonly called green algae, so that the remaining Chlorophyta represented a more nearly monophyletic group. Among the green algae, the Charophyta are sister to the land plants but are themselves a paraphyletic collection leading to multiple taxonomic schemes with various classes and orders being promoted to phylum status (Laurin-Lemay 2012) (Figure 3.2). Land plants are a sister group to the Zygnematales/Desmidiales lineages (Guiry 2013) which are included in the Charophyta *sensu lato*. Two species of charophytes have been associated with neurotransmitters; of these, *Micrasterias denticulata*, a member of the Desmidiales, would presumably be the closest relative of land plants (Schiechl et al. 2008; Beilby et al. 2015b).

Unlike the freshwater charophyte species, the remaining algae associated with neurotransmitters are all multicellular marine algae, that is, seaweeds (Table 3.1). Many of these species are large enough to contribute substantial structure and productivity to shallow water benthic communities.

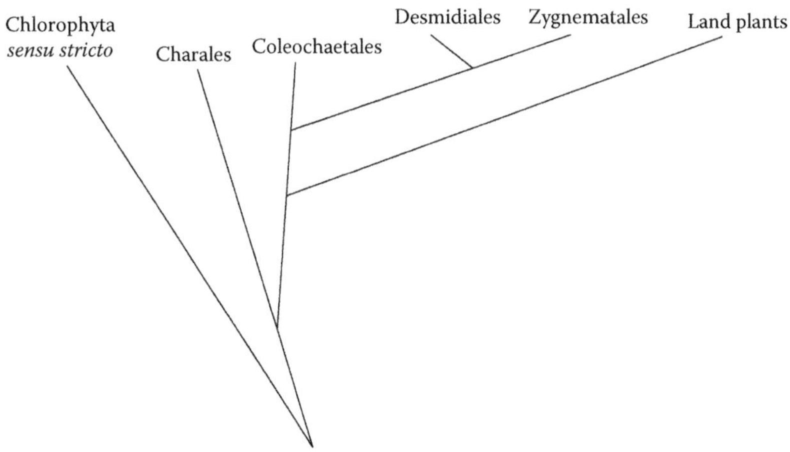


**FIGURE 3.2** Diversity of algae formerly included in the Phylum Chlorophyta and their descendant land plants. Chlorophyta *sensu stricta* are still regarded as a Phylum today, with the former members of Class Charophyceae (shown here as Charales, Coleochaetales, Desmidiales, and Zygnematales) split out and referred to as a single paraphyletic Phylum Charophyta or with the orders (-ales) indicated in the figure represented at various taxonomic levels. Branch lengths are not drawn to scale of genetic differences. (Based on Laurin-Lemay, S. et al., *Curr. Biol.,* 22(15), R593–R594, 2012; Guiry, M.D. et al., *Algae,* 28, 1–29, 2013.)

**TABLE 3.1**
**Algae That Produce Acetylcholine, Dopamine, Epinephrine, Histamine, Norepinephrine, and Serotonin**

| Species | Location | Neurotransmitter | Reference |
|---|---|---|---|
| **PHYLUM CHAROPHYTA** | | | |
| *Chara australis* | Sydney, Australia | Serotonin | M.J. Beilby et al. (2015) |
| *Micrasterias denticulata* | Unknown | Acetylcholine | Schiechl et al. (2008) |
| **PHYLUM CHLOROPHYTA** | | | |
| *Caulerpa cupressoides* | Pacheco Beach, Cumbuco, Brazil | Histamine | de Alencar et al. (2011) |
| *Caulerpa mexicana* | Pacheco Beach, Cumbuco, Brazil | Histamine | de Alencar et al. (2011) |
| *Caulerpa prolifera* | Pacheco Beach, Cumbuco, Brazil | Histamine | de Alencar et al. (2011) |
| *Caulerpa sertularioides* | Pacheco Beach, Cumbuco, Brazil | Histamine | de Alencar et al. (2011) |
| *Ulvaria obscura* | San Juan Island, Washington, USA | Dopamine | Tocher and Meeuse (1966) |
| **PHYLUM OCHROPHYTA** | | | |
| *Desmarestia aculeata* | Helgoland, Germany | Phenethylamine | Steiner and Hartmann (1968) |
| *Dictyopteris delicatula* | Pacheco Beach, Cumbuco, Brazil | Histamine | de Alencar et al. (2011) |
| *Lobophora variegata* | Pacheco Beach, Cumbuco, Brazil | Histamine | de Alencar et al. (2011) |
| *Sargassum tenerrimum* | Okha, India | Acetylcholine | Gupta et al. (2013) |
| **PHYLUM RHODOPHYTA** | | | |
| *Botryocladia occidentalis* | Pacheco Beach, Cumbuco, Brazil | Histamine | de Alencar et al. (2011) |
| *Bryothamniuon seaforthii* | Pacheco Beach, Cumbuco, Brazil | Histamine | de Alencar et al. (2011) |
| *Bryothamnion triquertum* | Pacheco Beach, Cumbuco, Brazil | Histamine | de Alencar et al. (2011) |
| *Ceramium rubrum* | Helgoland, Germany | Phenethylamine | Steiner and Hartmann (1968) |
| *Cryptonemia crenulata* | Pacheco Beach, Cumbuco, Brazil | Histamine | de Alencar et al. (2011) |

*(Continued)*

**TABLE 3.1 (*Continued*)**
**Algae That Produce Acetylcholine, Dopamine, Epinephrine, Histamine, Norepinephrine, and Serotonin**

| Species | Location | Neurotransmitter | Reference |
|---|---|---|---|
| *Cystoclonium purpureum* | Helgoland, Germany | Phenethylamine | Steiner and Hartmann (1968) |
| *Furcellaria lumbricalis* | Kimmeridge, Dorset, England; Koster Fjord, Sweden | Histamine | Andersson and Bohlin (1984); Barwell (1979, 1989) |
| *Gelidium crinale* | Şile, Black Sea, Turkey | Phenethylamine | Percot et al. (2009) |
| *Gracilaria bursa-pastoris* | Soğandere, Aegean Sea, Turkey | Phenethylamine | Percot et al. (2009) |
| *Halymenia floesii* | Kepez, Aegean Sea, Turkey | Phenethylamine | Percot et al. (2009) |
| *Laurencia obtusata* | Bembridge, Isle of Wight, England | Acetylcholine | Barwell (1980) |
| *Phyllophora crispa* | Şile, Black Sea, Turkey | Phenethylamine | Percot et al. (2009) |
| *Polyides rotunda* | Helgoland, Germany | Phenethylamine | Steiner and Hartmann (1968) |
| *Polysiphonia morrowii* | Kepez, Aegean Sea, Turkey | Phenethylamine | Percot et al. (2009) |
| *Polysiphonia tripinnata* | Şile, Black Sea, Turkey | Phenethylamine | Percot et al. (2009) |
| *Polysiphonia urceolata* | Helgoland, Germany | Phenethylamine | Steiner and Hartmann (1968) |

Kelps (Ochrophyta; Phaeophyceae, Laminariales) form highly complex and productive habitats for a variety of others plants and animals (Dayton 1985). Macroalgae may limit phytoplankton blooms via competition for nutrients (e.g., Fong et al. 1993), thus serving as a buffer against excessive eutrophication. Tropical macroalgae also stabilize sediments via rhizoids (e.g., seen in *Caulerpa* spp.; Reise 2002) or by consolidating substrata in reef systems (e.g., seen in coralline red algae; Goreau 1963). In contrast to these ecosystem services, several marine seaweeds may form harmful blooms (Nelson and Lee 2001; Piñón-Gimate et al. 2008) or overgrow in eutrophic conditions, leading to localized anoxia (Solidoro et al. 1997).

## 3.2 DISTRIBUTION OF POTENTIAL NEUROTRANSMITTERS IN MARINE AND FRESHWATER ALGAE

Acetylcholine (ACh), dopamine, serotonin, histamine, and phenethylamine occur in marine and freshwater algae (Table 3.1). In a recent literature search, we found no reports of algae producing adrenaline or noradrenaline. This may be a result of algae not producing these compounds or a lack of studies on the topic, as literature on the production of neurotransmitters by algae other than the charophytes is limited.

Based on the available data, there are no discernable trends in the production of different neurotransmitters among major macroalgal taxa. Histamine is the most widely reported compound taxonomically and occurs in a higher proportion of tropical than temperate macroalgae. It is reported from one temperate European red alga, *Furcellaria lumbricalis* (Barwell 1979; Andersson and Bohlin 1984), and numerous tropical algae from Brazil (de Alencar et al. 2011). Six green (Phylum Chlorophyta), 31 brown (Phylum Ochrophyta, Class Phaeophyceae) and 38 red (Phylum Rhodophyta) macroalgal species surveyed in England did not contain detectable levels of histamine (Barwell 1979). In contrast, histamine was found in five out of six green macroalgal species, four out of four red macroalgal species, and two out of three brown macroalgal species from Brazil (de Alencar et al. 2011). To our knowledge, there are no reports of histamine from charophytes.

Acetylcholine has been identified in one species each in the charophytes, ochrophytes, and rhodophytes (Table 3.1) using either chromatography or NMR spectroscopy. Phenethylamine (phenylethylamine) has only been reported from brown and red macroalgae (Table 3.1). It was absent from seven green algal species surveyed in England and Turkey (Steiner and Hartmann 1968;

Percot et al. 2009). In the same surveys, only 1 of 16 brown algae contained the compound, whereas it was present in 12 of 22 red algae. Serotonin and dopamine are less widely distributed among algae. Serotonin has only been reported from the charophyte *Chara australis* (Beilby et al. 2015b), and dopamine has only been reported from the green alga *Ulvaria obscura* (Tocher and Meeuse 1966b). Dopamine does not occur in detectable concentrations in two related green algal species in Washington, USA (Van Alstyne et al. 2006).

## 3.3 PHYSIOLOGICAL AND ECOLOGICAL FUNCTIONS OF ALGAL NEUROTRANSMITTERS

### 3.3.1 Acetylcholine: Possible Electrical Signal Transmission in Charophytes

Virtually all eukaryotic cells maintain a membrane potential across their plasma membranes. The ion gradients responsible for these potentials are established by electrogenic pumps driven by ATP hydrolysis. In plant cells, the main pumps are $H^+$-ATPases that move protons out of the cell creating at once an electrochemical gradient and a pH gradient (Chasan and Schroeder 1992). The resting membrane potential (RMP) of animal cells is determined largely by passive diffusion of potassium ions ($K^+$) out of the cell and is, therefore, sensitive to changes in external $K^+$ concentration. Animal nerve cells have RMPs of about –70 mV (inside relative to external medium). Outward diffusion of $K^+$ also contributes to the RMPs of plant cells in conjunction with outward proton pumping; thus, plant cells are sensitive to both external $K^+$ concentration and external pH. The RMP of most plant cells ranges between −120 and −180 mV. The electrochemical and pH gradients associated with the RMP facilitate co-transport processes across the membrane and in certain cells the altering of the membrane potential is important for transmitting signals in the form of action potentials (Hartmann and Gupta 1989; Roshchina 2001).

Electrically excitable cells of animal and plant cells are characterized by the presence of voltage-gated ion channels on their plasma membranes. When opened, these populations of ion channels can generate an action potential (AP), an *all-or-none* depolarizing pulse that rapidly changes the membrane voltage. In animal nerve cells, APs are often generated by the opening of voltage-gated cation (sodium or calcium) channels when a stimulus causes membrane depolarization that exceeds a certain threshold voltage. The resting potential is then reestablished when $K^+$ exits the cell through delayed opening of voltage-gated $K^+$ (rectifier) channels. During the transmission of electrical impulses in nerve cells, depolarization proceeds down the axon until reaching its terminus, where neurotransmitters are stored in synaptic vesicles. For the signal to be transmitted to an adjacent neuron, voltage-gated $Ca^{2+}$ channels are opened in axon terminals causing an inward $Ca^{2+}$ flux. The intracellular $Ca^{2+}$ concentration shift facilitates the exocytosis of neurotransmitters from vesicles into the synaptic cleft, where they diffuse from the pre-synaptic neuron to receptors on the post-synaptic neuron. The binding of the neurotransmitter initiates either an excitatory or inhibitory response in the post-synaptic cell. The neurotransmitter is then either enzymatically broken down (in the case of ACh) or moved back into the pre-synaptic terminals.

Charophytes, and possibly other algae, may transmit electrical signals using a similar system. The generation of action potentials in charophytes has been particularly well studied because they produce large, multinucleate cells that are amenable to experimental methods typically used in neurobiology, such as perfusing individual membranes and patch clamping (Beilby 2016). In fact, the first measurements of action potentials were made in *Nitella mucronata*, a freshwater charophyte (Umrath 1930).

There are several differences in the production of action potentials in animals and charophytes. For example, the transmission of electrical signals in charophytes is about 1000 times slower than in animal nerve cells (Dettbarn 1962). Another difference is the mechanism that causes depolarization. In *Chara corallina*, and *Nitellopsis obtusa*, depolarization results when $Cl^-$ exits cells through ion channels, rather than $Na^+$ entering cells (Beilby and Coster 1979; Shiina and Tazawa 1987). In *C. corallina*, exogenous ACh increases $Cl^-$ channel activation and the addition of nicotine to the

ACh solution, but not muscarine, further increases the probability of the channels opening (Gong and Bisson 2002). The addition of ACh also slows the rate of repolarization of the action potential in *N. obtusa* and *C. corallina*, (Gong and Bisson 2002; Kisnierienė et al. 2009), but it does not affect the rate of depolarization in *C. corallina* (Gong and Bisson 2002).

In animals, two types of cholinergic receptors (receptors that respond to ACh) occur (Sargent 1993). Nicotinic receptors, which respond to nicotine as well as ACh, occur in neurons and skeletal muscles. Muscarinic receptors, which respond to muscarine, a mushroom natural product, are found in the brain and other tissues (Caulfield 1993). Gong and Bisson (2002) suggest that the charophyte ACh receptor is sufficiently different from animal nicotinic receptors that it must have evolved independently. They argue that the differences in the receptors include the fact that charophyte receptors are anion channels and animal receptors are cation channels; in charophytes, the receptor binding site is on the inside (vacuolar side) of the membrane, whereas in animals, it is external; and, in animals, ACh and nicotine bind to a single site, whereas in charophytes the two ligands appear to bind to different sites. However, whether ACh actually binds to a receptor protein has recently been questioned. Work on *Chara braunii* internodal cells suggests that ACh does not bind to the channels. Rather, it is hydrolyzed by acetylcholinesterase to form hydrogen ions ($H^+$), choline, and acetate (Fillafer and Schneider 2016). The resulting hydrogen ions are then responsible for triggering the depolarization of the membrane.

Do similar transmission systems exist in other algal phyla? Well-characterized cholinergic systems only exist for the charophytes; they have not been described for other algal phyla. Nonetheless, there is evidence that similar systems may exist. ACh occurs in at least one brown and one red algal species (Barwell 1980; Gupta et al. 2013) and cholinesterase has been reported in four species of green algae, one brown alga, and five red algae (Gupta et al. 1998), as well as in *Nitella obtusa* (Dettbarn 1962). While the occurrence of these compounds alone does not prove the existence of cholinergic transmission systems, they are suggestive of their presence. Given the diverse evolutionary history of the major algal phyla, similarities and differences in the functioning of these compounds in different algal phyla may lead to a better understanding of the evolution of electrical signaling systems

### 3.3.2 Serotonin: Circadian Patterns in Charophytes

Serotonin in charophytes may be involved in alleviating the effects of temporally varying oxidative stresses or in regulating growth. The compound can be a precursor of melatonin, a strong antioxidant that may be involved in setting or stabilizing circadian patterns (Beilby et al. 2015a, 2015b; Beilby 2016). Its concentrations in *C. australis* undergo daily patterns that differ in the summer and winter (Beilby et al. 2015b). When *C. australis* were maintained on a summer light cycle (12/12 light/dark), serotonin concentrations peaked at 11 a.m. and were approximately 400 ng $g^{-1}$. Small peaks also occurred at 3 a.m. and 8 a.m. The large midday serotonin peaks in the summer corresponded with peaks in melatonin and preceded peaks in auxin. When plants were maintained on a winter light cycle (9/15 light/dark), the 11 a.m. peak was not present, but small peaks still occurred at about 3 a.m. and 8 p.m. Midday auxin peaks were also absent, and the melatonin peaks were reduced. The lower melatonin concentrations in the evening and in the winter may occur because reductions in photosynthesis at these times result in a reduced production of reactive oxygen species (ROS), reducing the need for high concentrations of antioxidants.

### 3.3.3 Histamine: Life History and Tissue Variability in *Furcellaria lumbricalis*

The function of histamine in marine algae is not known, but its concentrations in the red alga *F. lumbricalis* differ among parts of the thallus and among life history stages (Barwell 1989). Like many red seaweeds, *F. lumbricalis* produces three distinct macroscopic stages, an asexual tetrasporophyte, a male gametophyte, and a female gametophyte. Average histamine concentrations ranged from 60 to 1190 μg $g^{-1}$ of algae. Although they varied among different parts of the algal thalli, they were generally higher in male gametophytes than in female gametophytes or tetrasporophytes. In male gametophytes, concentrations were highest in the outer branchlets (ramuli) and lowest in the middle of the thallus. In female gametophytes, concentrations were lowest in the branchlets and highest in the bases of the algae. In tetrasporophytes, they were also lowest in the branchlets but highest in unbranched parts of the thallus.

### 3.3.4 Phenethylamine

Phenethylamine and related compounds produced by terrestrial plants can be feeding deterrents (Harley and Thorsteinson 1967) and are a component of blends of volatile compounds that serve as oviposition cues in butterflies (Nishida 2014). In macroalgae, phenethylamine has been hypothesized to be a feeding deterrent (Smith 1977), but no definitive assays have been conducted to determine whether this is the case.

### 3.3.5 Dopamine: Production and Ecological Functions in *Ulvaria obscura*

Dopamine is only known to be produced by one alga, *U. obscura* (Figure 3.3a–c; formerly *Monostroma fuscum*). The compound was initially identified in the 1960s in *U. obscura* from Washington State, USA (Tocher and Craigie 1966a). A phenolase enzyme that oxidizes dopamine to form dopamine quinone was also identified from *U. obscura* at the same time (Tocher and Meeuse 1966b). Subsequent work on the production of dopamine by *U. obscura* has focused on its ecological, rather than physiological, relevance to the alga (Van Alstyne et al. 2015) and to the dynamics of *green tides*, large growths or accumulations of green seaweeds (Figure 3.3a) that are becoming increasingly abundant worldwide (Charlier 2008; Ye et al. 2011; Smetacek and Zingone 2013).

#### 3.3.5.1 Basic Structure of *Ulvaria* Thalli

Catecholamines have been found in many plants, but few specifically produce dopamine at high concentration, and the organs or cells where such production occurs varies by species (Kulma and Szopa 2007). Thus, the relatively high concentrations of dopamine found in *U. obscura* make it a potential model system for studying plant catecholamine synthesis at the cellular level. In contrast to the various cell types that form the complex multilayer tissues of higher plants, the thalli of *U. obscura* have monostromatic blades consisting of a single layer of tall, closely packed vegetative cells. As shown in Figure 3.4a, the cell walls of the alga are thick at each end but are thinner laterally. A large central vacuole occupies much of the cell volume, displacing organelles such as chloroplasts to more peripheral locations within the cortical cytosol. Each chloroplast contains one to three pyrenoids and multiple starch grains that vary in number, size, and position. The single nucleus is positioned nearer to one pole, and the perinuclear cytosol contains many dictyosomes (Golgi bodies) and associated vesicles.

#### 3.3.5.2 Histochemical Localization of Dopamine in *Ulvaria* Thalli

Indirect immunofluorescence experiments using anti-dopamine monoclonal antibodies on thin sections cut from plastic-embedded *U. obscura* thalli show strong immunoreactivity associated with chloroplast thylakoids in stromal areas adjacent to starch grains and pyrenoidal starch sheathes (Figure 3.4b–e). In whole-mount specimens, we occasionally found anti-dopamine staining of vesicles located in the cortical cytosol between the chloroplasts and plasma membrane (Ridgway, unpublished data). Such vesicles are rarely seen in sections of plastic-embedded specimens, however, suggesting that most are lost during sample processing steps. This conclusion is supported by our attempts to localize dopamine using the glyoxylic acid histofluorescence technique (De Le Torre 1980), which requires incubation of samples in a buffered glyoxylic acid solution followed by a desiccation step and heating to 80°C–90°C. This processing releases significant amounts of dopamine across the cell walls primarily from cortical cytosol vesicles, but also from chloroplasts, creating a haze of fluorescent product (Figure 3.4f). Quenching cell wall fluorescence with Toluidine Blue

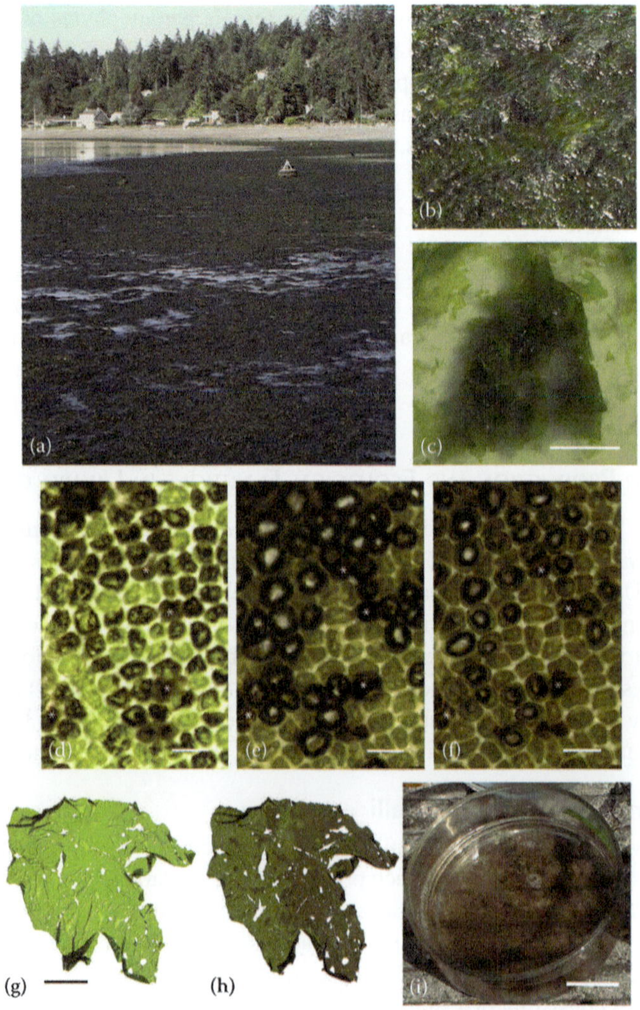


**FIGURE 3.3** (a) A moderate bloom of ulvoid green macroalgae revealed during low tide at Pleasant Beach, Bainbridge Island (Kitsap County, Washington, USA) on July 9, 2009. (b) A closer view of the algal bloom in (a) showing the dense accumulation (up to 12 cm thick in some areas) of various ulvoid algal species in the mid-intertidal zone of the beach. (c) A solitary specimen of *U. obscura* in about 8 cm of seawater in the lower intertidal zone of Pleasant Beach on the same day as (a) and (b). Measure bar in (c) = 10 cm. (d–f) Solubility of heat-induced dopamine oxidation products formed within cells of *U. obscura*. (d) Bright field microscopy of cells within a whole-mounted thallus sample that was blotted to remove excess seawater and then incubated in a 95°C oven for 5 min. Brown or black oxidation products are seen within vesicles of many of the cells. For reference, four cells that had undergone dopamine release prior to field collection of the thallus sample are denoted (asterisks). (e, f) The same thallus region as in (d) photographed after 1 min exposure (e) and 3 min exposure (f) to filtered seawater demonstrating the solubility of many of the oxidation products. Reference cells marked (asterisks) are the same as in (d). All measure bars = 20 μm. (g–h) *U. obscura* thallus prior to (g) and after (h) 3 h of desiccation on an overcast day. The brown coloration is due to the formation of melanins from dopamine. Measure bar = 5 cm. (i) Bowl containing filtered seawater that had been added to desiccated *U. obscura*. The *U. obscura* were removed 5 min after the seawater was added and the reddish coloration developed within 30 –60 min. Measure bar = 5 cm.

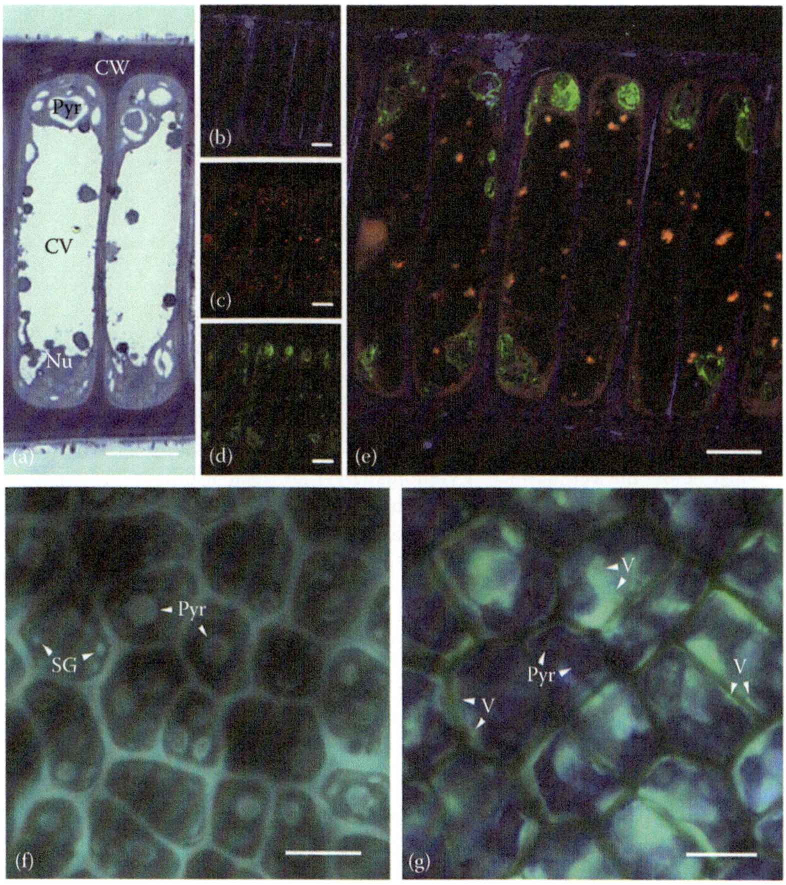


**FIGURE 3.4** (a–e) Cell structure and immunohistochemical localization of dopamine in *U. obscura*. (a) Thallus cross-sectional view using brightfield microscopy showing key structural features of cells as stained by Stevenel's Blue. The thallus consists of a single layer of elongated vegetative cells with relatively thick cell walls (CW) at the surface. Chloroplasts are located within the cortical cytosol at the poles of the cell; these contain pyrenoids (Pyr) that are readily identified by their shell-like starch sheathes, as well as adjacent ellipsoidal starch grains (seen as white voids). A central vacuole (CV) occupies most of the volume of the cell, relegating the cytosol to a thin layer between the tonoplast and cell membrane except in the region of the nucleus (N) where rough endoplasmic reticulum and Golgi bodies predominate. (b–e) Thallus cross-sectional view using epifluorescence microscopy. The color channels shown include: (b; blue channel) CalcoFluor White staining of cell wall cellulose, (c; red channel) chlorophyll autofluorescence, (d; green channel) dopamine immunofluorescence staining via a commercial monoclonal primary antibody (Gemacbio S.A., Saint-Jean d'Illac., France) generated against dopamine conjugated to bovine serum albumen via glutaraldehyde, and (e) a merged image combining the three channels at higher magnification. Note the strong immunofluorescence signals associated with stromal regions around pyrenoids and starch grains. All measure bars = 10 μm. (f–i) Whole-mount preparations of *U. obscura* thalli. (f) The typical distribution of dopamine using the glyoxylic acid (GA) histofluorescence method shows staining (bluish white) associated with cell walls, pyrenoids (Pyr), and starch granules (SG), plus significant haze due to released dopamine. (g) When cell wall fluorescence is blocked by applying 0.05% Toluidine Blue O prior to glyoxylic acid staining, some thalli show strong histofluorescence of vesicles (V) clustered within cortical regions of the cytosol just below the plasma membrane, and lighter staining associated with pyrenoids (Pyr) and starch granules (not labeled). *(Continued)*

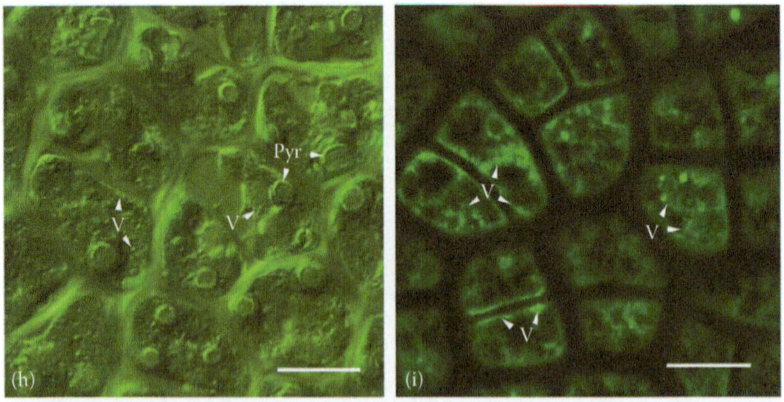


**FIGURE 3.4 (Continued)** (h) Differential interference contrast microscopy of living, non-stained cells at a focal plane just below the cell surface. Pyrenoids (Pyr) are seen within chloroplasts and clusters of small vesicles (V) are apparent in the cytosol. (i) Spinning disk confocal fluorescence microscopy of living cells photographed at a focal plane just below the cell surface showing the distribution of acidic vesicles (V) stained with the fluorescent probe LysoSensor Green DND-189 (Molecular Probes, Eugene, OR). All measure bars = 10 µm.

O dye improves contrast and enables dopamine histofluorescence to be localized in any remaining vesicles (Figure 3.4g). Cortical cytosolic vesicles can often be seen in freshly collected thallus specimens in the focal plane just below the cell wall (Figure 3.4h). Vesicles in this plane can be stained in living thallus samples using the pH-sensitive dye LysoSensor Green DND-189 (Molecular Probes, Eugene, Oregon, USA), indicating the lumen of these putative dopamine-containing vesicles is acidic (Figure 3.4i) (Ridgway, unpublished data).

#### 3.3.5.3 Proposed Cellular Pathways for Dopamine Storage and Release

Taken together, these dopamine localization studies suggest possible routes for dopamine synthesis, transport, storage, and release by *U. obscura* (summarized in Figure 3.5). The main site of dopamine synthesis appears to be in thylakoid membranes, particularly in areas of the chloroplast stroma near pyrenoidal sheathes and surrounding starch grains. The chloroplasts do not seem to be the main site for dopamine storage, however. Instead, dopamine appears to be transported across the chloroplast membranes to the cytosol where it is either (1) directly taken up by acidic vesicles derived from the endoplasmic reticulum (ER)-to-Golgi pathway, which are then stored in the cortical cytosol before release or (2) moved into the cortical ER and later enclosed in acidic vesicles budded from the ER membrane. The latter pathway is of interest because it suggests a route by which dopamine might be moved throughout the cell for storage while being protected from the more oxidative environment of the cytosol. In addition to acidic vesicles, further dopamine storage may be associated with cytosolic invaginations into the central vacuole. These invaginations involve the tonoplast and resemble those described for phenol-storing cells of banana roots (Beckman and Mueller 1970; Mueller and Beckman 1974). Storage in the acidic vesicles and/or cortical endoplasmic reticulum would facilitate regulated release of dopamine through plasma membrane signaling events, whereas loss of membrane integrity would result in unregulated (catastrophic) release of dopamine.

#### 3.3.5.4 Comparisons with Dopamine Production in Animals and Higher Plants

It is interesting to compare the proposed pathways for dopamine synthesis and release in *U. obscura* with other cellular systems known to involve catecholamines. In the dopaminergic neurons of animals, for example, dopamine is synthesized from tyrosine in a two-step process occurring in the cytosol (Meiser et al. 2013; Segura-Aguilar et al. 2014). In the rate-limiting first step, tyrosine is hydroxylated to *L*-dihydroxyphenylalanine (*L*-DOPA) in an oxygen-requiring reaction catalyzed

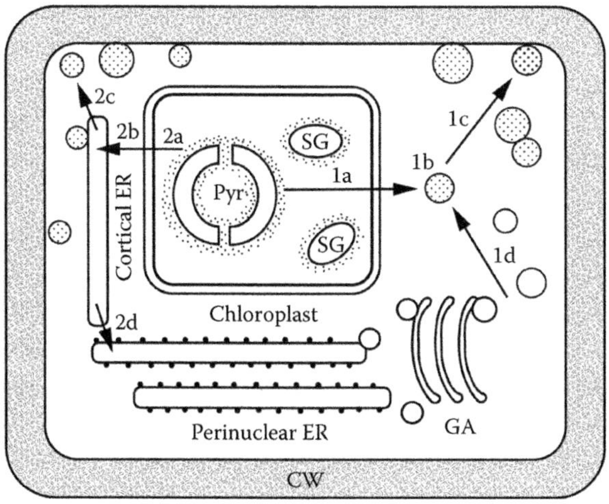


**FIGURE 3.5** Proposed cellular pathways for dopamine synthesis, storage, and release in *U. obscura*. Histochemical evidence suggests dopamine is produced in chloroplast stromal regions adjacent to starch grains (SG) and pyrenoidal (Pyr) starch sheathes (stippled areas). Dopamine appears to be transported to the cytosol where it is taken up and stored in vesicles derived from two possible pathways. In Pathway 1, dopamine is transported across chloroplast membranes (1a) to the cytosol and is uploaded into acidic vesicles (1b) derived from the endomembrane system (1d), which includes the perinuclear endoplasmic reticulum (ER) and Golgi apparatus (GA). The dopamine-containing vesicles, which vary in size due to fusion events, migrate to the cortical regions of the cytosol (1c) adjacent to the plasma membrane and cell wall (CW). In Pathway 2, dopamine transported across the chloroplast membranes (2a) to the cytosol is taken up by the cortical endoplasmic reticulum (2b) and loaded into acidic vesicles produced by the ER membrane; these vesicles are then distributed within the cortical cytosol (2c). Since the cortical ER can directly communicate with the perinuclear ER (2d), convergence between the two pathways may occur. Regardless of the pathway, this model infers that dopamine release could be either a regulated (exocytotic) process or, if plasma membrane integrity is lost, a non-regulated process.

by tyrosine hydroxylase (TH). In the second step, L-DOPA is decarboxylated to dopamine in a reaction catalyzed by aromatic amino acid decarboxylase (AADC). Dopamine does not accumulate in the cytosol, however, since TH and AADC form a complex with vesicular monoaminergic transporter 2 (VMAT2), a membrane protein that efficiently moves dopamine into synaptic vesicles (Cartier et al. 2010). The uptake by synaptic vesicles removes free dopamine from the potentially oxidizing environment of the cytosol (pH ~7.4) and into the reducing environment (pH 2.0–2.4) of the vesicle interior (Guillot and Miller 2009). The low pH is the result of VMAT2 being coupled to a $H^+$-ATPase on the vesicular membrane that creates a proton gradient to drive dopamine uptake (Segura-Aguilar et al. 2014). Dopamine is stored in synaptic vesicles within presynaptic terminals until released to the synaptic cleft via calcium-dependent exocytosis. Although the site of dopamine synthesis differs from that found in *U. obscura*, the uptake of dopamine into acidic vesicles for storage and regulated release may be similar.

Dopamine synthesis, storage, and release in higher plants differs in several key aspects from that just described for animal neurons but provides clues to how these stages are carried out by *U. obscura*. Some plant cells, such as those of Scotch broom (*Cytisus scoparius*), resemble animal cells in using TH to convert tyrosine to *L*-DOPA and AADC (also known as DOPA decarboxylase) to convert *L*-DOPA to dopamine (Tocher and Tocher 1972). In other plants, such as red banana (*Musa sapientum*), an alternate pathway is utilized in which tyrosine is decarboxylated to tyramine in a reaction catalyzed by tyrosine decarboxylase followed by tyramine hydroxylation to produce dopamine in a reaction catalyzed by monophenol hydroxylase (Nagatsu et al. 1972; Kulma and Szopa 2007). The route that gives rise to tyrosine in both cases is the shikimate pathway, a metabolic pathway localized to plastids and therefore not found in animals. In this pathway, phosphoenolpyruvate (from glycolysis) and erythrose-4-phosphate (from the oxidative pentose phosphate pathway) are brought together and converted in a series of enzyme-catalyzed steps first to shikimate, then to chorismate, and then to the aromatic amino acids tyrosine and phenylalanine (Soares et al. 2014). Most phenylpropanoids, including alkaloids, flavonoids, phenolic acids, and lignin are produced in biosynthesis pathways that branch from the shikimate pathway (Cheynier et al. 2013). The cellular compartment for production of these phenylpropanoids shifts away from the plastids in many cases. In *U. obscura* and a few other notable plants, such as banana (Kanazawa and Sakakibara 2000), dopamine is the main end-product whereas in other plants, such as the opium poppy (*Panaver sominifer*), it is merely a key intermediate metabolite in a complex pathway leading to diverse end-products (Facchini and De Luca 1995; Facchini 2001). Storage of dopamine in at least one poppy species, *Papaver bracteatum*, has been localized to small (<3 μm diam.) round *vacuoles* that differed from vacuoles in which the morphinan alkaloids, thebaine and morphine, were localized (Kutchen et al. 1986).

### 3.3.6 Ecological Functions

#### 3.3.6.1 Bloom Formation and Feeding Deterrence

*U. obscura* is one of several species of green seaweeds in the Order Ulvales that commonly occurs in green tides (Figure 3.3a) in Washington State, USA (Nelson et al. 2003b, 2008). Typically, it is found in deeper waters and is more abundant in the winter and spring than *Ulva lactuca*, the other abundant ulvoid in many blooms (Nelson et al. 2003b). *U. obscura* is not as well adapted for growing in intertidal environments as *U. lactuca*. In the intertidal zone, it grows more slowly than *U. lactuca*, but growth rates of the two species are similar in the subtidal zone (Nelson et al. 2008). It dries more quickly than *U. lactuca* and, for a given amount of drying, has a lower photosynthetic yield (Nelson et al. 2010). It also has lower growth rates in hyposaline waters (Nelson et al. 2008).

The ability of *U. obscura* to dominate subtidal blooms despite not being able to outgrow *U. lactuca* is due to its ability to resist grazing (Nelson et al. 2008). When given a choice of *U. obscura* versus other ulvoid species in choice assays, herbivorous snails (*Littorina sitkana*), crustaceans (*Idotea wosnesenskii*), and urchins (*Strongylocentrotus droebachiensis*) eat significantly less *U. obscura* (Van Alstyne et al. 2006; Nelson et al. 2008). A related urchin (*Strongylocentrotus purpuratus*) has lower absorption efficiencies when fed *U. obscura* (reported as *M. fuscum*) as compared to other common algal species (Vadas 1977). Using a bioassay-guided fractionation, Van Alstyne and coworkers (2006) demonstrated that in *U. obscura*, dopamine is the primary feeding deterrent towards *S. droebachiensis*. It also deters feeding by *I. wosnesenskii* and *L. sitkana* when incorporated into artificial diets.

The degree of protection provided by dopamine may vary spatially and temporally. In *U. obscura* from the Salish Sea, dopamine concentrations vary seasonally with peaks occurring in mid-summer (Figure 3.6). Concentrations differ among islands but tend to be consistent among sites on a given island. Inter-annual variation also occurs. Dopamine concentrations measured over a three-year period were consistently higher in 2006 than in 2007 and 2008. Within sites, dopamine concentrations in *U. obscura* tend to be consistent among depths (Figure 3.7), suggesting that seasonal, inter-annual, and among-island variability is not caused by either light or temperature, as light levels and temperatures should be higher at the shallow sites than the deep sites.

#### 3.3.6.2 Release of Dopamine in Seawater

Intertidal and shallow subtidal *U. obscura* can become stranded on beaches during low tides, where the algal thalli can become desiccated. Desiccated *U. obscura* often develop brown-colored patches on the surface of the thallus (Figure 3.3g–h). This discoloration is also seen at the microscopic level when *U. obscura* is heated briefly in the lab. Within minutes, damaged *U. obscura* cells produce

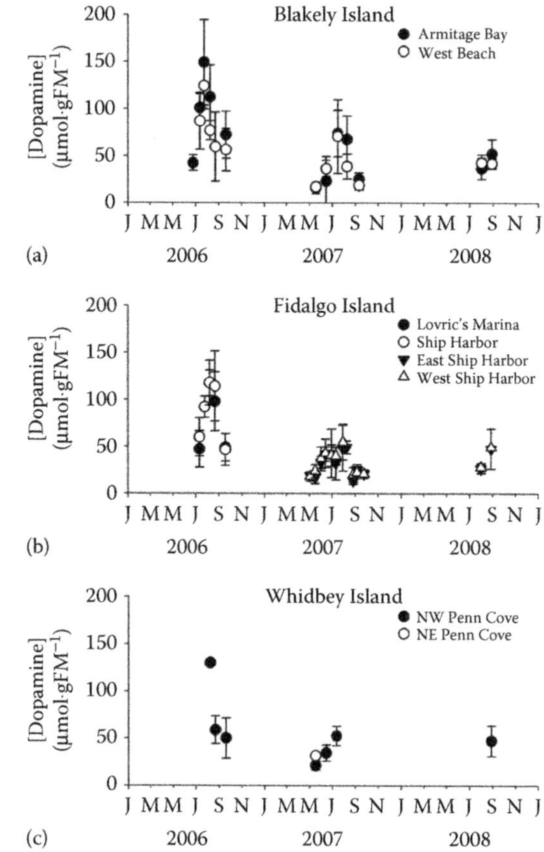


**FIGURE 3.6** Concentrations of dopamine (µmol dopamine g fresh mass$^{-1}$) in *U. obscura* from sites on three islands in the Salish Sea, Washington, USA. At all sites, dopamine concentrations peak in the mid-summer despite high levels of inter-island and inter-annual variability. The three islands are within 40 km of one another, and the sites on the islands are located less than 5 km away from one another. (a) Blakely Island, (b) Fidalgo Island, (c) Whidbey Island. Algae were collected during the summer growing season by hand during low tides and dopamine measurements were made by HPLC as described in Van Alstyne et al. (2006). Data are means ± 1 SD (N = 1–24).

colored, water-soluble oxidation products (Figure 3.3d–f) that are released into seawater. These oxidation products are likely to be the same products (Figure 3.8) that are produced from dopamine in animal cells at synapses (Meiser et al. 2013; Tan and Chen 1992). In animals, dopamine is first oxidized to form dopamine quinone, which results in the release of ROS (Figure 3.8). This reaction can be catalyzed by enzymes, such as the phenolase enzyme produced by *U. obscura* (Tocher and Meeuse 1966b), or by metals. Dopamine quinone is then cyclized to form aminochrome (2,3 dihydroindole-5,6 quinone). Following several additional reactions, the quinones polymerize to produce melanins (also referred to as neuromelanins). UV spectra of both commercially obtained dopamine-HCl and exudates released by *U. obscura* in a phosphate buffer solution containing sodium meta-periodate (which accelerates the oxidation of dopamine to aminochrome) show very similar changes in spectra over time (Figure 3.9). This provides strong evidence that the discoloration produced by *U. obscura* exudates in seawater is due to the oxidation of dopamine.

Several lines of evidence point to the occurrence of large-scale releases of dopamine by *U. obscura* blooms when stranded algae are rehydrated. On beaches where large biomasses of *U. obscura* occur,

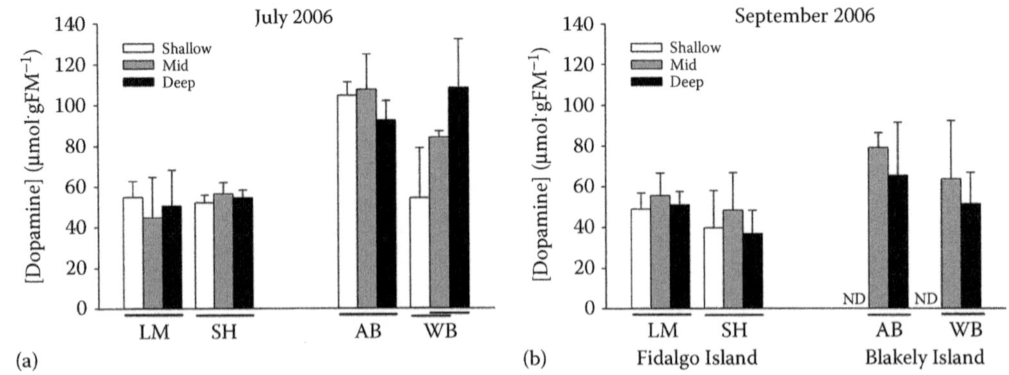


**FIGURE 3.7** Concentrations of dopamine (µmol dopamine g fresh mass$^{-1}$) in *U. obscura* from two sites on Fidalgo Island (LM: Lovric's Marina, SH: Ship Harbor) and two sites on Blakely Island (AB: Armitage Bay, WB: West Beach) along 50 m transect lines running from the mid intertidal zone (shallow) to the subtidal zone (deep). Data are means + 1 SD (N = 2–17). Lines below the graph connect data that are not significantly different from one another ($P > 0.05$ from *post hoc* Tukey's tests following analyses of variance). With the exception of the West Beach site in July 2006, dopamine concentrations did not differ significantly with depth. ND: no data (*U. obscura* was not found in shallow algal collections at the Blakely Island sites in September). Dopamine measurements were made as described in Van Alstyne et al. (2006).

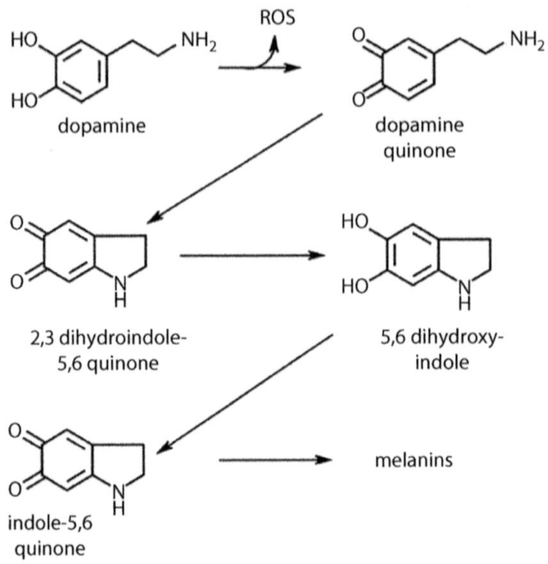


**FIGURE 3.8** Reactions that are proposed to occur when dopamine is in an aqueous solution. (From Tan, T.C. and Chen, Y., *Int. J. Chem. Kinet.*, 24, 1023–1034, 1992.)

discolored water can be observed at the edge of an incoming tide that is rehydrating desiccated algae (Nelson et al. 2003a). Similar discolorations are observed when desiccated *U. obscura* is rehydrated with seawater in laboratory experiments (Figure 3.3i). Experimental work that simulated *U. obscura* being stranded at low tide then rehydrated by an incoming tide found that seawater dopamine concentrations following rehydration were positively correlated with solar radiation during emersion and could be as high as 1340 µM (Van Alstyne et al. 2011). The loss of dopamine was due to the algae becoming desiccated; it was not impacted by UV or visible light levels, water temperature, or exudates from neighboring *U. obscura* (Van Alstyne et al. 2013).

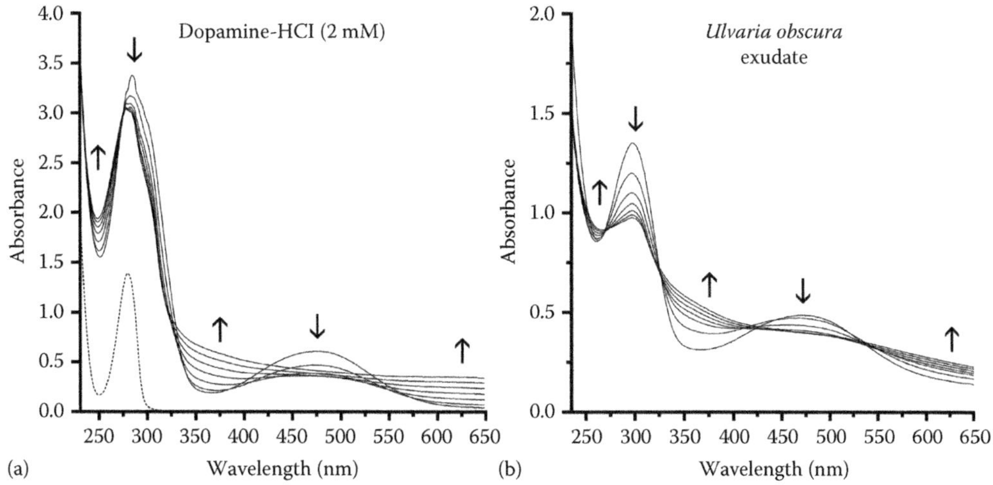


**FIGURE 3.9** UV-visible spectra of oxidation products seen after the addition of sodium meta-periodate to either (a) a 2 mM dopamine-HCl sample in phosphate buffer (pH 7.4) or (b) a sample of *U. obscura* exudate in the same phosphate buffer. In both cases, the solid lines are overlaid spectra taken at different time points (corresponding to 2, 5, 10, 15, 20, 25, and 30 min) after periodate addition. These spectra reveal a similar time evolution of the two peaks ($\lambda_{max}$ = 300 and 475 nm) associated with production of 2,3 dihydroindole-5,6 quinone (also known as aminochrome). These peaks gradually decay (downward arrows) as the oxidation continues, giving rise to products that increase in absorbance throughout the visible spectrum (upward arrows) as melanins are formed. For reference, in (a) a single spectrum of a 2 mM dopamine-HCl sample in distilled water (dashed line) is provided showing the characteristic peak absorption at 280 nm.

### 3.3.6.3 Waterborne Dopamine as a Toxin and Allelochemical

Ulvoid algae can affect the survival and growth of nearby benthic and planktonic organisms (Magre 1974; Johnson and Welsh 1985; Wang et al. 2009; Tang and Gobler 2011). Exudates of *U. obscura* inhibit the germination and growth of embryos of the brown alga *Fucus distichus* and the growth of *U. obscura* and *U. lactuca*. They also inhibit the growth of benthic microalgae (Nelson et al. 2003a). Dopamine in seawater at concentrations similar to those found in exudates inhibits the germination of *F. distichus* embryos and slows the growth of *U. lactuca* (Van Alstyne et al. 2014); however, the specific mechanisms underlying these effects are not known. Because dopamine in seawater rapidly oxidizes to form quinones, which bind to DNA (Graham et al. 1978) and inactivate proteins (Stokes et al. 1999), the quinones generated by dopamine released from *U. obscura* may be the toxic agents. Dopamine oxidation also produces ROS, which cause damage to lipids, protein, and DNA (Halliwell and Gutteridge 1989). Desiccated *U. obscura* releases ROS at a rate of about 0.3 μmol $g^{-1}$ of algae when rehydrated (van Hees and Van Alstyne 2013). These ROS may be byproducts of photosynthesis, cellular respiration, and oxidative stress responses, in addition to being reaction products of dopamine oxidation.

*U. obscura* exudates also affect the survival and development of invertebrate larvae. Nelson and coworkers (2003a) found that *U. obscura* exudates caused 100% mortality in oyster (*Crassostrea gigas*) larvae using densities of algae typically found in blooms (14 g fresh mass per L). Using a range of dilutions of *U. obscura* exudates, it was estimated that half of the *C. gigas* larvae would fail to develop properly to the D-hinge stage within 24 h after being exposed to exudates from *U. obscura* at densities of 0.21 g $l^{-1}$ (Nelson and Gregg 2013). In similar experiments, Rivera Vázquez et al. (2017) found that two-hour-old exudates from *U. obscura* at threefold higher densities did not affect the germination of *C. gigas* zygotes or increase larval mortality after 25 min and one h of exposure, respectively. However, exposure to exudates did increase shell lengths and widths of *C. gigas* larvae at the straight-hinged veliger stage. The differences in the results of these studies may be due to differences in the exposure times to the exudates, the ages or concentrations of the exudates, or to differences in the amounts of toxins being produced by *U. obscura* from different sites or collected at different times of the year.

The effects of *U. obscura* exudates and dopamine on invertebrate larvae differ among species and stages. Dopamine decreases the survival of Dungeness crab (*Metacarcinus magister*) zoeae but does not affect the survival of shore crab (*Cancer oregonensis*) megalopae and juveniles (Van Alstyne et al. 2014). Ten and 100 μM dopamine has no effect on the fertilization success of sand dollar (*Dendraster excentricus*) eggs, but exudates of *U. obscura* reduce it, suggesting that the exudates contain other toxins (Rivera Vázquez et al. 2017). The same concentrations of dopamine have no impact on the frequency of development to the gastrula stage; however, when larvae are exposed to 10 and 100 μM dopamine, they develop smaller archenterons (primitive guts). Dopamine is also responsible for regulating arm length, which affects foraging efficiency in the pluteus stage of sea urchins (*Stronglylocentrotus purpuratus*) (Adams et al. 2011). A week after six-armed *D. excentricus* plutei were exposed to 10 and 100 μM dopamine for an hour, they had shorter arms than control larvae suggesting that dopamine released by *U. obscura* may alter the development of planktonic invertebrate larvae in nearby waters (Rivera Vázquez et al. 2017).

#### 3.3.6.4 The Benefits and Costs of Producing Dopamine

The protection that dopamine provides against a taxonomically wide range of herbivores likely contributes to *U. obscura*'s ability to form large blooms. However, it may also limit the distribution of the alga. By causing damage to *U. obscura*'s tissues when the alga is desiccated, dopamine could limit *U. obscura*'s distribution in the intertidal zone to times of the year when solar irradiances are reduced at low tide or to lower on the shore where it is exposed for shorter periods of time when the tide is out (Nelson et al. 2003b). This hypothesis assumes that the presence of dopamine either causes tissues to become damaged more easily when they are desiccated or that the release of dopamine further damages desiccated tissues. It is also possible that *U. obscura* is more susceptible to desiccation for reasons unrelated to dopamine and the release of dopamine is a response to fatally stressed cells losing membrane integrity. Because ulvoid algae have diffuse growth (cells throughout the thallus can divide), loss of a desiccated part of the thallus does not necessarily cause the death of the entire individual, but it is associated with losses of tissue mass and poorer survival rates (Van Alstyne et al. 2013). Using nitrogen to produce dopamine is also likely to be costly to *U. obscura* because nitrogen is generally a limiting nutrient to algal growth in waters where the alga is found (Nelson et al. 2003a).

## 3.4 CONCLUSION

Many marine and freshwater algae produce compounds that are important neurotransmitters in animals. In most cases, the functions of these compounds are unknown and studies are needed to determine their basic physiological and/or ecological roles. Similarities in neurotransmitters among such evolutionarily disparate groups as animals, terrestrial plants, green and red algae, and brown algae suggest evolutionary convergences in the production of these compounds, though the synthetic pathways used to produce them and their functions may differ among major taxonomic groups. Our current understanding of the synthesis and use of neurotransmitters by algae is limited and even for the best-studied systems critical information is still missing. For example, although ACh-like signal transmission in the charophytes has been studied since the 1930s, to our knowledge, ACh has yet to be identified from this group.

Algae may be good model systems for studying the production and function of neurotransmitters. Because of their large, readily accessible cells, charophytes have been used as models for studying electrical signal transduction. The production of dopamine by *U. obscura* is another potential model system. Recent studies have demonstrated many parallels in *U. obscura* and animals in terms of how dopamine is synthesized, stored, and released, and in the subsequent conversion of dopamine to quinones and melanins, processes that may be important human cognition and neurological diseases (e.g., Meiser et al. 2013; D'Esposito and Postle 2015). Because *U. obscura* is abundant, easily maintained and manipulated in the laboratory, and produces copious amounts of dopamine, it may be a model system for understanding these processes in all organisms.

## ACKNOWLEDGMENTS

This work was funded in part by a grant from the National Science Foundation (IOS-00100768).

## REFERENCES

Adams, D.K., M.A. Sewell, R.C. Angerer, and L.M. Angerer. 2011. Rapid adaptation to food availability by a dopamine-mediated morphogenetic response. *Nature Comm.* 2:592.

Adl, S.M., A.G. Simpson, C.E. Lane, J. Lukeš, D. Bass, S.S. Bowser, M.W. Brown et al. 2012. The revised classification of eukaryotes. *J. Eukaryot. Microbiol.* 59:429–514.

Andersson, L. and L. Bohlin. 1984. Studies of Swedish marine organisms. III: Procedure for the isolation of the bioactive principle, histamine, from the red alga *Furcellaria umbricalis* (Huds.) Lamour. *Acta Pharm. Suec.* 21:373–376.

Barwell, C.J. 1979. Occurrence of histamine in the red alga *Furcellaria lumbricalis* (Huds.) Lamour. *Bot. Mar.* 1:341–344.

Barwell, C.J. 1980. Acetylcholine in the red alga *Laurencia obtusa* (Huds.) Lamour. *Bot. Mar.* 23:63–64.

Barwell, C.J. 1989. Distribution of histamine in the thallus of *Furcellaria lumbricalis*. *J. Appl. Phycol.* 1:341–344.

Beckman, C.H. and W.C. Mueller. 1970. Distribution of phenols in specialized cells of banana roots. *Phytopathology* 60:79–82.

Beilby, M.J. 2016. Multi-scale characean experimental system: From electrophysiology of membrane transporters to cell-to-cell connectivity, cytoplasmic streaming and auxin metabolism. *Front. Plant Sci.* 7:1052.

Beilby, M.J. and H.G.L. Coster. 1979. The action potential in *Chara corallina*. III. The Hodgkin-Huxley parameters for the plasmalemma. *Funct. Plant Biol.* 6:337–353.

Beilby, M.J., S. Al Khazaaly, and M.A. Bisson. 2015a. Salinity-induced noise in membrane potential of Characeae *Chara australis*: Effect of exogenous melatonin. *J. Membrane Biol.* 248:93–102.

Beilby, M.J., C.E. Turi, T.C. Baker, F.J. Tymm, and S.J. Murch. 2015b. Circadian changes in endogenous concentrations of indole-3-acetic acid, melatonin, serotonin, abscisic acid and jasmonic acid in Characeae (*Chara australis* Brown). *Plant Signal. Behav.* 10:e1082697.

Cartier, E.A., L.A. Parra, T.B. Baust, M. Quiroz, G. Salazar, Y. Faundez, L. Egana, and G.E. Torres. 2010. A biochemical and functional complex involving dopamine synthesis and transport into synaptic vesicles. *J. Biol. Chem.* 151: 957–966.

Caulfield, M.P. 1993. Muscarinic receptors—characterization, coupling and function. *Pharmacol. Therapeut.* 58:319–379.

Charlier, R.H. 2008. How Brittany and Florida coasts cope with green tides. *Int. J. Environ. Stud.* 65:191–208.

Chasan, R. and J.I. Schroeder. 1992. Excitation in plant membrane biology. *Plant Cell* 4:1180–1188.

Cheynier, V., G. Comte, K.M. Davies, V. Lattanzio, and S. Martens. 2013. Plant phenolics: Recent advances on their biosynthesis, genetics, and ecophysiology. *Plant Physiol. Biochem.* 72:1–20.

Dayton, P.K. 1985. Ecology of kelp communities. *Ann. Rev. Ecol. Syst.* 16:215–245.

de Alencar, D.B., K.M. dos Santos Pires-Cavalcante, M.B. de Sousa, F.A. Viana, and S. Saker-Sampaio. 2011. Aminas biogênicas em macroalgas marinhas do Estado do Ceará, Brasil. *Revista Ciência Agron.* 42:349–353.

del Campo, J., M.E. Sieracki, R. Molestina, P. Keeling, R. Massana, and I. Ruiz-Trillo. 2014. The others: Our biased perspective of eukaryotic genomes. *Trends Ecol. Evol.* 29: 252–259.

De Le Torre, J.C. 1980. An improved approach to histofluorescence using the SPG method for tissue monoamines. *J. Neurosci. Methods* 3:1–5.

D'Esposito, M. and B.R. Postle. 2015. The cognitive neuroscience of working memory. *Ann. Rev. Psychol.* 66:115–142.

Dettbarn, W.D. 1962. Acetylcholinesterase activity in *Nitella*. *Nature* 194:1175–1176.

Facchini, P.J. 2001. Alkaloid synthesis in plants: Biochemistry, cell biology, molecular regulation, and metabolic engineering applications. *Annu. Rev. Plant Physiol.* 52:29–66.

Facchini, P.J. and V. De Luca. 1995. Phloem-specific expression of tyrosine/dopa-decarboxylase genes and the biosynthesis of isoquinoline alkaloids in opium poppy. *Plant Cell* 7:1811–1821.

Fillafer, C. and M.F. Schneider. 2016. On the excitation of action potentials by protons and its potential implications for cholinergic transmission. *Protoplasma* 253:357–365.

Fong, P., R.M. Donohoe, and J.B. Zedler. 1993. Competition with macroalgae and benthic cyanobacterial mats limits phytoplankton abundance in experimental microcosms. *Mar. Ecol. Prog. Ser.* 100:97–102.

Gong, X.-Q. and M.A. Bisson. 2002. Acetylcholine-activated $Cl^-$ channel in the *Chara* tonoplast. *J. Membrane Biol.* 188: 107–113.

Goreau, T.F. 1963. Calcium carbonate deposition by coralline algae and corals in relation to their roles as reef-builders. *Ann. NY Acad. Sci.* 109:127–167.

Graham, D., S. Tiffany, W.J. Bell, and W. Gutknecht. 1978. Autoxidation versus covalent binding of quinones as the mechanism of toxicity of dopamine, 6-hydroxydopamine, and related compounds toward C1300 neuroblastoma cells *in vitro*. *Mol. Pharmacol.* 14:644–653.

Graham, J.E., L.W. Wilcox, and L.E. Graham. 2008. *Algae*, 2nd ed. San Francisco, CA: Benjamin-Cummings.

Guillot, T.S. and G.W. Miller. 2009. Protective actions of the vesicular monoamine transporter 2 (VMAT2) in monoaminergic neurons. *Mol. Neurobiol.* 39: 149–170.

Guiry, M.D. 2013. Taxonomy and nomenclature of the Conjugatophyceae (= Zygnematophyceae). *Algae* 28:1–29.

Gupta, A., M.R. Vijayaraghavan, and R. Gupta. 1998. The presence of cholinesterase in marine algae. *Phytochemistry* 49:1875–1877.

Gupta, V., R.S. Thakur, C.R.K. Reddy, and B. Jha. 2013. Central metabolic processes of marine macrophytic algae revealed from NMR-based metabolome analysis. *RSC Adv.* 3:7037–7047.

Halliwell, B. and J. Gutteridge. 1989. *Free Radicals in Biology and Medicine*. Oxford, UK: Oxford University Press.

Harley, K.L.S. and A.J. Thorsteinson. 1967. The influence of plant chemicals on the feeding behavior, development, and survival of the two-striped grasshopper, *Melanoplus bivittatus* (Say), Acrididae: Orthoptera. *Can. J. Zool.* 45:305–319.

Hartmann, E. and R. Gupta. 1989. Acetylcholine as a signaling system in plants. In *Plant Biology: Second Messengers in Plant Growth and Development*, vol. 6, (Eds.) Boss, W. and Gupta, R. New York: Alan R. Liss. pp. 257–288.

Johnson, D.A. and B.L. Welsh. 1985. Detrimental effects of *Ulva lactuca* (L.) exudates and low oxygen on estuarine crab larvae. *J. Exp. Mar. Biol. Ecol.* 86:73–83.

Kanazawa, K. and H. Sakakibara. 2000. High content of dopamine, a strong antioxidant, in Canvendish banana. *J. Agric. Food Chem.* 48:844–848.

Kisnierienė, V., V. Sakalauskas, A. Pleskačiauskas, V. Yurin, and O. Rukšėnas. 2009. The combined effect of $Cd_2^+$ and ACh on action potentials of *Nitellopsis obtusa* cells. *Open Life Sci.* 4:343–350.

Kulma, A. and J. Szopa. 2007. Catecholamines are active compounds in plants. *Plant Sci.* 172: 433–440.

Kutchen, T.M., M. Rush, and C.J. Coscia. 1986. Subcellular localization of alkaloids and dopamine in different vacuolar compartments of *Papaver bracteatum*. *Plant Physiol.* 81:161–166.

Laurin-Lemay, S., H. Brinkmann, and H. Philippe. 2012. Origin of land plants revisited in the light of sequence contamination and missing data. *Curr. Biol.* 22(15):R593–R594.

Lee, R.E. 1999. *Phycology*, 3rd ed. Cambridge, UK: Cambridge University Press.

Magre, E.J. 1974. *Ulva lactuca* L. negatively affects *Balanus balanoides* (L.) (Cirripedia Thoracica) in tidepools. *Crustaceana* 27:231–234.

Meiser, J., D. Weindl, and K. Hiller. 2013. Complexity of dopamine metabolism. *Cell Comm. Signal.* 11: 34.

Mueller, W.C. and C.H. Beckman. 1974. Ultrastructure of the phenol-storing cells in the roots of banana. *Physiol. Plant Pathol.* 4:187–190.

Nagatsu, I., Y. Sudo, and T. Nagatsu. 1972. Tyrosine hydroxylation in the banana plant. *Enzymologia* 43: 25–31.

Nelson, T.A. and B.A. Gregg. 2013. Determination of $EC_{50}$ for normal oyster larval development in extracts from bloom-forming green seaweeds. *Nautilus* 127:156–159.

Nelson, T.A., K. Haberlin, A.V. Nelson, H. Ribarich, R. Hotchkiss, K.L. Van Alstyne, L. Buckingham, D.J. Simunds, and K. Fredrickson. 2008. Ecological and physiological controls of species composition in green macroalgal blooms. *Ecology* 89:1287–1298.

Nelson, T.A. and A. Lee. 2001. A manipulative experiment demonstrates that ulvoid algal blooms reduce eelgrass shoot density. *Aquat. Bot.* 71:149–154.

Nelson, T.A., D.J. Lee, and B.C. Smith. 2003a. Are 'green tides' harmful algal blooms? Toxic properties of water-soluble extracts from two bloom-forming macroalgae, *Ulva fenestrata* and *Ulvaria obscura* (Ulvophyceae). *J. Phycol.* 39:874–879.

Nelson, T.A., A.V. Nelson, and M. Tjoelker. 2003b. Seasonal and spatial patterns of "green tides" (ulvoid algal blooms) and related water quality parameters in the coastal waters of Washington state, USA. *Bot. Mar.* 46:263–275.

Nelson, T.A., J. Olson, L. Imhoff, and A.V. Nelson. 2010. Aerial exposure and desiccation tolerances are correlated to species composition in "green tides" of the Salish Sea (northeastern Pacific). *Bot. Mar.* 53:103–111.

Nishida, R. 2014. Chemical ecology of insect–plant interactions: Ecological significance of plant secondary metabolites. *Biosci. Biotechnol. Biochem.* 78:1–13.

Percot, A., A. Yalcın, V. Aysel, H. Erdugan, B. Dural, and K.C. Güven. 2009. β-Phenylethylamine content in marine algae around Turkish coasts. *Bot. Mar.* 52:87–90.

Piñón-Gimate, A., E. Serviere-Zaragoza, M.J. Ochoa-Izaguirre, F. Páez-Osuna. 2008. Species composition and seasonal changes in macroalgal blooms in lagoons along the southeastern Gulf of California. *Bot. Mar.* 51:112–123.

Reise, K. 2002. Sediment mediated species interactions in coastal waters. *J. Sea Res.* 48:127–141.

Ridgway, R.L. unpublished data Rivera Vázquez, Y., K.L. Van Alstyne, and B.L. Bingham. 2017. Exudates of the green alga *Ulvaria obscura* (Kützing) affect larval development of *Dendraster excentricus* (Eschscholtz) and *Crassostrea gigas* (Thunberg). *Mar. Biol.* 164: 194.

Roshchinań, V.V. 2001. *Neurotransmitters in Plant Life*. Enfield, NH: Science Publishers, Inc.

Sargent, P.B. 1993. The diversity of neuronal nicotinic acetylcholine receptors. *Ann. Rev. Neurosci.* 16:403–443.

Schiechl, G., M. Himmelsbach, W. Buchberger, H.H. Kerschbaum, and U. Lütz-Meindl. 2008. Identification of acetylcholine and impact of cholinomimetic drugs on cell differentiation and growth in the unicellular green alga *Micrasterias denticulata. Plant Sci.* 175:262–266.

Segura-Aguilar, J., I. Paris, P. Munoz, E. Ferrari, L. Zecca, and F.A. Zucca. 2014. Protective and toxic roles of dopamine in Parkinson's disease. *J. Neurochem.* 129: 898–915.

Shiina, T. and M. Tazawa. 1987. $Ca^{2+}$-activated $Cl^-$ channel in plasmalemma of *Nitellopsis obtusa. J. Membrane Biol.* 99:137–146.

Smetacek, V. and A. Zingone. 2013. Green and golden seaweed tides on the rise. *Nature* 504:84–88.

Smith, T.A. 1977. Phenylethylamine and related compounds in plants. *Phytochemistry* 16:9–16.

Soares, A.R., R. Marchiosi, R.D.C. Barbarosa de Lima, W. Dantas dos Santos, and O. Ferrarese-Filho. 2014. The role of L-DOPA in plants. *Plant Signal. Behav.* 9:e28275.

Solidoro, C., V.E. Brando, C. Dejak, D. Franco, R. Pastres, and G. Pecenik. 1997. Long-term simulations of population dynamics of *Ulvar.* in the lagoon of Venice. *Ecol. Model.* 102:259–272.

Steiner, M. and T. Hartmann. 1968. The occurrence and distribution of volatile amines in marine algae. *Planta* 79:113–121.

Stokes, A., T. Hastings, and K. Vrana. 1999. Cytotoxic and genotoxic potential of dopamine. *J. Neurosci. Res.* 55:659–665.

Tan, T.C. and Y. Chen. 1992. Enzymatic oxidation of dopamine by polyphenol oxidase. *Int. J. Chem. Kinet.* 24:1023–1034.

Tang, Y.Z. and C.J. Gobler. 2011. The green macroalga, *Ulva lactuca*, inhibits the growth of seven common harmful algal bloom species via allelopathy. *Harmful Algae* 10:480–488.

Tocher, R.D. and J.S. Craigie. 1966a. Enzymes of marine algae: II. Isolation and identification of 3 hydroxytyramine as the phenolase substrate in *Monostroma fuscum. Can. J. Bot.* 44:605–608.

Tocher, R.D. and B.J.D. Meeuse. 1966b. Enzymes of marine algae: I. Studies of phenolase in the *green alga, Monostroma fuscum. Can. J. Bot.* 44:551–561.

Tocher, R.D. and S.C. Tocher. 1972. Dopa decarboxylase in *Cytisus scoparius. Phytochemistry* 11: 1661–1667.

Umrath, K. 1930. Untersuchungen über plasma und plasmaströmung an characeen. *Protoplasma* 9:576–597.

Vadas, R.L. 1977. Preferential feeding: An optimization strategy in urchins. *Ecol. Monogr.* 47:337–371.

Van Alstyne, K.L., K.J. Anderson, D.H. van Hees, and S.-A. Gifford. 2013. Dopamine release by *Ulvaria obscura* (Chlorophyta): Environmental triggers and impacts on the photosynthesis, growth, and survival of the releaser. *J. Phycol.* 49:719–727.

Van Alstyne, K.L., K.J. Anderson, A.K. Winans, and S.-A. Gifford. 2011. Dopamine release by the green alga *Ulvaria obscura* after simulated immersion by incoming tides. *Mar. Biol.* 158:2087–2094.

Van Alstyne, K.L., E.L. Harvey, and M. Cataldo. 2014. Effects of dopamine, a compound released by the green-tide macroalga *Ulvaria obscura* (Chlorophyta), on marine algae and invertebrate larvae and juveniles. *Phycologia* 53:195–202.

Van Alstyne, K.L., A.V. Nelson, J.R. Vyvyan, and D.A. Cancilla. 2006. Dopamine functions as an antiherbivore defense in the temperate green alga *Ulvaria obscura. Oecologia* 148:304–311.

Van Alstyne, K.L., T.A. Nelson, and R.L. Ridgway. 2015. Environmental chemistry and chemical ecology of "green tide" seaweed blooms. *Int. Comp. Biol.* 55:518–532.

van Hees, D.H., and K.L. Van Alstyne. 2013. Effects of emersion, temperature, dopamine, and hypoxia on extracellular oxidant accumulations surrounding the bloom-forming seaweeds *Ulva lactuca* and *Ulvaria obscura. J. Exp. Mar. Biol. Ecol.* 448:207–213.

Wang, Y., B. Zhou, and X.X. Tang. 2009. Effects of two species of macroalgae-Ulva pertusa and Gracilaria lemaneiformis-on growth of Heterosigma akashiwo (Raphidophyceae). *J. Appl. Phycol.* 21:375–385.

Williams, T.A. 2014. Evolution: Rooting the eukaryotic tree of life. *Curr. Biol.* 24:R151–R152.

Ye, N.H., X.W. Zhang, Y.Z. Mao, C.W. Liang, D. Xu, J. Zou, Z.M. Zhuang, and Q.Y. Wang. 2011. 'Green tides' are overwhelming the coastline of our blue planet: Taking the world's largest example. *Ecol. Res.* 26:477–485.

# *Section II*

## *Role of Neurotransmitters in Regulation of Growth and Development*

# 4 Melatonin and Serotonin in Plant Morphogenesis and Development

*Lauren A.E. Erland and Praveen K. Saxena*

## CONTENTS



## 4.1 INTRODUCTION

Melatonin (*N*-acetyl-5-methoxytryptamine) and serotonin (5-hydroxytryptamine) are best known for the roles they play in the vertebrate nervous system. Melatonin has been referred to as the chemical expression of darkness due to its important role in maintaining circadian rhythms with increasing levels indicating nighttime and decreasing levels in the morning signaling the arrival of daybreak (Klein and Weller 1970; Reiter 1993; Brzezinski 1997). Melatonin is produced primarily by the pineal gland in humans, where it was first isolated and identified; however, diverse extra-pineal sources have been identified in recent years including the reproductive and gastrointestinal systems (Lerner et al. 1958; Venegas et al. 2012; Acuña-Castroviejo et al. 2014). Since then, melatonin has also been recognized as an important regulator for diverse other rhythms in animals including migration and reproduction (Duncan et al. 1990; Cavallo and Ritschel 1996; Arendt 1998). Further, significant interest in this compound has identified roles for melatonin in mitigating diverse disease processes (Petrie et al. 1993; Szczepanik 2007; Arendt et al. 2008; Rios et al. 2010; Lanfumey et al. 2013; Ma et al. 2015; Reiter et al. 2016). Serotonin, by comparison, is a neurotransmitter across the nervous system, and is recognized as a *happy* molecule, with low levels being associated with depression and anxiety disorders. As a result, drugs or medicinal plants that have hallucinogenic or other mind-altering results have been found to have significant effects on serotonin signaling and levels in the central nervous system. In plants, serotonin was first identified in the 1950s in the medicinal plant cowhage (*Mucuna pruriens* L.) about a decade after its discovery in vertebrates (Bowden et al. 1954). Melatonin by contrast was not identified in plants until the mid-1990s, being described in two separate reports on seeds and edible plants (Dubbels et al. 1995; Hattori et al. 1995). By comparison, melatonin was first described in the pigmented skin of frogs in the early 1900s, and it was isolated from the bovine pineal gland and chemically characterized in the 1950s (Lerner et al. 1958, 1959).

The biosynthetic pathway for melatonin and serotonin in plants is similar to that present in humans, with one major exception (Murch et al. 2000). The first intermediate in the pathway in humans is 5-hydroxytryptophan (5-HTP), which is produced via hydroxylation of tryptophan by tryptophan hydroxylase (TPH), while in plants tryptophan is decarboxylated to form the amino acid tryptamine by tryptophan decarboxylase (TDC) (Simonneaux and Ribelayga 2003; Kang et al. 2007b). In both cases, biosynthesis then proceeds via serotonin in reactions catalyzed by aromatic amino acid decarboxylase (AADC) in humans and tryptamine 5-hydroxylase (T5H) in plants (Kang et al. 2007a, 2008). This is then followed by acetylation of serotonin by serotonin-*N*-acetyltransferase (SNAT) to form *N*-acetylserotonin (NAS), followed by methylation by acetylserotonin-*O*-methyltransferase (ASMT, also known as hydroxyindole-*O*-methyltransferase; HIOMT) to form melatonin (Kang et al. 2011; Park et al. 2014). As is often the case in plants, however, there have been several alternative biosynthetic routes that have been identified to date including direct conversion of tryptamine to NAS by SNAT, conversion of serotonin to melatonin via the alternative intermediate 5-methoxytryptamine (5-MT) and the involvement of the phenylpropanoid biosynthetic enzyme caffeic acid-*O*-methyltransferase (COMT; Figure 4.1) (Byeon et al. 2014a; Lee et al. 2014; Byeon et al. 2015; Back et al. 2016; Tan et al. 2016). Indeed, the importance of COMT in melatonin biosynthesis has been highlighted by recent studies, for example in *Escherichia coli*, expression of ASMT alone led to poor yields for melatonin, while cotransformation with COMT led to high levels of melatonin production (Byeon and Back 2016a).

As compared with other organisms, plants are unique in that their cells possess the potential for indeterminate growth. This is due to the capacity for differentiation, de-differentiation or re-differentiation of post-embryonic cells, a process referred to as totipotency. A strict balance of plant growth regulating substances and their associated signaling pathways govern the expression of totipotency and subsequent growth and development of tissues and organs. Auxins and cytokinins were the first classes of plant growth regulators that were identified as important in determining the morphogenetic outcomes of plant cells, exerting their effects via antagonistic interactions (Skoog and Miller 1957). For instance, the ratio of auxin to cytokinin is a major determinant in the final growth pattern and has been applied in *in vitro* culture systems where high auxin leads to increased root growth, while high cytokinin leads to increased shoot production. A balance between the two leads to undifferentiated cellular growth referred to as callus (Figure 4.2). Since this discovery, many new classes of plant growth regulators, which work to fine tune plant growth, have been identified with the major classes being recognized as: auxins, cytokinins, gibberellins, jasmonic

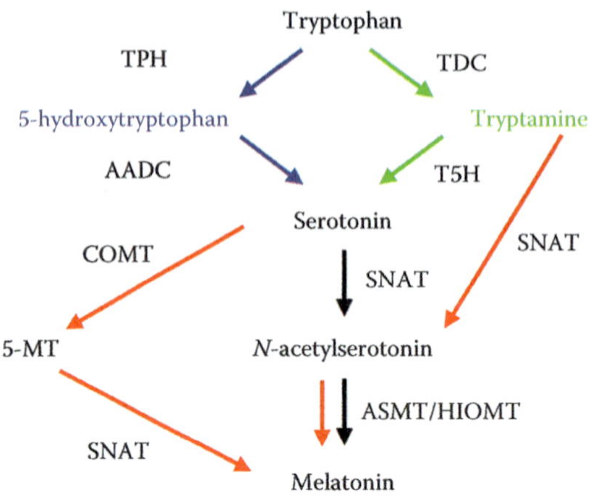


**FIGURE 4.1** Biosynthetic routes for melatonin and serotonin in plants. Green—plant specific pathway; blue—animal specific pathway; red-alternative pathway. AADC—arylalkylamine *N*-acetyltransferase); ASMT—acetylserotonin-*O*-methyltransferase; COMT—caffeic acid *O*-methyltransferase; HIOMT—hydroxyindole-*O*-methyltransferase; 5-MT—5-methoxytryptamine; SNAT—serotonin *N*-acetyltransferase; T5H—tryptamine 5-hydroxylase; TDC—tryptophan decarboxylase; TPH—tryptophan hydroxylase.

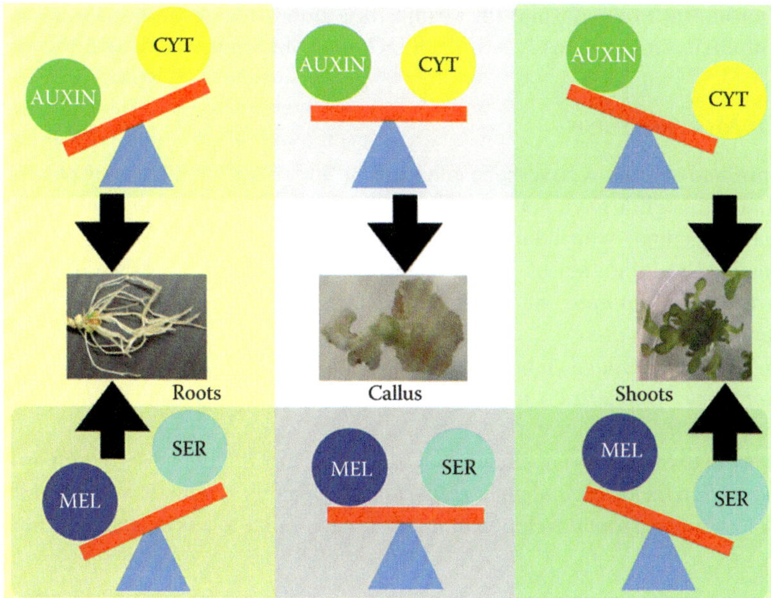


**FIGURE 4.2** Overview of the melatonin/serotonin balance and its relation to that observed for auxin and cytokinins. Aux—auxin; Cyt—cytokinin; Mel—melatonin; Ser—serotonin.

acid, abscisic acid, brassinosteroids, and ethylene (Mandava 1988; Amasino 2005; Woodward and Bartel 2005; Bari and Jones 2009; Werner and Schmülling 2009; Enders and Strader 2015; Gantait et al. 2015). There exists, however, an ever-increasing interest toward nontraditional classes of plant growth regulators and their roles in directing morphogenesis and plant development. One such class is the indoleamine neurotransmitters serotonin and melatonin, whose discovery in plants is relatively recent (Erland et al. 2015). Though their classification has been difficult, they have recently been proposed as a novel class of plant growth regulator (Ramakrishna et al. 2011; Erland et al. 2015, 2016). The rationale for new classification will be discussed in this chapter and is based on the diverse roles these compounds have been found to play in plants, with special focus on *in vitro* plant tissue culture experiments, as they have provided the best evidence for this proposal.

## 4.2 EFFECTS OF MELATONIN AND SEROTONIN ON PLANT GROWTH

Melatonin and serotonin have been found to have diverse effects on plant growth and morphogenesis. In general, melatonin has been found to have a positive effect on root development, while serotonin appears to support shoot growth. However, despite increasing interest in indoleamines in the current plant biology literature, much remains unknown about the physiology of these processes. For instance, rather than acting as a universal regulator (i.e., auxin) melatonin and serotonin appear to act as gatekeepers, with their effects largely occurring through mediation of other signaling pathways as will be discussed in the following sections. Furthermore, the balance that seems to exist between serotonin and melatonin during morphogenesis is of particular interest, as it parallels that which has been observed between auxins and cytokinins. Given melatonin is a metabolite of serotonin, they are very closely related biosynthetically, allowing for a remarkable capacity for very fine control of the pathway. Despite the apparent importance of both compounds in plant processes, the majority of literature has focused solely on melatonin. In spite of this, literature does exist on serotonin. Furthermore, some information can indirectly be drawn from studies on melatonin that utilized transgenic plants, as these studies neglected to quantify serotonin levels, despite the strong likelihood that serotonin levels would vary in response to modification of SNAT, ASMT, TDC, or ASMT/HIOMT levels.

### 4.2.1 Root Morphogenesis

Both melatonin and serotonin have been found to be important mediators of plant root development, potentially though a phytohormone-like action. For example, both compounds function in a dose-dependent manner with relatively low concentrations having positive effects, while high concentrations lead to inhibition of growth. These effects have been observed in diverse species including *Arabidopsis thaliana*, St. John's wort (*Hypericum perforatum* L.), mimosa (*Mimosa pudica* L.), lupin (*Lupinus albus* L.), cucumber (*Cucumis sativus* L.), mung bean (*Vigna radiata* L.), sunflower (*Helianthus annuus* L.), barley (*Hordeum vulgare* L.), rice (*Oryza sativa*), pomegranate (*Punica granatum* cv "Wonderful"), tomato (*Solanum lycopersicum* L.), and walnut (*Juglans regia* L.) among others (Table 4.1).

The first system to explore the application of exogenous melatonin and serotonin on root development was observed in the medicinal plant *H. perforatum*. In this system, it was found that melatonin was capable of promoting root development in a manner similar to auxin, and treatment of plantlets

**TABLE 4.1**
**Effects of Melatonin and Serotonin on Plant Root Growth and Differentiation**

| Effect | Melatonin? | Serotonin? | Species | Reference |
|---|---|---|---|---|
| Increased lateral root number/ formation | × | × | *A. thaliana*; *C. sativus*; *L. albus*; *O. sativa* | Arnao and Hernández-Ruiz (2007); Pelagio-Flores et al. (2011); Pelagio-Flores et al. (2012); Koyama et al. (2013); Zhang et al. (2013b); Zhang et al. (2014); Liang et al. (2017) |
| Increased adventitious root number/ formation | × | × | *A. thaliana*; *L. albus*; *O. sativa*; *Prunus* spp.; *S. lycopersicum*; *Withania somnifera* | Arnao and Hernández-Ruiz (2007); Pelagio-Flores et al. (2011); Park and Back (2012); Pelagio-Flores et al. (2012); Sarropoulou et al. (2012b); Byeon et al. (2014b); Adil et al. (2015); Wen et al. (2016) |
| Increased root number | × | | *H. perforatum*; *Brassica juncea*; *Prunus* spp.; *P. granatum* | Murch et al. (2000); Murch et al. (2001); Murch and Saxena (2002); Chen et al. (2009); Sarropoulou et al. (2012a); Sarropoulou et al. (2012b); Sarrou et al. (2014) |
| Increased root elongation/ length | × | × | *H. vulgare*; *H. annuus*; *V. radiata*; *Prunus* spp.; *L. albus*; *O. sativa*; *A. thaliana*; *S. lycopersicum* | Csaba and Pal (1982); Arnao and Hernández-Ruiz (2007); Park and Back (2012); Sarropoulou et al. (2012a); Sarropoulou et al. (2012b); Szafrańska et al. (2012); Bajwa et al. (2014); Mukherjee et al. (2014); Wen et al. (2016) |
| Inhibition primary root growth | × | × | *A. thaliana*; *B. juncea*; *O. sativa*; *L. albus* | Hernández-Ruiz et al. (2004); Chen et al. (2009); Pelagio-Flores et al. (2011); Park and Back (2012); Pelagio-Flores et al. (2012); Bajwa et al. (2014); Wang et al. (2016); Pelagio-Flores et al. (2011) |
| Induction of rooting | | × | *Juglans* spp. | Gatineau et al. (1997) |

with auxin transport and action inhibitors 2,3,5-triiodobenzoic acid (TIBA) and p-chlorophenoxy-isobutyric acid (PCIB) led to both a decrease in endogenous melatonin content as well as decreases in root production. The effect described earlier could not be fully recovered with auxin treatment, suggesting an important role for melatonin (Murch et al. 2001). The use of inhibitors in *in vitro* systems to study the effects of the indoleamines on plant morphogenesis was continued in mimosa where both serotonin and melatonin treatment led to an increase in root production, though the effects appeared to be more significant for melatonin treatment (Ramakrishna et al. 2009). It is also interesting to note that the effect of serotonin treatment may be more difficult to understand as increased supply of serotonin may function both directly while also increasing endogenous melatonin levels as it is a precursor for melatonin. Given both the upstream enzyme TDC and downstream enzyme SNAT have been shown to be important regulatory steps in the pathway (Kang et al. 2007b, 2008; Byeon and Back 2016b), it would be interesting to see a greater number of morphogenetic studies quantify endogenous levels of these two indoleamines after treatment. Overall, it appears that melatonin in particular acts via a unique melatonin-specific mechanism versus acting directly as an auxin (although melatonin does have auxin-like activity and may accomplish some of its effects through modulation of auxin action). For instance, a report in maize utilized the classical auxin assays: coleoptile elongation while also monitoring ACC (1-aminocyclopropane-1-carboxylic acid) synthase gene expression (an important ethylene biosynthetic enzyme closely regulated by auxin) and found that melatonin did not have comparable effects to auxin (Kim et al. 2016).

To date, the system that has best characterized the effects and mechanisms of melatonin and serotonin treatments on root system architecture involve the model plant *Arabidopsis*. Pelagio-Flores et al. (2011) utilized *A. thaliana* auxin mutants, which were knockouts for auxin transport (AUX1), action/signaling (DR5), biosynthesis (BA3) and localization and ubiquitination (AXR 1, 2, and 4), and found that serotonin functioned via repression of auxin activity in primary and adventitious roots as well as lateral root primordia. Serotonin was also found to promote growth of pre-existing lateral root primordia via this inhibition, leading to maturation and development of lateral roots. Additionally, though this action was independent of AUX 1 and AXR 4 it was found to require AXR 1 and 2, suggesting that the primary mechanism of serotonin in this capacity is as an auxin inhibitor (Pelagio-Flores et al. 2011).

In comparison, melatonin was found to regulate root architecture through auxin-independent mechanisms (Pelagio-Flores et al. 2012). This study found that melatonin enhanced lateral and adventitious root growth but did not activate auxin-inducible gene expression suggesting that it did not require auxin signaling. A subsequent study supported these results, finding that melatonin suppressed primary root growth via inhibition of root meristem size, which was attributed to modification of auxin transport, particularly PIN1, 3 and 7 and down-regulation of auxin biosynthesis with YUC1, 2, 5, 6, and TAR2 being decreased, though three other YUC transcripts were upregulated (Wang et al. 2016). In contrast, a study in rice (*O. sativa*) seedlings also found that melatonin promoted lateral root growth (length and number) but inhibited length of embryonic roots. RNA-seq experiments found that this was due to upregulation of genes associated with auxin signaling by melatonin treatment (Liang et al. 2017). Another transcriptomics study also found that treatment of *Arabidopsis* with melatonin generally led to down-regulation of transcripts associated with auxin signaling as well as down-regulation of pathways associated with cell wall synthesis, while also simultaneously enhancing expression of calcium signals and stress-related hormone pathways such as salicylic acid (SA), ethylene, jasmonic acid (JA), and abscisic acid (ABA) (Weeda et al. 2014). The auxin-dependent action of melatonin in promoting root growth was also found to be important in tomato (*S. lycopersicum*). A study by Wen et al. (2016) found that melatonin was capable of enhancing adventitious root formation in de-rooted tomato seedlings. This study is unique in that it also examined the involvement of nitric oxide (NO) signaling in this effect. The authors found that treatment with exogenous melatonin (12.5–75 μM) enhanced production of adventitious roots, and this effect was associated with the accumulation of NO in tissues. The accumulation of indole-3-acetic acid (IAA) levels in the hypocotyl as well as the observation that exogenous application of NO treatment could enhance endogenous melatonin production in de-rooted seedlings reaffirms further the idea that interactions between the two pathways is occurring. Through the application of NO scavengers and gene expression analysis, it was additionally found that NO functioned downstream of melatonin via down-regulation of *S*-nitrosoglutathione reductase (GSNOR) and that melatonin mediated expression of several auxin transport and signaling genes (Wen et al. 2016).

Overall, melatonin and serotonin appear to play an important role in mediating plant root growth. In general, melatonin treatment appears to have a stronger effect on root growth than does serotonin treatment. Additionally, it appears in many cases that melatonin treatment inhibits primary root growth, and instead promotes branching by increasing the number of lateral and adventitious roots, while also promoting root elongation. This effect may be due to changes in endogenous concentrations as studies using transgenic systems such as rice and *Arabidopsis* have observed this type of response after exogenous application (Arnao and Hernández-Ruiz 2007; Chen et al. 2009; Park and Back 2012; Sarropoulou et al. 2012a, 2012b; Szafrańska et al. 2012; Zhang et al. 2013b; Byeon and Back 2014; Mukherjee et al. 2014; Sarrou et al. 2014; Erland et al. 2015). In addition to the interaction with auxin, several other potential mechanisms for promotion of root growth by melatonin have been reported. For example, in several cherry rootstocks (*Prunus cerasus* L., *P. avium* × *P. cerasus*, *P. cerasus* × *P. canescens*, and *P. avium* × *P. mahaleb*) melatonin action was not only suggested to be similar to the effects of auxin but also to enhance total carbohydrate and proline content, while also enhancing chlorophyll (a and b) levels (Sarropoulou et al. 2012a, 2012b). Similar to the above, improvement of photosynthetic pigment levels by melatonin was associated with improved lateral rooting in cucumber (Zhang et al. 2013b). Furthermore, a similar mechanism was also suggested in maize (*Zea mays* L.) where 10 μM of melatonin enhanced overall root biomass, although at higher concentrations (0.1–1 mM) melatonin showed a distinct inhibitory effect. In treated maize seedlings, enhanced root biomass was associated with enhanced leaf growth and modified carbohydrate metabolism. It was found that melatonin treatment was associated with accelerated nighttime starch metabolism, increased sucrose transport, and enhanced hexokinase activity as well as promoting photosynthetic activity (Zhao et al. 2015a).

Though this article does not address the significant body of knowledge examining the important roles melatonin and serotonin play in mitigating plant stress, this modification of root branching is likely of particular value to plants that are growing under low nutrient or drought conditions. Thus, it is possible that the role of melatonin and serotonin in mediating growth patterns could be related to their apparent original role in evolution as stress-mitigating compounds (Tan et al. 2012, 2014; Manchester et al. 2015). For an in-depth review of the role of melatonin and serotonin in plant stress mitigation and survival, readers are referred to several recent reviews (Akula and Ravishankar 2014; Arnao and Hernández-Ruiz 2014; Hardeland 2016). Taken as a whole, there is still a need for further research into the role that melatonin and serotonin play in plant root system architecture, particularly as mechanisms appear to be species dependent.

### 4.2.2 Shoot Morphogenesis

Interestingly, despite the relative wealth of mechanistic studies into the role indoleamines play during root differentiation, there are relatively fewer investigations into the role they play in shoot morphogenesis (Table 4.2). This is likely due to the fact that melatonin and serotonin were originally thought to act as auxins, which in *in vitro* studies is most closely associated with root production. Hypocotyl and coleoptile elongation assays, classical systems for the study and screening of auxins, as well as the discovery of new auxins and homologs, have provided some contradictory information on this point. For instance, the most recent study by Kim et al. (2016) discussed in earlier sections of this review concluded that in maize melatonin does not function as an auxin, as it did not show significant homology to IAA activity in the maize hypocotyl elongation assays. In contrast, earlier studies using lupin hypocotyls (Hernández-Ruiz et al. 2004; Arnao and Hernández-Ruiz 2007), oat, wheat, canary grass (*Phalaris canariensis* L.), and barley

**TABLE 4.2**
**Effects of Melatonin and Serotonin on Plant Aerial Growth and Differentiation**

| Effect | Melatonin? | Serotonin? | Species | Reference |
|---|---|---|---|---|
| Increased shoot production | × | × | *P. granatum*; *Vaccinium corymbosum*; *M. pudica*; *H. perforatum*; *Coffea canephora* | Murch et al. (2001); Litwinczuk and Wadas-Boron (2009); Ramakrishna et al. (2009); Ramakrishna et al. (2011); Sarrou et al. (2014) |
| Increased hypocotyl elongation | × | × | *H. annuus*; *L. albus* | Hernández-Ruiz et al. (2004); Mukherjee et al. (2014) |
| Increased cotyledon expansion | × | | *L. albus* | Hernández-Ruiz and Arnao (2008) |
| Increased coleoptile elongation | × | | *P. canariensis*; *Avena sativa*; *Triticum aestivum*; *H. vulgare* | Hernández-Ruiz et al. (2005) |
| Promotion of germination | × | × | *C. sativus*; *Hippeastrum hybridium* | Roshchina (2001); Posmyk et al. (2009) |
| Modification of stem/leaf branching patterns | × | | *O. sativa*; *S. lycopersicum* | Okazaki et al. (2010); Wang et al. (2013) |
| Inhibition seedling growth | × | | *Zea mays* | Zhao et al. (2015a) |
| Increased seedling growth | × | | *O. sativa*; *L. albus*; *Brassica olereacea rubrum*; *Z. mays*; *G. max* | Hernández-Ruiz et al. (2004); Hernández-Ruiz et al. (2005); Hernández-Ruiz and Arnao (2008); Posmyk et al. (2008); Byeon and Back (2014); Wei et al. (2015); Zhao et al. (2015a) |

(Hernández-Ruiz et al. 2005) found similar effects for melatonin as compared with auxin. In lupin, melatonin was again found to promote hypocotyl elongation, with the effect being most significant in de-rooted hypocotyls leading the authors to suggest it could be an auxinic hormone (Hernández-Ruiz et al. 2004). Specifically, in the coleoptile elongation assays melatonin showed effects of 10%–55% of that observed for coleoptile elongation induced by IAA, while showing similar primary root growth inhibition as observed after IAA treatment. Additionally, the authors quantified IAA and melatonin levels in the untreated tissues and found similar levels and therefore concluded that these results supported prior suggestions that melatonin may be acting as an auxinic hormone. Follow-up studies in cotyledons also found that melatonin was capable of promoting expansion of lupin cotyledons similarly to that observed in response to IAA treatment.

Serotonin has been found to have a positive effect on shoot growth and multiplication *in vitro*. The same keystone paper that identified melatonin as an important promoter of root morphogenesis in *H. perforatum* also found that serotonin was an important regulator of shoot production in these *in vitro* cultures, with serotonin promoting increased shoot production. Treatment with several mammalian serotonin inhibitors, including p-chlorophenylalanine (p-CPA), d-amphetamine, fluoxetine (Prozac), methylphenidate (Ritalin), and hydralazine were found to inhibit this effect (Murch et al. 2001). These results were again suggested to be related to modification of auxin activity (Murch et al. 2001; Murch and Saxena 2004). Interestingly, melatonin and serotonin appear to be important in directing the growth effects of the synthetic substituted phenylurea thidiazuron (TDZ) (Murch and Saxena 2004; Jones et al. 2007). This synthetic growth regulator was originally developed as a cotton defoliant but has since been identified to have pleiotropic effects in morphogenesis, which range from promotion of callus production to intact seedling regeneration and somatic embryo production. TDZ is often referred to as a cytokinin for simplicity; however, due to its diverse range of effects, it also appears to have auxin-like activity, making its classification more complex (Saxena et al. 1992; Murthy et al. 1995; Murthy et al. 1998; Erland and Mahmoud 2014). Etiolated hypocotyls of *H. perforatum* treated with TDZ generally showed increased shoot production; however, co-treatment with p-CPA, Ritalin, Prozac, or amphetamine reduced this shoot production by up to 50%. Use of radiolabeled tryptophan further showed that serotonin production was enhanced by treatment with TDZ when inhibitors were absent (Murch and Saxena 2004). A study in *Echinacea purpurea* (L.) further demonstrated a connection between TDZ activity and the indoleamines. Treatment of echinacea leaf explants with TDZ resulted in increased endogenous melatonin and serotonin levels and resulted in a mixture of shoot organogenesis and somatic embryo production. From these studies, the authors concluded that serotonin and melatonin may function as a downstream signal for TDZ in association with calcium and auxin signaling (Jones et al. 2007). Serotonin and melatonin have also been found to promote shoot multiplication and organogenesis in mimosa. This study found that exogenous treatment with serotonin could increase percent of explants with shoots by up to 65%, while also increasing the total number of shoots produced from a single shoot (up to 22 shoots per explant), shoot height, and total biomass. Melatonin was slightly reduced, but still had a significant effect on shoot production with a 55% increase and an increase by 10 shoots per explant as compared to the control. Interestingly, this study not only examined the effects of the serotonin inhibitors p-CPA and Prozac, but also examined the effects of several calcium inhibitors, as well as calcium supplementation, and it was found that calcium played an important role during this response. For instance, co-application of calcium with melatonin or serotonin could further increase the percent of explants with shoots by up to 20%, while increasing the number of shoots produced by up to an additional 7 shoots. Additionally, treatment with calcium inhibitors or chelators almost completely reversed the promotable effect of serotonin or melatonin treatment and suggests that melatonin and serotonin are not only important mediators of morphogenetic outcomes, but also that their effect is dependent upon calcium signaling (Ramakrishna et al. 2009).

Both melatonin and serotonin have also been shown to have diverse effects on early seedling germination and growth. These studies encompass both exogenous application and studies examining the modified growth patterns of transgenic rice, tomato, and *Arabidopsis* lines, which have knockouts or up-regulation of genes in the indoleamine biosynthetic pathway. For example, experiments using *Micro-Tom* tomatoes (overexpression of the last biosynthetic enzyme in melatonin biosynthesis, HIOMT, or its mammalian analog AANAT (arylalkylamine *N*-acetyltransferase), individuals were found to have a modified branching structure, suggesting a role for melatonin in mediating growth patterns. It is interesting, however, to note that although the authors also found decreased auxin content in these mutants, serotonin levels were not determined. Given serotonin has been found to have diverse effects on shoot production, it is possible that these mutants may be responding to decreased serotonin levels resulting from the accelerated conversion of serotonin (via NAS) to melatonin in tissues. In rice, which were transformed to overproduce TDC (the first step in the biosynthetic pathway), an 11 to 25-fold increase in serotonin was observed depending on tissue type. Although melatonin was not quantified in this study, increased serotonin content was associated with shorter plants, again suggesting a role in growth patterning (Kang et al. 2007b). Indeed, the importance of T5H was highlighted in another study in rice, which found that overexpression of T5H coupled with supplementation with tryptamine could increase serotonin content, and supplementation was associated with modified root growth patterns (Kang et al. 2007a).

In soybean (*Glycine max*), melatonin pretreatment of seeds also had a positive effect on seedling growth leading to increased plant height, leaf size, and improved pod and seed number (Wei et al. 2015). While in maize, exogenous melatonin treatment led to improved seedling growth, which was attributed to enhanced sucrose metabolism. In particular, melatonin treatment was found to increase hexokinase activity, phloem transport of sucrose, and photosynthesis (Zhao et al. 2015a). At high doses, melatonin was also found to be capable of enhancing leaf area and slowing leaf senescence in *Arabidopsis*, while at low concentrations root growth was promoted. This is interesting, as at high doses it is possible that melatonin may be having other regulatory effects on the indoleamine pathway, though this has not been investigated (Hernández et al. 2015).

Melatonin and serotonin treatment have also been associated with less conventional means of morphogenesis, such as somatic embryogenesis. Somatic embryo production was found to be promoted in response to melatonin and serotonin treatment in coffee (*Coffea canephoora* Pierre ex Froehn.) This is of special interest due to the connection between TDZ (a potent inducer of somatic embryogenesis) and melatonin and serotonin suggested by Murch et al. (2004) and Jones et al. (2007) (see previous paragraph). If TDZ functions even partially by modifying the melatonin–serotonin balance, this may lead to an interesting new role for melatonin and serotonin.

## 4.3 MECHANISMS AND SIGNALS

Though initial investigations may have focused on discovering a simple explanation for the action of melatonin and serotonin, it has become clear that their mechanisms and functions are much more nuanced. Though melatonin and serotonin treatment may appear similar to auxin or cytokinin effects, they do not work to mimic the functions of these growth regulators. Instead, it seems more likely that they act to modulate these pathways and others, acting as gatekeepers for plant signals leading to diverse signals. Some of the plant signals proposed to be modulated by melatonin and serotonin include: calcium, cAMP, phytohormones (including auxin, JA, GA, and ABA among others), sucrose/carbon, nitrogen, phytochrome, and reactive oxygen species (ROS). Melatonin and serotonin have also been extensively shown to effect gene expression patterns in tissues with both targeted expression analyses and large-scale RNA-Seq/trancriptomics studies having been undertaken (Zhang et al. 2013a; Weeda et al. 2014; Hu et al. 2016). Several large-scale metabolomics studies have also helped to elucidate the broad effects of melatonin treatment across both primary and specialized metabolism (Qian et al. 2015; Zhao et al. 2015a, 2015b). It is noteworthy that the majority of these studies have focused on the mechanisms of melatonin treatment to combat diverse stresses including cold, salinity, metal, and heat among others. Therefore, there still exists a need for further studies specifically examining serotonin and examining the mechanisms of melatonin and serotonin in normal growth and developmental processes in the absence of an applied stress.

One of the best-defined mechanisms of melatonin and serotonin action is their role as potent antioxidants both directly and via modulation of other plant antioxidant systems such as the ascorbate-glutathione cycle and superoxide dismutase and peroxidase activity (Tan et al. 2002, 2015; Rosen et al. 2006; Reiter et al. 2007; Balzer and Hardeland 2014; Bajwa et al. 2015). In addition to acting to detoxify ROS and other reactive species, including reactive nitrogen species, quenching of these species may also have important impacts on signaling networks, as ROS are increasingly acknowledged as important signaling molecules and not just as toxic compounds (Mittler et al. 2011).

There are also limited studies that exist examining the localization of melatonin and serotonin in plant tissues. Many studies use broad exogenous treatment, for example, foliar sprays or liquid treatment of the roots, but the actual location of action is unclear from these reports. A better understanding of the location of action and location of biosynthesis of melatonin will assist in better understanding their functions. Serotonin has been found to localize to vascular tissues, such as the xylem and phloem of cultured explants of coffee, and recent reports have found that melatonin can undergo long-distance transport via the phloem (Ramakrishna et al. 2011; Li et al. 2017). In animals, melatonin is known to be synthesized by the mitochondria in extra-pineal locations such as the reproductive tract or gastrointestinal system (Acuña-Castroviejo et al. 2014). Due to this and the fact that the chloroplast and mitochondria have both evolved from some of the most ancient life forms on Earth serving as locations for high levels of ROS, it has been extensively hypothesized that melatonin and serotonin would be synthesized in the chloroplast and/or mitochondria of plants. Some of the first reports supporting this came in 2014 and examined localization of the biosynthetic enzymes SNAT and ASMT in rice and *Arabidopsis* where it was found to be localized to the chloroplast, while enzymes cloned from animals showed differential localization to the cytoplasm (Byeon et al. 2013, 2014b; Lee et al. 2014). Interestingly, the enzyme catalyzing an alternative route, COMT, was found to localize to the cytoplasm (Lee et al. 2014). A more recent study also demonstrated that melatonin could be synthesized by isolated chloroplasts, further supporting this hypothesis. Despite these monumental efforts, much more work is still required to demonstrate the location of serotonin biosynthesis, as these chloroplasts were fed NAS, and the aforementioned localization studies are for conversion downstream of serotonin (Zheng et al. 2017).

Despite the wealth of studies now available examining phyto-melatonin and serotonin, the search for a receptor for these compounds is still elusive. To date, a single protein has been found in complex with melatonin, a quinone reductase-like receptor cloned from *H. perforatum*. Unfortunately, downstream signaling of this has not be characterized, and only a crystal structure is currently available (Sliwiak et al. 2016). No receptor has been found for phytoserotonin, which brings about the possibility that these compounds may function in a manner other than that of a traditional phytohormone, with a nontraditional receptor.

### 4.3.1 Applications and Implications in Conclusion

As melatonin and serotonin are increasingly acknowledged as important plant growth regulators, their application to modern agricultural and biotechnology continue to expand as well. In addition to opening new avenues for the direct use of these compounds, from anti-browning, or anti-stress agents or promoters of plant biomass accumulation and enhanced yield, understanding the important functions of these compounds in plants will also help direct current strategies. This is particularly true as their action as gatekeepers of other processes becomes clear and may influence the way in which broader plant signaling and phytohormone networks are manipulated and viewed. Melatonin and to a lesser extent serotonin have been suggested to have many beneficial effects in modern plant systems, including toward the development of new technologies such as cryopreservation and cryobanks where they serve as potent protective molecules (Zhao et al. 2011; Uchendu et al. 2013, 2014; Popova et al. 2016). Additionally, they may be applied to micropropagation and plant tissue culture strategies to enhance morphogenesis in recalcitrant or endangered species due to their ability to mediate growth outcomes. In summary, though there is a significant body of information accumulating, the majority of these studies focus on the functions and mechanisms, of melatonin especially, in response to diverse biotic and abiotic stresses. There still exists a need for further studies focusing specifically on the roles these compounds play under unstressed conditions and through plant morphogenesis and development as well as studies targeted toward serotonin action in order to start untangling the individual roles of these two important molecules.

## REFERENCES

Acuña-Castroviejo D, Escames G, Venegas C et al. 2014. Extrapineal melatonin: Sources, regulation, and potential functions. *Cell Mol Life Sci* 71:2997–3025.

Adil M, Abbasi BH, Khan T. 2015. Interactive effects of melatonin and light on growth parameters and biochemical markers in adventitious roots of *Withania somnifera* L. *Plant Cell Tiss Organ Cult* 123:405–412.

Akula R, Ravishankar GA. 2011. Influence of abiotic stress signals on secondary metabolites in plants. Plant Signal Behav 6:1720–1731.

Amasino R. 2005. 1955: Kinetin Arrives. The 50th anniversary of a new plant hormone. *Plant Physiol* 138:1177–1184.

Arendt J. 1998. Melatonin and the pineal gland: Influence on mammalian seasonal and circadian physiology. *Rev Reprod* 3:13–22.

Arendt J, Van Someren EJ, Appleton R, Skene DJ, Akerstedt T. 2008. Clinical update: Melatonin and sleep disorders. *Clin Med* 8:381–383.

Arnao MB, Hernández-Ruiz J. 2007. Melatonin promotes adventitious- and lateral root regeneration in etiolated hypocotyls of *Lupinus albus* L. *J Pineal Res* 42:147–152.

Arnao MB, Hernández-Ruiz J. 2014. Melatonin: Plant growth regulator and/or biostimulator during stress? *Trend Plant Sci* 19:789–797.

Back K, Tan DX, Reiter RJ. 2016. Melatonin biosynthesis in plants: Multiple pathways catalyze tryptophan to melatonin in the cytoplasm or chloroplasts. *J Pineal Res* 61:426–437.

Bajwa VS, Shukla MR, Sherif SM, Murch SJ, Saxena PK. 2014. Role of melatonin in alleviating cold stress in *Arabidopsis thaliana*. *J Pineal Res* 56:238–245.

Bajwa VS, Shukla MR, Sherif SM, Murch SJ, Saxena PK. 2015. Identification and characterization of serotonin as an anti-browning compound of apple and pear. *Postharvest Biol Technol* 110:183–189.

Balzer I, Hardeland R. 2014. Melatonin in algae and higher plants—Possible new roles as a phytohormone and antioxidant. *Bot Acta* 109:180–183.

Bari R, Jones JDG. 2009. Role of plant hormones in plant defence responses. *Plant Mol Biol* 69:473–488.

Bowden K, Brown BG, Batty JE. 1954. 5-Hydroxytryptamine: Its occurrence in cowhage. *Nature* 174:925–926.

Brzezinski A. 1997. Melatonin in humans. *New Engl J Med* 336:186–195.

Byeon Y, Back K. 2014. An increase in melatonin in transgenic rice causes pleiotropic phenotypes, including enhanced seedling growth, delayed flowering, and low grain yield. *J Pineal Res* 56:408–414.

Byeon Y, Back K. 2016a. Melatonin production in *Escherichia coli* by dual expression of serotonin N-acetyltransferase and caffeic acid O-methyltransferase. *App Microbiol Biotechnol* 100:6683.

Byeon Y, Back K. 2016b. Low melatonin production by suppression of either serotonin N-acetyltransferase or N-acetylserotonin methyltransferase in rice causes seedling growth retardation with yield penalty, abiotic stress susceptibility, and enhanced coleoptile growth under anoxic conditions. *J Pineal Res* 60:348–359.

Byeon Y, Choi GH, Lee HY, Back K. 2015. Melatonin biosynthesis requires N-acetylserotonin methyltransferase activity of caffeic acid O-methyltransferase in rice. *J Exp Bot* 66:6917–6925.

Byeon Y, Lee HY, Lee K et al. 2013. Cellular localization and kinetics of the rice melatonin biosynthetic enzymes SNAT and ASMT. *J Pineal Res* 56:107–114.

Byeon Y, Lee HY, Lee K, Back K. 2014a. Caffeic acid *O*-methyltransferase is involved in the synthesis of melatonin by methylating *N*-acetylserotonin in *Arabidopsis*. *J Pineal Res* 57:219–227.

Byeon Y, Lee HY, Lee K, Back K. 2014b. A rice chloroplast transit peptide sequence does not alter the cytoplasmic localization of sheep serotonin *N*-acetyltransferase expressed in transgenic rice plants. *J Pineal Res* 57:147–154.

Cavallo A, Ritschel WA. 1996. Pharmacokinetics of melatonin in human sexual maturation. *J Clin Endocrinol Metab* 81:1882–1886.

Chen Q, Qi W, Reiter RJ et al. 2009. Exogenously applied melatonin stimulates root growth and raises endogenous indoleacetic acid in roots of etiolated seedlings of *Brassica juncea*. *J Plant Physiol* 166:324–328.

Csaba G, Pal K. 1982. Effects of insulin, triiodothyronine, and serotonin on plant seed development. *Protoplasma* 110:20–22.

Dubbels R, Reiter RJ, Klenke E et al. 1995. Melatonin in edible plants identified by radioimmunoassay and by high performance liquid chromatography-mass spectrometry. *J Pineal Res* 18:28–31.

Duncan MJ, Fang JM, Dubocovich ML. 1990. Effects of melatonin agonists and antagonists on reproduction and body-eight in the Siberian hamster. *J Pineal Res* 9:231–242.

Enders TA, Strader LC. 2015. Auxin activity: Past, present, and future. *Am J Bot* 102:180–196.

Erland L, Mahmoud SS. 2014. An efficient method for regeneration of lavandin (*Lavandula* × *intermedia* cv. 'Grosso'). *In Vitro Cell Dev Biol—Plant* 50:646–654.

Erland L, Murch SJ, Reiter RJ, Saxena PK. 2015. A new balancing act: The many roles of melatonin and serotonin in plant growth and development. *Plant Signal Behav* 10:e1096469–15.

Erland LAE, Turi CE, Saxena PK. 2016. Serotonin: An ancient molecule and an important regulator of plant processes. *Biotechnol Adv* 34:1347–1361.

Gantait S, Sinniah UR, Ali MN, Sahu NC. 2015. Gibberellins—A multifaceted hormone in plant growth regulatory network. *Curr Protein Pept Sci* 16:406–412.

Gatineau F, Fouché JG, Kevers C et al. 1997. Quantitative variations of indolyl compounds including IAA, IAA-aspartate and serotonin in walnut microcuttings during root induction. *Biol Plant* 39:131–137.

Hardeland R. 2016. Melatonin in plants—Diversity of levels and multiplicity of functions. *Front Plant Sci* 7:198.

Hattori A, Migitaka H, Iigo M et al. 1995. Identification of melatonin in plants and its effects on plasma melatonin levels and binding to melatonin receptors in vertebrates. *Biochem Mol Biol Int* 35:627–634.

Hernández IG, Gomez FJV, Cerutti S et al. 2015. Melatonin in *Arabidopsis thaliana* acts as plant growth regulator at low concentrations and preserves seed viability at high concentrations. *Plant Physiol Biochem* 94:191–196.

Hernández-Ruiz J, Arnao MB. 2008. Melatonin stimulates the expansion of etiolated lupin cotyledons. *Plant Growth Regul* 55:29–34.
Hernández-Ruiz J, Cano A, Arnao MB. 2004. Melatonin: A growth-stimulating compound present in lupin tissues. *Planta* 220:140–144.
Hernández-Ruiz J, Cano A, Arnao MB. 2005. Melatonin acts as a growth-stimulating compound in some monocot species. *J Pineal Res* 39:137–142.
Hu W, Kong H, Guo Y et al. 2016. Comparative physiological and transcriptomic analyses reveal the actions of melatonin in the delay of postharvest physiological deterioration of cassava. *Front Plant Sci* 7:138–12.
Jones AMP, Cao J, O'Brien R, Murch SJ, Saxena PK. 2007. The mode of action of thidiazuron: Auxins, indoleamines, and ion channels in the regeneration of *Echinacea purpurea* L. *Plant Cell Rep* 26:1481–1490.
Kang K, Kang S, Lee K et al. 2008. Enzymatic features of serotonin biosynthetic enzymes and serotonin biosynthesis in plants. *Plant Signal Behav* 3:389–390.
Kang K, Kong K, Park S et al. 2011. Molecular cloning of a plant N-acetylserotonin methyltransferase and its expression characteristics in rice. *J Pineal Res* 50:304–309.
Kang S, Kang K, Lee K, Back K. 2007a. Characterization of tryptamine 5-hydroxylase and serotonin synthesis in rice plants. *Plant Cell Rep* 26:2009–2015.
Kang S, Kang K, Lee K, Back K. 2007b. Characterization of rice tryptophan decarboxylases and their direct involvement in serotonin biosynthesis in transgenic rice. *Planta* 227:263–272.
Kim M, Seo H, Park C, Park WJ. 2016. Examination of the auxin hypothesis of phytomelatonin action in classical auxin assay systems in maize. *J Plant Physiol* 190:67–71.
Klein DC, Weller JL. 1970. Indole metabolism in the pineal gland: A circadian rhythm in *N*-acetyltransferase. *Science* 169:1093–1095.
Koyama FC, Carvalho TLG, Alves E et al. 2013. The structurally related auxin and melatonin tryptophan-derivatives and their roles in *Arabidopsis thaliana* and in the human malaria parasite *Plasmodium falciparum*. *J Eukaryot Microbiol* 60:646–651.
Lanfumey L, Mongeau R, Hamon M. 2013. Biological rhythms and melatonin in mood disorders and their treatments. *Pharmacol Ther* 138:176–184.
Lee HY, Byeon Y, Lee K et al. 2014. Cloning of *Arabidopsis* serotonin N-acetyltransferase and its role with caffeic acid O-methyltransferase in the biosynthesis of melatonin in vitro despite their different subcellular localizations. *J Pineal Res* 57:418–426.
Lerner AB, Case JD, Mori W, Wright MR. 1959. Melatonin in peripheral nerve. *Nature* 183:1821.
Lerner AB, Case JD, Takahashi Y, Lee TH, Mori W. 1958. Isolation of melatonin, the pineal gland factor that lightens melanocytes. *J Am Chem Soc* 80:2587.
Li H, Chang J, Zheng J, Dong Y, Liu Q, Yang X, Wei C, Zhang Y, Ma J, Zhang X. 2017. Local melatonin application induces cold tolerance in distant organs of *Citrullus lanatus* L. via long distance transport. *Sci Rep* 7:40858.
Liang C, Li A, Yu H et al. 2017. Melatonin regulates root architecture by modulating auxin response in rice. *Front Plant Sci* 8:89–12.
Litwinczuk W, Wadas-Boron M. 2009. Development of highbush blueberry (*Vaccinium corymbosum* hort. non L.) in vitro shoot cultures under the influence of melatonin. *Acta Sci Pol* 8:3–12.
Ma Z, Yang Y, Fan C, Han J, Wang D, Di S, Hu W, Liu D, Li X, Reiter RJ, Yan X. 2015. Melatonin as a potential anticarcinogen for non-small-cell lung cancer. *Oncotarget* 19:1–18.
Manchester LC, Coto-Montes A, Boga JA et al. 2015. Melatonin: An ancient molecule that makes oxygen metabolically tolerable. *J Pineal Res* 59:403–419.
Mandava NB. 1988. Plant growth-promoting brassinosteroids. *Ann Rev Plant Physiol Plant Mol Biol* 39:23–52.
Mittler R, Vanderauwera S, Suzuki N et al. 2011. ROS signaling: The new wave? *Trend Plant Sci* 16:300–309.
Mukherjee S, David A, Yadav S et al. 2014. Salt stress-induced seedling growth inhibition coincides with differential distribution of serotonin and melatonin in sunflower seedling roots and cotyledons. *Physiol Plant* 152:714–728.
Murch SJ, Campbell SSB, Saxena PK. 2001. The role of serotonin and melatonin in plant morphogenesis: Regulation of auxin-induced root organogenesis in in vitro-cultured explants of St. John's wort (*Hypericum perforatum* L.). *In Vitro Cell Dev Biol—Plant* 37:786–793.
Murch SJ, Krishnaraj S, Saxena PK. 2000. Tryptophan is a precursor for melatonin and serotonin biosynthesis in in vitro regenerated St. John's wort (*Hypericum perforatum* L. cv. Anthos) plants. *Plant Cell Rep* 19:698–704.
Murch SJ, Saxena PK. 2002. Melatonin: A potential regulator of plant growth and development? *In Vitro Cell Dev Biol—Plant* 38:531–536.

Murch SJ, Saxena PK. 2004. Role of indoleamines in regulation of morphogenesis in in vitro cultures of St. John's wort (*Hypericum perforatum* L.). *Acta Hortic* 629:425–432.

Murthy B, Murch SJ, Saxena PK. 1995. Thidiazuron-induced somatic embryogenesis in intact seedlings of peanut (*Arachis hypogaea*): Endogenous growth regulator levels and significance of cotyledons. *Physiol Plant* 94:268–276.

Murthy BNS, Murch SJ, Saxena PK. 1998. Thidiazuron: A potent regulator of in vitro plant morphogenesis. *In Vitro Cell Dev Biol—Plant* 34:267–275.

Okazaki M, Higuchi K, Aouini A, Ezura H. 2010. Lowering intercellular melatonin levels by transgenic analysis of indoleamine 2,3-dioxygenase from rice in tomato plants. *J Pineal Res* 49:239–247.

Park S, Back K. 2012. Melatonin promotes seminal root elongation and root growth in transgenic rice after germination. *J Pineal Res* 53:385–389.

Park S, Byeon Y, Lee HY et al. 2014. Cloning and characterization of a serotonin N-acetyltransferase from a gymnosperm, loblolly pine (*Pinus taeda*). *J Pineal Res* 57:348–355.

Pelagio-Flores R, Muñoz Parra E, Ortíz-Castro R, López-Bucio J. 2012. Melatonin regulates *Arabidopsis* root system architecture likely acting independently of auxin signaling. *J Pineal Res* 53:279–288.

Pelagio-Flores R, Ortíz-Castro R, Méndez-Bravo A, Macías-Rodríguez L, López-Bucio J. 2011. Serotonin, a tryptophan-derived signal conserved in plants and animals, regulates root system architecture probably acting as a natural auxin inhibitor in *Arabidopsis thaliana*. *Plant Cell Physiol* 52:490–508.

Petrie K, Dawson AG, Thompson L, Brook R. 1993. A double-blind trial of melatonin as a treatment for jet lag in international cabin crew. *Biol Psychiatry* 33:526–530.

Popova E, Kim HH, Saxena PK et al. 2016. Frozen beauty: The cryobiotechnology of orchid diversity. *Biotechnol Adv* 34:380–403.

Posmyk MM, Bałabusta M, Wieczorek M et al. 2009. Melatonin applied to cucumber (*Cucumis sativus* L.) seeds improves germination during chilling stress. *J Pineal Res* 46:214–223.

Posmyk MM, Kuran H, Marciniak K, Janas KM. 2008. Presowing seed treatment with melatonin protects red cabbage seedlings against toxic copper ion concentrations. *J Pineal Res* 45:24–31.

Qian Y, Tan D-X, Reiter RJ, Shi H. 2015. Comparative metabolomic analysis highlights the involvement of sugars and glycerol in melatonin-mediated innate immunity against bacterial pathogen in *Arabidopsis*. *Sci Rep* 5:15815.

Ramakrishna A, Giridhar P, Jobin M, Paulose CS, Ravishankar GA. 2011. Indoleamines and calcium enhance somatic embryogenesis in *Coffea canephora* P ex Fr. *Plant Cell Tiss Organ Cult* 108:267–278.

Ramakrishna A, Giridhar P, Ravishankar GA. 2009. Indoleamines and calcium channels influence morphogenesis in in vitro cultures of *Mimosa pudica* L. *Plant Signal Behav* 4:1136–1141.

Ramakrishna A, Giridhar P, Ravishankar GA. 2011. Phytoserotonin: A review. *Plant Signal Behav* 6:800–809.

Reiter RJ. 1993. The melatonin rhythm: Both a clock and a calendar. *Experientia* 49:654–664.

Reiter RJ, Mayo JC, Tan DX et al. 2016. Melatonin as an antioxidant: Under promises but over delivers. *J Pineal Res* 61:253–278.

Reiter RJ, Tan D, Terron MP, Flores LJ. 2007. Melatonin and its metabolites: New findings regarding their production and their radical scavenging actions. *Acta Biochim Pol* 54:1–9.

Rios ERV, Venâncio ET, Rocha NFM et al. 2010. Melatonin: Pharmacological aspects and clinical trends. *Int J Neurosci* 120:583–590.

Rosen J, Than NN, Koch D et al. 2006. Interactions of melatonin and its metabolites with the ABTS cation radical: Extension of the radical scavenger cascade and formation of a novel class of oxidation products, C2-substituted 3-indolinones. *J Pineal Res* 41:374–381.

Roshchina VV. 2001. *Neurotransmitters in Plant Life*. Enfield, NH: Science Publishers, Inc.

Sarropoulou V, Dimassi-Theriou K, Therios I, Koukourikou-Petridou M. 2012a. Melatonin enhances root regeneration, photosynthetic pigments, biomass, total carbohydrates and proline content in the cherry rootstock PHL-C (*Prunus avium* × *Prunus cerasus*). *Plant Physiol Biochem* 61:162–168.

Sarropoulou VN, Therios IN, Dimassi-Theriou KN. 2012b. Melatonin promotes adventitious root regeneration in in vitro shoot tip explants of the commercial sweet cherry rootstocks CAB-6P (*Prunus cerasus* L.), Gisela 6 (*P. cerasus* × *P. canescens*), and MxM 60 (*P. avium* × *P. mahaleb*). *J Pineal Res* 52:38–46.

Sarrou E, Therios I, Dimassi-Theriou K. 2014. Melatonin and other factors that promote rooting and sprouting of shoot cuttings in *Punica granatum* cv. Wonderful. *Turk J Bot* 38:293–301.

Saxena PK, Malik KA, Gill R. 1992. Induction by thidiazuron of somatic embryogenesis in intact seedlings of peanut. *Planta* 187:421–424.

Simonneaux V, Ribelayga C. 2003. Generation of the melatonin endocrine message in mammals: A review of the complex regulation of melatonin synthesis by norepinephrine, peptides, and other pineal transmitters. *Pharmacol Rev* 55:325–395.

Skoog F, Miller CO. 1957. Chemical regulation of growth and organ formation in plant tissues cultured in vitro. *Symp Soc Exp Biol* 11:118–130.

Sliwiak J, Dauter Z, Jaskolski M. 2016. Crystal structure of Hyp-1, a *Hypericum perforatum* PR-10 protein, in complex with melatonin. *Front Plant Sci* 7:352–10.

Szafrańska K, Glińska S, Janas KM. 2012. Ameliorative effect of melatonin on meristematic cells of chilled and re-warmed *Vigna radiata* roots. *Biol Plant* 57:91–96.

Szczepanik M. 2007. Melatonin and its influence on immune system. *J Physiol Pharmacol* 58 Suppl 6:115–124.

Tan DX, Hardeland R, Back K, Manchester LC, Alatorre-Jimenez MA, Reiter RJ. 2016. On the significance of an alternate pathway of melatonin synthesis via 5-methoxytryptamine: Comparisons across species. *J Pineal Res* 61:21–40.

Tan DX, Manchester LC, Esteban-Zubero E, Zhou Z, Reiter RJ. 2015. Melatonin as a potent and inducible endogenous antioxidant: Synthesis and metabolism. *Molecules* 20:18886–18906.

Tan DX, Manchester LC, Liu X, Rosales-Corral SA, Acuna-Castroviejo D, Reiter RJ. 2012. Mitochondria and chloroplasts as the original sites of melatonin synthesis: A hypothesis related to melatonin's primary function and evolution in eukaryotes. *J Pineal Res* 54:127–138.

Tan DX, Reiter RJ, Manchester LC, Yan MT, El-Sawi M, Sainz RM, Mayo JC, Kohen R, Allegra MC, Hardeland R. 2002. Chemical and physical properties and potential mechanisms: Melatonin as a broad-spectrum antioxidant and free radical scavenger. *Curr Top Med Chem* 2:181–197.

Tan DX, Zheng X, Kong J, Manchester LC, Hardeland R, Kim SJ, Xu X, Reiter RJ. 2014. Fundamental issues related to the origin of melatonin and melatonin isomers during evolution: Relation to their biological functions. *Int J Mol Sci* 15:15858–15890.

Uchendu EE, Shukla MR, Reed BM, Saxena PK. 2013. Melatonin enhances the recovery of cryopreserved shoot tips of American elm (*Ulmus americana* L.). *J Pineal Res* 55:435–442.

Uchendu EE, Shukla MR, Reed BM, Saxena PK. 2014. An efficient method for cryopreservation of St. John's wort and tobacco: Role of melatonin. *Acta Hortic* 1039:233–241.

Venegas C, García JA, Escames G et al. 2012. Extrapineal melatonin: Analysis of its subcellular distribution and daily fluctuations. *J Pineal Res* 52:217–227.

Wang L, Zhao Y, Reiter RJ et al. 2013. Changes in melatonin levels in transgenic "Micro-Tom" tomato over-expressing ovine AANAT and ovine HIOMT genes. *J Pineal Res* 56:134–142.

Wang Q, An B, Wei Y et al. 2016. Melatonin regulates root meristem by repressing auxin synthesis and polar auxin transport in *Arabidopsis*. *Front Plant Sci* 7:1–11.

Weeda S, Zhang N, Zhao X et al. 2014. *Arabidopsis* transcriptome analysis reveals key roles of melatonin in plant defense systems. *PLoS One* 9:e93462.

Wei W, Li QT, Chu YN et al. 2015. Melatonin enhances plant growth and abiotic stress tolerance in soybean plants. *J Exp Bot* 66:695–707.

Wen D, Gong B, Sun S et al. 2016. Promoting roles of melatonin in adventitious root development of *Solanum lycopersicum* L. by regulating auxin and nitric oxide signaling. *Front Plant Sci* 7:787.

Werner T, Schmülling T. 2009. Cytokinin action in plant development. *Curr Opin Plant Biol* 12:527–538.

Woodward AW, Bartel B. 2005. Auxin: Regulation, action, and interaction. *Ann Bot* 95:707–735.

Zhang HJ, Zhang N, Yang RC et al. 2014. Melatonin promotes seed germination under high salinity by regulating antioxidant systems, ABA and GA4 interaction in cucumber (*Cucumis sativus* L.). *J Pineal Res* 57:269–279.

Zhang N, Zhang HJ, Zhao B et al. 2013a. The RNA-seq approach to discriminate gene expression profiles in response to melatonin on cucumber lateral root formation. *J Pineal Res* 56:39–50.

Zhang N, Zhao B, Zhang HJ, Weeda S. 2013b. Melatonin promotes water-stress tolerance, lateral root formation, and seed germination in cucumber (*Cucumis sativus* L.). *J Pineal Res* 54:15–23.

Zhao H, Su T, Huo L et al. 2015a. Unveiling the mechanism of melatonin impacts on maize seedling growth: Sugar metabolism as a case. *J Pineal Res* 59:255–266.

Zhao H, Xu L, Su T et al. 2015b. Melatonin regulates carbohydrate metabolism and defenses against *Pseudomonas syringae* pv. tomato DC3000 infection in *Arabidopsis thaliana*. *J Pineal Res* 59:109–119.

Zhao Y, Qi L-W, Wang W-M et al. 2011. Melatonin improves the survival of cryopreserved callus of *Rhodiola crenulata*. *J Pineal Res* 50:83–88.

Zheng X, Tan DX, Allan AC et al. 2017. Chloroplastic biosynthesis of melatonin and its involvement in protection of plants from salt stress. *Sci Rep* 7:41236.

# 5 The Multi-Regulatory Properties of Melatonin in Plants

*Marino B. Arnao and Josefa Hernández-Ruiz*

## CONTENTS



## 5.1 MELATONIN IN PLANTS

Melatonin (*N*-acetyl-5-methoxytryptamine) was discovered in 1958 in the bovine pineal gland (Lerner et al., 1958). Since then it has become one of the most studied biological molecules, and its role has been much studied in mammals, birds, amphibians, reptiles, and fish. Melatonin has many physiological roles in animals (Maronde and Stehle, 2007; Pandi-Perumal et al., 2008; Jan et al., 2009; Hardeland et al., 2012), influencing circadian rhythms, mood, sleep, body temperature, locomotor activity, food intake, retina physiology, sexual behavior, seasonal reproduction, and the immune system (Carrillo-Vico et al., 2013). Melatonin acts as a signal of darkness, providing information to the brain and peripheral organs and serving as an endogenous synchronizer for physiological rhythms (e.g., sleep–wake cycles, seasonal reproduction, and endocrine release cycles). Alterations in rhythmic-melatonin production have been associated with many disorders (Maronde and Stehle, 2007; Wilhelmsen et al., 2011; Hardeland, 2012) such as Alzheimer's and Parkinson's syndrome (Srinivasan et al., 2005), glaucoma, multiple sclerosis, depression, insomnia, chronic fatigue syndrome, schizophrenia, anxiety, metabolic syndrome, osteoporosis, and certain types of cancer (Seely et al., 2012; Di Bella et al., 2013).

Possibly the first identification of endogenous melatonin in a higher plant was made in 1993 by van Tassel and co-workers (van Tassel and O'Neill, 1993). The authors detected melatonin by radioimmunoassay (RIA) in the Japanese morning glory (*Pharbitis nil*), although these results were not published extensively until 1995 (van Tassel et al., 1995). It was also in 1995 that two papers were published simultaneously, demonstrating the presence of melatonin and establishing its content in some higher plants. In Germany, Dubbels and co-workers used RIA and GC-MS to measure the melatonin levels in extracts of *Nicotiana tabacum* and in five edible plants (Dubbels et al., 1995). Two months later the publication of the Japanese group appeared, which quantified the presence of melatonin in extracts of a large number of edible plants using RIA and HPLC with fluorescence detection (Hattori et al., 1995). Also in 1995, a Czech research group identified the presence of melatonin in *Chenopodium rubrum* by LC-MS/MS (Kolar et al., 1995). Since the first articles published in 1995, no more than 35 articles were published on *Plant melatonin* until 2005. Since 2006, the publications on melatonin in plants have shown an exponential increase (about 90 in 2015/2016). Currently, research on plant melatonin is one of the most active fields with high rates of publications, as regards aspects of the biochemistry and physiology of plants and related with food science and technology.

In the following sections, we present an overview of the most relevant aspects of this interesting molecule: its levels in plants, its biosynthesis, and its physiological actions, focusing on the most novel and interesting aspects related with higher plants.

## 5.2 MELATONIN LEVELS IN PLANTS

In plants, melatonin levels have been shown to vary from a few picograms up to micrograms per gram of material analyzed. One of the problems for its the reliable measurement is the difficulty involved in its extraction and recovery from plants (Reiter et al., 2001; Kolar and Machackova, 2005; Cao et al., 2006; Pape and Lüning, 2006; Hardeland et al., 2007). Generally, melatonin is extracted from liquid-nitrogen-treated plant tissue using organic solvents, such as methanol, chloroform, or ethyl acetate. However, when aqueous extraction is used, variable or low recovery rates have been reported (Burkhardt et al., 2001; Reiter et al., 2005; Pape and Lüning, 2006). Also, direct sample extraction procedures (without the homogenization of fresh tissues) using only organic solvent are recommendable. The parameter *extraction efficiency* is a decisive factor in the global estimation of melatonin content in plants. Thus, the reliability of the melatonin level data will depend, among others factors, on the recovery rate. According to our data, the inclusion of an ultrasonic treatment in the direct sample extraction procedure results in different levels of efficiency (2%–20%), depending on the sample, although the inclusion of this treatment in the extraction procedure may lead to a significant increase in phytomelatonin extraction (Cao et al., 2006; Arnao and Hernández-Ruiz, 2009a). Other possible problems, such as false-negative and false-positive results, overestimations, losses by destruction, co-elution in chromatographic methods, cross-reactivities in immunological procedures, and so forth, have been described in the determination of melatonin in complex matrices (Hardeland et al., 2007). Analysis by liquid chromatography and identification by mass spectrometry (LC-MS/MS) are the most recommended techniques for the detection and quantification of melatonin. In this respect, LC-MS/MS with positive electrospray ionization and multiple-reaction-monitoring is widely used. The innovative technique of liquid chromatography with time-of-flight/mass spectrometry (LC-TOF/MS) has also been applied to melatonin detection in plants. However, less specific but very sensitive techniques such as LC with electrochemical or fluorometric detection are also excellent if accompanied by identification by LC-MS/MS (Kolar and Machackova, 2005; Arnao and Hernández-Ruiz, 2006; Cao et al., 2006; Hernández-Ruiz, and Arnao, 2008a; Arnao and Hernández-Ruiz 2009a; Garcia-Parrilla et al., 2009; Paredes et al., 2009; Huang and Mazza, 2011a; Tan et al., 2012; Gómez et al., 2013). Due to cross-reactivity, immunological techniques such as RIA and ELISA present serious problems in plants, unlike in animals (van Tassel and O'Neill, 2001; Pape and Lüning, 2006; Hardeland et al., 2007).

For example, in a study of seven herbs used in Thai traditional medicine as sleeping aids, the authors used two different techniques for melatonin quantification: ELISA and liquid chromatography with fluorimetric detection (Padumanonda et al., 2014). Table 5.1 shows the melatonin content of these herbs obtained by both methods. As can be seen, only in one case (*Sesbania sesban*) were both measurements similar, while in all other cases the differences were notable. In two cases (*Momordica charantia* and *Senna tora*), even, melatonin was not detected by ELISA. Although many studies on melatonin in plants have applied immunodetection techniques (mostly ELISA), the use of LC-MS/MS to obtain more reliable data, avoiding cross-reactivity, is always recommended.

**TABLE 5.1**
**Melatonin Content of Thai Herbs[a] Determined by ELISA and LC**

| Herb Extract | ELISA[b] | LC[b] |
|---|---|---|
| *Piper nigrum* | 865 | 1092 |
| *Baccaurea ramiflora* | 76.7 | 43.2 |
| *Sesbania glandiflora* | 43.7 | 26.3 |
| *Momordica charantia* | nd | 21.4 |
| *Senna tora* | nd | 10.5 |
| *Sesbania sesban* | 7.3 | 8.7 |
| *Moringa oleifera* | nd | nd |

[a] Adapted from Padumanonda, T. et al., *DARU J. Pharm. Sci.*, 22, 1–6, 2014.
[b] Expressed as ng of melatonin/g dry weight.
nd: not detected.

Tables 5.2 and 5.3 present data on the melatonin content of fruits and vegetables and of seeds/nuts and medicinal plants, respectively. Among fruits (Table 5.2), goji and jujube berries show the highest melatonin content (Chen et al., 2003). Apples and cherries also have a high content. While several varieties of *Vitis vinifera* present a considerable melatonin content, there is a high degree of dispersion in the data, as in most plant data. In general, tropical fruits have a low melatonin content.

**TABLE 5.2**
**Content of Melatonin in Some Fruits and Vegetables**

| Common Name/*Species* | Melatonin Content[a] | Reference |
|---|---|---|
| **Fruits** | | |
| Goji berry/*Lycium barbarum* L. | 530 | Chen et al. (2003) |
| Jujube/*Ziziphus jujube* Lam. | 256 | Chen et al. (2003) |
| Apple/*Malus domestica* Borkh. | 134 | Lei et al. (2013) |
| Sweet cherry/*Prunus avium* L. | 120 | Zhao et al. (2013) |
| Grape/*V. vinifera* L. | 18 | Vitalini et al. (2011) |
| Strawberry/*Fragaria × ananassa* Duch. | 12 | Hattori et al. (1995) |
| Pomegranate/*Punica granatum* L. | 5 | Mena et al. (2012) |
| Fig/*Ficus carica* L. | 4 | Zohar et al. (2011) |
| Orange/*Citrus × sinensis* L. | 3 | Fernández-Pachón et al. (2014) |
| Mango/*Mangifera indica* L. | 0.7 | Johns et al. (2013) |
| Pineapple/*Ananas comosus* L. | 0.3 | Sae-Teaw et al. (2013) |
| **Vegetables** | | |
| Tomato/*Solanum lycopersicum* L. | 114 | Sturtz et al. (2011) |
| Pepper (bell)/*Capsicum annuum* L. | 42 | Huang and Mazza (2011) |
| Kohlrabi/*Brassica napus napobrassica* | 7.8 | Pasko et al. (2014) |
| Carrot/*Daucus carota* Hoffm. | 0.05 | Hattori et al. (1995) |
| Onion/*Allium cepa* L. | 0.03 | Hattori et al. (1995) |
| Cucumber/*Cucumis sativum* L. | 0.02 | Hattori et al. (1995) |
| Beetroot/*Beta vulgaris* L. | 0.002 | Dubbels et al. (1995) |

[a] Expressed as ng of melatonin/g fresh weight.

**TABLE 5.3**
**Content of Melatonin in Some Seeds/Nuts and Medicinal Herbs**

| Common Name/*Species* | Melatonin Content[a] | Reference |
|---|---|---|
| **Seed/Nuts** | | |
| White mustard/*Brassica hirta* L. | 189 | Manchester et al. (2000) |
| Black mustard/*Brassica nigra* L. | 129 | Manchester et al. (2000) |
| Fenugreek/*Trigonella foenum-graecum* L. | 43 | Manchester et al. (2000) |
| Almond/*Prunus amygdalus* Batsch. | 39 | Manchester et al. (2000) |
| Sunflower/*Helianthus annuus* L. | 29 | Manchester et al. (2000) |
| Alfalfa/*Medicago sativa* L. | 16 | Manchester et al. (2000) |
| Cardamom/*Elettaria cardamomum* L. | 15 | Manchester et al. (2000) |
| Flax/*Linum usitatissimum* L. | 12 | Manchester et al. (2000) |
| Anise/*Pimpinella anisum* L. | 7 | Manchester et al. (2000) |
| Coriander/*Coriandrum sativum* L. | 7 | Manchester et al. (2000) |
| Lupin/*Lupinus albus* L. | 3.8 | Hernández-Ruiz and Arnao (2008a) |
| Walnut/*Juglans regia* L. | 3.5 | Reiter et al. (2005) |
| Peanut/*Arachis hypogaea* L. | 2.3 | Blask et al. (2004) |
| **Medicinal Herbs** | | |
| Tea/*Camellia sinensis* L. | 0.4 | Blask et al. (2004) |
| Coffee beans/*Coffea* sp. | 6,500 | Ramakrishna et al. (2012) |
| Thyme/*Thymus vulgaris* L. | 38,000 | Stege et al. (2010) |
| Chinese liquorice/*Glycyrrhiza uralensis* Fisch. | 34,000 | Afreen et al. (2006) |
| Sage/*Salvia officinalis* L. | 29,000 | Stege et al. (2010) |
| St. John's wort/*Hypericum perforatum* L. | 23,000 | Murch and Saxena (2006) |
| Peppermint/*Mentha piperita* | 19,500 | Chen et al. (2003) |
| Feverfew/*T. parthenium* L. | 1,700 | Murch et al. (1997) |
| Purslane/*Portulaca oleracea* L. | 19 | Simopoulos et al. (2005) |
| Laurel/*Laurus nobilis* L. | 8 | Zohar et al. (2011) |

[a] Expressed as ng of melatonin/g dry weight.

Among vegetables (Table 5.2), the Solanaceae family (tomatoes and peppers) have a high melatonin content, as do *Brassica* species. In the case of seeds/nuts (Table 5.3), mustard seeds have high melatonin levels. In aromatic/medicinal plants, thyme, liquorice root, sage and St. John's wort plants have very high levels (Table 5.3), even higher than coffee beans. Generally, Chinese medicinal herbs have the highest melatonin content, but studies of medicinal plants of non-Chinese origin are scarce. Other medicinal plants such as feverfew (*Tanacetum parthenium*) have also been seen to be rich in melatonin.

## 5.3 BIOSYNTHESIS OF MELATONIN

Chemically, melatonin (*N*-acetyl-5-methoxytryptamine) is an indolic compound derived from serotonin (5-hydroxytryptamine) (Figure 5.1). Both biogenic amines are synthesized from the amino acid tryptophan in an extensively studied biosynthetic pathway in both animals and plants (Reiter, 1991; Arnao and Hernández-Ruiz, 2006, 2014a, 2015a, 2015b; Tan et al., 2015; Back et al., 2016; Nawaz et al., 2016). In plants, tryptophan is converted into tryptamine by tryptophan decarboxylase (TDC) (Figure 5.1). Tryptamine is then converted into 5-hydroxytryptamine (commonly known as

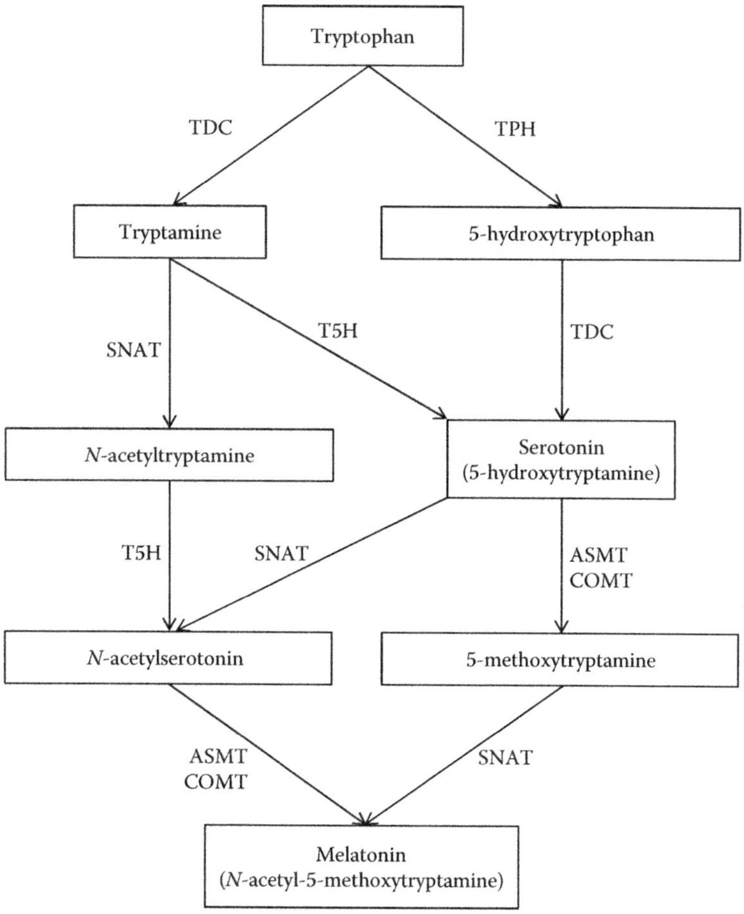


**FIGURE 5.1** Biosynthetic pathway of melatonin in plants. The enzymes of the respective steps are: T5H, tryptophan 5-hydroxylase; TDC, tryptophan decarboxylase; SNAT, serotonin *N*-acetyltransferase; ASMT, acetylserotonin methyltransferase and COMT, caffeic acid *O*-methyltransferase. (From Arnao, M.B. and Hernández-Ruiz, J., *J. Pineal Res.*, 59, 133–150, 2015a. With permission.)

serotonin) by tryptamine 5-hydroxylase (T5H), an enzyme that has been only characterized in rice, and which possibly acts with a large number of substrates, although this has not been studied in depth (Kang et al., 2007; Fujiwara et al., 2010; Park et al., 2012, 2013b). The *N*-acetylation of serotonin is catalyzed by the enzyme serotonin *N*-acetyltransferase (SNAT) (Ferry et al., 2000; Byeon et al., 2015b, 2016). *N*-acetylserotonin is then methylated by acetylserotonin-*O*-methyltransferase (ASMT), a hydroxyindole-*O*-methyltransferase that generates melatonin. In plants, the methylation of *N*-acetylserotonin can also be made by a caffeic acid *O*-methyltransferase (COMT), an enzyme that can act on a broad diversity of substrates including caffeic acid and quercetin (Byeon et al., 2014a, 2015a; Lee et al., 2014b). Serotonin can also be transformed into 5-methoxytryptamine by ASMT (and by COMT), and then generate melatonin through the action of SNAT. Also, melatonin can be generated through the formation of *N*-acetyltryptamine, which is converted into *N*-acetylserotonin. Finally, serotonin can be formed from 5-hydroxytryptophan, after the action of tryptophan hydroxylase (TPH) and TDC, the latter step occurring mainly in animals but also in plants to a lesser extent.

To resume, melatonin in plants can be synthesized in many ways, the most relevant being the sequence: tryptophan → tryptamine → serotonin → *N*-acetylserotonin → melatonin. All the named enzymes have been detected and characterized in rice and *Arabidopsis*, except TPH, which is well-known in animals but not in plants. Nevertheless, some authors have proposed that T5H can act as a hydroxylase with low substrate specificity and is capable of acting in all the hydroxylation steps described (Arnao and Hernández-Ruiz, 2014a, 2014b). This same broad substrate specificity can also be attributed to SNAT, ASMT, and COMT enzymes.

As regards to regulation of melatonin biosynthesis, TDC and T5H transcript expression levels are regulated by light: high levels of expression occur under constant light and low levels under dark conditions (Park et al., 2012, 2013b; Byeon et al., 2014b). Although the SNAT enzyme is localized in chloroplasts, its activity is inhibited by chlorophylls (Byeon et al., 2013b). ASMT seems to be the rate-limiting enzyme in plants, and its mRNA level is higher in the dark than under continuous light conditions. Interestingly, TDC, SNAT, and ASMT activities also seem to be regulated by high temperature, as is the ancestral cyanobacterial SNAT gene (Byeon et al., 2013a; Byeon and Back, 2014b). However, more studies on the conditions that determine the alternative pathways are necessary to better understand melatonin biosynthesis in plants.

## 5.4 MELATONIN AND ABIOTIC STRESS

One of the most studied aspects related with melatonin has been its possible role as a protective agent against abiotic stress situations in plants. Melatonin acts as an effective free radical scavenger against hazardous reactive molecules, both reactive oxygen and reactive nitrogen species (ROS/RNS), such as hydroxyl radical, superoxide anion, singlet oxygen, hydrogen peroxide, hypochlorous acid, nitric oxide, peroxynitrite anion, peroxynitrous acid, and lipid peroxyl radicals, among others. The excellent properties of melatonin as a natural antioxidant against ROS/RNS and the absence of pro-oxidant effects have been the subject of a great deal of research (Reiter et al., 2000, 2014; Tan et al., 2000; Teixeira et al., 2003; Fischer et al., 2013; Arnao and Hernández-Ruiz, 2015b). Table 5.4 represents a compilation of most of the studies that demonstrate the protective role that melatonin performs in stress situations in plants. In general, cold, heat, salinity, drought, and chemical toxicity are countered or mitigated by the presence of melatonin. The protective role of melatonin in cold conditions and/or freezing tissues has been demonstrated. In general, melatonin protects plants against cold/freezing conditions, preserving the tissues and mitigating associated harmful effects. Higher plant survival, shoot and root growth, and sucrose and proline levels have been observed in melatonin-treated plants, together with a lower presence of ROS/RNS, lipid membrane peroxidation, and cell damage (Table 5.4). For example, the presence of melatonin in both pre-culture and regrowth media enhanced the growth of frozen shoot explants of American elm (*Ulmus americana*), demonstrating the usefulness of melatonin for the long-term storage of germplasm for plant cell culture (Uchendu et al., 2013). Also, in a recent study with soybean (*Glycine max*) plants, seeds imbibed with melatonin optimized parameters such as seedling growth, leaf size, plant height, biomass and pod, and seed number. Melatonin treatment improved the salt and drought tolerance of plants, demonstrating the significant potential of melatonin for improving field crops (Wei et al., 2015).

Melatonin also improves the induced-senescence process, and it has been reported that barley leaves treated with melatonin delay dark-induced senescence in a concentration-dependent manner, slowing down chlorophyll loss in detached leaves (Arnao and Hernández-Ruiz, 2009c). This protective effect was confirmed in other species such as apple, cucumber, rice, pea, ryegrass, and *Arabidopsis* (Table 5.4) and has been related with the protective effect that melatonin has on photosynthetic pigments (chlorophylls and carotenoids), enhancing photosynthetic efficiency in chloroplastic photosystems and alleviating the photoinhibition caused by abiotic stressors (Arnao and Hernández-Ruiz, 2014a, 2015a). In a recent paper, melatonin-treated tomato seedlings showed an increase in the effective quantum yield of Photosystem II, the photochemical quenching coefficient and the proportion of *open* PSII centers that control plants under saline conditions. In this way, damage to the photosynthetic electron transport chain in Photosystem II was mitigated. In addition,

**TABLE 5.4**
**Effects of Melatonin in Abiotic Stress Responses**

| Abiotic Stressor | Plant Species | Melatonin Treatment (μM) | Effects Observed | Reference |
|---|---|---|---|---|
| Cold | *Arabidopsis* | 10–30 | ↑ Fresh weight, shoot height, and primary roots | Bajwa et al. (2014) |
| Cold | *Arabidopsis* | 20 | ↑ Plant survival | Shi and Chan (2014) |
| Cold, salt, drought | *Arabidopsis* | 50 | ↑ Sucrose, survival rate | Shi et al. (2015c) |
| Cold | Cucumber | 25–100 | ↑ Germination, ↓ membrane peroxidation | Posmyk et al. (2009) |
| Cold | Cucumber | 50–500 | ↑ GSH pool, ↓ ROS burst | Balabusta et al. (2016) |
| Cold | *Rhodiola crenulata* | 0.1 | ↑ Cryopreservation of callus | Zhao et al. (2011) |
| Cold | Mung bean | 20 | ↑ Root growth, ↓ membrane peroxidation | Szafranska et al. (2013) |
| Cold | American elm | 0.1–0.5 | ↑ Regrowth frozen shoots | Uchendu et al. (2013) |
| Cold | Watermelon | 150 | ↑ Photosynthesis, ↓ cold-related microRNA | Li et al. (2016b) |
| Cold, salt, drought | Bermudagrass | 20–100 | ↑ Fresh weight, osmoregulation, ↓ ROS burst, cell damage | Shi et al. (2015b); Fan et al. (2015) |
| Cold, salt, drought | Rice | — | Changes in MEL biosynthesis enzymes transcripts, ↑ TDC, ASMT | Wei et al. (2016) |
| Cold | Wheat | 1 mM | ↑ Redox balance, Chls, osmoregulation, ↓ ROS burst | Turk et al. (2014) |
| Cold | Cabbage | 10–1000 | ↑ Anthocyanins, proline, redox balance, ↓ ROS burst | Zhang et al. (2016b) |
| Cold, drought | Barley | 1 mM | ↑ Photosynthesis efficiency, ABA, water content, ↓ ROS burst | Li et al. (2016c) |
| Heat | *Phacelia* | 0.3–90 | ↑ Germination | Tiryaki and Keles (2012) |
| Heat | *Arabidopsis* | 5–20 | ↑ Thermotolerance | Shi et al. (2015e) |
| Heat | Tomato | 10 | ↑ Thermotolerance and cell protection | Xu et al. (2016) |
| Metal-Cu | Pea | 5 | ↑ Plant survival | Tan et al. (2007b) |
| Metal-Cu | Red cabbage | 1–100 | ↑ Fresh weight, germination, ↓ membrane peroxidation | Posmyk et al. (2008) |
| Metal-Cd | Tomato | 25–500 | ↑ Cd tolerance, phytochelatins, ATPase activity | Hasan et al. (2015) |
| Metal-Pb | Tobacco | 0.2 | ↑ Cell culture growth, ↓ mortality cells, ROS burst | Kobylinska and Posmyk (2016) |
| Oxidative | *Arabidopsis* | 5–10 | ↑ Plant survival, autophagy, ↓ oxidized proteins | Wang et al. (2015) |
| Oxidative | *Pisum sativum* | 50–200 | ↑ Photosynthesis efficiency, pigments, water content, ↓ ROS burst | Szafranska et al. (2016) |
| Salinity | *Malus* | 0.1 | ↑ Shoot height, leaf number, chlorophylls, ↓ electrolyte leakage | Li et al. (2012) |

*(Continued)*

**TABLE 5.4 (*Continued*)**
**Effects of Melatonin in Abiotic Stress Responses**

| Abiotic Stressor | Plant Species | Melatonin Treatment (μM) | Effects Observed | Reference |
|---|---|---|---|---|
| Salinity | *Malus* | 0.1 | ↑ Shoot height, $K^+$ channels, $K^+$ level, ↓ ROS burst | Li et al. (2016a) |
| Salinity, drought | Soybean | 50–100 | ↑ Seedling growth, leaf size, biomass, seed yield | Wei et al. (2015) |
| Salinity | Citrus | 1 | ↑ Osmoregulation, Chls, ↓ ROS burst, membrane peroxidation | Kostopoulou et al. (2015) |
| Salinity | Sunflower | 15 | ↑ Root and hypocotyl growth, antioxidant potential | Mukherjee et al. (2014) |
| Salinity | Cucumber | 1 | ↑ Germination, $GA_4$, ↓ ROS burst, membrane peroxidation, ABA | Zhang et al. (2014a) |
| Salinity | *Vicia faba* | 100–500 | ↑ Plant height, RWC, photosynthetic pigments, osmolites, phenolic | Dawood and El-Awadi (2015) |
| Salinity | Tomato | 50–150 | ↑ Photosynthesis, PSII efficiency, D1 protein turnover, ↓ ROS burst | Zhou et al. (2016b) |
| Alkalinity | Tomato | 0.25–1 | ↑ Seedling growth, photosynthesis, ion homeostasis, ↓ ROS burst | Liu et al. (2015) |
| Drought | Cucumber | 100 | ↑ Germination, root growth | Zhang et al. (2013) |
| Drought | Grape | 0.05–0.2 | ↑ Seedling growth, osmoregulation, photosynthesis, ↓ ROS burst | Meng et al. (2014) |
| Drought | *Malus* | 100 | ↑ Water status, Chls, photosynthesis efficiency, ↓ ROS burst | Li et al. (2015) |
| Leaf-senescence | Barley | 0.01–1 | ↓ Senescence, ↑ Chls | Arnao and Hernández-Ruiz (2009c) |
| Leaf-senescence | *Malus* | 10 mM | ↓ Senescence, ROS burst, ↑ Chls, photosynthesis efficiency | Wang et al. (2012); Wang et al. (2013) |
| Leaf-senescence | *Arabidopsis* | 20–125 | ↓ Senescence, ROS burst, ↑ Chls, photosynthesis efficiency | Shi et al. (2015d) |
| Leaf-senescence | Rice | 10–20 | ↓ Senescence, ROS burst, cell death, ↑ Chls | Liang et al. (2015) |
| Leaf-senescence | Perennial ryegrass | 20–100 | ↓ Senescence, ROS burst, ↑ Chls, photosynthesis efficiency | Zhang et al. (2016a) |
| Postharvest | Peach | 100 | ↓ Senescence, weight loss, respiration rate, ROS burst, ↑ firmness | Gao et al. (2016) |
| Postharvest | Cassava tuber | 100 | ↓ Senescence, ROS burst, starch degradation, ↑ shelf life | Hu et al. (2016) |
| Postharvest | Tomato | 50 | ↑ Fruit ripening, anthocyanins | Sun et al. (2015); Sun et al. (2016) |

↑, Increased content or increased action; ↓, Decreased content or decreased action.

**TABLE 5.5**
**Effects of Abiotic Stress on the Endogenous Melatonin Levels**

| Plant Species | Abiotic Stressor | Increased Level of Melatonin versus Control | Reference |
|---|---|---|---|
| Lupin | Zn, NaCl, $H_2O_2$, cold, drought | 1.5–12 times | Arnao and Hernández-Ruiz (2013a) |
| Barley | Zn, NaCl, $H_2O_2$ | 6 times | Arnao and Hernández-Ruiz (2009b) |
| Barley | Drought, cold | 2 times | Li et al. (2016c) |
| Tomato | Field-growth conditions | 10 times | Arnao and Hernández-Ruiz (2013b) |
| Grape | Field-growth conditions | 15 times | Boccalandro et al. (2011) |
| Cherry | Field-growth conditions | 10 times | Zhao et al. (2013) |
| Sunflower | NaCl | 2–6 times | Mukherjee et al. (2014) |
| Bermudagrass | NaCl, drought, cold | 2–3 times | Shi et al. (2015b); Fan et al. (2015) |
| *Arabidopsis* | Cold | 2 times | Shi and Chan (2014) |
| | Heat | 2–5 times | Shi et al. (2015e) |
| | NaCl, drought, cold | 3–6 times | Shi et al. (2015c) |
| | Fe-deficiency | 6 times | Zhou et al. (2016a) |
| Rice | Cd | 6 times | Byeon et al. (2015c) |
| Tomato | Cd | 2 times | Hasan et al. (2015) |
| Ryegrass | Darkness | 2 times | Zhang et al. (2016a) |
| *Vitis* | NaCl | 5.5 times | Jiao et al. (2016) |
| | Osmotic | 1.5 times | Jiao et al. (2016) |

melatonin pretreatment facilitated the repair of PSII by maintaining the availability of D1 protein that was otherwise reduced by salinity (Zhou et al., 2016b).

Recent data have clearly demonstrated that endogenous melatonin levels change with environmental conditions. These data suggested that melatonin is accumulated in the plant tissues as a protective molecule in response to different environmental abiotic stressors, such as cold, ultraviolet (UV) radiation, the light-dark cycle, chemical agents, water deficit, and so on. Table 5.5 shows some of the most relevant papers that demonstrated that endogenous melatonin levels were increased in the face of different abiotic stressors. The global influence of environmental factors on the melatonin levels of plant organs was clearly demonstrated in tomato and lupin plants by Arnao and co-workers (Arnao and Hernández-Ruiz, 2013a, 2013b). In both cases, plants grown in field conditions contained 10-fold (in tomato) and 3-fold (in lupin) the melatonin content of plants grown in artificial conditions (culture chambers). The influence of environmental factors on melatonin levels was later seen in water hyacinth plants (Tan et al., 2007a), grape berry skin (Boccalandro et al., 2011), and cherry fruits (Zhao et al., 2013). Salinity, cold, drought, and heavy metals have been the abiotic environmental agents most frequently studied as inductors of melatonin biosynthesis in plants (Table 5.5). More recently, an interesting study provided evidence to support that melatonin can increase the tolerance of plants to iron deficiency. Melatonin treatment increased the resistance of plants to iron deficiency by enhancing remobilization of iron from cell walls of *Arabidopsis*, in a process dependent on the production of polyamine-induced nitric oxide (NO) under iron-deficient conditions, thereby alleviating chlorosis (Zhou et al., 2016a).

Figure 5.2 show a general scheme of melatonin's action as a positive effector on several physiological processes. Abiotic and biotic stressors (see the following) provoke an increase in endogenous melatonin levels through the up-regulation of melatonin biosynthetic genes. Both abiotic and biotic stress effects are mediated by an oxidation burst, ROS induction being the first cellular signal. It is known that stressors act as negative effectors in many cellular and physiological processes such as photosynthesis and membrane integrity (Figure 5.2 Panel A). Endogenous melatonin can change the expression of many genes and regulation factors that attenuate or reverse the

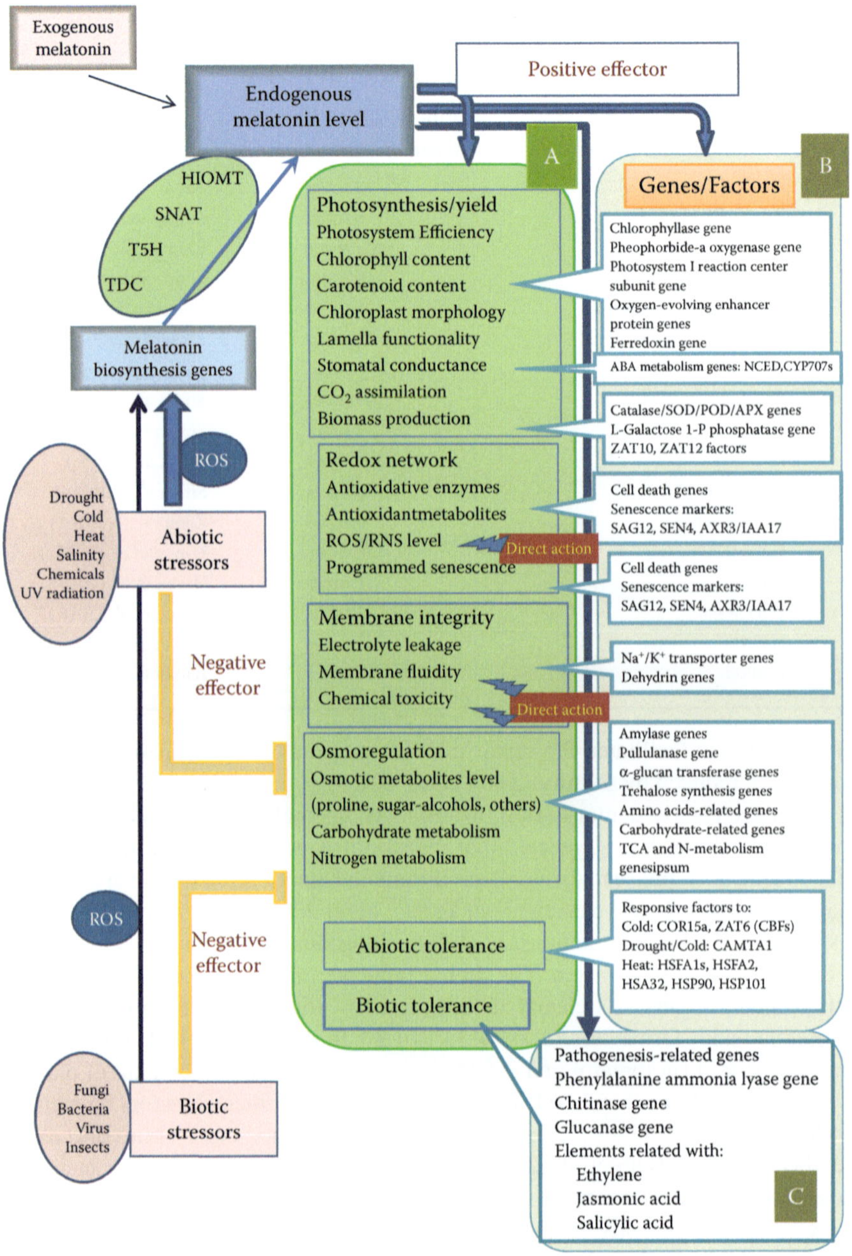


**FIGURE 5.2** Scheme of melatonin action as a positive effector in several physiological processes. Abiotic and biotic stressors provoke an increase in endogenous melatonin levels through the up-regulation of melatonin biosynthetic genes. Both abiotic and biotic stress effects are mediated by an oxidation burst, ROS induction being the first cellular signal. Stressors act as negative effectors in many cellular and physiological processes such as photosynthesis, membrane integrity, and so on (Panel A). Endogenous melatonin can change the expression of many genes and regulation factors that attenuate or reverse the negative effects of biotic/abiotic stressors on physiological processes, acting as a positive effector against stress (Panels B and C). Also, melatonin can act directly as a free radical scavenger (direct antioxidant) on ROS/RNS, lipid peroxides and toxic chemicals, controlling relevant aspects such as membrane integrity and the proper functioning of the redox network. (From Arnao, M.B. and Hernández-Ruiz, J., *J. Pineal Res.*, 59, 133–150, 2015a. With permission.)

negative effects of biotic/abiotic stressors on physiological processes, acting as a positive effector against stress (Figure 5.2 Panels B and C). Also, melatonin can act directly as free radical scavenger (direct antioxidant) of reactive oxygen and reactive nitrogen species, lipid peroxides, and toxic chemicals, controlling aspects such as membrane integrity and the proper functioning of the redox network (Arnao and Hernández-Ruiz, 2015b).

A broad variety of transgenic plants expressing ectopic genes or isogenes related with melatonin biosynthesis have been used. Table 5.6 shows most of the studies on melatonin with

**TABLE 5.6**
**Melatonin-Related Transgenic Plants**

| Modified Plant | Ectopic Enzyme Gene | Response[a] | Reference |
|---|---|---|---|
| OE[b] in rice | Human SNAT | ↑ Melatonin<br>↑ Cold resistance | Kang et al. (2010) |
| OE in rice | Sheep SNAT | ↑ Melatonin<br>↑ Herbicide resistance | Park et al. (2013a) |
| OE in rice | Sheep SNAT | ↑ Melatonin<br>↑ Root growth | Park and Back (2012) |
| OE in rice | Sheep SNAT | ↑ Melatonin<br>↑ Robustness<br>↑ Biomass, ↓ Yield | Byeon and Back (2014a) |
| OE in rice | Rice T5H isogenes | ↓ Melatonin | Park et al. (2012) |
| SE[c] in rice | Rice T5H isogenes | ↑ Melatonin | Park et al. (2013c) |
| OE in rice | Rice ASMT isogenes | ↑ Melatonin | Park et al. (2013a) |
| OE in rice | Rice TDC isogenes | ↑ Melatonin | Byeon et al. (2014b) |
| OE in rice | Rice COMT | ↑ Melatonin | Byeon et al. (2015a) |
| SE in rice | Rice COMT | ↓ Melatonin | |
| SE in rice | Rice SNAT/ASMT | ↓ Melatonin<br>↓ Cold resistance<br>↓ Growth, ↓ Seed yield<br>↑ Senescence | Byeon and Back (2016) |
| OE in tomato | Sheep SNAT/ASMT | ↑ Melatonin<br>↑ Drought resistance | Wang et al. (2014) |
| OE in tomato | *Chlamydomonas* SNAT | ↑ Melatonin | Okazaki et al. (2009) |
| OE in tomato | Rice Indoleamine 2,3-Dioxygenase | ↓ Melatonin<br>↓ Branching | Okazaki et al. (2010) |
| OE in tomato | Tomato ASMT | ↑ Thermotolerance<br>↑ Cell protection | Xu et al. (2016) |
| OE in *Nicotiana* | Human SNAT/ASMT | ↑ Melatonin<br>↑ Resistance to UV-B | Zhang et al. (2012) |
| OE in *Arabidopsis* | Apple ASMT | ↑ Melatonin<br>↑ Drought resistance<br>↑ Biomass, ↑ Roots | Zuo et al. (2014) |
| OE in switchgrass | Sheep SNAT/ASMT | ↑ Biomass, ↑ Roots<br>Differential gene ontology | Yuan et al. (2016a)<br>Yuan et al. (2016b) |
| SE in *Arabidopsis* | *Arabidopsis* SNAT | ↓ Melatonin<br>↓ Salicylic acid, ↓ Defense genes<br>↑ Pathogen susceptibility | Lee et al. (2015) |

[a] ↑, Increased content or increased action; ↓, Decreased content or decreased action.
[b] OE: Overexpressed gene.
[c] SE: Suppressed gene expression.

physiological objectives using transgenic plants. The most commonly used ectopic genes were those which codify SNAT and ASMT enzymes (Figure 5.1) from different sources including human and sheep. Some studies have been made with other melatonin biosynthesis genes, such as T5H and TDC. Rice plants have been the most widely used receptors of these ectopic genes, and many rice transgenic plants have been obtained to study the melatonin metabolism (Table 5.6). Tomato and *Nicotiana sylvestris* plants have also been used. In general as a response, SNAT/ASMT overexpressing plants showed a net increase in endogenous melatonin levels, while other related indoles also changed its levels in transgenic plants compared with the levels in untransformed plants. Some phenotypic changes compared with wild-type plants are to be expected. For example, the increased melatonin level in overexpressing rice plants was related with greater resistance to butafenacil, a singlet oxygen-generating herbicide (Park et al., 2013c). Also, the overexpression of apple-ASMT in *Arabidopsis* resulted in a higher endogenous melatonin level, low ROS content, increased biomass and, in general, a greater tolerance to drought compared to wild type plants (Zuo et al., 2014). A higher resistance to abiotic stress conditions such as drought, low or high temperatures, and UV can be observed in melatonin-rich overexpressing plants *versus* wild-type plants. These results demonstrate that endogenous melatonin, even at very low concentrations, is critical as a first line of defense against stressors (Table 5.5 and Figure 5.2).

## 5.5 MELATONIN AND BIOTIC STRESS

In animals, melatonin has important immunomodulatory, antioxidant, anti-inflammatory, and neuroprotective effects, suggesting that this methoxyindole is good therapeutic alternative for fighting bacterial, viral, and parasitic infections (Bonilla et al., 2004). During sepsis, melatonin has been reported to block the overproduction of pro-inflammatory cytokines and increase interleukin-10 levels. Melatonin administration also augmented the weight of the spleen in endotoxin-septic rats, counteracted sepsis-induced apoptosis in the spleen, and neutralized inflammatory infiltration in different tissues (Carrillo-Vico et al., 2013). *In vitro* studies have demonstrated that melatonin is effective against multidrug-resistant Gram-positive and Gram-negative bacteria, carbapenem-resistant *Pseudomonas aeruginosa*, *Acinetobacter bau-mannii*, and methicillin-resistant *Staphylococcus aureus* (Tekbas et al., 2008). Therefore, treatment with melatonin was neuroprotective to hippocampal neurons of rats infected with *Klebsiella pneumonia*. These anti-inflammatory and anti-oxidative effects of melatonin make this indole an alternative for preventing the neurocognitive damage produced by *K. pneumonia* (Wu et al., 2011). Also, its use as an *in vivo* antibiotic has been suggested, for example, in human newborns suffering from septicemia (Gitto et al., 2001).

Venezuelan equine encephalomyelitis (VEE) is an important human and equine disease caused by the VEE virus. ROS have been implicated in the dissemination of the responsible virus, and their deleterious effects may be diminished by melatonin treatment. The administration of melatonin significantly decreased the virus level in blood and brain compared with the levels seen in infected control mice (Bonilla et al., 2004). Lastly, the role of melatonin has been studied in diseases caused by protozoan parasites such as *Plasmodium* spp., *Entamoeba histolytica*, *Trypanosoma cruzi*, and *Toxoplasma gondii*, among others (Vielma et al., 2014).

In agriculture, pathogenic diseases are responsible for major production and economic losses in crops, and particular attention has been paid recently to the action(s) of melatonin on plant biotic stress. Table 5.7 show a list of the papers related with the positive effect of melatonin on innate plant immunity. In the first paper related with plant-pathogen infection, treating trees through their roots by irrigation with melatonin at different concentrations improved resistance of *Malus prunifolia* against the fungus *Diplocarpon mali* (Marssonina apple blotch). At 20 days, melatonin-treated

**TABLE 5.7**
**Effects of Melatonin in Biotic Stress Responses**

| Plant Species | Biotic Stressor | Melatonin Treatment (μM) | Effects Observed | Reference |
|---|---|---|---|---|
| *M. prunifolia* | *D. mali* | 50–500 | ↑ Resistance to fungal infection<br>↓ Leaf lesions, cell death<br>↓ Pathogen expansion | Yin et al. (2013) |
| *Arabidopsis* and tobacco | *P. syringae* DC3000 | 10 | ↑ Defense-related genes<br>↑ Resistance (10-fold vs. mock) | Lee et al. (2014a) |
| *Arabidopsis* | *Pseudomonas syringae* DC3000 | SE of SNAT | ↓ Melatonin (50%), -SA<br>↓ Defense-related genes<br>↓ Resistance to infection | Lee et al. (2015) |
| *Arabidopsis* | *Pseudomonas syringae* DC3000 | 20 | ↑ NO and melatonin<br>↑ Defense-related genes<br>↑ Resistance | Shi et al. (2015a) |
| *Arabidopsis* and tobacco | *Pseudomonas syringae* DC3000 | 1 | ↑ MAP kinases cascade | Lee and Back (2016) |
| *Arabidopsis* | *Pseudomonas syringae* DC3000 | 50 | ↑ CBF/DREB1 (stress factors)<br>↑ CCA1 (internal clock factors)<br>↑ Defense-related genes | Shi et al. (2015a); Shi et al. (2016) |
| *Lupinus albus* | *Penicillium* spp. | 20–70 | ↑ Resistance to fungal infection | Arnao and Hernández-Ruiz (2015a) |
| Rice | *Xanthomonas oryzae*, *Xoo Magnaporthe oryzae*, *blast fungus* | — | Changes in melatonin biosynthesis enzymes transcripts | Wei et al. (2016) |
| *Vitis labruscana* | Endophytic bacteria | — | ↑ Melatonin in roots<br>↑ Plant height, Chls<br>↑ Root length, lateral roots<br>↓ ROS burst | Jiao et al. (2016) |

↑, Increased content or increased action.
↓, Decreased content or decreased action.

apple trees showed a lower number of damaged leaves, a higher chlorophyll content, a more efficient Photosystem II, and a less defoliation than infected-untreated trees. Melatonin contributed to greater resistance to fungal infection, reduced lesions, inhibited pathogen expansion, and generally alleviated disease damage (Yin et al., 2013). Also, different concentrations of melatonin showed growth inhibition activity against several plant fungal pathogens such as *Alternaria* spp., *Botrytis* spp., and *Fusarium* spp. growing in standard media. Melatonin decreased the rate of infection in plant-pathogen attacks by *Penicillium* spp. in non-sterilized *Lupinus albus* seeds (Arnao and Hernández-Ruiz, 2015a).

The most widely used model in plant-bacterial pathogen interaction studies is *Arabidopsis/Pseudomonas syringae* (Table 5.7). In general, the application of melatonin to *Arabidopsis* (also tobacco) induces pathogenesis-related genes, which further supports the idea that melatonin may be a defense-signaling molecule in plants against pathogens. Melatonin up-regulates pathogen-related, salicylic acid (SA)- and ethylene-dependent genes, an effect that was suppressed in mutants defective in SA and ethylene signaling. Also, melatonin increased nitric oxide (NO) and SA-related genes, accompanied by reduced susceptibility to the pathogen, leading to an increase in both melatonin and NO. SNAT knockout mutants did not only exhibit reduced levels of melatonin but also lower levels of SA, along with greater susceptibility to the pathogen (Lee et al., 2015) (Tables 5.6 and 5.7). Recently, it has been found that mitogen-activated protein kinase (MAPK) signaling through diverse MAPK kinase cascades is also required for melatonin-mediated innate immunity in plants (Lee and Back, 2016). The data obtained to date show that melatonin acts upstream of the defense gene signaling pathway, inducing the biosynthesis of SA, jasmonic acid (JA), NO, and ethylene, which, together, elicit disease resistance in a well-known co-action (Zhu and Lee, 2015). The emergence of an oxidative burst during the early stages of the plant-pathogen interaction seems to increase endogenous melatonin levels (Figure 5.2). In conclusion, melatonin seems to be involved in innate plant immunity against fungal and bacterial pathogens via a SA and NO-dependent pathway. However, no studies about plant viruses and melatonin have been published to date.

Interestingly, endophytes (e.g., soil microbes) form symbiotic relationships with plants. These symbiotic organisms are important in defending their hosts against pathogens and may also promote the growth of their host plants *via* nitrogen fixation, phosphorus solubilization, and the enhancement of plant hormone levels. The occurrence of melatonin biosynthesis in endophytic bacteria, in *Vitis* sp. (Table 5.7) provides evidence for a novel form of communication between beneficial endophytes and host plants *via* melatonin (Jiao et al., 2016). In this sense, high endogenous melatonin levels have recently been detected in the soil fungi *Trichoderma* spp., which may be related with the proposed function of these fungi as of root/stem growth promoters, plant-pathogen resistance inducers, and as alleviating bioagents in plants against various environmental stressors, such as heavy metals, radiation, acid and alkaline soils, and extreme temperatures, among others (Liu et al., 2016b).

## 5.6 MELATONIN AND CIRCADIAN RHYTHMS

The role of melatonin as a regulator of light-dark cycles has been clearly established in mammals (Reiter, 1993; De la Iglesia et al., 2000; Reppert and Weaver, 2002). In plants, the circadian oscillator is able to adjust the phase of a variety of biological processes, such as gene and metabolic regulation and protein stability, to coincide with daily and/or seasonal cycles. Thus, circadian regulation increases photosynthesis and growth rates, and may affect flowering and seed yield in crops, and biotic/abiotic stress responses (Dodd et al., 2005, 2014; Atkins and Dodd, 2014; Dixon et al., 2014).

An oscillating behavior in melatonin levels has been observed in the plant species studied. Figure 5.3 shows changes in melatonin levels during a photoperiodic cycle of 12 h light/12 h darkness (Arnao and Hernández-Ruiz, 2015a). The data obtained for lupin and barley are very similar to those for sweet cherry and apple leaves, in which two peaks of melatonin appeared (Figure 5.3, Panels F and G). In both cases, melatonin peaks were higher in roots than in leaves or cotyledons. The results of the studies with *Chenopodium* (Kolar et al., 1997), *Eichhornia* (Tan et al., 2007a), *Vitis* (Boccalandro et al., 2011), *Malus* (Zuo et al., 2014), *Prunus* (Zhao et al., 2013), *Lupinus*, *Hordeum* (Arnao and Hernández-Ruiz, 2015a) and *Arabidopsis* (Shi et al., 2016) indicate that the time of day at which the sample is taken seems to be relevant for melatonin levels in plant tissues, since, depending on the time of day or night, differences of several orders of magnitude are recorded. Whatever the case, a circadian rhythm of melatonin also seems to exist in plants.

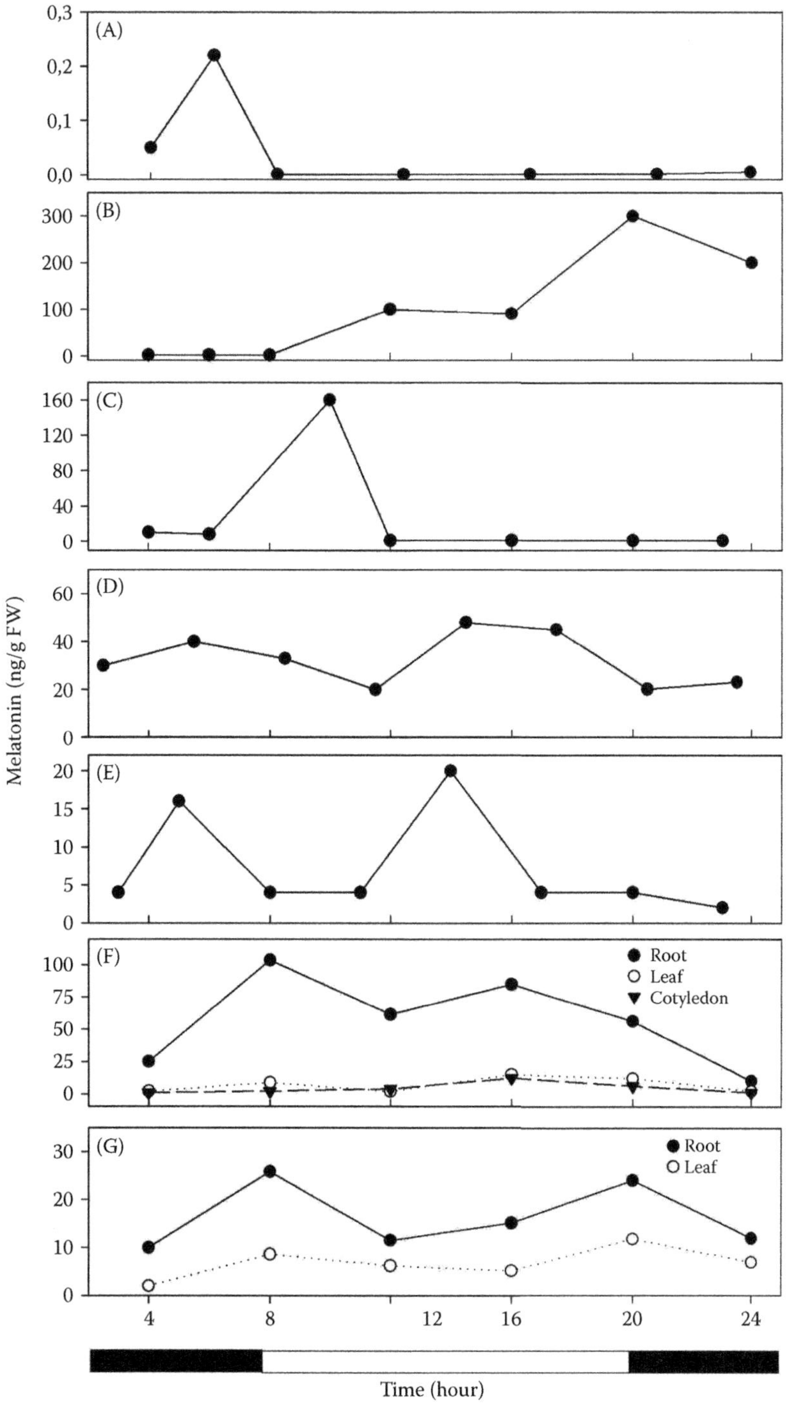


**FIGURE 5.3** Changes in melatonin level during a photoperiodic cycle of 12 h light/12 h darkness. (A) Chenopodium, (B) Hyacinth, (C) Grapevine, (D) Apple, (E) Cherry, (F) Lupin and (G) Barley. (From Arnao, M.B. and Hernández-Ruiz, J., *J. Pineal Res.*, 59, 133–150, 2015a. With permission.)

## 5.7 MELATONIN AND PLANT HORMONES

Table 5.8 shows the effect that some melatonin treatments have on endogenous plant hormones. In the following we discuss the most relevant data on the relationship of melatonin with the principal plant hormones, such as auxin, gibberellins, abscisic acid, and ethylene.

### 5.7.1 AUXIN

The case of *Brassica juncea*, which showed increased levels of endogenous indole-3-acetic acid (IAA) after melatonin treatment, was the first described (Table 5.8). In general, melatonin treatment induces a slight increase in endogenous IAA as can also be observed in *Brassica oleracea* and tomato plants. However, in transgenic plants that over-produce melatonin, a substantial decrease in IAA levels has been reported. In all the cases described in Table 5.8, the phenotypic effects clearly

**TABLE 5.8**
**Effects of Melatonin Related with Plant Hormones**

| Plant Hormone | Plant Species | Melatonin Treatment or Over-Accumulation | Effects Observed | Reference |
|---|---|---|---|---|
| Auxin, IAA | *B. juncea* roots | 0.01–0.5 μM, 24 h | ↑ IAA<br>↑ Root growth | Chen et al. (2009) |
| | Tomato de-rooted plants | 12.5–100 μM, 6 days | ↑ IAA<br>↑ Adventitious roots | Wen et al. (2016) |
| | Tomato overexpressing sheep SNAT/ASMT[a] | ↑ up to 6 times | ↓ IAA up to-7-times<br>↓ Apical dominance<br>Branching phenotype<br>↑ Tolerance to drought | Wang et al. (2014) |
| | *A. thaliana* overexpressing apple ASMT | ↑ up to 4 times | ↓ IAA up to 1.4-times<br>↑ Lateral roots, biomass<br>↑ Tolerance to drought | Zuo et al. (2014) |
| | *Arabidopsis* seedlings | 1 mM, 16 h | ↑ IAA-amino synthase genes<br>↑ Auxin-carrier proteins | Weeda et al. (2014) |
| | *Arabidopsis* roots | 600 μM | ↓ IAA<br>↓ Root meristem sizes | Wang et al. (2016) |
| Gibberellins (GAs) | Cucumber seedlings | 10–500 μM, 24 h | ↑ GAs<br>↑ GAs biosynthesis genes<br>↑ Germination rate | Zhang et al. (2014a)<br>Zhang et al. (2014b) |
| Abscisic acid (ABA) | Cucumber seedlings | 10–500 μM, 24 h | ↓ ABA<br>↓ ABA biosynthesis genes<br>↑ ABA catabolism genes | Zhang et al. (2014a) |
| | Apple leaves | 10–500 μM, 24 h | ↓ ABA<br>↓ ABA biosynthesis genes<br>↑ ABA catabolism genes | Li et al. (2015) |
| | Barley leaves | 1 mM, 5 days | ↑ ABA<br>↑ Photosynthesis efficiency<br>↑ Tolerance to drought, cold | Li et al. (2016c) |

*(Continued)*

**TABLE 5.8 (*Continued*)**
**Effects of Melatonin Related with Plant Hormones**

| Plant Hormone | Plant Species | Melatonin Treatment or Over-Accumulation | Effects Observed | Reference |
|---|---|---|---|---|
| Ethylene | Tomato fruits | 50 μM, 2 h | ↑ Ethylene, climacteric peak<br>Changes in flavors, lycopene, softening<br>↑ ACC synthase genes, anthocyanins<br>↑ Ethylene receptor genes | Sun et al. (2015)<br>Sun et al. (2016) |
| | Tomato plants | 0.1 mM | ↑ Ascorbic acid, lycopene, Ca, P<br>↑ Quality and yield of tomato fruits | Liu et al. (2016a) |
| | Lupin seedlings | 0.01–1 mM, 1 h | ↓ Ethylene rate | Arnao and Hernández-Ruiz (2007a) |
| | *Arabidopsis* seedlings | 1 mM, 16 h | ↑ ACC synthase genes | Weeda et al. (2014) |
| | Cassava roots | 100 μM, 2 h | ↓ Postharvest deterioration<br>↑ Shelf life of tuberous | Hu et al. (2016) |
| | Peach fruits | 0.1 mM, 10 min | ↓ Postharvest senescence<br>↑ Weight, firmness, ascorbic acid | Gao et al. (2016) |
| | Strawberry fruits | 100 μM, 5 min | ↓ Postharvest senescence<br>↑ ATP, antioxidants, shelf life | Aghdam and Fard (2017) |

[a] SNAT: Serotonin *N*-acetyltransferase; ASMT: Acetylserotonin methyltransferase.
↑, Increased content or increased action.
↓, Decreased content or decreased action.

resemble auxinic responses, promoting root growth, rhizogenesis, including a reduced apical dominance, with high degree of branching in the case of tomato overexpressing sheep SNAT/ASMT (Zuo et al., 2014). This suggests that melatonin does not replace IAA in the apical dominance function, but that both take part in the same physiological action.

The growth-promoting activity of melatonin is one of the more specific auxinic roles exerted by melatonin. Table 5.9 shows the many studies in which exogenous melatonin has been seen to change the growth pattern in different species. In all the studies presented, melatonin induced growth in aerial parts and promoted or inhibited growth in roots, depending on the concentration assayed. Generally, growth inhibition only occurs at high melatonin concentrations (100 μM), although in one recent study using *Zea mays*, melatonin was seen to have no effect on growth promoting/inhibiting activities in the conditions assayed, which contrasted with the findings of a previous paper, in which leaves and roots of *Z. mays* were seen to respond in the expected way to melatonin treatment (Zhao et al., 2015). Nevertheless, in the presence of NaCl, melatonin partially restored NaCl-inhibited growth in maize coleoptiles and roots (Kim et al., 2016) (see data in Table 5.9). Generally, the growth-promoting effect of melatonin is high when a stress condition affects plant

**TABLE 5.9**
**Effect of Exogenous Melatonin on Growth in Different Organs and Species**

| Species | Organ | Melatonin Treatment on Growth | Maximum Growth Response (vs without Melatonin) | Reference |
|---|---|---|---|---|
| *Lupinus albus* | Hypocotyl (sections) | 1.0–10 μM | ↑ Up to 3.0 times | Hernández-Ruiz et al. (2004) |
| | Hypocotyl (derooted) | 0.1–10 μM | ↑ Up to 3.5 times | |
| | Cotyledon | 0.1–1.0 μM | ↑ Up to 1.2 times | Hernández-Ruiz and Arnao (2008b) |
| *Phalaris canariensis* | Coleoptile (sections) | 0.1–10 μM | ↑ Up to 3.6 times | Hernández-Ruiz et al. (2005) |
| | Roots (intact) | 0.1 nM–100 μM | ↓ Up to 0.4 times | |
| *Triticum aestivum* | Coleoptile (sections) | 10–100 μM | ↑ Up to 3.5 times | |
| | Roots (intact) | 10 nM–1 μM | ↑ Up to 1.4 times | |
| | | 10–100 μM | ↓ Up to 0.7 times | |
| *Hordeum vulgare* | Coleoptile (sections) | 10 nM–1 μM | ↑ Up to 3.3 times | |
| | | 10–100 μM | ↓ Up to 0.8 times | |
| *Avena sativa* | Coleoptile (sections) | 10 nM | ↑ Up to 1.2 times | |
| | | 0.1 μM | ↓ Up to 0.3 times | |
| | Roots (intact) | 0.1 nM–100 μM | ↓ Up to 0.7 times | |
| *B. juncea* | Roots (intact) | 0.1 μM | ↑ Up to 1.2 times | Chen et al. (2009) |
| | | 100 μM | ↓ Up to 0.6 times | |
| *B. oleracea* | Whole plant | 1–10 μM | ↑ Up to 1.2 times | Posmyk et al. (2008) |
| | Whole plant in cooper | 10 μM | ↑ Up to 1.5 times | |
| *Helianthus annuus* | Hypocotyl | 5–30 μM | ↑ Up to 2.2 times | Mukherjee et al. (2014) |
| | Hypocotyl with NaCl | 15 μM | ↑ Up to 1.3 times | |
| | Primary root | 15–30 μM | ↑ Up to 1.8 times | |
| | Primary root with NaCl | 15 μM | ↑ Up to 1.2 times | |
| *Oryza sativa* | Seminal root (wild type) | 0.5–5 μM | ↑ Up to 1.4 times | Park and Back (2012) |
| | Seminal root (transgenic line) | — | ↑ Up to 1.5 times | |
| *A. thaliana* | Lateral roots | 100–300 μM | ↑ Up to 3.2 times | Pelagio-Flores et al. (2012) |
| | Whole plant | 10–100 μM | ↑ Up to 1.4 times | Bajwa et al. (2014) |
| | Whole plant at 4°C | 30 μM | ↑ Up to 1.7 times | |
| | Primary roots | 10–40 μM | ↑ Up to 1.3 times | |
| | Primary roots at 4°C | 30 μM | ↑ Up to 1.4 times | |
| | Primary roots | 0.1 μM | ↑ Up to 1.1 times | Hernández et al. (2015) |
| *G. max* | Leaf area | 50–100 μM | ↑ Up to 1.3 times | Wei et al. (2015) |
| *Cynodon dactylon* | Whole plant | 4–100 μM | Not increases in growth | Shi et al. (2015b) |
| | Whole plant at 4°C | 20–100 μM | ↑ Up to 1.4 times | |
| | Whole plant in drought | 20–100 μM | ↑ Up to 1.4 times | |
| | Whole plant at 4°C | 20–100 μM | ↑ Up to 1.5 times | |
| *Prunus avium* × *Prunus cerasus* | Primary root | 0.1–1 μM | ↑ Up to 2.5 times | Sarropoulou et al. (2012a) |
| *Cucumis sativus* | Shoot in osmotic stress | 100–500 μM | ↑ Up to 1.7 times | Zhang et al. (2013) |
| | Root in osmotic stress | 100–500 μM | ↑ Up to 3.0 times | |

*(Continued)*

**TABLE 5.9 (*Continued*)**
**Effect of Exogenous Melatonin on Growth in Different Organs and Species**

| Species | Organ | Melatonin Treatment on Growth | Maximum Growth Response (vs without Melatonin) | Reference |
|---|---|---|---|---|
| *Punica granatum* | Adventitious roots | 5 µM | ↑ Up to 1.4 times | Sarrou et al. (2014) |
| *Solanum lycopersicum* | Shoot with cadmium | 25–500 µM | ↑ Up to 2.4 times | Hasan et al. (2015) |
| | Root with cadmium | 25–500 µM | ↑ Up to 3.3 times | |
| *Malus hupehensis* | Whole plants | 0.1 µM | Not increases in growth | Li et al. (2016a) |
| | Whole plants in NaCl | 0.1 µM | ↑ Up to 1.2 times | |
| | Whole plants in K deficiency | 0.1 µM | ↑ Up to 1.3 times | |
| | Whole plants in all-nutrient deficiency | 0.1 µM | ↑ Up to 1.7 times | |
| *Z. mays* | Leaves | 1–10 µM | ↑ Up to 1.2 times | Zhao et al. (2015) |
| | Roots | 100–1000 µM | ↓ Up to 0.5 times | |
| | | 10 µM | ↑ Up to 1.2 times | |
| | | 100–1000 µM | ↓ Up to 0.6 times | |
| | Coleoptiles | 0.1–100 µM | Not increases in growth | Kim et al. (2016) |
| | Coleoptiles with NaCl | 10 µM | ↑ Up to 1.3 times | |
| | Roots | 1–100 µM | Not increases in growth | |
| | Roots with NaCl | 10 µM | ↑ Up to 1.1 times | |

development, as can be seen in the case of *Helianthus* and *Z. mays* grown in saline conditions, and in *Arabidopsis* and *Cynodon* exposed to cold-stress, among others (Table 5.9). Melatonin induced a three- to four-fold increase in growth compared to control plants in the aerial tissues of *Lupinus*, *Phalaris*, *Triticum*, *Hordeum*, *Arabidopsis*, and *Cucumis*, and a less pronounced increase in others (Table 5.9).

The rhizogenic activity of melatonin has also been described. In the pioneering study by Arnao group's (Arnao and Hernández-Ruiz, 2007b), where the strong induction of lupin roots by melatonin was compared with the effect of IAA, it was seen that melatonin induced the activity of root primordia from pericycle cells and clearly affected the appearance of both adventitious and lateral roots, modifying the pattern of distribution, the time-course, the number and length of adventitious roots, and the number of new lateral roots. Figures 5.4 and 5.5 depict representative pictures of the rooting activity of melatonin.

Since this study, many other species have confirmed that melatonin promotes the generation of lateral and adventitious roots. For example, in *Arabidopsis thaliana*, melatonin was seen to increase the appearance of adventitious roots twofold and the appearance of lateral roots by up to threefold but to have no effect on root hair density (Pelagio-Flores et al., 2012). Also, in three transgenic lines of *A. thaliana*, which over-produce melatonin, a high number of lateral roots (compared with wild-type) was induced (Zuo et al., 2014). Similar rhizogenetic activity was obtained in sweet cherries (Sarropoulou et al., 2012a, 2012b), cucumber (Zhang et al., 2013, 2014b), pomegranate (Sarrou et al., 2014) and tomato plants (Wen et al., 2016).

A very novel aspect also related to the auxinic activity of melatonin is its possible role in the gravitropic response of roots. But the only evidence concerning the possible participation of melatonin in tropic responses was that provided recently by Arnao and co-workers. This preliminary study

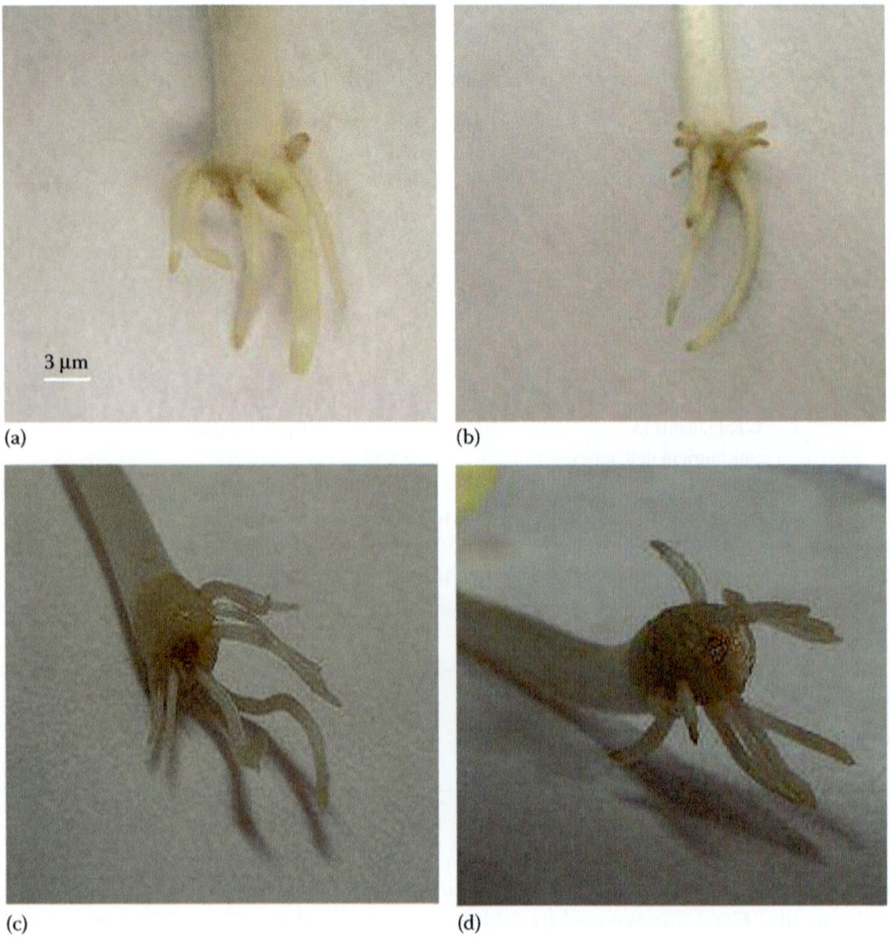


**FIGURE 5.4** Representative images of the root regeneration induced by melatonin or IAA in 6 day-old etiolated lupin after 8 days of treatment at 25°C in darkness. (a) Adventitious root induction by 10 μM melatonin; (b) idem by 1 μM IAA; (c) Lateral root induction by 1 μM melatonin; (d) idem by 1 μM IAA. (From Arnao, M.B. and Hernández-Ruiz, J., *J. Pineal Res.*, 42, 147–152, 2007b. With permission.)

showed the effect that an exogenous imbalance of IAA or melatonin has on the tropic response of lupin roots. The authors suggested that the disruption (through IAA- or melatonin-enriched agar blocks) of the natural level of auxin (or melatonin) in lupin roots gave rise to an artificial internal gradient that provoked a high lateral growth of roots, with the consequent loss of verticality. Once the auxinic gradient was normalized on both sides of the root, vertical growth was reestablished, guided by the force of gravity (Arnao and Hernández-Ruiz, 2017).

Although IAA and melatonin share tryptophan as a precursor in their respective biosynthetic pathways, to date no conclusive data have been presented to confirm metabolic inter-conversion between melatonin and IAA in plants. Also, the expression pattern of auxin-related genes exhibited minimal changes in melatonin-treated *Arabidopsis* plants with respect to untreated plants. Thus, only one IAA-amino synthase was up-regulated with no change in the expression of auxin biosynthesis genes. More significant was the down-regulation of several auxin-influx carrier proteins (AUX1/LAX) in response to melatonin. AUX1 regulates lateral root development, root gravitropism,

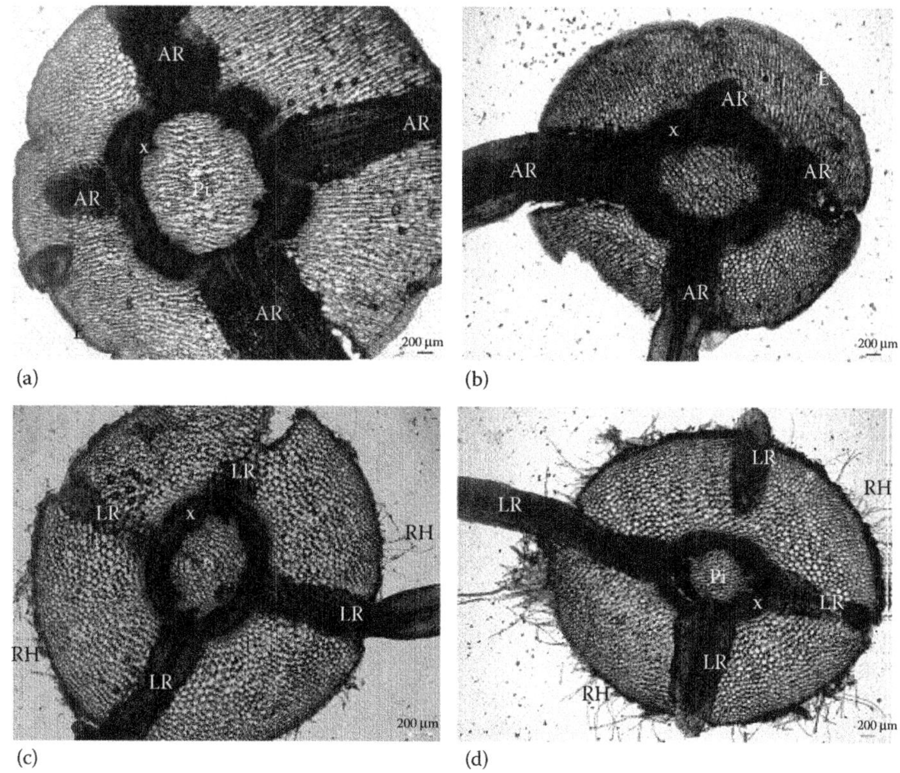


**FIGURE 5.5** Light micrographs of transverse sections of 6-day-old etiolated lupin after 8 days of treatment at 25°C in darkness. (a) Adventitious root induction by 10 μM melatonin; (b) idem by 1 μM IAA; (c) lateral root induction by 1 μM melatonin; (d) idem by 1 μM IAA. AR, adventitious root; C, cortex; E, epidermis; Pi, pith; RH, root hairs; LR, lateral root; X, xylem. (From Arnao, M.B. and Hernández-Ruiz, J., *J. Pineal Res.*, 42, 147–152, 2007b. With permission.)

root hairs, and leaf phyllotaxy (Swarup and Péret, 2012), suggesting that melatonin can interfere in the action of auxin through changes in auxin carriers that modify local IAA gradients (Weeda et al., 2014; Arnao and Hernández-Ruiz, 2015a; Wang et al., 2016). Curiously, in lupin roots and some monocot roots, a melatonin gradient similar to the IAA gradient has been described, suggesting that both gradients co-participate in plant growth (Hernández-Ruiz et al., 2004; Hernández-Ruiz and Arnao, 2008a). Recently, Wang and co-workers demonstrated that melatonin negatively regulated auxin biosynthesis and the auxin response in *Arabidopsis*. The auxin polar transporter inhibitor 2,3,5-triiodobenzoic acid (TIBA) did not enhance the melatonin-mediated reduction of root meristem size, indicating that polar auxin transport may be necessary for the regulation of root meristem size by melatonin treatment (Wang et al., 2016).

### 5.7.2 Gibberellins and Abscisic Acid

In the case of gibberellins (GA) and abscisic acid (ABA), the levels of both plant hormones were altered by melatonin treatment (Table 5.8). Thus, melatonin up-regulated GA biosynthesis genes such as GA20ox and GA3ox in cucumber seedlings in saline conditions, contributing to a high level of activates GAs as $GA_4$, promoting salt-inhibited germination process (Zhang et al., 2014a, 2014b). Also, melatonin treatment induced the up-regulation of ABA catabolism genes (two CYP707 monooxygenases) and the down-regulation of 9-cis-epoxycarotenoid dioxygenase (NCED), a key enzyme in ABA biosynthesis, which resulted in a rapid decrease in ABA levels during seed germination under salt stress (Zhang et al., 2014a). Similar data were obtained in apple leaves in drought conditions, where melatonin pre-treatment halved the ABA content through regulation of the same ABA biosynthesis and catabolism enzymes as mentioned earlier (Li et al., 2015).

Using wild-type and an ABA-deficient mutant of barley, the effect of melatonin treatment (either foliarly or rhizospherically) in a process of drought priming-induced cold tolerance was investigated. All stressing treatments significantly increased the levels of endogenous ABA and melatonin in wild-type barley leaves (no ABA increase occurred in the mutant), suggesting that melatonin increased drought priming-induced ABA accumulation under cold stress. Thus, the application of melatonin resulted in a higher ABA concentration in the drought-primed plants when exposed to cold stress, which could help to maintain a better plant water status, modulating subcellular antioxidant systems and ABA levels in barley. In addition, both drought priming and exogenous melatonin application increased the endogenous melatonin concentration in barley mutant, indicating that the response was ABA independent. Also, exogenous application of melatonin significantly enhanced chlorophyll content of the primed plants under cold stress. Together with the positive effects on electron transport and Rubisco activity, both drought priming and exogenous application of melatonin could alleviate the cold stress-induced photosynthetic limitation in barley (Li et al., 2016c).

### 5.7.3 Ethylene

In the case of ethylene (Table 5.8), the effect of exogenous melatonin on ethylene metabolism, postharvest ripening and the quality of tomato fruit showed that tomatoes treated with 50 μM melatonin for 2 h manifested substantial changes in their fruit ripening parameters, such as lycopene levels, fruit softening, flavor, and ethylene-signaling and biosynthesis enzymes with respect to untreated tomatoes (Sun et al., 2015). Exogenous melatonin slightly increased ethylene generation and the subsequent timing of the climacteric peak through the up-regulation of 1-aminocyclopropane-1-carboxylic acid (ACC) synthase expression. Also, the ethylene receptor genes, NR and ETR4, and the transducing elements, EIL1, EIL3, and ERF2, were up-regulated by melatonin.

By contrast, melatonin produced a strong inhibition (up to 65% in roots, compared with a control) in the rate of ethylene production in etiolated lupin seedlings (Arnao and Hernández-Ruiz, 2007a). This inhibitory effect on ethylene production manifested itself in vegetative tissues as a regulation by auxin, in which IAA was able to induce ACC synthase expression, while blocking the induction of ACC oxidase expression by ethylene, in agreement with the model proposed by Kang's group in mungbean hypocotyls (Kim et al., 2001). Also, in *Arabidopsis* melatonin-treated plants, two ACC synthases were up-regulated, one auxin-inducible according to the previous model (Weeda et al., 2014). Possibly, this opposite effect was due to differences in the auxin-mediated response between vegetative and reproductive tissues.

Exogenous melatonin treatment induced the up-regulation of genes related with lycopene biosynthesis, aroma/flavor, cell wall structure, and aquaporins in tomatoes, leading to the conclusion that melatonin promotes postharvest tomato fruit ripening through increased ethylene production and signaling (Sun et al., 2015). In a recent differential proteomic analysis of tomato fruits, 241 proteins involved in several ripening-related pathways, including cell wall metabolism, oxidative phosphorylation, carbohydrate, flavonoid biosynthesis, and fatty acid metabolism, were significantly influenced by melatonin. Also, eight proteins related to anthocyanin accumulation during fruit ripening were increased, suggesting that the exogenous application of melatonin positively regulates fruit ripening, while negatively regulating fruit senescence (Sun et al., 2016). Also in tomato plants, melatonin-treated plants (in seeds) have much higher yields as well as higher ascorbic acid, lycopene, and Ca levels, while the content of N, Mg, Cu, Zn, Fe, and Mn decrease. By contrast, plants irrigated weekly with melatonin-supplemented nutrient solutions showed significant improvements in their contents of soluble solids, ascorbic acid, lycopene, citric acid, and P when compared with control plants that received only a standard solution. Also, the higher sucrose and glucose contents observed indicated that melatonin increased the fruit sugar content. As tomato fruits develop over time, levels of soluble galactose increase, thereby stimulating ethylene production and, subsequently, promoting the ripening. Also, melatonin-treated plants have significantly higher levels of citric acid. The flavor of fruits is optimal when there are high concentrations of both sugars and organic acids. In general, melatonin increased both fruit yield (up to 13%) and quality in tomato fruits (Liu et al., 2016a).

In peach fruits, melatonin treatment effectively slowed the senescence in the two peach cultivars assayed, as indicated by reduced weight loss, decay incidence, and respiration rate, while the firmness, total soluble solids, and ascorbic acid contents were maintained (Table 5.8). Similarly, the postharvest application of melatonin delayed senescence and maintained the quality of peach fruit, which may be attributed to its capacity to mediate antioxidative actions, suggesting that melatonin treatment may be a promising method for delaying senescence and maintaining fruit quality of postharvest peach fruit (Gao et al., 2016). Also in melatonin-treated strawberry, improvements in postharvest decay indicators and high ATP contents, antioxidant activity, and polyphenols level have been described (Aghdam and Fard, 2017).

With regard to salicylic acid (SA) and jasmonic acid (JA), both plant regulators are involved in biotic stress responses, as mentioned earlier (see *Melatonin and biotic stress* section). Also, in an extensive genetic functional analysis of *Arabidopsis* treated with exogenous melatonin, most genes in the ABA, SA, JA, and ethylene pathways were up-regulated, confirming that melatonin alters the expression of stress-response genes involved in all the steps of the pathway, from receptors to transcription factors. These results confirm the critical roles of melatonin in defense against both biotic and abiotic stress in plants (Weeda et al., 2014; Lee et al., 2015). By contrast, in a recent work, the treatment of rice leaves with different plant hormones (IAA, $GA_3$, JA, SA, and ethylene, at a fixed concentration) did not induce melatonin biosynthesis and only a low level was measured in response to ABA, suggesting that melatonin biosynthesis is not linked to plant hormone responses (Byeon et al., 2015c).

## 5.8 SUMMARY

Melatonin, a classical animal hormone, was discovered in plants in 1995 and was soon seen to have multiple possible roles in plant physiology. Its protective function against biotic and abiotic stressors is now fully accepted and may be very relevant for application in agronomic techniques to improve crop yield. Its implication in photosynthesis efficiency, $CO_2$ assimilation, the carbohydrates/lipid/nitrogen metabolisms, and osmoregulation appears clear although some important aspects remain to be understood. Melatonin shows some activities similar to classical auxin. Its role as a modulator of growth and rooting has led to it being considered a plant hormone. Nevertheless, the fact that its receptor in plants has not been identified and the scarce data concerning its mechanism of action means that we are still at the beginning of this exciting road.

## REFERENCES

Afreen F., Zobayed S.M.A., and Kozai T. 2006. Melatonin in *Glycyrrhiza uralensis*: Response of plant roots to spectral quality of light and UV-B radiation. *Journal of Pineal Research* 41:108–115.

Aghdam M.S. and Fard J.R. 2017. Melatonin treatment attenuates postharvest decay and maintains nutritional quality of strawberry fruits (*Fragaria* × *anannasa* cv. Selva) by enhancing GABA shunt activity. *Food Chemistry* 221:1650–1657.

Arnao M.B. and Hernández-Ruiz J. 2006. The physiological function of melatonin in plants. *Plant Signaling & Behavior* 1:89–95.

Arnao M.B. and Hernández-Ruiz J. 2007a. Inhibition of ACC oxidase activity by melatonin and IAA in etiolated lupin hypocotyls. In *Advances in Plant Ethylene Research*, (Eds.) Ramina, A., Chang, C., Giovannoni, J., Klee, H., Perata, P., and Woltering, E., pp. 101–103. Dordrecht, the Netherlands: Springer.

Arnao M.B. and Hernández-Ruiz J. 2007b. Melatonin promotes adventitious- and lateral root regeneration in etiolated hypocotyls of *Lupinus albus* L. *Journal of Pineal Research* 42:147–152.
Arnao M.B. and Hernández-Ruiz J. 2009a. Assessment of different sample processing procedures applied to the determination of melatonin in plants. *Phytochemical Analysis* 20:14–18.
Arnao M.B. and Hernández-Ruiz J. 2009b. Chemical stress by different agents affects the melatonin content of barley roots. *Journal of Pineal Research* 46:295–299.
Arnao M.B. and Hernández-Ruiz J. 2009c. Protective effect of melatonin against chlorophyll degradation during the senescence of barley leaves. *Journal of Pineal Research* 46:58–63.
Arnao M.B. and Hernández-Ruiz J. 2013a. Growth conditions determine different melatonin levels in *Lupinus albus* L. *Journal of Pineal Research* 55:149–155.
Arnao M.B. and Hernández-Ruiz J. 2013b. Growth conditions influence the melatonin content of tomato plants. *Food Chemistry* 138:1212–1214.
Arnao M.B. and Hernández-Ruiz J. 2014a. Melatonin: Plant growth regulator and/or biostimulator during stress? *Trends in Plant Science* 19:789–797.
Arnao M.B. and Hernández-Ruiz J. 2014b. Melatonin: Possible role as light-protector in plants. In *UV Radiation: Properties, Effects, and Applications*, (Ed.) Radosevich, J.A., pp. 79–92. New York: Nova Science Publishers.
Arnao M.B. and Hernández-Ruiz J. 2015a. Functions of melatonin in plants: A review. *Journal of Pineal Research* 59:133–150.
Arnao M.B. and Hernández-Ruiz J. 2015b. Melatonin: Synthesis from tryptophan and its role in higher plants. In *Amino Acids in Higher Plants*, (Ed.) D' Mello, J.P.F., pp. 390–435. Boston, MA: CAB Intern.
Arnao M.B. and Hernández-Ruiz J. 2017. Growth activity, rooting capacity and gravitropism: Three auxinic precepts fulfilled by melatonin. *Acta Physiologiae Plantarum* 39:127.
Atkins K.A. and Dodd A.N. 2014. Circadian regulation of chloroplasts. *Current Opinion in Plant Biology* 21:43–50.
Back K., Tan D.X., and Reiter R.J. 2016. Melatonin biosynthesis in plants: Multiple pathways catalyze tryptophan to melatonin in the cytoplasm or chloroplasts. *Journal of Pineal Research* 61:426–437.
Bajwa V.S., Shukla M.R., Sherif S.M., Murch S.J., and Saxena P.K. 2014. Role of melatonin in alleviating cold stress in *Arabidopsis thaliana*. *Journal of Pineal Research* 56:238–245.
Balabusta M., Szafranska K., and Posmyk M.M. 2016. Exogenous melatonin improves antioxidant defense in cucumber seeds germinated under chilling stress. *Frontiers in Plant Science* 7:575.
Blask D.E., Dauchy R.T., Sauer L.A., and Krause J.A. 2004. Melatonin uptake and growth prevention in rat hepatoma 7288CTC in response to dietary melatonin: Melatonin receptor-mediated inhibition of tumor linoleic acid metabolism to the growth signaling molecule 13-hydroxyoctadecadienoic acid and the potential role of phytomelatonin. *Carcinogenesis* 25:951–960.
Boccalandro H.E., Gonzalez C.V., Wunderlin D.A., and Silva M.F. 2011. Melatonin levels, determined by LC-ESI-MS/MS, fluctuate during the day/night cycle in *Vitis vinifera cv Malbec:* Evidence of its anti-oxidant role in fruits. *Journal of Pineal Research* 51:226–232.
Bonilla E., Valero N., Chacín-Bonilla L., and Medina-Leendertz S. 2004. Melatonin and viral infections. *Journal of Pineal Research* 36:73–79.
Burkhardt S., Tan D.X., Manchester L.C., Hardeland R., and Reiter R.J. 2001. Detection and quantification of the antioxidant melatonin in montmorency and balaton tart cherries (*Prunus cerasus*). *Journal of Agricultural and Food Chemistry* 49:4898–4902.
Byeon Y. and Back K. 2014a. An increase in melatonin in transgenic rice causes pleiotropic phenotypes, including enhanced seedling growth, delayed flowering, and low grain yield. *Journal of Pineal Research* 56:408–414.
Byeon Y. and Back K. 2014b. Melatonin synthesis in rice seedlings in vivo is enhanced at high temperatures and under dark conditions due to increased serotonin N-acetyltransferase and N-acetylserotonin methyltransferase activities. *Journal of Pineal Research* 56:189–195.
Byeon Y. and Back K. 2016. Low melatonin production by suppression of either serotonin N-acetyltransferase or N-acetylserotonin methyltransferase in rice causes seedling growth retardation with yield penalty, abiotic stress susceptibility, and enhanced coleoptile growth under anoxic conditions. *Journal of Pineal Research* 60:348–359.
Byeon Y., Choi G.H., Lee H.Y., and Back K. 2015a. Melatonin biosynthesis requires N-acetylserotonin methyltransferase activity of caffeic acid O-methyltransferase in rice. *Journal of Experimental Botany* 66:6917–6925.
Byeon Y., Lee H.Y., and Back K. 2015b. Chloroplastic and cytoplasmic overexpression of sheep serotonin N-acetyltransferase in transgenic rice plants is associated with low melatonin production despite high enzyme activity. *Journal of Pineal Research* 58:461–469.

Byeon Y., Lee H.Y., and Back K. 2016. Cloning and characterization of the serotonin N–acetyltransferase–2 gene (SNAT2) in rice (*Oryza sativa*). *Journal of Pineal Research* 61:198–207.
Byeon Y., Lee H.Y., Hwang O.J. et al. 2015c. Coordinated regulation of melatonin synthesis and degradation genes in rice leaves in response to cadmium treatment. *Journal of Pineal Research* 58:470–478.
Byeon Y., Lee H.Y., Lee K., and Back K. 2014a. Caffeic acid O-methyltransferase is involved in the synthesis of melatonin by methylating N-acetylserotonin in *Arabidopsis*. *Journal of Pineal Research* 57:219–227.
Byeon Y., Lee K., Park Y.I., Park S., and Back K. 2013a. Molecular cloning and functional analysis of serotonin N-acetyltransferase from the cyanobacterium *Synechocystis* sp. PCC 6803. *Journal of Pineal Research* 55:371–376.
Byeon Y., Park S., Kim Y.S., and Back K. 2013b. Microarray analysis of genes differentially expressed in melatonin-rich transgenic rice expressing a sheep serotonin N-acetyltransferase. *Journal of Pineal Research* 55:357–363.
Byeon Y., Park S., Lee H.Y., Kim Y.S., and Back K. 2014b. Elevated production of melatonin in transgenic rice seeds expressing rice tryptophan decarboxylase. *Journal of Pineal Research* 56:275–282.
Cao J., Murch S.J., O'Brien R., and Saxena P.K. 2006. Rapid method to accurate analysis of melatonin, serotonin and auxin in plant samples using liquid chromatography-tandem mass spectrometry. *Journal of Chromatography A* 1134:333–337.
Carrillo-Vico A., Lardone P.J., Alvarez-Sanchez N., Rodríguez-Rodriguez A., and Guerrero J.M. 2013. Melatonin: Buffering the immune system. *International Journal of Molecular Sciences* 14:8638–8683.
Chen G., Huo Y., Tan D.X. et al. 2003. Melatonin in chinese medicinal herbs. *Life Science* 73:19–26.
Chen Q., Qi W.B., Reiter R.J., Wei W., and Wang B.M. 2009. Exogenously applied melatonin stimulates root growth and raises endogenous IAA in roots of etiolated seedling of *Brassica juncea*. *Journal of Plant Physiology* 166:324–328.
Dawood M.G. and El-Awadi M.E. 2015. Alleviation of salinity stress on *Vicia faba* L. plants via seed priming with melatonin. *Acta Biologica Colombiana* 20:223–235.
De la Iglesia H.O., Meyer J., Carpino A.J., and Schwartz W.J. 2000. Antiphase oscillation of the left and right suprachiasmatic nuclei. *Science* 290:799–801.
Di Bella G., Mascia F., Gualano L., and Di Bella L. 2013. Melatonin anticancer effects: Review. *International Journal of Molecular Sciences* 14:2410–2430.
Dixon L.E., Hodge S.K., van Ooijen G. et al. 2014. Light and circadian regulation of clock components aids flexible responses to environmental signals. *New Phytologist* 203:568–577.
Dodd A.N., Dalchau N., Gardner M.J., Baek S.J., and Webb A.A. 2014. The circadian clock has transient plasticity of period and is required for timing of nocturnal processes in *Arabidopsis*. *New Phytologist* 201:168–179.
Dodd A.N., Salathia N., Hall A. et al. 2005. Plant circadian clocks increase photosynthesis, growth, survival, and competitive advantage. *Science* 309:633.
Dubbels R., Reiter R.J., Klenke E. et al. 1995. Melatonin in edible plants identified by radioimmunoassay and by HPLC-MS. *Journal of Pineal Research* 18:28–31.
Fan J., Hu Z., Xie Y. et al. 2015. Alleviation of cold damage to photosystem II and metabolisms by melatonin in Bermudagrass. *Frontiers in Plant Science* 6:925.
Fernández-Pachón M.S., Medina S., Herrero-Martín G. et al. 2014. Alcoholic fermentation induces melatonin synthesis in orange juice. *Journal of Pineal Research* 56:31–38.
Ferry G., Loynel A., Kucharczyk N. et al. 2000. Substrate specificity and inhibition studies of human serotonin-N-acetyltransferase. *Journal of Biological Chemistry* 275:8794–8805.
Fischer T.W., Kleszcynski K., Hardkop L.H., Kruse N., and Zillikens D. 2013. Melatonin enhances antioxidative enzyme gene expression (CAT, GPx, SOD), prevents their UVR-induced depletion, and protects against the formation of DNA damage in ex vivo human skin. *Journal of Pineal Research* 54:303–312.
Fujiwara T., Maisonneuve S., Isshiki M. et al. 2010. *Sekiguchi* lesion gene encodes a cytochrome P450 monooxygenase that catalyzes conversion of tryptamine to serotonin in rice. *Journal of Biological Chemistry* 285:11308–11313.
Gao H., Zhang Z.K., Chai H.K. et al. 2016. Melatonin treatment delays postharvest senescence and regulates reactive oxygen species metabolism in peach fruit. *Postharvest Biology and Technology* 118:103–110.
Garcia-Parrilla M.C., Cantos E., and Troncoso A.M. 2009. Analysis of melatonin in foods. *Journal of Food Composition and Analysis* 22:177–183.
Gitto E., Karbownik M., and Reiter R.J. 2001. Effects of melatonin treatment in septic newborns. *Pediatric Research* 50:756–760.

Gómez F.J.V., Hernández I.G., Martinez L.D., Silva M.F., and Cerutti S. 2013. Analytical tools for elucidating the biological role of melatonin in plants by liquid chromatography-tandem mass spectrometry. *Electrophoresis* 34:1749–1756.

Hardeland R. 2012. Melatonin in aging and disease. Multiple consequences of reduced secretion, options and limits of treatment. *Aging & Disease* 3:194–225.

Hardeland R., Madrid J.A., Tan D.X., and Reiter R.J. 2012. Melatonin, the circadian multioscillator system and health: The need for detailed analysis of peripheral melatonin signal. *Journal of Pineal Research* 52:139–166.

Hardeland R., Pandi-Perumal S., and Poeggeler B. 2007. Melatonin in plants: Focus on a vertebrate night hormone with cytoprotective properties. *Functional Plant Science & Biotechnology* 1:32–45.

Hasan M., Ahammed G.J., Yin L. et al. 2015. Melatonin mitigates cadmium phytotoxicity through modulation of phytochelatins biosynthesis, vacuolar sequestration and antioxidant potential in *Solanum lycopersicum* L. *Frontiers in Plant Science* 6:601.

Hattori A., Migitaka H., Iigo M. et al. 1995. Identification of melatonin in plants and its effects on plasma melatonin levels and binding to melatonin receptors in vertebrates. *Biochemistry and Molecular Biology International* 35:627–634.

Hernández I.G., Gomez F.J.V., Cerutti S., Arana M.V., and Silva M.F. 2015. Melatonin in *Arabidopsis thaliana* acts as plant growth regulator at low concentrations and preserves seed viability at high concentrations. *Plant Physiology and Biochemistry* 94:191–196.

Hernández-Ruiz J. and Arnao M.B. 2008a. Distribution of melatonin in different zones of lupin and barley plants at different ages in the presence and absence of light. *Journal of Agricultural and Food Chemistry* 56:10567–10573.

Hernández-Ruiz J. and Arnao M.B. 2008b. Melatonin stimulates the expansion of etiolated lupin cotyledons. *Plant Growth Regulation* 55:29–34.

Hernández-Ruiz J., Cano A., and Arnao M.B. 2004. Melatonin: Growth-stimulating compound present in lupin tissues. *Planta* 220:140–144.

Hernández-Ruiz J., Cano A., and Arnao M.B. 2005. Melatonin acts as a growth-stimulating compound in some monocot species. *Journal of Pineal Research* 39:137–142.

Hu W., Kong H., Guo Y. et al. 2016. Comparative physiological and transcriptomic analyses reveal the actions of melatonin in the delay of postharvest physiological deterioration of cassava. *Frontiers in Plant Science* 7:736.

Huang X. and Mazza G. 2011a. Application of LC and LC-MS to the analysis of melatonin and serotonin in edible plants. *Critical Review in Food Science and Nutrition* 51:269–284.

Huang X. and Mazza G. 2011b. Simultaneous analysis of serotonin, melatonin, piceid and resveratrol in fruits using liquid chromatography tandem mass spectrometry. *Journal of Chromatography A* 1218:3890–3899.

Jan J.E., Reiter R.J., Wasdell M.B., and Bax M. 2009. The role of the thalamus in sleep, pineal melatonin production, and circadian rhythm sleep disorders. *Journal of Pineal Research* 46:1–7.

Jiao J., Ma Y., Chen S. et al. 2016. Melatonin-producing endophytic bacteria from grapevine roots promote the abiotic stress-induced production of endogenous melatonin in their hosts. *Frontiers in Plant Science* 7:1387.

Johns N.P., Johns J., Porasupthana S., Plaimee P., and Sae-Teaw M. 2013. Dietary intake of melatonin from tropical fruit altered urinary excretion of 6-sulfatoxymelatonin in healthy volunteers. *Journal of Agricultural and Food Chemistry* 61:913–919.

Kang K., Lee K., Park S., Kim Y.S., and Back K. 2010. Enhanced production of melatonin by ectopic overexpression of human serotonin *N*-acetyltransferase plays a role in cold resistance in transgenic rice seedlings. *Journal of Pineal Research* 49:176–182.

Kang S., Kang K., Lee K., and Back K. 2007. Characterization of tryptamine 5-hydroxylase and serotonin synthesis in rice plants. *Plant Cell Reports* 26:2009–2015.

Kim J.E., Kim W.T., and Kang B.G. 2001. IAA and $N^6$-Benzyladenine inhibit ethylene-regulated expression of ACC oxidase and ACC synthase genes in mungbean hypocotyls. *Plant and Cell Physiology* 42:1056–1061.

Kim M., Seo H., Park C., and Park W.J. 2016. Examination of the auxin hypothesis of phytomelatonin action in classical auxin assay systems in maize. *Journal of Plant Physiology* 190:67–71.

Kobylinska A. and Posmyk M.M. 2016. Melatonin restricts Pb-induced PCD by enhancing BI-1 expression in tobacco suspension cells. *Biometals* 29:1059–1074.

Kolar J. and Machackova I. 2005. Melatonin in higher plants: Occurrence and possible functions. *Journal of Pineal Research* 39:333–341.

Kolar J., Machackova I., Eder J. et al. 1997. Melatonin: Occurrence and daily rhythm in *Chenopodium rubrum*. *Phytochemistry* 44:1407–1413.

Kolar J., Machackova I., Illnerova H. et al. 1995. Melatonin in higher plant determined by radioimmunoassay and liquid chromatography-mass spectrometry. *Biological Rhythm Research* 26:406–409.

Kostopoulou Z., Therios I., Roumeliotis E., Kanellis A.K., and Molassiotis A. 2015. Melatonin combined with ascorbic acid provides salt adaptation in *Citrus aurantium* L. seedlings. *Plant Physiology and Biochemistry* 86:155–165.

Lee H.Y. and Back K. 2016. Mitogen-activated protein kinase pathways are required for melatonin-mediated defense responses in plants. *Journal of Pineal Research* 60:327–335.

Lee H.Y., Byeon Y., and Back K. 2014a. Melatonin as a signal molecule triggering defense responses against pathogen attack in *Arabidopsis* and tobacco. *Journal of Pineal Research* 57:262–268.

Lee H.Y., Byeon Y., Lee K., Lee H.J., and Back K. 2014b. Cloning of *Arabidopsis* serotonin N-acetyltransferase and its role with caffeic acid O-methyltransferase in the biosynthesis of melatonin *in vitro* despite their different subcellular localizations. *Journal of Pineal Research* 57:418–426.

Lee H.Y., Byeon Y., Tan D.X., Reiter R.J., and Back K. 2015. *Arabidopsis* serotonin N-acetyltransferase knockout mutant plants exhibit decreased melatonin and salicylic acid levels resulting in susceptibility to an avirulent pathogen. *Journal of Pineal Research* 58:291–299.

Lei Q., Wang L., Tan D.X. et al. 2013. Identification of genes for melatonin synthetic enzymes in Red Fuji apple (*Malus domestica Borkh.cv.Red*) and their expression and melatonin production during fruit development. *Journal of Pineal Research* 55:443–451.

Lerner A.B., Case J.D., Takahashi Y., Lee T.H., and Mori W. 1958. Isolation of melatonin, a pineal factor that lightens melanocytes. *Journal of American Chemical Society* 80:2587.

Li C., Liang B., Chang C. et al. 2016a. Exogenous melatonin improved potassium content in *Malus* under different stress conditions. *Journal of Pineal Research* 61:218–229.

Li C., Liang D., Chang C., Jia D., and Ma F. 2015. Melatonin mediates the regulation of ABA metabolism, free-radical scavenging, and stomatal behavior in two *Malus* species under drought stress. *Journal of Experimental Botany* 66:669–680.

Li C., Wang P., Wei Z. et al. 2012. The mitigation effects of exogenous melatonin on salinity-induced stress in *Malus hupehensis*. *Journal of Pineal Research* 53:298–306.

Li H., Dong Y., Chang J. et al. 2016b. High-throughput microRNA and mRNA sequencing reveals that microRNAs may be involved in melatonin-mediated cold tolerance in *Citrullus lanatus* L. *Frontiers in Plant Science* 7:1231.

Li X., Tan D.X., Jiang D., and Liu F. 2016c. Melatonin enhances cold tolerance in drought-primed wild-type and abscisic acid-deficient mutant barley. *Journal of Pineal Research* 61:328–339.

Liang C., Zheng G., Li W. et al. 2015. Melatonin delays leaf senescence and enhances salt stress tolerance in rice. *Journal of Pineal Research* 59:91–101.

Liu J., Zhang R., Sun Y. et al. 2016a. The beneficial effects of exogenous melatonin on tomato fruit properties. *Scientia Horticulturae* 207:14–20.

Liu N., Jin Z., Wang S. et al. 2015. Sodic alkaline stress mitigation with exogenous melatonin involves reactive oxygen metabolism and ion homeostasis in tomato. *Scientia Horticulturae* 181:18–25.

Liu T., Zhao F., Liu Z. et al. 2016b. Identification of melatonin in *Trichoderma* spp. and detection of melatonin content under controlled-stress growth conditions from *T. asperellum*. *Journal of Basic Microbiology* 56:843.

Manchester L.C., Tan D.X., Reiter R.J. et al. 2000. High levels of melatonin in the seeds of edible plants. Possible function in germ tissue protection. *Life Science* 67:3023–3029.

Maronde E. and Stehle J.H. 2007. The mammalian pineal gland: Known facts, unknown facets. *Trends Endocrinology & Metabolism* 18:142–149.

Mena P., Gil-Izquierdo A., Moreno D.A., Martí N., and García-Viguera C. 2012. Assessment of the melatonin production in pomegranate wines. *LWT—Food Science and Technology* 47:13–18.

Meng J.F., Xu T.F., Wang Z.Z. et al. 2014. The ameliorative effects of exogenous melatonin on grape cuttings under water-deficient stress: Antioxidant metabolites, leaf anatomy, and chloroplast morphology. *Journal of Pineal Research* 57:200–212.

Mukherjee S., David A., Yadav S., Baluska F., and Bhatla S.C. 2014. Salt stress-induced seedling growth inhibition coincides with differential distribution of serotonin and melatonin in sunflower seedling roots and cotyledons. *Physiologia Plantarum* 152:714–728.

Murch S.J. and Saxena P.K. 2006. A melatonin-rich germplasm line of St. John's wort (*Hypericum perforatum* L.). *Journal of Pineal Research* 41:284–287.

Murch S.J., Simmons C.B., and Saxena P.K. 1997. Melatonin in feverfew and other medicinal plants. *Lancet* 350:1598–1599.

Nawaz M.A., Huang Y., Bie Z. et al. 2016. Melatonin: Current status and future perspectives in plant science. *Frontiers in Plant Science* 6:1230.

Okazaki M., Higuchi K., Aouini A., and Ezura H. 2010. Lowering intercellular melatonin levels by transgenic analysis of indoleamine 2,3-dioxygenase from rice in tomato plants. *Journal of Pineal Research* 49:239–247.

Okazaki M., Higuchi K., Hanawa Y., Shiraiwa Y., and Ezura H. 2009. Cloning and characterization of a *Chlamydomonas reinhardtii* cDNA arylalkylamine N-acetyltransferase and its use in the genetic engineering of melatonin content in the Micro-Tom tomato. *Journal of Pineal Research* 46:373–382.

Padumanonda T., Johns J., Sangkasat A., and Tiyaworanant S. 2014. Determination of melatonin content in traditional Thai herbal remedies used as sleeping aids. *DARU Journal of Pharmaceutical Sciences* 22:1–6.

Pandi-Perumal S.R., Trakht I., Srinivasan V. et al. 2008. Physiological effects of melatonin: Role of melatonin receptors and signal transduction pathways. *Progress in Neurobiology* 85:335–353.

Pape C., Lüning K. 2006. Quantification of melatonin in phototrophic organisms. *Journal of Pineal Research* 41:157–165.

Paredes S.D., Korkmaz A., Manchester L.C., Tan D.X., and Reiter R.J. 2009. Phytomelatonin: A review. *Journal of Experimental Botany* 60:57–69.

Park S. and Back K. 2012. Melatonin promotes seminal root elongation and root growth in transgenic rice after germination. *Journal of Pineal Research* 53:385–389.

Park S., Byeon Y., and Back K. 2013a. Functional analyses of three ASMT gene family members in rice plants. *Journal of Pineal Research* 55:409–415.

Park S., Byeon Y., and Back K. 2013b. Transcriptional suppression of tryptamine 5-hydroxylase, a terminal serotonin biosynthetic gene, induces melatonin biosynthesis in rice (*Oryza sativa* L.). *Journal of Pineal Research* 55:131–137.

Park S., Lee D.E., Jang H. et al. 2013c. Melatonin-rich transgenic rice plants exhibit resistance to herbicide-induced oxidative stress. *Journal of Pineal Research* 54:258–263.

Park S., Lee K., Kim Y.S., and Back K. 2012. Tryptamine 5-hydroxylase-deficient *Sekiguchi* rice induces synthesis of 5-hydroxytryptophan and *N*-acetyltryptamine but decreases melatonin biosynthesis during senescence process of detached leaves. *Journal of Pineal Research* 52:211–216.

Pasko P., Sulkowska-Ziaja K., Muszynska B., and Zagrodzki P. 2014. Serotonin, melatonin, and certain indole derivatives profiles in rutabaga and kohlrabi seeds, sprouts, bulbs, and roots. *LWT—Food Science and Technology* 59:740–745.

Pelagio-Flores R., Muñoz-Parra E., Ortiz-Castro R., and Lopez-Bucio J. 2012. Melatonin regulates *Arabidopsis* root system architecture likely acting independently of auxin signaling. *Journal of Pineal Research* 53:279–288.

Posmyk M.M., Balabusta M., Wieczorek M., Sliwinska E., and Janas K.M. 2009. Melatonin applied to cucumber (*Cucumis sativus* L.) seeds improves germination during chilling stress. *Journal of Pineal Research* 46:214–223.

Posmyk M.M., Kuran H., Marciniak K., and Janas K.M. 2008. Presowing seed treatment with melatonin protects red cabbage seedlings against toxic copper ion concentrations. *Journal of Pineal Research* 45:24–31.

Ramakrishna A., Giridhar P., Sankar K., and Ravishankar G. 2012. Melatonin and serotonin profiles in beans of *Coffea* species. *Journal of Pineal Research* 52:470–476.

Reiter R., Manchester L., and Tan D. 2005. Melatonin in walnuts: Influence on levels of melatonin and total antioxidant capacity of blood. *Nutrition* 21:920–924.

Reiter R.J. 1991. Pineal melatonin: Cell biology of its synthesis and of its physiological interactions. *Endocrine Reviews* 12:151–180.

Reiter R.J. 1993. The melatonin rhythm: Both a clock and a calendar. *Experientia* 49:654–664.

Reiter R.J., Tan D.X., Burkhardt S., and Manchester L.C. 2001. Melatonin in plants. *Nutrition Reviews* 59:286–290.

Reiter R.J., Tan D.X., and Galano A. 2014. Melatonin: Exceeding expectations. *Physiology (Bethesda)* 56:371–381.

Reiter R.J., Tan D.X., Qi W. et al. 2000. Pharmacology and physiology of melatonin in the reduction of oxidative stress *in vivo*. *Biological Signals & Receptors* 9:160–171.

Reppert S.M. and Weaver D.R. 2002. Coordination of circadian timing in mammals. *Nature* 418:935–941.

Sae-Teaw M., Johns J., Johns N.P., and Subongkot S. 2013. Serum melatonin levels and antioxidant capacities after consumption of pineapple, orange, or banana by healthy male volunteers. *Journal of Pineal Research* 55:58–64.

Sarropoulou V.N., Dimassi-Theriou K.N., Therios I.N., and Koukourikou-Petridou M. 2012a. Melatonin enhances root regeneration, photosynthetic pigments, biomass, total carbohydrates and proline content in the cherry rootstock PHL-C (*Prunus avium* × *Prunus cerasus*). *Plant Physiology and Biochemistry* 61:162–168.

Sarropoulou V.N., Therios I.N., and Dimassi-Theriou K.N. 2012b. Melatonin promotes adventitious root regeneration in in vitro shoot tip explants of the commercial sweet cherry rootstocks CAB-6P (*Prunus cerasus* L.), Gisela 6 (*P. cerasus* × *P. canescens*), and MxM 60 (*P. avium* × *P.mahaleb*). *Journal of Pineal Research* 52:38–46.

Sarrou E., Therios I.N., and Dimassi-Theriou K.N. 2014. Melatonin and other factors that promote rooting and sprouting of shoot cuttings in *Punica granatum* cv. Wonderful. *Turkist Journal Botany* 38:293–301.

Seely D., Wu P., Fritz H. et al. 2012. Melatonin as adjuvant cancer care with and without chemotherapy: A systematic review and meta-analysis of randomized trials. *Integrative Cancer Therapies* 11:293–303.

Shi H. and Chan Z. 2014. The cysteine2/histidine2-type transcription factor zinc finger of arabidopsis thaliana 6-activated c-repeat-binding factor pathway is essential for melatonin-mediated freezing stress resistance in *Arabidopsis. Journal of Pineal Research* 57:185–191.

Shi H., Chen Y., Tan D.X. et al. 2015a. Melatonin induces nitric oxide and the potential mechanisms relate to innate immunity against bacterial pathogen infection in *Arabidopsis. Journal of Pineal Research* 59:102–108.

Shi H., Jiang C., Ye T. et al. 2015b. Comparative physiological, metabolomic, and transcriptomic analyses reveal mechanisms of improved abiotic stress resistance in bermudagrass [*Cynodon dactylon* (L). Pers.] by exogenous melatonin. *Journal of Experimental Botany* 66:681–694.

Shi H., Qian Y., Tan D.X., Reiter R.J., and He C. 2015c. Melatonin induces the transcripts of CBF/DREB1s and their involvement in both abiotic and biotic stresses in *Arabidopsis. Journal of Pineal Research* 59:334–342.

Shi H., Reiter R.J., Tan D.X., and Chan Z. 2015d. Indole-3-Acetic Acid Inducible 17 positively modulates natural leaf senescence through melatonin-mediated pathway in *Arabidopsis. Journal of Pineal Research* 58:26–33.

Shi H., Tan D.X., Reiter R.J. et al. 2015e. Melatonin induces class A1 heat shock factors (HSFA1s) and their possible involvement of thermotolerance in *Arabidopsis. Journal of Pineal Research* 58:335–342.

Shi H., Wei Y., and He C. 2016. Melatonin-induced CBF/DREB1s are essential for diurnal change of disease resistance and CCA1 expression in *Arabidopsis. Plant Physiology and Biochemistry* 100:150–155.

Simopoulos A.P., Tan D.X., Manchester L.C., and Reiter R.J. 2005. Purslane: A plant source of omega-3 fatty acids and melatonin. *Journal of Pineal Research* 39:331–332.

Srinivasan V., Pandi-Perumal S.R., Maestroni G.J.M. et al. 2005. Role of melatonin in neurodegenerative diseases. *Neurotoxicologic Research* 7:293–318.

Stege P.W., Sombra L.L., Messina G., Martinez L.D., and Silva M.F. 2010. Determination of melatonin in wine and plant extracts by capillary electrochromatography with immobilized carboxylic multi-walled carbon nanotubes as stationary phase. *Electrophoresis* 31:2242–2248.

Sturtz M., Cerezo A., Cantos-Villar E., and Garcia-Parrilla M. 2011. Determination of the melatonin content of different varieties of tomatoes (*Lycopersicon esculentum*) and strawberries (*Fragaria ananassa*). *Food Chemistry* 127:1329–1334.

Sun Q., Zhang N., Wang J. et al. 2016. A label-free differential proteomics analysis reveals the effect of melatonin on promoting fruit ripening and anthocyanin accumulation upon postharvest in tomato. *Journal of Pineal Research* 61:138–153.

Sun Q.Q., Zhang N., Wang J. et al. 2015. Melatonin promotes ripening and improves quality of tomato fruit during postharvest life. *Journal of Experimental Botany* 66: 657.

Swarup R., Péret B. 2012. AUX/LAX family of auxin influx carriers-an overview. *Frontiers in Plant Science* 3:225.

Szafranska K., Glinska S., and Janas K.M. 2013. Ameliorative effect of melatonin on meristematic cells of chilled and re-warmed *Vigna radiata* roots. *Biologia Plantarum* 57:91–96.

Szafranska K., Reiter R.J., and Posmyk M.M. 2016. Melatonin application to *Pisum sativum* L. seeds positively influences the function of the photosynthetic apparatus in growing seedlings during paraquat-induced oxidative stress. *Frontiers in Plant Science* 7:1663.

Tan D.X., Hardeland R., Manchester L.C. et al. 2012. Functional roles of melatonin in plants, and perspectives in nutritional and agricultural science. *Journal of Experimental Botany* 63:577–597.

Tan D.X., Manchester C.L., Esteban-Zubero E., Zhou Z., and Reiter J.R. 2015. Melatonin as a potent and inducible endogenous antioxidant: Synthesis and metabolism. *Molecules* 20:18886–18906.

Tan D.X., Manchester L.C., Di Mascio P. et al. 2007a. Novel rhythms of N-acetyl-N-formyl-5-methoxykynuramine and its precursor melatonin in water hyacinth: Importance for phytoremediation. *The FASEB Journal* 21:1724–1729.

Tan D.X., Manchester L.C., Helton P., and Reiter R.J. 2007b. Phytoremediative capacity of plants enriched with melatonin. *Plant Signaling & Behavior* 2:514–516.

Tan D.X., Manchester L.C., Reiter R.J. et al. 2000. Significance of melatonin in antioxidative defense system: Reactions and products. *Biological Signals & Receptors* 9:137–159.

Teixeira A., Morfim M.P., de Cordova C.A.S. et al. 2003. Melatonin protects against pro-oxidant enzymes and reduces lipid peroxidation in distinct membranes induced by the hydroxyl and ascorbyl radicals and by peroxynitrite. *Journal of Pineal Research* 35:262–268.

Tekbas O.F., Ogur R., Korkmaz A., Kilic A., and Reiter R.J. 2008. Melatonin as an antibiotic: New insights into the actions of this ubiquitous molecule. *Journal of Pineal Research* 44:222–226.

Tiryaki I. and Keles H. 2012. Reversal of the inhibitory effect of light and high temperature on germination of *Phacelia tanacetifolia* seeds by melatonin. *Journal of Pineal Research* 52:332–339.

Turk H., Erdal S., Genisel M. et al. 2014. The regulatory effect of melatonin on physiological, biochemical and molecular parameters in cold-stressed wheat seedlings. *Plant Growth Regulation* 74:139–152.

Uchendu E.E., Shukla M.R., Reed B.M., and Saxena P.K. 2013. Melatonin enhances the recovery of cryopreserved shoot tips of American elm (*Ulmus americana* L.). *Journal of Pineal Research* 55:435–442.

van Tassel D. and O'Neill S. 2001. Putative regulatory molecules in plants: Evaluating melatonin. *Journal of Pineal Research* 31:1–7.

van Tassel D.L. and O'Neill S.D. 1993. Melatonin: Identification of a potential dark signal in plants. *Plant Physiology* 102:659.

van Tassel D.L., Roberts N., and O'Neill S.D. 1995. Melatonin from higher plants: Isolation and identification of N-acetyl-5-methoxytryptamine. *Plant Physiology* 108:101.

Vielma J.R., Bonilla E., Chacín-Bonilla L. et al. 2014. Effects of melatonin on oxidative stress, and resistance to bacterial, parasitic, and viral infections: A review. *Acta Tropica* 137:31–38.

Vitalini S., Gardana C., Zanzotto A. et al. 2011. The presence of melatonin in grapevine *(Vitis vinifera* L.) berry tissues. *Journal of Pineal Research* 51:331–337.

Wang L., Zhao Y., Reiter R.J. et al. 2014. Changes in melatonin levels in transgenic Micro-Tom tomato over-expressing ovine AANAT and ovine HIOMT genes. *Journal of Pineal Research* 56:134–142.

Wang P., Sun X., Li C. et al. 2013. Long-term exogenous application of melatonin delays drought-induced leaf senescence in apple. *Journal of Pineal Research* 54:292–302.

Wang P., Sun X., Wang N., Tan D.X., and Ma F. 2015. Melatonin enhances the occurrence of autophagy induced by oxidative stress in *Arabidopsis* seedlings. *Journal of Pineal Research* 58:479–489.

Wang P., Yin L., Liang D. et al. 2012. Delayed senescence of apple leaves by exogenous melatonin treatment: Toward regulating the ascorbate-glutathione cycle. *Journal of Pineal Research* 53:11–20.

Wang Q., An B., Wei Y. et al. 2016. Melatonin regulates root meristem by repressing auxin synthesis and polar auxin transport in *Arabidopsis*. *Frontiers in Plant Science* 7:1882.

Weeda S., Zhang N., Zhao X. et al. 2014. *Arabidopsis* transcriptome analysis reveals key roles of melatonin in plant defense systems. *PLoS One* 9:e93462.

Wei W., Li Q., Chu Y.-N. et al. 2015. Melatonin enhances plant growth and abiotic stress tolerance in soybean plants. *Journal of Experimental Botany* 66:695–707.

Wei Y., Zeng H., Hu W. et al. 2016. Comparative transcriptional profiling of melatonin synthesis and catabolic genes indicates the possible role of melatonin in developmental and stress responses in rice. *Frontiers in Plant Science* 7:676.

Wen D., Gong B., Sun S. et al. 2016. Promoting roles of melatonin in adventitious root development of *Solanum lycopersicum* L. by regulating auxin and nitric oxide signaling. *Frontiers in Plant Science* 7:718.

Wilhelmsen M., Amiriam I., Reiter R.J., Rosenberg J., and Gögenur I. 2011. Analgesic effects of melatonin: A review of current evidence from experimental and clinical studies. *Journal of Pineal Research* 51:270–277.

Wu U.I., Mai F.D., Sheu J.N. et al. 2011. Melatonin inhibits microglial activation, reduces pro-inflammatory cytokine levels, and rescues hippocampal neurons of adult rats with acute *Klebsiella pneumoniae* meningitis. *Journal of Pineal Research* 50:159–170.

Xu W., Cai S.Y., Zhang Y. et al. 2016. Melatonin enhances thermotolerance by promoting cellular protein protection in tomato plants. *Journal of Pineal Research* 61:457–469.

Yin L., Wang P., Li M. et al. 2013. Exogenous melatonin improves *Malus* resistance to *Marssonina* apple blotch. *Journal of Pineal Research* 54:426–434.

Yuan S., Guan C., Liu S. et al. 2016a. Comparative transcriptomic analyses of differentially expressed genes in transgenic melatonin biosynthesis ovine HIOMT gene in switchgrass. *Frontiers in Plant Science* 7:1613.
Yuan S., Huang Y., Liu S. et al. 2016b. RNA-seq analysis of overexpressing ovine AANAT gene of melatonin biosynthesis in switchgrass. *Frontiers in Plant Science* 7:1289.
Zhang H.J., Zhang N., Yang R.C. et al. 2014a. Melatonin promotes seed germination under high salinity by regulating antioxidant systems, ABA and GA4 interaction in cucumber (*Cucumis sativus* L.). *Journal of Pineal Research* 57:269–279.
Zhang J., Li H., Xu B., Li J., and Huang B. 2016a. Exogenous melatonin suppresses dark-induced leaf senescence by activating the superoxide dismutase-catalase antioxidant pathway and down-regulating chlorophyll degradation in excised leaves of perennial ryegrass (*Lolium perenne* L.). *Frontiers in Plant Science* 7:1500.
Zhang L.J., Jia J., Xu Y. et al. 2012. Production of transgenic *Nicotiana sylvestris* plants expressing melatonin synthetase genes and their effect on UV-B-induced DNA damage. *In Vitro Cellular & Developmental Biology-Plant* 48:275–282.
Zhang N., Sun Q., Li H. et al. 2016b. Melatonin improved anthocyanin accumulation by regulating gene expressions and resulted in high reactive oxygen species scavenging capacity in cabbage. *Frontiers in Plant Science* 7:197.
Zhang N., Zhang H.J., Zhao B. et al. 2014b. The RNA-seq approach to discriminate gene expression profiles in response to melatonin on cucumber lateral root formation. *Journal of Pineal Research* 56:39–50.
Zhang N., Zhao B., Zhang H.J. et al. 2013. Melatonin promotes water-stress tolerance, lateral root formation, and seed germination in cucumber (*Cucumis sativus* L.). *Journal of Pineal Research* 54:15–23.
Zhao H., Su T., Huo L. et al. 2015. Unveiling the mechanism of melatonin impacts on maize seedling growth: Sugar metabolism as a case. *Journal of Pineal Research* 59:255–266.
Zhao Y., Qi L.W., Wang W.M., Saxena P.K., and Liu C.Z. 2011. Melatonin improves the survival of cryopreserved callus of *Rhodiola crenulata*. *Journal of Pineal Research* 50:83–88.
Zhao Y., Tan D.X., Lei Q. et al. 2013. Melatonin and its potential biological functions in the fruits of sweet cherry. *Journal of Pineal Research* 55:79–88.
Zhou C., Liu Z., Zhu L. et al. 2016a. Exogenous melatonin improves plant iron deficiency tolerance via increased accumulation of polyamine-mediated nitric oxide. *International Journal of Molecular Sciences* 17:1777.
Zhou X., Zhao H., Cao K. et al. 2016b. Beneficial roles of melatonin on redox regulation of photosynthetic electron transport and synthesis of D1 protein in tomato seedlings under salt stress. *Frontiers in Plant Science* 7:1823.
Zhu Z., Lee B. 2015. Friends or foes: New insights in jasmonate and ethylene co-actions. *Plant and Cell Physiology* 56:414–420.
Zohar R., Izhaki I., Koplovich A., and Ben-Shlomo R. 2011. Phytomelatonin in the leaves and fruits of wild perennial plants. *Phytochemistry Letters* 4:222–226.
Zuo B., Zheng X., He P. et al. 2014. Overexpression of MzASMT improves melatonin production and enhances drought tolerance in transgenic *Arabidopsis thaliana* plants. *Journal of Pineal Research* 57:408–417.

# 6 Serotonin and Melatonin in Root Morphogenesis
## *Functions and Mechanisms*

*Ramón Pelagio-Flores, Jesús Salvador López-Bucio, and José López-Bucio*

**CONTENTS**



## 6.1 INTRODUCTION

Plants synthesize several neurotransmitters that are important for growth, development, and adaptation to environmental stress. A myriad of physiological responses are orchestrated by the non-protein amino acid gamma-aminobutyric acid, L-glutamate, and the indoleamines serotonin (5-hydroxytryptamine) and melatonin (*N*-acetyl-5-methoxytryptamine), including adaptation to acid pH, extreme temperatures, salinity, mechanical damage, metabolism, and defense (Bouche and Fromm 2004; Ramesh et al. 2015; Pelagio-Flores et al. 2016; Kan et al. 2017). All aforementioned molecules mediate changes in root architecture, adaptation, and survival.

The action of indoleamines is different regarding other neurotransmitters. High concentrations of L-glutamate (mM range) elicit a specific sequence of changes in root morphogenesis in *Arabidopsis* caused by the inhibition of cell division at the root tip followed by an increased formation of lateral roots (Walch-Liu et al. 2006; Forde et al. 2013, 2014; López-Bucio et al. 2018). In contrast, low (10 to 160 μM) concentrations of serotonin stimulate lateral root development, but higher doses repress both primary and lateral root growth (Pelagio-Flores et al. 2011). Melatonin also inhibits primary root growth and increases lateral and adventitious root formation in *Arabidopsis* and crop plants, albeit at much higher concentrations than serotonin (Pelagio-Flores et al. 2012; Wang et al. 2016; Liang et al. 2017). This is somewhat surprising, since serotonin acts as the precursor of melatonin in a biosynthetic pathway that uses the amino acid L-tryptophan as the first substrate, a property that is also shared with indole-3-acetic acid (IAA), the most abundant auxin in plants (Back et al. 2016). The aims of this chapter are to provide the reader with the most recent information from the field of indoleamine research and discuss their possible mechanisms of action, particularly regarding the cross talk with the phytohormone pathways that orchestrate root morphogenesis.

## 6.2 ROOT GROWTH AND PATTERNING

In plants, two main parts can be identified, the shoot and the root. The shoot harbors the photosynthetic and reproductive tissues, whereas the root grows belowground to support the shoot and is directly involved in taking up and providing water and nutrients from soil. Roots sustain the plant microbiome via releasing nutritional compounds for bacterial and fungal nutrition (Badri and Vivanco 2009; Baetz and Martinoia 2014).

In most plant species, including *Arabidopsis*, the growth of the primary root is indeterminate due to its outstanding ability to produce new cells at the root apical meristem (Shishkova et al. 2008). Following germination, the primary root produces lateral roots that extend the exploratory potential. Lateral roots originate from a specific tissue, the pericycle, and once emerged from the primary root, they behave as signal organizing centers, which perceive environmental stimuli (Ruíz-Herrera et al. 2015; Shahzad and Amtmann 2017). Besides lateral roots, many crops and horticultural species develop adventitious roots that assist lateral roots when foraging for nutrients (Martínez de la Cruz et al. 2015; Steffens and Rasmussen 2016). Adventitious roots may develop from the hypocotyl, stem, or leaves, and from different tissues, such as the pericycle, mesophyll, parenchyma, cambium, protoxylem, and epidermis. Although the formation of lateral or adventitious roots may have common genetic elements, there are important differences at the level of genes and proteins involved in these processes (Bellini et al. 2014).

Auxins positively control lateral root initiation and growth, and its signaling pathway is influenced by interactions with other phytohormones (Woodward and Bartel 2005). Cytokinin and abscisic acid negatively regulate lateral root formation, whereas brassinosteroids induce root branching via cross talk with auxin biosynthesis and/or signaling (De Smet et al. 2003; Laplaze et al. 2007). On the other hand, ethylene and jasmonic acid may have positive or negative roles during lateral root formation, growth, and positioning along the primary root (Negi et al. 2005; Raya-González et al. 2012). Although every phytohormone may inhibit primary root growth when supplied at high concentrations, their functions for lateral root development is not always stimulating, and different environmental factors may shape root architecture. For instance, a stress-induced response, consisting of reduced primary root growth and increased formation of lateral roots, has been reported for plants exposed to several abiotic stress conditions, including low phosphate deficiency or exposure to aluminium (Al), rare earth elements lanthanum and gadolinium, or chromium (Cr), and it was postulated that this adaptive trait redirects plant growth to diminish stress exposure (López-Bucio et al. 2002; Potters et al. 2009; Ruíz-Herrera et al. 2012; Ruíz-Herrera and López-Bucio 2013; Martínez-Trujillo et al. 2014).

Sucrose, glucose, and vitamins are important structural, metabolic, and regulatory components for root function (Raya-González et al. 2017; Ayala-Rodríguez et al. 2017). They are tightly linked to the activation of the Target of Rapamycin (TOR) kinase, which is a master regulatory protein for root growth and activation of cell division in meristems (Xiong et al. 2013). Two recent reports demonstrated the critical role of vitamin B9 (folate polyglutamate) for root meristem maintenance related to auxin gradients at the root tip (Srivastava et al. 2011; Reyes-Hernández et al. 2014). The *Arabidopsis* MEDIATOR complex, a transcriptional multiproteinic complex ubiquitous in eukaryotes, links sugar availability with auxin-response and cell niche gene expression (Raya-González et al. 2017).

Several kinds of volatiles and diffusible compounds from plant or microbial origin, orchestrate root growth (Gutiérrez-Luna et al. 2010; Ortiz-Castro et al. 2008, 2009, 2014). Currently, the study of the mechanisms by which plants coordinate root morphogenesis to signal perception is a very active area of research where auxins or auxin signal mimics play critical roles (Ortiz-Castro et al. 2011; Muñoz-Parra et al. 2017). Since all aforementioned factors reprogram root architecture, most of them would function through the same or similar mechanisms, but this is not the case. Indeed, sugars influence auxin biosynthesis, transport, and/or response (Raya-González et al. 2017), whereas other stimuli, including fungal volatiles, may rather interfere with, or may specifically regulate the auxin signaling pathway in a tissue specific manner (Garnica-Vergara et al. 2016).

## 6.3 AUXIN IS A CRITICAL PLAYER FOR ROOT MORPHOGENESIS

Auxin acts as a morphogen instructing cell fate and controlling the pool of stem cells placed at the root apex (Heidstra and Sabatini 2014). A complex network of interactions between ethylene and cytokinins coordinates many developmental and environmental inputs by converging on the regulation of genes involved in auxin signaling and/or transport. Therefore, the finely regulated interactions between hormones ultimately modulate the distribution and the perception of auxin (Sozzani and Iyer-Pascuzzi 2014; Pacifici et al. 2015).

The signal transduction pathway leading to auxin-mediated gene activation is well-known. The auxin receptors belong into the F-box TRANSPORT INHIBITOR RESPONSE 1/AUXIN SIGNALING F-BOX PROTEIN (TIR1/AFB) family of proteins. Downstream of the receptors, the Auxin/INDOLE-3-ACETIC ACID (Aux/IAA) transcriptional repressors and the AUXIN RESPONSE FACTOR (ARF) transcription factors mediate cell responses. Auxin promotes the interaction between TIR1/AFB and Aux/IAA proteins, resulting in degradation of the Aux/IAAs and the release of ARFs to bind to the DNA-elements in the promoters of genes that are induced by auxins (Lavy and Estelle 2016).

More than two decades of study in the mechanisms of auxin perception have provided specific probes for accessing and exploring the auxin signaling machinery, and transgenic plants expressing reporter genes under the control of defined promoter elements such as *DR5* or *BA3* makes possible to study IAA localization and/or response in plant tissues through utilizing the most commonly used genetic constructs *DR5:GUS*, *DR5:GFP* or *BA3:GUS* (Ulmasov et al. 1997; Oono et al. 1998; Ottenschläger et al. 2003).

*DR5:GFP* is mainly expressed in columella cell layers at the root tip (Ulmasov et al. 1997; Ottenschläger et al. 2003), whereas the *BA3:GUS* line showed that the majority of IAA-induced GUS expression was localized in the elongation zone (Oono et al. 1998). The combined use of auxin-regulated gene reporter constructs makes possible to characterize the functions of compounds with auxin activity as well as to evaluate the auxin response under environmental and nutritional situations and are valuable tools toward clarifying the *in planta* functions of neurotransmitters.

## 6.4 SEROTONIN AND MELATONIN IN ROOT GROWTH AND DEVELOPMENT

Accumulating information supports a role for serotonin and melatonin as bioactive plant growth modulators in tissue culture explants and intact plants. Serotonin increased the number of roots from nodal segments in *Mimosa pudica* L. (Ramakrishna et al. 2009) and from explants from different varieties of cherries (Sarropoulou et al. 2012a, 2012b). In a similar manner to auxins and other molecules from bacterial and fungal origin, serotonin inhibits primary root growth while improving lateral root patterning. In barley seedlings, serotonin showed higher activity than IAA and the synthetic auxins naphthylacetic acid and trichloro-phenoxyacetic acid (Csaba and Pal 1982). In *Arabidopsis thaliana*, this indoleamine stimulated lateral root development at low concentrations, while at higher doses repressed lateral root growth, primary root growth, and root hair development but stimulated adventitious root formation (Pelagio-Flores et al. 2012).

Murch and coworkers (2001) showed an increase in *de novo* root formation in *Hypericum perforatum* L. that correlated with high levels of melatonin, while in *Lupinus albus* melatonin promoted adventitious and lateral root formation in a similar manner to IAA, allowing the authors to propose an auxinic mechanism for this molecule (Arnao and Hernández Ruiz 2007; Hernández-Ruiz and Arnao 2008). Melatonin supplementation to etiolated seedlings of wild leaf mustard (*Brassica juncea*) had a transient stimulatory effect on root growth only observed in young seedlings and at low concentrations (0.1 μM), but higher concentrations (100 μM) repressed growth (Chen et al. 2009). As in the *Lupinus* reports, during the promotion of lateral root growth in cucumber (*Cucumis sativus*) (Zhang et al. 2013), and in *Punica granatum* (Sarrou et al. 2014),

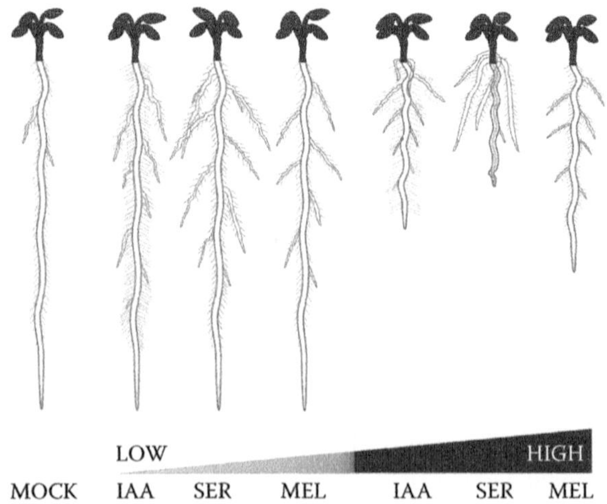


**FIGURE 6.1** The configuration of *Arabidopsis* root architecture in response to auxin, serotonin, and melatonin. Auxin, serotonin, and melatonin are indicated as IAA, SER and MEL, respectively. At low concentrations, all three molecules sustain normal primary root growth and induce root branching, while at high concentrations, an impaired primary root growth and lateral or adventitious root formation hallmark the specific responses to indoleamines.

the effect of melatonin was mostly attributed to an auxinic mechanism, although any convincing functional evidence to support this conclusion was missing.

Most reports have failed to establish a correlation between serotonin and melatonin levels with any biological activity. To solve this problem and compare the morphogenetic effects of both indoleamines in the same plant species and growth conditions, Pelagio-Flores et al. (2011, 2012), tested their effects in the model plant *A. thaliana*. In these assays, melatonin stimulated lateral and adventitious root formation without significantly affect the primary root growth, and these effects differed from serotonin, its proposed biosynthetic precursor. Later on, Wang et al. (2016) showed that high melatonin concentrations (1 mM) regulate root meristem by repressing auxin synthesis and polar auxin transport in *Arabidopsis*, thus both serotonin and melatonin can repress root growth albeit at very different concentrations (Figure 6.1).

### 6.4.1 Serotonin and Melatonin Perception

Pharmacological studies in several plant species led some authors to suggest that either serotonin or melatonin, or both, could act as agonists of the auxin signaling pathway (Arnao and Hernández-Ruiz 2014; Wen et al. 2016; Arnao and Hernández-Ruiz 2017). However, this possibility is controversial, since evidence from the fields of development, genetics, and molecular biology have provided contradictory information to support that the phenotypical and/or morphological responses to these indoleamines are due to an auxinic role in plant cells.

The proposed auxinic role of serotonin and melatonin is not supported in *Arabidopsis* or maize (Pelagio-Flores et al. 2011; Pelagio-Flores et al. 2012; Kim et al. 2016; Pelagio-Flores et al. 2016; Wang et al. 2016). Moreover, these indoleamines not only fail to act in concert, since the primary root growth inhibition by serotonin cannot be explained via its conversion into melatonin, but just by its interaction with reactive oxygen species (ROS), nitric oxide, ethylene, and jasmonic acid (Weeda et al. 2014; Pelagio-Flores et al. 2016; Wen et al. 2016). Pelagio-Flores et al. (2016) tested the role of ROS and jasmonic acid-ethylene signaling in root architecture reconfiguration by serotonin. In such report, an *A. thaliana* mutant defective at the RADICAL-INDUCED CELL DEATH1 (RCD1) locus, which was resistant to paraquat-induced ROS accumulation in primary roots showed decreased inhibition or root growth in response to serotonin. Moreover, several jasmonic acid- and ethylene-related mutants including *coronatine insensitive 1* (*coi1*), *jasmonic acid resistant 1* (*jar1*), *ethylene resistant 1* (*etr1*), *ethylene insensitive 2* (*ein2*) and *ein3* showed tolerance to serotonin in the inhibition of primary root growth and ROS redistribution within the root tip when compared with wild-type seedlings. In competence assays, $AgNO_3$, a well-known blocker of ethylene action, normalized root growth when supplemented with serotonin, whereas roots of *ethylene overproducer 3* (*eto3*) mutants were oversensitive to serotonin. Comparison of ROS levels in wild type (WT), *etr1*, *jar1* and *rcd1* primary root tips using the ROS-specific probe 2′,7′-dichlorofluorescein diacetate and confocal imaging showed that serotonin inhibition of primary root growth likely occurs independently of its conversion into melatonin. All these physiological, molecular, and genetic evidences strongly support the hypothesis that indoleamines affect root growth and development in an auxin-independent and ROS-dependent manner.

In rice transgenic plants expressing sheep serotonin *N*-acetyltransferase, increased melatonin production enhanced primary and seminal root growth (Park and Back 2012). Interestingly, follow-up studies in these plants revealed the differential expression of 464 genes, among the up-regulated genes transcription factors from the leucine-rich repeat and zinc-finger families were represented, whereas suppressed genes included jasmonate and senescence-associated genes (Byeon et al. 2013). The absence of auxin-regulated genes in the transcriptomes supports the idea that this neurotransmitter mediates its physiological responses likely acting independently or interfering with auxin.

RNA-seq analyses were conducted to explore the mechanism of melatonin-induced lateral root formation in cucumber under salt stress. The transcriptomes showed a total of 317 differentially expressed genes between control and melatonin treatment, from which 121 genes were up-regulated and 196 down-regulated, with 77 genes regulated in a melatonin-dependent manner. Interestingly, auxin-related genes expected to be over-represented showed minimal expression in the melatonin treatment, allowing the authors to conclude that melatonin modulates root development mostly by regulating the reactive oxygen species system and transcription factors (Zhang et al. 2014). In *Arabidopsis*, treatment with a low (100 pM) melatonin concentration significantly affected the expression of 81 genes (51 down-regulated and 30 up-regulated), while a high (1 mM) concentration increased the number of regulated genes to 1308 (566 up-regulated and 742 down-regulated). Genes related to the auxin pathway were mostly repressed, but instead a large number of genes were involved in plant stress and defense responses, which belong into the abscisic acid, ethylene, salicylic acid, and jasmonic acid pathways (Weeda et al. 2014).

## 6.5 CONCLUSION

Serotonin, melatonin, and IAA are all biosynthesized from the amino acid L-tryptophan. These molecules not only share some structural similarity but are ubiquitous to plants, being produced by a great variety of species belonging to different families and in response to environmental stress (Paredes et al. 2009; Mukherjee et al. 2014). The regulation of root architecture, a complex trait similarly influenced by indoleamines has provided very valuable information on their function and molecular mechanisms of action, particularly regarding phytohormone cross talk.

Pharmacological application and/or endogenous accumulation of serotonin or melatonin via transgenic approaches have shown a strong biological activity of both indoleamines modulating primary root growth and increasing root branching in several plant species. From these studies, serotonin is by far the most active compound, repressing root growth and promoting lateral and adventitious root formation, and thus, the primary root growth inhibition by serotonin cannot be explained via its conversion into melatonin (Pelagio-Flores et al. 2016). Although a recent report (Wen et al. 2016) showed that melatonin increased expression of PIN auxin transporters in *Solanum lycopersicum*, no correlation with increased auxin levels in roots could be found, which suggest that melatonin may fine-tune auxin redistribution in particular tissues, growth conditions, and/or in a dose-dependent manner. High melatonin concentrations may instead repress auxin biosynthesis and/or interfere with auxin perception, affecting in this way root meristem functioning, which requires auxin (Weeda et al. 2014; Wang et al. 2016).

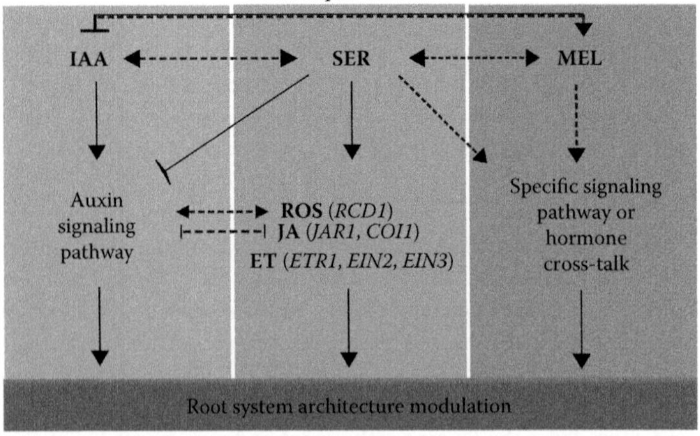


**FIGURE 6.2** Indoleamine regulation of root morphogenesis and its interactions with phytohormones and second messengers. In *Arabidopsis*, serotonin and melatonin suppress the auxin signaling pathway and cross talk with jasmonic acid, ethylene, and reactive oxygen species. These interactions explain the attributes of serotonin and melatonin in root morphogenesis but also may help explain the induction of defense and adaptation to environmental stress conferred to plants. SER, MEL, and IAA represent serotonin, melatonin, and auxin, respectively.

Genetic and transcriptomic approaches indicate that serotonin changes ROS distribution at the meristem and elongation regions of the primary root, and its bioactivity involves components of the jasmonic acid-ethylene signaling pathways (Weeda et al. 2014; Pelagio-Flores et al. 2016). This may influence its properties in modulating root morphogenesis, defense, and senescence-related programs as well as to confer stress tolerance (Figure 6.2).

Ongoing research in this topic will show the ubiquity of functions of indoleamines in crops and horticultural species, while model plants will be helpful toward identifying the receptors, kinases/phosphatases, and transcription factors mediating cell responses to these neurotransmitters.

## ACKNOWLEDGMENTS

The authors appreciate the financial support from the Consejo Nacional de Ciencia y Tecnología (CONACYT, México, Grant 177775), the Consejo de la Investigación Científica UMSNH (grant 2.26), and the Marcos Moshinsky foundation for research on plant neurotransmitters.

## REFERENCES

Arnao, M. B. and J. Hernández-Ruiz. 2017. Growth activity, rooting capacity, and tropism: Three auxinic precepts fulfilled by melatonin. *Acta Physiologiae Plantarum* 39:127.

Arnao, M. B. and J. Hernández-Ruiz. 2014. Melatonin: Plant growth regulator and/or biostimulator during stress? *Trends in Plant Science* 19:789–797.

Arnao, M. B. and J. Hernández-Ruiz. 2007. Melatonin promotes adventitious- and lateral root regeneration in etiolated hypocotyls of *Lupinus albus* L. *Journal of Pineal Research* 42:147–152.

Ayala-Rodríguez, J. A., Barrera-Ortiz, S., Ruiz-Herrera, L. F. and J. López-Bucio. 2017. Folic acid orchestrates root development linking cell elongation with auxin response and acts independently of the TARGET OF RAPAMYCIN signaling in *Arabidopsis thaliana*. *Plant Science* 264:168–178.

Back, K., Tan, D. X. and R. J. Reiter. 2016. Melatonin biosynthesis in plants: Multiple pathways catalyze tryptophan to melatonin in the cytoplasm or chloroplasts. *Journal of Pineal Research* 61:426–437.

Badri, D. and J. Vivanco. 2009. Regulation and function of root exudates. *Plant Cell and Environment* 32:666–681.

Baetz, U. and E. Martinoia. 2014. Root exudates: The hidden part of plant defense. *Trends in Plant Science* 19:90–97.

Bellini, C., Pacurar, D. I. and I. Perrone. 2014. Adventitious roots and lateral roots: Similarities and differences. *Annual Reviews of Plant Biology* 65:639–666.

Bouche, N. and H. Fromm. 2004. GABA in plants: Just a metabolite? *Trends in Plant Science* 9:110–115.

Byeon, Y., Park, S., Kim, Y. S. and K. W. Back. 2013. Microarray analysis of genes differentially expressed in melatonin-rich transgenic rice expressing a sheep serotonin N-acetyltransferase. *Journal of Pineal Research* 55:357–363.

Chen, Q., Qi, W. B., Reiter, R. J., Wei, W. and B. M. Wang. 2009. Exogenously applied melatonin stimulates root growth and raises endogenous indoleacetic acid in roots of etiolated seedlings of *Brassica juncea*. *Journal of Plant Physiology* 166:324–328.

Csaba, G. and K. Pal. 1982. Effect of insulin triiodothyronine and serotonin on plant seed development. *Protoplasma* 110:20–22.

De Smet, I., Signora, L., Beeckman, T., Inzé, D., Foyer, C. H. and H. Zhang. 2003. An abscisic acid-sensitive checkpoint in lateral root development of *Arabidopsis*. *Plant Journal* 33:543–555.

Forde, B. G. 2014. Glutamate signalling in roots. *Journal of Experimental Botany* 65:779–787.

Forde, B. G., Cutler, S. R., Zaman, N. and P. J. Krysan. 2013. Glutamate signalling via a MEKK1 kinase-dependent pathway induces changes in *Arabidopsis* root architecture. *Plant Journal* 75:1–10.

Garnica-Vergara, A., Barrera-Ortiz, S., Muñoz-Parra, E. et al. 2016. The volatile 6-n-pentyl-2H-pyran-2-one from *Trichoderma atroviride* regulates *Arabidopsis* root morphogenesis via auxin signaling and ETHYLENE INSENSITIVE 2 functioning. *New Phytologist* 209:1496–1512.

Gutiérrez-Luna, F. M., López-Bucio, J., Altamirano-Hernández, J., Valencia-Cantero, E., Reyes de la Cruz, H. and L. Macías-Rodríguez. 2010. Plant growth-promoting rhizobacteria modulate root-system architecture in *Arabidopsis thaliana* through volatile organic compound emission. *Symbiosis* 51:75–83.

Heidstra, R. and S. Sabatini. 2014. Plant and animal stem cells: Similar yet different. *Nature Reviews Molecular and Cellular Biology* 15:301–312.

Hernández-Ruiz, J. and M. B. Arnao. 2008. Melatonin stimulates the expansion of etiolated lupin cotyledons. *Plant Growth Regulation* 55:29–34.

Kan, C. C., Chung, T. Y., Wu, H. Y., Juo, Y. A. and M. H. Hsieh. 2017. Exogenous glutamate rapidly induces the expression of genes involved in metabolism and defense responses in rice roots. *BMC Genomics* 18:186.

Kim, M., Seo, H., Park, C. and W. J. Park. 2016. Examination of the auxin hypothesis of phytomelatonin action in classical auxin assay systems in maize. *Journal of Plant Physiology* 190:67–71.

Laplaze, L., Benkova, E., Casimiro, I. et al. 2007. Cytokinins act directly on lateral root founder cells to inhibit root initiation. *Plant Cell* 19:3889–3900.

Lavy, M. and M. Estelle. 2016. Mechanisms of auxin signaling. *Development* 143:3226–3229.

Liang, C., Li, A., Yu, H. et al. 2017. Melatonin regulates root architecture by modulating auxin response in rice. *Frontiers in Plant Science* 8:134.

López-Bucio, J., Hernández-Abreu, E., Sánchez-Calderón, L., Nieto-Jacobo, M. F., Simpson, J. and L. Herrera-Estrella. 2002. Phosphate availability alters architecture and causes changes in hormone sensitivity in the *Arabidopsis* root system. *Plant Physiology* 129:244–256.

López-Bucio, J. S., Raya-González, J., Ravelo-Ortega, G. et al. 2018. Mitogen activated protein kinase 6 and MAP kinase phosphatase 1 are involved in the response of *Arabidopsis* roots to L-glutamate. *Plant Molecular Biology*. doi:10.1007/s11103-018-0699-8.

Martínez de la Cruz, E., García-Ramírez, E., Vázquez-Ramos, J. M., Reyes de la Cruz, H. and J. López-Bucio. 2015. Auxins differentially regulate root system architecture and cell cycle protein levels in maize seedlings. *Journal of Plant Physiology* 176:147–156.

Martínez-Trujillo, M., Méndez-Bravo, A., Ortiz-Castro, R. et al. 2014. Chromate alters root system architecture and activates expression of genes involved in iron homeostasis and signaling in *Arabidopsis thaliana*. *Plant Molecular Biology* 86:35–50.

Mukherjee, S., Anisha, D., Sunita, Y., Baluška, F. and S. Chander. 2014. Salt stress-induced seedling growth inhibition coincides with differential distribution of serotonin and melatonin in sunflower seedling roots and cotyledons. *Physiologia Plantarum* 152:714–728.

Muñoz-Parra, E., Pelagio-Flores, R., Raya-González, J., Salmerón-Barrera, G., Ruiz-Herrera, L. F., Valencia-Cantero, E. and J. López-Bucio. 2017. Plant-plant interactions influence developmental phase transitions, grain productivity and root system architecture in *Arabidopsis* via auxin and PFT1/MED25 signaling. *Plant Cell and Environment* 40:1887–1899.

Murch, S. J., Campbell, S. B. and P. K. Saxena. 2001. The role of serotonin and melatonin in plant morphogenesis: Regulation of auxin-induced root organogenesis in in vitro-cultured explants of St. John's wort (*Hypericum perforatum* L.). *In Vitro Cell and Developmental Biology of Plants* 37:786–793.

Negi, S., Ivanchenko, M. G. and G. K. Muday. 2005. Ethylene regulates lateral root formation and auxin transport in *Arabidopsis thaliana*. *Plant Journal* 55:175–187.

Oono, Y., Chen, Q., Overvoorde, P. J., Köhler, C. and A. Theologis. 1998. Age mutants of *Arabidopsis* exhibit altered auxin-regulated gene expression. *Plant Cell* 10:1649–1662.

Ortíz-Castro, R., Pelagio-Flores, R., Méndez-Bravo, A., Ruíz-Herrera, L. F., Campos-García, J. and J. López-Bucio. 2014. Pyocyanin, a virulence factor produced by *Pseudomonas aeruginosa*, alters root development through reactive oxygen species and ethylene signaling in *Arabidopsis*. *Molecular Plant-Microbe Interactions* 27:364–378.

Ortíz-Castro, R., Díaz-Pérez, C., Martínez-Trujillo, M., del Río, R., Campos-García, J. and J. López-Bucio. 2011. Transkingdom signaling based on bacterial cyclodipeptides with auxin activity in plants. *Proceedings of the National Academy of Sciences United States of America* 108:7253–7258.

Ortíz-Castro, R., Contreras-Cornejo, H. A., Macías-Rodríguez, L. and J. López-Bucio. 2009. The role of microbial signals in plant growth and development. *Plant Signaling and Behavior* 4:701–712.

Ortíz-Castro, R., Martínez-Trujillo, M. and J. López-Bucio. 2008. *N*-acyl-*L*-homoserine lactones: A class of bacterial quorum-sensing signals alter post-embryonic root development in *Arabidopsis thaliana*. *Plant Cell and Environment* 31:1497–1509.

Ottenschläger, I., Wolff, P., Wolverton, C., Bhalerao, R. P., Sandberg, G., Ishikawa, H., Evans, M. and K. Palme. 2003. Gravity-regulated differential auxin transport from columella to lateral root cap cells. *Proceedings of the National Academy of Sciences United States of America* 100:2987–2991.

Pacifici, E., Polverari, L. and S. Sabatini. 2015. Plant hormone cross-talk: The pivot of root growth. *Journal of Experimental Botany* 66:1113–1121.

Paredes, S. D., Korkmaz, A., Manchester, L. C., Tan, D. X. and R. J. Reiter. 2009. Phytomelatonin: A review. *Journal of Experimental Botany* 60:57–69.

Park, S. and K. Back. 2012. Melatonin promotes seminal root elongation and root growth in transgenic rice after germination. *Journal of Pineal Research* 53:385–389.

Pelagio-Flores, R., Ruiz-Herrera, L. F. and J. López-Bucio. 2016. Serotonin modulates *Arabidopsis* root growth via changes in reactive oxygen species and jasmonic acid-ethylene signaling. *Physiologia Plantarum* 158:92–105.

Pelagio-Flores, R., Muñoz-Parra, E., Ortiz-Castro, R. and J. López-Bucio. 2012. Melatonin regulates *Arabidopsis* root system architecture likely acting independently of auxin signaling. *Journal of Pineal Research* 53:279–288.

Pelagio-Flores, R., Ortíz-Castro, R., Méndez-Bravo, A., Macías-Rodríguez, L. and J. López-Bucio. 2011. Serotonin, a tryptophan-derived signal conserved in plants and animals, regulates root system architecture probably acting as a natural auxin inhibitor in *Arabidopsis thaliana*. *Plant and Cell Physiology* 52:490–508.

Potters, G., Pasternak, T. P., Guisez, Y. and M. Jansen. 2009. Different stresses, similar morphogenic responses: Integrating a plethora of pathways. *Plant Cell and Environment* 32:158–169.

Ramakrishna, A., Giridhar, P. and G. A. Ravishankar. 2009. Indoleamines and calcium channels influence morphogenesis in in vitro cultures of *Mimosa pudica* L. *Plant Signaling and Behavior* 12:1–6.

Ramesh, S., Tyerman, S. D., Xu, B. et al. 2015. GABA signalling modulates plant growth by directly regulating the activity of plant-specific anion transporters. *Nature Communications* 6:7879.

Raya-González, J., López-Bucio, J. S., Prado-Rodríguez, J. C., Ruiz-Herrera, L. F., Guevara-García, A. A. and J. López-Bucio. 2017. The MEDIATOR genes MED12 and MED13 control *Arabidopsis* root system configuration influencing sugar and auxin responses. *Plant Molecular Biology* 95:141–156.

Raya-González, J., Pelagio-Flores, R. and J. López-Bucio. 2012. The jasmonate receptor COI1 plays a role in jasmonate-induced lateral root formation and lateral root positioning in *Arabidopsis thaliana*. *Journal of Plant Physiology* 169:1348–1358.

Reyes-Hernández, B. J., Srivastava, A. C., Ugartechea-Chirino, Y. et al. 2014. The root indeterminacy-to-determinacy developmental switch is operated through a folate-dependent pathway in *Arabidopsis thaliana*. *New Phytologist* 202:1223–1236.

Ruíz-Herrera, L. F., Shane, M. W. and J. López-Bucio. 2015. Nutritional regulation of root development. *Wiley Interdisciplinary Reviews: Developmental Biology* 4:431–443.

Ruíz-Herrera, L. F. and J. López-Bucio. 2013. Aluminum induces low phosphate adaptive responses and modulates primary and lateral root growth by differentially affecting auxin signaling in *Arabidopsis* seedlings. *Plant and Soil* 371:593–609.

Ruíz-Herrera, L. F., Sánchez-Calderón, L., Herrera-Estrella, L. and J. López-Bucio. 2012. Rare earth elements lanthanum and gadolinium induce phosphate-deficiency responses in *Arabidopsis thaliana* seedlings. *Plant and Soil* 353:231–247.

Sarrou, E., Therios, I. and K. Dimassi-Theriou. 2014. Melatonin and other factors that promote rooting and sprouting of shoot cuttings in *Punica granatum* cv. Wonderful. *Turkish Journal of Botany* 38:293–301.
Sarropoulou, V., Dimassi-Theriou, K., Therios, I. and M. Koukourikou-Petridou. 2012a. Melatonin enhances root regeneration, photosynthetic pigments, biomass, total carbohydrates and proline content in the cherry rootstock PHL-C (*Prunus avium* × *Prunus cerasus*). *Plant Physiology and Biochemistry* 61:162–168.
Sarropoulou, V. N., Therios, I. N. and K. N. Dimassi-Theriou. 2012b. Melatonin promotes adventitious root regeneration in in vitro shoot tip explants of the commercial sweet cherry rootstocks CAB-6P (*Prunus cerasus* L.), Gisela 6 (*P. cerasus* × *P. canescens*), and MxM 60 (*P. avium* × *P. mahaleb*). *Journal of Pineal Research* 52:38–46.
Shahzad, Z. and A. Amtmann. 2017. Food for thought: How nutrients regulate root system architecture. *Current Opinion in Plant Biology* 39:80–87.
Shishkova, S., Rost, T. L. and J. G. Dubrovsky. 2008. Determinate root growth and meristem maintenance in angiosperms. *Annals of Botany* 101:319–340.
Sozzani, R. and A. Iyer-Pascuzzi. 2014. Postembryonic control of root meristem growth and development. *Current Opinion in Plant Biology* 17:7–12.
Srivastava, A. C., Ramos-Parra, P. A., Bedair, M. et al. 2011. The folylpolyglutamate synthetase plastidial isoform is required for postembryonic root development in *Arabidopsis*. *Plant Physiology* 155:1237–1251.
Steffens, B. and A. Rasmussen. 2016. The physiology of adventitious roots. *Plant Physiology* 170:603–617.
Ulmasov, T., Murfett, J., Hagen, G. and T. J. Guilfoyle. 1997. Aux/IAA proteins repress expression of reporter genes containing natural and highly active synthetic auxin response elements. *Plant Cell* 9:1963–1971.
Walch-Liu, P., Liu, L., Remans, T., Tester, M. and B. G. Forde. 2006. Evidence that L-glutamate can act as an exogenous signal to modulate root growth and branching in *Arabidopsis thaliana*. *Plant and Cell Physiology* 47:1045–1057.
Wang, Q., An, B., Wei, Y. et al. 2016. Melatonin regulates root meristem by repressing auxin synthesis and polar auxin transport in *Arabidopsis*. *Frontiers in Plant Science* 7:1882.
Weeda, S., Zhang, N., Zhao, X. et al. 2014. *Arabidopsis* transcriptome analysis reveals key roles of melatonin in plant defense systems. *PLoS One* 9:e93462–e93462.
Wen, D., Gong, B., Sun, S. et al. 2016. Promoting roles of melatonin in adventitious root development of *Solanum lycopersicum* L. by regulating auxin and nitric oxide signaling. *Frontiers in Plant Science* 7:718.
Woodward, A. W. and B. Bartel. 2005. Auxin: Regulation, action, and interaction. *Annals of Botany* 95:707–735.
Xiong, Y., McCormack, M., Li, L., Hall, Q., Xiang, C. and J. Sheen. 2013. Glucose-TOR signalling reprograms the transcriptome and activates meristems. *Nature* 496:181–186.
Zhang, N., Zhang, H. J., Zhao, B. et al. 2014. The RNA-seq approach to discriminate gene expression profiles in response to melatonin on cucumber lateral root formation. *Journal of Pineal Research* 56:39–50.
Zhang, N., Zhao, B., Zhang, H. J. et al. 2013. Melatonin promotes water-stress tolerance, lateral root formation, and seed germination in cucumber (*Cucumis sativus* L.) *Journal of Pineal Research* 54:15–23.

# 7 Acetylcholine as a Regulator of Differentiation and Development in Tomato

*Kiran Bamel and Rajendra Gupta*

## CONTENTS



## 7.1 INTRODUCTION

Acetylcholine (ACh) is present in all living organisms ranging from microbes, lower plants (such as algae, fungi, bryophytes, pteridophytes) gymnosperms, angiosperms, and animals. It is reported to be present in the nervous as well as non-nervous tissues of animals (Roshchina, 2001). The participation and importance of ACh in non-neuronal functions of animals has also been known for many decades. However, due to the overwhelming importance of ACh in the nervous system where it is the most important neurotransmitter, the non-nervous roles did not receive much attention in the beginning. Whittaker (1963), for the first time, emphasized the widespread presence of ACh in non-nervous tissues in animals and cautioned against ascribing an exclusive neuronal role to ACh. This caution was echoed by Silver (1974) and, subsequently, Sastry and Sadavongvivad (1979) presented the first comprehensive review on non-neuronal roles of ACh in animals. The non-neuronal aspect of ACh is being talked about and gaining considerable attention recently (Small et al., 1996; Grisaru et al., 1999; Lauder and Schambra, 1999; Wessler et al., 1999, 2008). The increased interest in non-nervous functions of ACh can be attributed to enhanced awareness about presence of ACh in (a) the non-nervous tissues of animals with a well-defined nervous system, (b) animals lacking a nervous system or having poorly organized nervous systems, and (c) plants. The presence in plants of several other neurotransmitters, agonists, and antagonists and their effects on plant growth and development have also increased the curiosity regarding the non-classical roles of neurotransmitters in general (Roshchina, 2001; Murch, 2006). The Cholinesterase enzyme (ChE) responsible for the breakdown of acetylcholine is also universally present in plants (Gupta and Gupta, 1997). ACh is also present in non-nervous living organisms (Kawashima et al., 2007). The presence of ACh

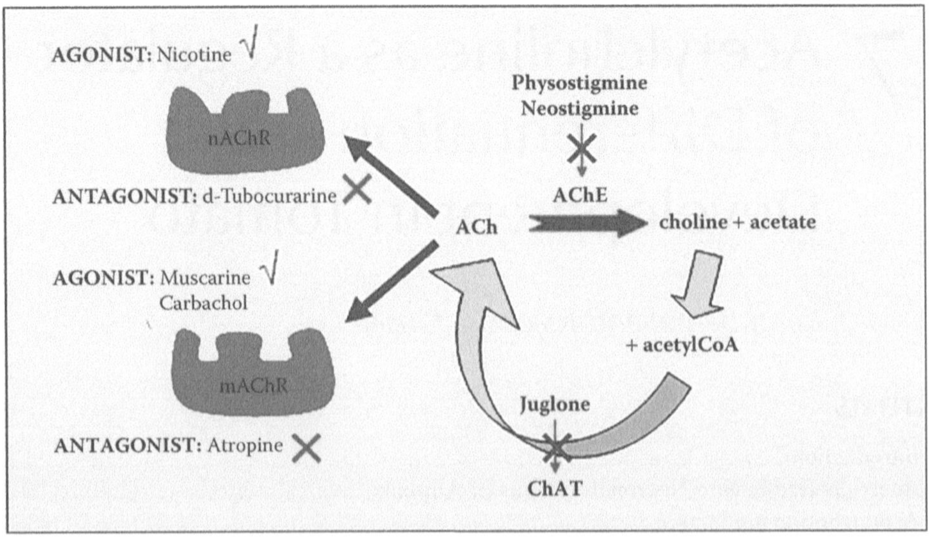


**FIGURE 7.1** Components of ACh system.

and ChE in plants have generated enormous interest in plant biologists to know about the functions of ACh systems in this group of living organisms that lack nervous system. This curiosity has been the driving force behind the emergence of *plant neurobiology* as a new field in plant biology (Baluska et al., 2006; Stahlberg, 2006). Since the present chapter explores the possible role of ACh in morphogenesis in a plant, it is essential to summarize the current status of non-neuronal roles of ACh in living organisms, in general, with special reference to morphogenic role. An overview of involvement of various components of the ACh system in signal transduction in nerves has also been presented briefly (Figure 7.1).

## 7.2 ACETYLCHOLINE IN NON-NEURONAL SYSTEMS OF ANIMALS

The occurrence of ACh in bacteria, algae, protozoa, and tubellaria suggests an extremely early appearance of ACh in the evolutionary process (Wessler et al., 1999). The choline acetyl transferase (ChAT) protein can be regarded as a basic cellular component, as it is expressed in a majority of human cells. Nicotinic and muscarinic acetylcholine receptors (AChRs) are widely expressed in non-neuronal tissue, for example endothelial cells, epithelial cells, placenta, and circulating blood cells. A receptor responding to both nicotinic and muscarinic agonists has been reported in insects (Lapied et al., 1990) and bovine (Shirvan et al., 1991) neurosecretory cells. Besides this chimeric receptor, all the subtypes of muscarinic receptor (Grando et al., 1995) and few subunits of nicotinic receptor complex (Grando et al., 1996) are expressed in human keratinocytes. The action of acetylcholinesterase (AChE) on differentiation and proliferation has been demonstrated in so many experimental systems that they must be considered firmly established functions of AChE.

The cholinergic components affect the motility of two photosynthetic bacteria, namely *Rhodospirillum rubrum* and *Thiospirillum jenense* (Sastry and Sadavongvivad, 1979). ACh and AChE are present in cellular slime molds *Physarum polycephalum* and *Physarella oblonga* influencing the protoplasmic streaming (Hoitink and Dijk, 1966). The gill plates of some fishes contain ACh, AChE, and ChAT (Sastry and Sadavongvivad, 1979). Antagonists of AChR act as inhibitors of ACh-caused cell movements in sea urchin (Lauder and Schambra, 1999). Cell movements occurring during gastrulation and postgastrulation stages are regulated by ACh (Gustafson and Toneby, 1970).

Both AChE and butyrylcholinesterase (BChE) share amino-acid sequence homology with two adhesion proteins, glutactin (Olson et al., 1990) and neurotactin (Barthalay et al., 1990; De la Escalarn et al., 1990) found in *Drosophila*. Besides the enzyme activity, particular glyco-decoration of ChEs may be relevant for their role as a new class of adhesive molecule. The HNK-1 carbohydrate epitope of ChE, which has been linked to cell differentiation and migration processes (Jungalwala, 1994) is a common motif of many cell adhesion molecules.

Cholinesterases indeed can influence cell growth and neurite differentiation (Layer, 1990; Layer et al., 1993). Lipton and Kater (1989) showed that ACh is secreted by some growing axons and thereby inhibits outgrowth of neurites. ACh released from growth cones inhibits neurite extension while ChE attracts and directs neurites by creating ACh-free spaces (Layer, 1991). Early appearance of cholinergic neurons and their receptors in embryonic rat brain and transient expression of nicotinic receptor in optic nerve fiber suggests that ACh could play a role in neurogenesis and development of the retinotechtal projection (Lauder, 1993).

Certain neurotransmitters regulate regeneration, for example catecholamines produced by hydrozoan larvae can act extracellularly to regulate metamorphosis and play a role in regeneration in *Planaria* (Lauder, 1993). The early presence of cholinergic system in mesenchyme and developing cartilage of chick limb and myotomes also supports the hypothesis that ACh has morphogenic roles (Willbold and Layer, 1994).

ChEs have been implicated in the control of proliferation and differentiation of some haematopoietic cell systems. AChE and BChE genes are amplified in leukemias and platelet disorders (Soreq and Zakut, 1990; Soreq et al., 1992; Zakut et al., 1992). High levels of ChE activity have been reported in primary brain and ovarian tumors (Small et al., 1995). Small cell lung carcinomas possess muscarinic acetylcholine receptor (mAChR) (Williams and Lennon, 1991) as well as nicotinic acetylcholine receptor (nAChR) (Codignola et al., 1994), which when stimulated inhibit or stimulate cell proliferation, respectively. The non-catalytic functions of AChE, which affect cell proliferation and differentiation, are partly due to the three C-terminal variants of AChE, namely AChE-S, AChE-R, and AChE-E. Structural functions of AChE variants can explain their proliferative and developmental roles in blood and bone, as well as retinal and neuronal cells (Grisaru et al., 1999).

Lauder (1993) suggested a role of ACh in cellular differentiation. Two possibilities have been indicated by Small et al. (1995) by which ChE influences cellular differentiation. First, by regulating the local concentration of ACh, which may influence cellular differentiation *via* nicotinic or muscarinic receptors. Second, a non-catalytic mechanism may be involved. AChE acts as a postmitotic differentiation marker in most vertebrates (Puelles et al., 1987; Hanneman et al., 1988; Moody and Stein, 1988; Layer, 1991). In chick, AChE and BChE are expressed in motoneuronal cell bodies and in dermomyotomes, thus, representing the very first sign of muscle differentiation (Layer, 1991). The appearance of AChE in early events in the development of a variety of embryonic tissues (Drews, 1975) and long before the start of synaptic activity supports the view that cholinesterases may be directly involved in embryogenesis, independent of their roles in the fully developed nervous system.

## 7.3 ACETYLCHOLINE IN PLANTS

In plants, ACh is involved in several physiological phenomenons (Fluck and Jaffe, 1974b; Hartmann and Gupta, 1989; Tretyn and Kendrick, 1991; Roshchina and Vikhlyantsev, 2009). ACh regulates seed germination (Kostir et al., 1965), hypocotyl elongation (Lawson et al., 1978), promotes pollen tube elongation in *Lilium longiflorum* (Tezuka et al., 2007), and growth of certain intact or isolated organs of plants. ACh increases the size of stem segments in wheat (Tretyn et al., 1990), soybean (Mukherjee, 1980), and cucumber (Verbeke and Vendrick, 1977). It also helps in emergence and elongation of lateral roots of *Raphanus sativus* (Sugiyama and Tezuka, 2011). Sansebastiano et al. (2014) have reported that in tomato ACh affects cell elongation by affecting the transcription of expansin gene *LeEXPA2*. AChE also confers heat tolerance (Yamamoto et al., 2011) and affects differentiation and proliferation in many experimental systems (Wessler et al., 1999). ACh can stimulate growth by interacting with different hormones, such as auxins, gibberellins, and ethylene or through the changes in ion fluxes (Lawson et al., 1978). In a majority of cases, ACh-induced growth changes are related to the quality of light exposures (Tretyn and Kendrick, 1991). Dekhuijzen (1973) reported the stimulatory effect of ACh on the growth of wheat seedlings. Jones (1986) observed that ACh caused retardation of growth and mineral uptake in *Glycine max*. ACh induced an increase in dry weight of the root as well as their $K^+$ and $Ca^{++}$ content. On the contrary, $K^+$ and $Ca^{++}$ contents and dry weight of the aerial parts decreased. Later, Jones and Stutte (1988) also observed that ACh and neostigmine (Nst) inhibited 1-aminocyclopropane-1-carboxylic acid (ACC)-mediated ethylene evolution from soybean leaf discs. Atropine, an antagonist of ACh action, had a promoting effect. Jones and Stutte (1986) also studied the effect of red light on ethylene evolution and found that red light mimicked the ACh effect. Nst and red light in combination with ACh partially overcame the atropine-induced response.

The meristematic and differentiating plant tissues also have higher ACh (Jaffe, 1970; Evans, 1972; Fluck and Jaffe, 1974a; Mukherjee, 1980; Hoshino, 1983), as well as high AChE activity (Raineri and Modensi, 1986). ACh and AChE have also been related to morphogenic processes. Tyagi (1982) studied the effect of ACh and two inhibitors of AChE, *viz*. neostigmine and physostigmine on *in vitro* androgenesis in *Datura innoxia* and showed that ACh as well as anti-ChE promote haploid formation. In lichen, *Parmelia caperata*, ChE activity was found in symbionts, alga, and fungus. During the early morphogenesis of soredia, the activity was associated with algal proliferation. After completion of differentiation, both the inner dividing algal layer and an outer fungal hyphae envelop showed ChE activity, suggesting the involvement of cholinergic system in regulating morphogenic processes, such as, cell division, oriented tip growth, and membrane interactions of the symbionts (Raineri and Modensi, 1986). In *Narcissus confusus*, the content of an anti-ChE compound galanthamine increased with tissue differentiation (Selleś et al., 1999).

It is evident that ACh, as well as enzymes of its metabolism, are present at several non-neural locations in animals as well as plants. Effects of exogenous applications of ACh and several chemicals that affect the cholinergic response in nerves also effects several other physiological phenomena in different organisms. ACh seems to be one of the most ancient chemical signals used by living organisms for regulation of growth, development, morphogenesis, and coordination of movements. It seems that during the course of evolution, several cell types in different organisms use ACh for various purposes. There is compelling evidence for involvement of ACh in differentiation and morphogenesis in many animals, but there is a dearth of research on plant systems. Earlier, there are few reports of the role of ACh and other chemicals affecting the cholinergic system on growth of intact or excised parts of plants, but a lack of studies on differentiation and morphogenesis leaves a gap in our understanding of the ACh effects in general. The widespread occurrence of the components of the ACh system in plants calls for investigation on morphogenic role of ACh in plants. It is possible that we may be on the threshold of discovering a common and widespread non-neuronal function of ACh.

There are generally five accepted classical plant hormones *viz*. auxins, cytokinins, gibberellins, ethylene, and abscisic acid. The list has been expanded in the last two decades and now includes oligosaccharins, jasmonates, systemins, salicylic acid, and brassinosteroids. Other chemicals listed as growth regulators but short of the status of hormone include polyamines, triacontanol, turgorin, and lunularic acid (Salisbury, 1992; Davies, 1995; Crozier et al., 2000; Taiz and Zeiger, 2002). Acetylcholine is not listed under plant growth hormone or regulators in any standard plant physiology textbook, treatise or review, although hundreds of research papers on its presence, metabolism, and physiological effects have been published in the last three decades. Since nature observes parsimony in resource utilization, it would be difficult to understand the reason for the presence of ACh and enzymes of its metabolism in plants unless ACh were not a participant in some physiological process(es). Although, scores of research papers have been published on the physiological response to ACh, only a few of these try to address the questions listed earlier. Most of the papers deal with exogenous application of ACh and one or more chemicals that influence its breakdown or action. Most of the papers do not present experimental evidence employing essential controls, such as use of breakdown products or precursors of ACh. Bamel and coworkers have focused on morphogenic effects of exogenous application of ACh, its breakdown products, inhibitors of synthesis and breakdown, and agonists and antagonists of ACh receptors (Bamel et al., 2007; Bamel et al., 2015; Bamel et al., 2016).

To understand the effect of plant growth regulators and plant hormones on morphogenesis and differentiation, tissue culture techniques have been widely employed (Street, 1977; Razdan, 1993). Bamel et al. (2007, 2015, 2016) performed experiments on leaf explants of tomato cultured *in vitro* to study the effects of ACh as a putative plant hormone as it belongs to the family Solanaceae, and members of this family provide a model system for different types of experimentations concerning physiology, growth, nutrition, and morphogenesis (Narayanaswamy, 1997). Tomato leaf explants are known to have excellent regenerative and morphogenic potential *in vitro*, and protocols for *in vitro* culture and morphogenesis in tomato (Kartha et al., 1976, Gunay and Rao, 1980), culture, and maintenance of cell lines as cell suspension cultures, or calli, are well-established (Sink and Reynolds, 1986). Tomato leaf explants contain high ChE activity as compared to cotyledons, stems, and roots (Figure 7.3). Plants of the tomato family Solanaceae have the highest concentration of ChE among hundreds of plants tested thus far (Gupta and Gupta, 1997). Several plants of the tomato family Solanaceae contain chemicals that affect the cholinergic response in animals, for example anti-ChE: solanine, solanidine, and α-chaconine; AChR-agonists: nicotine and nor-nicotine; and AChR-antagonists: atropine and scopolamine (Roshchina, 2001). The widespread occurrence of the components of ACh system in plants and its involvement in morphogenesis and differentiation calls for investigation on morphogenic role of ACh in plants.

The tomato leaf explants are known to exhibit rhizogenesis when cultured on a nutrient medium supplemented with IAA in micromolar or lesser concentrations. The explants undergo caulogenesis when grown on a medium supplemented with cytokinins whereas a combination of auxins and cytokinins induces only rhizogenesis, only caulogenesis or both rhizogenesis and caulogenesis depending on their relative concentrations. The classical response is rhizogenesis if the auxin to cytokinin ratio is high and caulogenesis if the ratio is low (Skoog and Miller, 1957). It is well-known that the response to exogenous application of hormones may not be dependent on concentration alone but also on sensitivity of tissue to the hormone, endogenous level of hormone, uptake and metabolism of exogenously applied hormone, mechanism of action of the hormone, glycosylation, and hydrolysis of the hormones.

The aforementioned factors may, in turn, be influenced by different parameters, such as the physiological stage of the donor plant, its nutritional, hormonal, as well as environmental conditioning and the spatial gradient of hormone in plant (Tran, 1981). Therefore, to obtain reproducible results, the seedlings were grown *in vitro* under a well-controlled environment, and for each set of the experiment, the seedlings were taken from the same batch of cultures. To avoid the differences in the physiological stage of the explant, 30-day-old seedlings were used and for each culture only the first and the second leaves from cotyledons were selected as explants.

## 7.4 ACETYLCHOLINE REGULATES MORPHOGENESIS IN TOMATO

### 7.4.1 Acetylcholine Induces Root Formation in *In Vitro* Cultured Leaf Explants

Exogenous application of ACh enhanced rooting in a dose-dependent manner in excised leaf explants of *Lycopersicon esculentum* (Bamel et al., 2007). The roots emerged from the midrib of explants. There was no rooting on control explants, whereas ACh ($10^{-7}$ M or higher) induced rooting. The maximum response to ACh was induction of rooting in about 15%–20% explants (Figure 7.2a–d). Hartmann (1978) has demonstrated that as high as 90% of the exogenously supplied ACh is hydrolysed before its uptake by the ChE situated at the cell surface. Therefore, tests for ChE were performed. AChE activity was decreased from a high value of 578 pmol ATChI hydrolysed/mg protein/s in controls to 138 in leaves cultured on $10^{-4}$ M Neostigmine (76% inhibition). This level of ChE would be enough to bring down the concentration of ACh in culture medium from $10^{-3}$ M to $10^{-4}$ M in three days. Consequently, the concentration of ACh available to the explants over a culture

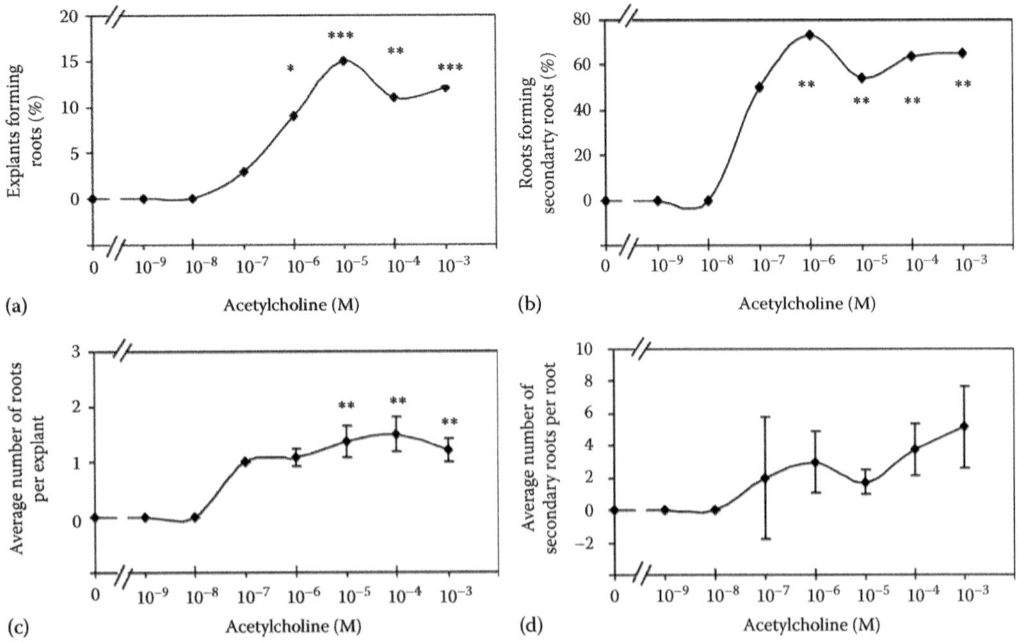


**FIGURE 7.2** (a–d) Effect of ACh on morphogenic responses of leaf explants of tomato, *Lycopersicon esculentum* var. Pusa Ruby. ACh ($10^{-9}$ to $10^{-3}$ M) was provided continuously in the culture medium from the first day of culture for 30 days. Leaf explants used here were excised from 30-day-old seedlings raised *in vitro* on MS basal medium. The data presented here are an average of three experiments with 144 explants per treatment. Error bars show ±S.E.M. * Denotes significant differences between treatment and control ($P \leq 0.05$); ** Highly significant ($P \leq 0.01$); *** Very highly significant ($P \leq 0.001$). (a) Percentage of leaf explants forming roots. (b) Percentage of roots forming secondary roots. (c) Average number of roots formed per responding explant. (d) Average number of secondary roots developed per root. (Adapted from Bamel, K. et al., *Life Sci.*, 80, 2393–2396, 2007. With permission.)

period of four to five weeks may be several magnitudes lower than what was supplied initially. This may be one of the reasons for low response to ACh. However, other factors such as its uptake, metabolism inside the cells, and sensitivity of the tissue will also have to be taken into consideration for further analysis. When the hydrolytic products of ACh, namely, choline ($10^{-9}$–$10^{-3}$ M) and acetate ($10^{-9}$–$10^{-3}$ M) were exogenously supplied to the cultures, it was observed that these chemicals did not elicit any rhizogenic response in the explants, indicating that ACh has a specific effect on rhizogenesis.

The effect of anti-ChE chemicals, neostigmine (Nst), and physostigmine (Pst), on morphogenic response of tomato leaf explants was also studied. It was believed that, being inhibitors of ChE, these two chemicals would increase the endogenous titer of ACh and hence aid in rooting of explants. Tretyn et al. (1987) reported that changes in ACh levels in response to different light exposures are related to change in ChE levels. Riov and Jaffe (1973) have also documented that inhibitors of ChE (growth retardants related to (+)-limonene) enhance the endogenous ACh level and have effects similar to ACh in inhibiting the secondary root growth of bean. An endogenous anti-ChE compound, galanthamine, has been reported in *N confuses*, where its levels are related to differentiation status of explants (Selleś et al., 1999).

Nst augmentation also caused rooting in leaf explants (Bamel et al., 2007). The optimum response was seen at $10^{-4}$ M Nst for primary root and $10^{-3}$ M for secondary roots (Figure 7.3). Supplementation of Nst in different combinations with ACh to the control enhanced the effect of ACh. Sixty-four percent explants rooted on $10^{-4}$ M ACh+$10^{-3}$ M Nst when only 14% formed roots on ACh alone. An interesting observation was the inhibition of callusing on leaf explants on ACh and Nst. Callusing is otherwise considered an unwanted but frequent feature in plant tissue cultures. Although levels of

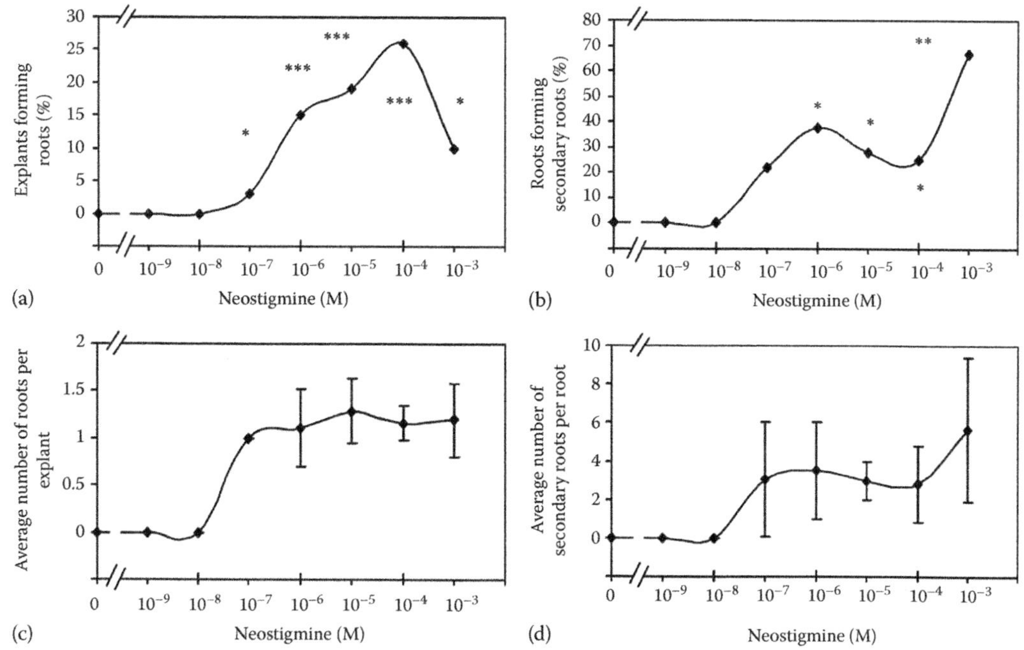


**FIGURE 7.3** (a–d) Effect of neostigmine on morphogenic responses of leaf explants of tomato, *Lycopersicon esculentum* var. Pusa Ruby. Neostigmine ($10^{-9}$ to $10^{-3}$ M) was provided continuously in the culture medium from the first day of culture for 30 days. Leaf explants used here were excised from 30-day-old seedlings raised *in vitro* on MS basal medium. The data presented here are an average of three experiments with 144 explants per treatment. Error bars show±S.E.M. * Denotes significant differences between treatment and control ($P \leq 0.05$); ** Highly significant ($P \leq 0.01$); *** Very highly significant ($P \leq 0.001$). (a) Percentage of leaf explants forming roots. (b) Percentage of roots forming secondary roots. (c) Average number of roots formed per responding explant. (d) Average number of secondary roots developed per root. (Adapted from Bamel, K. et al., *Life Sci.*, 80, 2393–2396, 2007. With permission.)

ACh in explants treated with neostigmine is not known, it can be safely assumed that inhibition of AChE by its inhibitor Nst or Pst would result in elevation of endogenous levels of ACh. In experiments on rhizogenesis, as well as caulogenesis, Nst and Pst not only mimicked the action of ACh, but their effects were found to be more and more pronounced as compared to exogenously applied ACh.

Effect of exogenous application of ACh on rooting in intact seedlings has been reported earlier. Jaffe (1970, 1972) found that ACh caused a decrease in secondary root formation in mung bean seedlings. Inhibition of ChE activity in bean roots by application of several growth retardants was also accompanied by inhibition of secondary root formation and supported the hypothesis that ACh may be an endogenous growth retardant. Later, Kasturi and Vasantharajan (1976) also presented evidence for inhibition of secondary root formation in pea seedlings due to inhibition of ChE activity caused by pesticides. However, Kasemir and Mohr (1972) and Gupta (1983) failed to observe any effect of ACh on the growth of primary roots or formation of secondary roots.

The promotive/inductive effect of ACh on rooting of tomato leaf explants was the first observation of this nature recorded for any plant. Of the two inhibitors, Nst was found to be more effective on rhizogenesis (Bamel et al., 2007). In animals, physostigmine is known to be a potent inhibitor of ChEs (Silver, 1974), and it prolongs the physiological effects of acetylcholine. However, ChE activity in tomato (Kavita, 2002) as well as in other plants (Roshchina, 2001) is relatively insensitive to Pst as compared to Nst. In wheat seeds, ChE activity was not inhibited by Pst even at high concentrations ($10^{-5}$ and $10^{-4}$ M) tried by Molyneux and McKinlay (1989). Based on this result, it seems that physostigmine is not very effective against plant ChEs. Later, Sugiyama and Tezuka (2011) also showed that the root growth in *R. sativus* is promoted by ACh and neostigmine.

### 7.4.2 Acetylcholine Suppresses Shoot and Callus Formation

The effect of various chemicals affecting the cholinergic system was studied and compared to the response of leaf explants reared on shoot regeneration medium (SRM), that is Murashige and Skoogs' medium (Murashige and Skoogs, 1962) supplemented with $1.7 \times 10^{-6}$ M IAA + $1.3 \times 10^{-5}$ M BA (Bamel et al., 2016). ACh affected the morphogenic potential of leaf explants of tomato by reducing caulogenesis (Figures 7.4 and 7.5). ACh ($10^{-4}$ M) fortified media supported shoot formation only in 55% explants in comparison to about 83% response in control.

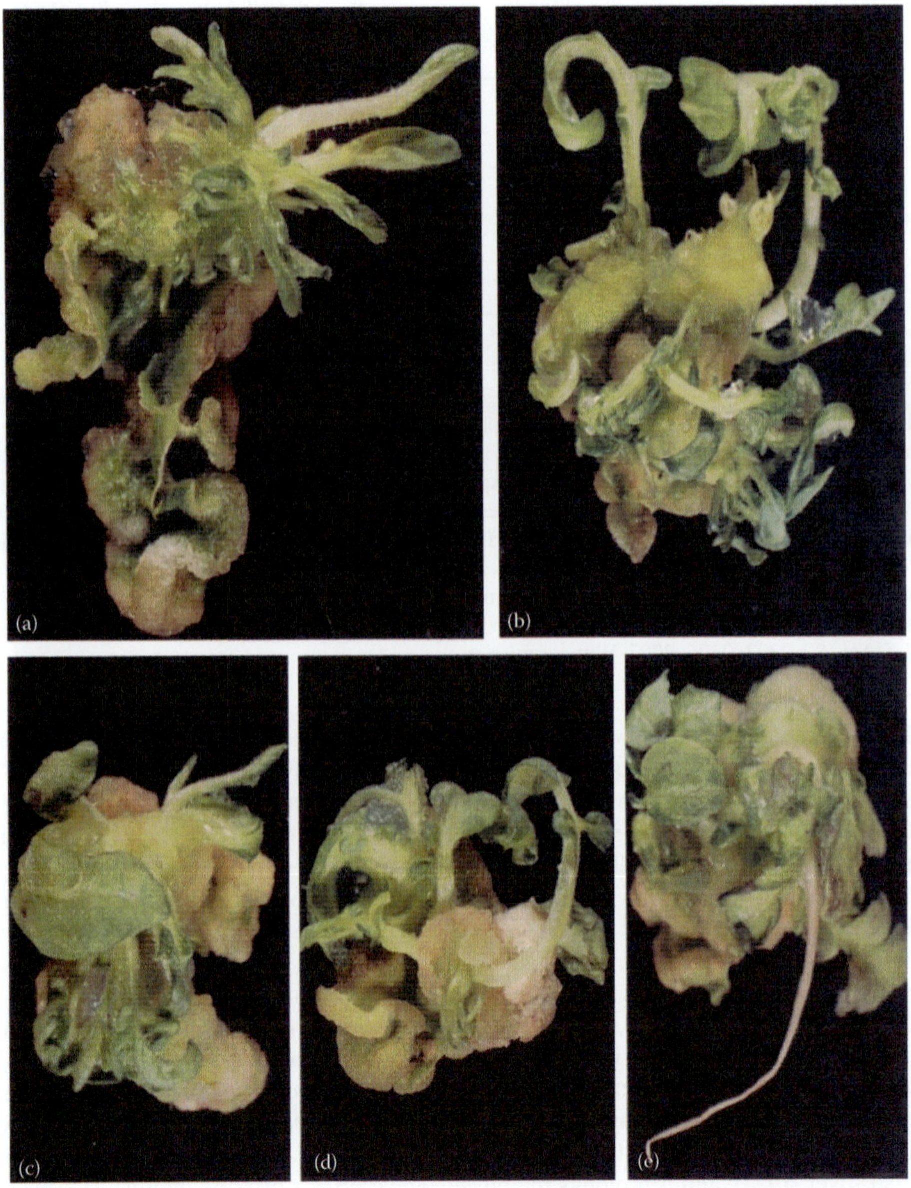


**FIGURE 7.4** Morphogenic response of *Lycopersicon esculentum* leaf explants, excised from 30-day-old *in vitro* raised seedlings and cultured for 30 days on acetylcholine-adjuvanted SRM medium containing $1.3 \times 10^{-5}$ M BA and $1.7 \times 10^{-6}$ M IAA, under 16-hour photoperiod. (a) Leaf explants on SRM (control). b–e: Explants cultured on SRM containing different levels of ACh, (b): $10^{-9}$, (c): $10^{-6}$, (d): $10^{-5}$; (e): $10^{-4}$ M. (Adapted from Bamel, K. et al., *Plant Signal. Behav.*, 11, e1187355, 2016. With permission.)

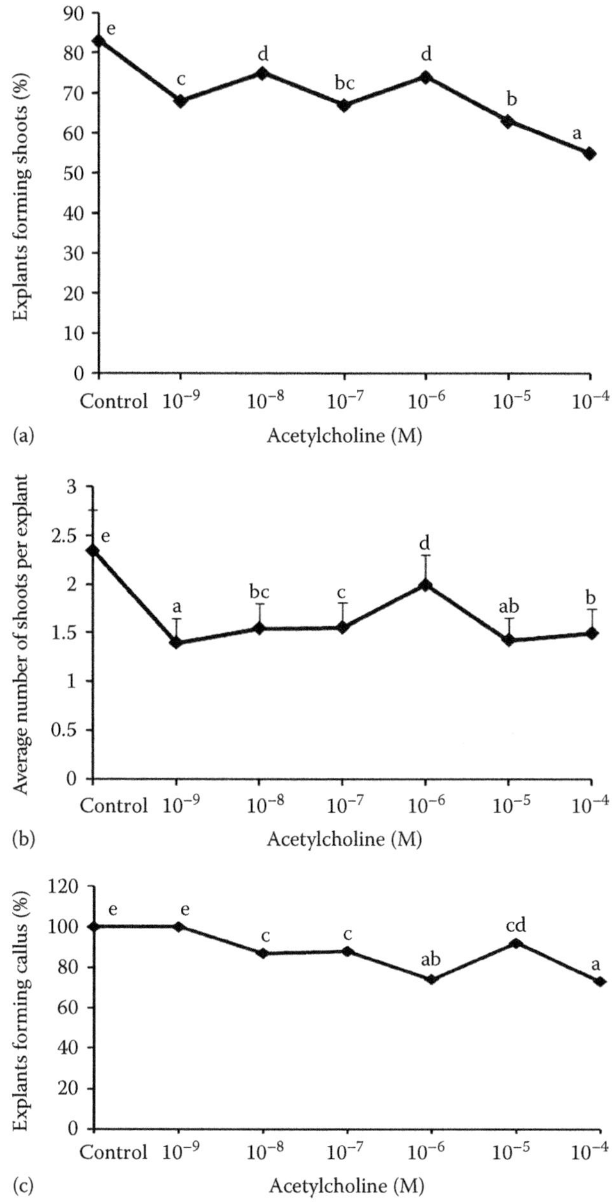


**FIGURE 7.5** Caulogenic response of *Lycopersicon esculentum* leaf explants, excised from 30-day-old *in vitro* raised seedlings and cultured for 30 days on acetylcholine-adjuvanted SRM medium containing $1.3 \times 10^{-5}$ M BA and $1.7 \times 10^{-6}$ M IAA, under 16-hour photoperiod after 30 days of culture. (a) explants forming shoots (%), (b) average number of shoots per explants, (c) explants forming callus (%). Values followed by the same letters are not significantly different ($P \leq 0.05$) by Tukey post hoc test using Statistical Package for Social Sciences (SPSS) software. (Adapted from Bamel, K. et al., *Plant Signal. Behav.*, 11, e1187355, 2016. With permission.)

Less number of cultures organised callus on ACh-supplemented media, and the callus mass was also reduced (Figures 7.6. and 7.7). A good amount of greenish-brown callus was formed on the controls, whereas on ACh-supplemented media only moderate amounts of callus, though its color and texture remained the same. The breakdown products of ACh, namely choline and acetate, did not affect the morphogenic response. The number of shoots formed and callus mass

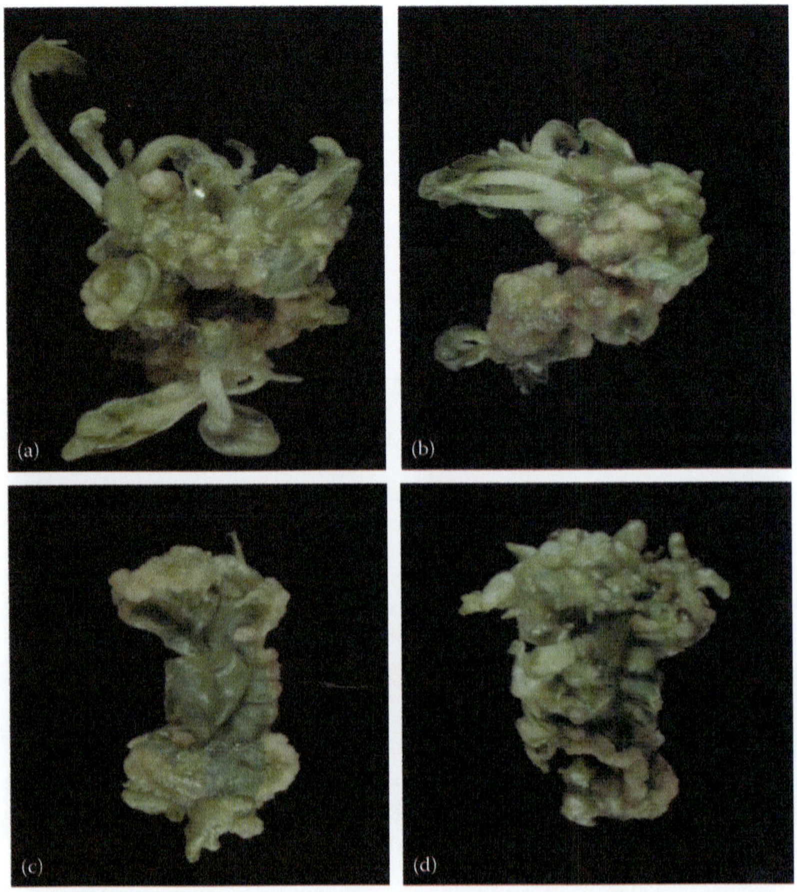


**FIGURE 7.6** Morphogenic response of *Lycopersicon esculentum* leaf explants, excised from 30-day-old *in vitro* raised seedlings and cultured for 30 days on neostigmine-adjuvanted SRM medium containing $1.3 \times 10^{-5}$ M BA and $1.7 \times 10^{-6}$ M IAA, under 16-hour photoperiod. (a) Leaf explants on SRM (control). (b–d) Explants cultured on SRM containing different levels of Neostigmine, (b): $10^{-9}$, (c): $10^{-8}$, (d): $10^{-5}$ M. (Adapted from Bamel, K. et al., *Plant Signal. Behav.*, 11, e1187355, 2016. With permission.)

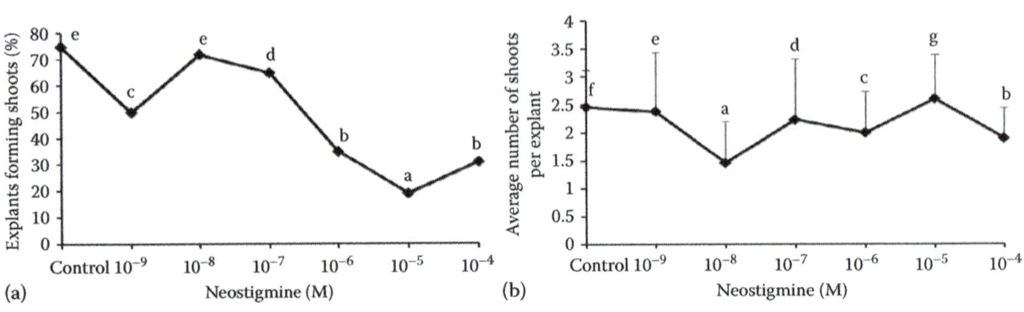


**FIGURE 7.7** Caulogenic response of *Lycopersicon esculentum* leaf explants, excised from 30-day-old *in vitro* raised seedlings and cultured for 30 days on neostigmine-adjuvanted SRM medium containing $1.3 \times 10^{-5}$ M BA and $1.7 \times 10^{-6}$ M IAA, under 16-hour photoperiod after 30 days of culture. (a) Explants forming shoots (%), (b) average number of shoots per explants. Values followed by the same letters are not significantly different ($P \leq 0.05$) by Tukey post hoc test using Statistical Package for Social Sciences (SPSS) software. (Adapted from Bamel, K. et al., *Plant Signal. Behav.*, 11, e1187355, 2016. With permission.)

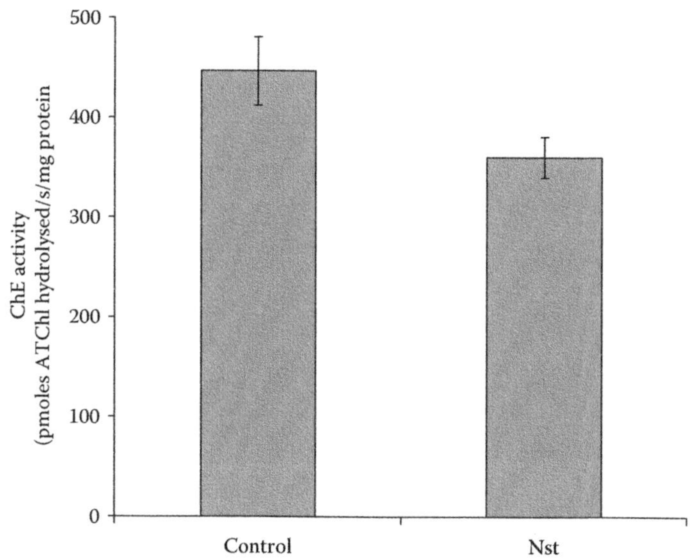


**FIGURE 7.8** ChE activity of extracts of *in vitro*-raised cultures of *Lycopersicon esculentum* on shoot regeneration medium supplemented with $10^{-5}$ M Nst. Error bars show SEM; $P \leq 0.05$. (Adapted from Bamel, K. et al., *Plant Signal. Behav.*, 11, e1187355, 2016. With permission.)

were comparable to the control. Neostigmine affected callogenesis as well as caulogenesis (Figures 7.6 and 7.7). Even at $10^{-9}$ M Nst and $10^{-8}$ M Nst the decline in the amounts of callus was visible. Physostigmine also reduced morphogenesis but though the decline in caulogenesis was clearly discernible, the response did not show a typical dose-response pattern. Nst reduced the specific activity of ChE in the cultures. The specific activity of ChE decreased to 360.5 units in the cultures raised on Nst-supplemented medium, whereas it was 446.5 units on the control (Figure 7.8). There was slight decline in the caulogenic response when different levels of acetylcholine were added along with neostigmine ($10^{-5}$ M) to the SRM (Figures 7.9 and 7.10). The maximum inhibition caused by ACh was about 33% (at $10^{-4}$ M) over the control. Callus amount was also reduced on all ACh concentrations compared to the control. The sizes of the explants also increased. Presently it is not known whether ACh inhibits cell division per se or promotes cell differentiation in tomato leaf cultures. But the meristematic zone of root and shoot tips highest concentrations of ACh and ChE in plants (Jaffe 1970; Riov and Jaffe, 1973). In animals also ChE activity is known to influence cellular differentiation animals by regulating local ACh concentrations. Small et al. (1995) have also indicated that lowering of ACh levels and increase in ChE levels are associated with carcinogenic activity in several organs in animal systems. This observation is similar to the inhibitory effect of ACh on callus formation in tomato leaf explants. Earlier workers on the contrary have shown that ACh acting through nAChR is known to promote carcinogenic activity (Williams and Lennon, 1991) while through mAChR, it inhibits cell division (Codignola et al., 1994). The ChE inhibitors, neostigmine, and physostigmine not only mimicked the action of ACh, but their effects were also more pronounced as compared to exogenously applied ACh (Bamel et al., 2016). As much of the ACh may be hydrolysed by ChE enzyme in leaf explants, the exogenous application of anti-ChE would ensure a continuously higher ACh concentration inside cells. Similar antitumor activity of an anti-ChE (galanthamine) is reported in animal tissues (Wenigner et al., 1995) and in *N confuses* (Selleś et al., 1999). Galanthamine changes the plant growth reminiscent of 2,4–D toxicity in cultures of *Artemesia tridentata* (Turi et al., 2014).

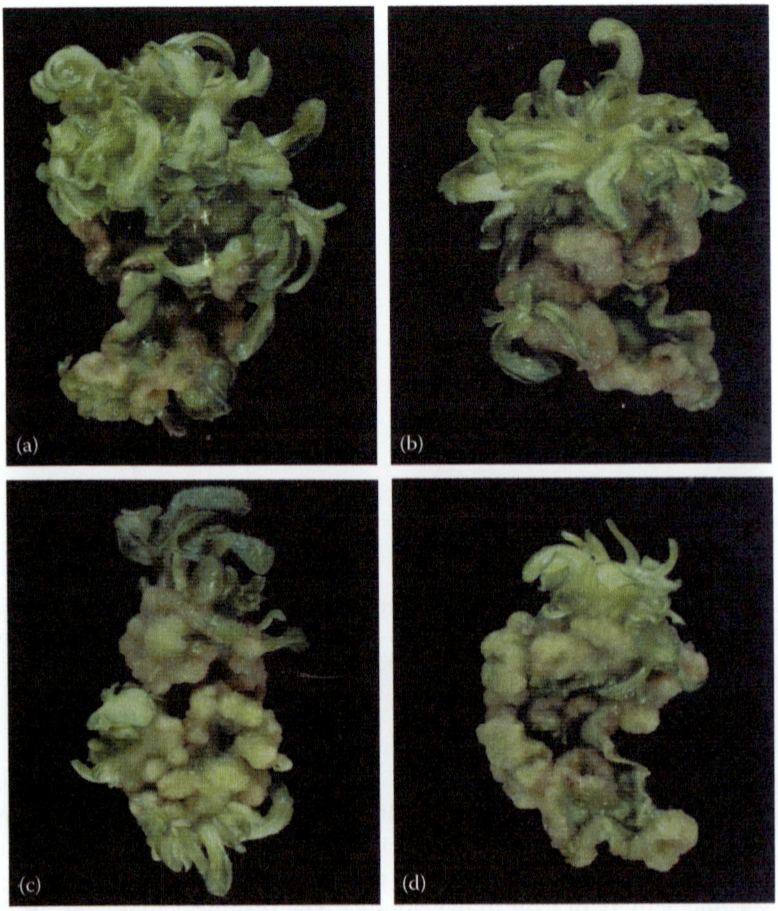


**FIGURE 7.9** Morphogenic response of *Lycopersicon esculentum* leaf explants, excised from 30-day-old *in vitro*-raised seedlings and cultured for 20 days on ACh+ Nst-adjuvanted SRM medium, under 16-hour photoperiod. (a) Leaf explants on SRM (control). (b) Explants cultured on SRM containing $10^{-5}$ M Nst, (c–d) Explants cultured on SRM containing $10^{-5}$ M Nst and different levels of ACh, (c) $10^{-8}$ M ACh, (d) $10^{-7}$ M ACh. (Adapted from Bamel, K. et al., *Plant Signal. Behav.*, 11, e1187355, 2016. With permission.)

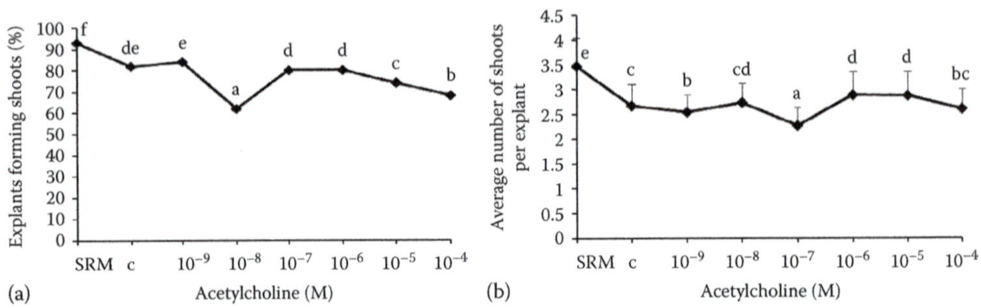


**FIGURE 7.10** Effect of ACh in the presence of $10^{-5}$ M Nst on the caulogenic response of *Lycopersicon esculentum* leaf explants excised from 30-day-old *in vitro*-raised seedlings after 20 days of culture. (a) Explants forming shoots (%), (b) average number of shoots per explant. Values followed by the same letters are not significantly different ($P \leq 0.05$) by Tukey post hoc test using Statistical Package for Social Sciences (SPSS) software. (Adapted from Bamel, K. et al., *Plant Signal. Behav.*, 11, e1187355, 2016. With permission.)

### 7.4.3 Agonist of Nicotinic Acetylcholine Receptor Mimic the Response of Acetylcholine

Nicotine is an alkaloid of tobacco (*Nicotiana tabacum* L. and *Nicotiana rustica* L.). The pharmacological action of nicotine is because it can mimic the action of ACh (Taylor 2001). Nicotine is also known to affect differentiation in plants (Tabata et al., 1971; Miller et al., 1983; Peters et al., 1974a), but its mechanism of action is not known. The presence of ACh receptors or its binding sites in plants has been reported, though indirectly, in several systems. ACh receptors belonging to two different classes and similar to mAChR and nAChR of animals have been identified in wheat protoplasts (Tretyn et al., 1990). Lukasiewicz et al. 1997 studied the effect of ACh and chemicals affecting ACh system on flowering in *Pharbitis nil*. They showed that ACh, inhibitors of ChE, and agonists for the receptors of ACh stimulate flowering when applied to the cotyledons of 5-day-old seedlings of *P. nil* during 24 h inductive night preceded by 72 h of darkness and then 24 h of low intensity light, terminated by 15 min impulse of far-red light. They concluded that ACh controlled flowering of short day plants. Nicotine also improves germination, lengths of radicle and plumule, and seedling vigor in maize cv. Diara but has contradictory effect on the same characteristics in rice cv. Saket 4 (Rizvi et al., 1989a). The researchers also showed that nicotine application to maize cv. Diara seeds increased solubilization of stored starch by increasing amylase activity at 7.5 mM level (Rizvi et al., 1989b). It affects several other plant processes, such as protoplast swelling (Tretyn et al., 1990), seed germination (Hartmann and Gupta 1989), secondary root formation (Peters et al., 1974b), ACC-mediated ethylene evolution (Jones and Stutte, 1988), and hypocotyl elongation (Hoshino, 1983). nAChR are also reported to be present in respiratory epithelial cells, endothelial cells, placenta, blood cells, keratinocytes, most immune cells (Skok, 2007; Kawashima et al., 2012), and cell organelle mitochondria (Gergalova et al., 2014). In these cells they are mediating different non-neuronal functions. Interestingly, several chemicals that interact with the animal AChRs are present in plants, particularly members of Solanaceae, the tomato family. These chemicals include nAChR agonist-nicotine (Hoffman and Taylor, 2001). The effect of nicotine on root formation in tobacco has been shown earlier (Tabata et al., 1971; Miller et al., 1983; Peters et al., 1974a, b). It is the plant where it is found in higher concentration. Differentiation of shoot and root primordia in callus cultures of tobacco were closely related with the nicotine levels, and it was reasoned that the higher differentiation in the high alkaloid genotypes may be because of higher nicotine synthesis (Miller et al., 1983). Nicotine also induce roots in *Phaseolus vulgaris* cultures (Peters et al., 1974b). Increase in internal nicotine concentrations by ethylene treatment also increases the adventitious rooting in the cultures (Wang et al., 1994). There was an indirect but positive correlation between the nicotine levels and adventitious root formation in callus tissues of *N. tabacum*. Its most interesting fact is that morphogenic response to nicotine has been shown by plants belonging to the Solanaceae and the Fabaceae where high concentration of acetylcholine-hydrolysing enzyme cholinesterase (ChE) has been found (Gupta and Gupta 1997). It would be worthwhile to study the effect of nicotine in plants with low ChE levels. The amount of nicotine in cultures should also be measured, as it is unstable and metabolized in biological systems. The possible involvement of nicotinic binding sites or nAChR in the morphogenic response elicited by ACh on explants of tomato was studied (Bamel et al., 2015). For this, nicotine as an agonist and d-tubocurarine as an antagonist of the nicotinic binding sites or nAChR that compete with ACh for the receptor binding sites were fortified to the basal medium and their effect on rhizogenesis was studied. It was observed that exogenous supplementation of nicotine effectively stimulates of rooting in *L. esculentum* leaf explants (Figure 7.11). At $10^{-7}$ M level, 64% explants formed roots (optimum response) in comparison to the control where only 20.5% explants developed roots (Figure 7.12) within 30 days of culture. Nicotine also induced secondary roots on all concentrations, whereas no secondary roots were formed on the control (Figure 7.12). The increase in secondary roots per root on Nicotine supplemented media was dramatic and as per expectation d-tubocurarine, an antagonist of nicotinic acetylcholine receptor (nAChR) in animals, reduced the rooting potential. The roots were not branched. The effective levels of d-tubocurarine level were $10^{-8}$ to $10^{-4}$ M where

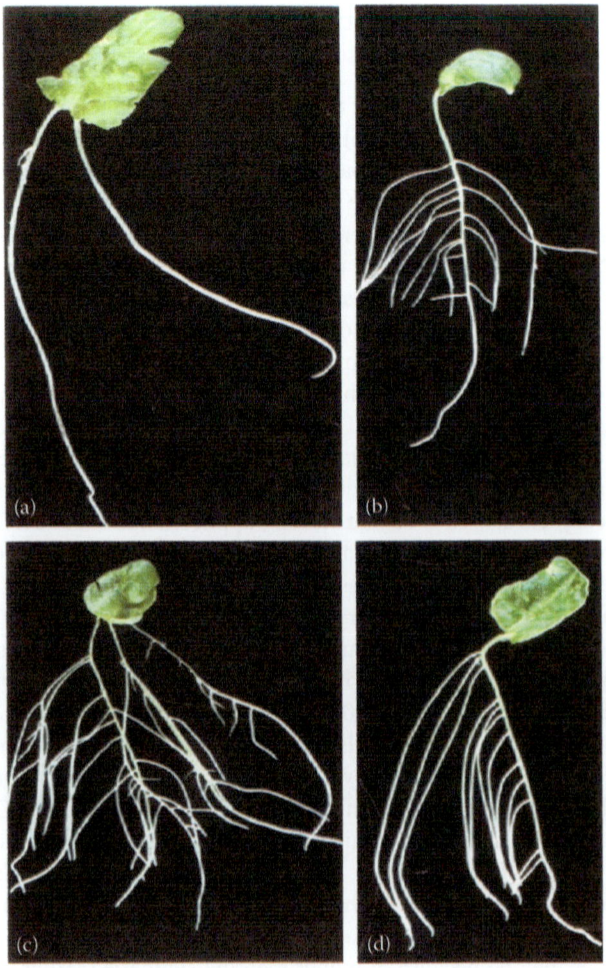


**FIGURE 7.11** (a–d) Morphogenic response of leaf explants of *Lycopersicon esculentum* var. Pusa Ruby excised from 30-day-old seedling cultured on MS medium supplemented with nicotine after 30 days of culture. (a) Leaf explants on MS basal. (b–d) Leaf explants on MS medium containing different levels of nicotine, (b) $10^{-8}$ M, (c) $10^{-7}$ M, (d) $10^{-3}$ M. (Adapted from Bamel, K. et al., *Int. Immunopharmacol.*, 29, 231–234, 2015. With permission.)

decline in rhizogenic capacity was observed (Figure 7.13). The response was lowest at $10^{-4}$ M, which supported rooting in 52% cultures. Though tubocurarine reduced root formation, it did not largely affect the secondary root formation. There is no reason to believe that the nAChR or the nicotine binding sites in leaf explants of tomato may be identical (chemically and structurally) to the ones found in animal nerves. It is interesting to note that while nicotine mimicked the action of ACh completely in tomato (Figure 7.12), d-tubocurarine reduced the percentage of rooting explants (Figure 7.13a), but the average number of roots formed per responding explant were same as in control (Figure 7.13b) in tomato leaves. There are two possibilities for the lack of a clear-cut response to d-tubocurarine: different structures of nicotinic binding sites or nAChR of tomato and animals or lack of uptake of d-tubocurarine molecules by leaf explants in tissue culture. Therefore, to conclude several depolarizing and receptor blocking agents must be tried before a conclusion can be drawn about the physiological nature of nicotinic binding sites or AChRs in tomato vis-a-vis the animals (Bamel et al., 2015). Earlier, it was also shown that nicotine inhibits secondary wall formation, cell growth, and differentiation in the unicellular green alga *Micrasterias denticulata* (Schiechl et al., 2008). Since ACh is ubiquitously present in all plants (Kawashima et al., 2007) and affects several physiological functions,

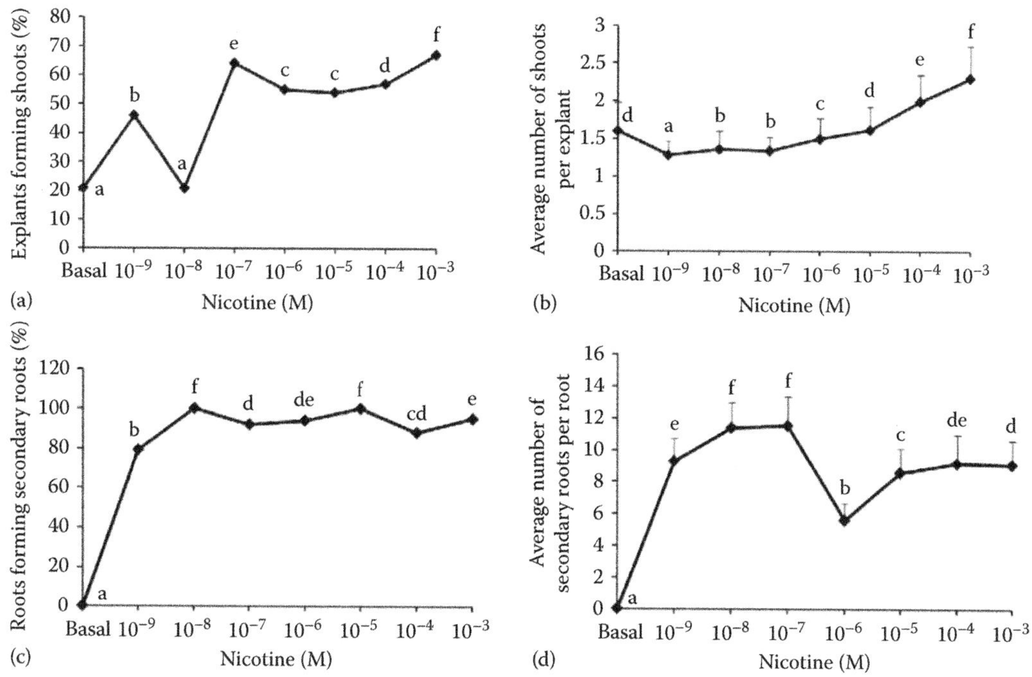


**FIGURE 7.12** (a–d) Effect of nicotine on morphogenic responses of leaf explants excised from *in vitro*-raised 30-day-old seedlings of *Lycopersicon esculentum* var. Pusa Ruby, after 30 days of culture. (a) Percentage of leaf explants forming roots. (b) Percentage of roots forming secondary roots. (c) Average number of roots formed per responding explant. (d) Average number of secondary roots formed per root. Values followed by the same letters are not significantly different ($P \leq 0.05$) by Tukey post hoc test using Statistical Package for Social Sciences (SPSS) software. (Adapted from Bamel, K. et al., *Int. Immunopharmacol.*, 29, 231–234, 2015. With permission.)

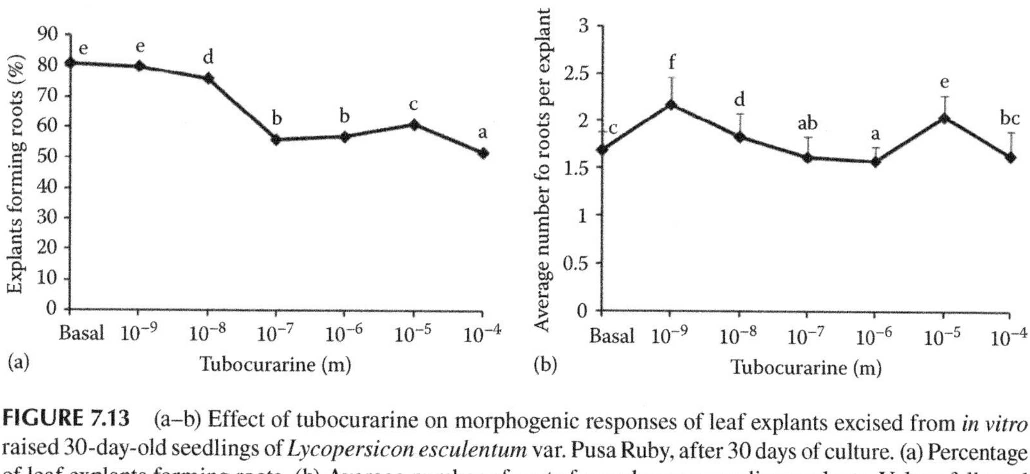


**FIGURE 7.13** (a–b) Effect of tubocurarine on morphogenic responses of leaf explants excised from *in vitro* raised 30-day-old seedlings of *Lycopersicon esculentum* var. Pusa Ruby, after 30 days of culture. (a) Percentage of leaf explants forming roots. (b) Average number of roots formed per responding explants. Values followed by the same letters are not significantly different ($P \leq 0.05$) by Tukey post hoc test using Statistical Package for Social Sciences (SPSS) Software. (Adapted from Bamel, K. et al., *Int. Immunopharmacol.*, 29, 231–234, 2015. With permission.)

the ability of nicotine to simulate the action of ACh in tomato leaves suggests that (1) nicotinic binding sites or nAChR is also present in plants and (2) morphogenic action of ACh (Bamel et al., 2007) and nicotine (Tabata et al., 1971; Miller et al., 1983; Peters et al., 1974 a, b; Bamel et al., 2015) may be through the same molecular mechanism as operative in the animal nerves.

#### 7.4.3.1 Choline Acetyl Transferase Inhibitor Juglone Antagonizes the Effect of Acetylcholine

Juglone, the inhibitor of ACh-synthesizing enzyme ChAT, also supports that ACh regulates rhizogenesis, caulogenesis, and callogenesis, as its supplementation to the controls induces a response that is opposite that of ACh (Bamel et al., unpublished). This is as per expectations, because if the ACh system is operating in the tomato leaf responses, then addition of Juglone will bring down the endogenous ACh levels and the hence the opposite effects.

Bamel and coworkers (2007, 2015, 2016) have shown that acetylcholine may be a natural plant growth regulator. It is stated that it induces rooting and suppresses shoot and callus formation of tomato leaf explants *in vitro* because (a) ACh, but not its breakdown products choline and acetate, can affect the response; (b) application of neostigmine, an inhibitor of enzyme acetylcholinesterase, which would elevate the endogenous ACh level, mimics the ACh effect; (c) neostigmine, when used in combination with ACh, potentiates the ACh effect (Bamel et al., 2007); and (d) nicotine and tubocurarine, the agonist and the antagonist for the nAChR in animal nerves, consolidates the action of ACh (Bamel et al., 2015). Thus, on the basis of the experimental evidences, it is proposed that ACh is a natural plant hormone, which is a regulator of differentiation in tomato leaf explants. However, extensive investigations are necessary to unravel the molecular mechanism of action of acetylcholine. In view of the recent surge of interest on the non-neuronal role of ACh in animals as well as plants, particularly on its role in differentiation and morphogenesis, we may be on the threshold of a new common role for ACh in plants and animals.

## 7.5 SUMMARY AND CONCLUSION

Acetylcholine not only performs the special function of neurotransmission in animals as ascribed to them since ages, but it is also involved in several other non-neuronal roles both in plants and animals. The involvement of acetylcholine and other components of acetylcholine system-like inhibitors of ChE and ChAT also consolidate this fact. The presence of receptor of ACh in plants is also evidenced by the simulation of ACh functions by the agonists and opposite effects of the antagonists. The most interesting findings of the recent years is the involvement of ACh in morphogenesis and differentiation in plants. Acetylcholine suppresses callusing and promotes rooting and inhibits shoot formation in leaf explants of tomato. ACh also induces/promotes rooting in stem cuttings *in vitro*. Breakdown products of ACh do not affect rooting. Inhibitors of ACh hydrolysis, neostigmine, and physostigmine also mimic ACh action. Treatment of explants with neostigmine and physostigmine was found to inhibit acetylcholinesterase activity in tomato leaves. Therefore, it is assumed that effects of neostigmine and physostigmine result from elevation of endogenous level of ACh. Juglone, an inhibitor of ACh synthesis, causes effects that are opposite of the effects of exogenous application of ACh (unpublished data). Nicotine, an agonist of the n-AChR receptor, mimics the action of the ACh effect on promotion of root formation as well as shoot inhibition. D-tubocurarine, an antagonist of n-AChR receptor, did not have any clear-cut antagonist action of ACh either on rhizogenesis or on caulogenesis. Further work is required to conclude anything on this aspect. The effect of inhibitors of ACh metabolism strongly supports the role of ACh as a natural regulator of morphogenesis. Enormous scientific data are already available on factors affecting differentiation in plant tissue culture, which has led to the current scientific consensus that (1) auxins are chiefly responsible for rooting; (2) rooting is affected by high auxin/cytokinin ratio, while shoot formation requires low auxin/cytokinin ratio; and (3) high IAA/GA ratio promotes rooting. If ACh is a natural regulator of differentiation in tomato leaf cultures, the relationship of ACh-controlled effects with the classical effects of auxins, cytokinins, gibberellins, and other growth regulators needs to be investigated.

## REFERENCES

Baluska F, Barlow PW, Mancuso S, Volkmann D. 2006. Preface. In: F. Baluska, S. Mancuso, D. Volkmann (Eds.). *Communications in Plants- Neuronal Aspects of Plant Life*. Berlin, Germany: Springer.

Bamel K, Gupta R, Gupta SC. 2016. Acetylcholine suppresses shoot formation and callusing in leaf explants of in vitro raised seedlings of tomato, *Lycopersicon esculentum* Miller var. Pusa Ruby. *Plant Signal. Behav.*, 11(6): e1187355.

Bamel K, Gupta R, Gupta SC. 2015. Nicotine promotes rooting in leaf explants of in vitro raised seedlings of tomato, *Lycopersicon esculentum* var Pusa Ruby. *Int Immunopharmacol.*, 29:231–234; PMID:26363975; doi:10.1016/j.intimp.2015.09.001.

Bamel K, Gupta SC, Gupta R. 2007. Acetylcholine causes rooting in leaf explants of in vitro raised tomato (*Lycopersicon esculentum* Miller var Pusa Ruby) seedlings. *Life Sci.*, 80:2393–2396; PMID:17328922; doi:10.1016/j.lfs.2007.01.039.

Barthalay Y, Hipeau-Jacquotte R, de la Escalera S, Jimenez F, Piovant M. 1990. Drosophila neurotactin mediates heterophilic cell adhesion. *EMBO J.*, 9:3603–3609.

Codignola A, Tarroni P, Cattaneo MG, Vicentini, LM, Clementi F, Sher E. 1994. Serotonin release and cell proliferation are under the control of bungarotoxin sensitive nicotinic receptors in small-cell lung carcinoma cell lines. *FEBS Lett.*, 342:286–290.

Crozier A, Kamiya Y, Bishop G, Yokota T. 2000. Biosynthesis of hormones and elicitor molecules. In: B. B. Buchanan, W. Gruissem, and R. L. Jones (Eds). *Biochemistry and Molecular Biology of Plants*. American Society of Plant Physiologists, Rockville, MD, pp. 850–928.

Davies PJ. 1995. The plant hormones: Their nature, occurrence and functions. In: P. J. Davies (Ed.). *Plant Hormones: Physiology, Biochemistry and Molecular Biology*. Kluwer Academic Publishers, Dordrecht, the Netherlands, pp. 1–12.

De la Escalera S, Bockamp EO, Moya F, Piovant M, Jiménez F. 1990. Characterization and gene cloning of neurotactin, a Drosophila transmembrane protein related to cholinesterases. *EMBO J.*, 9:3593–3601.

Dekhuijzen HM. 1973. The effect of acetylcholine on the growth inhibition by CCC in wheat seedlings. *Planta*, 111:149–156.

Drews U. 1975. Cholinesterase in embryonic development. *Prog. Histochem. Cytochem.*, 7:1–52. doi:10.4161/psb.28645.

Evans ML. 1972. Promotion of cell elongation in Avena coleoptiles by acetylcholine. *Plant Physiol.*, 50:414–416.

Fluck RA, Jaffe MJ. 1974a. Cholinesterases from plant tissues. III. Distribution and subcellular localization in Phaseolus aureus Roxb. *Plant Physiol.*, 53:752–758.

Fluck RA, Jaffe MJ. 1974b. The acetylcholine system in plants. *Curr. Adv. Plant Sci.*, 5:1–22.

Gergalova G, Lykhmus O, Komisarenko S, Skok M. 2014. α7 nicotinic acetylcholine receptor control cytochrome c release from isolated mitochondria through kinase mediated pathways. *Int. J. Biochem. Cell Biol.*, 49:26–31.

Grando SA, Horton RM, Pereira EFR, Diethelm-Okito BM, George PM, Albuquerque EX, Conti-Fine BM. 1995. A nicotinic acetylcholine receptor regulating cell adhesion and motility is expressed in human keratinocytes. *J. Invest. Dermatol.*, 105:774–781.

Grando SA, Hortron RM, Mauro TM, Kist DA, Lee TX, Dahl MV. 1996. Activation of keratinocyte nicotinic cholinergic receptors stimulates calcium influx and enhances cell differentiation. *J. Invest. Dermatol.*, 107:412–418.

Grisaru D, Sternfeld M, Eldor A, Glick D, Soreq H. 1999. Structural roles of acetylcholinesterase variants in biology and pathology. *Eur. J. Biochem.*, 264:672–686; PMID:10491113; doi:10.1046/j.1432-1327.1999.00693.x.

Gunay AL, Rao PS. 1980. In vitro propagation of hybrid tomato plants (Lycopersicon esculentum Mill.) using hypocotyls and cotyledon explants. *Ann. Bot.*, 45:205–207.

Gupta A and Gupta R. 1997. A survey of plants for presence of cholinesterase activity. *Phytochemistry*, 46: 827–31. doi:10.1016/S0031-9422(97)00393-2.

Gupta R. 1983. Studies on cholinesterase in Bengal gram and probable role of acetylcholine in plants. Ph.D. thesis, University of Delhi, New Delhi, India.

Gustafson T, Toneby M. 1970. On the role of serotonin and acetylcholine in sea urchin morphogenesis. *Exp. Cell Res.*, 62:452–462.

Hanneman E, Trevarrow B, Metcalfe WK, Kimmel CB, Westerfield M. 1988. Segmental pattern of development of the hindbrain and spinal cord of the zebrafish embryo. *Development*, 103:49–58.

Hartmann E and Gupta R. 1989. Acetylcholine as a signaling system in plants. In: W. Boss and D. J. Morre (Eds). *Second Messengers in Plant Growth and Development*. Alan R, Liss, New York, 57–88.

Hartmann E. 1978. Uptake of acetylcholine by bean hypocotyl hooks. *Z. Pflanzenphysiol.*, 86:303–311.
Hoffman BB, Taylor P. 2001. Neurotransmission: The autonomic and somatic motor\nervous systems, In: J.G. Hardman, L.E. Limbird (Eds.), *Goodman and Gilman's The Pharmacological Basis of Therapeutics*, McGraw Hill, Medical Publishing Division, New York, pp. 115–153.
Hoitink AWJH, Van Dijk G. 1966. The influence of neurohumoral transmitter substances on protoplasmic streaming in the myxomycetes *Physarella oblonga. J. Cell Physiol.*, 67:133–140.
Hoshino T. 1983. Effects of acetylcholine on the growth of Vigna seedling. *Plant Cell Physiol.*, 24:551–556.
Jaffe MJ. 1972. Acetylcholine as a native metabolic regulator of phytochrome-mediated processes in bean roots. In: V. C. Runeckles and T. S. Tso (Eds.). *Recent Advances in Phytochemistry*, Vol. 5, Structural and Functional Aspects of Phytochemistry. Academic Press, New York, 81–104.
Jaffe MJ. 1970. Evidence for the regulation of phytochrome-mediated processes in bean roots by the neurohumor, acetylcholine. *Plant Physiol.*, 46:768–777.
Jones RA, and Stutte CA. 1986. Acetylcholine and red light influence on ethylene evolution from soybean leaf tissues. *Ann. Bot.*, 57:897–900.
Jones RA and Stutte CA. 1988. Inhibition of ACC-mediated ethylene evolution from soybean leaf tissue by acetylcholine. *Phyton* 48:107–113.
Jones RA. 1986. Modification of membrane associated processes in vegetative soybeans (Glycine max L. Merr.) by acetylcholine. Dissertation Abstracts International, B. Sciences and Engineering. 46:3297B.
Jungalwala FB. 1994. Expression and biological functions of sulfoglucuronyl glycolipids (SGGLs) in the nervous system-a review. *Neurochem. Res.*, 19:945–957.
Kartha KK, Gamborg OL, Skyluk LP, Constabel F. 1976. Morphogenic investigations on in vitro leaf cultures of tomato (Lycopersicon esculentum Mill. Starfire) and high-frequency plant regeneration. *Z. Pflanzenphysiol.*, 77:292–301.
Kasemir H and Mohr H. 1972. Involvement of acetylcholine in phytochrome-mediated processes. *Plant Physiol.*, 49:453–454.
Kasturi R and Vasantharajan VN. 1976. Properties of acetylcholinesterases from Pisum sativum. *Phytochemistry*, 15:1345–1347.
Kavita P. 2002. Preliminary characterization of a cholinesterase from leaves of tomato-Lycopersicon esculentum Miller var. Pusa Gaurav. M. Phil. Thesis, University of Delhi, Delhi, India.
Kawashima K, Fujii T, Moriwaki Y, Misawa H. 2012. Critical roles of acetylcholine and the muscarinic and nicotinic acetylcholine receptors in the regulation of immune function. *Life sci.*, 91(21–22):1027–1032.
Kawashima K, Misawa H, Moriwaki Y, Fujii YX, Fujii T, Horiuchi Y, Yamada T, Imanaka T, Kamekura M. 2007. Ubiquitous expression of acetylcholine and its biological functions in life forms without nervous systems. *Plant Sci.*, 80:2206–2209.
Koštir J, Klenha J, Jiraček V. 1965. The effect of choline and acetylcholine on seed germination in agricultural plants. *Rost Vyroba* 12:1239–1279.
Lapied B, Le Corrone H, Hue, B. 1990. Sensitive nicotinic and mixed nicotinic-muscarinic receptors in insect neurosecretory cells. *Brain Res.*, 533:132–136.
Lauder JM and Schambra UB. 1999. Morphogenetic roles of acetylcholine. *Env. Health Persp.*, 107:65–69; doi:10.1289/ehp.99107s165.
Lauder JM. 1993. Neurotransmitter as growth regulatory signals: Role of receptors and second messengers. *Trends Neurosci.*, 16:233–240.
Lawson VR, Brady RM, Campbell A, Knox BG, Knox GD, Walls RL. 1978. Interaction of acetylcholine chloride with IAA, GA3, and red light in the growth of excised apical coleoptile segments. *Bull. Torr. Bot. Club.*, 105:187–191; doi:10.2307/2484113.
Layer PG, Weikert T, Alber R. 1993. Cholinesterases regulate neurite growth of chick nerve cells in vitro by means of a non-enzymatic mechanism. *Cell Tissue Res.*, 273:219–226.
Layer PG. 1990. Cholinesterases preceding major tracts in vertebrate neurogenesis. *Bioessays*, 12:415–420.
Layer PG. 1991. Cholinesterases during development of avian nervous system. *Molec. Cell. Neurobiol.*, 11: 7–33.
Lipton AS and Kater SB. 1989. Neurotransmitter regulation of neuronal outgrowth, plasticity and survival. *Trends Neurosci.*, 12:265–270.
Lukasiewicz RH, Tretyn A, Cymerski M, Kopcewicz J. 1997. The effects of exogenous acetylcholine and other cholinergic agents on photoperiodic flower induction of Pharbitis nil. *Acta Soc. Bot. Pol.* 66:47–54.
Miller RD, Collins GB, Davis DL. 1983. Effects of nicotine precursors on nicotine content in callus cultures of burley tobacco alkaloid lines. *Crop Sci.*, 23:561–565.
Molyneux DE, McKinlay RG. 1989. Observations on germination and acetylcholinesterase activity in wheat seeds. *Ann. Bot.*, 63:81–86.

Moody SA, Stein DB.1988. The development of acetylcholinesterase activity in the embryonic nervous system of the frog, Xenopus laevis. *Dev. Brain Res.*, 39:225–232.

Mukherjee I. 1980. The effect of acetylcholine on hypocotyl elongation in soybean. *Plant Cell Physiol.*, 21:1657–1660; PMID:25385982; doi:10.1093/pcp/21.8.1657.

Murashige T and Skoog F. 1962. A revised medium for rapid growth and bioassays with tobacco tissue cultures. *Physiol. Plant*, 15:473–497; doi:10.1111/j.1399-3054.1962.tb08052.x.

Murch SJ. 2006. Neurotransmitters, neuroregulators and neurotoxins in plants. In: F. Baluska, S. Mancuso, D. Volkmann (Eds.). *Communications in Plants- Neuronal Aspects of Plant Life*. Springer, Berlin, Germany, pp. 137–151.

Narayanaswamy S. 1977. Regeneration of plants from tissue cultures. In: J. Reinert and Y. P. S. Bajaj (Eds.). *Applied and Fundamental Aspects of Plant Cell Tissue and Organ Culture*. Springer-Verlag, Berlin, Germany, pp. 176–206.

Olson PF, Fessler LI, Nelson RE, Sterne RE, Campbell AG, Fessler JH. 1990. Glutactin, a novel Drosophila basement membrane-related glycoprotein with sequence similarity to serine esterases. *EMBO J.*, 9: 1219–1227.

Peters JE, Wu PHL, Sharp WR, Paddock EF. 1974a. Rooting and the metabolism of nicotine in tobacco callus cultures. *Physiol. Plant.*, 31:97–100.

Peters JE, Crocoma OJ, Sharp WR. 1974b. Exogenous alkaloid control of growth and root morphogenesis in Phaseolus vulgaris tissue cultures. *Am. J. Bot.* 61:10.

Puelles L, Amat JA, Martinez-de-la-Torre M. 1987. Segment-related mosaic neurogenetic pattern in the forebrain and mesencephalon of early chicken embryos. 1. Topography of AChE-positive neuroblasts up to stage HH 18. *J. Comp. Neurol.*, 266:247–268.

Raineri M, Modenesi P. 1986. Preliminary evidence for a cholinergic-like system in lichen morphogenesis. *Histochem. J.*, 18:647–657.

Razdan MK. 1993. *An Introduction to Plant Tissue Culture*. Oxford & IBH, Publishing, New Delhi, India, p. 398.

Riov J and Jaffe, MJ. 1973. A cholinesterase from bean roots and its inhibition by plant growth retardants. *Experientia* 29:264–265.

Rizvi SJH, Mishra GP, Rizvi V. 1989a. Allelopathic effects of nicotine on maize. I. Its possible importance in crop rotation. *Plant Soil.* 116:289–291.

Rizvi SJH, Mishra GP, Rizvi V. 1989b. Allelopathic effects of nicotine on maize. II. Some aspects of its mechanism of action. *Plant Soil.* 116:292–293.

Roshchina VV and Vikhlyantsev IM. 2009. Mechanism of chemo-signaling in allelopathy: Role of ion channels and cytoskeleton in development of plant microspores. *Allelopathy J.*, 23:25–36.

Roshchina VV. 2001. *Neurotransmitters in Plant Life*. Science Publishers, Enfield, NH, p. 283.

Salisbury FB and Ross CW. 1992. *Plant Physiology Hormones and Growth Regulators: Cytokinins, Ethylene, Abscisic Acid, and Other Compounds*. Wadsworth, Belmont, MA, pp. 382–407.

Sansebastiano GD, Fornaciari S, Barozzi F, Piro G, Arru L. 2014. New insights on plant cell elongation: a role for acetylcholine. *Int. J. Mol. Sci.*, 15:4565–4582; PMID:24642879; doi:10.3390/ijms15034565.

Sastry BVR and Sadavongvivad C. 1979. Cholinergic systems in non-nervous tissues. *Pharmacol. Rev.* 30:65–131.

Schiechl G, Himmelsbach G, Buchburger W, Kerschbaum HH, Lutz-Meindl U. 2008. Identification of acetylcholine and impact of cholinomimetic drugs on cell differentiation and growth in the unicellular green alga Micrasterias denticulata. *Plant Sci.* 175:262–266.

Selleś M, Viladomat F, Bastida J, Codina CI. 1999. Callus induction, somatic embryogenesis and organogenesis in Narcissus confusus: Correlation between the state of differentiation and the content of galanthamine and related alkaloids. *Plant Cell Rep.* 18:646–651.

Shirvan MH, Pollard HB, Heldman, E. 1991. Mixed nicotinic and muscarinic features of cholinergic receptor coupled to secretin in bovine chromaffin cells. *Proceedings of the National Academy of Sciences (USA)*, 88:4860–4864.

Silver A. 1974. The biology of cholinesterases. *North-Holland Research Monographs, Frontiers of Biology*, Vol. 36, North-Holland Publishing, Amsterdam, the Netherlands, p. 596.

Sink KC, Reynolds JF. 1986. Tomato (Lycopersicon esculentum L.). In: Y. P. S. Bajaj (Ed.). *Biotechnology in Agriculture and Forestry*, Vol. 2. Springer-Verlag, Berlin, Germany, pp. 319–344.

Skok MV. 2007. Non-neuronal nicotinic acetylcholine receptors: Cholinergic regulation of the immune processes. *Neurophysiology* 39(4–5):264–271.

Skoog F, Miller CO. 1957. Chemical regulation of growth and organ formation in plant tissue cultured in vitro. *Symp. Soc. Exp. Biol.*, 11:118–131.

Small DH, Michaelson S, Sberna G. 1996. Non-classical actions of cholinesterases: Role in cellular differentiation, tumorigenesis and Alzheimer's disease. *Neurochem. Int.*, 28:453–483; PMID:8792327; doi:10.1016/0197-0186(95)00099-2.

Small DH, Reed G, Whitefield B, Nurcombe V. 1995. Cholinergic regulation of neurite outgrowth from isolated chick sympathetic neurons in culture. *J. Neurosci.*, 15:144–151.
Soreq H, Gnatt A, Loewenstein Y, Neville LF. 1992. Excavations into the active-site gorge cholinesterases. *Trends Biochem. Sci.*, 17:353–358.
Soreq H, and Zakut H. 1990. Amplification of butrylcholinesterase and acetylcholinesterase genes in normal and tumor tissues: Putative relationship to organophosphorus poisoning. *Pharm. Res.*, 7:1–7.
Stahlberg R. 2006. Historical overview on plant neurobiology. *Plant Signal. Behav.*, 1:6–8; PMID:19521469; doi:10.4161/psb.1.1.2278.
Street HE. 1977. *Plant Tissue and Cell Culture*, 2nd ed. Blackwell Scientific Publishing, Oxford, p. 614.
Sugiyama K and Tezuka T. 2011. Acetylcholine promotes the emergence of lateral roots of Raphanus sativus. *Plant Signal. Behav.*, 6:1545–1553; PMID:21900743; doi:10.4161/psb.6.10.16876.
Tabata M, Yamamoto H, Hiraoka N, Marumoto Y, Konoshima M. 1971. Regulation of nicotine production in tobacco tissue culture by plant growth regulators. *Phytochemistry* 10:723–729.
Taiz L, Zeiger E. 2002. *Plant Physiology*, 3rd ed. Sinauer Associates, Sunderland, MA, p. 690.
Taylor P. 2001. Agents acting at the neuromuscular junction and autonomic ganglia. In: J.G. Hardman, L.E. Limbird (Eds.). *Goodman and Gilman's The Pharmacological Basis of Therapeutic*s, 10th ed. McGraw-Hill Medical Publishing Division, New York, pp. 193–213.
Tezuka T, Akita I, Yoshino N, Suzuki Y. 2007. Regulation of self-incompatibility by acetylcholine and cAMP in Lilium longiflorum. *J. Plant Physiol.* 164:878–885; PMID:16882455; doi:10.1016/j.jplph.2006.05.013.
Tran Than KM. 1981. Control of morphogenesis in in vitro cultures. *Ann. Rev. Plant Physiol.*, 32:291–311.
Tretyn A, Bossen ME, Kendrick RE. 1990. The influence of acetylcholine on the swelling of wheat (Triticum aestivum L.) protoplasts. *J. Plant Physiol.*, 136:24–29.
Tretyn A and Kendrick RE. 1991. Acetylcholine in plants: Presence, metabolism and mechanism of action. *Bot. Rev.*, 57:33–73; doi:10.1007/BF02858764.
Tretyn A and Kendrick RE. 1990. Induction of unrolling by phytochrome and acetylcholine in etiolated wheat seedlings. *Photochem. Photobiol.*, 52:123–129.
Tretyn A. 1987. Influence of red light and acetylcholine on $^{43}Ca^{2+}$ uptake by oat coleoptile cells. *Cell Biol. Int. Rep.*, 11:887–896.
Turi CE, Axwik KE, Smith A, Jones AMP, Saxena PK, Murch SJ. 2014. Galanthamine, an anticholinesterase drug, effects plant growth and development in Artimisia tridentata Nutt. via modulation of auxin and neurotransmitter signaling. *Plant Signal. Behav.*, 9: e28654.
Tyagi, AK. 1982. Induction of embryo from pollen grains and selection of variants from haploid cells of Datura innoxia Mill. Ph.D. Thesis, University of Delhi, New Delhi, India.
Verbeke M and Vendrick JC. 1977. Are acetylcholine-like cotyledon-factors involved in the growth of the cucumber hypocotyl? *Z Pflanzenphysiol.*, 83:335–340; doi:10.1016/S0044-328X(77)80157-8.
Wang HL, Ho HH, Su JC. 1994. Effect of ethylene on adventitious root formation and nicotine content of tobacco callus tissues. *Bot. Bull. Acad. Sin.*, 35:217–222.
Wenigner B, Italiano LJP, Bastida J, Bergonon S, Codina C, Lobstein A, Anton R. 1995. Cytotoxic activity of Amaryllidaceae alkaloids. *Planta Med.*, 61:77–79.
Wessler I, Kirkpatrick CJ, Racke K. 1999. The cholinergic "pitfall": Acetylcholine, a universal cell molecule in biological systems, including humans. *Clin. Exp. Phramacol. Physiol.*, 26:198–205; doi:10.1046/j.1440-1681.1999.03016.x.
Wessler I and Kirkpatrick CJ. 2008. Acetylcholine beyond neurons: The non-neuronal cholinergic system in humans. *Br. J. Pharmacol.*, 154:1558–1571; PMID:18500366; doi:10.1038/bjp.2008.185.
Whittaker VP. 1963. Identification of acetylcholine and related esters of biological origin. In: G. B. Koelle (Ed.). *Handbook of Experimental Pharm*acology, Vol. 15. Springer-Verlag, Berlin, Germany, pp. 1–39.
Willbold E and Layer PG. 1994. Butyrylcholinesterase regulates laminar retinogenesis of the chick embryo in vitro. *Eur. J. Cell Biol.*, 64:192–199.
Williams CL and Lennon VA. 1991. Activation of muscarinic acetylcholine receptors inhibits cell cycle progression of small-cell lung carcinoma. *Cell Regul.*, 2:373–381.
Yamamoto K, Sakamoto H, Momonoki YS. 2011. Maize acetylcholinesterase is a positive regulator of heat tolerance in plants. *J. Plant Physiol.*, 168:1987–1992; PMID:21757255; doi:10.1016/j.jplph.2011.06.001.
Zakut H, Lapidot-Lifson Y, Beeri R, Ballin A, Soreq H. 1992. In vivo gene amplification in non-cancerous cells: Cholinesterase genes and oncogenes amplify in thrombocytopenia associated with Lupus erythematosus. *Mutat. Res.*, 276:275–284.

# Section III

## Cellular Location of Neurotransmitters

### Their Reception and Signaling

# 8 Tools for Microanalysis of the Neurotransmitter Location in Plant Cells

*Victoria V. Roshchina*

**CONTENTS**



## 8.1 INTRODUCTION

The location of neurotransmitters within living cells is an important problem for biologists. This is related, mainly, to the histochemical methods based on the use of special dyes or reagents making the compounds visible. The arsenal of the approaches is applied by the plant biologists, too, using both assays from the neurological field and new possibilities originating from the plant features. Today, we may consider histochemical determination of acetylcholine basing on the color reaction of the enzyme of its hydrolysis named cholinesterase and fluorescent reactions for neurotransmitters, mainly, biogenic amines.

## 8.2 COLOR REACTIONS FOR CHOLINESTERASE AS MARKER OF ACETYLCHOLINE IN THE CELL

Reactions for cholinesterase determination include the reaction of substrate hydrolysis and then staining with reagent, besides, variants with inhibitors of cholinesterase are done in order to show specificity of the esterase. First experiments on the cholinesterase location in plant cells were based on the ferricyanide/ferrocyanide reaction of Karnovsky–Roots (Karnovsky and Roots 1964) where product was red-brown Hettchet's pigment (copper ferrocyanide). By electron microscopy, the optical density of the product was analyzed (Fluck and Jaffe 1974; Maheshvary and Gupta 1982; Bednarska and Tretyn 1989; Gorska-Brylass et al. 1990; Bednarska 1992). This method was used for the determination cholinesterase activity in pollen and pistil (Bednarska and Tretyn 1989; Bednarska 1992). Cholinesterase/acetylcholinesterase, an enzyme-hydrolyzed acetylcholine, may be a sensor of acetylcholine on the cell surface, because it was found in the cell wall and plasmalemma as well as in intercellular space (Fluck and Jaffe 1974; Bednarska and Tretyn 1989; Bednarska 1992).

Color reactions for cholinesterases were originated from the experiments related to the animal nervous system. Recent experiments for similar determination of the cholinesterase activity with thiocholine as substrate were based on the ferricyanide/ferrocyanide reaction of Karnovsky-Roots

(Karnovsky and Roots 1964) where the product was red-brown Hettchet's pigment (copper ferrocyanide) and was well seen on isolated enzymes from electric eel (Budantsev and Roshchina 2005, 2007). First histochemical research in animals was also based on the color reactions with some azo compounds (Menten et al. 1944; Nachlas and Seligman 1949). Some authors applied azo dye Fast Red TR for determination of spot (band) of cholinesterase reacting with β-naphthyl acetate as substrate in gel-electrophoresis (Harris et al. 1962).

First attempts to apply color methods to the study of the cholinesterase location in plant cells were done on pollen and showed the colored reaction in plasmalemma and first drop from the microspore (Roshchina et al. 1994; Roshchina 1999, 2001, 2007). This method was also spread to animals—worm *Planaria tigrina* and isolated organelles from plant objects (Roshchina 2014; Roshchina and Svirst 2016). Recently basing on the method with application Fast Red TR dye (Kakariari et al. 2000), histochemical staining was also used for plant cells, for example of the trichome of *Hibiscus rosa-sinensis* (Roshchina and Svirst 2016). Recently azo dyes (Figure 8.1) were specially used for determination of cholinesterase in plant cells and isolated organelles (Roshchina and Svirst 2016; Roshchina 2017, 2018).

Main mechanisms of the dyes staining are seen on Figure 8.1. Fast Red TR salt and β-naphthyl acetate as substrate (Roshchina and Svirst 2016) transfer of the colorless object to red color. Color reaction red analogue of Ellman reagent named 2,2-dithio-bis-(p-phenyleneazo)-bis-(1-oxy-8-chlorine-3,6)-disulfur acid in the form of sodium salt was also used (Roshchina et al. 1994). In this case, red reagent binding with thiocholine, forming as a result of the acetylthiocholine hydrolysis, forms blue product (Roshchina 2001, 2007). This method was also applied for extracts from plant objects, especially from pollen (Roshchina et al. 1994; Roshchina and Semenova 1995; Roshchina 2001, 2007). In 1999, the first drop of secretion from pollen of knight star *Hippeastrum hybridum* was stained (Roshchina 1999) using this dye, and the main color was seen in the plasmalemma of the cell (Roshchina 2007). The blue coloration is well-seen in the plasmatic membrane that is under the cell wall (Figure 8.2). An earlier location of the cholinesterase in plasmalemma was demonstrated with electron microscopy (Fluck and Jaffe 1974; Maheshwary and Gupta 1982; Gorska-Brylass et al. 1990; Bednarska 1992).

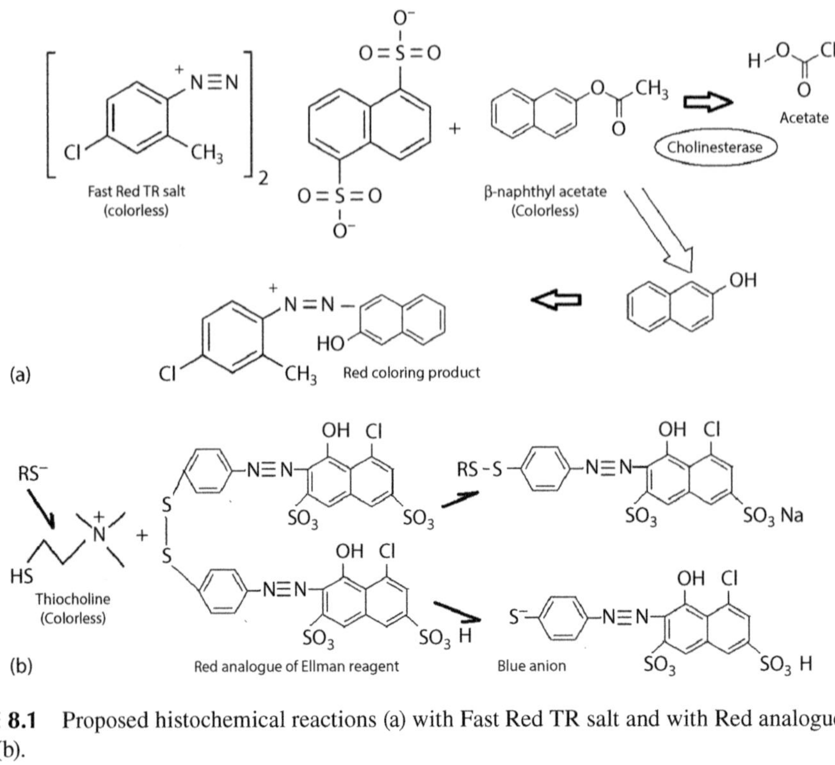


**FIGURE 8.1** Proposed histochemical reactions (a) with Fast Red TR salt and with Red analogue of Ellman reagent (b).

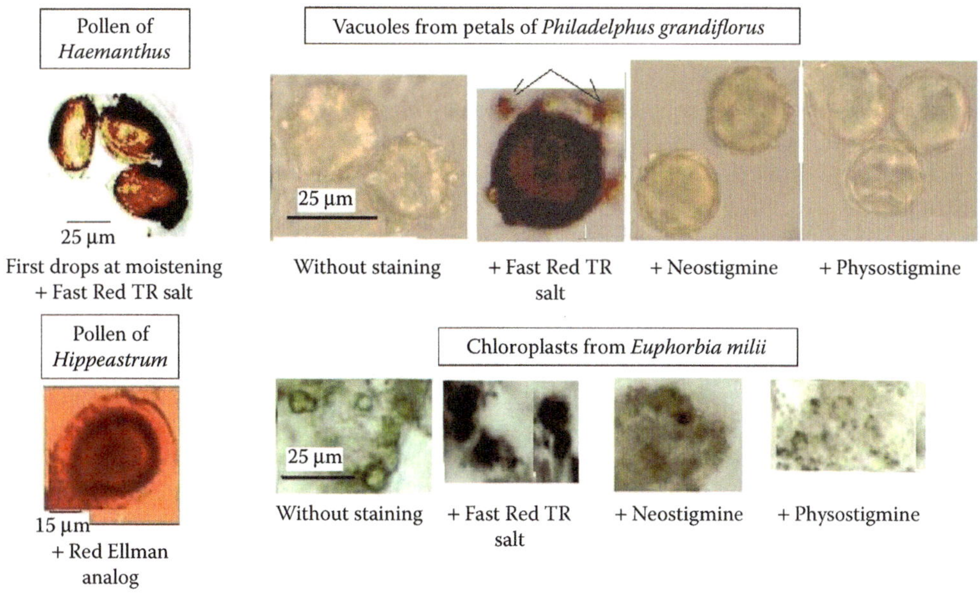


**FIGURE 8.2** Examples of the cell staining with azo dyes at the reactions for cholinesterase activity.

Dye Fast Red TR salt was also used for a model system—pollen from Katharine blood lily *Haemanthus katharinae* (Roshchina 2018). Figure 8.2 demonstrates the red coloration of first drops after the moistening of the pollen grains. The evidence of the cholinesterase activity confirms by the decrease in rates of the hydrolysis of acetylcholine or acethylthiocholine after addition of the inhibitors neostigmine or physostigmine. Today by both azo dyes Fast Red TR salt and Red analogue of Ellman reagent various plant objects were analyzed showing red and/or blue coloring after the incubation with substrates β-naphthyl acetate or acethylthiocholine (Table 8.1). To our regret, only the first dye is an industrial reagent.

**TABLE 8.1**
**The Samples Which Are Clear Coloring in Red or Blue after the Staining with Azo Dyes Fast Red TR (a) or Red Analogue of Ellman Reagent (b), Relatively After the Hydrolysis of the Substrate Used**

| Object | Colored Target | Object | Colored Target |
|---|---|---|---|
| Worm *Planaria tigrina* | Muscle[b] | Pollen of *Hippeastrum hybridum* | Pollen tube[b] |
| Bracts of *Humulus lupulus* | Veins[a,b] | Pollen of *Epiphyllium hybridum* | Pollen tube[b] |
| 2-day-old seedlings of *Raphanus sativus* | Growing tip[a,b] | Vacuole of petals from *Begonia acetosa* | Tonoplast[a] |
| Leaf secretory hairs of *Hibiscus rosa-sinensis* | Hairs[a,b] | Vacuole of *Euphorbia milii* | Tonoplast[b] |
| Trap of *Utricularia* sp. | Trap[b] | Nuclei of *Dolichothele albescens* flower petals | Outer membrane[b] |
| Secretory hairs of *Drosera capensis* | Head of secretory hair[b] | Chloroplasts of *Euphorbia milii* | Outer membrane[b] |

*Source:* Roshchina (1999, 2017); Roshchina and Svirst (2016); Roshchina, V.V., *Neurotransmitters in Plant Life*, Science Publishers, Enfield, NH, 283 p, 2001; *Model Systems to Study Excretory Function of Higher Plants*, Springer, Dordrecht, the Netherlands, 2014.

As we can see, main colored structures such as muscle or contracting organelles or trap of carnivorous plants have motility. At a neuro-muscle junction, acetylcholine triggers a contraction, and enzyme acetylcholinesterase breaks its down, which allows a neuron to return to the resting state after activation. Therefore, cholinesterases play an important role in controlling acetylcholine-induced contractions of muscle, for example of planarians. The finding of contractile elements, acetylcholine, and cholinesterase in any cell not only in animal, but in plants and microorganisms (Roshchina 2016), requires us to study a location of a cholinesterase activity in cells and their connection with motile systems. In swelling pollen after moistening, one could see that main color of blue product of the reaction on cholinesterase activity is concentrating in plasmalemma (Figure 8.2). The connection of motility with cholinesterase activity is discussed in animal physiology (Boettger and McClintock 2012).

The sensitivity of plant contractile elements to acetylcholine and the connection with cholinesterase activity has been demonstrated earlier as in phloem transport (Yang et al. 2007) and chloroplasts (Roshchina and Mukhin 1984; Gorska Brylass et al. 1990). Data about pollen cholinesterase activity were summed in monograph (Roshchina 2001).

Possible involvement of cholinesterase, the enzyme that hydrolyzed acetylcholine, in signal processes of plants has been studied by the histochemical analysis of its presence in secreting cells and individual organelles, and also experiments on the influence of exogenous cholinesterase from different plant species on the growth of reaction have been carried out (Roshchina 2017, 2018). For research on whole cells, plants with visible secretion, firefly *Jacobinia pauciflora* Lindau and Katharine blood lily *H. katherinae*, whose pistils and pollen actively secrete as well as hop *Humulus lupulus* L., having secretory glands in the female flowers bracts, were selected. In order to isolate and examine organelles of hothouse, models were used such as bracts and leaves of milkweed mile (*Euphorbia milii* Desmoulins), white petals of mock orange *Philadelphus grandiflorus* Wild, and *Kalanchoe blossfeldiana* var Simone L. By vital histochemical staining methods and biochemical analysis with reagent Fast Red TR, Ellman reagent, and an analogue of Ellman reagent dithio-bis-(p-phenyleneazo)-bis-(1-oxy-8-chloro-3,6) sodium disulfate in intact secretory cells, cholinesterase activity has been found that was concentrated mainly in the plasmalemma, free space of cell, and cell walls. By histochemical methods, the presence of this enzyme in the outer membranes of organelles isolated (nuclei, vacuoles, and chloroplasts) has been shown. Inhibitors of cholinesterase neostigmine, physostigmine, and berberine reduced or completely blocked the hydrolysis of acetylcholine cells or organelles.

## 8.3 FLUORESCENT REACTIONS FOR CELLULAR OBSERVATION OF LOCATION OF BIOGENIC AMINES AND THEIR POSSIBLE RECEPTORS

Fluorescence as a parameter to monitor and measure both inter and intracellular contacts with participation of neurotransmitters is considered appropriate for such analyses in plant cells (Roshchina et al. 2012, 2014, 2016; Roshchina and Yashin 2014; Roshchina 2016). The study of location and binding of the neurotransmitters in plant cells needs to have special methods and approaches. In the following we shall consider perspectives in the use of fluorescence to investigate some known neurotransmitters and binding of antineurotransmitter compounds in plant cells.

Approaches to study a neurotransmitter or its reception in cells using fluorescence may be visible emission based on emerging reaction products, fluorescent labels, or fluorescent model cellular systems known for the biogenic amines determination in animal tissues. Plant physiologists may observe several fluorescent reactions: (1) with neurotransmitter or its agonist, (2) with antagonist of neurotransmitter binding with its receptor.

### 8.3.1 Fluorescent Reactions with Neurotransmitters

#### 8.3.1.1 Acetylcholine

Today, the main ways to determine acetylcholine in plant cells by the fluorescent method is to analyze the presence of enzymes of the hydrolysis named as cholinesterase that is determined with the red analogue of Ellman reagent for the fluorescent product with thiocholine (Roshchina et al. 1994; Roshchina 2001) or the binding with its agonist muscarine (Roshchina 2008). More information was received when fluorescing antagonists of acetylcholine d-tubocurarine (acting on nicotinic type of animal cholinoreceptor) or atropine (binding with muscarine type of cholinoreceptor) were used (Roshchina 2004, 2008). Both drugs bound with plasmatic membranes, fluorescing in blue or blue green at range 415–460 nm as well berberine and chelerythrine, natural inhibitors of cholinesterase, emitted in yellow or green yellow (510–540 nm) (Roshchina 2007).

#### 8.3.1.2 Biogenic Amines

Main fluorescent reactions for plant cells may be characterized based on animal assays for biogenic amines with special histochemical reagents on biogenic amines, mainly aldehydes, that leads to the formation of fluorescent products. Thus, fluorescent indoleamines were formed from serotonin with the maximum of the emission at 520–527 nm in animal neuronal cells while catecholamines with aldehydes forming fluorescent products with the emission maximum at 460–475 nm were registered (Falck 1962). Mechanisms of the reactions for catecholamines with glyoxylic acid were considered by some authors (Björklund et al. 1972) as well as reaction for histamine with o-phthalic aldehyde (Cross et al. 1971). Kimura (1968) was the first who applied this achievement to study serotonin and catecholamines histochemically, and later Barwell (1979) determined histamine in the red alga of *Furcellaria* by similar approach. Later, subcellular localization of alkaloids and dopamine in different vacuolar compartments of *Papaver bracteatum* was studied using fluorescent method with glyoxylic acid and gas chromatography (Kutchan et al. 1986). Today, the three main fluorescent reactions for catecholamines, serotonin, and histamine are applied to plant cells (Figure 8.3). The reactions on catecholamines occur after the treatment with glyoxylic acid (an emission maximum of the product with catecholamines 470–475 nm) in plant cells (Roshchina et al. 2012, 2014, 2015b). Figure 8.3 demonstrates the fluorescent rings in blue on pollen of *Clivia miniata*. Recently, fluorescent reaction of dopamine with resorcinol was also represented for animal cells (Lin et al. 2015), but it is not used in plant cells yet. As for serotonin determination only formaldehyde giving blue or blue-green emission at excitation by light 360–380 nm or 390–425 nm as observed for pollen of *Dolichotele albescens*. Here under the last actinic light, one can see green-yellowish fluorescent nucleus and apertures from which the pollen tube enters during development. This shows the accumulation of serotonin in the sites of pollen. o-Phthalic aldehyde as a reagent for histamine induced blue or green (in high concentrations about $10^{-3}$ M) lightening of cells as a whole, like of pollens from various species or vegetative microspores from horsetail *Equisetum arvense* (Figure 8.3). If histamine is present, the fluorescence may be seen within the cell (in nuclei and chloroplasts of horsetail for example) or excretes out as a halo (Figure 8.3).

Due to the use of the fluorescent reactions, the presence of catecholamines, serotonin, and histamine was established in several vegetative tissues and pollen from 25 plant species (Roshchina 2014; Roshchina and Yashin 2014; Roshchina et al. 2014). Localization of catecholamines and histamine was seen within cells analyzed, in particular the neurotransmitters may be accumulated in organelles—fluorescence was marked in nuclei and chloroplasts of some objects (Roshchina et al. 2014).

This approach has also been applied to isolated organelles from various plants (Roshchina 2016; Roshchina et al. 2016). In the publications the external images of the isolated organelles (nuclei, vacuoles, chloroplasts) and their fluorescence spectra after staining with reagents to catecholamines

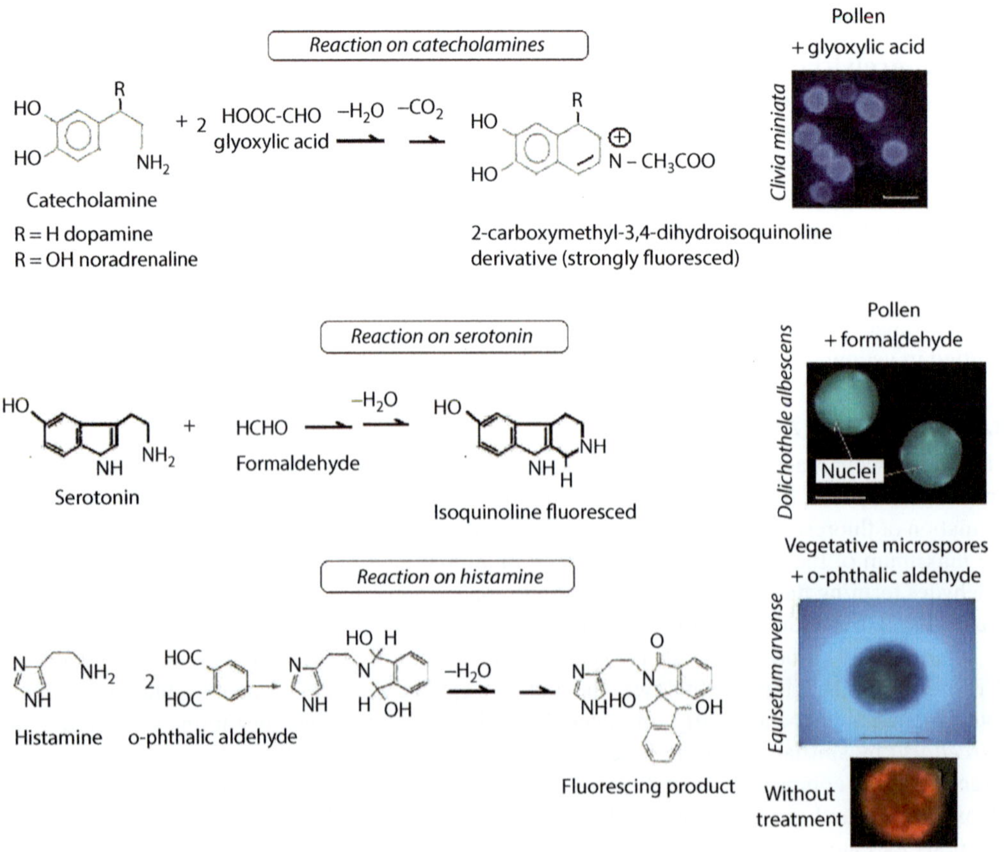


**FIGURE 8.3** Detection of the presence and localization of biogenic amines in intact cell based on the emission of the products of their reactions. Left—Possible mechanisms of fluorescent reaction. Right—views under the luminescence microscope (photos) after treatment of cells of *Clivia miniata* (bar = 50 μm), *Dolichothele albescens* (bar = 20 μm) and *Equisetum arvense* (bar = 20 μm) with glyoxylic acid, formaldehyde, and o-phthalic aldehyde, relatively.

and histamine as well as fluorescent antagonists of neurotransmitters were analyzed. The presence of biogenic amines was determined as increase of the emission from products forming in histochemical reactions with glyoxylic acid or ortho-phthalic aldehyde in blue (460–475 nm) or blue-green (510–520 nm) spectral region. Some examples of the effects of reagents for biogenic amines are on Figure 8.4. Nuclei, vacuoles, and chloroplasts of represented plants have marked fluorescent reactions on catecholamines and histamine, while reactions on serotonin were weak. As seen on Figure 8.4, all studied isolated organelles, in different degrees, gave a blue or blue-green emission with glyoxylic acid and o-phthalic aldehyde that was dependent on the amount of neurotransmitter. Moreover, we also need to remember about possibilities of biogenic amines to conjugate with flavonoids and other compounds (Roshchina 2001) that may influence the emission.

Control samples (without any treatment) had no autofluorescence, and only after exposure to the reagents for catecholamines or histamine the fluorescence of 460–475 nm appears, which indicates the presence of these neurotransmitters in the nuclei (possibly on the surface). Earlier similar effects were in the intact pollen and vegetative microspores of horsetail *E. arvense*, and similar histochemical staining led to the arising of nuclei blue fluorescence absent in untreated cells (Roshchina and Yashin 2014; Roshchina et al. 2015).

According to the aforementioned algorithm, experiments with isolated vacuoles were carried out using four plant species *Beta vulgaris* var. rubra, *Epiphyllum hybridum*, *E. milii*, *Saintpaulia*

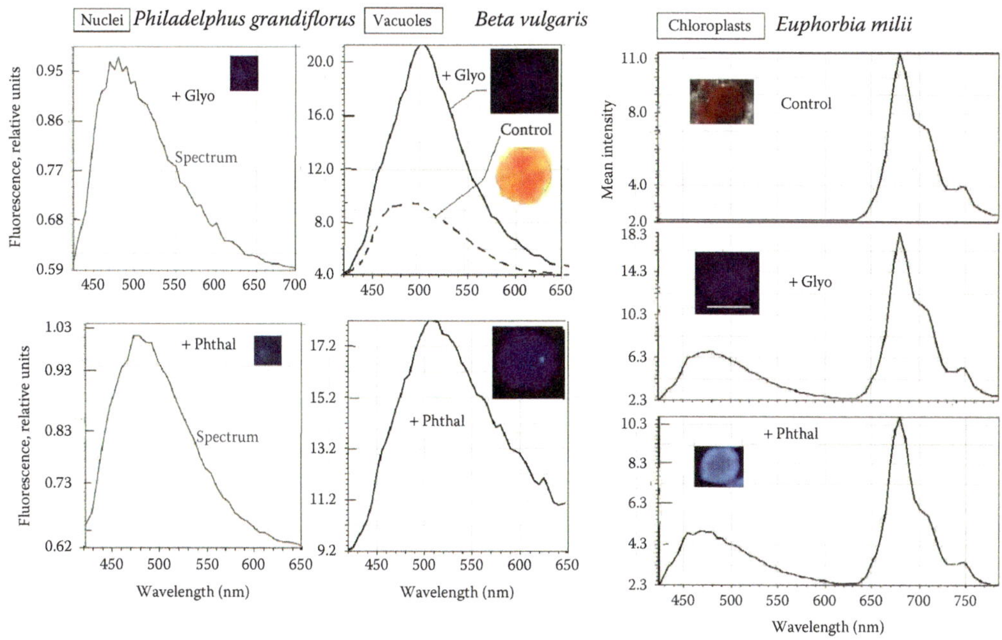


**FIGURE 8.4** Fluorescence and the fluorescence spectra observed and recorded by confocal microscope after the treatment of isolated organelles with reagents for catecholamines—glyoxylic acid (glyo) and reagent for histamine-o-phthalic aldehyde (phthal). Left—reactions with nuclei (diameter 4–6 μm), from cell of white petals of *Philadelphus grandiflorus* (control organelles without treatment have no any emission), Middle—reactions with vacuoles from roots of *Beta vulgaris* var. rubra (diameter 25 μm), control red-colored organelle having weak emission in blue without treatment/and the emission spectrum shown by broken line as control), Right—reactions with chloroplasts(diameter 8–10 μm), from *Euphorbia milii* (control organelles without treatment have no marked emission in blue-green region).

*ionantha* (Roshchina 2016; Roshchina et al. 2016). The staining for the detection of the presence of catecholamines with glyoxylic acid or the presence of histamine with o-phthalic aldehyde induced all the samples to fluoresce at 460–500 nm. Unlike the nuclei and vacuoles in the fluorescence spectra of control sample of organelle, there was a noticeable area of 675–680 nm, due to the presence of chlorophyll, and in many plastides no measurable emission in blue-green spectral region (Figure 8.4). Treatment with glyoxylic acid or ortho-phthalic aldehyde resulted in the maximum 460–470 nm appearance in the fluorescence spectrum indicating the presence of catecholamines and histamine in the chloroplasts. Their location may be inside or, rather, on the surface because there is the reduction in the height of the maximum of 675–680 nm (peculiar to thylakoids placed inside the plastids). Furthermore, the surface binding of histamine specifies the kind of optical slice of chloroplast—most emission is seen on the external membrane (Roshchina 2016; Roshchina et al. 2016). Thus, binding of biogenic amines may occur on the surface of organelles, and then the appropriate question about their reception is actual.

More expressive picture can be illustrated for vacuoles from *E. hybridum* and *E. milii*. Autofluorescence from all samples of vacuoles are quite weak in this area of the spectrum; therefore, the enhanced lightening after the treatment with glyoxylic acid or o-phthalic aldehyde belong to products of the reactions with the reagents. The intensity of the luminescence at 460–475 nm increased, as seen in the spectra of the samples stained, that related to the most significant (10 times) concentration of catecholamines than that of histamine present in the vacuoles. Moreover, in vacuoles *E. hybridum* noticeable lightening was on the surface of cells after treatment with ortho-phthalic aldehyde, while the emission after the staining with glyoxylic acid was observed within the organelles, too, as shown on the optical slice of laser-scanning confocal microscope (Figure 8.4).

In early work on isolated vacuoles from cell culture of *P. bracteatum* the presence of catecholamines has also been shown by the fluorescence method with glyoxylic acid, and then using chromatography HPLC the biogenic amines were identified, mainly, as dopamine.

No doubt that the observation of biogenic amines in vacuoles confirms the known store and toxin-neutralizing (for example high concentrations of amines) cellular functions of these organelles. Vacuoles are a depot of many oxidants and antioxidants that play a key role in redox homeostasis of plant cells (Andreev 2012). Catecholamines are involved in redox reactions, including formation of alkaloids (Roshchina 1991, 2001), so their presence in vacuoles is, probably, a necessary part of the metabolism of the plant cell.

Form of the biogenic amines in plant cell is not clear yet. Monograph (Roshchina 2001) contains information about the presence of dopamine in the form of glycosides in seeds of *Entada pursaetha* and occurrence of histamine (depending on the type of object and body) in the form *N*-acetylhistamine, *N*,*N*-dimethyltilhistamine, cinnamoylhistamine, or feruloylhistamine as well.

### 8.3.2 Fluorescent Reactions with Agonists and Antagonists of Neurotransmitters

The studies of the location neurotransmitters may be analyzed using fluorescent agonists and antagonists of neurotransmitters that are used in experiments on plant cells (Table 8.2). All of them fluoresce

**TABLE 8.2**
**Fluorescing Agonists and Antagonists of Neurotransmitters**

| Fluorescent Agent (Agonist or Antagonist) | Fluorescence (Color/Maxim in nm) | Cellular Component Target of Fluorescent Staining |
|---|---|---|
| **Antiacetylcholine Agent** | **Pure Compound** | |
| Muscarine (agonist) | Blue/415 | ChR in Plm |
| Atropine (antagonist) | Blue/415 | ChR in Plm |
| d-Tubocurarine (antagonist) | Blue/415 | ChR, Plm, Vac, Chl |
| **Anticatecholamine Agent** | | |
| 6,7-diOHATN (agonist) | Blue/465 | AdR (DR), Plm |
| BODIPY-dopamine (agonist) | Green/518 | AdR (DR), Plm, Nu |
| Yohimbine (antagonist) | Blue/455 | AdR, Plm, Nu |
| Labetalol (antagonist) | Blue/460 | AdR, Plm |
| Prazosin (antagonist) | Blue/450 | AdR, Plm |
| **Antiserotonin Agent** | | |
| BODIPY-5HT (agonist) | Green/518 | SR, Plm, Nu |
| Kur-14 (antagonist) | Green/520 | SR, Plm |
| Inmecarb (antagonist) | Blue/460 | SR, Plm |
| Antihistamine agent | | |
| Azulene (antagonist) | Blue/415 | HR in Plm |

*Source:* Roshchina, V.V., Luminescent cell analysis in allelopathy, in *Cell Diagnostics. Images, Biophysical and Biochemical Processes in Allelopathy*, Roshchina, V.V. and Narwal, S.S. (Eds.), pp. 103–115, Science Publisher, Enfield, NH, 2007; New tendency and perspectives in evolutionary considerations of neurotransmitters in microbial, plant, and animal cells, in *Advances in Experimental Medicine and Biology*, Series: Microbial Endocrinology: Interkingdom Signaling in Infectious Disease and Health, Lyte, M. (Ed.), vol. 874, pp. 25–77, Springer, New York, 2016; Roshchina, V.V. et al., *Dokl. Russ. Acad. Sci.*, 393, 832–835, 2003; *Biochem. (Moscow) Suppl. Ser. A Membr. Cell Biol.*, 6, 105–112, 2012; *Biochem. (Moscow) Suppl. Ser. A Membr. Cell Biol.*, 10, 233–239, 2016.

ChR, AdR, DR, SR, HR—cholinoreceptor, adrenoreceptor, dopamine receptor, serotonin receptor, histamine receptor, relatively; Plm—plasmalemma, Nu—nuclei, Vac—vacuoles, Chl—chloroplasts.

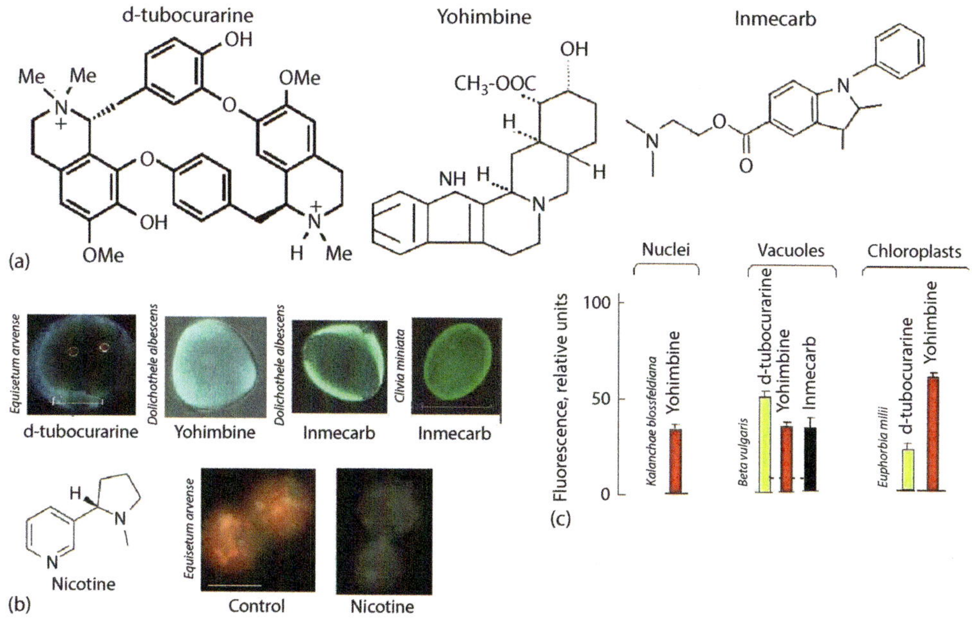


**FIGURE 8.5** Fluorescence of antagonists (a) and agonist (b) in model plant cells and isolated organelles (c). a and b images of *Equisetum arvense* vegetative microspores with d-tubocurarine and pollen of *Dolichothele albescens* done by confocal microscope Leica TSL -5, other images—by luminescence microscope Leica 6000. c—the fluorescence intensity values of ROI from isolated organelles done by confocal microscope Leica TSL -5.

in blue or blue-green (Roshchina 2008). Besides, we can mark that fluorescent reagents studied only on animals yet. The fluorescent tracer of dopamine or fluorescein isothiocyanate-dopamine (FITC-DA) was synthesized through conjugation of the isothiocyanate reactive group (–N=C=S) of FITC to the primary amine group of the neurotransmitter for living zebrafish (Lin et al. 2015), but in plant cells this reagent is not used yet as well as BODIPY-acetylcholine or 4,4-Difluoro-4-bora-3a,4a-diaza-s-indacene-acetylcholine (Bezuglov et al. 2004).

On intact plant cells, the reagents represented on Table 8.2 usually bounded on the surface (Figure 8.5). Using luminescence and confocal microscopy, the data have been received that main emission of d-tubocurarine and yohimbine (Roshchina 2004; Roshchina et al. 2012; Roshchina 2014) as well as antagonists of serotonin inmecarb and Kur-14 were seen in the plasmalemma layer (Roshchina et al. 2012; Roshchina 2014, 2016). This shows the possible presence of cholinoreceptors, adrenoreceptors, and serotonin receptors as sites for acetylcholine and catecholamines in plasmalemma. This picture was observed on some pollens and vegetative microspores of horsetail (Figure 8.5a and b). Antagonist of acetylcholine a-tubocurarine fluoresces in blue binding on the surface of vegetative microspore of *E. arvense* (rings show chloroplasts emitted in red), while pollens from *D. albescens* and *C. miniata* treated with antagonist of catecholamines yohimbine or antagonists of serotonin inmecarb have blue or green fluorescence on the surface layer (cell wall and plasmalemma sites). Binding of blue-green fluoresced yohimbine with vegetative microspores of *E. arvense* was shown on images and the emission spectra received by confocal microscopy (Roshchina et al. 2013). The emission was seen in the cell wall and especially in plasmalemma. The agonist of acetylcholine nicotine itself has no fluorescence in the visible spectral region, but when it was added to vegetative microspore of *E. arvense*, the cell began to fluoresce in blue-green, unlike the emission in red in control. A mechanism of the agonist action is not clear, but some metabolic processes may take place on the cellular surface that mask all emissions of the cell as a whole. Unfortunately, we don't know yet about the existence of fluorescent (in the visible spectral region) antagonists of histamine, except azulene examined on vegetative microspores of horsetail (Roshchina et al. 2012, 2013). Azulene modulated the germination of the microspores, while chamazulene-containing extracts from some species in experiments with pollens of *H. hybridum* and *Vallota speciosa* inhibited their development (Roshchina et al. 2011). Both synthetic and natural azulenes penetrated into the cells, and after that, the nucleus and chloroplasts fluoresced in blue-green (Roshchina et al. 2011a, 2013).

Intracellular binding of the reagents was also studied on isolated organelles (Roshchina 2016; Roshchina et al. 2016). Among studied objects, nuclei, vacuoles, and chloroplasts bounded yohimbine that fluoresces the intense emission visible at 490–520 nm after adding yohimbine 10-5 M, which may relate to the possible presence of the adrenoreceptors on the surface of the nucleus (Figure 8.5c). Yohimbine has been on the surface of the *E. milii* vacuoles, and its emission is almost 100 times greater than the intensity of the autofluorescence of the organelles (Roshchina et al. 2016). Fluorescence of other samples was seen after the treatment of d-tubocurarine that shows the presence of a nicotinic-type cholinoreceptor like in animal cells. A similar emission was peculiar after the treatment of inmecarb, antagonist of serotonin, that bound appropriate receptors. Vacuoles from roots of *B. vulgaris* var. rubra bounded d-tubocurarine and inmecarb, too, while chloroplasts from leaves of *E. milii*—only d-tubocurarine. Earlier, the optical slices previously isolated nuclei of white petals from mock orange *P. grandiflorus* using laser-scanning confocal microscopy has shown that fluorescent agonist of dopamine called BODIPY-dopamine also binds to the surface of these organelles in the site of possible dopamine receptors as well (Roshchina 2005). Of course there is the question of the possible reception of neurotransmitters within the cells and intracellular signaling with their participation. Experiments similar to those done with nuclei and vacuoles were carried out on isolated chloroplasts (Roshchina 2016; Roshchina et al. 2016). The sensitivity of isolated chloroplasts to biogenic amines and their reception has previously been investigated based on photochemical reactions (electronic transport and related photophosphorylation) (Roshchina and Mukhin 1984; Roshchina 1990b). Binding constants for catecholamines, histamine, acetylcholine, as well as their agonists and antagonists, estimated on the rate of photophosphorylation, were about $10^{-9} - 10^{-8}$ M, but the effects of serotonin have not been found at all.

## 8.4 CONCLUSION

The use of color and fluorescent reactions at the studies of the neurotransmitters in plant cells opens new possibilities for the intercellular and intracellular signaling. Using reagents on cholinesterase activity (marker of the presence of acetylcholine) or reagents for catecholamines, serotonin, and histamine one could estimate the location of neurotransmitters. Moreover, fluorescent agonists and antagonists binding in the cell permit us to know the presence of the appropriate receptor for the neurotransmitter.

## REFERENCES


Andreev, I.M. 2012. Role of vacuole in redox-homeostasis of plant cells. *Russian Journal of Plant Physiology* 59(5): 660–667.

Barwell, C.J. 1979. The occurrence of histamine in the red alga of *Furcellaria lumbricalis* Lamour. *Botanica Marina* 22: 399–401.

Bednarska, E. 1992. The localization of nonspecific esterase and cholinesterase activity in germinating pollen and in pollen tube of *Vicia faba* L. The effect of actinomicyn D and cycloheximide. *Biologia Plantarum* 34: 229–240.

Bednarska, E., Tretyn, A. 1989. Ultrastructural localization of acetylcholinesterase activity in the stigma of *Pharbitis nil* L. *Cellular Biology International Reports* 13: 275–281.

Bezuglov, V.V., Gretskaya, N.M., Esipov, S.S. et al. 2004. Fluorescent lipophilic analogs of serotonin, dopamine and acetylcholine: Synthesis, mass-spectrometry and biological activity. *Bioorganic Chemistry (Bioorg Khim, Russia)* 30(5): 512–519.

Björklund, A., Lindvall, O., Svensson, L.A. 1972. Mechanisms of fluorophore formation in the histochemical glyoxylic acid method for monoamines. *Histochemie* 32(2): 113–131.

Boettger, S.A., McClintock, J.B. 2012. Acetyl cholinesterase activity and muscle contraction in the sea urchin *Lytechinus variegatus* (Lamarck) following chronic phosphate exposure. *Environmental Toxicology* 27(4): 193–201.
Budantsev, A.Y., Roshchina, V.V. 2005. Testing of cholinesterase activity to acetylcholinesterase. *Plant Resources (Russia)* 41:131–138.
Budantsev, A.Y., Roshchina, V.V. 2007. Cholinesterase activity as a biosensor reaction for natural allelochemicals: Pesticides and pharmaceuticals. In *Cell Diagnostics. Images, Biophysical and Biochemical Processes in Allelopathy*, Roshchina, V.V. and Narwal, S.S. (Eds.), pp. 127–145. Enfield, NH: Science Publisher.
Cross, S.A.M., Even, S.W.B., Rost, F.W.D. 1971. A study of the methods available for the cytochemical localization of histamine by fluorescence induced with o-phthalaldehyde or acetaldehyde. *HistochemicalJournal* 3(6): 471–476.
Falck, B. 1962. Observation on the possibilities of the cellular localization of monoamines by a fluorescence method. *Acta Physiologica Scandinavica* 56(197): 5–25.
Fluck, R.A., Jaffe, M.J. 1974. Cholinesterases from plant tissues VI. Distribution and subcellular localization in *Phaseolus aureus* Roxb. *Plant Physiology* 53:752–758.
Gorska-Brylass, A., Rascio, N., Mariani, P. 1990. Cytochemical localization of cholinesterase in the thylakoids of chloroplasts of *Marchantia polymorpha. Cell Biology International Reports* 14: 208.
Harris H., Hopkinson D.A.., Robson E.B. 1962 Two-dimensional electrophoresis of pseudocholinesterase components in normal human serum. *Nature.* 196 ( 4861), 1296–1298
Kakariari, E., Georgalaki, M., Kalantzopoulos, G., Tsakalidou, E. 2000. Purification and characterization of an intracellular esterase from *Propionibacterium freudenreichii* ssp. *freudenreichii* ITG 14. *Le Lait* 80(5): 491–501.
Karnovsky, M. and Roots, L.A. 1964. A direct colouring method for cholinesterases. *Journal of Histochemistry and Cytochemistry* 12 : 219–25.
Kimura, M. 1968. Fluorescence histochemical study on serotonin and catecholamine in some plants. *The Japanese Journal of Pharmacology* 18: 162–168.
Kutchan, T.M., Rush, M., Coscia, C.J. 1986. Subcellular localization of alkaloids and dopamine in different vacuolar compartment of *Papaver bracteatum. Plant Physiology* 81(1): 161–166.
Lin, H.J., Lu, H.H., Liu, K.M., Chau, C.M., Hsieh, Y.Z., Li, Y.K., Liau, I. 2015. Toward live cell imaging of dopamine neurotransmission with fluorescent neurotransmitter analogues. *Chemical Communications* 51: 14080–14083.
Maheshwary, S.C., Gupta, R., Gharyal, P.K. 1982. Cholinesterase in plants. In: *Recent Developments in Plant Sciences: S.M Sircar Memorial Volume.* (Ed., S.P.Sen). pp. 145-160. Today and Tomorrows printers and publishers. New Delhi.
Menten, M.L., Junge, J., Green, M.H. 1944. A coupling histochemical azo dye test for alkaline phosphatase in the kidney. *Journal of Biological Chemistry* 153: 471–477.
Nachlas, M.M., Seligman, A.M. 1949. The histochemical demonstration of esterase. *Journal of the National Cancer Institute* 9: 415–425.
Roshchina, V.V. 1988. Characterization of pea chloroplast cholinesterase: Effect of inhibitors of animal enzymes. *Photosynthetica* 22(1): 20–26.
Roshchina, V.V. 1990a. Biomediators in chloroplasts of higher plants. 3. Effect of dopamine on photochemical activity. *Photosynthetica* 24(1): 117–121.
Roshchina. V.V. 1990b. Biomediators in chloroplasts of higher plants. 4. Reception by photosynthetic membranes. *Photosynthetica* 24: 539–549.
Roshchina, V.V. 1991. *Biomediators in Plants. Acetylcholine and Biogenic Amines.* Pushchino, Russia: Biological Center of USSR Academy of Sciences, 192 pp.
Roshchina, V.V. 1999a. Mechanisms of cell-cell communication. In *Allelopathy Update*, Narwal, S.S. (Ed.), vol. 2, pp. 3–25. Enfield, NH: Science Publishers.
Roshchina, V.V. 1999b. Chemosignalization at pollen. *Uspekhi Sovremennoi Biologii (Trends in Modern Biology, Russia)* 119: 557–566.
Roshchina, V.V. 2001. *Neurotransmitters in Plant Life.* Enfield, NH: Science Publishers, 283 p.
Roshchina, V.V. 2004. Cellular models to study the allelopathic mechanisms. *Allelopathy Journal* 13(1): 3–16.
Roshchina, V.V. 2005. Allelochemicals as fluorescent markers, dyes and probes. *Allelopathy Journal* 16(1): 31–46.
Roshchina, V.V. 2007. Luminescent cell analysis in allelopathy. In *Cell Diagnostics. Images, Biophysical and Biochemical Processes in Allelopathy*, Roshchina, V.V. and Narwal, S.S. (Eds.), pp. 103–115. Enfield, NH: Science Publisher.

Roshchina, V.V. 2008. *Fluorescing World of Plant Secreting Cells*. Enfield, NH: Science Publishers, p. 338.
Roshchina, V.V. 2014. *Model Systems to Study Excretory Function of Higher Plants*. Dordrecht, the Netherlands: Springer.
Roshchina, V.V. 2016. New tendency and perspectives in evolutionary considerations of neurotransmitters in microbial, plant, and animal cells. In *Advances in Experimental Medicine and Biology*, Series: *Microbial Endocrinology: Interkingdom Signaling in Infectious Disease and Health*, Lyte, M. (Ed.), vol. 874, pp. 25–77. New York: Springer.
Roshchina, V.V. 2017. Serotonin in Cells: Fluorescent Tools and Models to Study. In: *Serotonin and Melatonin (Serotonin and Melatonin: Their Functional Role in Plants, Food, Phytomedicine, and Human Health*. Eds Gokare A. Ravishankar, Akula Ramakrishna, CRC Press, Boca Raton, 2017 Eds. Rachavendra and Ramakrishna A, pp.47-60, CRC Press, Boca Raton
Roshchina, V.V. 2018. Cholinesterase in secreting cells and isolated organelles. *Biological Membranes (Russia)* 35 (2): 143–149.
Roshchina, V.V., Bezuglov, V.V., Markova, L.N. et al. 2003. Interaction of living cells with fluorescent derivatives of biogenic amines. *Doklady Russian Academy of Sciences* 393(6): 832–835.
Roshchina, V.V., Melnikova, E.V. 1998a. Chemosensory reactions at the interaction pollen-pistil. *Biology Bulletin* 6: 678–685.
Roshchina, V.V., Melnikova, E.V. 1998b. Allelopathy and plant generative cells. Participation of acetylcholine and histamine in a signalling at the interactions of pollen and pistil. *Allelopathy Journal* 15(2): 171–182.
Roshchina, V.V., Melnikova, E.V., Kovaleva, L.V., Spiridonov, N.A. 1994. Cholinesterase of pollen grains. *Doklady Biological Sciences* 337: 424–427.
Roshchina, V.V., Mukhin, E.N. 1984. Acetylcholinesterase activity in chloroplasts of higher plants. *Doklady Russian Academy of Sciences* 278(3): 754–757.
Roshchina, V.V. and Semenova, M.N. 1995. Neurotransmitter systems in plants. Cholinesterase in excreta from flowers and secretory cells of vegetative organs in some species. In: *Proceedings of the Plant Growth Regulation Society of America*. (Eds. D.Greene and G.Cutler) 22: Annual Meeting, July18-20, 1995, pp. 353–57. Minneapolis: Fritz C.D.
Roshchina, V. V. and Shvirst N.E. 2016. Cholinesterase in contractile structures of plants and animals: histochemical experiments with azocompounds. In: *Biological Motility*. Ed.Udaltsov S.N. Materials of International Symposium. 12-14 May 2016. Pushchino.Pushchino: Synchrobook, 2016. P.198–202.
Roshchina, V.V., Yashin, V.A. 2014. Neurotransmitters catecholamines and histamine in allelopathy: Plant cells as models in fluorescence microscopy. *Allelopathy Journal* 34(1):1–16.
Roshchina, V.V., Yashin, V.A., Kuchin, A.V. 2015a. Fluorescent analysis for bioindication of ozone on unicellular models. *Journal of Fluorescence* 25(3): 595–601. doi:10.1007/s10895-015-1540-2.
Roshchina, V.V., Yashin, V.A, Kuchin, A.V. 2015b. Fluorescence in the study of neurotransmitters in plant cells and their reception. In *Reception and Intracellular Signaling. Proceedings of the International Conference*, May 25–28, Zinchenko, V.P. and Berezhnov, A.V. (Eds.), vol. 1, pp. 364–369. Pushchino, Russia: Fix-Print.
Roshchina, V.V., Yashin, V.A., Kuchin, A.V. 2016. Fluorescence of neurotransmitters and their reception in plant cell. *Biochemistry (Moscow), Supplement Series A: Membrane and Cell Biology*, 10(3): 233–239. doi:10.1007/s10895-015-1540-2.
Roshchina, V.V., Yashin, V.A., Kuchin, A.V., Kulakov, V.I. 2014. Fluorescent analysis of catecholamines and histamine in plant single cells. *International Journal of Biochemistry Photon* 195:344–351.
Roshchina, V.V., Yashin, V.A., Vikhlyantsev, I.M. 2012. Fluorescence of plant microspores as biosensors. *Biochemistry (Moscow), Supplement series A: Membrane and Cell Biology* 6(1): 105–112.
Roshchina, V.V., Yashin, V.A., Yashina, A.V., Goltyaev, M.V. 2011a. Colored allelochemicals in the modeling of cell-cell allelopathic interactions. *Allelopathy Journal* 28: 1–12.
Roshchina, V.V., Yashina, A.V., Yashin, V.A., Goltyaev, M.V. 2011b. Fluorescence of biologically active compounds in plant secretory cells. In *Research Methods in Plant Science*, Narwal, S.S., Pavlovic, P., and John, J. (Eds.), vol. 2, pp. 3–25. Houston, TX: Studium Press.
Roshchina, V.V., Yashin, V.A., Yashina, A.V., Goltyaev, M.V. 2013. Microscopy for modeling of cell-cell allelopathic interactions. In *Allelopathy. Current Trends and Future Applications*, Cheema, Z.A., Farooq, M., and Wahid, A. (Eds.), pp. 407–427. Berlin, Germany: Springer.
Yang, C.J., Zhai, Z.X., Guo, Y.H., Gao, P. 2007. Effects of acetylcholine, cytochalasin B and amiprophos-methyl on phloem transport in radish (*Raphanus sativus*). *Journal of Integrative Plant Biology* 49(4): 550–555.

# 9 Inter- and Intracellular Signaling in Plant Cells with Participation of Neurotransmitters (Biomediators)

*Victoria V. Roshchina*

## CONTENTS



## 9.1 INTRODUCTION

Contacts (microorganism-microorganism, microorganism-plant, microorganism-animal, plant-animal, plant-plant, and animal-animal) in Nature may occur via signaling by bioactive compounds (Csaba and Muller 1996). Compounds known as neurotransmitters/neuromediators. may play role of exogenous and intragenous signals. Neurotransmitters acetylcholine, dopamine, noradrenaline (norepinephrine), adrenaline (epinephrine), serotonin, and histamine found in all living organisms have been considered as ancient components of irritability showing characteristics of chemosignals and metabolic regulators (Roshchina 1991, 2010, 2016a). In any case, the compounds found in every organism can be called biomediators or biotransmitters (Roshchina 1989) rather than neurotransmitters/neuromediators. But the term *neurotransmitters* is widespread today, and we also use it for better understanding the reader audience (Roshchina 2010, 2016a).

Now it is known that acetylcholine and biogenic amines (catecholmines dopamine, noradrenaline and adrenaline, serotonin, histamine) can also play an important role in the signal system of plants and microorganisms. As hydrophilic substances, they do not penetrate through the plasmatic membrane and act outside, but similar signaling, perhaps, occurs in the interior cell between organelles.

This chapter concentrates on the new information about neurotransmitters with respect to intercellular and intracellular signaling in plants. The signaling with neurotransmitters will be considered (1) as possible signaling outside the cell (intercellular signaling with the compounds excreted by any cells—from plant cell of the same plant individual or other plant species to animal and microorganisms), and (2) as intracellular signaling (within plant cell).

## 9.2 APPROACHES AND METHODOLOGY

In a dependence on the aim of study and following approaches applied in neurobiology one can to use (1) Methods of molecular biology; (2) Pharmacological analysis; (3) Inhibitory analysis; and (4) Radiometric, electrophysiological, fluorescent, and histochemical technique. In first direction structure of proteins (receptors and enzymes participated in cholinergic and aminergic systems), mechanisms of ligand-protein binding, genes coding of the enzymes have been investigated. In animals, receptors for neurotransmitters were studied and some genes coded are known, but similar information for plants is absent, yet only the *Arabidopsis* model (Muralidharan et al. 2013) for analysis of inclusion of similar brain proteins in DNA binding complex was represented (Lu et al. 1992). Pharmacological approach is characterized by use of specific ligands (mainly drugs) binding with appropriate receptors in mammalians. Based on the methods, effects of the compounds on various processes sensitive to neurotransmitters and an occurrence of their receptors are studied. The nature of reception is analyzed. Spreading of the signal of the neurotransmitter from the cell surface to interior is studied with specific inhibitors of systems related to secondary messengers. Participation of some cellular compartments in the spreading of primary signals and location of intracellular receptors and enzymes of cholinergic and aminergic systems are studied by radiometric, fluorescent, or histochemical reactions. At all approaches, one needs to choose a suitable model or model system including sensitive plant species or individual processes.

## 9.3 MODEL PROCESSES SENSITIVE TO NEUROTRANSMITTERS

There are more evidences that neurotransmitters are multifunctional substances participating in various living processes that play a universal role as signal and regulatory compounds (Roshchina 1991, 2001, 2010, 2016a; Momonoki and Momonoki 1991, Momonoki et al. 1997; Wessler et al. 2001; Baluska et al. 2005, 2006; Brenner et al. 2006; Kulma and Szopa 2007; Ramakrishna et al. 2011; Erland et al. 2016). The processes selectively sensitive to neurotransmitters may serve as models for analysis of reception and spreading the chemosignal from the surface into cell interior (Figure 9.1). Among established and hypothetical functions of the compounds in plants, first of all is known participation in signaling. For example, acetylcholine, itself, like in neural systems, changes membrane potential it is able to induce action potential, spreading promoted within accepting cell, final result of signaling may be realized in regulation of growth reaction (Koštiř et al. 1965; Bamel et al. 2015, 2016), changes in fertilization and development (Roshchina 1991, 2001). In any case, the surface enzyme cholinesterase decomposes such active agent preventing the long irritation, or it may prevent parasitic invasion. Moreover, the acetylcholine-hydrolizing enzyme regulates stomata opening, recognition, and fertilization. Most hypothetical, although attractive, is the proposed universal role of cholinesterase as producer of acetic acid regulating cellular pH (Borodyuk 1990). Yamamoto with co-workers (2016) have shown that acetylcholinesterase overexpression causes an enhanced gravitropic response in rice seedlings and suggested

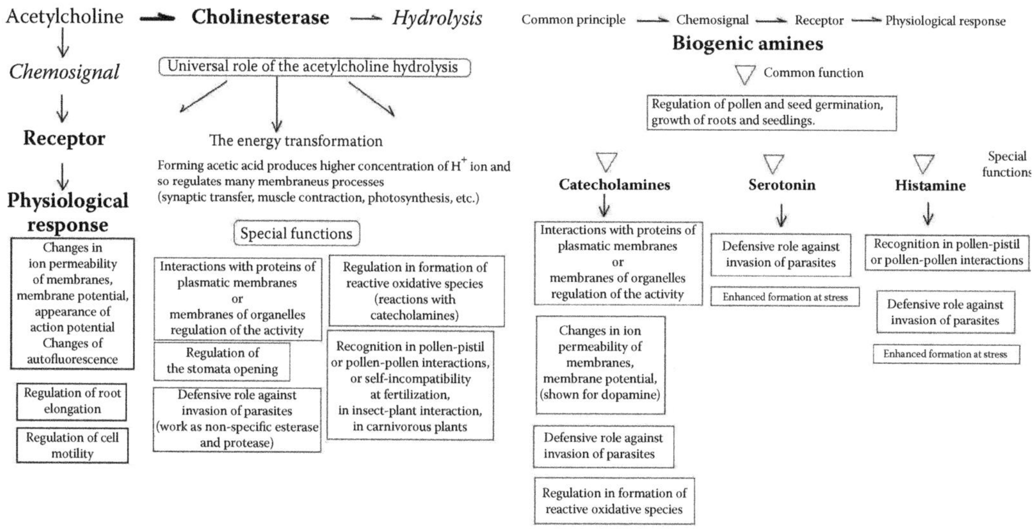


**FIGURE 9.1** Common scheme of plant sensitive reactions to neurotransmitters and functions of the tandem acetylcholine-cholinesterase and biogenic amines. (From Koštiř, J. et al., *Rostlinnà Vŷroba. Rocnik*, 11, 1239–1280, 1965; Fluck, R.A. and Jaffe, M.J., The acetylcholine system in plants, In: Smith, E. (Ed.), *Current Advances in Plant Sciences*, Vol. 5, Science Engineering, Medical and Data, Oxford, UK, pp. 1–22, 1974a; Fluck, R.A. and Jaffe, M.J., *Phytochemistry*, 13, 2475–2480, 1974b; Maheshwary, S.C. et al., Cholinesterase in plants, In: Sen, S.P. (Ed.), *Recent Developments in Plant Sciences: S.M. Sircar Memorial Volume*, Today and Tomorrows Printers and Publishers, New Delhi, India, pp. 145–160, 1982; Grosse, W., *Phytochemistry*, 21, 819–822, 1982; Hartmann, E. and Gupta, R., Acetylcholine as a signaling systems in plants, In: Boss, W.F. and Morve, D.I. (Eds.), *Second Messengers in Plant Growth and Development*, Allan R. Liss, New York, pp. 257–287, 1989; Borodyuk, N.R., *Vestnik Selkhozyastvennoi Nauki*, 6, 87–95, 1990; Tretyn, A. and Kendrick, R.E., *Bot. Rev.*, 57, 33–73, 1991; Roshchina, V.V., *Biomediators in Plants. Acetylcholine and Biogenic Amines*, Biological Center of USSR Academy of Sciences, Pushchino, Russia, 192 p, 1991; Roshchina, V.V., *Biol. Nauk.*, 9, 124–129, 1992; Roshchina, V.V., *Neurotransmitters in Plant Life*, Science Publishers, Enfield, NH, 2001; Roshchina, V.V., Evolutionary considerations of neurotransmitters in microbial, plant and animal cells, In Lyte, M. and Freestone, P.P.E. (Eds.), *Microbial Endocrinology: Interkingdom Signaling in Infectious Disease and Health*, Springer, New York, pp. 17–52, 2010; Roshchina, V.V., New tendency and perspectives in evolutionary considerations of neurotransmitters in microbial, plant, and animal cells, In: Lyte, M. (Ed), *Advances in Experimental Medicine and Biology*, Vol. 874. Series: Microbial Endocrinology: Interkingdom Signaling in Infectious Disease and Health, Springer International Publishing, New York, Chapter 2, pp. 25–77, 2016a; Kuklin, A.I. and Conger, B.V., *J. Plant Growth Regul.*, 14, 91–97, 1995; Kulma, A. and Szopa, J., *Plant Sci.*, 172, 433–440, 2007; Ramakrishna et al. 2011; Sugiyama, K. and Tezuka, T., *Plant Signal Behav.*, 6, 1545–1553, 2011; Ramakrishna 2011; Mukherjee, S. et al., *Physiol. Plant*, 152, 714–728, 2014; Di Sansebastiano, G.P.D. et al., *Int. J. Mol. Sci.*, 15, 4565–4582, 2014; Bamel, K. et al., *Int. Immunopharmacol.*, 29, 231–234, 2015; Bamel, K. et al., *Plant Signal Behav.*, 11, e1187355, 2016.)

that the function of the rice enzyme relates to hypothesized that the plant acetylcholinesterase regulates asymmetric distribution of hormones and substrates due to gravity stimulus. Common features of biogenic amines are in regulation of the growth and development as well as in defensive reactions at stress (Kuklin and Conger 1995; Kulma and Szopa 2007). Many of aforementioned reactions serve as model processes for analysis of neurotransmitters' effects (Roshchina 2014).

## 9.4 EXCRETIONS AND SECRETIONS AS MEDIA FOR CELL–CELL OR ORGANELLE–ORGANELLE CONTACTS WITH NEUROTRANSMITTERS

Neurotransmitters and some enzymes of their synthesis and catabolism appear to be met in cellular excretions and secretions where they serve as chemosignals and regulators of many cellular responses. As medium may be water, oil, and gases or their mixtures. But it is especially of interest that first drop of pollen grain after moistening (Roshchina 1999b), excretions of pollen (Bednarska 1992), and pistil stigma (Bednarska and Tretyn 1989) contain cholinesterase, enzyme catalyzes the acetylcholine hydrolysis. The enzyme is also the indicator or marker of the presence of acetylcholine in a cell itself. Moreover, it was cloning for maize *Zea mays* and identified as a mixture of disulfide- and noncovalently linked 88-kD homodimers consisting of 42- to 44-kD polypeptides (Sagane et al. 2005).

Possible involvement of cholinesterase, the enzyme that hydrolyzed acetylcholine, in signal processes of plants has been studied by the histochemical analysis of its presence in secreting cells, and experiments on the influence of exogenous cholinesterase from different plant species on the growth reaction have been carried out (Roshchina 2017, 2018). Whole cells with visible secretion (such as firefly *Jacobinia pauciflora* Lindau, Katharine blood lily *Haemanthus katharinae*, whose pistils and pollen actively secrete as well as hop *Humulus lupulus* L., having secretory glands in the female flowers bracts) were analyzed. In intact secretory cells, cholinesterase activity has been found that was concentrated mainly in the plasmalemma, free space of cell, and cell walls. Inhibitors of cholinesterase neostigmine, physostigmine, and berberine reduced or completely blocked the hydrolysis of acetylcholine. On the models of single intact cells—vegetative microspores from *Equisetum arvense* and pollen of *Corylus avellana*—addition of foreign exogenous cholinesterase (isolated from leaves of *Pisum satium* or from pollen and pistils of *Hippeastrum hybridum*) inhibited their germination (Roshchina 2017). On the contrary, the treatment of the cells with inhibitors of cholinesterase stimulated the process (Roshchina 2018). This confirmed supposion about participation of cholinesterases in signaling in order to stimulate or depress the cell development.

Acetylcholine (Barwell 1980) and biogenic amines such as histamine (Barwell 1979, 1989) or dopamine (van Alstyne et al. 2014) are also found in the water-living plants and their excretions. A lot of marine algae contain cholinesterase (Gupta et al. 1998, 2001). At last, smells of some representatives of Araceae family contain histamine, up to 1 μL/ml, in particular in river-lived devil's tongue or snake palm *Hydrosme rivieri* (Roshchina and Roshchina 1993). The functions are difficult to separate, and in any case signaling is known as first. After that, metabolic reactions occur as physiological responses on the signal. At intercellular contacts before the signaling, interactions in excretions (contained neurotransmitters and enzymes of their metabolism) from surfaces of both cell-donor and cell-acceptor take place.

## 9.5 LOOK ON COMMON MECHANISMS OF INTERCELLULAR AND INTRACELLULAR SIGNALING IN PLANTS

Cell–cell signaling by neurotransmitter (biomediator) or/and enzymes of their metabolism constantly takes place in one and the same organism or in biocenosis between other organisms (plant-plant, plant-animal, plant-microorganism). The process begins via the excretion from cell-donor into the medium surrounded and then perception (reception) of the compound by cell-acceptor, independently on the same organism or another one. Excretions of the compounds and enzymes of their metabolism take place in the biocenosis relationships. Secreting enzyme cholinesterase, catalyzing the acetylcholine hydrolysis, has plural functions in plant life, participating in fertilization, protein-protein interactions, recognition, etc. (Roshchina 2001; Malik 2014).

The main model of cell–cell contact with participation of neurotransmitters originated from synapses of cells in the mammalian nervous system where the excretion of neurotransmitter from cell-donor occurs and the compound interacts with cell-acceptor. Reception of neurotransmitters by sensitive plant surfaces, mainly plasmalemma and membranes of organelles, occurs via the binding with special sites (sensors-receptors), and received information is transported into the cell (from outside plant cell) or organelle (from cytoplasm) through transducins (G-proteins) and via systems of secondary messengers. Common ways of such signaling in animals was represented completely in fundamental books edited by Cassimeris with co-authors (2011), Buchanan et al. 2015 and Plopper with co-authors (2015). Secondary messengers of cyclic AMP and GMP, inositol triphosphate, $Ca^{2+}$ ions are also found in plant cells and within cellular organelles. Enzymes of deamination of biogenic amines (aminooxidases) are also included in redox reactions of plant cells. Some results of experiments demonstrate participation of secondary messengers, such as inositol triphosphate, $Ca^{2+}$ ions, and cAMP in the transfer of the information (received from neurotransmitter outside plant cell) within the cell.

After the finding of cholinesterase and biogenic amines as well some receptors in plant organelles, in particular in chloroplasts (Roshchina and Mukhin 1984; Roshchina 1986, 1990a, 1990b), the common principle of the signaling also has been considered for organelle-organelle interactions (Roshchina 1991, 2001).

Thus, communication by extracellular signals usually involves several steps as follows:

1. Synthesis > release of the neurotransmitter (biomediator) by the cell-donor or organelle-donor of the secretion
2. Transport of the signal to the target cell (cell-acceptor of the signal or organelle-acceptor)
3. Detection of the signal by an appropriate receptor protein (cholinoreceptor for acetylcholine or special receptors for biogenic amines)
4. Triggered by the receptor-signal complex including cascade of reactions of secondary messengers
5. A change in cellular metabolism influencing functioning and development
6. Removal of the neurotransmitter, which often terminates the cellular response

Signaling mechanism at the cellular level is similar with known for mammalians and includes following processes: (1) the binding of aforementioned substances with appropriate receptors on the plasmatic membrane (intercellular signaling) or on the outer membrane of organelle (intracellular signaling); (2) the conformation changes of the excited receptors transfer into the cell or organelle interior; and (3) the signal transfer through cytoplasm or the organelle interior occurs by electric changes, perhaps, the action potential spreading or by systems of secondary messengers (cyclic AMP, or GMP, calcium ions, etc.). A special role belongs to G-proteins (heterotrimeric protein with Gα, Gβ/Gγ subunits) and constitutes one of the most important components of the cell signaling cascade (Tuteja 2009). G Protein Coupled Receptors (GPCRs) perceive many extracellular signals, transduce them to heterotrimeric G proteins and further transduce these signals intracellularly to appropriate downstream effectors, thereby playing an important role in various signaling pathways. GPCRs exist as a superfamily of integral membrane protein receptors that contain seven transmembrane α-helical regions, which bind to a wide range of ligands. The end of the signaling is realized in various metabolic responses. More known data are concerned with cAMP (Gehring 2010) or inositol triphosphate (Stevenson et al. 2000) in plants.

Cell–cell contacts are met between plant cell and animal cell or between cells of plants and microorganisms where neurotransmitters or enzymes of their metabolism are excreted from the cell surface. The key mechanisms by which the neuromediators' signal acts consist in changing the membrane permeability, the membrane potential, and the associated processes such as the cascade of the regulatory processes with participation of secondary messengers. In the following, we shall consider some examples that deal with the determination of neurotransmitters, or their metabolic enzymes, in cells and cellular excretions of plants.

## 9.6 NEUROTRANSMITTERS IN CELL EXCRETIONS

The level of acetylcholine is controlled by enzyme cholinesterase, which catalyzes the hydrolysis of the compound (Massoulie et al. 1993). Significance of the enzyme in signaling at fertilization and breeding is remarkable. For example, cholinesterase activity was found in first drop of excretion from pollen grain after moistening (Roshchina 1999b). Moreover, lack of cholinesterase in pollen (Kovaleva and Roshchina 1997) or lower activities of acetylcholinesterase and choline acetyltransferase in pistils (Tezuka et al. 2007) were associated with self-incompatibility. Cholinesterase/

acetylcholinesterase, be a sensor of acetylcholine on the cell surface, because it was found in the cell wall and plasmalemma as well as in the intercellular space (Fluck and Jaffe 1974c; Bednarska and Tretyn 1989; Bednarska 1992; Roshchina et al. 1994, 2001).

A special role in plant excretions belongs to acetylcholine as substrate and cholinesterase as hydrolyzing enzyme. The evidence of the cholinesterase activity confirms by the decrease in rates of the hydrolysis of acetylcholine or acethylthiocholine after addition of the inhibitors neostigmine or physostigmine. Main experiments for determination of the location of activity done by electron microscopy with histochemical method of Koelle (1963) or with color reactions were based on the ferricyanide/ferrocyanide reaction of Karnovsky-Roots (Karnovsky and Roots 1964) where the product was red-brown Hettchet's pigment (copper ferrocyanide). The enzyme has been found in the excretions of roots (Fluck and Jaffe 1974a, 1974b), nectar, and secretions on the pistil stigma (Roshchina and Semenova 1995). Red analogue of Ellman reagent named 2,2-dithio-bis-(p-phenyleneazo)-bis-(1-oxy-8-chlorine-3,6) -disulfur acid in form of sodium salt (red reagent binding with thiocholine as the product of the acetylthiocholine hydrolysis became blue) was also used for the analysis of pollen excretions and histochemical staining on the cholinesterase activity (Roshchina et al. 1994; Roshchina 2001, 2007). Histochemical reaction for the enzyme activity, in which red color product arises from colorless reagents, was also carried out with Fast Red TR salt and β-naphthyl acetate as substrate (Roshchina and Shvirst 2016). The example is on Figure 9.2 for pollen from Katharine blood lily *H. katharinae*. Another search for the enzyme activity, in which red color product arises from colorless reagents, was also carried out with Fast Red TR salt and β-naphthyl acetateas substrate (Roshchina and Shvirst 2016; Roshchina 2018).

In intact secretory cells, cholinesterase activity has been found to be concentrated mainly in plasmalemma, free space of cell, and cell wall.

Cholinesterase finding both in free space and excretions of plant cells makes possible involvement of the enzyme in intercellular signaling. For example, it is confirmed by some effects on germination of the unicellular models such as vegetative microspores of horsetail *E. arvense* and pollen of *Corylus avellana* (Roshchina 2017a, 2018). In the experiments it has been shown the stimulation of the process by acetylcholine and inhibitors of cholinesterase neostigmine, physostigmine, and berberine. The concentration of acetylcholine (range $10^{-7}$–$10^{-4}$ M) encourages this process in vegetative microspores.

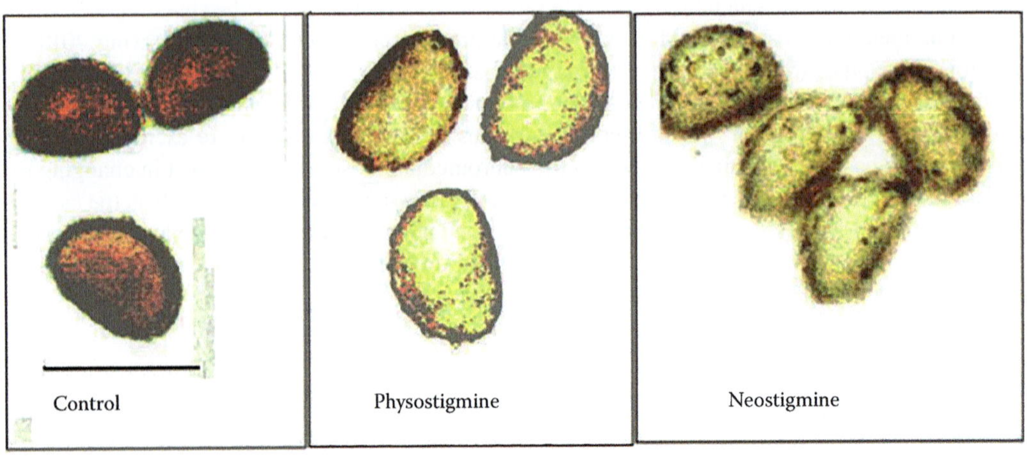


**FIGURE 9.2** The staining of pollen from *Haemanthus katherinae* Baker on the cholinesterase activity with Red analogue of Ellman Reagent. Blue coloration of control cells exposed in acetylthiocholine $10^{-3}$ M during 1 h (most intensive on the surface of the pollen grain) contrasted weak coloration after preliminary treatment with the cholinesterase inhibitors physostigmine or neostigmine $5 \times 10^{-5}$ M. Bar = 50 μm.

The most noticeable effect is marked for neostigmine in the concentration concentration $10^{-4}$ M, while for physostigmine at $10^{-5}$ M. All used compound stimulated the pollen germination by 30%–150% in a comparison with a control. Parallel experiments with acetylcholine also showed significant (up to 100%) stimulation of the process. It should mark that this value may be higher for the stored samples of vegetative microspores collected eailer. For pollen *C. avellana*, similar sensitivity in the germination showed, mainly, for inhibitors neostigmine and physostigmine, whereas the effect of acetylcholine itself was little different from the control. It follows from this the acetylcholine hydrolysis catalyzed by cholinesterase eliminates acetylcholine, neutralizing its effects. Thus, the cholinesterase inhibition leads to absence of hydrolysis of acetylcholine, which is accumulated and increases the number of germinated microspores. The data were correlated with a fact that foreign cholinesterases isolated from leaves of pea *P. sativum* or pollen of *H. hybridum* led to depression of the model cells' germination (Roshchina 2017). Stimulation of the pollen tube growth on the pistil stigma of flower *Hippeastrum hybridum* by neostigmine was observed after the treatment of this pistil with the inhibitor that accelerated fertilization (Roshchina and Melnikova 1998a, b). Interactions such as microorganism-plant or plant-animal are possible due to the fact that all participants contain acetylcholine and cholinesterase. Cholinesterase from salivary glands of insects (Bridges 1972; Miles 1972) can participate in the relationship with acetylcholine excreted from plant cell or vice versa, cholinesterase released from plant interacts with acetylcholine of insect, although direct studies have not been conducted. However, changes of plant growth induced by parasite attack may depend on the cholinesterase activity of saliva.

For determination of biogenic amines in cells and their excretions, fluorescent methods were used after the histochemical staining of the cells with glyoxylic acid (the emission maximum at 470–480 nm) or *o*-phthalic aldehyde and formaldehyde (emission at 460 nm at high concentrations) as shown in papers (Kimura 1968; Kutchan et al. 1986; Roshchina 2014). Figure 9.3 represents the photo where excretion of catecholamines or histamine is seen as fluorescent halo around the cells.

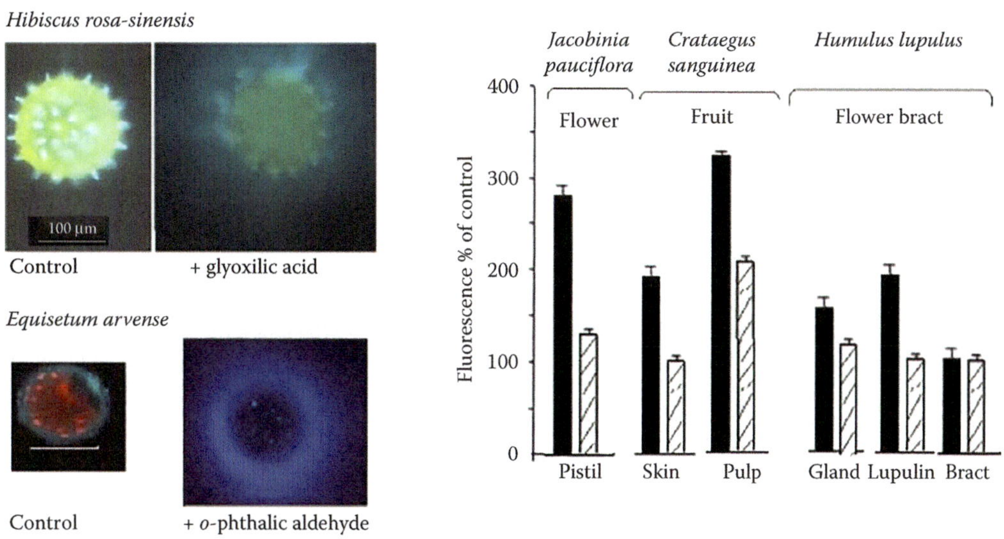


**FIGURE 9.3** The occurrence of catecholamines and histamine in some plant organs based on the fluorescence increase after the staining with glyoxylic and *o*-phthalic aldehyde. Left—the staining of pollen from *Hibiscus rosa-sinensis* (bar—100 µm) with glyoxylic acid, reagent for catecholamines, and vegetative microspores of *Equisetum arvense* (bar = 20 µm) with *o*-phthalic aldehyde, reagent for the determination of histamine. Common view of microspores before (control) and after the treatment with the reagents under UV-light (360–405 nm) of luminescent microscope. Blue fluorescence of secretion is seen as blue holo or green-yellow emission. Right—the level of the fluorescence at 460 or 480 nm after the treatment with glyoxylic acid (dark rectangles) or *o*-phthalic aldehyde (white rectangles) in some plant species.

Pollen from 25 species as well as leaf and stem stinging hairs of *Urtica dioica* and vegetative microspores of horsetail *E. arvense* were analyzed (Roshchina et al. 2014; Roshchina 2016b). The cells and their excretions being excited by UV-light may fluoresce in blue-green after staining with glyoxylic acid or in blue/blue-green or yellow after the treatment with *o*-phthalic aldehyde. The reagents penetrated into the cells and stained DNA-containing organelles such as nuclei and chloroplasts. The attempt to estimate the occurrence and content of catecholamines and histamine by the fluorescent method in individual plant cells may be useful for express-diagnostics of these biogenic amines as biologically active compounds in ecosystems and drugs in medicinal plants.

The occurrence of catecholamines and histamine in pollen excretions may be a cause that influences the germination of foreign pollen (pollen allelopathy). For example, enriched in catecholamines and histamine ($10^{-5}$–$10^{-3}$ M) pollen of *Populus balsamifera* stimulates in mixtures the germination of pollen *C. avellana* (Roshchina and Melnikova 1996; Roshchina and Yashin 2014). Due to the histamine presence, perhaps, pollen of *Anthriscus sylvestris* was able to stimulate the germination of pollen *Knautia arvensis* (Roshchina et al. 2009b).

One of the examples that show quantitative data of the occurrence of catecholamines and histamine in secreting cells of pistil, pollen, and glands is given on Figure 9.5.

## 9.7 RECEPTION OF NEUROTRANSMITTERS

Reception both on the plasmatic membranes in cell-acceptor occurs when any transmitter acts outside and on the outer membrane of an organelle received from the neuromediator within the cell. It is known that some of molecules carry signals over long distances, whereas others act locally to convey information between neighboring cells. Moreover, signaling molecules differ in their mode of action on their target cells. Like animals, there are plant surface receptors for neurotransmitters. Cholinoreceptors in mammals have been distinguished in two main types with subtypes, depending on the sensitivity of the reactions to agonists of acetylcholine, such as nicotine or muscarine, and are called nicotinic or muscarinic receptors, respectively. For acetylcholine in plant cells, pharmacological analysis with its agonists nicotine or muscarine showed the presence of nicotinic or muscarinic types of cholinoreceptors or receptors with mixed characteristics (Roshchina 2001). Nicotinic receptors of animals were characterized by molecular biologists (Piccioto et al. 1998; Corringer et al. 2000), including some ionic channels related to the receptors (Pichon 1993). The dopamine receptors have been classified into two groups, the D1-like and D2-like dopamine receptors, respectively, based on molecular biology and pharmacological studies as well the gene structures of the receptors (D'Souza 2015). The gene structures of these two classes of receptors are dissimilar with respect to the organization of their coding and regulatory regions. Biogenic amines also have receptors called adrenoreceptors (Lanier and Limbird 1997), serotonin receptors (Davis et al. 2002), and histamine receptors (Parsons and Ganellin 2006), which are described in literature devoted to mammalians. In animals, there is a known receptor part (this class of ligand binding ion channels called Cys-loop receptors protein) for acetylcholine, serotonin, and GABA that plays a major role in fast synaptic transmission and also known in prokaryotes.

As for plants themselves, today molecular biologists (Mukherjee 2015) by physico-chemical and sequence analysis try to study only one of the receptor structures from alga *Chlamydomonas reinhardtii* and indicate that it is an anionic receptor and Cys-loop receptors protein with some domains similar to neurotransmitter-gated ion-channel ligand-binding domains of animals. A model of this protein was generated using homology modeling based on a nicotinic acetylcholine receptor of *Torpedo marmorata*. Using methods of molecular biology in plant expression of the mammalian serotonin receptor in plant and amphibian cells was shown (Beljelarskaya and Sutton 2003). Moreover, transformed potato *Solanum tuberosum* was created with a cDNA encoding human dopamine receptor HD1 (Skirycz et al. 2005). Expression of human dopamine receptor results in altered tuber carbon metabolism. Thus, although our knowledge is small yet, we see that incompatibility is absent between protein receptors for species from various kingdoms.

Up to now, studies of reception in plants is carried out like in animal physiology where the main approach to analysis of reception consists of the application of pharmacological analysis—the comparison of the cellular responses on neurotransmitters and their agonists (imitators) or antagonists (acting as blockers). Orienting on animal similar postulates the studies are carried out, too, although we have small information yet, one should consider them here. In plants, mainly pharmacological analysis was used. Received data concerning cholinoreceptors were summed in publications of Hartman and Gupta (1989), Tretyn and Kendrick (1991), and Roshchina (1991, 2001). Sensitive reactions to the acetylcholine agonists and antagonists were uptake of $C^{14}$-acetate, root pressure, release of ethylene, development of embryos, unrolling of leaves, pistil stigma autofluorescence, fertilization after the pistil treatment, and water uptake by dry pollen. Concerning receptors to catecholamines and serotonin some information was represented in monographs (Roshchina 1991, 2001), in reviews of Kulma and Szopa (2007) and Roshchina (2016b) as well about histaminic receptors in plants—in monograph (Roshchina 2001). Model physiological processes sensitive to agonists and antagonists of catecholamines were (1) changes in membrane potential, (2) cytoplasm movement, (3) stimulation of flowering, and (4) pollen germination, while those sensitive to serotonin and histamine were seed and pollen germination. Using fluorescence features of some agonists and antagonists in small concentrations $10^{-7}$–$10^{-5}$ M (Roshchina 2008, 2014, 2016a, 2016b; Roshchina et al. 2012) some experiments on the cellular level were carried out, and it was shown on intact cells from various species that the target of the fluorescent agents was plasmatic membranes. Some signaling molecules are able to cross the plasma membranes and bind to intracellular receptors in the cytoplasm or organelles. At higher concentrations dopamine, for example, penetrates into animal cells (in form liposome as well) and binds with actin (Shubina et al. 2009).

As for organelles, sensitive reactions in plant cells to endogenous acetylcholine and catecholamines as well to their agonists and antagonists were non-cyclic photophosphorylation and $Na^+/K^+$ or $Ca^{2+}/Mg^{2+}$ efflux in chloroplasts (Roshchina 1990c). Serotonin effects were absent in isolated plastids. Recently by fluorescent methods, the presence of catecholamines and histamine in nuclei and vacuoles and binding of their agonists and antagonists on the outer membranes of the organelles was shown (Roshchina et al. 2016). In this connection, there is a very relevant question about the form of biogenic amines within plant cells. There is information about the presence of dopamine in the form of glycosides in the seeds of *Entada pursaetha* and the occurrence of histamine (depending on the species of object and organ) in the form N-acetylhistamine, N,N-dimethylhistamine, cinnamoylhistamine, or feruloylhistamine as well (Ponchet et al. 1982; Hikino et al. 1983; Roshchina 2001). Serotonin is met in green coffee beans in glycosylated forms (Servillo et al. 2016).

## 9.8 SPREADING OF RECEIVED SIGNAL

In animal cells, first response on the event was realized in the changes in ion permeability and membrane potential or of the inclusion of cascade of secondary messengers—cyclic AMP or GMP (regulation of key protein phosphorylation), $Ca^{2+}$-ions (regulation of membrane states and metabolic activity), inositol-3-phosphate (regulation of the metabolism of lipid, etc.), and others. Data also shows possibilities to a signal spreading throughout cellular contractile systems (Roshchina 2005, 2006b; Roshchina and Vikhlyantsev 2009; Roshchina et al. 2012).

### 9.8.1 Electric Signaling (Changes in Ion Permeability and Membrane Potential of Whole Cells)

Jaffe (1970) was the first who has shown that acetylcholine is capable of increasing the ionic permeability of plant membranes and inducing changes in membrane potential. The depolarizing effect of acetylcholine was also demonstrated in other experiments with arising and spreading of action potentials in the cells-acceptors of the mediator (see references in reviews of Hartmann and Gupta

[1989]; Tretyn and Kendrick [1991]; and in monographs of Roshchina [1991, 2001]). According to Hartmann and Gupta (1989), the ion basis of action potentials in plant cells is given by $K^+$, $Cl^-$, and $Ca^{2+}$ fluxes through the relative channels. As a whole, acetylcholine induced the shifts in membrane potential (mainly, decrease) due to the changes in ion permeability of plasmalemma, while its antagonists α-bungarotoxin and atropine as well as antagonist of serotonin trifluoperasine strongly decreased $Na^+$ efflux (Roshchina 2001). For acetylcholine, there are two types of acetylcholine receptor—nicotinic (receptors respond to nicotine) and muscarinic (sensitive to muscarine). Recently, it was established that muscarinic and nicotinic acetylcholine receptors are involved in the regulation of stomata function—the opening and closing movement—in the plants *Vicia faba* and *P. sativum* (Wang et al. 1998, 1999a, 2000). Leng et al. (2000) showed the regulation role of acetylcholine and its antagonists in inward rectified $K^+$ channels from *V. faba* guard cells. The location of the muscarinic receptor was shown in plasmatic membranes and chloroplast membranes (Meng et al. 2001), and cholinesterase activity was found in the cells (Wang et al. 1999b). The germination of plant microspores such as vegetative microspores of horsetail *E. arvense* or pollen (generative microspores) of knight's star *H. hybridum* was blocked by the antagonists of acetylcholine, which are linked with nicotinic cholinoreceptors and $Na^+$/ $K^{++}$ ion channels (Roshchina and Vikhlyantsev 2009). Main experimental data for multicellular tissues are shown in Table 9.1.

Analysis of the Table 9.1 presented the data showing the effects of acetylcholine on the ionic permeability of plants and membrane potential indicates that most often acetylcholine induces depolarization—a decrease in membrane resting potential and a change of the potassium permeability. In plants belonging to Characeae family, the $K^+$ efflux increased after activation on $K^+$ channels after the treatment with acetylcholine in concentration $10^{-3}$ M, and action potential arose (Kisnieriene et al. 2012). The changes in permeability of the plasmalemma induced by exogenous acetylcholine are reminiscent of the phenomena proceeding in synapses of animal cells, that is when coming in contact with the postsynaptic membrane, acetylcholine increases the permeability for $Na^+$ and $K^+$ ions and biogenic amines for $Ca^{2+}$, thus generating the action potential, which propagates through the organism (Roshchina 2001; Zherelova et al. 2014, 2016). Trifluoroperazine, as calmodulin blocker, also decreased $Ca^{2+}$ efflux. Antagonist of noradrenaline yohimbine stimulated $K^+$ and $Mg^{2+}$ efflux. Besides, the effects on plasmalemma permeability, the red pigment betacyanin efflux was also observed in variants with histamine, serotonin, dopamine, and trifluoroperazine (Roshchina 2001). This showed on the additional action on the tonoplast permeability. Unlike multicellular animals, the transmission of the signal from cell to cell in multicellular plants proceeds not through synaptic contacts but via continuous cytoplasmic strands (Zherelova et al. 2014). Consequently, in plant cells, acetylcholine and other mediators are involved in the transmission of the signal inside the cell but not between cells via intercellular contacts. Biogenic amines are also capable of changing the membrane potential. The effect of the neurotransmitter dopamine on the ionic channels of the electrically excitable plasma membrane of *Chara corallina* giant cells was investigated (Zherelova et al. 2014). The method of voltage clamping on the plasma membrane with registration of ionic currents under normal conditions and in the presence of dopamine was used. It was shown that the influence of dopamine on the ionic currents of the cells depended on the dose and time of exposure. The influence of dopamine on the membrane structure was reversible; after removal of the neurotransmitter, a complete restoration of the amplitude and the development kinetics of the current was observed. A 24h-long incubation of *C. corallina* cells in the presence of dopamine (2 mM) produced a drop in the resting potential of the cells, decreased the membrane resistance, and stopped the movement of the cytoplasm. The effect of the haloperidol, a dopamine D-receptor antagonist on the function of ionic channels of the electrically excitable plasma membrane and on the cytoskeleton of *C. corallina* cells was investigated recently (Zherelova et al. 2016). Haloperidol was shown to block plasmalemmal $Na^{2+}$-channels. Apart from $Na^{2+}$-current reduction, the presence of the antagonist of dopamine slowed the kinetics of both activation and inactivation of ionic channels The influence of haloperidol on the membrane structure was reversible. After removal of the compound, both the amplitude and the kinetics of the current development were seen to be completely restored. Haloperidol had no effect on $Ca^{2+}$-activated chloride channels.

**TABLE 9.1**
**Changes in Membrane Potential and Ionic Permeability Induced by Neurotransmitters in Whole Cells of Plants**

| Neurotransmitter-Biomediator | Plant | Organ | Ion Permeability | Membrane Potential (MP) | Reference |
|---|---|---|---|---|---|
| Acetylcholine | *Phaseolus aureus* | Roots | $H^+$ permeability increases | Mp decreases (depolarization) | Jaffe (1970) |
| | *Phaseolus vulgaris* | Etiolated hypocotyles | $K^+$ permeability increases | MP decreases (slight depolarization on blue light) | Hartmann (1977) |
| | *Nitella* | Cells | $K^+$ permeability increases | MP decreases from 136 to 123 mV (depolarization) | Yurin et al. (1979a, 1979b) |
| | *Avena spp.* | Coleoptyles | $Ca^{+2}$ and $H^+$ permeability increases | | Tretyn (1987) |
| | *Beta vulgaris* var. *rubra* | Root disks | $Na^+/K^+$ effkux decreased/increased, relatively | | Roshchina (2001) |
| | *Nitella flexilis* | Cells | $K^+$ permeability increases | | Kisnieriene et al. (2012) |
| | *Nitellopsis obtusa* | Cells | $K^+$ ion channels activation | Generation of action potential | Kisnieriene et al. (2009) |
| | *Vicia faba* L. | Guard cell protoplasts | $K^+$ ion channel regulated stomata opening and closing | | Madhavan et al. (1995) |
| | *Nicotiana glauca* Graham | Guard cell protoplasts | $K^+$ ion channel regulated stomata opening and closing | | Madhavan et al. (1995) |
| Dopamine | *Chara carollina* | Cells | | Changes in MP | Zherelova et al. (2014, 2016) |
| Dopamine | *Beta vulgaris* var. *rubra* | Root disks | $Na^+/K^+$ and $Ca^{2+}$ efflux decreased | | Roshchina (2001) |
| Adrenaline | *Nitella syncarpa* | Cells | | Shifts in membrane potential | Oniani et al. (1973, 1974) |
| Noradrenaline | *Nitella syncarpa* | Cells | | Shifts in membrane potential | Oniani et al. (1973, 1974) |
| Serotonin | *Beta vulgaris* var. *rubra* | Root disks | $Na^+$ transport depression | | Pickles and Sutcliffe (1955) |
| Serotonin | *Nitella syncarpa* | Cells | | Shifts in membrane potential | Oniani et al. (1973, 1977) |
| Serotonin | *Beta vulgaris* var. *rubra* | Root disks | $Na^+/K^+$ and $Ca^{2+}$ efflux decreased | | Roshchina (2001) |
| Histamine | *Beta vulgaris* var. *rubra* | Root disks | $Na^+/K^+$ and $Ca^{2+}$ efflux decreased | | Roshchina (2001) |

Besides data shown on multicellular systems, there are experiments on single cells—vegetative microspores of spore-bearing plants and pollen from seed-bearing plants. Input and output of ions or pigment (from vacuole) in plasmalemma and chloroplasts were also studied (Roshchina 1990c, 2001, 2009b). The water uptake by dry microspores (2 mg in 5 mL) of *E. arvense* has been blocked by antagonists of acetylcholine (acting on nicotinic cholinoreceptors and connected ion channels), bungarotoxin, and d-tubocurarine (Roshchina 2009b) as well as tetraethylammonium (blocker of $K^+$-ion channels) and verapamil (blocker of $Ca^{2+}$-ion channels).

Although the main research concentrates on algae or unicellular models, the application of electrophysiological technique to multicellular plants, especially to woody plants (Gurovich et al. 2012) may show signaling in the long transporting systems.

### 9.8.2 Spreading Signal within a Cell or Organelle via the Systems of Secondary Messengers

Reception of neurotransmitters by sensitive plant surfaces, mainly plasmalemma and membranes of organelles, occurs via the binding with special sites (sensors-receptors), and received information is transported into cell (from outside plant cell) or organelle (from cytoplasm) through transducins (G-proteins) and via systems of secondary messengers. G-proteins are found in plants and may be transductors from different signals (Taddese et al. 2014). Participation of G-proteins in the stimulation by noradrenaline and adrenaline water-pumping root activity in *Z. mays* was shown with inhibitor of G-protein activity guanosine thiodiphosphate and stimulator of the activity by guanosine thiotriphosphate (Zholkevich et al. 2007b). Secondary mediators participate both in the transmission of external signals into the cellular interior and may strongly strengthen the original signal. To secondary messengers of cyclic AMP and GMP, inositol triphosphate, $Ca^{2+}$ ions belong, and they are also found in plant cells and within cellular organelles (Stevenson et al. 2000). Common scheme of transduction of chemosignals earlier described for animals looks as follows:

**Neurotransmitter→ receptor→ G-proteins→ secondary messengers (for example cAMP)→ protein kinases→ phosphorylation of proteins→ protein phospatases→ dephosphorylation of proteins**

Each molecule of receptor, combining with signaling molecule, activates many molecules of adenylate cyclase, for example, which, in its turn, catalyzes the formation of a number of molecules of cAMP as secondary messenger. At the end, the strengthening of initial signals ~$10^7$–$10^8$-fold occurs when it spreads along all chains of the signal transmission from receptor to observed cellular response. Thus, few signaling molecules of mediators or hormones can change functional or metabolic activity of all cells. Similar mechanisms are described in a monograph with small experimental data related to neurotransmitters (Roshchina 2001).

Transduction of the signal in plant cells from exogenous neurotransmitter via secondary messengers is not studied widely and analyzed by the help of specific inhibitors and stimulators (Figure 9.4). The experiments to show a direct participation of the cAMP system in intracellular signaling on pollen germination of *H. hybridum* (suitable model and process to the analysis) was demonstrated by use of agents acting on different sites of the system such as activator of adenylate cyclase (*adenylyl cyclase*) forscolin (diterpene from plant *Coleus*), activator of the intracellular cyclic AMP synthesis dibutyryl cAMP, inhibitors of phosphodiesterase theophylline, and isobutylmethylxanthine (Roshchina and Melnikova 1998a, 1998b). Forscolin and dibutyryl cAMP stimulated the germination of *H. hybridum* pollen *in vitro* where as theophylline and isobutylmethylxanthine inhibited the process, and exogenous cAMP and cGMP practically had no effect. Therefore, endogenous cyclic nucleotides can participate in the transduction of the information from plasmalemma to nucleus. Stimulating effects of acetylcholine, histamine,

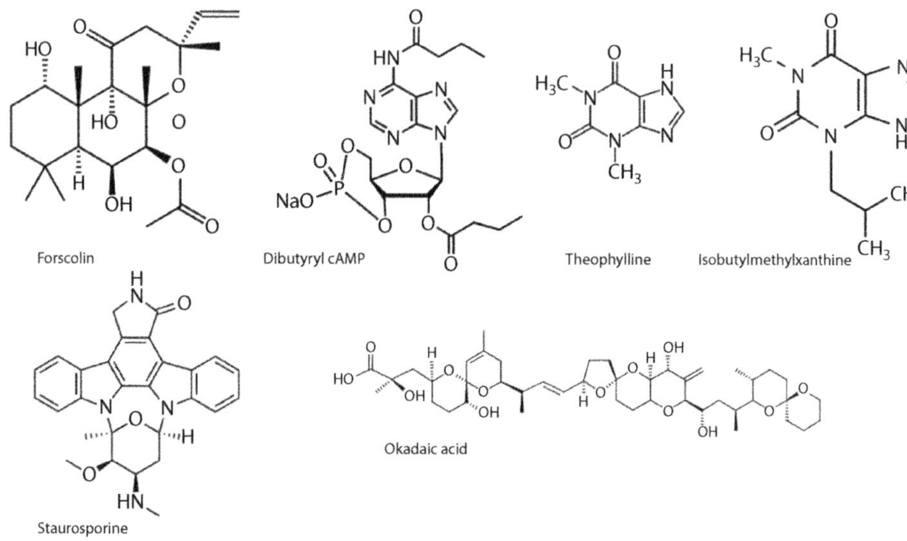


**FIGURE 9.4** Compounds used at the study of signaling with secondary messengers.

and serotonin were not observed if the pollen was preliminarily treated by theophylline and isobutylmethylxanthine. Thus, these compounds can act on the pollen germination as the final reply to exogenous neurotransmitter via messenger system of cAMP. Forskolin and theophylline were also used in a series of experiments on unicellular vegetative (*E. arvense*) microspores and generative microspores (*H. hybridum* pollen) when chemical signal transduction from the cell surface to organelles was studied with neurotransmitters acetylcholine, dopamine, and serotonin, their agonists, and antagonists, $Na^+$, $K^+$, and $Ca^{2+}$ channel blockers (Roshchina 2006a). Both types of microspores exposed to neurotransmitters, their agonists, forskolin, and theophylline demonstrated growth activation, while neurotransmitter antagonists and ion channel blockers inhibited this process. No stimulating effects of neurotransmitters were observed for cells pretreated with the antagonists and ion channel blockers. Pathways of chemical signal transduction from the cell surface to organelles are discussed.

Mechanisms of intercellular signaling were also studied with forskolin (stimulator adenylate cyclase) and theophylline (inhibitor of phosphodiesterase) using pigmented model cells where model process to analysis of signaling with participation of secondary messengers is excretion of pigments and other low-molecular metabolites from the cells. Colored objects able to excrete pigments are: (1) unicellular system—vegetative spores and generative spores (pollen); (2) multicellular systems of plants and animals including flower petal cells. Through membranes of the cell sensitive to neurotransmitters secretory products or metabolites may release out cell that well-seen in pigmented cells that has been shown on model of blue petals from *Saintpaulia* ionantha flower (Roshchina and Yashin 2013) or root disks of *Beta vulgaris* var. *rubra* (Roshchina 2001). Under the influence of external chemosignal (passing from plasmalemma to organelles through the cytoplasm) secretory products can go outside and from vacuoles. On such plant objects, one can visually highlight pigments from cells and quantitatively analyze both the output of the products and spectral characteristics of these compounds. It can be expected to consider changes in absorption and fluorescence of intact samples of their own and their secrets in solution under the influence of various factors. Pigmented cells are proposed to be models that excrete pigments and by this mode one permits to observe and estimate the process. The quantity of the secretion pigment released appears to be analyzed by spectral methods.

In root discs from *Beta vulgaris* var. *rubra* one could see the betacyanin efflux stimulated by acetylcholine, serotonin, and histamine (Roshchina 2001). The petal cells of blossoming *S. ionantha* Wendl.

released anthocyanins and proteins (including the enzyme cholinesterase) to the outer environment after the treatment with acetylcholine and its agonists nicotine, muscarine, arecoline, and quinuclidinyl benzilate (Roshchina and Yashin 2013). The secretion was largely obstructed by the antagonists of acetylcholine d-tubocurarine, tetraethylammonium, atropine, and platyphylline that indicates the participation of mixed-type cholinoreceptors (both nicotinic and muscarinic) in the secretory process. Since theophylline and forskolin, agents affecting the synthesis and catabolism of cyclic AMP, stimulated the output of secretory products, it is expected to yield intracellular signaling with secondary messenger cyclic AMP from plasmalemma to the vacuoles containing anthocyanins. The sensitivity of the petal cells to acetylcholine as well as to its agonists and antagonists permitted us to use the object as a model for studies of plant cholinergic system. Acetylcholine and its agonists nicotine, arecoline, and quinuclidinyl benzylate ($10^{-7}$–$10^{-6}$ M) induced stimulation of the process while antagonists d-tubocurarine, atropine, and platyphylline inhibited (Roshchina and Yashin 2013; Roshchina 2014). Forskolin and theophylline acting on the metabolism of secondary messenger cAMP also stimulated the excretion of the pigments and parts of proteins. But if the cells were preliminary treated by antagonists d-tubocurarine, atropine, and platyphylline, the process decreased in different degrees that showed one of the ways of signaling by acetylcholine from plasmalemma to cellular organelles with the help of secondary messenger cAMP. Among proteins excreted, cholinesterase activity was observed. The experiments were also carried out with other anthocyanin-containing petals from various species: cacti *Epiphyllum hybridum* (L.) and *Hibiscus rosa-cinensis* (L.), and from bracts of milkweed *Euphorbia milii* Desmoul, scales of red onion *Allium cepa* L., or containing pigments in leaf oil gland and ducts as St. John's wort *Hypericum perforatum* L., and bracts of hop *H. lupulus* L. (Roshchina et al. 2013). Unlike acetylcholine, biogenic amines such as dopamine, serotonin, and histamine did not demonstrate verifiable results on these model systems. As for other secondary messengers, their participation in the signaling by exogenous neurotransmitters on intact cells were not studied yet.

Participation of protein kinases or protein phosphatases in the intracellular signaling when catecholamines served as chemosignal was studied with special inhibitors staurosporine (antibiotic AM-2282 or STS, alkaloid) isolated from the *Streptomyces staurosporeus*) and okadaic acid (toxin, polyketide, polyether derivatives of C38 fatty acid, produced by several species of dinoflagellates), respectively (Figure 9.4). The participation of protein kinases and protein phosphatases in the signal transfer from exogenous catecholamines on the exudation and water-pumping ability of roots from *Z. mays* were used. These agents blocked the transfer signal from adrenaline into root cells based on the exudation intensity (Zholkevich et al. 2007a, 2007b).

Direct information of the participation ions of $Ca^{2+}$ in muscarine-acetylcholine-receptor mediated acetylcholine signal transduction in guard cells of *Vicia faba* was received in experiments with Fluo-3 fluorescence (Meng et al. 2004).

### 9.8.3 Signal Spreading via Cellular Contractile Systems

Involvement of contractile components in chemical signal transduction from the cell surface to the organelles was studied using anticontractile agents (Figure 9.5) such as (1) binding with actin of microfilaments—cytochalasin B (an inhibitor of actin polymerization in microfilaments) and phalloidin (binds with actin of microfilaments and hinders its depolymerization) and (2) binding with tubulin-containing microtubules—alkaloids colchicine and vinblastine (inhibitors of tubulin polymerization in microtubules) or taxol (stabilizer of microtubules depolymerization by suppressing that disrupts the normal process of dynamic reorganization of the network of microtubules, which is important for cellular functions at the stage of mitosis and cell cycle interphases). Besides above-mentioned reagents, latrunculins A and B are used as inhibitors of G-actin-polymerization, toxins from marine sponges (for example from genus *Latrunculia*) or phalloidin, bicyclic heptapeptide from fungi *Amanita phalloides*, that binds with F-actin and stops is depolymerization (Figure 9.5).

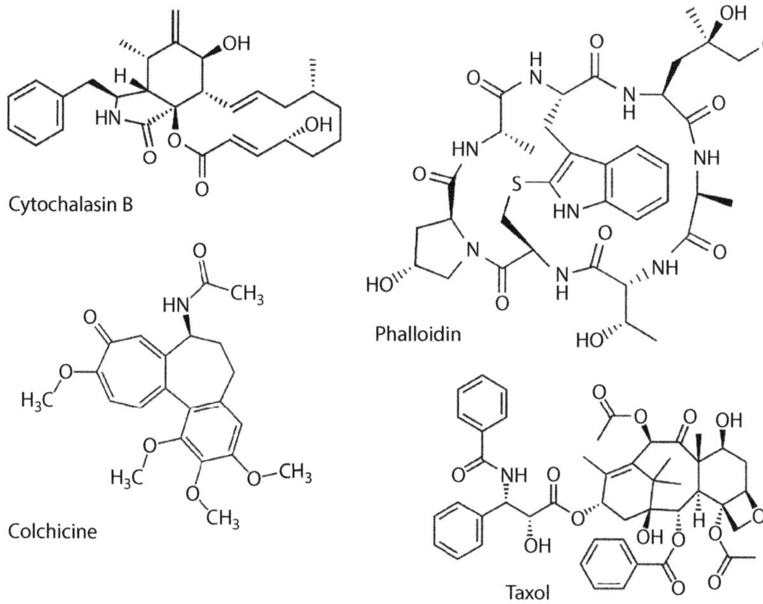


**FIGURE 9.5** Anticontractile compounds.

The compounds were tested on unicellular systems -vegetative microspores of field horsetail *E. arvense* and generative microspores (pollen) of knight's star *H. hybridum* (Roshchina 2005). It is known that the germination of pollen, as evaluated by the emergence and growth of the pollen tube, depends on the activity of the contractile proteins actin and myosin. Experiments included the observation of the microspores germination rate after the preliminary treatment with cytochalasin B (an inhibitor of actin polymerization in microfilaments), colchicine, and vinblastine (inhibitors of tubulin polymerization in microtubules) decreased the microspores germination (Roshchina 2005, 2006b).

The addition of acetylcholine to microspores pretreated with cytochalasin B or colchicine not only did not stimulate their germination but also notably decelerated their development. If these cells were first treated with ion channel blockers and then with blockers of contractile protein polymerization, the rate of inhibition of microspore germination was in many cases similar to the conditions of blocking only the ion channels. Apparently, blocking of ion channels or inhibition of polymerization of contractile proteins suppresses intracellular transduction of exogenous signal. A close relationship between the cytoskeleton and ion channels cannot be excluded. Exogenous dopamine and serotonin, usually stimulated the microspores germination as chemical signals, after the preliminary treatment with cytochalasin B (an inhibitor of actin polymerization in microfilaments), colchicines, and vinblastine (inhibitors of tubulin polymerization in microtubules) had no similar effect on the microspores germination. Both types of the treated microspores demonstrated suppressed development, particularly, after cytochalasin B treatment. Pretreatment with ion channel blockers and then by anticontractile agents (cytochalasin B or colchicine) either had no effect or increased the inhibition of microspore growth. Possible role of contractile proteins in chemical signaling during microspore development was seen due to the absence of the stimulation of microspore development by biogenic amines after pretreatment of cells with anticontractile agents for actin and tubulin. Actin binding inhibited growth to a greater extent than tubulin binding. The interaction of neurotransmitters or peroxides with the plasma membrane is realized through the contact with putative receptors. Binding of anticontractile agents is likely to occur in the same plasmalemma regions, although they also penetrate inside the cell and directly interact with the corresponding proteins of the motor (motile) structures.

Contractile proteins may participate in phloem transport of assimilates in *Raphanus sativus,* while acetylcholine in some concentrations stimulates the phloem transport of carbohydrates and the photosynthesis products while cytochalasin B, inhibitor of actin in microfilaments, inhibits it (Yang et al. 2007). Intensity of the exudation of the cutter roots of *Z. mays* was also stimulated by acetylcholine ($5 \times 10^{-5}$ M), but the anticontractile agents cytochalasin B acting on actin or colchicines acting on tubulin inhibited the process (Zholkevich et al. 2003).

Chemical signaling can be mediated by contractile proteins similar to the receptor-transducin interactions (realized through secondary messenger's calcium, cyclic AMP, or GMP, etc.) or receptor-controlled opening of ion channels. It is also possible that contractile proteins or their analogues are associated with or integrated into transducins and ion channels, which is the subject of future research.

One of the stimulatory mechanisms by biogenic amines, in particular catecholamines, may be due to the formation of reactive oxygen species such as free radicals and peroxides (in our experiments with hydrogen peroxide and *tert*-butyl peroxide), which also stimulated the microspores stimulation (Roshchina 2005, 2006). The microspore pretreatment with cytochalasin B and colchicine followed by the addition of the peroxides decreased the germination rate. At the same time, increased blue fluorescence was observed in certain cell regions (along the cell wall and around nuclei and chloroplasts) where the corresponding contractile proteins could be localized. The involvement of actin and tubulin in chemical signal transduction from the cell surface to the nucleus is proposed.

Participation of contractile structures in the water uptake is also proposed. Dry microspores of *E. arvense* at the presence of antagonists-blocker of ion channels such as bungarotoxin and d-tubocurarine inhibited the process (Roshchina 2009b).

Haloperidol, antagonist of dopamine, inhibited cytoplasm motion in cells of *C. corallina* related to microfilamentary complex (Zherelova et al. 2016). Once haloperidol was removed from cell by washing solution, cytoplasmic motion was restored.

Catecholamines also stimulated root exudation and the root-pumping activity acting via contractile proteins (Zholkevich et al. 2003, 2007a, 2007b). This was confirmed with the use of various anticontractile compounds such as colchicines binding with tubulin of microtubules and taxol (stabilizer of microtubules depolymerization by suppressing that disrupts the normal process of dynamic reorganization of the network of microtubules, which is important for cellular functions at the stage of mitosis and cell cycle interphases) or phalloidin (binds with actin of microfilaments and hinders its depolymerization). The analysis of the events showed participation of G-proteins, protein kinases, and protein phosphatases in the stimulation effects. Chemosignaling processes on plant microspores with participation of dopamine also linked with its redox effects on contractile stuctures of the plant cells because high concentrations of dopamine (~ mM), acting as a strong oxidant-allelochemical (Roshchina 2005, 2006a), decreased the root growth and cell viability in *Glycine max* (Guidotti et al. 2013). A defense function for catecholamines in the plant cell, related to their high amounts and ability to form toxic dopachrome, has also been considered in the literature (Roshchina 1991, 2001; Szopa et al. 2001; Kulma and Szopa 2007).

Contractile effects of acetylcholine connected with membrane ion permeability were also observed in the regulation of the stomata function—the opening and closing movement in plants such as *V. faba* and *P. sativum* (Wang et al. 1998, 1999a, 2000). It was established that muscarinic and nicotinic acetylcholine receptors are involved in the event. A regulatory role for acetylcholine and its antagonists in inward rectified K+ channels from guard cells protoplasts from leaf stomata of *V. faba* was found (Leng et al. 2000). $Ca^{2+}$ and Ca-related systems were found to participate in acetylcholine-regulated signal transduction during stomata opening and closing (Wang et al. 2003a; Meng et al. 2004). Smolinski and Gorska-Brylass (1994) have also shown cholinesterase activity in stomata of *Marchantia*.

### 9.8.4 Vesicular and Transporter Ways of the External Signal

An important problem to be considered is how exogenous acetylcholine and biogenic amines signal to plant cells, because their water-soluble molecules cannot penetrate through plasmatic membranes. They are too hydrophilic in order to pass across lipid layers of membranes. Their transport into and out, like in animal cell, may occur via (1) free diffusion through the plasma membrane (slowly); (2) exocytosis, resulting from fusion of a secretory granule with the plasma membrane; sorting of proteins in the secretory system also occur; and (3) fleeting release from a granule through a transient pore without full fusion or the release through a specialized plasmalemma molecule such as the mediatophore. Synaptic vesicles in animals are also essential for regulation of this type of release. They fuse with the plasma membrane only late after activity and seems to be involved in calcium sequestration and extrusion. Cellular mechanisms of the excretion may be active and passive. Active one is exocytosis. Now the concept of four fundamental mechanisms of membrane fusion in eukaryotic cells from yeast to mammalian neurons are considered. According to that, there are endosome-derived and Golgi-derived vesicular pathways to plasmatic membrane. Hydrophilic substances such as neurotransmitters and water-insoluble particles are are transported via the secretory vesicles or secretory granules by exocytosis or the rapture of the plasmatic membrane. The secretion of the large granules is supposed to connect with the formation of a fusion pore between the interior of a granule and the extracellular space. A novel special group of proteins-neurotransporters through plasmatic membrane has been also found (Schloss et al. 1992).

One of the mechanisms of internal transport of information may be also occur via special trasporters of neurotransmitters, which can transfer the compounds into and out of cells in animal nervous system (Blakely and Edwards 2012). For example, in rat striatum, dopamine transporters are phosphorylated on N-terminal serines (Foster et al. 2002). But the same mechanism for plants is only proposed. Here the vesicular system as a transport system is more understandable today.

In plant and animal cells, most hormones and mediators are localized in special vesicles—which differ from other similar structures by a coating consisting of fibrillar clathrin, with a molecular mass of 180 kDa, linked with polypeptide 35 kDa—and form a cover (Coleman et al. 1988). Coated vesicles transport the substances to organelles via endo- and exocytosis. They can transfer substances between organelles and membrane recycling. One can think these vesicles to be met in any cell and to have a universal meaning. In membrane, hormone, mediator, or other ligand are binding with special receptor that induces the arising of vesicles which gemmate from membrane. Then in cell the translocation and transformation of these vesicles take place. When the vesicle is transported to a membrane and fuses with it, the vesicle puts off the cover and liberates from clathrin so that the content of the vesicle can be released. In this process, ATPase with molecular mass 70 kDA takes part. Biological meaning of the formation of coated vesicles is unclear yet. Clathrinic cover is supposed to prevent lysis of these structures and their contents. Perhaps, the clathrinic coat promotes the coupling of vesicles with cytoskeleton following their transport in the cell. The translocation of clathrinic vesicles within cells is promoted due to the organization of the cytoskeleton, including contractile systems of microtubules and microfilaments. Microtubules are observed not only in cytoplasm, but also inside organelles, in particular

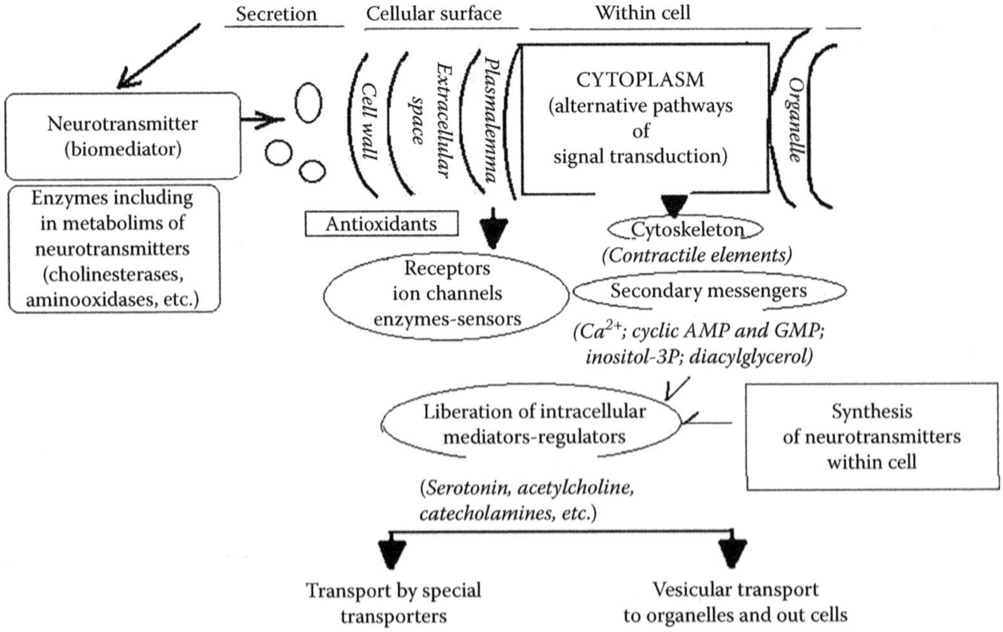


**FIGURE 9.6** Possible alternative ways of the signal transfer induced by exogenous neurotransmitter.

in chloroplasts (Vaughan and Wilson 1981) that permit these vesicles to move within the plastid too. Since acetylcholine and catecholamines are found in chloroplasts (Roshchina 1989) intracellular mediation also may take place. As a whole, the information transfer from exogenous neurotransmitter looks like as seen on Figure 9.6.

## 9.9 SIGNALING IN ORGANELLES

Intracellular fast regulation of physiological processes with using metabolites such as neurotransmitters of the excitation impulse from plasmalemma to organelles, have first arisen in unicellular organisms, where plasmalemma itself received signal from environment. Intracellular signalization as evolutional development of the adaptations and type of transmission of information appears to store from unicellular to multicellular organisms (Roshchina 2010, 2016a; Roshchina et al. 2016). Could the signaling occur between organelles within a cell? This question has been actual from the works of Buznikov done on animal embryos lack of a nervous system (see monographs 1967, 1987, 1990) and of Roshchina on chloroplasts isolated from various plant species (Roshchina 1989, 1990a, 1990b, 1993, see also monographs 1991, 2001). The intracellular signaling with participation nuclei from rats also was considered by Tretyak and Arkhipova (1992) based on influence of catecholamines and serotonin on the RNA-synthesizing capacity of isolated nuclei and chromatin of brain and rat liver (Arkhipova et al. 1988). The approaches to understanding the event should include the information similarly known for whole cells and related to (1) the presence of neurotransmitters, their receptors, and enzymic components of cholinergic and aminergic systems in separate cell compartments; (2) changes in ion permeability and membrane potential in isolated organelles; (3) the presence of possible systems of secondary messengers within organelles; and (4) the presence of contractile systems and others transport systems. In the following, all these questions (if it is possible to nowdays) will be considered.

### 9.9.1 Cholinesterase and Receptors of Acetylcholine

Cholinesterase activity of organelles may be a marker of the presence of acetylcholine and its participation in the information transfer and regulation the reactions occur within the cellular compartments. The cholinesterase activity is located within the cell too—as has been shown first in root cell cytoplasm and plasmalemma of *Phaseolus aureus* (Fluck and Jaffe 1974c, 1976), nucleus of *P. sativum* (Maheshwary et al. 1982), and chloroplasts (Roshchina and Mukhin 1984, 1986, 1987; Gorska-Brylass et al. 1990; Gorska-Brylass and Smolinski 1992). Perhaps, the contacts between organelles include the enzyme. When the hydrolysis of acetylcholine by the chloroplasts of pea *P. sativum* and common nettle *U. dioica* was analyzed (Roshchina and Mukhin 1984, 1987a), it has been shown that the highest hydrolyzing activity is concentrated in chloroplasts and is inhibited by specific inhibitors of animal cholinesterases neostigmine and physostigmine. The cholinesterase activity has been found in fractions of outer membranes and thylakoids (Roshchina 1989). Moreover, the activity of the enzyme in thylakoids was approximately seven-fold higher than in chloroplast envelope. The concentration curves of the rate of hydrolysis of cholinic esters in a dependence on substrate show that the chloroplast cholinesterase hydrolyzes acetylcholine with a higher rate than butyrylcholine, and the excess of substrate to depress the cholinesterase activity (Roshchina and Mukhin 1984).

In other cellular compartments, excluding the plasmalemma region and free space of the cell, the cholinesterase activity was not studied yet. Color reactions with azo dyes (Roshchina 2001; Roshchina and Shvirst 2016) permitted us to use this approach on isolated nuclei, vacuoles, and chloroplasts (Roshchina 2017a, 2018). By histochemical methods, the presence of this enzyme in the outer membranes of organelles isolated (nuclei, vacuoles, and chloroplasts) has been shown. Inhibitors of cholinesterase neostigmine, physostigmine, and berberine reduced or completely blocked the hydrolysis of acetylcholine cells or organelles.

Some examples of color reactions for organelles from white petals of *Philadelphus grandiflorus* is shown on Figure 9.7. The intensive blue color on the periphery of the outer membranes of nuclei and vacuoles was seen. The color intensity was decreased after the preliminary treatment with the inhibitors of cholinesterase—neostigmine and physostigmine then addition of the substrate acetylthiocholine. Using azodyes Fast Red TR and Red Analogue of Ellman reagent (Roshchina and Shvirst 2016) may observe cholinesterase in cells and tissues of animal (*Planaria* worm) and various plant cells. As we see from Table 9.2 the cholinesterase activity was observed by histochemical methods in nuclei, vacuoles, and chloroplasts. Using suitable models, one can see the reactions on cholinesterase activity on both colored and colorless objects.

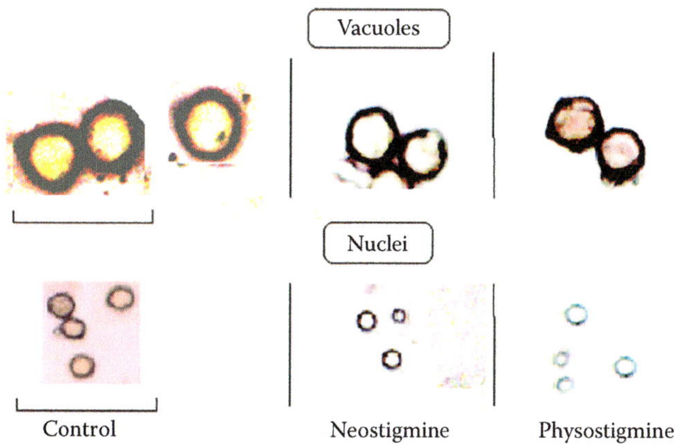


**FIGURE 9.7** Staining of vacuoles and nuclei from petals *Philadelphus grandiflorus* with Red Ellman reagent after the 60 min exposure in $10^{-3}$ M acetylthiocholine. Red Ellman reagent colored vacuoles and nuclei in blue on the outer membrane.

**TABLE 9.2**
**Occurrence of Cholinesterase in Some Plant Organelles**

| Object | Part of Plant or Organelle | Dye Colored | Inhibition Neostigmine | Inhibition Physostigmine |
|---|---|---|---|---|
| *Kalanhoe blossfeldiana* (petals) | Nuclei | Fast Red TR | Yes | Yes |
| *Philadelphus grandiflorus* (petals) | Nuclei | Red Ellman reagent | Yes | Yes |
| *Philadelphus grandiflorus* (petals) | Vacuoles | Red Ellman reagent | Inhibited, but not completely | Inhibited, but not completely |
| *Philadelphus grandiflorus* (petals) | Vacuoles | Fast Red TR | No | No |
| *Euphorbia milii* (leaves) | Vacuoles | Fast Red TR | Yes | Yes |
| *Euphorbia milii* (pink bracts) | Vacuoles | Red Ellman reagent | Yes | Yes |
| *Euphorbia milii* (leaves) | Chloroplasts | Fast Red TR | Yes | Yes |
| *Euphorbia milii* (leaves) | Chloroplasts | Red Ellman reagent | Yes | Yes |

*Source:* Roshchina and Shvirst 2016 and unpublished data.

### 9.9.2 Biogenic Amines and Their Reception

The use of fluorescent reagents for the histochemical detection of catecholamines or histamine, as well as luminescent antagonists of the intracellular neurotransmitters revealed that they can bind to certain cellular compartments (Roshchina et al. 2003, 2014; Roshchina 2016b, 2017). The subjects of the consideration were (1) application of reagents forming fluorescent products 2016a, b, (for catecholamines—glyoxylic acid, formaldehyde- or for histamine *ortho*-phthalic aldehyde) to show the presence and binding of the compounds in cells, (2) binding of their fluorescent agonists and antagonists with cell. After the treatment with glyoxylic acid (a reagent used for the detection of catecholamines), blue fluorescence with maximum at 460–475 nm was visualized in nuclei and chloroplasts (in control preparations no emission in this spectral region was recorded), as well as an intense fluorescence, exceeding the control level, in the vacuoles. After the exposure to *ortho*-phthalic aldehyde (a reagent used for the histamine detection), blue emission was more noticeable in nuclei and chloroplasts, which correlates with previously observed effects on intact cells, such as pollen and vegetative microspores. A comparison of the intensities of the biogenic amine-related emission in various organelles showed that the greatest emission was in vacuoles and the weakest, in chloroplasts (see Table 9.3). Thus, on the surface, and possibly within the organelle, fluorescence could demonstrate the presence of biogenic amines. Antagonists of the neurotransmitters (d-tubocurarine for acetylcholine; yohimbine for dopamine; norepinephrine and inmecarb for serotonin), which fluoresce in the blue and blue-green region and usually bind with the plasmalemma of intact cells, also interacted with the membranes of the organelles studied. The staining with reagents on biogenic amines leads to the appearance blue or blue-green emission on the surface and excretions of intact cells as well in some DNA-containing organelles within cells, for example nuclei (Figure 9.8). The difference between autofluorescence and histochemically induced fluorescence may reflect the occurrence and amount of biogenic amines in the cells studied. Fluorescence intensity depended on the object; most effect was recorded for yohimbine in the outer membrane of the nucleus, vacuoles, and chloroplasts. After the treatment of isolated cellular organelles with glyoxylic acid, blue emission with maximum 460–475 nm was seen in nuclei and chloroplasts (in control variants in this spectral region the noticeable emission was absent) and very expressive fluorescence (more than twenty times as compared to control) in the vacuoles. After exposure to *ortho*-phthalic

**TABLE 9.3**
**The Receptor-Dependent Reactions on Neurotransmitters in Isolated Plant Cell Organelles. Ion Permeability (IP) and Membrane Potential (MP)**

| Neurotransmitter/ Biomediator | Organelle | Receptor-Sensitive Reaction (According Binding Agonists and Antagonists) |
|---|---|---|
| Acetylcholine | Mitochondria | Uptake of oxygen (stimulation) |
| | Chloroplasts | Ion permeability and membrane potential (decreased MP, stimulates the efflux of ions $Na^+$ and $K^+$ from intact organelles, stimulates the uptake of $H^+$-ions independently on the radiation both by intact plastids and thylakoids). Stimulation of non-cyclic photophosphorylation |
| | Vacuoles | Ion permeability for $Cl^-$ (decreased MP) |
| Dopamine | Chloroplasts | Non-cyclic photophosphorylation |
| Noradrenaline | Chloroplasts | Ion permeability (stimulates efflux of $Ca^{2+}$ and $Mg^{2+}$ ions from intact chloroplasts). Stimulation of non-cyclic photophosphorylation |
| Adrenaline | Chloroplasts | Ion permeability (stimulates efflux of $Ca^{2+}$ and $Mg^{2+}$ ions from intact chloroplasts). Stimulation of non-cyclic photophosphorylation |
| Serotonin | Chloroplasts | Ion permeability (stimulates efflux of $Ca^{2+}$ and $Mg^{2+}$ ions from intact chloroplasts). |
| Histamine | Chloroplasts | Stimulation of non-cyclic photophosphorylation |

*Source:* Yunghans and Jaffe 1972; Roshchina 1989, 1990, 2001.

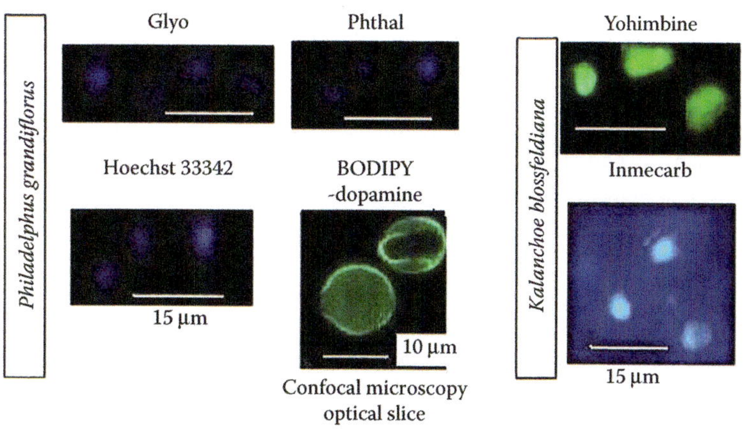


**FIGURE 9.8** Fluorescent analysis of the presence of biogenic amines and their antagonists' binding in the isolated nuclei. (From Roshchina et al. 2016a; V.V., *J. Fluorescence*, 26, 1029–1043, 2016b.)

aldehyde, blue emission was more noticeable in nuclei and chloroplasts. Fluorescent agonists (muscarine, 6,7-diOHATN, BODIPY-dopamine or BODIPY-5HT) or antagonists (d-tubocurarine for acetylcholine, yohimbine for dopamine and norepinephrine, inmecarb for serotonin) of neurotransmitters that bounded with animal receptors fluorescent in blue (460–480 nm) or blue-green (490–530 nm) and usually are bounded with external membrane of the isolated organelles studied (Roshchina et al. 2005; Roshchina 2016b). Indirect information about the presence of serotonin in chloroplasts is also seen from the occurrence of enzyme chloroplast-encoded serotonin *N*-acetyltransferase here (Byeon et al. 2015). The mammalian serotonin receptor is included in plant cells by molecular biology methods (Beljelarskaya and Sutton 2003).

The observations of Yunghan and Jaffe (1972) indicated that acetylcholine stimulates the uptake of oxygen by whole tissues and isolated mitochondria of secondary roots of *P. aureus* but had no effect on the synthesis of ATP. In vacuoles, the arising of a microaction potential may take place because there is a decrease in the membrane potential related to a chloride channel in the tonoplast (vacuolar membrane) of *C. corallina*, which also responds to acetylcholine (Gong and Bisson 2002). Changes in membrane permeability and photoreactions were registered for ions in isolated chloroplasts (Roshchina 1989, 1990c). Sensitive to acetylcholine, its agonists and antagonists reactions were $Na^+$ (in lesser degree $K^+$) efflux from intact chloroplasts and non-cyclic photophosphorylation in thylakoids (partly broken plastids). Saturation for the concentration curves was observed as well as selectivity to some agonists and antagonists (see details in monographs of Roshchina 2001). This type of reception was similar to nicotinic cholinoreceptor of animals, although in some cases features like in muscarinic cholinoreceptor were also observed. Based on the concept of universal mechanism of intracellular chemical signalization from plasmalemma to organelles, the reactions of chloroplasts on mediators were considered (Roshchina 1989, 1990a) according to symbiotic theory of the plastid origination, where chloroplast is the descendant of cyanobacteria, which has been caught by eucaryotic cell using endocytosis that led to somewhat symbiosis of two earlier independent organisms. Possible participation of neurotransmitters in intracellular signaling confirms by some facts, such as effects of low ($10^{-10}$–$10^{-7}$ M) concentrations of exogenous mediators, the presence of endogenous acetylcholine and biogenic amines in chloroplasts, and, at last, the finding of the enzyme cholinesterase in plastids.

One could imagine that the signaling into an organelle to be similar with the receiving exogenous signal by plasmalemma and transfer into a cell interior (See aforementioned sections) and transfer into the organelle interior. Participation of secondary messengers in the signal transfer into organelle is also possible, for cAMP and enzyme of its synthesis adenylate cyclase (adenylyl cyclase) that are present in chloroplasts of *Nicotiana tabacum* (Witters et al. 2005). Earlier adenylate cyclase was found in fractions of chloroplasts (Brown et al. 1989) as well as in nucleus, mitochondria, microsomes, tonoplast (Brown and Newton 1981; Ladror and Zielinskii 1989). By immunocytochemical technique, the enzyme activity has shown to be predominantly located at the intermembrane space of the chloroplast envelope when use of adenylyl imidodiphosphate as a substrate. Intact isolated chloroplasts were submitted to the same cytoenzymological procedure and revealed stromal adenylyl cyclase activity. A family (10 human genes) of enzymes called *adenylyl cyclases* that synthesize cAMP from ATP can concentrate in the *nucleus* (Pollard et al. 2017). Moreover, in chloroplasts fractions have found guanylate cyclase (Brown et al. 1989). In vacuoles of *B. vulgaris* var. *rubra* also contain cAMP and adenyllyl cyclase/adenylyl cyclase (Rykun 2013; Rykun et al. 2015; Staehelin 2015).

Universal inclusion of calcium is possible too. Noradrenaline, adrenaline, and serotonin stimulate $Ca^{2+}$ efflux from intact chloroplasts (Roshchina 1989a, 1989b). Actin and myosin are also met in chloroplasts (Ohnishi 1964) and in animal nuclei (De Lanerolle and Serebryannyy 2011); therefore, their participation in the traffic of signals from endogenous neurotransmitters into the intraorganelle space is supposed too. Similar opinion was in the review devoted interactions between endomembrane systems and actin cytoskeleton (Wang and Hussey 2015). We have no the similar data about nuclei from plants, but actin and myosin are found in mouse nuclei (De Lanerolle and Serebryanyy 2011) as well as in chloroplasts (Ohnishi 1964).

## 9.10 SIGNALING AT CELL–CELL INTERACTIONS IN THE SAME ORGANISM AND BETWEEN ORGANISMS

Neurotransmitters regulate their own metabolic processes within a cell, and the relationships (allelopathy) between neighbors in biocenosis, they may serve as attractants or repellents as well as oxidative agents (biogenic amines). Interactions between microorganism–microorganism,

microorganism–animal (human) and microorganism–plant play essential roles in the environment, and thus neurotransmitter compounds should be considered universal agents of irritation in this living relationship. Key role of neurotransmitters excreted in plant fertilization belongs to contacts between pollen and pistil at fertilization or pollen–pollen relations on pistil stigma in allelopathic competition (Roshchina 1999; Roshchina 2001, 2014, 2016a).

### 9.10.1 Cell–Cell Interaction at Fertilization

At plant fertilization, generative male microspore (named pollen) is added on the surface of the pistil by wind, insects, or by the human hand in the selection practice. Foreign pollen, which can be carried by wind or insect on the surface of pistil stigma, interacted with pollen of its own sp. that led to the antagonistic or favorable effects on the pollen germination. The first attempt to analyze direct pollen–pistil interaction was when the spreading of the excitation wave (electric impulse) along the pistil stigma was measured with electrodes. Thus, the action potential was first registered in the reproductive organs of *Ipomoea* flower, when pollen was added on pistil stigma (Sinyukhin and Britikov 1967). Primary reactions of contacting cells, both pistil and pollen, at fertilization may be not only changes in electric potential, but also in their autofluorescence that reflects the sexual recognition *own-self species*. Chemical relations between different species in a competition between pollen of own and foreign species usually take place when added on the same pistil. In the latter cases, *pollen–pistil* and *pollen–pollen* interactions will be analyzed in connection with a fertility and pollen allelopathy. Pollen–pollen contacts have been described earlier (Roshchina et al. 2009a). In the following we shall consider modeling in system pollen–pistil.

Pollen allelopathy, known as pollen–pollen competition on the pistil surface, plays a major role in normal plant fertilization in biocenosis (Roshchina et al. 2009a, 2009b).

At the contacts, *pollen–pistil* recognition of specific signal–stimuli in plant excretions by acceptor cells takes place. Pistil or pollen secretion is a primary medium for a chemosignal spreading from pollen surface to plasmalemma (Roshchina et al. 1998). Secretion of the recipient pollen cover may serve as a recognizing and transporting liquid like olfactory slime in animals for chemosignal. Acetylcholine and cholinesterase, enzymes that hydrolyse this compound, are found in plant secretions, in particular from those as pollen and pistil of some species (Bednarska and Tretyn 1989; Roshchina et al. 1994; Roshchina 1999, 2001a, 2001b, 2007a; Roshchina and Semenova 2005). The secretions of pollen contain neurotransmitters acetylcholine and histamine (Marquardt and Vogg 1952), catecholamines (Roshchina and Melnikova 1999), as well as the enzymes cholinesterase and hydrolyzing acetylcholine (Bednarska 1992; Roshchina et al. 1994).

The treatment of the pistil stigma with acetylcholine and histamine as well as their agonists (imitators) and antagonists (binding with membrane receptors in the same part of the membrane and regulating their activity) also induced the shifts in the fluorescence spectra and intensity (Roshchina and Melnikova 1998a, 1988b; Roshchina 2001). The addition of acetylcholine on the pistil stigma almost doubled the fluorescence intensity in maximum 475 nm and shoulders 530 nm. However, after the preliminary treatment with its antagonist d-tubocurarine, which blocks animal nicotinic cholinoreceptor, the effects were not observed. A similar picture was for antagonist atropine (Roshchina 2001). It showed the presence of the receptors of acetylcholine on the stigma surface. Moreover, this response was absent if the pistil stigma was preliminarily treated with antagonists of acetylcholine or histamine such as d-tubocurarine or clemastin (tavegyl) that links receptors on the cellular surface. Thus, autofluorescence of the cells at contacts of pollen and pistil appears to be a biosensor reaction for neurotransmitters and the antineurotransmitter substances. Another surface sensor on the pistil is enzyme cholinesterase, which is present in the stigma excretions (Roshchina and Semenova 1995). The treatment of the pistil stigma with neostigmine, inhibitor of cholinesterase also makes smooth the fluorescence spectra of the structure. Therefore, it was shown that acetylcholine and histamine induced the changes in the pistil stigma fluorescence, like is observed at pollen addition (Roshchina 2001).

Addition of acetylcholine and its antagonists binding with receptors or anticholinesterase agents on pistil surface before pollination led to the significant changes in the formation of fruits and seeds (Roshchina and Melnikova 1998a): (1) time of the fertilization (the closing cells after pollination) in control was about one day after addition of pollen, acetylcholine, and its antagonists atropine and d-tubocurarine retarded the process up to four to six days, and the beginning of the fruit formation up to eight-nine days; (2) formed fruits and seeds were different because inhibitors of cholinesterase stimulated about twofold the fruit weight, while antagonists such as α-bungarotoxin and d-tubocurarine decreased. Analysis with special inhibitors of natural and artificial origins, which affect receptors, ion channels, and contractile systems in both animal and plant cells has been carried out (Roshchina 2007; Roshchina and Vikhlyantsev 2009). The negative effects on fertilization (reduction in fruits and lack of seeds) were also observed. Ugly fruits without seeds were received after the treatment with α-bungarotoxin antagonist of acetylcholine acting as the blocker of $Na^+$ channels and nicotinic type of cholinoreceptor α-bungarotoxin, tetraethylammonium that blocks of $K^+$ channel binding with nicotinic type of cholinoreceptor, verapamil, calcium channel blocker; cytochalasin B, inhibitor of actin polymerization, and alkaloid papaverin, antispazmatic drug relaxing contractile systems. Unlike these compounds, colchicine, inhibitor of tubulin polymerization, did not depress the fruit and seed development. Therefore, the pistil has cholinoreceptors, and one way of the acetylcholine signaling within cell may include contractile proteins, more really, actin.

### 9.10.2 Plant Cell Interactions with Various Organisms

Neurotransmitters may participate in intercellular contacts between organisms of various kingdoms, for example, the insect secretions induced the bioelectric reactions in potato (Volkov and Haack 1995). There are direct evidences related to neurotransmitters, mainly connected with water inhabitants. One can mark that increased dopamine content in some algae, in particular *Ulvaria obscura,* has led to the consideration of the neurotransmitter as a feeding deterrent (van Alstyne et al. 2006, 2013, 2014). The confirmation of dopamine production acting as a defense mechanism against grazers was done from experiments with isopods, snails, and sea urchin eating the agar-based foods contained exogenous dopamine. Damaged algae were also found to release a water-soluble reddish-black substance (dopachrome) that inhibits the development of brown algal embryos, reduced the rates of macroalgal and epiphyte growth, and caused increase mortality in oyster larvae (Nelson et al. 2003). Yamamoto with co-workers (1999) have studied roles of dopamine and serotonin in larval attachment of the barnacle *Balanus amphitrite.* This is a novel ecological role for a catecholamine. Histamine released by host alga of sea urchin *Holopneustes purpurascens* was able to induce settlement of larvae (Swanson et al. 2004). A defense function of catecholamines for the plant cells themselves has also been considered in the literature (Roshchina 1991, 2001; Szopa et al. 2001; Kulma and Szopa 2007).

Moreover, in a result of this connection biogenic amines may be regulators of the insect development. One example of such communications is the leaves of *Urtica* genus which leaves serve as the site of the butterfly development. Image of fluorescent products as a reaction on the presence of dopamine and noradrenaline on egg input of butterfly *Vanessa* spp. was seen on the leaf of stinging nettle *Urtica urens* (Roshchina 2016c). Both plants and insects contain the neurotransmitters, but in the latter the concentration is higher. Figure 9.9 illustrates the fluorescent reaction on histamine in the same object. In ultraviolet or violet light, leaves of *U. urens* are weakly emitted, and after the treatment with *o*-phthalic aldehyde—a reagent for histamine—we see the bright fluorescence of the stinging hairs emerging off the leaf and the grena with eggs of butterfly. Eggs of the butterfly may need acetylcholine and biogenic amines as development and growth regulators. That is why the insects prefer neurotransmitter-enriched *Urtica* species. Regulatory roles of biogenic amines were also seen in the reproductive behavior insects on plants such as rice (Yamane and Miyatake 2010; Yamane 2013). The development of the insects depends on dopamine in western tarnished plant bug *Lygus hesperus* (Brent et al. 2016), dopamine and serotonin in the cricket, *Gryllus bimaculatus* (Woodring

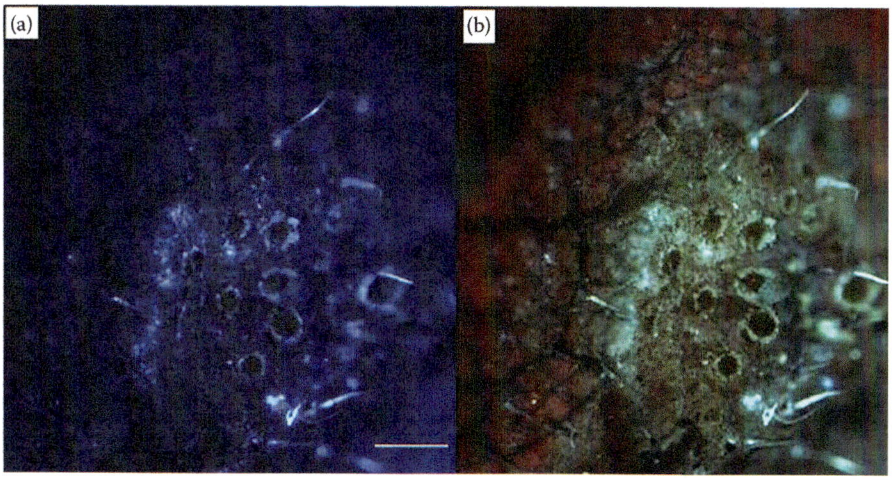


**FIGURE 9.9** Image of fluorescent products as a reaction on the presence of histamine on grena with open eggs of butterfly *Vanessa* spp. seen on the leaf of stinging nettle *Urtica urens*. Bar—100 μm. a and b—excitation by light 360–380 nm and 400–420 nm, respectively.

and Hoffmann 1994). Dopamine and noradrenaline were released in the salivary glands and brain of the tick *Boophilus microplus* (Megaw and Robertson 1974). Cones of *Drosofilla* excrete acetylcholine during the development (Yao et al. 2000).

Special interest is in the carnivorous plants' slime, which contains cholinesterase such as in genus *Utricularia* and *Drosera* (Roshchina and Semenova 1995; Roshchina 2001). Recognition of the insect in this case may be reflecting not only mechanical irritation, but a sense to acetylcholine from the animal.

Interactions of plants with bacteria or fungi also may carry out through neurotransmitters and enzymes of their metabolism (Lyte and Ernst 1992; Oleskin et al. 1998; Oleskin 2012). It may have an essential role in food products where microbial infection induces accumulation bacterial catecholamines or histamine (Silla Santos 1996; Ekici and Coscun 2002; Ekici et al. 2006). Extremly important are the microorganisms–plant interactions for food and pharmaceutical practice, because high concentrations of biogenic amines and histamine contained in the organisms from both kingdoms may act as toxins (Bodmer et al. 1999; Ekici and Coskun 2002; Ekici et al. 2006). Leaf wilting and heat stress, perhaps, have relation to the interaction with microorganisms (Momonoki and Momonoki 1991, 1992, 1993a). In defense mechanisms against fungal infections, enhanced synthesis of serotonin is known for rice (Ishihara et al. 2008, Kang et al. 2009).

## 9.11 METABOLIC SIGNALING IN THE DEVELOPMENT AND AT STRESS

Changes in quality and quantity of neurotransmitters appear also to be considered as metabolic signaling during the development (phase of the maturing) and at stress (a defense adaptation). In this case, inhabitors of biocenosis, including insects and herbivaries that use the compounds as food and regulators of their metamorphosis, react on the alterations. Neurotransmitters/biomediators of plants serve as carriers of important ecological information like it is supposed for pollen (Geodakyan 1978), which is often enriched in the substances, especially biogenic amines (Roshchina 2014; Roshchina and Yashin 2014; Roshchina et al. 2014). Analysis of received information for 50 years (Roshchina 2001) shows that catecholamines and serotonin are accumulated in fruits at maturing of some species, for example banana as demonstrated with fluorescent method (Kimura 1968). Barwell (1979, 1989) connected the algae development with different amounts of histamine. At different stages of development in *Papaver somniferum*, the amount of dopamine is markedly changed (Kamo and Mahlberg 1984). Similar event takes place in economical species for serotonin and its derivative melatonin, animal hormone (Murch et al. 2009, 2010). Keep in mind that animals may regulate metabolism with the neurotransmitters by eating the plants; their sense percepts the compounds. Their food role may be peculiar to microorganisms, both fungi and bacteria (Strakhovskaya et al. 1993; Oleskin et al. 1998) living on the plants too.

Stress also impresses in the accumulation of biogenic amines. Parasitic infection induced serotonin accumulation in rice intensively studied by Kang with co-workers (2009). In potato, catecholamines are involved in plant responses toward biotic and abiotic stresses (high salt condition, changes in light and dark, cold, etc.) due to the decrease in their catabolism (Swiedrych et al. 2004). Heat stress response of plants (*Macroptilum*, maize, rice) may also occur through the enchanced acetylcholinesterase activity that decreased the amount of acetylcholine (Momonoki and Momonoki 1992, 1993b; Yamamoto et al. 2011, 2012). Increased salt concentration enhanced biogenic amines and histamine in germinating microspores of horsetail *E. arvense* (Roshchina and Yashin 2014). Analogical effects were seen in the object and pollen from various species under the high concentration of ozone in air (Roshchina et al. 2014, 2015a, 2015b). The role of neurotransmitters in the defense and adaptation of plants could be of scientific interest in the future.

## 9.12 PERSPECTIVES IN FOLLOWING STUDIES

1. The studies of character of cellular reactions at signaling, including electric, growth, and motile responses
2. Search of new models suitable for the observation for the cellular reactions
3. Inhibitory (pharmacological) analysis of the receptors presence and possibly their genetic identification
4. Analysis of the neurotransmitters (biomediators) participation in symbiotic or parasitic relations of microorganisms within plant or animal organisms that is of ecological and medicinal importance for practice

## 9.13 CONCLUSION

The compounds known as neurotransmitters or biomediators in plant cells participate both in signaling outside cell named intercellular signaling (with the compounds and enzymes of their metabolism, as established for cholinesterase excreted by any cells—from plant cell of the same individual plant or other plant species to animal and microorganisms) and in intracellular signaling (within plant cell). It includes transfer of external signals by secondary messengers, contractile, vesicular systems, or direct influence of intracellular neurotransmitters as intracellular regulators on the organelles. Neurotransmitters regulate their own metabolic processes within a cell and the relationships (allelopathy) between neighbors within biocenosis where serve as attractants or repellents as well as oxidative agents (mainly biogenic amines). Interactions between microorganism–microorganism, microorganism–animal (human), and microorganism–plant play essential roles in the environment, and thus neurotransmitter compounds should be considered as universal agents of irritation in this living relationship.

## REFERENCES


Arkhipova, L.V., Tretyak, T.M., Ozolin, O. 1988. The influence of catecholamines and serotonin on RNA-synthesizing capacity of isolated nuclei and chromatin of brain and rat liver. *Biochemistry* (USSR) 53: 1078–1081.

Baluska, F., Hlavacka, A., Mancuso, S., Barlow, P.W. 2006. Neurobiological view of plants and their body plan. In: Baluska, F., Mancuso, S., Volkmann, D. (Eds.), *Communication in Plants – Neuronal Aspects of Plant Life*. Springer, Berlin, Germany, pp. 19–35.

Baluska, F, Volkmann., D., Menzel, D. 2005. Plant synapses: Actin-based domains for cell-to-cell communication. *Trends Plant Sci* 10: 106–111.

Bamel, K., Gupta, R., Gupta, S.C. 2015. Nicotine promotes rooting in leaf explants of in vitro raised seedlings of tomato, *Lycopersicon esculentum* var Pusa Ruby. *Int Immunopharmacol* 29: 231–234. doi:10.1016/j.intimp.2015.09.0012.

Bamel, K., Gupta, R., Gupta, S.C. 2016. Acetylcholine suppresses shoot formation and callusing in leaf explants of in vitro raised seedlings of tomato, Lycopersicon esculentum Miller var. Pusa Ruby. *Plant Signal Behav* 11 (6): e1187355. doi:10.1080/15592324.2016.1187355.

Barwell, C.J. 1979. The occurrence of histamine in the red alga of *Furcellaria lumbricalis* Lamour. *BotanicaMarina* 22: 399–401.

Barwell, C.J. 1989. Distribution of histamine in the thallus of Furcellaria lumbricalis. *J Appl Phycol* 1: 341–344.

Barwell, C.J. 1980. Acetylcholine in the red algae *Laurencia obtusa* (Huds.) Lamour. *Botanica Marina* 23: 63–64.

Bednarska, E. 1992. The localization of nonspecific esterase and cholinesterase activity in germinating pollen and in pollen tube of Vicia faba L. The effect of actinomicyn D and cycloheximide. *Biologia Plantarum* 34: 229–240.

Bednarska, E., Tretyn, A. 1989. Ultrastructural localization of acetylcholinesterase activity in the stigma of *Pharbitis nil* L. *Cell Biol Int* 13: 275–281.

Beljelarskaya, S.N. and Sutton, F. 2003. Expression of the mammalian serotonin receptor in plant and amphibian cells. *Mol Biol* 37 (3): 387–391. doi:10.1023/A:1024283126198.

Brent, C., Miyasaki, K., Vuong, C. et al. 2016. Regulatory roles of biogenic amines and juvenile hormone in the reproductive behavior of the western tarnished plant bug (*Lygus hesperus*). *J Comp Physiol B* 186: 169–179.

Blakely, R.D. and Edwards, R.H. 2012. Vesicular and plasma membrane transporters for neurotransmitters. *Cold Spring Harb Perspect Biol* 4 (2). doi:10.1101/cshperspect.a005595.

Bodmer, S., Imark, C., Kneubühl, M. 1999. Biogenic amines in foods: Histamine and food processing. *Inflamm Res* 48: 296–300.

Borodyuk, N.R. 1990. Role of the acetylcholine hydrolysis for the vital biological systems. *Vestnik Selkhozyastvennoi Nauki* 6: 87–95.

Bridges, R.G. 1972. Choline metabolism in insects. In: Treherne, J.E., Berridge, M.J., Wigglesworth, V.B. (Eds.), *Advances in Insect Physiology*, vol. 9. Academic Press, London, UK, pp. 51–111.

Brenner, E.D., Stahlberg, R., Mancuso, S. et al. 2006. Plant neurobiology: An integrated view of plant signaling. *Trends Plant Sci* 11: 413–419.

Brown, E.G. and Newton, R.P. 1981. Cyclic AMP and higher plants. *Phytochemistry* 20: 2453–2463.

Brown, E.G., Newton, R.P., Evans, D.E., Walton, T.J., Younis, L.M. and Vaughan, J.M. 1989. Influence of light on cyclic nucleotide metabolism in plants, effect of dibutyryl cyclic nucleotides on chloroplast components. *Phytochemistry* 28: 2559–2563.

Buchanan, B.B., Gruissen, W., Jones, R.L. (Eds.). 2015. *Biochemistry and Molecular Biology of Plants*, 2nd ed. Wiley-Blackwell, Chichester, UK.

Buznikov, G.A. 1967. *Low Molecular Regulators of Embryonal Development*. Nauka, Moscow, Russia.

Buznikov, G.A. 1987. *Neurotransmitters in Embryogenesis*. Nauka, Moscow, Russia.

Buznikov, G.A. 1990. *Neurotransmitters in Embryogenesis*. Harwood Academic Press, Chur, Switzerland.

Buznikov, G.A., Shmukler, Y.B. 1981. Possible role of "prenervous" neurotransmitters in cellular interactions of early embryogenesis: A hypothesis. *Neurochem Res* 6 (1): 55–68. doi:10.1007/BF00963906.

Buznikov, G.A., Shmukler, Y.B., Lauder, J.M. 1996. From oocyte to neuron: Do neurotransmitters function in the same way throughout development? *Cell Mol Neurobiol* 16 (5): 532–559.

Buznikov, G.A., Shmukler, Y.B., Lauder, J.M. 1999. Changes in the physiological roles of neurotransmitters during individual development. *Neurosci Behav Physiol* 29: 11–21. doi:10.1007/BF02461353.

Byeon, Y., Lee, H.Y., Choi, D.W. et al. 2015. Chloroplast-encoded serotonin *N*-acetyltransferase in the red alga *Pyropia yezoensis*: Gene transition to the nucleus from chloroplasts. *J Exp Bot* 66 (3): 709–717.

Cassimeris, L., Lindappa, V.R., Plopper, G. (Eds.). 2011. *Lewin's Cells*. Jones and Bartlett Publshing, Boston, MA.

Coleman, J., Evans, D., Hawes, C. 1988. Plant coated vesicles. *Plant Cell and Environment* 11: 669–684.

Corringer, P.J., Le Novère, N., Changeux, J.P. 2000. Nicotinic receptors at the amino acidic level. *Ann Rev Pharmacol Toxicol* 40 (1): 431–458.

Csaba, G. and Muller, W.E.G. 1996. *Signalling Mechanisms in Protozoa and Invertebrates*. Springer, Berlin, Germany.

Davis, K.L., Charney, D., Coyle, J.T., Nemeroff, C. (Eds.). 2002. *Neuropsychopharmacology*. The fifth generation of progress. Lippincott Williams and Wilkinsm, Philadelphia, PA.

De Lanerolle, P. and Serebryannyy, L. 2011. Nuclear actin and myosins: Life without filaments. *Nature Cell Biology* 13: 1282–1288. doi:10.1038/ncb2364.

Di Sansebastiano, G.P.D., Fonaciari, S., Barozzi, F. et al. 2014. New insights on plant cell elongation: Role for acetylcholine. *Int J Mol Sci* 15: 4565–4582. doi:10.3390/ijms15034565.

D'Souza, U.M. 2015. Gene and promoter structures of the dopamine receptors. In: K.A. Neve (Ed.), *The Dopamine Receptors*, 2nd ed. Humana Press, a part of Springer Science+Business Media, LLC 2010, New York.

Ekici, K., Coskun, H., Tarakci, Z. et al. 2006. The contribution of herbs to the accumulation of histamine in "otlu" cheese. *J Food Biochem* 30: 362–371.

Ekici, K. and Coşkun, H. 2002. Histamine content of some commercial vegetable pickles. *Proceedings of ICNP-2002*, Trabzon, Turkey, pp. 162–164.

Erland, L.A.E., Turi, C.E., Saxena, P.K. 2016. Serotonin: An ancient molecule and an important regulator of plant processes. *Biotechnol Adv* 34 (8): 1347–1361.

Fluck, R.A., Leber, P.A., Lieser, J.D. et al. 2000. Choline conjugates of auxins. I. Direct evidence for the hydrolysis of choline-auxin conjugates by pea cholinesterase. *Plant Physiol Biochem* 38: 301–308.

Fluck, R.A. and Jaffe, M.J. 1974a. The acetylcholine system in plants. In: Smith, E. (Ed.), *Current Advances in Plant Sciences*, Vol. 5. Science Engineering, Medical and Data, Oxford, UK, pp. 1–22.

Fluck, R.A. and Jaffe, M.J. 1974b. The distribution of cholinesterases in plant species. *Phytochemistry* 13: 2475–2480.

Fluck, R.A. and Jaffe, M.J. 1974c. Cholinesterases from plant tissues VI. Distribution and subcellular localization in *Phaseolus aureus* Roxb. *Plant Physiol* 53: 752–758.

Fluck, R.A. and Jaffe, M.J. 1976. The acetylcholine system in plants. In: Smith, H. (Ed.), *Commentaries in Plant Science*. Pergamon Press, Oxford, UK, pp. 119–136.

Foster, J.D., Pananusorn, B., Vaughan, R.A. 2002. Dopamine transporters are phosphorylated on N-terminal serines in rat striatum. *J Biol Chem* 277: 25178–25186.

Geodakyan, V.A. 1978. Pollen amount as transmitter of ecological information and regulator of evolutional plant plastificity. *J General Biol* (USSR) 39 (5): 743–753.

Gehring, C. 2010. Adenyl cyclases and cAMP in plant signaling—past and present. *Cell Commun Signal*. doi:10.1186/1478-811X-8-15.

Gomes, B.R., de Cássia Siqueira-Soares, R., Dantas dos Santos, W. et al. 2014. The effects of dopamine on antioxidant enzymes activities and reactive oxygen species levels in soybean roots. *Plant Signal Behav* 9 (12): e977704.

Gong, X.O. and Bisson, M.A. 2002. Acetylcholine-activated Cl- channel in the *Chara* tonoplast. *J Membr Biol* 188: 107–113.

Gorska-Brylass, A., Rascio, N., Mariani, P. 1990. Cytochemical localization of cholinesterase in the thylakoids of chloroplasts of *Marchantia polymorpha*. *Cell Biol Int* 14 (abstr. Suppl): 208.

Gorska-Brylass, A. and Smolinski D.J. 1992. Ultrastructural localization of acetylcholinesterase activity in stomata of *Marchantia polymorpha* L. W: L. MeGias-MeGias et al. (Eds.), *EUREM 92, Seria Electron Microscopy*, vol. 3. Universidad de Granada, Spain, pp. 439–440.

Grosse, W. 1982. Function of serotonin in seeds of walnuts. *Phytochemistry* 21: 819–822.

Gupta, A. and Gupta, R. 1997. A survey of plants for presence of cholinesterase activity. *Phytochemistry* 46: 827–831.

Gupta, A., Thakur, S.S., Uniyal, P.L., Gupta, R. 2001. A survey of Bryophytes for presence of cholinesterase activity. *Am J Bot* 88: 2133–2135.

Gupta, A., Vijayaraghavan, M.R., Gupta, R. 1998. The presence of cholinesterase in marine algae. *Phytochemistry* 49 (7): 1875–1877.

Gurovich, L.A. 2012. Electrophysiology of woody plants. In: Oraii, S. (Ed.), *Electrophysiology–From Plants to Heart*. InTech, Rijeka, Croatia, pp. 1–25.

Guidotti, B.B., Gomes, B.R., Siqueira-Soares, R.D. et al. 2013. The effects of dopamine on root growth and enzyme activity in soybean seedlings. *Plant Signal Behav* 8: e25477.

Hartmann, E. and Gupta, R. 1989. Acetylcholine as a signaling systems in plants. In: Boss, W.F. and Morve, D.I. (Eds.), *Second Messengers in Plant Growth and Development*. New York: Allan R. Liss, pp. 257–287.

Hikino, H., Ogata, M., Konno, C. 1983. Structure of feruloylhistamine a hypotensive principle of Ephedra roots. *Planta Med* 48: 108–109.

Ishihara, A., Hashimoto, Y., Tanaka, C. et al. 2008. The tryptophan pathway is involved in the defense responses of rice against pathogenic infection via serotonin production. *Plant J* 54 (3): 481–495.

Iyer, L.M., Aravind, L., Coon, S.L. et al. 2004. Evolution of cell–cell signaling in animals: Did late horizontal gene transfer from bacteria have a role? *Trends Genet* 20 (7): 292–299.

Jaffe, M.J. 1970. Evidence for the regulation of phytochrome mediated processes in bean roots by the neurohumor, acetylcholine. *Plant Physiol* 46: 768–777.

Kamo, K.K. and Mahlberg, P.G. 1984. Dopamine biosynthesis at different stages of plant development in *Papaver somniferum*. *J Nat Prod* 47: 682–686.

Kang, K., Kim, Y.S., Park, S., Back, K. 2009. Senescence-induced serotonin biosynthesis and its role indelaying senescence in rice leaves. *Plant Physiol* 150 (3): 1380–1393.

Karnovsky, M. and Roots, L.A. 1964. A direct colouring method for cholinesterases. *J Histochem Cytochem* 12: 219–225.

Kimura, M. 1968. Fluorescence histochemical study on serotonin and catecholamine in some plants. *Jap J Pharmacol* 18: 162–168.

Kisnieriene, V., Sakalauskas, V., Pleskačiauskas, A., Yurin, V., Rukšėnas, O. 2009. The combined effect of $Cd^{2+}$ and ACh on action potentials of *Nitellopsis obtusa* cells. *Cent Eur J Biol* 4 (3): 343–350.

Kisnieriene, V., Ditchenko, T.I., Kudryashov, A.P. et al. 2012. The effect of acetylcholine on Characeae $K^+$ channels at rest and during action potential generation. *Cent Eur J Biol* 7 (6): 1066–1075.

Koelle, G.B. 1963. Cytological distributions and physiological functions of cholinesterases. *Handb Exp Pharmacol* I5 (l): 187–298.

Koštiř, J., Kleňha, J., Jirāŝek, V. 1965. Vliv cholinu a acetylcholinu na Kliceni semen hospodarsky dulezitych plodin (The influence of choline and acetylcholine on the germination of seeds of economically important plants.) *Rostlinnà Vŷroba.Rocnik* 11 (28): 1239–1280.

Kovaleva, L.V. and Roshchina, V.V. 1997. Does cholinesterase participate in the intercellular interaction in pollen-pistil system? *Biol Plant* 39 (2): 207–213.

Kuklin, A.I. and Conger, B.V. 1995. Catecholamines in plants. *J Plant Growth Regul* 14: 91–97.

Kulma, A. and Szopa, J. 2007. Catecholamines are active compounds in plant. *Plant Sci* 172: 433–440.

Kutchan, T.M., Rush, M., Coscia, C.J. 1986. Subcellular localization of alkaloids and dopamine in different vacuolar compartment of *Papaver bracteatum*. *Plant Physiol* 81 (1): 161–166.

Ladror, U.S. and Zielinski, R.E. 1989. Protein kinase activities in tonoplast and plasmalemma membranes from corn roots. *Plant Physiology*. 89: 151–158.

Lanier, L.L. and Limbird, L. (Eds) 1997. *α2-Adrenergic Receptors. Structure, Function and Theuropeutic Implication*. Harwood Academic Publishers, Amsterdam, the Netherlands.

Lembeck, F. and Skofitsch, G. 1984. Distribution of serotonin in *Juglans regia* seeds during ontogenetic development and germination. *Zeischrift der Pflanzenphysiologie* 114: 349–353.

Leng, Q., Hua, B., Guo., Y., Lou, C. 2000. Regulating role of acetylcholine and its antagonists in inward rectified K+ channels from guard cells protoplasts of *Vicia faba*. *Sci Chin Ser C* 43 (2): 217–224.

Lu, G., DeLisle, A.J., de Vetten, N.C., Ferl, R.J. 1992. Brain proteins in plants: An *Arabidopsis* homolog to neurotransmitter pathway activators is part of a DNA binding complex. *Proc Natl Acad Sci USA* 89: 11490–11494.

Lyte, M. and Ernst, S. 1992. Catecholamine induced growth of gram-negative bacteria. *Life Sci* 50: 203–212.

Madhavan, S., Sarath, G., Lee, B.H., Pegden, R.S. 1995. Guard cell protoplasts contain acetylcholinesterase activity. *Plant Sci* 109: 119–127.

Maheshwary, S.C., Gupta, R., Gharyal, P.K. 1982. Cholinesterase in plants. In: Sen, S.P. (Ed.), *Recent Developments in Plant Sciences: S.M. Sircar Memorial Volume*. Today and Tomorrows Printers and Publishers, New Delhi, India, pp. 145–160.

Malik, S. 2014. Cholinesterase in animals and *Arabidopsis thaliana*. *IJIRD* 3 (2): 280–281.

Marquardt, P. and Vogg, G. 1952. Pharmakologische und chemische Untersuchungen uber Wirkstoffe in Bienenpollen. *Arzneimittel forschung* 21 (353): 267–271.

Massoulie, Y., Pezzementi, L., Bon, S., Krejci, E., Vallette, F. 1993. Molecular and cellular biology of cholinesterases. *Prog Neurobiol* 41: 31–91.

Megaw, W.J. and Robertson, H.A. 1974. Dopamine and noradrenaline in the salivary glands and brain of the tick *Boophilus microplus*: effect of reserpine. *Cell Mol Life Sci* 30: 1261–1262.

Meng, F., Liu, X., Zhang, S., Lou, C. 2001. Localization of muscarinic acetylcholine receptor in plant guard cells. *Chin Sci Bull* 46: 586–589.

Meng, F., Miao, L., Zhang, S., Lou, C. 2004. $Ca^{2+}$ is involved in muscarine acetylcholine receptor-mediated acetylcholine signal transduction in guard cells of *Vicia faba*. *Chin Sci Bull* 49 (5): 471–475.

Miles, P.W. 1972. The saliva of hemiptera. In: Treherne, J.E., Berridge, M.J., Wigglesworth, V.B. (Eds.), *Advances in Insect Physiology*, vol. 9. Academic Press, London, UK, pp. 183–256.

Momonoki, Y.S. and Momonoki, T. 1991. Changes in acetylcholine levels following leaf wilting and leaf recovery by heat in plant cultivars. *Nippon Sakumotsu Gakkai Kiji.Jpn J Crop Sci* 60: 283–290.

Momonoki, Y.S. and Momonoki, T. 1992. The influence of heat stress on acetylcholine content and its hydrolyzing activity in *Macroptilium atropurpureum* cv. Siratro. *Jpn J Crop Sci* 61: 112–118.

Momonoki, Y.S. and Momonoki, T. 1993a. Changes in acetylcholine-hydrolyzing activity in heat-stressed plant cultivars. *Jpn J Crop Sci* 62: 438–448.

Momonoki, Y.S. and Momonoki, T. 1993b. Histochemical localization of acetylcholinesterase in leguminous plant, siratro (*Macroptilium atropurpureum*). *Jpn J Crop Sci* 62: 571–576.

Momonoki, Y.S. 1997. Asymmetric distribution of acetylcholinesterase in gravistimulated maize seedlings. *Plant Physiol* 114: 47–53.

Mukherjee A. 2015. Computational analysis of a cys-loop ligand gated ion channel from the green alga *Chlamydomonas reinhardtii*. *Mol Biol* 49 (5): 742–754.

Mukherjee, S., David, A., Yadav, S., Baluška, F., Bhatla, S.C. 2014. Salt-induced seedling growth inhibition coincides with differential distribution of serotonin and melatonin in sunflower seedling roots and cotyledons. *Physiol Plant* 152 (4): 714–728.

Muralidharan, M., Buss, K., Larrimore, K.E. et al. 2013. The *Arabidopsis thaliana* ortholog of a purported maize cholinesterase gene encodes a GDSL-lipase. *Plant Mol Biol* 81 (6): 565–576.

Murch, S.J., Alan, A.R., Ca, J., Saxena, P.K. 2009. Melatonin and serotonin in flowers and fruits of *Datura metel* L. *J Pineal Res* 47: 277–283.

Murch, S.J., Hall, B.A., Le, C.H., Saxena, P.K. 2010. Changes in the levels of indoleamine phytocemicals during veraison and ripening of wine grapes. *J Pineal Res* 49: 95–100.

Nelson, T., Lee, D., Smith, B. 2003. Are 'green tides' harmful algal blooms? Toxic properties of water-soluble extracts from two bloom-forming macroalgae *Ulva fenestrate* and *Ulvaria obscura* (Ulvophyceae). *J Phycol* 39: 874–879.

Ohnishi, J. 1964. Le changement de volume du chloroplaste, accompagne de photophosphorylation et les proteins ressemblantes a l'active et a la myosine extractes du chloroplasts. *J Biochem* (*Tokyo*) 55: 494–503.

Oleskin, A.V., Kirovskaya, T.A., Botvinko, I.V., Lysak, L.V. 1998. Effects of serotonin (5-hydroxytryptamine) on the growth and differentiation of microorganisms. *Microbiology* (*Russia*) 67: 305–312.

Oleskin, A.V. 2012. *Biopolytics: The Political Potential of the Life Sciences.* Nova Science Publishers, New York.

Oniani, D.A., Vorob'ev, L.N., Kudrin, A.N. 1973. Influnce of noradrenaline, adrenaline and izadrin on the motility of protoplasm of cells of *Nitella syncarpa*. In: Yankyavichus, K.K. (Ed.), *Characean Algae and Their Use in the Studies of the Biological Celluar Processes.* Institute of Botany of Lit SSR, Vilnus, Lithuania, pp. 423–432.

Oniani, D.A., Vorob'ev, L.N., Kudrin, A.N. 1974. Influence of noradrenaline on the motility of protoplasm of cells of *Nitella*. *Soobshcheniya Grusinian SSR* (*Proceedings of Grusinian SSR*) 73: 457–460.

Oniani, D.A., Kudrin, A.N., Lomsadze, B.A., Vorob'ev, L.N. 1977. The study of bioelectric potential and rate of the protoplasm movement in cells of *Nitella syncarpa* under adrenaline and noradrenaline. *Soobshcheniya Grusinian SSR* (*Proceedings of Grusinian SSR*) 86: 457–460.

Parsons, M.S. and Ganellin, C. 2006. Histamine and its receptors. *Br J Pharmacol* 147: S127–S135.

Picciotto, M.R., Zoli, M., Rimondini, R. et al. 1998. Acetylcholine receptors containing the β2 subunit are involved in the reinforcing properties of nicotine. *Nature* 391 (6663): 173–177.

Pichon, Y. (Ed.). 1993. *Comparative Molecular Neurobiology*. Birkhäuser, Basel, Switzerland.

Pickles, V.R. and Sutcliffe, J.F. 1955. The effects of 5-hydroxytryptamine, indole-3-acetic acid and some other substances, on pigment effusion, sodium uptake, and potassium efflux, by slices of red beet-root in vitro. *Biochim Biophys Acta* 17: 244–251.

Plopper, G., Sharp, D., Sikorski, E. (Eds.). 2015. *Lewin's Cells*, 3rd ed. Jones and Barrtlett Learning, Burlington, MA.

Pollard, T.D.J., Earnshaw, W.C., Lippincott-Schwartz, J., Johnson, G.T. 2017. *Cell Biology*, 3rd ed. Elsevier, Philadelphia, PA.

Ponchet, M., Martin-Tanguy, J., Marais, A., Martin, C. 1982. Hydroxycinnamoyl acid amides and aromatic amines in the influorescences of some Araceae species. *Phytochemistry* 21: 2865–2869.

Ramakrishna, A., Giridhar, P., Ravishankar, G.A. 2011. Phytoserotonin. *Plant Signal Behav* 6 (6): 800–809.

Roshchina, V.V. 1988. Characterization of pea chloroplast cholinesterase: Effect of inhibitors of animal enzymes. *Photosynthetica* 22: 20–26.

Roshchina, V.V. 1989. Biomediators in chloroplasts of higher plants. 1. The interaction with photosynthetic-membranes. *Photosynthetica* 23: 197–206.

Roshchina, V.V. 1990a. Biomediators in chloroplasts of higher plants. 2. The acetylcholine-hydrolyzing proteins. *Photosynthetica* 24: 110–116.

Roshchina, V.V. 1990b. Biomediators in chloroplasts of higher plants. 3. Effect of dopamine on photochemical activity. *Photosynthetica* 24: 117–121.

Roshchina, V.V. 1990c. Biomediators in chloroplasts of higher plants. 4. Reception by photosyn- thetic membranes. *Photosynthetica* 24: 539–549.

Roshchina, V.V. 1991. *Biomediators in Plants. Acetylcholine and Biogenic Amines.* Biological Center of USSR Academy of Sciences, Pushchino, Russia, 192 pp.

Roshchina, V.V. 1992. The action of neurotransmitters on the seed germination. *Biol Nauk* 9: 124–129.
Roshchina, V.V. 1999a. Mechanisms of cell-cell communication. In: Narwal, S.S. (Ed.), *Allelopathy Update*, Vol. 2. Science Publishers, Enfield, NH, pp. 3–25.
Roshchina, V.V. 1999b. Chemosignalization at pollen. *Uspekhi Sovremennoi Biologii (Trends in Modern Biology, Russia)* 119: 557–566.
Roshchina, V.V. 2001. *Neurotransmitters in Plant Life*. Science Publishers, Enfield, NH.
Roshchina, V.V. 2003. Autofluorescence of plant secreting cells as a biosensor and bioindicator reaction. *J Fluoresc* 13 (5): 403–420.
Roshchina, V.V. 2004. Cellular models to study the allelopathic mechanisms. *Allelopathy J* 13 (1): 3–16.
Roshchina, V.V. 2005. Contractile proteins in chemical signal transduction in plant microspores. *Biol Bull Ser Biol* 3: 281–286.
Roshchina, V.V. 2006a. Chemosignaling in plant microspore cells. *Biol Bull Ser Biol* 33: 414–420.
Roshchina, V.V. 2006b. Plant microspores as biosensors. *Trends Mod Biol* 126 (3): 262–274.
Roshchina, V.V. 2007. Cellular models as biosensors. In Roshchina, V.V. and Narwal, S.S. (Eds.), *Cell Diagnostics: Images Biophysical and Biochemical Processes in Allelopathy*. Science Publisher, Plymouth, UK, pp. 5–22.
Roshchina, V.V. 2008. *Fluorescing World of Plant Secreting Cells*. Science Publishers, Enfield, NH.
Roshchina, V.V. 2009a. Effects of proteins oxidants and antioxidants on germination of plant microspores. *Allelopathy J.* 23 (1): 37–50
Roshchina, V.V. 2009b. Acetylcholine and biogenic amines in nonsinaptic signal systems. In: Zinchenko, V.P., Kolesnikov, S.S., Berezhnov, A.V. (Ed.), *Reception and intracellular signaling. Materials of International Conference*, June 2–4. Institute of Cell Biophysics, ONTI Pushchino, Pushchino, Russia, pp. 694–698.
Roshchina, V.V. 2010. Evolutionary considerations of neurotransmitters in microbial, plant and animal cells. In Lyte, M. and Freestone, P.P.E. (Eds.), *Microbial Endocrinology: Interkingdom Signaling in Infectious Disease and Health*. Springer, New York, pp. 17–52.
Roshchina, V.V. 2014. *Model Systems to Study Excretory Function of Higher Plants*. Springer, Dordrecht, the Netherlands.
Roshchina, V.V. 2016a. New tendency and perspectives in evolutionary considerations of neurotransmitters in microbial, plant, and animal cells. In: Lyte, M. (Ed), *Advances in Experimental Medicine and Biology*, Vol. 874. Series: Microbal Endocrinology: Interkingdom Signaling in Infectious Disease and Health. Springer International Publishing, New York, Chapter 2, pp. 25–77. doi:10.1007/978-3-319-20215-02.
Roshchina, V.V. 2016b. The fluorescence methods to study neurotransmitters (biomediators) in plant cells. *J Fluorescence* 26 (3): 1029–1043. doi:10.1007/s10895-016-1791-6.
Roshchina, V.V. 2016c. Compounds earlier known as neurotransmitters are met and function in every living organism. Atlas of Science Another view of Science, February 2, 2016.
Roshchina, V.V. 2017. Serotonin in cells: Fluorescent tools and models to study. In: Ravishankar, G.A. and Ramakrishna, A. (Eds.), *Serotonin and Melatonin: Their Functional Role in Plants, Food, Phytomedicine, and Human Health*. CRC Press, Boca Raton, FL, pp. 47–60.
Roshchina, V.V. 2018. Cholinesterase in secreting cells and isolated organelles. *Biological Membranes (Russia)* 35(2): 143–149.
Roshchina, V.V. and Melnikova, E.V. 1996. Microspectrofluorometry: A new technique to study pollen allelopathy. *Allelopathy J* 3: 51–58.
Roshchina, V.V. and Melnikova, E.V. 1998a. Allelopathy and plant reproductive cells: Participation of acetylcholine and histamine in signalling in the interactions of pollen and pistil. *Allelopathy J* 5: 171–182.
Roshchina, V.V. and Melnikova, E.V. 1998b. Pollen-pistil interaction: Response to chemical signals. *Biol Bull* 25: 557–563.
Roshchina, V.V. and Mukhin, E.N. 1984. The acetylcholinesterase activity in chloroplasts of higher plants. *Doklady AN SSSR* 278: 754–757.
Roshchina, V.V. and Mukhin, E.N. 1985a. Acetylcholinesterase activity in chloroplasts and acetylcholine effects on photochemical reactions. *Photosynthetica* 19: 164–171.
Roshchina, V.V. and Mukhin, E.N. 1985b. Acetylcholine action on the photochemical reactions in chloroplasts. *Plant Sci* 42: 95–98.
Roshchina, V.V. and Mukhin, E.N. 1986. Acetylcholine, its role in plant life. *Uspekhi Sovremennoi Biologii (Trends in Modern Biology – Reviews, Russia)* 101: 265–274.
Roshchina, V.V. and Mukhin, E.N. 1987a. Acetylcholine-acetylcholinesterase system in isolated pea chloroplasts and regulation of some photosynthetic reactions. *Fiziologiya Rastenii (USSR)* 34: 67–73.
Roshchina, V.V. and Mukhin, E.N. 1987b. Acetylcholine changes photochemical reactions and stimulates $Na^+$ and $K^+$ efflux from chloroplasts. *Fiziologiya Rastenii (USSR)* 34: 907–911.

Roshchina, V.V. and Roshchina, V.D. 1993. *The Excretory Function of Higher Plants*. Springer, Berlin, Germany, 314 pp.
Roshchina, V.V. and Semenova, M.N. 1990. Plant cholinesterases: Activity and substrate-inhibitory specificity. *J Evo Biochem Physiol* 26: 644–651.
Roshchina, V.V. and Semenova, M.N. 1995. Neurotransmitter systems in plants. Cholinesterase in excreta from flowers and secretory cells of vegetative organs in some species. In: Greene, D. and Cutler, G. (Eds.), *Proceedings of the Plant Growth Regulation Society of America. 22: Annual Meeting*, July 18–20. Fritz C.D., Minneapolis, MN, pp. 353–357.
Roshchina, V.V. and Shvirst, N.E. 2016. Cholinesterase in contractile structures of plants and animals: Histochemical experiments with azocompounds. In: Udaltsov, S.N. (Ed.), *Biological Motility. Materials of International Symposium*, May 12–14. Synchrobook, Pushchino, Russia, pp. 198–202.
Roshchina, V.V, and Vikhlyantsev, I.M. 2009. Mechanisms of chemosignalling in allelopathy: Role of Ion channels and cytoskeleton in development of plant microspores. *Allelopathy J* 23 (1): 25–36.
Roshchina, V.V. and Yashin, V.A. 2013. Secreting cells of *Saintpaulia* as models in the study of plant cholinergic system. *Biological Membranes* (Russia) 23 (6): 454–461.
Roshchina, V.V. and Yashin, V.A. 2014. Neurotransmitters catecholamines and histamine in allelopathy: Plant cells as models in fluorescence microscopy. *Allelopathy J* 34 (1): 1–16.
Roshchina, V.V, Melnikova, E.V., Kovaleva, L.V., Spiridonov, N.A. 1994. Cholinesterase of pollen grains. *Dokl Biol Sci* 337: 424–427.
Roshchina, V.V., Bezuglov, V.V., Markova, L.N. et al. 2003. Interaction of living cells with fluorescent derivatives of biogenic amines. *Dokl Russian Acad Sci* 393 (6): 832–835.
Roshchina, V.V., Markova, L.N., Bezuglov, V.V., Buznikov, G.A., Shmukler, Y.B., Yashin, V.A., Sakharova, N.Y. 2005. Linkage of fluorescent derivatives of neurotransmitters with plant generative cells and animal embryos. In Zinchenko, V.P. (Ed.), *Materials of International Symposium: "Reception and Intracellular Signaling."* ONTI, Pushchino, Russia, pp. 399–402.
Roshchina, V.V., Popov, V.I., Novoselov, V.I., Melnikova, E.V., Gordon, R.Y., Peshenko, I.V., Fesenko, E.E. 1998. Transduction of chemosignal in pollen. *Cytologia (Russia)* 40: 964–971.
Roshchina, V.V., Yashin, V.A., Yashina, A.V., Gol'tyaev, M.V., Manokhina, I.A. 2009a. Microscopicobjects for the study of chemosignaling. In Zinchenko, V.P., Kolesnikov, S.S., Berezhnov, A.V. (Eds.), *Reception and Intracellular Signalling*. Biological Center of the Russion Acadamy of Science, Pushchino, Russia, pp. 699–703.
Roshchina, V.V., Yashina, A.V., Yashin V.A., Prizova, N.K. 2009b. Models to study pollen allelopathy. *Allelopathy J* 23 (1): 3–24.
Roshchina, V.V., Yashin, V.A., Shvirst, N.E., Prizova, N.K., Khaibullaeva, L.M., Kuchin, A.V. 2013. Secreting cells as models to study of the role of acetylcholine in signaling and communications of organisms. In Zinchenko, V.P. and Berezhnov, A.V. (Eds.), *International Conference "Reception and Intracellular Signalling"*, May 27–30, Biological Center of the Russion Acadamy of Science, EMA, Pushchino, Russia, vol. 2. pp. 790–795.
Roshchina, V.V., Yashin, V.A., Vikhlyantsev, I.M. 2012. Fluorescence of plant microspores as biosensors. *Membr Cell Biol* 6 (1): 105–112.
Roshchina, V.V., Yashin V.A., Kuchin A.V. 2015a. Microfluorescent analysis for bioindication of ozone on unicellular models. *Phys Wave Phenom* 23 (3): 1–7.
Roshchina, V., Yashin, V.A., Kuchin, A.V. 2015b. Fluorescent analysis for bioindication of ozone on unicellular models. *J Fluorescence* 25 (3): 595–601. doi:10.1007/s10895-015-1540-2.
Roshchina, V.V., Yashim, V.A., Kuchin, A.V., Kulakov, V.I. 2014. Fluorescent analysis of catecholamines and histamine in plant single cells. *Int J Biochem* 195: 344–351.
Rykun, O.V. 2013. Influence of biotic stress, calcium and hydrogen peroxide on the content of cAMP in vacuoles of roots beets in different periods of rest. *Vestnik Irkutsk State Agronomy Academy (GSHA)* 54: 20–26.
Rykun, O.V., Lomovatskaya, L.A., Romanenko, A.S. 2015. Plant vacuole as signaling compartment. In: Voinikov, V.K. (Ed.), *Factors of Plant Tolerantness to Extremal Natural and Technogenic Conditions. Materials of All-Russian Scientific Conference*, Irkutsjk, Russia, July 10–13, 2013. Direct Media, Moscow, Russia, pp. 222–224.
Sagane, Y., Nakagawa, T., Yamamoto, K., Michikawa, S., Oguri, S., Momonoki, Y.S. 2005. Molecular characterization of maize acetylcholinesterase. A novel enzyme family in the plant kingdom. *Plant Physiol* 138: 1359–1371.
Schloss, P., Mayser, W., Betz, H. 1992. Neurotransmitter transporters. A novel family of integral plasma membrane proteins. *FEBS Lett* 307: 76–80.
Schmukler, Y.B. and Nikishin, D.A. 2012. Transmitters in blastomere interaction. In Gowder, S. (Ed.), *Cell Interaction*. InTech, Rijeka, Croatia, pp. 31–65.

Segonzac, C. and Zipfel, C. 2011. Activation of plant pattern-recognition receptors by bacteria. *Curr Opin Microbiol* 14 (1): 54–61.

Servillo, L., Giovane, A., Casale, R., Cautela, D., D'Onofrio, N., Balestrieri, M.L., Castaldo, D. 2016. Glucosylated forms of serotonin and tryptophan in green coffee beans. *LWT-Food Sci Technol* 73: 117–122.

Shubina, V.S., Abramova, M.B., Lavrovskaya, V.P., Pavlik, L.L., Lezhnev, E.I., Moshkov, D.A. 2009. The influence of dopamine on the ultrastructure of BHK21 cells. *Cytology (Russia)* 51 (12): 996–1004.

Silla Santos, M.H. 1996. Biogenic amines: Their importance in foods. *Int J Food Micobiol* 29 (2–3): 213–231.

Sinyukhin, A.M. and Britikov, E.A. 1967. Action potential in the reproductive system of plants. *Nature* 215: 1278–1279.

Skirycz, A., Swiedrych, A., Szopa, J. 2005. Expression of human dopamine receptor in potato (*Solanum tuberosum*) results in altered tuber carbon metabolism. *BMC Plant Biol* 5: 1–15.

Smolinski, D.J. and Gorska-Brylass, A. 1994. Acetylcholinesterase activity in stomata of *Marchantia polymorpha* L. *Cell Biol Int* 18 (5): 539.

Staehelin, L.A. 2015. Membrane structure and membranous organelles. In: Buchanan, B.B., Gruissen, W., Jones, R.L. (Eds.) *Biochemistry and Molecular Biology of Plants*, 2nd ed. Wiley-Blackwell, Chichester, UK.

Stevenson, J.M., Perera, I.Y., Heilmann, I. et al. 2000. Inositol signaling and plant growth. *Trends Plant Sci* 5 (6): 252–258.

Strakhovskaya, M.G., Ivanova, E.V., Fraikin, G.Ya. 1993. Stimulatory effect of serotonin on the growth of the yeast *Candida guillermondii* and the bacterium *Streptococcus faecalis*. *Microbiology (Russia)* 62: 46–49.

Sugiyama, K. and Tezuka, T. 2011. Acetylcholine promotes the emergence and elongation of lateral roots of *Raphanus sativus*. *Plant Signal Behav* 6 (10): 1545–1553.

Swanson, R.L., Williamson, J.E., DeNys, R. et al. 2004. Induction of settlement of larvae of the sea urchin *Holopneustes purpurascens* by histamine from ahost alga. *Biol Bull* 206: 161–172.

Swiedrych, A., Kukuła, K.L., Skirycz, A., Szopa, J. 2004. The catecholamine biosynthesis route in potato is affected by stress. *Plant Physiol Biochem* 42: 593–600.

Szopa, J., Wilczynski, G., Fieh, O., Wencze, A., Willmitzer, L. 2001. Identification and quantification of catecholamines in potato plants (*Solanum tuberosum*) by GC-MS. *Phytochemistry* 58: 315–320.

Taddese, B., Upton, G.J.G., Bailey, G.R. et al. 2014. Do plants contain G protein-coupled receptors? *Plant Physiol* 164 (1): 287–307.

Tezuka, T., Akita, I., Yoshino, N., Suzuki, Y. 2007. Regulation of self-incompatibility by acetylcholine and cAMP in Lilium longiflorum. *J Plant Physiol* 164: 878–885.

Tretyak, T.M. and Arkhipova, L.V. 1992. Intracellular activity of mediators. *Uspekhi Sovremennoi Biologii (Trends Mod Biol)* 11: 265–272.

Tretyn, A. 1987. Influence of red light and acetylcholine on $^{43}Ca^{2+}$ -uptake by oat colepotile cells. *Cell Biol Int* 11: 887–896.

Tretyn, A. and Kendrick, R.E. 1991. Acetylcholine in plants: Presence, metabolism and mechanism of action. *Bot Rev* 57 (1): 33–73.

Tsavkelova, E.A., Botvinko, I.V., Kudrin, V.S., Oleskin, A.V. 2000. Detection of neurotransmitter amines in microorganisms using of high performance liquid chromatography. *Dokl Biochem* 372: 115–117 (in Russian issue 840–842).

Tuteja, N. 2009. Signaling through G protein coupled receptors. *Plant Signal Behav* 4 (10): 942–947.

van Alstyne, K.L., Anderson, K., Winans, A. 2011. Dopamine release by the green alga *Ulvaria obscura* after simulated immersion by incoming tides. *Mar Biol* 158: 2087–2094. doi:10.1007/s00227-011-1716-5.

van Alstyne, K.L., Anderson, K.J., van Hees, D.H., Gifford, S.A. 2013. Dopamine release by *Ulvaria obscura* (Chlorophyta): environmental triggers and impacts on photosynthesis growth and survival of the releaser. *J Phycol* doi:10.1111/jpy.12081.

van Alstyne, K.L., Harvey, E.L., Cataldo, M. 2014. Effects of dopamine a compound released by the greentide macroalga *Ulvaria obscura* (Chlorophyta), on marine algae and invertebrate larvae and juveniles. *Phycologia* 53 (2): 195–202.

van Alstyne, K.L., Nelson, A.V., Vyvyan, J.R., Cancilla, D.A. 2006. Dopamine functions as an antiherbivore defense in the temperate green alga *Ulvaria obscura*. *Oecologia* 148: 304–311.

Vaughn, K.C. and Wilson, K.J. 1981. Improved visualization of plastid fine structure: plastid microtubules. *Protoplasma* 108: 21–27.

Vaughn, K.C. and Wilson, K.J. 1984. Improved visualization of plastid fine structure: Plastid microtubules. *Protoplasma* 108: 21–27.

Volkov, A.G. and Haack, R.A. 1995. Insect induces bioelectrochemical signals in potato plants. *Bioelectrochemistry* 35: 55–60.

Wang, H., Wang, X., Zhang, S., Lou, C. 1998. Nicotinic acetylcholine receptor is involved in acetylcholine regulating of stomatal movement. *Sci Chin Ser C* 41: 650–656.

Wang, H., Wang, X., Lou, C. 1999a. Relationship between acetylcholine and stomatal movement. *Acta Bot Sin* 41: 171–175.

Wang, H., Wang, X., Zhang, S. 1999b. Extensive distribution of acetylcholinesterase in guard cells of Vicia faba. *Acta Bot Sin* 41: 364–367.

Wang, H., Wang, X., Zhang, S., Lou, C. 2000. Muscarinic acetylcholine receptor involved in acetylcholine regulating of stomatal function. *Chin Sci Bull* 45: 250–252.

Wang, H., Zhang, S., Wang, X., Lou, C. 2003a. Involvement of $Ca^{2+}$/CaM in the signal transduction of acetylcholine-regulating stomatal movement. *Chin Sci Bull* 48: 351354.

Wang, H., Zhang, S., Wang, X., Lou, C. 2003b. Role of acetylcholine on plant root-shoot signal transduction. *Chin Sci Bull* 48: 570–573.

Wang, P. and Hussey, P.J. 2015. Interactions between plant endomembrane systems and the actin cytoskeleton. *Front Plant Sci*. doi:10.3389/fpls.2015.00422.

Werle, E. and Raub, A. 1948. Über Vorkommen Bildung und Abbau biogener Amine bei Pflanzen unter besonderer Beruck-sichtigung des Histamins. *Biochem Z* 318: 538–553.

Wessler, I., Kilbinger, H., Bittinger, F., Kirkpatrick, C.J. 2001. The non-neuronal cholinergic system: The biological role of non-neuronal acetylcholine in plants and humans. *Jpn J Pharmacol* 85: 2–10.

Witters, E., Valcke, R., van Onckelen, H. 2005. Cytoenzymological analysis of adenylyl cyclase activity and 3':5'-cAMP immunolocalization in chloroplasts of *Nicotiana tabacum.New Phytol* 168 (1): 99–108.

Woodring, J. and Hoffmann, K.H. 1994. The effects of octopamine, dopamine and serotonin on juvenile hormone synthesis, in vitro. *J Insect Physiol* 40: 797–802.

Yamamoto, H., Shimizu, K., Tachibana, A., Fusetani, N. 1999. Roles of dopamine and serotonin in larval attachment of the barnacle *Balanus amphitrite*. *J Exp Zool* 284: 746–758.

Yamamoto, K., Sakamoto, H., Momonoki, Y.S. 2011. Maize acetylcholinesterase is a positive regulator of heat tolerance in plants. *J Plant Physiol* 168: 1987–1992.

Yamamoto, K. and Momonoki, Y.S. 2012. Tissue localization of maize acetylcholinesterase associated with heat tolerance in plants. *Plant Signal Behav* 7 (3): 301–305.

Yamamoto, K., Sakamoto, H., Momonoki, Y.S. 2016. Altered expression of acetylcholinesterase gene in rice results in enhancement or suppression of shoot gravitropism. *Plant Signal Behav* 11 (4). doi:10.1080/15592324.2016.1163464.

Yamane, T. 2013. Effects of the biogenic amines on female oviposition behavior in the rice leaf bug *Trigonotylus caelestialium* (Kirkaldy) (Heteroptera: Miridae). *Entomol News* 123: 161–167.

Yamane, T. and Miyatake, T. 2010. Reduced female mating receptivity and activation of oviposition in two *Callosobruchus* species due to injection of biogenic amines. *J Insect Physiol* 56: 271–276.

Yang, C.J., Zhai, Z.X., Guo, Y.H., Gao, P. Effects of acetylcholine, cytochalasin B and Amiprophosmethyl on phloem transport in radish (Raphanus sativus). *J Integr Plant Biol* 49 (4): 550–555.

Yao, W.-D., Rusch, J., Poo, M.M., Wu, C.F. 2000. Spontaneous acetylcholine secretion from developing growth cones of drosophila central neurons in culture: Effects of cAMP-pathway mutations. *J Neurosci* 20 (7): 2626–2637.

Yunghans, H. and Jaffe, M.J. 1972. Rapid respiratory changes due to red light or acetylcholine during the early events of phytochrome-mediated photomorphogenesis. *Plant Physiology* 49: 1–7.

Yurin, V.M., Ivanchenko, V.M., Galactionov, S.G. 1979a. *Regulation of Fuctions of Plant Cell Membranes*. Nauka I Technika, Minsk, Belarus.

Yurin, V.M., Gusev, V.V., Kudryashov, A.P. et al. 1979b. Testing of membrane-topic effect of physiologically active substances. I. Qualitative estimations. *Vestnik AN Belorussian SSR* 1: 97–100.

Zherelova, O.M., Kataev, A.A., Grischenko, V.M., Shtanchaev, R.S., Moshkov, D.A., Medvedev, B.I. 2014. Interaction of neuromediator dopamine with the ionic channels of *Chara corallina* cell plasmalemma. *Biomed Zh medline.ru* 15 (67): 834–846.

Zherelova, O.M., Kataev, A.A., Grischenko V.M., Shtanchaev R.S. 2016. Haloperidol modulates ionic transport of Chara coralline cells. *Cell Tissue Biol* 10 (6): 476–485

Zholkevich, V.N., Aniskin, D.N., Dustmamatov, A.G. 2003. On the stimulatory effect of neuromediators on the root pumping activity. *Dokl Biol Sci* 392: 419–421.

Zholkevich, V.N., Zhukovskaya, N.V., Popova, M.S. 2007a. Participation of proteinkinases and protein phospatases in signal transduction at the stimulatory effect of neuromediators on the root pumping activity. *Russian J Plant Physiol* 54 (4): 550–554.

Zholkevich, V.N., Zhukovskaya, N.V., Popova, M.S. 2007b. Stimulatory effects of adrenaline and noradrenaline on the root pumping activity and participation of G-proteins. *Russ J Plant Physiol* 54 (6): 885–892.

# 10 Neurotransmitters in Characean Electrical Signaling

*Vilma Kisnieriene, Indre Lapeikaite, and Vilmantas Pupkis*

## CONTENTS



## 10.1 INTRODUCTION

Plants possess many homologous molecules that are similar to the neurotransmitters of the animal nervous system such as glutamate, Gamma-Aminobutyric acid (GABA), serotonin, melatonin, dopamine, and acetylcholine (Roshchina 2001); therefore, it is reasonable to hypothesize that these compounds could be involved in signal transmission in addition to a possible metabolic role. Despite a long history of study, the plant neurotransmitters'-mediated system and its role in plants' signaling are not yet fully understood. In most investigations of plant neurotransmitters, detection and structural analysis of distinct components prevailed. Nevertheless, molecular studies alone do not provide information regarding physiological characteristics in the intact cells. Consequently, research of plant cholinergic, dopaminergic, glutamatergic, gabaergic system function, and coordination *in vivo* is needed. To date, there is no evidence about the accumulation mode of those chemicals and balance between their synthesis, diffusion, and hydrolysis in plants. It is unclear what role these compounds play in plant metabolism and signaling (Murch 2006). For example, the biological role of acetylcholine is focused mainly on its neurotransmitter function, although the presence of acetylcholine (ACh), cholineacetyltransferase, and acetylcholinesterase molecules in all forms of life support the idea that ACh, as a local cell molecule, could modulate important physiological functions at the very beginning of life (Horiuchi et al. 2003). Characterization of this neurotransmitter as a local mediator in the plant physiological response shed some light on the possible subcellular target of the acetylcholine signal, revealing it to be related to the vesicular transport (Di Sansebastiano et al. 2014).

Early events in the responses of plant cells to many physiological stimuli share common features, such as membrane depolarization and elevations in cytosolic $Ca^{2+}$. Membrane networks between plant cells may convey information about changing environments, monitor circadian cycles of light, and redirect plant growth. However, it is difficult to link experimentally chemical cell communication to electrical signaling in whole plants. If plant neurotransmitters may cause changes in membrane permeability similar to those found in the excitable membranes of animal cells, then they can interfere with electrical cellular signaling pathway in plants too. For such investigations, electrophysiological techniques could be useful to determine the effect of neurotransmitters on the activity of plant membrane ion transport systems and reveal participation in electrical signal transduction. The application of electrophysiological techniques in the combination with pharmacological data obtained from studies on animal cells has proven to be very powerful in the analysis of plant cell membrane processes. From the molecular studies, it is known that the ion channels that open directly in response to the neurotransmitters acetylcholine, serotonin, GABA, glycine, and others contain structurally similar subunits. Receptors for some neurotransmitters have been stored in evolution, although neurotransmitter-binding specificities and ion selectivity differ. Glutamate-gated ion channels are constructed from a distinct family of subunits and are thought to form tetramers, like the $K^+$ channels. Using electrophysiological methods neurotransmitters were shown to participate in regulation of ion permeability in various plants (Yurin et al. 1979; Yurin 1990).

The electrical measurements on intact or altered cells and cytoplasmic droplets using various experimental techniques unique to Characean plants form the basis for modern plant electrophysiology. Characean are aquatic plants, thus conventional electrical measurements can be carried out under their native external condition, aquatic solution (Figure 10.1).

Patch clamping in cytoplasmic droplets allows investigation of the tonoplast at one ion channel level, and voltage clamp technique allows all channel population investigations in steady state and after excitation. Two or more tandem cells are used to measure cell-to-cell electrical signaling by the perfusion and permeabilization techniques allowing access to both sides of the plasma membrane and tonoplast in a single cell context (Beilby 2016).

The contribution of Characean cells to the study of electrical signaling have been well documented in the book *The Physiology of Characean Cells* (Beilby and Casanova 2014). The present article intends to review the significance of Characean cells exclusively in the field of neurotransmitters' effect on plant ion transport and electrical signaling.

**FIGURE 10.1** Structure of *Nitellopsis obtusa* thallus and separated single mature intermodal cell (up to 15 cm in length and 1 mm in with). Insert: two cells between the node.

## 10.2 ALTERATIONS OF MEMBRANE POTENTIAL AS AN INDICATOR OF EARLY RESPONSE OF NEUROTRANSMITTERS

The potential differences in cell membranes are informative native electrochemical parameters to measure because they reflect ion transport across plasmalemma and tonoplast (vacuolar membrane), as well as ionic concentration differences between outside, cytoplasm, and vacuole. In spite of a fact that the electrical measurements do not distinguish between currents of different ions, the measurement of membrane potential remains a very useful tool in plant electrophysiology. Changes in membrane potential are vital means of signal cascades and transporter interaction, which enable the plant cell to maintain life-sustaining functions and respond to stress stimuli (Beilby et al. 2015b).

The regulation of membrane potential is known to constitute a signal in many cell types. There are many ion transporters in both plasma membrane and the tonoplast; moreover, new ones are being discovered in Characeae (Beilby 2016). Much electrophysiological information indicates the presence and action of voltage-gated channels in the plant plasma membrane. However, only a small number of transporter types dominate the membrane conductance and determine membrane potential. Whereas most animal cells in their resting stage are very close to the Nernst potential for $K^+$ ions, plant cells can obtain much higher values due to the operation not only of $K^+$ ion channels but also of an electrogenic $H^+$-ATPase-driven pumps. It is known that the membrane potential of Characean cells is composed of two components—passive diffusion potential and active potential generated by an electrogenic proton pump (Tsutsui et al. 1987). Under *normal* steady conditions, neutral pH and in light, the proton pump mainly controls the plasma membrane potential in plants. If the resting membrane potential (RP) varies in the range of −80 and −200 mV in plant cells (Hedrich 2012), RP potential of Characean *Nitellopsis obtusa* cells could be up to −255 mV. The membrane potential can change substantially depending on the pH and $K^+$ concentration of the outside medium. The cytoplasmic $K^+$ level is strictly controlled (80–100 mM) activity; that homeostasis is achieved by both the control of $K^+$ influx across the plasma membrane (PM) and by mobilizing $K^+$ from vacuolar reserves. Therefore, the diversity of bioelectrical responses in plant cells could be explained by different actions of exogenic factors on the $H^+$ pump and on passive channels activity. The highly negative membrane potential affects any passive ion movement. The proton pump generates an $H^+$ electrochemical gradient and provides a driving force for the rapid ion fluxes required for the uptake of various nutrients and amino acids. For example, the voltage dependence of $K^+$ conductance is well known in Characean cells. Thus, hyperpolarization of membrane potential (MP) may enhance $K^+$ uptake via voltage-gated $K^+$ inward rectifying channels or, alternatively, reduce $K^+$ efflux through outward $K^+$ channels. If MP becomes more negative than −300 mV, the background current is obscured by the inward rectifier current, $K^+_{irc}$, while if MP becomes more positive than −50 mV, the outward rectifier current, $K^+_{orc}$, predominates. The background current is thought to flow through non-selective MP-independent cation channels (Demidchik and Maathuis 2007), which supply micronutrients to the plant and contribute to signaling. The pump state and the underlying background state are the native states for the salt sensitive Characeae in their low salts lightly alkaline pond media. The inactivation of the pump brings the membrane potential to $E_{background}$, near −100 mV and to the excitation threshold. Measuring the membrane resistance (or conductance) is helpful for the determination of the transport processes affecting membrane potential. Changes in transmembrane potential difference of Characean cells could be used as the best test of the effect of various compounds and its combinations on the plasma membrane and the whole cell. Thus, changes in membrane potential can be used to reveal the instantaneous effect of neurotransmitters on the plant cell.

The first neurotransmitter discovered, acetylcholine is known to be a neurotransmitter at all autonomic ganglia, at many autonomically innervated organs, at the neuromuscular junction, and at many synapses in the central nervous system (CNS). Acetylcholine (ACh) is a phylogenetically ancient molecule, functioning as a local mediator as well as a neurotransmitter in almost all life forms on earth (Wessler et al. 2001). It has been further proposed that the primary mechanism of action of acetylcholine in plant cells is via the regulation of membrane permeability to $H^+$, $K^+$, $Na^+$ and $Ca^{2+}$ ions (Jaffe 1970; Hartmann 1977; Yurin et al. 1979; Roshchina 1991; Odjakova and Hadjiivanova 1997).

ACh may cause changes in membrane permeability; first it could be reflected in membrane potential and other basic electrical characteristics. Indeed, it was found that neurotransmitter ACh depolarizes membrane potential in Characean *N. obtusa* and *Nitella flexilis* (Kisnierienė et al. 2009). Changes in membrane potential and resistance after application of ACh indicate effect of ACh on ion transport systems involved in plant potential electrogenesis. It was demonstrated that 5 mM ACh activates $K^+$ ion channels in the resting state and found that ACh elicits depolarization of membrane potential immediately after application. 1 mM acetylcholine has depolarizing effect on the resting potential of *N. obtusa* cells (by $12.2 \pm 2.1$ mV from $-225.3 \pm 5.6$ mV) and causes a significant increase in membrane permeability (22.5%, $p < 0.002$, $n = 10$) (Kisnieriene et al. 2012). Based on our works, we propose that ACh increases background conductance. A rise in proton pump activity in 25 min counteracts this increase in background conductance, and the resting potential remains negative. The same effect of proton pump activation was shown after application of NaCl in Characean algae *Lamprothamnium* (Beilby and Shepherd 2006).

Additionally, 0,1 mM nicotine (ACh receptor agonist) caused strong depolarization of membrane potential, which started immediately after solution change. Depolarizing effect of nicotine depends on energetic state of the cell and resting potential in control conditions. Cells with very negative ($< -230$ mV) RP values depolarized by $52 \pm 7$ mV ($n = 5$) in 30 min after 0,1 mM nicotine application. The same concentration of nicotine caused RP depolarization and initiation of spontaneous repetitive action potential (AP) if membrane potential of the cells were about $-220$ mV ($n = 5$). In some cases, membrane potential depolarized up to $-60$ mV. However, the nicotine effect was reversible despite depolarization magnitude (Kisnieriene et al. 2018).

The acetylcholine is not a unique neurotransmitter affecting membrane potential of Characean. It is known that universal excitatory agent dopamine and other biogenic amines are present in a wide variety of organisms representing different evolutionary distant systematic groups (Roshchina 2016). In investigating the effect of dopamine on plant cells, it was established that a 24h-long incubation of *Chara corallina* cells in the presence of dopamine (2 mM) produced a drop in the resting potential of the cells, decreased the membrane resistance, and stopped the movement of the cytoplasm (Zherelova et al. 2014).

In our investigations of *N. obtusa* cells, we found a glutamate (Glu) effect on membrane potential (unpublished data). Notwithstanding, some ambiguity exists in interpreting the results of amino acid applications because of the existence of proton cotransport systems for amino acids, which themselves produce a transient depolarization of the membrane potential when an amino acid is applied externally (Kinraide and Etherton 1980).

Glu induces membrane depolarizations in higher plants. Dennison and Spalding (2010) demonstrated that glutamate can depolarize the plasma membrane of *Arabidopsis thaliana* root epidermal cells. This finding has led to the discovery of plant glutamate-activated cation channels (Demidchik et al. 2004). It was also shown that the amino acids Glu and glycine (Gly) trigger a large concomitant rise in membrane potential depolarization in *A. thaliana* plants (Qi et al. 2006). Mutations in four genes, all of which reduced glutamate receptor-like gene (GLR) expression in *A. thaliana* plants, caused reduced durations of surface potential changes in leafs (Mousavi et al. 2013).

## 10.3 EXPLORING CHARACEAE IN PLANT MELATONIN INVESTIGATIONS

The steady-state cytoplasmic concentrations of many substances give us important hints about transport systems at both the plasma membrane and tonoplast. For example, it was found that *Chara australis* produces endogenous melatonin in similar amounts to higher plants (Lazár et al. 2013). Characean is useful plant material for many biochemical assays where the seasonal and circadian nature of concentration of endogenous neurotransmitters could be shown. It was proved that melatonin (MEL) could be synthesized and taken up by Characean depending on the season.

Melatonin and serotonin (SER) concentrations were much smaller in winter cells, compared to summer cells, with only a weak correlation to IAA (indole-3-acetic acid) concentration changes. The large differences between summer and winter plants imply circadian rhythm or a role of serotonin/melatonin in seasonal development (Beilby et al. 2015a). The close synchronization between IAA and serotonin circadian cycling suggests IAA biosynthesis by the tryptamine pathway, which intersects with the serotonin/melatonin pathway (Beilby 2016).

Various studies have suggested specific physiological actions for melatonin in plants (Arnao and Hernández-Ruiz 2014). Melatonin is an amphipathic molecule that can easily diffuse through cell membranes into the cytoplasm and enter subcellular compartments (Zhang et al. 2015). It has been hypothesized that the role of melatonin in plants may be analogous to its function in mammals as a chemical signal of photoperiod, as a calmodulin binding factor, or as an antioxidant (Balzer and Hardeland 1996). MEL offers protection against reactive oxygen species (ROS) for chlorophylls as well as for the photosynthetic proteins in general. Beilby and co-workers exploited the antioxidant potential of melatonin to gain further insight into the electrophysiology of *C. australis* response to fatal salinity stress. Authors hypothesized that the salinity stress leads to an increase of ROS production in the cytoplasm, activating $H^+/OH^-$ channels, visible as salinity-induced noise in membrane potential. Cells treated for 24 h in 10 μM melatonin artificial pond water (APW) showed no increase in noise level in membrane potential upon transition from sorbitol to saline (Beilby et al. 2014). A direct interaction between MEL application and reduced production of ROS species at the cellular level was shown by using fluorescent probes and examining membrane potential. These findings have suggested that the natural circadian rhythms of tryptophan derivatives produced in *Chara* species may help to regulate ROS levels (Beilby et al. 2015a).

The upregulation of some anti-stress genes (vs cold, drought, osmotic stress) in melatonin-treated plants, and the induction of endogenous melatonin by these stressors, has confirmed a central role for melatonin as a signaling molecule in abiotic stress. Notwithstanding, currently, no melatonin receptor(s) or binding site(s) have been identified in plants, and animal-like melatonin receptor sequences have not been found so far in plant genomes (Reiter et al. 2015). Alternatively, since melatonin has a similar structure to auxin, it could possibly interact with the auxin receptor. Membrane potential alterations could link abiotic stress and different regulatory targets and pathways.

It was shown that an exogenous supply of SER and MEL to *Mimosa pudica* nodal explants exhibited shoot multiplication and proliferation, and it was further enhanced by calcium and calcium ionophore treatments thereby demonstrating the synergistic influence of SER/MEL and calcium treatment eliciting morphogenesis (Ramakrishna et al. 2009).

## 10.4 VARIABILITY OF ACTION POTENTIALS AS RESPONSE TO NEUROACTIVE COMPOUNDS

Plants respond to various stimuli in intracellular and extracellular environments by activation and propagation of fast electrical signals, that is action potentials (Davies 2004). Plasma membrane allows transmitting electrical signals in plants at the cell, tissue, and organ level for short and long distances (Hedrich et al. 2016). The first AP record in plants was acquired in Characean *Nitella mucronata* in 1930 (Umrath 1930). Characean are well-known as unique plants because of the generation of action potentials—large, transient changes in membrane potential (Beilby 2007). Characean, especially *C. corallina* and *N. obtusa* generate action potentials (AP) in response to mechanical stimulation, injury, or direct electrical stimulation (Beilby and Casanova 2014). In most experiments, the stimulus for an AP is a depolarising current. However, turgor change or chemical agents that cause depolarisation of membrane potential can also stimulate APs (Beilby 2007). The bioelectrical response of a Charophyte cell is rapid and highly sensitive to chemicals in environment (Beilby 2015b). Among other compounds, animal neurotransmitters involved in chemical transmission at synapses occur in plants (Murch 2006), and they could be involved in generation of Characean action potentials.

Action potentials have been investigated in Characean cells in an elaborate manner. It is known that fluxes through ion channels are responsible for action potential generation; thus, increased ion efflux accompanying the AP has been demonstrated in different charophytes (Beilby 2007). The AP is based on the activity of voltage-gated channels, which respond to (and cause) changes in membrane potential. Studies on Characean cells have also contributed to the exploration of P-type $H^+$ pumps (Mimura 1995) since the development of the membrane potential is linked to $H^+$ pump activation (Shimmen et al. 1994). Thus, the ion mechanism of the AP was intensely studied in giant Charophyta cells (Beilby and Casanova 2014). Action potentials involve a transient influx of $Ca^{2+}$ to the cytoplasm and effluxes of $K^+$ and $Cl^-$ outside (Lunevsky et al. 1983). Calcium is one of the most important signals within the cell and is implicated in a series of responses. The release of $Ca^{2+}$ from internal stores (in plants, mainly the vacuoles) or its entry through the plasma membrane is mediated by strictly controlled transporters, which themselves can be modulated by $Ca^{2+}_{cyt}$ in a feedforward or feedback manner (Tazawa and Kikuyama 2003). The regulation of transient $Ca^{2+}_{cyt}$ increase during an AP is very precise and occurs according to an all-or-none mechanism (Wacke and Thiel 2001). The changes in $Ca^{2+}$ homeostasis are reflected in AP generation (Beilby and Al Khazaaly 2016). This means that the AP is not entirely based on the time- and voltage-dependent activation properties of plasma membrane ion channels but on a complex signal transduction cascade. Calcium, when entering cells, not only activates $Ca^{2+}$-dependent (potential-controlled) chloride channels but suppresses the $H^+$-ATPase also. The chloride efflux and suppression of $H^+$-ATPase forms a depolarization phase to the peak level of the AP (Vodeneev et al. 2016). Furthermore, it should be noted that the cytoplasm of Characean is sandwiched between two membranes—plasmalemma and tonoplast. It was also established that both membranes, plasmalemma and tonoplast, are excitable. The excitation in the membranes proceeds simultaneously, accompanying each other so the action potentials at the plasmalemma and the tonoplast are coupled (Findlay 1970; Tazawa et al. 1976). In Characean, it was established that the increase in the cytoplasmic $Ca^{2+}$ during the action potential at the plasmalemma is the coupling factor that activates $Cl^-$ channels at the tonoplast (Kikuyama 1986). To date, ion transport systems involved in the generation of Characean action potentials have been established (Lunevsky et al. 1983; Thiel et al. 1997; Beilby 2007). It appears that the same ionic mechanism of AP generation applies also to higher plants (Fromm and Lautner 2007; Król et al. 2010).

In fact, the primary reaction of plants to neurotransmitters is often interrelated with changes in ion permeability of membranes (Roshchina 2001). For example, it has been shown that the primary mechanism of ACh action in plants was regulation of the membranes' permeability to protons (Jaffe 1970), $K^+$ (Roshchina, 2001), $Cl^-$ (Gong and Bisson 2002) and $Ca^{2+}$ (Tretyn and Kendrick 1991).

Analysis of *N. obtusa* membrane potentials and changes demonstrated in AP after application of ACh indicate effect of ACh on ion transport systems involved in AP generation (Figure 10.2).

Analysis of action potential characteristics in *N. obtusa* cells demonstrated that neurotransmitter ACh affects duration of AP irrespective of whether AP was spontaneous or electrically evoked (Kisnierienë and Sakalauskas 2007).

The pattern of single action potentials in APW and 5 mM ACh solutions was compared, and differences were found in the repolarization phase—ACh modulated repolarization process by increasing its duration. To evaluate velocity of AP repolarization, 50 s periods, starting from the peak of AP, were differentiated and dE/dt was plotted against voltage. The highest value of dE/dt was taken to evaluate prolongation of repolarization. The highest velocity of fast repolarization was $63.2 \pm 8.2$ mV/s in APW and $46.8 \pm 7.4$ mV/s in 5 mM ACh solution (Kisnierienė et al. 2009).

ACh could affect duration of repolarization, modulating activity of plasma membrane transport systems. If ACh acts as a signaling molecule in plants, it may cause changes in conductance of potassium channels and influence membrane permeability during AP generation. Gong and Bisson (2002) have demonstrated that ACh prolongs the repolarization phase of the action potential in another Characean *C. corallina*. They attributed this prolongation to the increased probability of the opening of the tonoplast chloride channels, which would tend to maintain the cell in a more

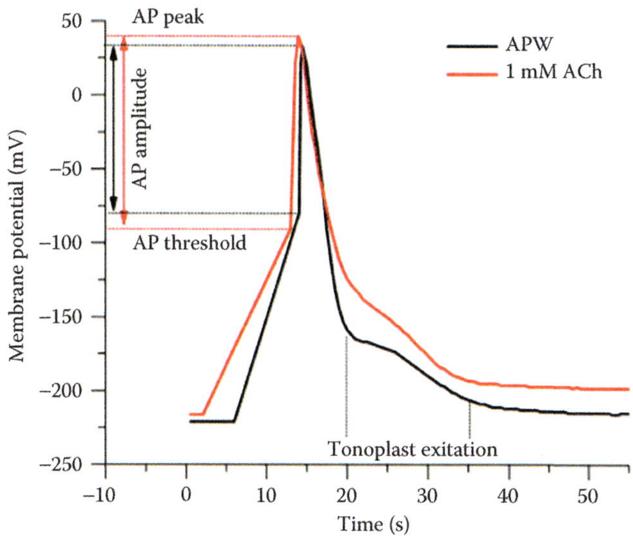


**FIGURE 10.2** Example of electrically stimulated action potential in *Nitellopsis obtusa* and its typical electrophysiological parameters. After exposure (1 h) to 1 mM ACh (acetylcholine) AP repolarization elongates and AP threshold hyperpolarizes. Since these AP registered using intercellular microelectrode impaled in vacuole, tonoplast excitation constituent of AP is prominent.

depolarized state. However, outward potassium ($K^+_{out}$) channels are the main transport system and $K^+_{out}$ current is the main current conditioning process of repolarization. If the $K^+_{out}$ channels at the plasma membrane were inhibited by ACh, the prolongation of AP could be seen. Results clearly showed ACh induced activation of $K^+$ channels and increase of $K^+_{out}$ and inward potassium ($K^+_{in}$) currents (Kisnieriene et al. 2012). So, reduction of repolarization rate could not be attributed to the inhibition of $K^+$ channels. Another transport system, which could prolong AP repolarization, is $H^+$-ATPase. By reducing activity of the proton pump, ACh could reduce flow of positive charges from the cell and prolong repolarization. Major prolongation of repolarization after inhibition of the proton pump by dicyclohexylcarbodiimide (DCCD) was noticed (Kisnierienė and Sakalauskas 2007). Micromolar concentrations of DCCD are sufficient to give complete and irreversible inhibition of proton pumping. $K^+$ channels proved to be insensitive to DCCD, a classic inhibitor of the electrogenic proton pump even at concentration 200 μM (Sokolik and Yurin, 1986). Since the application of DCCD is equivalent to inhibition of the $H^+$ pump current (Kishimoto et al. 1984), and if effect of ACh on repolarization is not dependent on inhibition of proton pump, we cannot rule out the possibility that inhibition of $H^+$-ATPase disturbed homeostasis of cell ions and cells became more sensitive to ACh. However, this inhibition allows a closer look at the different phases of the AP. For example, the effect of ACh on background conductance and increased depolarization could be counterbalanced by the activity of $H^+$-ATPase. Another possibility is that $K^+$ and $Cl^-$ inflow could be caused by the gradient and negative voltage generated by proton pump. It was found that ACh activates $K^+_{in}$ channels. If $Cl^-$ enters the cell by $2H^+/Cl^-$ cotransport system, DCCD indicates effect of ACh on $Cl^-$ efflux. Therefore, ACh increases duration of repolarization by modulating activity of $Cl^-$ channels (Kisnieriene et al. 2012).

It is generally accepted that certain electrical processes start at a particular value of membrane voltage. Excitation threshold of APs in every cell show voltage dependence.

We have demonstrated an enhanced excitability of *N. obtusa* cells after ACh application as cells started to generate AP at more negative voltages (unpublished data). Plant APs are caused by activity of $Ca^{2+}$-dependent $Cl^-$ channels in the plasma membrane and the tonoplast (Lunevsky et al. 1983). It was shown that $Ca^{2+}$-dependent $Cl^-$ channels opened most frequently between approximately −80 and −100 mV (Okihara et al. 1991). Our results in APW solution have confirmed these data.

It is possible that ACh has facilitated the opening of $Ca^{2+}$-dependent $Cl^-$ channels at more negative voltages and reduced excitation threshold after alteration of physical characteristics of the plasma membrane. ACh causing an increase in excitability could be explained by increased permeability to $Ca^{2+}$. It has been shown that ACh had increased intracellular $Ca^{2+}$ concentration (Tretyn 1987). It is possible that ACh accelerates accumulation of $Ca^{2+}{}_{cyt}$ required for $Ca^{2+}$-dependent $Cl^-$ channels activation, or there are acetylcholine receptors permeable to $Cl^-$ in the cell membrane or inside the cell (Gong and Bisson, 2002). Increased permeability to $Ca^{2+}$ after ACh application and the resulting rise of cytosolic $Ca^{2+}$ concentration could cause reduction of $Cl^-$ current latency. ACh-mediated prolongation of $Cl^-$ current inactivation could be determined by increasing time of the $Cl^-$ channels in open state. In this case, ACh could have inhibitory effect and could slow down rate of inactivation of $Cl^-$ channels participating in excitation (Gong and Bisson 2002). Otherwise, keeping longer increased intracellular $Ca^{2+}$ concentration could prolong time $Cl^-$ channels are in open state (Beilby and Shepherd 2006). The cytoplasmic concentration of $Ca^{2+}$ is maintained by two transport systems. It is proposed that the physiological role of $Ca^{2+}/H^+$ antiporters is to remove large amounts of $Ca^{2+}$ from the cytosol after $Ca^{2+}$ signal, while $Ca^{2+}$-ATPases maintain very a low level of $Ca^{2+}$ during resting state (Hirschi 2001). Consequently, prolongation of AP duration after DCCD application could be explained by longer time $Cl^-$ channels are in open state after increase of intracellular $Ca^{2+}$ concentration.

## 10.5 INVESTIGATIONS OF NEUROTRANSMITTERS IN PLANTS OBSERVING ACTION POTENTIAL TRAINS

Repetitive firing in *N. obtusa* cells when membrane potential reached a certain level of depolarization after application of various compounds was observed (Kisnierienė et al. 2009). As proposed by V. Shepherd and colleagues, this spontaneous firing resembles the onset of a critical instability where a time-ordered structure (repetitive firing) emerged after ramping currents to the excitation threshold (Shepherd et al. 2008). Namely, increased background conductance combined with loss of proton pump activity caused gradual depolarization of cell membrane potential to AP threshold. If resting membrane potential of *N. obtusa* cells ranges from −220 mV to −255 mV, it is far more negative than the AP threshold and depolarizing effect should be in high amplitude for spontaneous AP generation. Irrespective of the applied chemicals, the average time between spontaneous APs in *N. obtusa* cells was four to five minutes. This time is far shorter than it is required for a full repolarization after AP because of proton pump activity (Kisnierienė et al. 2009). However, it was reported that hyperpolarization or depolarization does not necessarily indicate the activity of $H^+$-pump, and if the $H^+$-pump activity was reduced, the resting potential could stay at the hyperpolarized level (Tsutsui and Ohkawa 2001). It was proposed that dynamics of membrane potential after electrically induced AP trains could be a better indicator of proton pump activity (Kisnieriene et al. (2018)). The repetitively triggered APs, their reproducibility, shape, and membrane potential dynamics after AP induction in APW and ACh solution were compared, and it was noticed that ACh caused membrane potential changes of dynamics after repetitive excitation in all investigated cells. Changes of membrane potential repolarization after repetitive (every 5 min) excitation of the cell were reproducible in the control solution and could be approximated by the exponent

$$y = y_0 + A_1 * \left(1 - \exp\left(\frac{-x}{t_1}\right)\right) \tag{10.1}$$

where:

$y_0$ is membrane potential before stimulation
$A_1$ describes range of membrane potential dynamics
$t_1$ is time constant

A significant decrease of the time constant from 5.5 ± 1.5 min in APW through 1.5 ± 0.65 min in 1 mM ACh solution down to 0.9 ± 0.35 min in 5mM ACh solution was observed. However, there was no significant difference between the time constant measures in APW and 0.1 mM ACh solution (Kisnierienė et al. 2009).

As mentioned earlier, depolarization of membrane potential to excitation threshold sets off trains of action potentials. In another investigation, it was found that MP of *Nitella* internodes was strongly dependent on the pH of the external medium. The slope of $E_m$ amounted to about 50 mV/pH, which was in the same order of magnitude (58 mV/pH in the pH range 4–5) found in *C. corallina*. These results indicated that the plasma membrane behaves like a $H^+$ electrode (Tazawa 2003). It was also found that application of a test medium acidified to pH ≤3.5 ± 0.2 with HCl resulted in immediate and repeatable triggering of APs within 1–2 s (Fillafer and Schneider 2016). Nevertheless, in another investigation it was shown that the pH 4.5 is the lower limit for excitability: the resting MP minus 100 mV, and excitation disappeared after about 60 min at this pH (Beilby 1984). It was proposed that protons are the main excitatory component of hydrolysates of ACh in *Chara braunii* (Fillafer and Schneider 2016). Dettbarn who found that Nitella is capable of hydrolysing acetylcholine up to 6.5 μmoles/g/h showed indirect possibility of acetylcholinesterase (AChE) activity in Characeae (Dettbarn 1962). The enzymatic activity of AChE has been shown to be altered by environmental contaminants such as metals. It was shown that $Cd^{2+}$ significantly inhibited AChE activity *in vitro*. Based on our data, we postulate AChE activity in *N. obtusa* cells as well (Kisnierienė et al. 2009). The high dose of ACh released inside the cell would have been expected to overcome a possible cholinesterase activity in the cell wall. The presence of acetylcholinesterase in a plant cell capable of propagating electric currents suggests the possibility of a fundamental similarity between the mechanism of bioelectrogenesis in plant and animal cells. We propose that amplitude and duration of AP and membrane potential dynamics triggered by repetitive stimulation have physiological implications. Repetitive action potentials could not only reflect the impact of various compounds, but they could accelerate death of the cells (Shepherd et al. 2008). Moreover, it was found that if the membrane potential of Characean cells reached more negative values, cells became more resistant to the toxic effect of various compounds (Kisnieriene et al. 2016). It has been proposed (Shabala et al. 2006) that frequency modulation, not only amplitude modulation, may be used for encoding environmental information in plants. It is possible that keeping the cell in a more depolarized state after the first AP in ACh solution has physiological implications. Probably the second AP causes accumulation of $Ca^{2+}$ sufficient to exert a physiological effect. For example, depolarization from a more positive membrane potential during the second AP generation consumes less energetic resources. It could be possible that two APs, separated by some critical time interval, have physiological meaning not only in *Dionaea*, (Trebacz et al. 1996; Volkov et al. 2008) but in other plant species too. Still, to access the sieve elements of higher plants for electrophysiological recordings is no easy task, given that this cell cable is embedded deep in the plant vascular tissue (Hedrich et al. 2016) and the data of AP transmission after neurotransmitters' application in plants are scanty. Already it was shown that 1 mM glutamic acid (Glu) or 5 mM GABA triggered APs, which propagated from leaf to leaf in barley. Glutamic acid and GABA induce APs not through membrane depolarization but presumably by binding to a putative receptor or to ligand-gated $Ca^{2+}$-conducting channels, respectively, followed by $Ca^{2+}$ induced activation of anion efflux (Felle and Zimmermann 2007). Injection of Glu solution at millimolar (200, 50, 5 mM) concentrations in the basal part of the stem evoked a series of APs in *Helianthus annuus* (Stolarz et al. 2010). Two glutamate receptor-like genes (GLR3.3 and 3.6) were identified as being involved in the propagation of electrical activity from the damaged to undamaged leaves (Hedrich et al. 2016).

## 10.6 VOLTAGE CLAMP APPROACH ON SINGLE CELL MODELS

The mechanisms that regulate the activity of neurotransmitter-gated ion channels are likely to be complex and responses can vary from cell to cell and from plant to plant. Information about ion channels and transporters can be assessed from both genomic investigations and electrophysiological characterizations of their activities. Voltage clamp techniques are used routinely in studies of voltage-sensitive ion channels, including measurements of their activation and inactivation dependence on voltage, kinetics of channel opening and closing, ion permeation, and channel inhibition by both organic and inorganic substances. The main value of the voltage clamp technique is that it allows one to measure the amount of ionic current crossing a membrane at any given voltage at a given time, and real-time processes in the intact cell can be investigated in detail using this method.

The electrophysiology of Characean is well explored using the voltage clamp technique. Standard intracellular electrophysiological methods and the voltage clamp technique are commonly used to determine the effect of various compounds on Characean electrical characteristics (Beilby 2016). It is very important and convenient that voltage and current data could be obtained from the simple single-cell system in various Characean species. Gained data can be applied as a common mechanism for comparable effects in higher plants. The simple morphology of the Characean offers a good experimental system to explore the two-electrode voltage clamp technique, which is much less invasive than the patch clamp technique because it does not damage the cell wall, membranes, and the cytosol. It is important because native cell walls may be a key element of many cell processes, for example in turgor regulation. Using Characean cells, very efficient voltage clamp methods that required only one microelectrode impalement for ion channels investigation was developed: membrane is clamped to the desired voltage by passing just enough current through the cell using Ag/AgCl wires (Lunevsky et al. 1983). Exploring Characean cells $K^+$, $Ca^{2+}$, and $Ca^{2+}$-activated $Cl^-$ currents can be registered separately in the steady state and after/during excitation. During long history of Characean investigations, it was proved that every transport system in the plasma membrane has very distinct MP dependencies and can be easily determined electrophysiologically (Berestovsky and Kataev 2005; Beilby and Casanova 2014).

Ion transport systems are often associated with many chemical compounds as an important early component of plant cell responses to specific stimulation. Yet little is known about the action of neurotransmitters on ion fluxes through the PM. Exploring Characean cells it was found that neurochemicals activate $K^+$, $Cl^-$, and $Ca^{2+}$ ion channels and enhance activation of $H^+$-ATPase. Voltage clamp method was used for the investigation of the activity of separate potassium ion transport systems. It is established that $K_{in}$ (inwardly) and $K_{out}$ (outwardly) channels opened at different membrane potential values: $K_{out}$ at −40 mV to −20 mV, $K_{in}$ at −150 mV to −180 mV (Sokolik and Yurin 1986). Membrane potential in this study was clamped at −40 mV for $K_{out}$ and −160 mV for $K_{in}$ investigations, and voltage–current characteristics were obtained by injecting short (30 ms) rectangular hyperpolarization or depolarization current pulses every 20 mV step from the clamped level. Effect of acetylcholine on $K^+$ channels was investigated in the *N. flexilis* cells. It was found that ACh increased conductance of both types—$K_{out}$ and $K_{in}$—rectifying $K^+$ channels (Kisnieriene et al. 2012). $K^+$ channels in the another Characean *C. corallina* were investigated after haloperidol, an antagonist of D2 dopamine receptors treatment, and all transient currents I–V curves show that haloperidol affects the plasma membrane conductance by blocking $K^+$ channels (Zherelova et al. 2016).

Using the voltage clamp method, the effect of ACh on $Cl^-$ conductance after excitation was proved. Increase of $Cl^-$ current in 5 mM ACh solution was observed (Kisnieriene et al. unpublished data). ACh determined increased efflux of negative charges during AP generation.

The method of clamping the plasma membrane potential with registration of ionic currents was also used to investigate the effect of neurotransmitter dopamine on the ionic channels of the electrically excitable plasma membrane of *C. corallina* cells. It was shown that the influence of dopamine on the ionic currents of the cells depended on the dose and time of exposure. The effect of dopamine on the membrane structure was reversible, after removal of the neurotransmitter, a complete restoration of the amplitude and the kinetics of the current was observed (Zherelova et al. 2014).

To reveal the mechanism underlying the selective effects of the chemical compounds onto calcium channels and $Ca^{2+}$-dependent chloride channels, the method of separating the calcium and the chloride components of the total current was proposed (Kataev et al. 2016). For example, it was shown that classical dopamine receptor antagonist haloperidol blocked $Ca^{2+}$ channels of the plasmalemma. In addition to bringing about a decrease in the amplitude of the calcium current, exposure to haloperidol decelerated the activation and inactivation of calcium channels. In the presence of haloperidol, the amplitudes of $Ca^{2+}$ currents were suppressed. Haloperidol did not affect $Ca^{2+}$-activated chloride channels. The authors suggest that $Ca^{2+}$ channels of *C. corallina* plasmalemma possess specific binding sites both for dopamine receptors and for their antagonists (Zherelova et al. 2016).

GABA is a major inhibitory neurotransmitter whose effect on Characean cells could be investigated using voltage clamp. GABA in CNS acts as a signal by regulating ion flow across cell membranes via two classes of receptors, the $GABA_A$ and $GABA_B$ (Bouché and Fromm 2004). Ionotropic $GABA_A$ receptors consist of multiple subunits that can assemble into a functional homomeric or heteromeric channel. GABA exerts its inhibitory effect in mature brain neurons by the activation of $Cl^-$ currents through $GABA_A$ receptor channels. This tends to hyperpolarize the membrane potential and inhibits excitability. In addition to acting on the ionotropic $GABA_A$ receptor, GABA is also an endogenous agonist of the $GABA_B$ receptor, which is a member of the large metabotropic G-protein coupled receptor superfamily (Lucas 2011).

Electrophysiological voltage clamp studies demonstrated that 5 μM GABA does not activate excitable $Cl^-$ currents in *C. corallina* as well as 160 μM nembutal (most effective agonist of the $GABA_A$ receptors) (Drinyaev et al. 2001). Recall for comparison that, according to published data, the concentrations of known and widely used blockers of $Ca^{2+}$-dependent $Cl^-$ current in plants such as A-9-C, ethacrynic acid, abscisic acid (ABA), and so on, necessary for complete blockage, is no less than 0.1 mM (Shiina and Tazawa 1987), and it is possible that too small of a concentration was applied in this case.

Some studies have suggested that avermectin (AVM, a drug used for treating diseases caused by parasitic roundworms) can promote GABA release and bind to its receptors to result in $Cl^-$ influx, hyperpolarization of the cell membrane, and inhibition of neural cells. It is proved that the effect of avermectin B1 on neurons of mammals is due to changes in $Cl^-$ current through the $GABA_A$ receptor/$Cl^-$ ionophore complex. In addition, AVMs can induce the opening of Glu-gated $Cl^-$ channels, increase membrane $Cl^-$ permeability, and reduce nerve conduction, which results in excitotoxicity, apoptosis, and necrosis (Chen et al. 2014). It was demonstrated in voltage clamp experiments with *C. corallina* that avermectin A1 (but not avermectins B1, A2, B2) regulate the $Ca^{2+}$-dependent $Cl^-$ currents. The low concentrations of avermectin A1 increase the $Cl^-$ currents in plant cell, while high concentrations of A1 inhibit them, thus resembling the effect of avermectin B1 on neurons of mammals. The $Ca^{2+}$-dependent $Cl^-$ current in these experiments was almost completely blocked at avermectin concentration of about 10 μM and avermectin A1 concentration of 10 nM (Drinyaev et al. 2001).

Results, obtained in Characean cells leads to the question: are the membrane transport systems of higher plants also affected by neurotransmitters?

To date, it is concluded that genes highly homologous to the animal GABA receptors are not present in the Arabidopsis genome (Bouché and Fromm 2004). Whole-cell voltage clamp experiments revealed that the influx of $Ca^{2+}$ increases in tobacco pollen tubes in response to exogenous GABA (Yu et al. 2014). Despite the lack of GABA receptor genes in plants, the amino acid sequences in the N-terminal domains of the putative GLRs in Arabidopsis have domains in common with $GABA_B$ receptors in animals; thus, it is tempting to consider that GABA could bind to this domain and modulate the activity of AtGLRs, in addition to their main ligands. Numerous reports over many years have shown that high levels of GABA accumulate rapidly in plant tissues exposed to a variety of different stresses. In response to environmental stresses, GABA production often increases so much that cellular levels of this non-protein amino acid exceed that of amino acids involved in protein synthesis (Kinnersley and Turano 2000). Other investigations have shown that stresses, including cold, heat, salinity, mechanical stress, etc. rapidly increase cellular levels of $Ca^{2+}$ and are interrelated to related electrical signaling (Reddy 2001; Huber and Bauerle 2016).

It was discovered that a family of plant anion channels, the aluminium ($Al^{3+}$)-activated Malate Transporters (ALMTs), are regulated by GABA, and this regulation has been proposed to transduce GABA metabolism into membrane signaling via an alteration of anion flux across cell membranes (Ramesh et al. 2015). GABA tightly regulates wheat root ALMT activity. As the inhibition of their activity can directly affect membrane potential with downstream physiological responses, it is likely that the different family members more broadly transduce GABA effects throughout plant tissues. In plants, the anion equilibrium potential is normally very positive so that when ALMT proteins are activated there is a depolarization observed. Plant action potentials are largely based on activation of voltage-dependent anion channels. Thus, GABA inhibition of ALMT will tend to hyperpolarize the membrane potential and decrease excitability, similar to the effect of GABA in animal neurons (Ramesh et al. 2017).

## 10.7 INVESTIGATIONS ON A SINGLE CHANNEL LEVEL WITH THE PATCH CLAMP TECHNIQUE

The patch clamp technique, in combination with advanced molecular cell biology and genomics methods, could provide a detailed picture of the plant membrane channels participating in electrical signaling. Recorded inward and outward currents using this method can be related to the data of macroscopic currents from algal cells under voltage clamp. The technique takes advantage of a high resistance seal (usually greater than 1 GΩ) formed between a microelectrode and a membrane thus providing an opportunity to investigate current dynamics of a single ion channel. Characean played a significant role in developing patch clamp investigations in plants. The first published single-channel recordings in plant cells have been performed in turgid cells of *N. obtusa* (Krawczyk 1978).

The presence of the cell wall poses some difficulties in applying the patch clamp technique to the plasma membrane. There are some solutions to this problem provided by various researchers. The cell wall may be removed using enzymes, perforated with a laser (De Boer et al. 1994), or removed surgically after plasmolysis (Laver 1991). Ion channels in both plant plasma membrane and tonoplast have been studied extensively, confirming the presence of certain ion channels and also investigating their characteristics and involvement in various functions (Tazawa et al. 1987; Thiel et al. 1997).

Coleman found a hyperpolarization-dependent $Cl^-$ channel in *C. australis* plasma membrane (Coleman 1986). $Ca^{2+}$-dependent $Cl^-$-permeable channel was found. The anion channel was activated by $Ca^{2+}$on the cytoplasmic side, and its opening was voltage dependent. The authors supposed that the anion channel is concerned with generation of action potentials at the plasma membrane (Okihara et al. 1991). The channel activity was inhibited by calmodulin antagonists. In addition, Okihara et al. (1993) demonstrated that activity of the $Ca^{2+}$-dependent $Cl^-$-permeable channel is regulated by calmodulin applied to the cytoplasmic side of the inside-out patch-clamp recording. The activity of the channel, which had been enhanced by $Ca^{2+}$, became low during the recording. However, the decrease of the activity could be stopped by applying calmodulin.

Homann and Thiel (1994) found a 40 pS $K^+$ channel in *Chara* that may play a role in conducting an outward current in the repolarization phase. They also observed two types of $Cl^-$ channels, which were active only during an action potential but not upon positive voltage steps. Laver has reported various kinds of $K^+$ channels in the plasma membrane, some of them showing sub-conductance states (Laver 1991).

There have been successful attempts to utilize the patch clamp technique for studying ion channels in the chloroplast thylakoid of Characean (Pottosin and Schoenknecht 1995) as well as in the intact chloroplast envelope (Pottosin 1992). To avoid any possible damage of the plasma membrane altogether while removing the cell wall, sometimes cytoplasmic droplets (Kamiya and Kuroda 1957) are used (Figure 10.3).

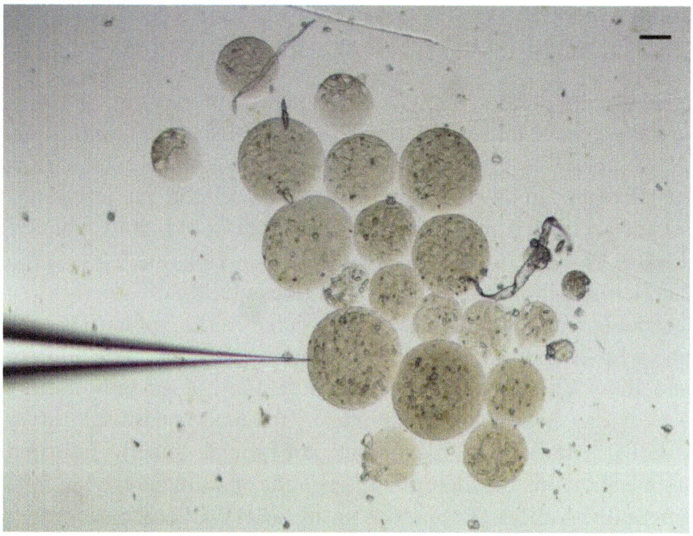

**FIGURE 10.3** A micrograph of cytoplasmic droplets isolated from internodal cells of *Nitellopsis obtusa*. A microelectrode is attached to one of the droplets. Horizontal bar represents 100 μm.

This technique provides an easy access to the tonoplast of the cell, which has been proven to be the source of the membrane of the cytoplasmic droplets (Sakano and Tazawa 1986).

Patch-clamp experiments have confirmed the presence of $Cl^-$ and $K^+$ channels in the Characean cell tonoplast (Tyerman and Findlay 1989). $Ca^{2+}$-modulated $Cl^-$ channels do exist in the membrane of *C. corallina* droplets (Berecki et al. 1999). Also high conductance $K^+$ channels were described by Lühring (1986) and a different kind of depolarization activated $K^+$ channels were observed by Pottosin and Andjus (Pottosin and Andjus 1994).

For some time, an interest has been taken in the effect of neuroactive compounds on the activity of ion channels of the Characeae. ACh could increase permeability of $Cl^-$ channels in plasmalemma and tonoplast. Patch clamp investigation of $Cl^-$ channels in *C. corallina* tonoplast demonstrated increased probability of open state, which would keep the cell more depolarized (Gong and Bisson 2002). It was found that an increase in $Cl^-$ concentration in the cytoplasm causes an increase in $Cl^-$ current (Beilby and Shepherd 2006). To activate tonoplast $Cl^-$ channel of Characeae, ACh must cross the plasma membrane and enter the cytoplasm. The effect of ACh in our experiments was reversible; thus, either ACh could not enter cytoplasm (then activation of $Cl^-$ channels is possible by the action of second messengers like $Ca^{2+}$) or ACh was hydrolyzed (Kisnierienė et al. 2009).

Another neuroactive compound Gamma-aminobutyric acid (GABA) affects ion channel activity in plant membranes and can be investigated by patch clamp method. There are indications that GABA receptors exist in plants or that plant glutamate receptors use GABA as a modulatory ligand (Bouché and Fromm 2004). To date, these investigations are scanty. Nevertheless, while the effect of GABA on the single ion channels of Characeae hasn't been investigated thoroughly, cytoplasmic droplets of *C. corallina* has been used as an alternative animal-free expression system for the $GABA_A$-receptor (and possibly other receptors) and research on the modulation and regulation of the $GABA_A$-receptor in connection with the effects of anesthetics (Duijn and Berecki 2001).

Despite substantial difficulties, the effects of neuroactive compounds are also being investigated in more complicated higher plant model systems using the patch clamp method. For example, the properties of the plant glutamate-activated cation currents were investigated in experiments of Arabidopsis root protoplasts and include voltage-independence, rapid activation upon changes in voltage and low selectivity among cations. The probability of observing glutamate-activated currents increased with increasing glutamate concentration (Demidchik et al. 2004).

## 10.8 THE ROLE OF NEUROTRANSMITTERS IN CHARACEAE CELL-TO-CELL ACTION POTENTIAL TRANSMISSION

Characean is very a useful system to study action potential propagation between plant cells. Two mechanisms of action potential transmission are well established in animals: one is the chemical transmission that takes place at synapses, and the other is known as electrotonic transmission, which occurs when currents can flow directly from cell-to-cell via gap junctions (Hille 2001). The communication between cells in plants is aided by the presence of highly dynamic structures called plasmodesmata (PD) under tight control of the plant. Gap junctions are the structures in animal cells equivalent to plasmodesmata in plants (Robards and Lucas 1990).

PD provide electrical connections between plant cells. Action potentials can propagate across the plasmodesmata, which supports their role in signaling in plants. The registration of electrical signal in adjacent cells of Characean after action potential induction in the first one is possible (Beilby 2016). The Characean cell size and organization made possible to arrange very illustrative experiments on PD. Characean experiments can provide general information about transport and signaling mechanisms through plant plasmodesmata because PD in *Chara* and higher plants are fundamentally similar (Brecknock et al. 2011). Characeae form primary PD containing endoplasmic reticulum but secondary PD were observed in some species of Characeae (Franceschi et al., 1994; Cook et al., 1997; Brecknock et al., 2011). However, while there has been some controversy over the presence of a desmotubule in *Chara* PD (Franceschi et al. 1994), it was shown that *Chara* PD contain desmotubules (Cook et al. 1997). It appears that *Chara* is a suitable model for studies of higher plant PD ultrastructure and function (Brecknock et al. 2011).

Long-distance nutrient transport in the Charales, apparently involving cytoplasmic streaming, is the functional equivalent of nutrient transport in the xylem and phloem in the parenchymatously constructed sporophytes of vascular embryophytes (Raven 2013). Ding et al. (1992) fed $NaH^{14}CO_3$ to a branchlet of *C. corallina* in an internode-branchlet complex and measured photoassimilates after 10 min in both the source branchlet and the sink internode, using thin-layer chromatography. The main photoassimilates transported were sucrose and amino acids. Transport was aided by downward concentration gradients of sucrose, serine, and glutamic acid between the cytoplasm. What is very interesting, cell-to-cell nutrient transport depends on the electrical properties of the cell. The stimulated cell may become isolated from the neighboring cells since the intercellular passage of substances through PD is inhibited when an AP is evoked. Beilby established that in the winter months the lateral internode cells exhibited low resting potential of ~−120 mV, (which is near excitation threshold), and restricted cell-to-cell communication. The exposure to excitation inhibitor $La^{3+}$ restored intercellular passage of substances as the action potential inhibited communication between nodes and internodes. In spring, the branch cells with more negative resting potential (~−210 mV) increased transport of 6-carboxyfluorcescein between internodes and adjacent nodes. As in winter cells, if cytoplasmic $Ca^{2+}$ was increased due to action potential or exposure to ionophore A23187, cell-to-cell transport was inhibited (Beilby 2016). Plasmodesmata not only facilitates transport of nutrients and photosynthates and ions across the cells but ensures steady signal transfer, permitting electrical coupling between adjacent cells.

Characean is a useful system for testing the mechanism of AP transmission because it was observed that an action potential generated in one internodal cell of *Nitella* or *Chara* trigger an action potential in a neighboring internodal cell. Sibaoka and Tabata investigated the AP transmission across the node in *C. braunii* using multiple electrodes in the two internodes and in one of the nodal cells. They found that the whole nodal cell did not produce an AP, although an area of the membrane bordering the nodal cell (end-membrane) could be excited. A stimulated internode produced greater electrotonic depolarization in the adjacent internode by a conducted AP, when the end-membrane excitation did not occur, so making the internode–internode transmission more likely (Sibaoka and Tabata 1981).

The fact that transmission does not occur in every instance supports the idea that the membranes within the plasmodesmata are not excitable, or that the membranes of the intervening nodal cells are not excitable (Spanswick 2012). It was suggested that an action potential would also be transmitted electrotonically to a separated internodal cell if cells were connected by a suitable solution bridge (Tabata 1990). Thus, the local current generated externally by one cell might be sufficient to depolarize the membrane at the junction of two cells even in the absence of a direct connection via plasmodesmata. For transmission of an action potential from one cell to another adjacent cell, the depolarizing membrane potential induced electrotonically in the adjacent cell by the action potential must exceed a critical threshold level (Tabata 1990).

AP measurements were performed in the tandem cell experiments after impalement of the two separate microelectrodes in two adjacent internodal cells of *N. obtusa* each 2 cm from the node (with appropriate reference electrodes in each compartment). Such type of electrophysiological measurements show real-time processes related to electrical communication between the two cells. Here we present membrane potential recording in the 2 adjacent cells after 1 mM acetylcholine treatment (Figure 10.4).

Usually when an AP is induced in the first cell, the neighboring cell shows only receptor potential. However, in some cases after receptor potential and electrical stimulation, the second cell fires spontaneous APs, after which the generation of AP in the first cell was recorded also. The membrane potential stays near excitation threshold level, which became more negative after ACh application. The action potential might release sufficient ions, or other molecules, into the apoplast to depolarize neighboring cells below the threshold for the generation of a separate action potential. When the AP in the first cell is electrically evoked again, in the second cell generation of AP without stimulation was recorded. It provides evidence on the electrical signal transduction through plasmodesmata.

To investigate the role of neurotransmitters as signaling molecules in the plant kingdom, we propose to use Characean cell as a model system. Specialized mechanisms of cell division that led to the evolution of plants and the position of Characean on the phylogenetic tree near the origin of land plants (Karol 2001) renew the interest in these algae and provide a new meaning for the investigations. Since Characeae species are close to higher plants and their value as relatively simple, easy-to-manipulate model system allows the formulation and verification of theories necessary to understand more complex level of functionality, adaptation, and information processing in higher plants and animals.

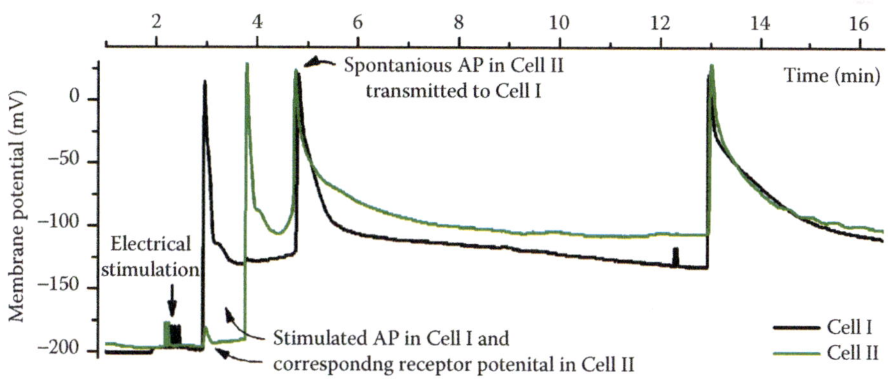


**FIGURE 10.4** Illustration of AP propagation after exposure to 1 mM ACh (acetylcholine) in two adjacent *Nitellopsis obtusa* cells. First, electrically stimulated AP in Cell I elicits receptor potential in Cell II (both indicated in arrows). After receptor potential and electrical stimulation, the second cell fires spontaneous AP (indicated with arrow), which elicits generation of AP in the Cell I. When AP in the first cell is electrically evoked again, in the second cell, generation of AP without stimulation was recorded also. Recorded using two intracellular microelectrodes impaled 2 cm from node.

## 10.9 SPECIFICITY OF CHARACEAN AND DIVERSE EXPERIMENTAL APPROACHES

Cells of Charophytes are well-characterized experimental systems to study a wide range of membrane transport phenomena. The results obtained from experiments using Characean cells have formed a basis for the studies of plant membrane functions (Shimmen et al. 1994). Basic knowledge on bioelectric phenomena of plant membranes has been greatly indebted to internodal cells of Charophytes. The main advantage offered by the Characean system is the possibility to use electrophysiology at the single cell level *in vivo*. A cell separated from the thallus can be considered as a single organism that maintains essential physiological characteristics for a long time (Shimmen et al. 1994). The usefulness of Characean in experiments is based on the ability of internodal cells to survive after isolation from the plant, on the simple morphology, on the large size and regular shape of these cells, facilitating creation of experimental systems on many different levels.

Characean internodal cells, being huge, have a typical plant-cell structure with a cellulose cell wall, plasma membrane bounding a relatively thin layer of cytoplasm containing moving multiple copies of the nucleus, mitochondria, protein bodies and statoliths, immobile chloroplasts, and in the center a huge vacuole enclosed by tonoplast. In most Characean cells, the vacuole occupies about 95% of the cell volume. Cortical actin filaments and microtubules are present along the plasma membrane and have an orientation that is predominantly transverse to the longitudinal axis in elongating internodes and random in non-elongating mature cells (Foissner and Wasteneys 2012).

Single cells are easily excised from the plants, and contents of cell compartments can be extracted and analyzed. Thus, the endogenous neurotransmitter content of cells and cell compartments can be measured; also, accurate doses of exogenous neurotransmitters can be easily (Beilby et al, 2015a).

Up to date, for Characean species there is no direct evidence that synaptic-like transmission (when the release of neurotransmitters from one cell induced action potential in another cell). There remains the possibility that the increase in channel-mediated fluxes of ions during an action potential could change the concentrations in the apoplast sufficiently to depolarize a neighboring cell below the threshold for an action potential.

## 10.10 CONCLUSION

Neurotransmitters affect all membrane transport systems involved in electrical signaling of Characean cells. Electrophysiological methods and electrical signaling pathway analysis of Characean open new opportunities for testing instantaneous impact of neurotransmitters on the intact living cell. Further research concerning plant membrane transport systems and regulation of electrical properties by neurotransmitters will lead us to understanding of their role in plant signaling as a whole. Since Characean are one of thoroughly investigated plant model system regarding electrical signaling, new experimental approaches or their combination could navigate toward unraveling this question. Undoubtedly, knowledge obtained using the Characean system helps us to understand more complex laws of functionality, adaptation, and information processing in higher plants and animals.

## REFERENCES

Arnao, M. B., and J. Hernández-Ruiz. 2014. Melatonin: Plant growth regulator and/or biostimulator during stress? *Trends in Plant Science* 19 (12): 789–797.

Balzer, I., and R. Hardeland. 1996. Melatonin in algae and higher plants-possible new roles as a phytohormone and antioxidant. *Plant Biology* 109 (3): 180–183.

Beilby, M. J. 1984. Current-voltage characteristics of the proton pump at chara plasmalemma: I. pH dependence. *Journal of Membrane Biology* 81 (2): 113–125.

Beilby, M. J., and V. A. Shepherd. 2006. The characteristics of $Ca^{2+}$-activated $Cl^-$ channels of the salt-tolerant charophyte lamprothamnium. *Plant, Cell & Environment* 29 (5): 764–777.
Beilby, M. J. 2007. Action potential in charophytes. *International Review of Cytology* 257 (7): 43–82.
Beilby, M. J., and M. T. Casanova. 2014. The Physiology of Characean Springer, pp. 205.
Beilby, M. J., S. Al Khazaaly, and M. A. Bisson. 2014. Salinity-induced noise in membrane potential of characeae chara australis: Effect of exogenous melatonin. *Journal of Membrane Biology* 248 (1): 93–102.
Beilby, M. J., C. E. Turi, T. C. Baker, F. J. M. Tymm, and S. J. Murch. 2015a. Circadian changes in endogenous concentrations of indole-3-acetic acid, melatonin, serotonin, abscisic acid and jasmonic acid in characeae (Chara australis brown). *Plant Signaling & Behavior*, 10 (11): e1082697.
Beilby M. J., C. E. Turi, and S. J. Murch. 2015b. *Rhythms in Plants: Dynamic Responses in a Dynamic Environment*, 2nd ed. pp. 343–367. Springer: Cham, Switzerland.
Beilby, M. J. 2016. Multi-scale characean experimental system: From electrophysiology of membrane transporters to cell-to-cell connectivity, cytoplasmic streaming and auxin metabolism. *Frontiers in Plant Science* 7: 1052.
Beilby, M. J., and S. Al Khazaaly. 2016. Re-modeling chara action potential: I. from thiel model of $Ca^{2+}$ transient to action potential form. *AIMS Biophysics* 3 (3): 431–449.
Berecki, G., Z. Varga, F. Van Iren, and B. Van Duijn. 1999. Anion channels in chara corallina tonoplast membrane: Calcium dependence and rectification. *Journal of Membrane Biology* 172 (2): 159–168.
Berestovsky, G. N., and A. A. Kataev. 2005. Voltage-gated calcium and $Ca^{2+}$-activated chloride channels and $Ca^{2+}$ transients: Voltage-clamp studies of perfused and intact cells of chara. *European Biophysics Journal* 34 (8): 973–986.
Bouché, N., and H. Fromm. 2004. GABA in plants: Just a metabolite? *Trends in Plant Science* 9 (3): 110–115.
Brecknock, S., T. P. Dibbayawan, M. Vesk, P. A. Vesk, C. Faulkner, D. A. Barton, and R. L. Overall. 2011. High resolution scanning electron microscopy of plasmodesmata. *Planta* 234 (4): 749–758.
Chen, L.-J., B.-H. Sun, Y. Cao, H.-D. Yao, J.-P. Qu, C. Liu, S.-W. Xu, and S. Li. 2014. The effects of avermectin on amino acid neurotransmitters and their receptors in the pigeon brain. *Pesticide Biochemistry and Physiology* 110: 13–19.
Coleman, H. A. 1986. Chloride currents in Chara—A patch-clamp study. *Journal of Membrane Biology* 93 (1): 55–61.
Cook, M. E., L. E. Graham, C. E. J. Botha, and C. A. Lavin. 1997. Comparative ultrastructure of plasmodesmata of Chara and selected bryophytes: Towards an elucidation of the evolutionary origin of plant plasmodesmata. *American Journal of Botany* 84: 1169–1178.
Davies, E. 2004. New functions for electrical signals in plants. *New Phytologist* 161 (3): 607–610.
De Boer, A. H., B. Van Duijn, P. Giesberg, L. Wegner, G. Obermeyer, K. Köhler, and K. W. Linz. 1994. Laser microsurgery: A versatile tool in plant (electro) physiology. *Protoplasma* 178 (1–2): 1–10.
Demidchik, V., P. A. Essah, and M. Tester. 2004. Glutamate activates cation currents in the plasma membrane of arabidopsis root cells. *Planta* 219 (1): 167–175.
Demidchik, V., and F. J. M. Maathuis. 2007. Physiological roles of nonselective cation channels in plants: From salt stress to signaling and development. *New Phytologist* 175 (3): 387–404.
Dennison, K. L., and E. P. Spalding. 2010. Glutamate-gated calcium fluxes in Arabidopsis. *Plant Physiology* 124 (4): 1511–1514.
Dettbarn, W. D. 1962. Acetylcholinesterase activity in Nitella. *Nature* 194:1175–1176.
Ding, D. Q., S. Amino, T. Mimura, K. Sakano, T. Nagata, and M. Tazawa. 1992. Quantitative analysis of intercellularly-transported photoassimilates in Chara corallina. *Journal of Experimental Botany* 43 (8): 1045–1051.
Di Sansebastiano, G. P., S. Fornaciari, F. Barozzi, G. Piro, and L. Arru. 2014. New insights on plant cell elongation: A role for acetylcholine. *International Journal of Molecular Sciences* 15 (3): 4565–4582.
Drinyaev, V. A., V. A. Mosin, E. B. Kruglyak, T. S. Sterlina, A. A. Kataev, G. N. Berestovsky, and Y. M. Kokoz. 2001. Effect of avermectins on $Ca^{2+}$-dependent $Cl^-$ currents in plasmalemma of chara corallina cells. *Journal of Membrane Biology* 182 (1): 71–79.
Duijn, B. van, and G. Berecki. 2001. The alga Chara corallina as an alternative for Xenopus oocytes. *NCA Newsletter*11: 6–8.
Felle, H. H., and M. R. Zimmermann. 2007. Systemic signaling in barley through action potentials. *Planta* 226 (1): 203–214.
Fillafer, C., and M. F. Schneider. 2016. On the excitation of action potentials by protons and its potential implications for cholinergic transmission. *Protoplasma* 253 (2): 357–365.

Findlay, G. P. 1970. Membrane electrical behaviour in Nitellopsis obtusa. *Australian Journal of Biological Sciences* 23 (4): 1033–1046.
Foissner, I., and G. O. Wasteneys. 2012. The Characean internodal cell as a model system for studying wound healing. *Journal of Microscopy* 247 (1): 10–22.
Franceschi, V. R., B. Ding, and W. J. Lucas. 1994. Mechanism of plasmodesmata formation in Characean algae in relation to evolution of intercellular communication in higher plants. *Planta* 192 (3): 347–358.
Fromm, J., and S. Lautner. 2007. Electrical signals and their physiological significance in plants. *Plant, Cell and Environment* 30 (3): 249–257.
Gong, X.-Q., and M. A. Bisson. 2002. Acetylcholine-activated $Cl^-$ channel in the Chara tonoplast. *Journal of Membrane Biology* 188 (2): 107–113.
Hartmann, E. 1977. Influence of acetylcholine and light on the bioelectric potential of bean (Phaseolus vulgaris L.) hypocotyl hook. *Plant and Cell Physiology* 18 (6): 1203–1207.
Hedrich, R. 2012. Ion channels in plants. *Physiological Reviews* 92 (4): 1777–1811.
Hedrich, R., V. Salvador-Recatalà, and I. Dreyer. 2016. Electrical wiring and long-distance plant communication. *Trends in Plant Science* 21 (5): 376–387.
Hille, B. 2001. *Ion Channel Excitable Membranes*. Sinauer Associates: Sunderland MA.
Hirschi, K. 2001. Vacuolar $H^+/Ca^{2+}$ transport: Who's directing the traffic? *Trends in Plant Science* 6 (3): 100–104.
Homann, U., and G. Thiel. 1994. $Cl^-$ and $K^+$ channel currents during the action potential in Chara. Simultaneous recording of membrane voltage and patch currents. *Journal of Membrane Biology* 141 (3): 297–309.
Horiuchi, Y., R. Kimura, N. Kato, T. Fujii, M. Seki, T. Endo, T. Kato, and K. Kawashima. 2003. Evolutional study on acetylcholine expression. *Life Sciences* 72 (15): 1745–1756.
Huber, A. E., and T. L. Bauerle. 2016. Long-distance plant signaling pathways in response to multiple stressors: The gap in knowledge. *Journal of Experimental Botany* 67 (7): 2063–2079.
Jaffe, M. J. 1970. Evidence for the regulation of phytochrome-mediated processes in bean roots by the neurohumor, acetylcholine. *Plant Physiology* 46 (6): 768–777.
Kamiya, N., and K. Kuroda. 1957. Cell operation in Nitella. I. Cell amputation and effusion of the endoplasm. *Proceedings of the Japan Academy* 33 (3): 149–152.
Karol, K. G. 2001. The closest living relatives of land plants. *Science* 294 (5550): 2351–2353.
Kataev, A., O. Zherelova, and V. Grishchenko. 2016. A characeae cells plasma membrane as a model for selection of bioactive compounds and drugs: Interaction of HAMLET-like complexes with ion channels of Chara corallina cells plasmalemma. *Journal of Membrane Biology* 249 (6): 801–811.
Kikuyama, M. 1986. Tonoplast action potential of Characeae. *Plant and Cell Physiology* 27 (8): 1461–1468.
Kinnersley, A. M., and F. J. Turano. 2000. Gamma aminobutyric acid (GABA) and plant responses to stress. *Critical Reviews in Plant Sciences* 19 (6): 479–509.
Kinraide, T. B., and B. U. D. Etherton. 1980. Electrical evidence for different mechanisms of uptake for basic, neutral, and acidic amino acids in oat coleoptiles. *Plant Physiology* 65 (6): 1085–1089.
Kishimoto, U., N. Kami-ike, Y. Takeuchi, and T. Ohkawa. 1984. A kinetic analysis of the electrogenic pump of Chara corallina. I. Inhibition of the pumpby DCCD. *Journal of Membrane Biology* 80: 175–183.
Kisnierienë, V., and V. Sakalauskas. 2007. The effect of aluminium on bioelectrical activity of the Nitellopsis obtusa cell membrane after $H^+$-ATPase inhibition. *Central European Journal of Biology* 2 (2): 222–232.
Kisnieriene V., I. Lapeikaite, V.Pupkis. 2018. Electrical signalling in *Nitellopsis obtusa*: potential biomarkers of biologically active compounds. *Functional Plant Biology*. 45(2): 132–142.
Kisnierienė, V., V. Sakalauskas, A. Pleskačiauskas, V. Yurin, and O. Rukšėnas. 2009. The combined effect of $Cd^{2+}$ and ACh on action potentials of Nitellopsis obtusa cells. *Central European Journal of Biology* 4 (3): 343–350.
Kisnieriene, V., T. I. Ditchenko, A. P. Kudryashov, V. Sakalauskas, V. M. Yurin, and O. Ruksenas. 2012. The effect of acetylcholine on Characeae $K^+$ channels at rest and during action potential generation. *Central European Journal of Biology* 7 (6): 1066–1075.
Kisnieriene V., I. Lapeikaite, and V. Pupkis. 2018. Electrical signalling in *Nitellopsis obtusa*: Potential biomarkers of biologically active compounds. *Functional Plant Biology* 45 (2): 132–142.
Kisnieriene, V., I. Lapeikaite, O. Sevriukova, and O. Ruksenas. 2016. The effects of $Ni^{2+}$ on electrical signaling of Nitellopsis obtusa cells. *Journal of Plant Research* 129 (3): 551–558.
Krawczyk, S. 1978. Ion channel formation in a living cell membrane. *Nature* 273: 56–57.
Król, E., H. Dziubińska, and K. Trębacz. 2010. What do plants need action potentials for? *Action Potential* 9: 1–26.
Laver, D. R. 1991. A surgical method for accessing the plasmamembrane of Chara Australis. *Protoplasma* 161 (2–3): 79–84.

Lazár, D., S. J. Murch, M. J. Beilby, and S. Al Khazaaly. 2013. Exogenous melatonin affects photosynthesis in Characeae Chara Australis. *Plant Signaling & Behavior* 8 (3): 1–5.
Lucas, A. 2011. GABA receptors and the immune system. Dissertation, pp. 1–23.
Lühring, H. 1986. Recording of single $K^+$ channels in the membrane of cytoplasmic drop of Chara Australis. *Protoplasma* 133 (1): 19–28.
Lunevsky, V. Z., O. M. Zherelova, I. Y. Vostrikov, and G. N. Berestovsky. 1983. Excitation of Characeae cell membranes as a result of activation of calcium and chloride channels. *Journal of Membrane Biology* 72 (1–2): 43–58.
Mimura, T. 1995. Physiological characteristics and regulation mechanisms of the $H^+$ pumps in the plasma membrane and tonoplast of Characean cells. *Journal of Plant Research* 108 (2): 249–256.
Mousavi, S. A. R., A. Chauvin, F. Pascaud, S. Kellenberger, and E. E. Farmer. 2013. GLUTAMATE RECEPTOR-LIKE genes mediate leaf-to-leaf wound signaling. *Nature* 500 (7463): 422–426.
Murch, S. J. 2006. Neurotransmitters, neuroregulators and neurotoxins in plants. *Communication in Plants: Neuronal Aspects of Plant Life* 137–151.
Odjakova, M., and C. Hadjiivanova. 1997. Animal neurotransmitter substances in plants. *Bulgarian Journal of Plant Physiol* 23: 94–102.
Okihara, K., T. A. Ohkawa, I. Tsutsui, and M. Kasai. 1991. A $Ca^{2+}$- and voltage-dependent Cl-sensitive anion channel in the Chara plasmalemma: A patch-clamp study. *Plant and Cell Physiology* 32 (5): 593–601.
Okihara, K., T. Ohkawa, and M. Kasa. 1993. Effects of calmodulin on $Ca^{2+}$-dependent $Cl^-$-sensitive anion channels in the Chara plasmalemma: A patch-clamp study. *Plant and Cell Physiology* 34 (1): 75–82.
Pottosin, I. I. 1992. Single channel recording in the chloroplast envelope. *FEBS Letters* 308 (1): 87–90.
Pottosin, I. I., and P. R. Andjus. 1994. Depolarization-activated $K^+$ channel in Chara droplets. *Plant Physiology* 106 (1): 313–319.
Pottosin, I. I., and G. Schoenknecht. 1995. Patch clamp study of the voltage-dependent anion channel in the thylakoid membrane. *Journal of Membrane Biology* 148 (2): 143–156.
Qi, Z., N. R. Stephens, and E. P. Spalding. 2006. Calcium entry mediated by GLR3.3, an Arabidopsis Glutamate receptor with a broad agonist profile. *Plant Physiology* 142 (3): 963–971.
Ramakrishna, A. P. Giridhar, and G. A. Ravishankar. 2009. Indoleamines and calcium channels influence morphogenesis in in vitro cultures of Mimosa Pudica L. *Plant Signaling & Behavior* 4 (12): 1136–1141.
Ramesh, S. A., S. D. Tyerman, B. Xu, J. Bose, S. Kaur, V. Conn, P. Domingos et al. 2015. GABA signaling modulates plant growth by directly regulating the activity of plant-specific anion transporters. *Nature Communications* 6: 7879.
Ramesh, S. A, S. D. Tyerman, M. Gilliham, and B. Xu. 2017. $\gamma$-aminobutyric acid (GABA) signaling in plants. *Cellular and Molecular Life Sciences* 2017. $\gamma$-Aminobutyric Acid (GABA) Signaling in Plants. *Cellular and Molecular Life Sciences* 74: 1577–1603.
Raven, J. A. 2013. Polar auxin transport in relation to long-distance transport of nutrients in the Charales. *Journal of Experimental Botany* 64 (1): 1–9.
Reddy, A. S. N. 2001. Calcium: Silver bullet in signaling. *Plant Science* 160 (3): 381–404.
Reiter, R. J., D. X. Tan, Z. Zhou, M. H. C. Cruz, L. Fuentes-Broto, and A. Galano. 2015. Phytomelatonin: Assisting plants to survive and thrive. *Molecules* 20 (4): 7396–7437.
Robards, A. W., and W. J. Lucas. 1990. Plasmodesmata. *Annual Review of Plant Physiology* 41 (1): 369–419.
Roshchina, V. V. 1991. Biomediators in plants. *Acetylcholine and Biogenic Amines in Plants.* Pushchino, Russia: Biological Center of USSR Academy of Sciences.
Roshchina, V. V. 2001. *Neurotransmitters in Plant Life.* Enfield, NH: Science Publishers.
Roshchina, V. V. 2016. New trends and perspectives in the evolution of neurotransmitters in microbial, plant, and animal cells. In *Microbial Endocrinology: Interkingdom Signaling in Infectious Disease and Health* (Ed.) M. Lyte. pp. 25–77. New York: Springer.
Sakano, K., and M. Tazawa. 1986. Tonoplast origin of the envelope membrane of cytoplasmic droplets prepared from Chara internodal cells. *Protoplasma* 131 (3): 247–249.
Shabala, S., L. Shabala, D. Gradmann, Z. Chen, I. Newman, and S. Mancuso. 2006. Oscillations in plant membrane transport: Model predictions, experimental validation, and physiological implications. *Journal of Experimental Botany* 57 (1): 171–184.
Shepherd, V. A., M. J. Beilby, S. Al Khazaaly, and T. Shimmen. 2008. Mechano-perception in Chara cells: The influence of salinity and calcium on touch-activated receptor potentials, action potentials and ion transport. *Plant, Cell and Environment* 31 (11): 1575–1591.
Shiina, T., and M. Tazawa. 1987. $Ca^{2+}$-Activated $Cl^-$ channel in plasmalemma of Nitellopsis obtusa. *Journal of Membrane Biology* 99 (2): 137–146.
Shimmen, T., T. Mimura, M. Kikuyama, and M. Tazawa. 1994. Characean cells as a tool for studying electrophysiological of plant cells characteristics. *Cell Structure and Function* 19 (5): 263–278.

Sibaoka, T., and T. Tabata. 1981. Electrotonic coupling between adjacent internodal cells of Chara braunii: Transmission of action potentials beyond the node. *Plant and Cell Physiology* 22 (3): 397–411.
Sokolik, A. I, and V. M. Yurin. 1986. Potassium channels in plasmalemma of Nitella cells at rest. *Membrane Biology* 9 (22): 9–22.
Spanswick, R. M. 2012. The role of plasmodesmata in the electrotonic transmission of action potentials. In *Plant Electrophysiology: Signaling and Responses*. pp. 233–247. Berlin, Germany: Springer.
Stolarz, M., E. Król, H. Dziubińska, and A. Kurenda. 2010. Glutamate induces series of action potentials and a decrease in circumnutation rate in Helianthus Annuus. *Physiologia Plantarum* 138 (3): 329–338.
Tabata, T. 1990. Electrotonic transmission of an action potential between two separated internodal cells of Chara through a bridge. *Plant and Cell Physiology* 31 (4): 513–518.
Tazawa, M., M. Kikuyama, and T. Shimmen. 1976. Electric characteristics and cytoplasmic streaming of Characeae cells lacking tonoplast. *Cell Structure and Function* 1 (2): 165–176.
Tazawa, M., T. Shimmen, and T. Mimura. 1987. Membrane control in the Characeae. *Annual Review of Plant Physiology* 38 (1): 95–117.
Tazawa, Masashi. 2003. Cell physiological aspects of the plasma membrane electrogenic $H^+$ Pump. *Journal of Plant Research* 116 (5): 419–442.
Tazawa, M., and M. Kikuyama. 2003. Is $Ca^{2+}$ release from internal stores involved in membrane excitation in Characean cells? *Plant and Cell Physiology* 44 (5): 518–526.
Thiel, G., U. Homann, and C. Plieth. 1997. Ion channel activity during the action potential in Chara: New insights with new techniques. *Journal of Experimental Botany* 48 (Special): 609–622.
Trebacz, K., M. B. Busch, Z. Hejnowicz, and A. Sievers. 1996. Cyclopiazonic acid disturbs the regulation of cytosolic calcium when repetitive action potentials are evoked in Dionaea Traps. *Planta* 198 (4): 623–626.
Tretyn, A. 1987. Influence of red light and Acetylcholine on Ca uptake by oat coleoptile cells. *Cell Biology International Reports* 11 (12): 887–896.
Tretyn, A., and R. E. Kendrick. 1991. Acetylcholine in plants: Presence, metabolism and mechanism of action. *Botanical Review* 57 (1): 33–73.
Tsutsui, I., T. Ohkawa, R. Nagai, and U. Kishimoto. 1987. Role of calcium ion in the excitability and electrogenic pump activity of the Chara corallina membrane: II. Effects of $La^{3+}$, EGTA, and calmodulin antagonists on the current-voltage relation. *Journal of Membrane Biology* 96 (1): 75–84.
Tsutsui, I., and T. Ohkawa. 2001. Regulation of the $H^+$ pump activity in the plasma membrane of internally perfused Chara corallina. *Plant & Cell Physiology* 42 (5): 531–537.
Tyerman, S. D., and G. P. Findlay. 1989. Current-Voltage curves of single $Cl^-$ channels which coexist with two types of $K^+$ channel in the tonoplast of Chara corallina. *Journal of Experimental Botany* 40 (1): 105–117.
Umrath, K. 1930. Untersuchungen über plasma und plasmaströmung an characeen. *Protoplasma* 9 (1): 576–597.
Vodeneev, V. A., L. A. Katicheva, and V. S. Sukhov. 2016. Electrical signals in higher plants: Mechanisms of generation and propagation. *Biophysics* 61 (3): 505–512.
Volkov, A. G., T. Adesina, V. S. Markin, and E. Jovanov. 2008. Kinetics and mechanism of Dionaea Muscipula trap closing. *Plant Physiology* 146 (2): 694–702.
Wacke, M., and G. Thiel. 2001. Electrically triggered all-or-none $Ca^{2+}$-liberation during action potential in the giant alga Chara. *Journal of General Physiology* 118 (1): 11–22.
Wessler, I., H. Kilbinger, F. Bittinger, and C. J. Kirkpatrick. 2001. The non-neuronal cholinergic system. The biological role of non-neuronal acetylcholine in plants and humans. *Japanese Journal of Pharmacology* 85 (1): 2–10.
Yu, G.-H., J. Zou, J. Feng, X.-B. Peng, J.-Y. Wu, Y.-L. Wu, R. Palanivelu, and M.-X. Sun. 2014. Exogenous γ-aminobutyric acid affects pollen tube growth via modulating putative $Ca^{2+}$-permeable membrane channels and is coupled to negative regulation on glutamate decarboxylase. *Journal of Experimental Botany* 65 (12): 3235–3248.
Yurin, V. M. 1990. Electrophysiological aspects of enobiology of plant cell. 3. Non-state ionic flow. Vestnik Akademii Nauk BSSR Serija. Biology *Biomedicinskij Zhurnal Medline.ru*. Biology 5: 7–10.
Yurin, V. M., Ivanchenko, V. M. and Galaktionov, S. G. 1979. *Regulation of the Functions of Plant Cell Membranes*. Minsk, Belarus: Nauka i Teknika.
Zhang, N., Q. Sun, H. Zhang, Y. Cao, S. Weeda, S. Ren, and Y. D. Guo. 2015. Roles of melatonin in abiotic stress resistance in plants. *Journal of Experimental Botany* 66 (3): 647–656.
Zherelova, O. M., A. A. Kataev, V. M. Grischenko, R. S. Shtanchaev, D. A. Moshkov, and B. I. Medvedev. 2014. Interaction of neuromediator dopamine with the ionic channels of Chara corallina cell plasmalemma. *Biomedicinskij Zhurnal Medline.ru*. 15 (67): 834–846.
Zherelova, O. M., A. A. Kataev, V. M. Grischenko, and R. S. Shtanchaev. 2016. Haloperidol modulates ion transport in Chara corallina cells. *Cell and Tissue Biology* 10 (6): 476–485.

# 11 Dopamine and Their Antagonist Modulates Ion Transport and Cytoplasmic Streaming in *Chara* Cells

*Anatolii A. Kataev, Olga M. Zherelova, V.M. Grischenko, and R.Sh. Shtanchaev*

## CONTENTS



## 11.1 INTRODUCTION

Classical neurotransmitters of synaptic transmission excitation, such as acetylcholine, catecholamines, serotonin, and histamine are present in plants, while components of the cholinergic and adrenergic regulation systems were discovered in plant cells. This makes quite plausible the idea of common principles of signaling and information transfer, in the form of electrical and chemical signals, in all living organisms.

Compounds acting as neuromediators or neurotransmitters in the animal central nervous system (CNS) were also detected in plants, where they perform regulatory and signaling functions (Roschina, 2001). Dopamine and other biogenic amines are universal excitatory agents present in a wide variety of organisms representing different evolutionary distant systematic groups. In particular, the effects of dopamine on ion permeability have been confirmed in several studies in plant objects (Roschina, 2001). Along with laboratory animals and target cell cultures, comprehensive research on the dopamine and their antagonists on ion channels of cell membranes can make use of *Chara corallina* alga as a convenient, readily available, and cost-efficient test model.

## 11.2 THE EFFECT OF DOPAMINE ON THE ELECTROEXCITABLE MEMBRANE AND THE CYTOSKELETON OF TARGET CELLS

The effect of neurotransmitter dopamine on the ionic channels of the electrically excitable plasma membrane of *C. corallina* giant cells was investigated (Zherelova et al., 2014). The plasmalemma of these cells features selective ion channels ($K^+$, $Ca^{2+}$, and $Cl^-$), the electrophysiological properties of which are in many respects similar to those of ion channels of animal cells (Tester, 1990; Pineros and Tester, 1997). The appropriateness of this cell model is justified by the fact that genes encoding ion channels of different families in plants exhibit considerable homology to their counterparts in animal cells (Ward et al., 2009).

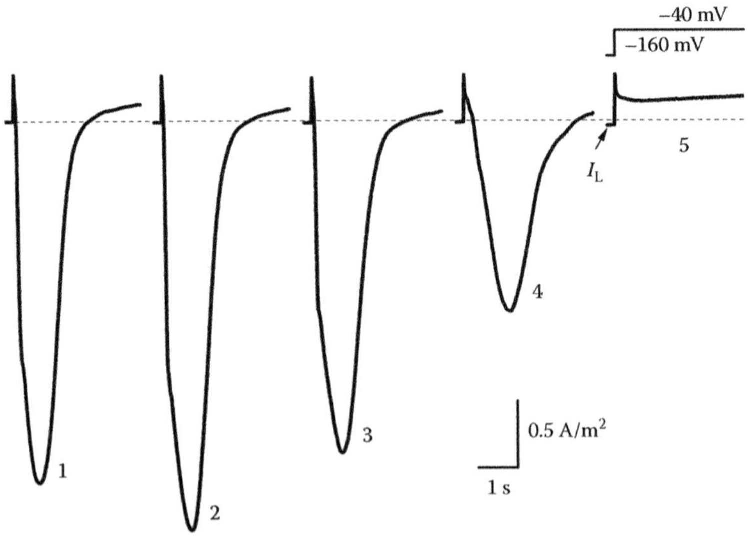


**FIGURE 11.1** Changes in the amplitude of the integral current in the plasmalemma of *Chara corallina* cells. 1-control; 2–5—in the presence of 25 mkM, 75 mkM, 125 mkM and 175 mM dopamine. Top—command voltage pulse on the membrane: from −160 to −40 mV. $I_L$—leakage current.

It is an important property of *C. corallina* cells that the cytoplasmic streaming can be observed visually (Tominaga et al., 1983; Kataev et al., 2012). In *Charophyta* cells, this motion is driven by interaction between the subcortical actin of internodes and endoplasmic myosin (Shimmen and Yokota, 2004).

The method of voltage clamping on the plasma membrane with registration of ionic currents under normal conditions and in the presence of dopamine was used. It was shown that the influence of dopamine on the ionic currents of the cells depended on the dose and time of exposure (Figure 11.1). The action of dopamine on the membrane structure was reversible, after removal of the neurotransmitter, a complete restoration of the amplitude and the development kinetics of the currents was observed (Figure 11.2). A 24h-long incubation of *C. corallina* cells in the presence of

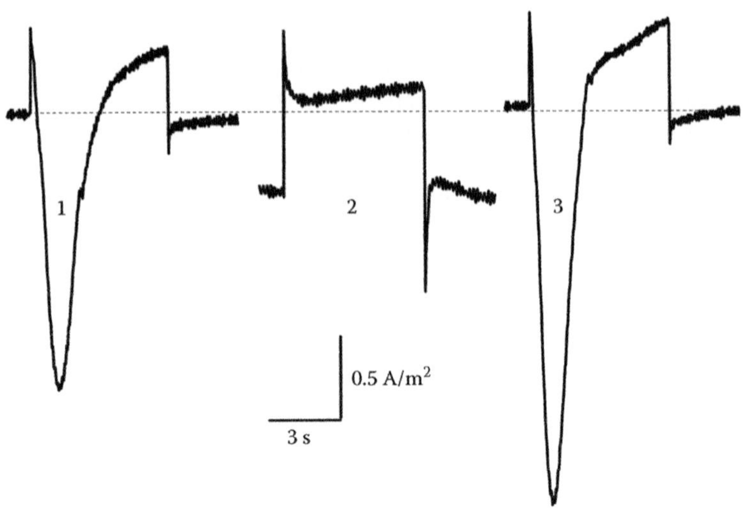


**FIGURE 11.2** Restoration of amplitude of integral current of a plasmalemma of *Chara corallina* after deleting dopamine from the solution washing a cell. (1)—monitoring, (2)—in the presence of 15 mM of dopamine, (3)—in 6 min after scrub of a cell from dopamine.

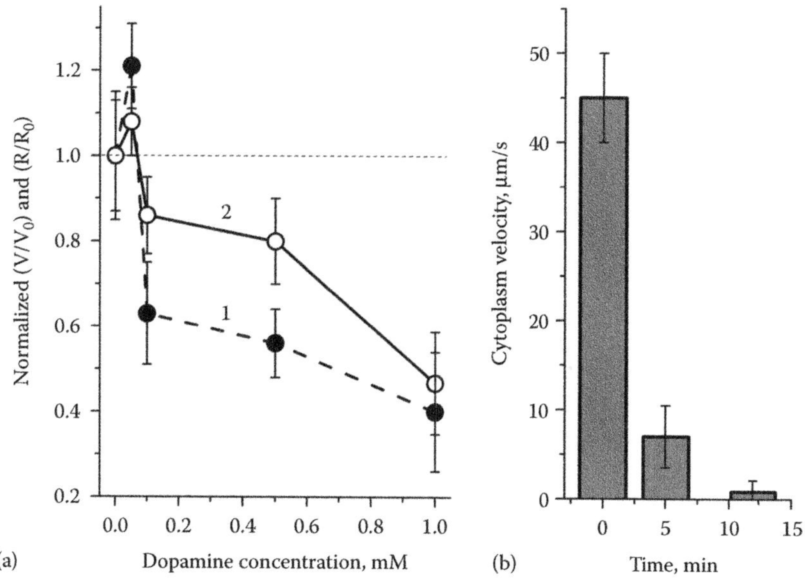


**FIGURE 11.3** Change of resting potential ($V_m$), membrane resistance ($R_m$), and cytoplasm velocity in the presence of dopamine. (a) Change $V_m$ and $R_m$ after 24 hours incubation in solution with dopamine in concentration 0; 0.05; 0.1; 0.5, and 1 mM. Normalized curves are given. 1 – $V_m$ the potential of rest of a plasmalemma, 2 – $R_m$ membrane resistance. (b) Change of speed of the movement of cytoplasm of cages of *Chara corallina* in the presence of dopamine 15 mM.

dopamine (2 mM) produced a drop in the resting potential of the cells, decreased the membrane resistance, and stopped the movement of the cytoplasm (Figure 11.3).

## 11.3 THE EFFECT OF HALOPERIDOL ON THE ION TRANSPORT OF TARGET CELLS

The progress that has been made in the understanding of the function of dopamine has largely being enabled by the discovery of their antagonists and by application of the latter in research and clinical practice; in particular, this concerns haloperidol, a butyrophenone derivative that blocks postsynaptic $D_2$ dopamine receptors. It is known that haloperidol interacts not only with dopamine receptors but also with ion channels of excitable membranes (Ogata et al., 1989; Akamine et al., 2002; Yang et al., 2005; Duncan et al., 2009) According to the modern understanding, $Ca^{2+}$ ions and ion channels are directly involved in the etiology of neurodegenerative diseases (Ehling et al., 2011; Ryazantseva et al., 2012). However, the mechanisms of its interaction with neuron membranes and ion transport systems are not sufficiently understood. For this reason, it is necessary to analyze in detail the effects of haloperidol on cellular systems of ion transport: calcium, potassium, and chloride channels. It is especially interesting to identify the mechanism by which haloperidol affects L-type $Ca^{2+}$ channels (Tarabova et al., 2009) and to determine how the cytoskeleton is involved in this process. Indeed, the behavior of actin, a key cytoskeletal protein, is regulated by intracellular $Ca^{2+}$ ions (Williamson and Ashley, 1982) and changes in response to dopamine, which causes polymerization of globular actin and formation of actin fibrils in the cell (Moshkov et al., 2010).

Giant electroexcitable cells of the freshwater alga *C. corallina* (6–12 cm long and 0.6–0.8 mm in diameter) possess a number of advantages for electrophysiological experiments: microelectrodes can be manipulated using light microscopy; electrophysiological parameters of the plasmalemma can be continuously observed both on the outer and the inner surface; $K^+$, $Ca^{2+}$, and $Ca^{2+}$-activated $Cl^-$ currents can be registered separately (Lunevsky et al., 1983; Kataev et al., 1984; Berestovsky and Kataev, 2005). Thus, the plasmalemma of *C. corallina* and other *Charophyta* cells represents a better model of plasma membrane than artificial lipid vesicles, and similar experimental models can be employed to investigate the mechanisms of interactions between biologically active compounds and their target cells and the corresponding transmembrane transport systems. In our previous works, we have already demonstrated the possibility of using this model in this kind of research (Drinyaev et al., 2001; Zherelova et al., 2009; Kataev et al., 2012; Kataev et al., 2016).

The influence of haloperidol on the electrophysiological characteristics of the plasmalemma of *C. corallina* cells was investigated by observing the time-dependent effects of its interaction with ion channels at different concentrations of the compound in the extracellular medium. Figure 11.4a illustrates the changes that exposure to haloperidol produces in the development of transient currents caused by membrane depolarization. It presents the currents observed before and after 10 μM haloperidol was applied externally on an isolated cell fragment (Figure 11.4a, curves 1 and 2–4, respectively). In the presence of haloperidol, the amplitudes of $Ca^{2+}$ and $Ca^{2+}$-dependent $Cl^-$ input currents diminished. It was shown previously that these currents are induced by activation of selective $Ca^{2+}$ and $Ca^{2+}$-activated $Cl^-$ channels in the plasmalemma (Lunevsky et al., 1983). In addition to diminishing the current, haloperidol delayed its activation and inactivation. Figure 11.4b shows the time-dependent change in the current amplitude (curve 1) and the increasing delay in the development of the peak current (curve 2) in the presence of 10 μM haloperidol.

To reveal the mechanism underlying the selective effects of the drug onto calcium channels and $Ca^{2+}$-dependent chloride channels, we separated the calcium and the chloride components of the total current. For this purpose, $Ca^{2+}$ concentration in the extracellular medium was increased to 20 mM, which accelerated inactivation of the $Ca^{2+}$ current and separated the calcium and chloride currents

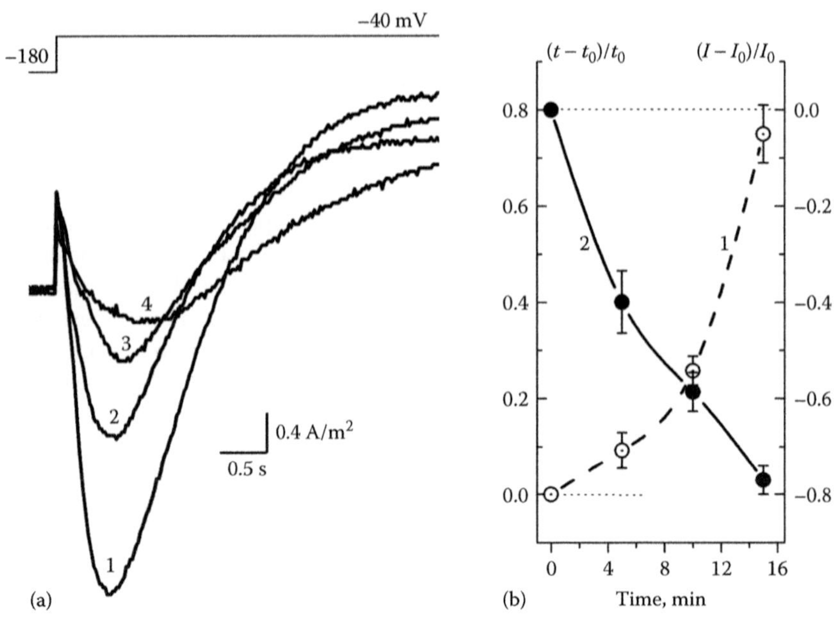


**FIGURE 11.4** Changes in the amplitude and kinetics of the integral current in the plasmalemma of *C. corallina* under the influence of haloperidol. (a): (1) control; (2), (3), and (4) 5, 10, and 15 min after haloperidol addition, respectively; top, command voltage pulse on the membrane: from −180 to −40 mV. (b) (1, *I*) Normalized curves of changes in the current amplitude and (2, *t*) the delay until the peak current value; vertical intervals show the error of the mean ($n = 12$). (b) The current amplitude and the delay until the peak current is reached as functions of time under exposure to 10 μM haloperidol.

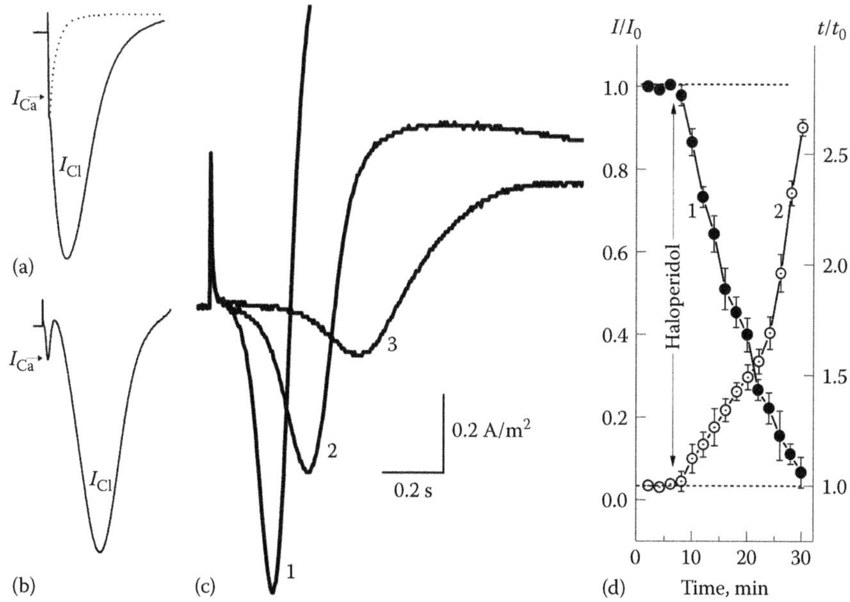


**FIGURE 11.5** Effect of haloperidol on calcium channels. (a) Total current in APW (at 0.5 mM $CaCl_2$). (b) Total current in APW containing 20 mM $CaCl_2$. (c) Time-dependent changes in the calcium component of the total current in the presence of 50 μM haloperidol: (1) control, (2) 15-min exposure, and (3) 30-min exposure. (d) Kinetics of changes (1) in the amplitude of the $Ca^{2+}$ current and (2) in the delay of the current peak moment; vertical intervals show the error of the mean ($n = 6$).

in time (Figure 11.5b). The $Ca^{2+}$ and $Cl^-$ components of the typical integral current developing in an artificial pond water (APW) solution with 0.5 mM $Ca^{2+}$ merge together (Figure 11.5a), whereas in the extracellular medium with a higher $Ca^{2+}$ concentration (20 mM) the current components separated due to more rapid inactivation of $Ca^{2+}$ channels (Figure 11.5b). The time-dependent effect of 50 μM haloperidol of the calcium component of the current is shown in Figure 11.5c: the drug blocks $Ca^{2+}$ channels and affects the kinetics of the current. The kinetics of haloperidol-induced inhibition of the $Ca^{2+}$ current depended on the duration of exposure. In the presence of 50 μM haloperidol, the $Ca^{2+}$-current was suppressed completely in 20–25 min (Figure 11.5d).

The effect of haloperidol-dependent inhibition of $Ca^{2+}$ current was reversible. After the drug was removed by ten exchanges of the external medium, the calcium current was partially restored (Figure 11.6a). In 25 min, the amplitude of the $Ca^{2+}$ current reached ~75% of the initial level; the kinetics of the current development were also restored (Figure 11.6a, curve 3). Figure 11.6b (curve 2) shows the relationship between the duration of cell exposure to 50 μM haloperidol and the current amplitude after drug withdrawal (curve 3), while the chart in Figure 11.6c shows the delay until the peak value $Ca^{2+}$ current as a function of time in the same experiments.

As it can be seen in Figure 11.4a, exposure to haloperidol led not only to suppression of the calcium current but also to a decrease in the amplitude of the $Ca^{2+}$-dependent chloride component of the total current. This could have been a consequence of a decrease in the calcium conductance of the membrane. To rule out the possibility that the development of the $Cl^-$ current is affected by calcium channels, it was necessary to activate $Cl^-$ channels directly by $Ca^{2+}$ ions introduced into the cell by means of cytoplasm perfusion with a $Ca^{2+}$-containing solution bypassing calcium channels. Thus, the behavior of $Cl^-$ channels could be compared between control cells and cells exposed to haloperidol to find out whether haloperidol directly affects $Cl^-$ channels. For this purpose, we performed experiments in internally perfused cells where the tonoplast had been removed (Kataev et al., 1984; Zherelova et al., 1987). In this setting, $Cl^-$ channels are activated by $Ca^{2+}$ ions and introduced directly into the

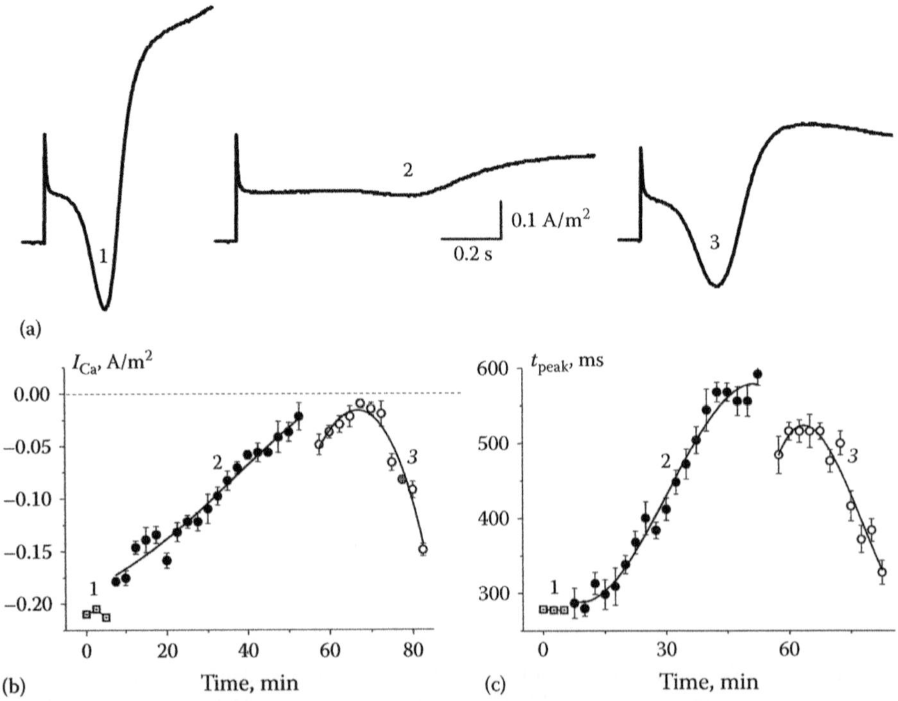


**FIGURE 11.6** Washing the cells from haloperidol. Restoration of the amplitude ($I_{Ca}$) and kinetics of $Ca^{2+}$ current. (a) $Ca^{2+}$ currents: (1) control, (2) in the presence of 50 μM haloperidol, and (3) after ten washes with haloperidol-free APW. (b) Kinetics of the $Ca^{2+}$ current inhibition and restoration after washing from haloperidol: (1) (squares), control; (2) (filled circles), in the presence of 50 μM haloperidol; and (3) (open circles), after washing from haloperidol. (c) Changes in the kinetics of the $Ca^{2+}$ current (a shift in the delay until the peak current and its reversal): curves (1)–(3) are the same as in (b). Each point was registered 2.5 min after the previous measurement; vertical intervals show the error of the mean ($n = 8$).

cell independently of the state of calcium channels. Figure 11.7a shows the control chloride current developing in response to an intracellular injection of $Ca^{2+}$ ions in the concentration of 0.2 mM, while Figure 11.7b shows activation of $Cl^-$ channels following 20-min exposure to 50 μM haloperidol present in the external APW medium. Obviously, neither the amplitude nor the kinetics of the chloride current changed; there was hardly any difference between the transient currents in control and in haloperidol-exposed cells. Intracellular administration of 10 μM haloperidol did not influence the development of the $Ca^{2+}$-dependent $Cl^-$ -current either. Therefore, we can conclude that haloperidol does not affect the state of $Ca^{2+}$-activated $Cl^-$ channels in the plasmalemma of *C. corallina*.

Figure 11.8 shows momentary current–voltage curves (I–V curves) obtained at different points of time: before (curve 1) and after addition of 10 μM haloperidol to the extracellular medium (curves 2 and 3). Current curves were obtained by applying sawtooth voltage with an amplitude of 270 mV and a duration of 30 ms. The capacitive current at the intersection of all momentary I–V curves corresponding to the reversal potential of the current ($E_R$) was $I_R = C(dV/dt) = 0.08$ μA. The growth rate of the sawtooth voltage was selected to be sufficiently high so as to prevent activation of $Ca^{2+}$ channels within this period. The fact that all momentary I–V curves share the same point of intersection ($E_R = -152$ mV) indicates that only one type of conductance changed in this setting: the conductance of $K^+$ channels. If another type of channel with a different reversal potential contributed to this process, the single $E_R$ point would shift and disintegrate. Taking into account that, in comparison to control, exposure to haloperidol leads to a stepwise decrease in the current induced by changing voltage (Figure 11.4, curves 1 and 4, respectively), it can be concluded that haloperidol affects the membrane conductance by blocking $K^+$ channels.

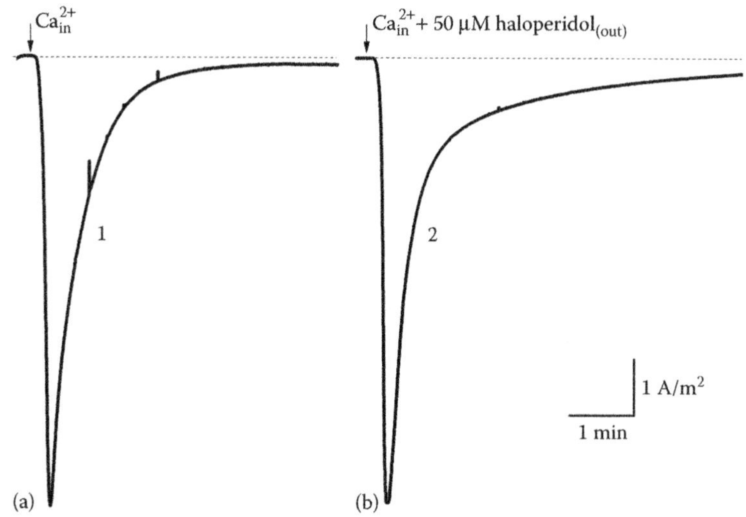


**FIGURE 11.7** Effect of haloperidol on $Ca^{2+}$-activated chloride channels in the plasmalemma of perfused cells with removed tonoplast. (a) (1) Control, a typical $Ca^{2+}$-activated chloride current induced by exchanging the calcium free EGTA-containing solution within the cell for a $Ca^{2+}$ solution; (b) (2) chloride current was induced by the same procedure following 20-min exposure to 50 μM haloperidol (HP) in the extracellular solution ($HP_{out}$). $V_m$ = −100 mV. Extracellular solution: APW containing 160 mM sucrose; solution for intracellular perfusion after tonoplast removal contained (mM): 10 KCl, 3 EGTA, 20 HEPES/Tris, 240 sucrose. Chloride channels were activated by exchanging the EGTA-containing intracellular solution for a 0.2 mM $CaCl_2$ solution (the moment of the exchange is indicated with an arrow). The plot shows a typical graph obtained in one of the repeated experiments ($n$ = 4).

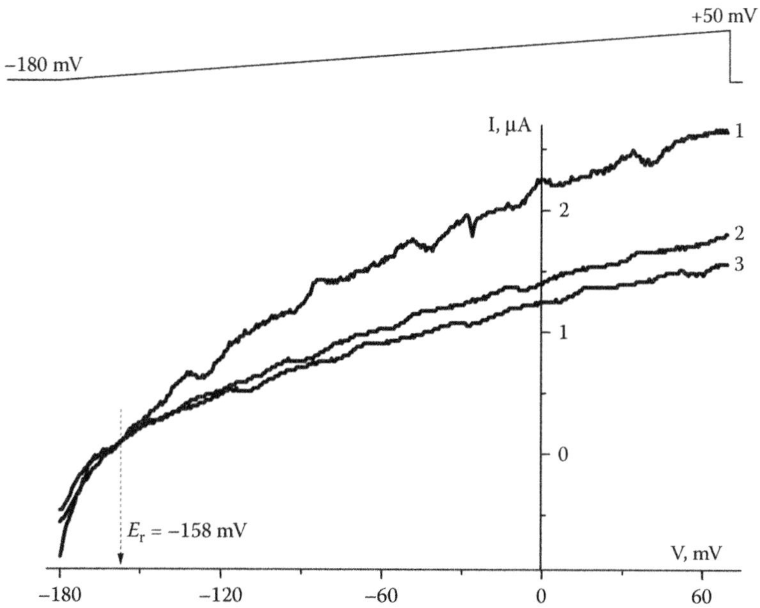


**FIGURE 11.8** Changes in the potassium conductance of the plasmalemma in the presence of haloperidol. Momentary I–V curves obtained by applying sawtooth voltage from −180 to +50 mV with a period of 30 ms were registered (1) before haloperidol addition and (2 and 3, respectively) 3 and 5 min after adding 50 μM haloperidol. The chart shows typical curves obtained in one of the repeated experiments ($n$ = 7).

## 11.4 EFFECT OF HALOPERIDOL ON THE CYTOSKELETON OF TARGET CELLS

In addition to changes in the electrophysiological characteristics in *C. corallina* cells, we also observed significant changes in the velocity of the cytoplasmic streaming, an important parameter characterizing the physiological state of a plant cell (Verchot-Lubicz, Goldstein, 2010). Using a previously developed technique (Kataev et al., 2012), we found that haloperidol not only slows down the cytoplasmic streaming in *C. corallina* cells but can also block it completely (Figure 11.9). After the drug was removed from the bathing solution, the cytoplasmic streaming was rapidly restored (Figure 11.9, curve 3). In control cells, the velocity of cytoplasmic streaming remained constant for 72 h. It should be pointed out that in presence of 60 μM haloperidol considerable change of resting potential ($V_m$) and membrane resistance ($R_m$) was observed (Figure 11.10).

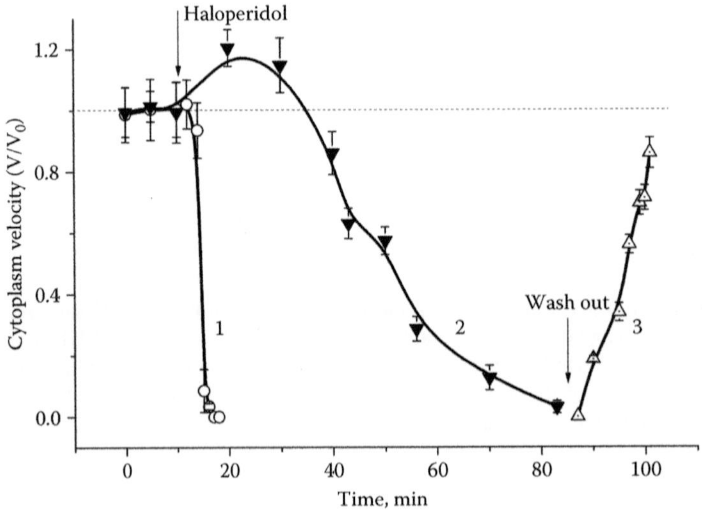


**FIGURE 11.9** Effect of different haloperidol concentrations on velocity of cytoplasmic streaming in *Chara corallina* cells. Exposure to (1) 0.25 and (2) 0.15 μM haloperidol; and (3) restoration of cytoplasmic streaming after haloperidol removal. Arrows indicate the moments of drug administration (HP) and the beginning of washing (wash) (n = 6).

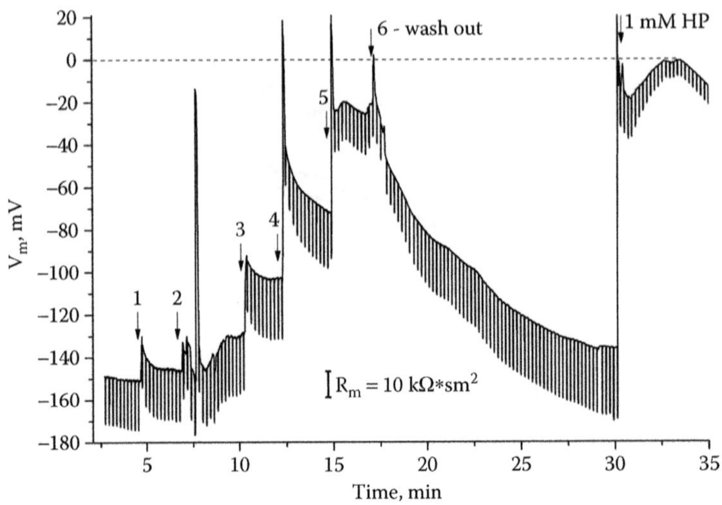


**FIGURE 11.10** Change of resting potential ($V_m$) and membrane resistance ($R_m$) in the presence of haloperidol. The arrows indicate the moments of introduction HP. 1-10, 2-20, 3-50, 4-140, 5-640 μM HP. 6-wash out. Vertical lines show the change of the $R_m$.

## 11.5 SUMMARY

In the present paper, which logically continues our research on the effects of neurotransmitters and their antagonists on the electroexcitable membrane and the cytoskeleton of *C. corallina* cells as a model target (Zherelova et al., 2014; Kataev et al., 2016), we have for the first time investigated the interactions of dopamine and haloperidol with ion channels of plasmalemma and with the cytosol of plant cells. The data obtained show that haloperidol affects both the plasmalemma and other structures of *C. corallina* cells in many ways, making it possible to consider its effects on ion transport systems in *Charophyta* cells and in animal cells as similar in many respects.

## REFERENCES

Akamine T, Nishimura Y, Ito K, Uji Y, Yamamoto T. Effects of haloperidol on $K^+$ currents in acutely isolated rat retinal ganglion cells. *Invest. Ophthalmol. Vis. Sci.* 2002;43:1257–1261.

Berestovsky GN, Kataev AA. Voltage-gated calcium and $Ca^{2+}$-activated chloride channels and $Ca^{2+}$ transients: voltage-clamp studies of perfused and intact cells of *Chara. Eur Biophys J.* 2005;34:973–986.

Drinyaev VA, Mosin VA, Kruglyak EB, Sterlina TS, Kataev AA, Berestovsky GN, Kokoz YM. Effect of Avermectins on $Ca^{2+}$-dependent $Cl^-$ currents in plasmalemma of *Chara corallina* cells. *J. Membr. Biol.* 2001;182:71–79.

Duncan CE, Schofield PR, Weickert CS. K channel interacting protein 3 expression and regulation by haloperidol in midbrain dopaminergic neurons. *Brain Res.* 2009;1304:1–13.

Ehling P, Bittner S, Budde T, Wiendl H, Meuth SG. Ion channels in autoimmune neurodegeneration. *FEBS Lett.* 2011;585:3836–3842.

Kataev AA., Zherelova OM., and Berestovsky GN. $Ca^{2+}$-induced activation and irreversible inactivation of chloride channels in the perfused plasmalemma of *Nitellopsis obtusa, Gen. Physiol. Biophys,* 1984;3:447–462.

Kataev AA, Andreeva-Kovalevskaya ZI, Solonin AS, Ternovsky VI. *Bacillus cereus* can attack the cell membranes of the alga *Chara corallina* by means of HlyII, *Biochim. Biophys. Acta.* 2012;1818:1235–1241.

Kataev A, Zherelova O, Grishchenko V. A Characeae cells plasma membrane as a model for selection of bioactive compounds and drugs: Interaction of HAMLET-like complexes with ion channels of *Chara corallina* cells plasmalemma. *J. Membrane Biol.* 2016;249:801–811.

Lunevsky VZ, Zherelova OM, Vostrikov IYa, Berestovsky GN. Excitation of Characeae cell membranes as a result of activation of calcium and chloride channels. *J. Membrane Biol.* 1983;72:43–58.

Moshkov DA, Abramova MB, Shubina VS, Lavrovskaya VL, Pavlik LL, Lezhnev EI. Effect of dopamine on viability of BHK-21 cells. *Bull. Exp. Biol. Med.* 2010;149(3):359–363.

Ogata N, Yoshii M, Narahashi T. Psychotropic drugs block voltage-gated ion channels in neuroblastoma cells. *Brain Res.* 1989;476:140–144.

Pineros M, Tester M. Calcium channels in higher plant cells: Selectivity, regulation and pharmacology. *J. Exp. Bot.* 1997;48:551–577.

Roshchina VV. *Neurotransmitters in Plant Life.* Enfield, NH: Scientific Publishers, 2001.

Ryazantseva MA, Mozhayeva GN, Kaznacheyeva EV. Calcium hypothesis of Alzheimer disease. *Usp. Fiziol. Nau.* 2012;43(4):59–72.

Shimmen T, Yokota E. Cytoplasmic streaming in plants. *Curr. Opin. Cell. Biol.* 2004;16:68–72.

Tarabova B, Novakova M, Lacinova L. Haloperidol moderately inhibits cardiovascular L-type calcium current. *Gen. Physiol. Biophys.* 2009;28:249–259.

Tester M. Plant ion channels: Whole-cell and single channel studies. Tansley Review no. 21. *New Phytol.* 1990;114:305–340.

Tominaga Y, Shimmen T, Tazawa M. Control of cytoplasmic streaming by extracellular $Ca^{2+}$ in permeabilized *Nitella* cells. *Protoplasma.* 1983;116:75–77.

Verchot-Lubicz J, Goldstein RE. Cytoplasmic streaming enables the distribution of molecules and vesicles in large plant cells. *Protoplasma.* 2010;240:99–107.

Ward JM, Maser P, Schroeder JI. Plant ion channels: Gene families, physiology, and functional genomics analyses. *Ann. Rev. Physiol.* 2009;71:59–82.

Williamson RE, Ashley CC. Free $Ca^{2+}$ and cytoplasmic streaming in the alga *Chara. Nature.* 1982;296:647–650.

Yang SB, Major F, Tietze LF, Rupnik M. Block of delayed-rectifier potassium channels by reduced haloperidol and related compounds in mouse cortical neurons. *J. Pharmacol. Exp. Ther.* 2005;315:352–362.

Zherelova O.M., Kataev AA, Berestovsky GN. The regulation of plasma membrane calcium channels of *Nitellopsis obtusa* by intracellular calcium. *Biofizika.* 1987;32(2):448–450.

Zherelova OM, Kataev AA, Grishchenko VM, Knyazeva EL, Permyakov SE, Permyakov EA. Interaction of antitumor alpha-lactalbumin-oleic acid complexes with artificial and natural membranes. *J. Bioenerg. Biomembr.* 2009;41:229–237.

Zherelova OM, Kataev AA, Grishchenko VM, Shtanchaev RSh, Moshkov DA, Medvedev BI. Interaction of neuromediator dopamine with the ionic channels of *Chara corallina* cell plasmalemma. *Biomed. Zh. Medline.ru.* 2014;15(67):834–846. (www.medline.ru.)

# 12 Possible Participation of Acetylcholine in Free-Radical Processes (Redox Reactions) in Living Cells

*Bogdan A. Kurchii*

## CONTENTS



## 12.1 INTRODUCTION

Acetylcholine as an unusual biologically active substance is presented in all living systems. First of all, it is the main neurotransmitter in the parasympathetic nervous system, as well as neurotransmitter performing neuromuscular transmission (Fröhlich, 2016). The compound is synthesized by choline acetyltransferase in the central nervous system (Govindasamy et al., 2004; Karczmar, 2007) and in a variety of non-neuronal cells, including epithelial and immune cells (Kawashima et al., 2012; Proskocil et al., 2004), lymphocytes (Kawashima and Fujii, 2000), and muscle and blood cells (Wessler et al., 1998; Wessler et al., 1999). Acetylcholine is contained in vesicles of the presynaptic nerve endings. Upon excitation it is released from these vesicles into the synaptic cleft, thereby affecting neurotransmission (Fisher and Wonnacott, 2012; Prado et al., 2002; Walker et al., 1996). Currently, general mechanisms for acetylcholine action as a neurotransmitter in animal bodies are well described (Albuquerque et al., 2009; Kruse et al., 2014; Lester, 2014; Myslivecek and Jakubik, 2016; Picciotto et al., 2012; Racké et al., 2006).

Today, we know that acetylcholine also occurs in plants and microorganisms participating in many signaling and regulatory reactions within organisms (Horiuchi et al., 2003; Roshchina, 2001).

All kingdoms contain a lot of redox agents that may interact with acetylcholine. How they influence on the neurotransmitter at the level of primary reactions is not known yet. Herein, a possible alternative mechanism for acetylcholine action on the molecular level is presented.

## 12.2 CHEMICAL PROPERTIES OF ACETYLCHOLINE: WHAT DO THEY TELL US?

The chemical properties of acetylcholine come from its molecular structure: it is a quaternary ammonium compound that contains a quaternary nitrogen atom, deprived of an electron. Also, according to the chemical structure, acetylcholine is an ester of choline and acetic acid.

Quaternary ammonium compounds are stable in acidic medium and are decomposed in alkaline medium (Smith, 2013). With the properties like these, one can explain the acidic environment in the synaptic vesicles (pH ~ 5.5) (Atluri and Ryan, 2006; Zhang et al., 2009), because in an alkaline medium acetylcholine cannot long exist as an integral structure and can degrade. These properties of acetylcholine were also confirmed *in vitro* (Wessler et al., 2015).

However well known, a decomposition reaction of quaternary ammonium salts in the Hofmann elimination reaction (also known as exhaustive methylation) in an alkaline medium can occur at elevated temperatures (Smith, 2013). This is a major obstacle to rapid and complete disintegration of quaternary ammonium compounds *in vivo* including acetylcholine. Confirmation of this chemical rule can be data about the presence of acetylcholine in the blood of various animal species (Kawashima and Fujii, 2000). This phenomenon is not fully understood because as follows from numerous reports mammalian neuronal tissues are alkalized. At the ordinal physiological state, pH of the nervous system is maintained near 7.4 (Mochizuki, 2005; Shiskin et al., 1937; Zhang et al., 2009). Nevertheless, neuronal pH may vary from 7.0 to 7.5–7.8 (Chesler, 1990, 2003). By these data, at least partially, we can explain published data why alkaline extracellular pH enhances excitability, while acidic extracellular pH depresses excitability (Chesler, 2003).

Maintaining intracellular pH homeostasis is essential for the functioning of the nervous system and is maintained by a family of membrane transport proteins, commonly called $Na^+/H^+$ exchangers (NHEs) or antiporters (Nha) (Orlowski and Grinstein, 2004; Yuan et al., 2014) that alkalize intracellular space.

Ion concentrations in the cytosol are: sodium ($Na^+$) 10 mM, potassium ($K^+$) 140 mM, calcium ($Ca^{2+}$) 4 μM, chloride ($Cl^-$) 4 mM. Ion concentrations in the extracellular space are: sodium ($Na^+$) 145 mM, potassium ($K^+$) 3 mM, calcium ($Ca^{2+}$) 1 mM, chloride ($Cl^-$) 110 mM (Fröhlich, 2016). Similar data were also reported in studies with different plants species. Plant cytosolic pH is regulated during the life cycle and can either increase or decrease. Under normal physiological conditions, the cytosolic pH is kept at 7.0–7.5 (Sakano, 2001), 7.2–7.5 (Felle, 2001), or 7.4 (Pratt et al., 2009). The cytosolic pH can be decreased to 6.7 to 6.8 when plants are exposed to abiotic stress (Felle, 2001). Altogether, plant pH levels differ from pH 4 (cell wall regions, vacuoles) to pH 7 (cytoplasm) up to pH 8.5 (chloroplast stroma during illumination). The concentration of $K^+$ is higher in the cytoplasm of plant cells whereas $Na^+$ and $Ca^{2+}$ are maintained in low concentration compared with the outside environment (Neuhaus, 2013). For instance, the cytosolic concentration of calcium in the plant is maintained in the range of 10–200 nM, whereas the concentration of $Ca^{2+}$ in the cell wall and an endoplasmic reticulum is 1–10 mM (Reddy and Reddy, 2004).

## 12.3 ABOUT ONE SPECIFIC CHEMICAL REACTION THAT IS INHERENT IN QUATERNARY AMMONIUM COMPOUNDS IN THE PLANTS

As stated earlier, quaternary ammonium compounds are decomposed *in vitro* at elevated temperatures and in alkaline solutions. However, this rule may be an exception in the living systems, which have a large number of chemically active substances, for example reductases, which compose various redox systems. Unfortunately, what type of primary and secondary reactions of acetylcholine occur in the cellular matrix is not yet fully understood.

Nevertheless, remarkable data were found for molecular mechanisms of paraquat action. Paraquat is commonly employed in agriculture as an herbicide (Lock and Wilks, 2010). Paraquat as a quaternary nitrogen compound is very soluble in water and a little in most organic solvents. It is stable in neutral and acidic media but easily hydrolyzed in alkaline media. *In vivo* paraquat

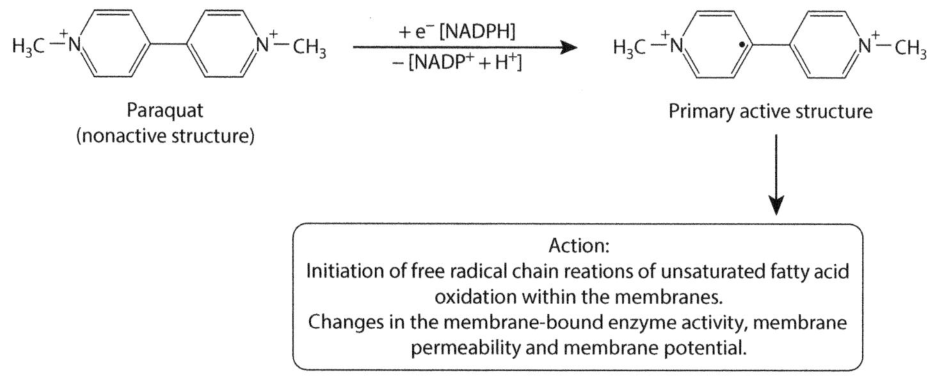


**FIGURE 12.1** Proposed mechanism for paraquat activation and mode of action. (Adapted from Bus, J.S. and Gibson, J.E., *Environ. Health Perspect.*, 55, 37–46, 1984; Clejan, L. and Cederbaum, A.I., *Biochem. Pharmacol.*, 38, 1779–1786, 1989. With permission.)

is undergoing a one-electron reduction (Bus and Gibson, 1984; Clejan and Cederbaum, 1989; Day et al., 1999). An important consequence of this reaction is the formation of monocation free radicals (Figure 12.1).

The subsequent reactions occur in accordance with the mechanisms of free radical reactions in biological systems (Bhattacharjee, 2014; Kehrer and Klotz, 2015).

A general conclusion from these studies is that quaternary nitrogen substances may or may not act through membrane receptors.

### 12.3.1 NADPH in Acetylcholine Activation

In the plasma membrane and cellular cytoplasm, redox reactions occur constantly (Pollak et al., 2007). In redox reactions (also known as oxidation-reduction reactions), one or more chemicals are oxidized (a process in which a molecule loses an electrons) while one or more chemicals are reduced (a process in which a molecule gains an electron).

The reducing agents are a particular interest. Among the reducing agents on the cell membranes are NADH-cytochrome P450 reductase and NADPH-cytochrome P450 reductase (Bedard and Krause, 2007; Dewald et al., 1979; Gan et al., 2009; Hart and Zhong, 2008; Wang et al., 1997).

NADPH-cytochrome P450 reductase is an enzyme that facilitates the transfer of electrons to numerous microsomal proteins (Strobel et al., 1995). It is found in all tissues including brain (Bergh and Strobel, 1992; Walther et al., 1986) and plants (Durst and Nelson, 1995).

NADPH is stable at alkaline pH and is decomposed in acid media (Lowry et al., 1961). The rate of degradation at the same condition was much faster in NADPH than in NADH (Wu et al., 1986).

NADH as a redox enzyme is present in the mammalian (Kitajima et al., 1981; Wilhelm and Hirrlinger, 2012, Winkler and Hirrlinger, 2015) and plant (Madyastha et al., 1993; Small and Wray, 1980; Soualmi and Champigny, 1986) cells. NADH-cytochrome *b* reductase and NADPH-cytochrome *c* reductase activities were also determined in the brain of rats (Takeshita et al., 1982). NADH, like NADPH, is stable at alkali solution (Rover et al., 1998; Wilhelm and Hirrlinger, 2011).

## 12.4 BY KNOWING THE CHEMICAL PROPERTIES OF A SUBSTANCE, IT IS POSSIBLE TO PREDICT THEIR BEHAVIOR IN BIOLOGICAL SYSTEMS

Based on the chemical properties of acetylcholine and taking into account all aforementioned research data, possible mechanisms of activation and mode of action for acetylcholine is shown in Figure 12.2.

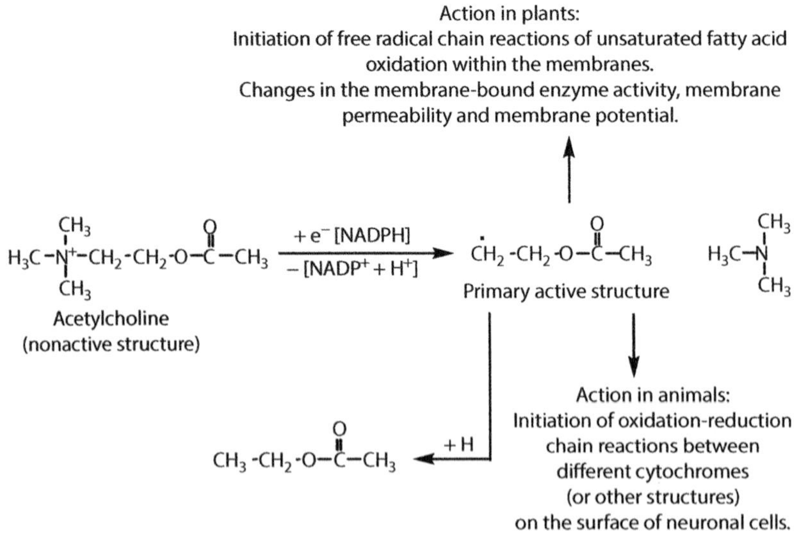


**FIGURE 12.2** Possible mechanism for activation and mode of action for acetylcholine at physiological concentration. (Adapted from Kurchii, B.A., *Ukr. Bioorg. Acta (Kiev)*, 14, 35–37, 2016. With permission.

In the reaction with reducing agents at the postsynaptic membrane, acetylcholine is decomposed to a neutral molecule trimethylamine and free radical of ethyl acetate. Simultaneously, acetylcholine in the synaptic cleft is also destroyed in a non-enzymatic reaction to form choline and acetic acid.

Physiological concentration means that the total number of free radicals of acetylcholine (primary active structure) in the synaptic clef is neutralized by antioxidant substances. This implies the existence of a balance between oxidants and antioxidants. As follows from this scheme, the reaction of acetylcholine with reductases (in this case this is NADPH) leads to decomposition of acetylcholine with free radical formation.

Nevertheless, the next question arises. How many acetylcholine molecules are released from synaptic vesicles and how many receptors are located on the membranes? It was calculated that each synaptic vesicle releases a quantum of 6.000–10.000 acetylcholine molecules (Kuffler and Yoshikami, 1975; Whittaker and Sheridan, 1965). Unfortunately, there is no information about the real quantity of acetylcholine molecules that are released by one neuron into the synaptic cleft and the real quantity of receptors at the postsynaptic membrane as well. It should also be noted that neuronal receptors are present on the presynaptic membrane (Schlicker and Feuerstein, 2016), which ensures possible reversible signal flow in the nervous system. One may conclude from this finding that free radical chain reactions of lipid oxidation within the membranes are not carried out during acetylcholine action. Hence, taking into account these evidences, I assumed that electric flow takes place on the neuronal cell surface. This means that oxidation-reduction chain reactions between different cytochromes (or other structures) occur on the neuronal cell surface. In this case, the subsequent events such as pump activation are consequents of oxidative reactions on the neuronal cell surface.

At the same time, the mechanisms of activation and action for non-quaternary ammonium compounds in the nervous systems are based on the free radical chain reactions of membrane lipid oxidation (Kurchii, 1996, 1998). Nevertheless, this is not a matter of this paper.

The mechanism of action of acetylcholine in a plant organism is somewhat different. First of all, this is due to the anatomical structure of tissues and the ultrastructure of plant cells. In addition, plant-specific cells that are similar to the neuronal animal cells in which acetylcholine is formed in large quantities is not identified. Hence, it can be formed within meristematic and differentiated cells.

Therefore, two variants of events are possible: the transformation of acetylcholine into a free radical state at the place of its formation or evolution from the cell and migration over long distances, where also it is transformed into a free radical state and initiates free radical chain reactions in the tissues. In the first and second case, free radical processes will spread to neighboring cells, which is a signal for the changes of various metabolic pathways.

A characteristic feature of these reactions will be slower changes in stationary physiological processes in comparison with the tissues of animal organisms.

## 12.5 CONCLUSION

The investigation of the mechanisms of biologically active substances at molecular levels including acetylcholine should be based on their chemical properties. Elementary chemical reactions also take place in the living systems. Furthermore, the chemical laws have not yet been canceled. Hence, if we know the chemical properties of the substance, we can anticipate its behavior in biological systems.

## REFERENCES


Albuquerque EX, Pereira EFR, Alkondon M, Rogers SW. Mammalian nicotinic acetylcholine receptors: From structure to function. *Physiol Rev.* 2009;89:73–120.

Atluri PP, Ryan TA. The kinetics of synaptic vesicle reacidification at hippocampal nerve terminals. *J Neurosci.* 2006;26:2313–2320.

Bedard K, Krause KH. The NOX family of ROS-generating NADPH oxidases: Physiology and pathophysiology. *Physiol Rev.* 2007;87:245–313.

Bergh AF, Strobel HW. Reconstitution of the brain mixed function oxidase system: Purification of NADPH-cytochrome P450 reductase and partial purification of cytochrome P450 from whole rat brain. *J Neurochem.* 1992;59:575–581.

Bhattacharjee S. Membrane lipid peroxidation and its conflict of interest: The two faces of oxidative stress. *Curr Sci.* 2014;107:1811–1823.

Bus JS, Gibson JE. Paraquat: Model for oxidant-initiated toxicity. *Environ Health Perspect.* 1984;55:37–46.

Chesler M. Regulation and modulation of pH in the brain. *Physiol Rev.* 2003;83:1183–1221.

Chesler M. The regulation and modulation of pH in the nervous system. *Progr Neurobiol.* 1990;34:401–427.

Clejan L, Cederbaum AI. Synergistic interactions between NADPH-cytochrome P-450 reductase, paraquat, and iron in the generation of active oxygen radicals. *Biochem Pharmacol.* 1989;38:1779–1786.

Day BJ, Patel M, Calavetta L, Chang L-Y, Stamler JS. A mechanism of paraquat toxicity involving nitric oxide synthase. *Proceedings of the National Academy of Sciences of United States of America* 1999;96:12760–12765.

Dewald B, Baggiotini M, Curnutte JT, Babior BM. Subcellular localization of the superoxide-forming enzyme in human neutrophils. *J Clin Invest.* 1979;63:21–29.

Durst F, Nelson DR. Diversity and evolution of plant P450 and P450-reductases. *Drug Metabol Drug Interact.* 1995;12:189–206.

Felle HH. pH: Signal and messenger in plant cells. *Plant Biol.* 2001;3:577–591.

Fisher SK, Wonnacott S. Acetylcholine. In: Brady ST, Siegel GJ, Albers RW, Price DL, Benjamins J, Fisher S, Hall A, Bazan N, Coyle J, Sisodia S (Eds). *Basic Neurochemistry: Principles of Molecular, Cellular, and Medical Neurobiology.* Amsterdam, the Netherlands: Academic Press Elsevier, 2012, pp. 258–282.

Fröhlich F. *Network Neuroscience.* Amsterdam, the Netherlands: Academic Press Elsevier, 2016.

Gan L, von Moltke LL, Trepanier LA, Harmatz JS, Greenblatt DJ, Court MH. Role of NADPH-cytochrome P450 reductase and cytochrome-$b_5$/NADH-$b_5$ reductase in variability of CYP3A activity in human liver microsomes. *Drug Metab Dispos.* 2009;37:90–96.

Govindasamy L, Pedersen B, Lian W, Kukar T, Gu Y, Jin S, Agbandje-McKenna M, Wu D, McKenna R. Structural insights and functional implications of choline acetyltransferase. *J Struct Biol.* 2004;148:226–235.

Hart SN, Zhong X-B. P450 oxidoreductase: Genetic polymorphisms and implications for drug metabolism and toxicity. *Expert Opin Drug Metab Toxicol.* 2008;4:439–452.

Horiuchi Y, Kimura R, Kato N, Fujii T, Seki M, Endo T, Kato T, Kawashima K. Evolutional study on acetylcholine expression. *Life Sci.* 2003;72:1745–1756.

Karczmar AG. *Exploring the Vertebrate Central Cholinergic Nervous System.* New York: Springer Science+Business Media, LLC, 2007.

Kawashima K, Fujii T, Moriwaki Y, Misawa H. Critical roles of acetylcholine and the muscarinic and nicotinic acetylcholine receptors in the regulation of immune function. *Life Sci.* 2012;91:1027–1032.

Kawashima K, Fujii T. Extraneuronal cholinergic system in lymphocytes. *Pharmacol Ther.* 2000;86:29–48.

Kehrer JP, Klotz L-O. Free radicals and related reactive species as mediators of tissue injury and disease: Implications for health. *Crit Rev Toxicol.* 2015;45:765–798.

Kitajima S, Yasukochi Y, Minakami S. Purification and properties of human erythrocyte membrane NADH-cytochrome $b_5$ reductase. *Archiv Biochem Biophys.* 1981;210:330–339.

Kruse AC, Kobilka BK, Gautam D, Sexton PM, Christopoulos A, Wess J. Muscarinic acetylcholine receptors: Novel opportunities for drug development. *Nat Rev Drug Discov.* 2014;13:549–560.

Kuffler SW, Yoshikami D. The number of transmitter molecules in a quantum: An estimate from iontophoretic application of acetylcholine at the neuromuscular synapse. *J Physiol* (*Lond*). 1975;251:465–482.

Kurchii BA. Acetylcholine: What happens when it interacts with reducing agents? *Ukr Bioorg Acta* (*Kiev*). 2016;14 (N 1):35–37 (*In English*).

Kurchii BA. Chemical structure of descriptors with an active hydrogen atom in certain bioregulators. *Ukr Biokhim Zh* (*Kiev*). 1996;68:3–13 (*In English*).

Kurchii BA. Kurchii BA. Using of descriptors to design of novel drugs and pesticides, and to predict the biological activity from the structure of chemicals which have not been bioassayed. In: Kurchii BA. (Ed.) What regulates the growth regulators? *Logos Publisher Kiev*, 1998, pp. 96–175 (*In English*).

Lester RAJ (Ed). *Nicotinic Receptors.* New York: Humana Press, 2014.

Lock EA, Wilks MF. Paraquat. In: Krieger R (Ed). *Hayes' Handbook of Pesticide Toxicology*, vol 1 & 2, 3rd ed. Amsterdam, the Netherlands: Elsevier, 2010, pp. 1771–1827.

Lowry OH, Passonneau JV, Rock MK. The stability of pyridine nucleotides. *J Biol Chem.* 1961;236:2756–2759.

Madyastha KM, Chary NK, Holla R, Karegowdar TB. Purification and partial characterization of microsomal NADH-cytochrome $b_5$ reductase from higher plant *Catharanthus roseus. Biochem Biophys Res Commun.* 1993;197:518–522.

Mochizuki M. Analysis of pH in blood plasma at steady state in vivo. *Yamagata Med J.* 2005;23:43–48.

Myslivecek J, Jakubik J (Eds). *Muscarinic Receptor: From Structure to Animal Models.* New York: Springer Science+Business Media, 2016.

Neuhaus G. Structure. In: Bresinsky A, Körner C, Kadereit JW, Neuhaus G, Sonnewald U (Eds). *Strasburger's Plant Sciences: Including Prokaryotes and Fungi.* Berlin, Germany: Springer-Verlag, 2013, pp. 11–235.

Orlowski J, Grinstein S. Diversity of the mammalian sodium/proton exchanger SLC9 gene family. *Pflugers Arch—Eur J Physiol.* 2004;447:549–565.

Picciotto MR, Higley MJ, Mineur YS. Acetylcholine as a neuromodulator: Cholinergic signaling shapes nervous system function and behavior. *Neuron.* 2012;76:116–129.

Pollak N, Dölle C, Ziegler M. The power to reduce: Pyridine nucleotides—Small molecules with a multitude of functions. *Biochem J.* 2007;402:205–218.

Prado MAM, Reis RAM, Prado VF, de Mello MC, Gomez MV, de Mello FG. Regulation of acetylcholine synthesis and storage. *Neurochem Int.* 2002;41:291–299.

Pratt J, Boisson AM, Gout E, Bligny R, Douce R, Aubert S. Phosphate (Pi) starvation effect on the cytosolic Pi concentration and Pi exchanges across the tonoplast in plant cells: An in vivo $^{31}$P-nuclear magnetic resonance study using methylphosphonate as a Pi analog. *Plant Physiol.* 2009;151:1646–1657.

Proskocil BJ, Sekhon HS, Jia Y, Savchenko V, Blakely RD, Lindstrom J, Spindel ER. Acetylcholine is an autocrine or paracrine hormone synthesized and secreted by airway bronchial epithelial cells. *Endocrinology.* 2004;145:2498–2506.

Racké K, Juergens UR, Matthiesen S. Control by cholinergic mechanisms. *Eur J Pharmacol.* 2006;533:57–68.

Reddy VS, Reddy ASN. Proteomics of calcium-signaling components in plants. *Phytochemistry.* 2004;65:1745–1776.

Roshchina VV. *Neurotransmitters in Plant Life.* Enfield, NH: Science Publisher, 2001.

Rover J Jr, Fernandes JCB, de Oliveira Neto G, Kubota LT, Katekawa E, Serrano SHP. Study of NADH stability using ultraviolet–visible spectrophotometric analysis and factorial design. *Anal Biochem.* 1998;260:50–55.

Sakano K. Metabolic regulation of pH in plant cells: Role of cytoplasmic pH in defense reaction and secondary metabolism. *Internat Rev Cytol.* 2001;206:1–44.

Schlicker E, Feuerstein T. Human presynaptic receptors. *Pharmacol Ther.* 2016. doi:10.1016/j.pharmthera.2016.11.005.

Shiskin C, Cape BA, Glasg MD. The pH of human blood plasma in respiratory and cardiac disease. *The Lancet.* 1937;5960:1191–1193.

Small IS, Wray JL. Ferrocyanide-activated NADH cytochrome *c* reductase from barley. *Plant Sci Lett.* 1980;18:389–393.

Smith MB. *March's Advanced Organic Chemistry: Reactions, Mechanisms, and Structure*, 7th ed. Hoboken, NJ: John Wiley & Sons, 2013.

Soualmi K, Champigny ML. Comparison of the NADH: Nitrate reductases from wheat shoots and roots. *J Plant Physiol.* 1986;125:35–45.

Strobel HW, Hodgson AV, Shen S. NADPH cytochrome P450 reductase and its structural and functional domains. In: Ortiz De Montellano PR (Ed). *Cytochrome P450: Structure, Mechanism, and Biochemistry*, 2nd ed. New York, Springer Science+Business Media, 1995, pp. 225–244.

Takeshita M, Miki M, Yubishui T. Cytochrome reductase activities in rat brain microsomes during development. *J Neurochem.* 1982;39:1047–1049.

Walker RJ, Brooks HL, Holden-Dye L. Evolution and overview of classical transmitter molecules and their receptors. *Parasitology.* 1996;113:S3–S33.

Walther B, Ghersi-Egea JF, Minn A, Siest G. Subcellular distribution of cytochrome P-450 in the brain. *Brain Res.* 1986;375:338–344.

Wang M, Roberts DL, Paschke R, Shea TM, Masters BSS, Kim J-JP. Three-dimensional structure of NADPH–cytochrome P450 reductase: Prototype for FMN- and FAD-containing enzymes. *Proceedings of the National Academy of Sciences of the United States of America.* 1997;94:8411–8416.

Wessler I, Kirkpatrick CJ, Racké K. Non-neuronal acetylcholine, a locally acting molecule, widely distributed in biological systems: Expression and function in humans. *Pharmacol Ther.* 1998;77:59–79.

Wessler I, Kirkpatrick CJ, Racké K. The cholinergic "pitfall": Acetylcholine, a universal cell molecule in biological systems, including humans. *Clin Exp Pharmacol Physiol.* 1999;26:198–205.

Wessler I, Michel-Schmidt R, Kirkpatrick CJ. pH-dependent hydrolysis of acetylcholine: Consequences for non-neuronal acetylcholine. *Int Immunopharmacol.* 2015;29:27–30.

Whittaker VP, Sheridan MN. The morphology and acetylcholine content of isolated cerebral cortical synaptic vesicles. *J Neurochem.* 1965;12:363–372.

Wilhelm F, Hirrlinger J. Multifunctional roles of NAD+ and NADH in astrocytes. *Neurochem Res.* 2012;37:2317–2325.

Wilhelm F, Hirrlinger J. The NAD+/NADH redox state in astrocytes: Independent control of the NAD+ and NADH content. *J Neurosci Res.* 2011,89:1956–1964.

Winkler U, Hirrlinger J. Crosstalk of signaling and metabolism mediated by the $NAD^+$/NADH redox state in brain cells. *Neurochem Res.* 2015;40:2394–2401.

Wu JT, Wu LH, Knight JA. Stability of NADPH: Effect of various factors on the kinetics of degradation. *Clin Chem.* 1986;32:314–319.

Yuan H, Shi Y, Sun D. Ion transporters in microglial function: New therapeutic targets for neuroinflammation in ischemic stroke? In: Chen J, Hu X, Stenzel-Poore M, Zhang JH (Eds). Immunological mechanisms and therapies in brain injuries and stroke. New York: Springer Science+Business Media, 2014, pp 121–134.

Zhang Q, Li Y, Tsien RW. The dynamic control of kiss-and-run and vesicular reuse probed with single nanoparticles. *Science.* 2009;323:1448–1453.

# 13 GABA/BABA Priming Causes Signaling of Defense Pathways Related to Abiotic Stress Tolerance in Plants

*K.C. Jisha, A.M. Shackira, and Jos T. Puthur*

## CONTENTS



## 13.1 INTRODUCTION

Abiotic stresses are the most significant factors leading to extensive and erratic loss in agricultural productivity across the world. Owing to the sessile nature of plants, they must cope with adverse environmental conditions, and therefore they must possess a variety of responses to cope with various environmental stresses (Gao et al. 2007). In plants, these stresses adversely affect the growth and productivity by triggering a series of morphological, biochemical, and molecular mechanisms (Wang et al. 2001). Even though the plants have developed specific mechanisms during the course of evolution to sense the subtle changes of growth conditions that trigger many signal transduction cascades (Gao et al. 2007), certain treatments activate stress responsive genes and finally leads to changes at the physiological and biochemical levels in plants.

The quick signaling for inducing the defense mechanism against the adverse environmental conditions requires a stress signal, which gets transduced like a neurotransmitter. Major neurotransmitters found in plants include acetylcholine, epinephrine, γ-aminobutyric acid (GABA), dopamine, levodopa, melatonin, serotonin, and so on (Fait et al. 2006). Later, it was discovered that GABA is largely and rapidly produced in plants in response to various biotic and abiotic stress factors. For example, when insects walk across a leaf, rapid accumulation of GABA was monitored within seconds (Scholz et al. 2015). In addition, β-aminobutyric acid (BABA) a close structural relative of GABA, is a potent inducer of acquired disease resistant in plants. However, its occurrence is rare in plants. BABA is well-known for conferring protection in different plant species against an exceptionally broad spectrum of stresses, including microbial pathogens, herbivores, and abiotic stresses (Jakab et al. 2001).

Hence, by identifying how plants use GABA/BABA as a stress signal, we will be equipped with a new tool to produce transgenic crops with increased tolerance toward adverse environmental conditions. This will further help in the global effort to breed more stress-resistant crops to fight food insecurity caused by the unfavorable climate conditions. Owing to the fact that most of the research on GABA/BABA as a signaling molecule was focused on mammalian cells, an attempt was made to summarize the available data on GABA/BABA as neurotransmitters for enhancing the abiotic stress tolerance potential in plants.

## 13.2 ENHANCING STRESS TOLERANCE IN PLANTS

In order to enhance the crop productivity under adverse environmental conditions, various efforts are underway to increase stress tolerance of crop plants. The basis of all these efforts is to improve the plant's acclimation capacity so that they can do better at the onset of stress. Even though substantial progress has been made on the aspect discussed earlier, breeding of plants for effective abiotic stress tolerance remains as a challenge even today. The research taking place worldwide is focused to design various strategies, such as breeding (conventional and mutation breeding), biotechnological (transgenic approach), and agronomical (priming approach) means either for the development of tolerant crops or for imparting tolerance to the available crop varieties against various stresses (Mir et al. 2012).

Traditional breeding methods had a major role in incorporating favorable genes to induce stress tolerance in crops. Although crop improvement was attained through conventional breeding methods, very few crops have been reported to show improved resistance against salt/drought stress at the field level (Srivastava 2012). Moreover, the conventional breeding approaches are highly time-consuming, cumbersome, and labor-intensive. In addition to this, undesirable characters often appear along with the desirable ones in conventional breeding programs (Kharkwal and Shu 2009). Transgenic methods mostly depend on the manipulation of genes that protect and maintain the function/structure of cellular components, which result in enhanced tolerance (Wang et al. 2003). The present efforts to enhance plant stress tolerance by gene manipulation have resulted in important achievements in many countries. However, the nature of the genetically complex mechanisms of abiotic stress tolerance, and the potential detrimental side effects, make this task extremely difficult and, moreover, the outcome may be unpredictable (Wang et al. 2003).

In contrast to the breeding and transgenic approaches, which solely depend upon the changes at the genome level, priming is an entirely different strategy where various kinds of regulatory molecules are used externally to strengthen the integral defense system of the plants (Ashraf et al. 2008). Nowadays, priming has emerged as a promising method for biotic and abiotic stress management whereby plants are trained through application of suitable priming agents (Goellner and Conrath 2008). Plants that have been exposed to stressful situations can *remember* these events and are able to defend themselves in a faster and better means upon renewed exposure to any stress in the future; this phenomenon is called priming. In other words, priming can be defined as the induction of a particular physiological state in plants by the treatment of natural and/or synthetic compounds to the seeds/seedlings before germination. Plants raised from primed seeds showed various advantages over the non-primed ones, such as sturdy and quick cellular defense response against abiotic stresses (Jisha et al. 2013).

Generally, the priming treatment is known to activate signaling components or transcription factors required for stress perception and tolerance. Since priming results in activation of cellular defense responses rather than the direct up-regulation of defense signaling cascades, the negative impact on plant fitness parameters such as growth rate and reproduction was found to be less significant (Jakab et al. 2005; Van Hulten et al. 2006; Jisha and Puthur 2014a, b; Vijayakumari and Puthur 2014). Priming for abiotic stress tolerance can be attained via application of various natural or chemical compounds. A priming agent may be stimuli that sensitize the plant's immune system for augmented activation against future stressed situation. Among different types of priming agents, γ-aminobutyric acid (GABA) and β-aminobutyric acid (BABA) constitute two powerful and widely used priming agents in plants, as it directly interferes with the plants' defense system, thereby providing increased resistance toward different abiotic stress factors.

## 13.3 GABA/BABA PRIMING IN PLANTS

GABA and BABA are non-protein amino acids that are used as priming agents for enhancing abiotic stress tolerance in plants. GABA and BABA constitute two potential plant priming hormones since it induces several defense-related signaling cascades in plants upon exposure to adverse environmental conditions. GABA is a naturally occurring, four-carbon non-protein amino acid (MW. 103.1 g/mol), which is found in all prokaryotic and eukaryotic organisms, and it acts mainly as an inhibitory neurotransmitter (Kinnersley and Turano 2000). GABA is a well-documented neurotransmitter in plants whose biosynthesis and mode of action are more or less elucidated.

On the other hand, BABA, an isomer of naturally occurring GABA, occurs very rarely in nature (Jakab et al. 2005; Mayer et al. 2006). So far BABA has been reported only in root exudates of tomato plants grown in solarized soils, in the xylem/phloem sap of *Eucalyptus regnans*, *Acacia*, and in some types of grapevine (Gamliel and Katan 1992; Barrado et al. 2009; Pfautsch et al. 2009). Interestingly, BABA, being similar to GABA in structure and also present naturally in plants, offers a great deal for enhancing the tolerance of plants toward various abiotic stresses. Thus, GABA and BABA trigger the defense system in order to equip the plant for the future exposure to different stress factors and to maintain a basal level of resistance. Figure 13.1 illustrates the chemical structures of GABA and BABA.

Interestingly, when these chemicals are applied exogenously, it seems to play a broad role in enhancing plant defense against both biotic stress, such as invasion of various bacterial, viral, and fungal pathogens, and abiotic stresses, such as drought, salt, and heat shock in plants (Cohen 2001; Jakab et al. 2001; Zimmerli et al. 2008). As a naturally occurring compound, GABA has many functions in plants. It can act as an osmolyte or help in the synthesis of osmolytes like proline during water deficit. It also serves as an amplifier of stress signals, inducer of ethylene production, and mediator for acquisition of minerals, and also has an anaplerotic role in the stress-related metabolism in plants (Kinnersley and Turano 2000). GABA, if applied superficially, exerts a stimulatory effect and acts as a growth-promoting substance in many plants (Kinnersley and Lin 2000; Roberts 2007). The beneficial effect of GABA priming on the plant growth and abiotic stress tolerance potential were proved recently in different plant species like *Lolium perenne* (Krishnan et al. 2013), *Oryza sativa* (Nayyar et al. 2014), black pepper (Vijayakumari and Puthur 2016), etc.

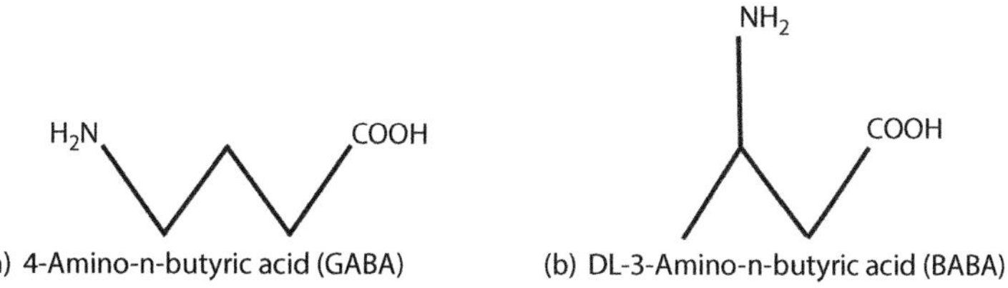


**FIGURE 13.1** Chemical structure of (a) GABA and (b) BABA.

Various enzymes are seen to get up-regulated under GABA treatment in plants. In sunflower, the expression of 1-Aminocyclopropane-1-carboxylic acid (ACC) synthase enzyme of ethylene metabolism was found to be altered by the exogenous application of GABA (Kathiresan et al. 1997), whereas in *Brassica napus*, GABA priming significantly increased the expression of mRNA for the BnNRT2 nitrate transporter in roots of plants (Beuve et al. 2004). In maize, GABA treatment significantly enhanced the activity of superoxide dismutase (SOD), peroxidases (POD), and catalases (CAT) (Tian et al. 2005). Similarly, in *Caragana intermedia*, GABA altered the expression of stress-related genes in roots subjected to NaCl stress (Shi et al. 2010). GABA priming was found to regulate nitrate uptake and metabolism in *Arabidopsis* seedlings by activating the enzymes concerned with nitrogen and carbon metabolism, including, nitrate reductase, glutamate synthase, glutamine synthetase, and phosphoenol pyruvate carboxylase (Barbosa et al. 2010).

BABA priming was also found to be one of the best priming method in various plants for biotic as well as abiotic stress (Jakab et al. 2001, 2005; Ton et al. 2005; Zimmerli et al. 2000, 2001, 2008; Cohen et al. 2007; Macarisin et al. 2009; Jisha and Puthur 2016a, b). BABA priming was found to enhance the drought and salt stress tolerance in *Arabidopsis* (Conrath et al. 2002; Jakab et al. 2005), and it exerts its action by priming plants to respond faster and stronger to future stresses. BABA-induced tolerance to osmotic stress is primarily based on priming for increased adaption responses rather than on the direct activation of these responses (Conrath et al. 2006). Several researchers reported the priming effect of BABA in increasing resistance to different microbial pathogens (Conrath et al. 2006; Slaughter et al. 2012; Pastor et al. 2014; Jisha and Puthur 2016a, b).

BABA-induced priming functions via the interaction of several hormones, such as salicylic acid (SA), abscisic acid (ABA), and ethylene (Jakab et al. 2005; Ton et al. 2005). Further mechanisms mediated by BABA are through the modulation of biochemical responses toward the stress. Sometimes it promotes the biosynthesis of secondary metabolites (phenols, anthocyanin, and phytoalexins) and elevates activity of enzymes associated with scavenging of active oxygen species, lignification, and plant secondary metabolism (Slaughter et al. 2008; Wu et al. 2010; Justyna and Ewa 2013). Moreover, activation of defense genes and accumulation of pathogenesis related (PR) proteins concerned with the antimicrobial activity were identified in many BABA-treated plants, such as tomato, peppers, potato, and so on (Cohen 1994; Hwang et al. 1997; Altamiranda et al. 2008; Šašek et al. 2011; Zhong et al. 2014).

Priming with BABA resulted in enhanced activities of polyphenoloxidase (PPO), phenylammonialyase (PAL), and peroxidase (POD) enzymes. BABA also induced the accumulation of pathogenesis-related (PR) proteins and deposition of callose and lignin, and thus reported to act through jasmonic acid, ethylene, and abscisic acid defense pathways (Cohen 1994; Hamiduzzaman et al. 2005). In tobacco plants, BABA induced proline accumulation and peroxidase activity under low-potassium stress and thereby enhanced the plant tolerance to low-potassium stress (Jiang et al. 2012). Whereas in wheat, BABA priming enhanced the drought-induced ABA accumulation in the leaves, and which further caused reduced water usage by closing of stomata (Du et al. 2012; Vijayakumari et al. 2016).

Moreover, BABA priming also reduced reactive oxygen species (ROS) production and lipid peroxidation by increasing the activity of antioxidant enzymes [superoxide dismutase (SOD), catalase (CAT), ascorbate peroxidase (APX), and glutathione reductase (GR)] (Rajaei and Mohamadi 2013). BABA-primed mustard plants exhibited increased ascorbate, anthocyanin, flavonoid, and calcium content in drought condition. Anthocyanin is a well-known modulator of stress signals, and in *Arabidopsis*, BABA stimulates anthocyanin biosynthesis by regulating the expression of CHS (chalcone synthase) and DFR (dihydroflavonol-4-reductase) enzymes (Wu et al. 2010). A list of plants primed with GABA/BABA for abiotic stress tolerance potential is summarized in Table 13.1.

**TABLE 13.1**
**The Specific GABA/BABA Priming Effect Related to Abiotic Stress Tolerance Potential in Various Plants**

| Sl No. | Plant | Priming Agent | Priming Against | Priming Effect | References |
|---|---|---|---|---|---|
| 1 | *Arabidopsis thaliana* (L.) Heynh. | BABA | Heat, NaCl | Accumulation of PR proteins | Jakab et al. (2001) |
| 2 | *Arabidopsis thaliana* (L.) Heynh. | BABA | Drought, NaCl | Enhanced ABA signaling | Jakab et al. (2005) |
| 3 | *Arabidopsis thaliana* (L.) Heynh. | BABA | Heat | Accumulation of ABA transcription factors | Zimmerli et al. (2008) |
| 4 | *Arabidopsis thaliana* (L.) Heynh. | BABA | Cadmium | The transcript levels of *GSH1* and *AtATM3* increased | Cao et al. (2009) |
| 5 | *Malus pumila* Mil. | BABA | Drought | Suppression of lignin biosynthesis | Macarisin et al. (2009) |
| 6 | *Caragana intermedia* Fabr. | BABA | NaCl | Reduced lipid peroxidation and regulate ethylene production | Shi et al. (2010) |
| 7 | *Malus × domestica* Borkh. nom. illeg. | BABA | Drought | Less water loss and reduced endogenous ABA | Tworkoski et al. (2011) |
| 8 | *Arabidopsis thaliana* (L.) Heynh. | BABA | Acid rain | Changes in protein profile and activate antioxidant system, salicylic acid, jasmonic acid, abscisic acid signaling pathways. | Liu et al. (2011) |
| 9 | *Triticum aestivum* L. | BABA | Drought | Increased antioxidant enzyme activities and reduced ROS | Du et al. (2012) |
| 10 | *Lolium perenne* L. | GABA | Drought | Increased RWC, peroxidase activity | Krishnan et al. (2013) |
| 11 | *Oryza sativa* L. | GABA | Heat | Increased leaf turgidity, antioxidant activity, osmoprotectant accumulation | Nayyar et al. (2014) |
| 12 | *Piper nigrum* L. | GABA | Drought | Increased antioxidant enzyme activity, accumulation of proline and sugar | Vijayakumari and Puthur (2016) |
| 13 | *Vigna radiate* (L.) R. Wilczek. | BABA | NaCl, drought | Increased photosynthesis, mitochondrial activities, enzymatic and non-enzymatic antioxidants | Jisha and Puthur (2016a) |
| 14 | *Zea mays* L. ssp. mays | GABA | NaCl | Increased photosynthesis, antioxidant activity, osmoprotectant accumulation | Wang et al. (2017) |

## 13.4 METABOLISM OF GABA/BABA

### 13.4.1 GABA Shunt

In plants, the γ-aminobutyric acid (GABA) shunt was first reported more than half a century ago in potato (*Solanum tuberosum*) tuber (Dent et al. 1947) and has been shown to be activated by light, during developmental phases. The GABA shunt involves three main reactions catalyzed by glutamate decarboxylase (GAD), GABA transaminase (GABA-T), and succinic semialdehyde

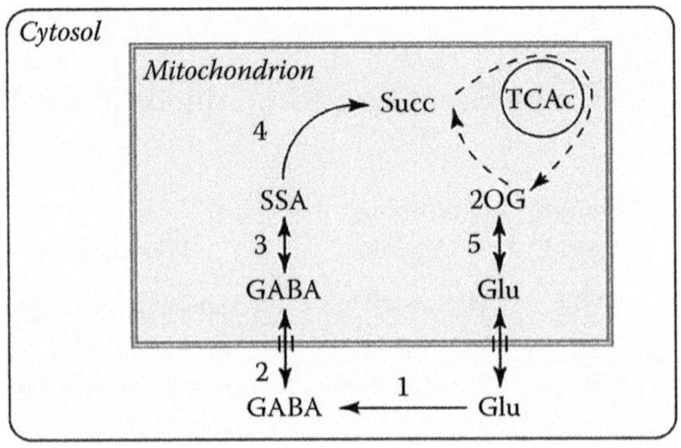


**FIGURE 13.2** Common depiction of the GABA shunt. (1) glutamate decarboxylase; (2) putative mitochondrial GABA transporter; (3) GABA transaminase; (4) succinic semialdehyde dehydrogenase; (5) glutamate dehydrogenase. Dotted arrows indicate unresolved metabolic routes. (TCAc, tricarboxylic acid cycle). (Data from Fait, A. et al., *Trends Plant Sci.*, 13, 14–19, 2008. With permission.)

dehydrogenase (SSADH), respectively. The pathway starts with the decarboxylation of glutamate (Glu) to produce GABA and $CO_2$ in the cytosol by the enzyme GAD. GABA is then presumably transported to the mitochondria by an as yet unidentified GABA transporter, where it is converted to succinic semialdehyde (SSA) by the enzyme GABA-T. Subsequently, SSA is converted either to succinate or 4-hydroxybutyrate (GHB) by the enzyme SSADH (Fait et al. 2008). Figure 13.2 represents the detailed pathway of GABA shunt.

### 13.4.2 BABA Metabolism

Recent reports of BABA's role in plants' metabolic processes opens up new lines of research in understanding the anabolic and catabolic pathway(s), as well as the exact role of BABA in defense. This might lead to novel approaches for defining defense mechanisms especially in the light of the wide effectiveness of this compound in defense priming. According to Takaishi et al. (2012), BABA could be synthesized from L-glutamate as proposed for the biosynthesis of the antibiotic incednine in bacteria. However, in order to gain more insight into the biosynthetic pathway of BABA, research is progressing with T-DNA insertion mutants of *Arabidopsis thaliana* (Thevenet et al. 2017). Since BABA is not metabolized in plants, it is thought to bind to cell wall proteins, making them more resistant to pathogen attack (Cohen and Gisi 1994; Ahmed et al. 2009). Such a mode of action could be occurring in the case of abiotic stress too.

## 13.5 SIGNALING CASCADES DURING GABA/BABA PRIMING

Priming causes many morphological, physiological, and biochemical changes in plants. The physiological state in which plants are able to respond more quickly is called the primed state of the plant (Beckers and Conrath 2007). Nowadays, priming has emerged as a promising approach in modern stress alleviation, as it protects plants against biotic as well as abiotic stresses without heavily affecting fitness (Van Hulten et al. 2006). The main benefit of a *primed state* is that it leads to the enhancement of cellular defense responses rather than the direct up-regulation of defense signaling cascades that would impart negative influences on plant growth (Heil 2002; Macarisin et al. 2009).

### 13.5.1 GABA Signaling Pathway

GABA synthesis is activated radically at the onset of stress, and it has been reported that GABA stimulates the genetic expression of various stress-related activities in plants (Lancien and Roberts 2006; Renault et al. 2010; Vijayakumari and Puthur 2016). In plants, GABA serves a dual role as a metabolite and/or as a signaling molecule (Bouché and Fromm 2004; Roberts 2007). According to Bao et al. (2015), GABA shunt is concerned with the regulation of homeostasis of metabolites in plants, which can positively affect the tolerance level of plants under abiotic stresses, so it is probable that GABA priming mediates significant changes in metabolome composition, by taking the role of a signaling molecule and also as a key metabolite during the stressful situation (Vijayakumari and Puthur 2016a).

Earlier GABA was found to be transported by the proline transporter 2 (ProT2), a quaternary transporter that mediates the influx of both GABA and proline with preference to the latter substrate (Britkreuz et al. 1999). The first high-affinity GABA-specific transporter in plants, GAT1 (AtGAT1-*A. thaliana* GABA transporter 1), was isolated thereafter from the cell membrane of *A. thaliana*. GAT1 is an $H^+$-driven, plasma membrane-localized transporter that recognizes GABA and several GABA-related compounds. In contrast to ProT2, GAT1 shows high affinity to GABA as compared to other substrates such as alanine and ß-alanine (Mayer et al. 2006). Another transporter of GABA was elucidated by the isolation of a GABA-permease (GABP), localized to the mitochondrial membrane (Michaeli et al. 2011).

In addition to the previous facts, recent reports suggested that GABA-based signaling in plants is mediated by ALMT (aluminium-activated malate transporters) as prime candidates. ALMT form a large multigenic anion channel family exclusive to plants with multiple physiological roles and discrete expression patterns (Dreyer et al. 2012). The concentrations of anions required to activate ALMT are commonly encountered within plant tissue, suggesting that ALMT are ordinarily active in cells or are at least primed for activation (Ramesh et al. 2015).

### 13.5.2 BABA Signaling Pathway

BABA is proved to induce systemic acquired resistance (SAR) in plants by the accumulation of pathogenesis-related (PR) proteins. According to Cao et al. (2009), BABA activates the ABA-dependent and salicylic acid-dependent defense mechanism in plants (Jakab et al. 2005). Studies have suggested that BABA increases the mRNA accumulation of ABA and ethylene early signaling intermediates in plants (Zimmerli et al. 2008). Moreover, the activation of ABA signaling further stimulates the cytosolic $Ca^{2+}$ level by inducing both $Ca^{2+}$ influx from the extracellular space and $Ca^{2+}$ release from intracellular space (Allen et al. 2000; Rajae and Mohamadi 2013).

Study of *Arabidopsis* mutants in BABA-IR led to the identification of the Impaired in BABA-induced Immunity 1 (IBI1) gene, encoding an aspartyl-tRNA synthetase (AspRS), which is thought to function as the BABA receptor in plants. R-BABA (active form of BABA) binds the L-asp binding domain of IBI1, thereby increasing the uncharged tRNA asp, which serves as a conserved signal for metabolic imbalance (Dong et al. 2000; Dever and Hinnebusch 2005; Wek et al. 2006; Li et al. 2013). The signaling cascades of the BABA-induced defense mechanism was detailed earlier by Schwarzenbacher et al. (2014).

Three mutants of *Arabidopsis* impaired in BABA-induced sterility (ibs) were identified, which are affected in distinct signaling routes of BABA-IR (Jakab et al. 2001; Ton et al. 2005). A mutation in the cyclin-dependent kinase-like gene IBS1 blocked BABA-induced priming of SA-dependent defenses, whereas ibs2 and ibs3 mutant plants are specifically affected in BABA-induced priming of cell wall defenses. The third mutant, ibs3, harbors a T-DNA insertion near the zeaxanthin epoxidase gene IBS3/ABA1/NPQ2 (nonphotochemical quencing 1), a gene involved in ABA biosynthesis. This finding confirmed the involvement of ABA in BABA-induced priming of cell wall defense (Ton and Mauch-Mani 2004).

BABA priming of abiotic stress resistance was found to be mainly mediated via ABA signaling cascades (Jakab et al. 2005; Zimmerli et al. 2008). However, SA-induced signaling was also involved in the protection of plants against abiotic stress (Kang and Saltveit 2002; Shakirova et al. 2003; Jakab et al. 2005). These reports imply that BABA acts at multiple levels to induce the plant for a faster and stronger activation of stress-specific defense mechanisms upon stress exposure when

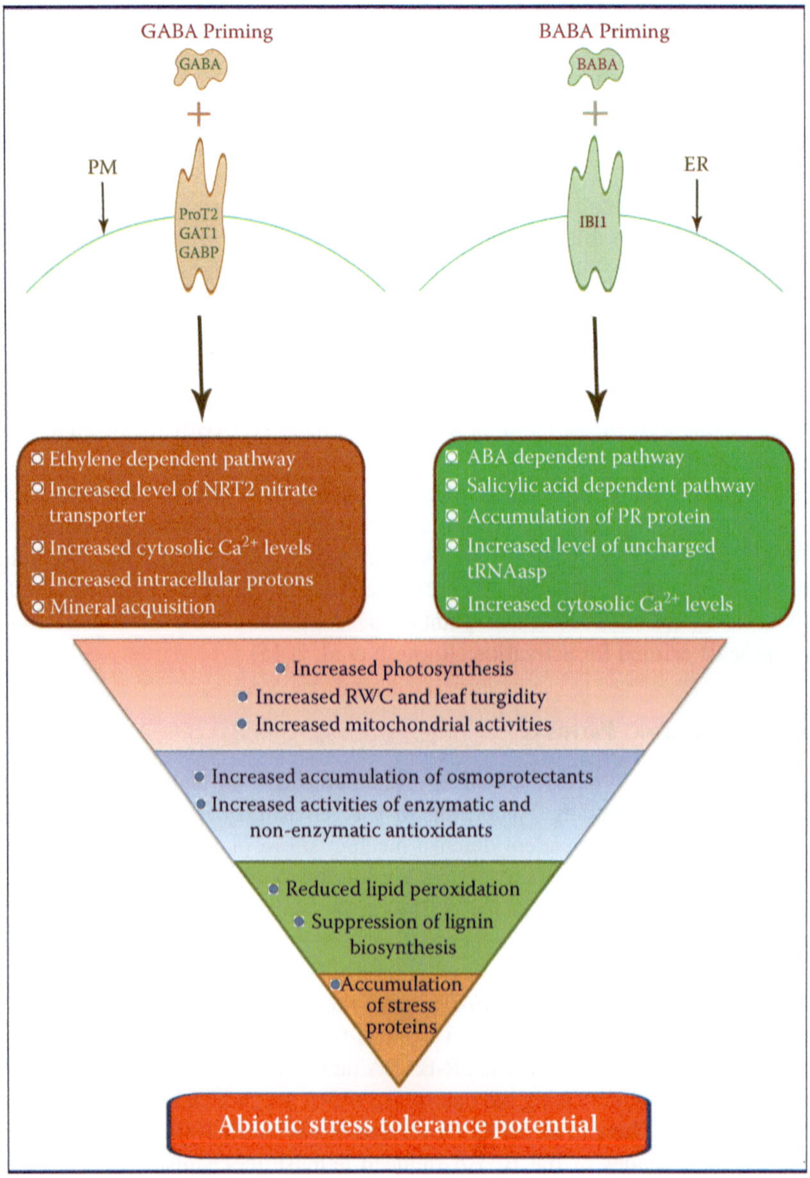


**FIGURE 13.3** Model of GABA/BABA induced signaling in plants. Priming of GABA/BABA induces different signaling pathways leading to changes in the expression level of several genes which ultimately imparts abiotic stress tolerance potential in plants. ProT2-proline transporter 2; GAT1-GABA transporter 1; GABP-GABA permease, IBI1- impaired in BABA-induced immunity, PM-plasma membrane; ER-endoplasmic reticulum membrane.

compared to the normal, genotypically determined response. Figure 13.3 illustrates the signaling pathway of GABA- and BABA-induced resistance in plants.

## 13.6 CONCLUSION

Nowadays GABA/BABA priming is receiving a wide acceptability due to their capacity to prepare the plants for abiotic stresses. It is well-known that the protective effect of non-protein amino acids like BABA and GABA on plants is due to induction of natural defense mechanisms against abiotic stresses, but at the same time not activating the complete defense mechanisms before the stress exposure. Even though the precise mechanism of GABA/BABA priming in plants is still a mystery, their role as signaling molecules during stress has more or less unraveled. As the complete elucidation of the signaling pathway in GABA/BABA priming-related abiotic stress tolerance in plants can offer much to the genetic engineering of crop plants to adapt with the adverse environmental conditions, more focus is needed in this field of research.

## REFERENCES


Ahmed, N., Abbasi, M. W., Shaukat, S. S., and Zaki, M. J. 2009. Induced systemic resistance in mung bean plant against root-knot nematode *Meloidogyne javanica* by DL-β-amino butyric acid. *Nematol Mediterr* 37:67–72.

Allen, A. M., Zhuo, J., and Mendelsohn, F. A. O. 2000. Localization and function of angiotensin At1 receptors. *Am J Hypertens* 13:31–38.

Altamiranda, E. A. G., Andreu, A. B., Daleo, G. R., and Olivieri, F. P. 2008. Effect of betaaminobutyric acid (BABA) on protection against *Phytophthora infestans* throughout the potato crop cycle. *Aust Plant Pathol* 37:421–427.

Ashraf, M., Athar, H. R., Harris, P. J. C., and Kwon, T. R. 2008. Some prospective strategies for improving crop salt tolerance. *Adv Agron* 97:45–110.

Bao, H., Chen, X., Lv, S. et al. 2015. Virus-induced gene silencing reveals control of reactive oxygen species accumulation and salt tolerance in tomato by γ-aminobutyric acid metabolic pathway. *Plant Cell Environ* 38:600–613

Barbosa, J. M., Singh, N. K., Cherry, J. H., and Locy, R. D. 2010. Nitrate uptake and utilization is modulated by exogenous γ-aminobutyric acid in *Arabidopsis thaliana* seedlings. *Plant Physiol Biochem* 48:443–450.

Barrado, E., Rodriguez, J. A., and Castrillejo, Y. 2009. Determination of primary amino acids in wines by high performance liquid magneto-chromatography. *Talanta* 78:672–675.

Beckers, G. J. M., and Conrath, U. 2007. Priming for stress resistance: From the lab to the field. *Curr Opin Plant Biol* 10:1–7.

Beuve, N., Rispail, N., Laine, P., Cliquet, J. B., Ourry, A., and Le Deunff, E. 2004. Putative role of γ-aminobutyric acid (GABA) as a long-distance signal in up-regulation of nitrate uptake in *Brassica napus* L. *Plant Cell Environ* 27:1035–1046.

Bouché, N., and Fromm, H. 2004. GABA in plants: Just a metabolite? *Trends Plant Sci* 9:110–115.

Britkreuz, K. E., Shelp, J. B., Fischer, W. N., Schwacke, R., and Rentsch, D. 1999. Identification and characterization of GABA, proline and quaternary ammonium compound transporters from *Arabidopsis thaliana. FEBS Lett* 450:280–284.

Cao, S. Q., Ren, G., Jiang, L. H., Yuan, B., and Ma, G. H. 2009. The role of β-aminobutyric acid in enhancing cadmium tolerance in *Arabidopsis thaliana. Russ J Plant Physiol* 56:575.

Cohen, Y. 1994. 3-Aminobutyric acid induces systemic resistance against *Peronospore tabacina. Physiol Mol Plant Pathol* 44:273–288.

Cohen, Y. 2001. The BABA story of induced resistance. *Phytoparasitica* 29:375–378.

Cohen, Y., and Gisi, U. 1994. Systemic translocation of $^{14}$C-DL-3-aminobutyric acid in tomato plants against *Phytophthora infestans. Physiol Mol Plant Pathol* 45:441–446.

Cohen, Y., Baider, A., Gotilieb, D., and Rubin, E. 2007. Control of *Bremia lectucae* in field-grown lettuce by DL-3-amino-n-butanoic acid (BABA). *Paper Presented at the 3rd QLIF Congress*, Hohenheim, Germany, pp. 20–23.

Conrath, U., Beckers, G. J. M., Flors, V. et al. 2006. Priming: Getting ready for battle. *Mol Plant Microbe Interact* 19:1062–1071.

Conrath, U., Pieterse, C. M. J., and Mauch-Mani, B. 2002. Priming in plant–pathogen interactions. *Trends Plant Sci* 7:210–216.

Dent, C. E., Stepka, W., and Steward, F. C. 1947. Detection of the free amino acids of plant cells by partition chromatography. *Nature* 160:682–683.

Dever, T. E., and Hinnebusch, A. G. 2005. GCN2 whets the appetite for amino acids. *Mol Cell* 18:141–142.

Dong, J. S., Qiu, H. F., Garcia-Barrio, M., Anderson, J., and Hinnebusch, A. G. 2000. Uncharged tRNA activates GCN2 by displacing the protein kinase moiety from a bipartite tRNA-Binding domain. *Mol Cell* 6:269–279.

Dreyer, I., Gomez-Porras, J. L., Riaño-Pachón, D. M., Hedrich, R., and Geiger, D. 2012. Molecular evolution of slow and quick anion channels (SLACs and QUACs/ALMTs). *Front Plant Sci* 3:263.

Du, Y., Wang, Z., Fan, J., Neil, C., Turner T. W., and Li, F. 2012. β-Aminobutyric acid increases abscisic acid accumulation and desiccation tolerance and decreases water use but fails to improve grain yield in two spring wheat cultivars under soil drying. *J Exp Bot* 63:4849–4860.

Fait, A., Yellin, A., and Fromm, H. 2006. GABA and GHB neurotransmitters in plants and animals. In *Communication in Plants: Neuronal Aspects of Plant Life* (Ed.) F. Baluska, S. Mancuso, and D. Volkmann, pp. 171–185, Springer: New York.

Fait, A., Fromm, H., Walter, D., Galili, G., and Fernie, A. R. 2008. Highway or byway: The metabolic role of the GABA shunt in plants. *Trends Plant Sci* 13:14–19.

Gamliel, A., and Katan, J. 1992. Influence of seed and root exudates on fluorescent *Pseudomonads* and fungi in solarized soil. *Phytopathol* 82:320–327.

Gao, J. P., Chao, D. Y., and Lin, H. X. 2007. Understanding abiotic stress tolerance mechanisms: Recent studies on stress response in rice. *J Integr Plant Biol* 49:742–750.

Goellner, K., and Conrath, U. 2008. Priming: It's all the world to induced disease resistance. *Eur J Plant Pathol* 121:233–242.

Hamiduzzaman, M. M., Jakab, G., Barnavon, L., Neuhaus, J. M., and Mauch-Mani, B. 2005. ß-amino butyric acid-induced resistance against downy mildew in grapevine acts through the potentiation of callose formation and JA signalling. *Mol Plant Micro Interact* 18:819–829.

Heil, M. 2002. Ecological costs of induced resistance. *Curr Opin Plant Biol* 5:345–350.

Hwang, B. K., Sunwoo, J. Y., Kim, Y. J., and Kim, B. S. 1997. Accumulation of beta-1,3-glucanase and chitinase isoforms, and salicylic acid in the DL-beta-amino-n-butyric acid-induced resistance response of pepper stems to *Phytophthora capsici. Physiol Mol Plant Pathol* 51:305–322.

Jakab, G., Cottier, V., Toquin, V. et al. 2001. D-Aminobutyric acid-induced resistance in plants. *Eur J Plant Pathol* 107:29–37.

Jakab, G., Ton, J., Flors, V., Zimmerli, L., Metraux, J. P., and Mauch-Mani, B. 2005. Enhancing *Arabidopsis* salt and drought stress tolerance by chemical priming for its abscisic acid responses. *Plant Physiol* 139:267–274.

Jiang, L., Yang, R. Z., Lu, Y. F., Cao, S. Q., Ci, L. K., and Zhang, J. J. 2012. β aminobutyric acid mediated tobacco tolerance to potassium deficiency. *Russ J Plant Physiol* 59:781–787.

Jisha, K. C., Vijayakumari, K., and Puthur, J. T. 2013. Seed priming for abiotic stress tolerance: An overview. *Acta Physiol Plant* 35:1381–1396.

Jisha, K. C., and Puthur, J. T. 2014a. Halopriming of seeds imparts tolerance to NaCl and PEG induced stress in Vigna *radiata* (L.) Wilczek varieties. *Physiol Mol Biol Plant* 20:303–312.

Jisha, K.C., and Puthur, J. T. 2014b. Seed halopriming outdo hydropriming in enhancing seedling vigour and osmotic stress tolerance potential of rice varieties. *J Crop Sci Biotechnol* 17:209–219.

Jisha, K.C., and Puthur, J. T. 2016a. Seed priming with BABA (β-amino butyric acid): A cost-effective method of abiotic stress tolerance in *Vigna radiata* (L.) Wilczek. *Protoplasma* 253:277–289.

Jisha, K.C., and Puthur, J. T. 2016b. Seed priming with beta-amino butyric acid (BABA) improves abiotic stress tolerance in rice seedlings. *Rice Science* 23:242–254.

Justyna, P. G., and Ewa, K. 2013. Induction of resistance against pathogens by β-aminobutyric acid. *Acta Physiol Plant* 35:1735–1748.

Kang, H. M., Saltveit, M. E. 2002. Chilling tolerance of maize, cucumber and rice seedling leaves and roots are differentially affected by salicylic acid. *Physiol Plant* 115:571–576.

Kathiresan, A., Tung, P., Chinnappa, C. C., and Reid, D. M. 1997. γ-Aminobutyric acid stimulates ethylene biosynthesis in sunflower. *Plant Physiol* 115:129–135.

Kharkwal, M. C., and Shu, Q. Y. 2009. The role of induced mutations in world food security. In *Induced Plant Mutations in the Genomics Era* (Ed.) Q. Y. Shu, pp. 33–38. Food and Agriculture Organization of the United Nations: Rome, Italy.

Kinnersley, A. M., and Lin, F. 2000. Receptor modifiers indicate that 4-aminobutyric acid (GABA) is a potential modulator of ion transport in plants. *Plant Growth Regul* 32:65–76.

Kinnersley, A. M., and Turano, F. J. 2000. Gamma-aminobutyric acid (GABA) and plant responses to stress. *Crit Rev Plant Sci* 19:479–509.

Krishnan, S., Laskowski, K., Shukla, V., and Merewitz, E. B. 2013. Mitigation of drought stress damage by exogenous application of non-protein amino acid γ-aminobutyric acid on perennial ryegrass. *J Amer Hort Sci* 138:358–366.

Lancien, M., and Roberts, M. R. 2006. Regulation of *Arabidopsis thaliana 14-3-3* gene expression by γ-aminobutyric acid. *Plant Cell Environ* 29:1430–1436.

Li, M. W., Auyeung, W. K., and Lam, H. M. 2013. The GCN2 homologue in *Arabidopsis thaliana* interacts with uncharged tRNA and uses *Arabidopsis* eIF2alpha molecules as direct substrates. *Plant Biol* 15:13–18.

Liu, T. W., Wu, F. H., Wang, W. H. et al. 2011. Effects of calcium on seed germination, seedling growth and photosynthesis of six forest tree species under simulated acid rain. *Tree Physiol* 31:402–413.

Macarisin, D., Wisniewski, M. E., Bassett, C., and Thannhauser, T. 2009. Proteomic analysis of β-aminobutyric acid priming and abscisic acid—induction of drought resistance in crabapple (*Malus pumila*): Effect on general metabolism, the phenylpropanoid pathway and cell wall enzymes. *Plant Cell Environ* 32:1612–1631.

Mayer, A., Eskandari, S., Grallath, S., and Rentsch, D. 2006. AtGAT1, a high affinity transporter for γ-aminobutyric acid in *Arabidopsis thaliana. J Biol Chem* 281:7197–7204.

Michaeli, S, Fait, A., Lagor, K. et al. 2011. A mitochondrial GABA permease connects the GABA shunt and the TCA cycle, and is essential for normal carbon metabolism. *Plant J* 67:485–498.

Mir, R. R., Zaman-Allah, M., Sreenivasulu, N., Trethowan, R., and Varshney, R. K. 2012. Integrated genomics, physiology and breeding approaches for improving drought tolerance in crops. *Theor Appl Genet* 125:625–645.

Nayyar, H., Kaur, R., Kaur, S., and Singh, R. 2014. γ-aminobutyric acid (GABA) imparts partial protection from heat stress injury to rice seedlings by improving leaf turgor and upregulating osmoprotectants and antioxidants. *J Plant Growth Regul* 33:408–419.

Pastor, V., Gamir, J., Camañes, G., Cerezo, M., Sánchez-Bel, P., and Flors, V. 2014. Disruption of the ammonium transporter AMT1.1 alters basal defences generating resistance against *Pseudomonas syringae* and *Plectosphaerella cucumerina. Front Plant Sci* 5:231.

Pfautsch, S., Gessler, A., Adams, M. A., and Rennenberg, H. 2009. Using amino-nitrogen pools and fluxes to identify contributions of understory *Acacia* spp. to overstory *Eucalyptus regnans* and stand nitrogen uptake in temperate Australia. *New Phytol* 183:1097–1113.

Rajaei, P., and Mohamadi, N. 2013. Effect of beta-aminobutyric acid (BABA) on enzymatic and non-enzymatic antioxidants of *Brassica napus* L. under drought. *Int J Biosci* 3:41–47.

Ramesh, S. A., Conn, V., Tyerman, S. D. et al. 2015. GABA signalling modulates plant growth by directly regulating the activity of plant-specific anion transporters. *Nat Commun* 6:7879.

Renault, H., Roussel, V., El Amrani, A. et al. 2010. The *Arabidopsis pop2-1* mutant reveals the involvement of GABA transaminase in salt stress tolerance. *BMC Plant Biol* 10:20.

Roberts, M. R. 2007. Does GABA act as a signal in plants? *Plant Signal Behav* 2:408–409.

Šašek, V., Nováková, M., Dobrev, P. I., Valentová, O., and Burketová, L. 2011. b-aminobutyric acid protects *Brassica napus* plants from infection by *Leptosphaeria maculans.* Resistance induction or a direct antifungal effect? *Eur J Plant Pathol* 133:279–289.

Scholz, S. S., Reichelt, M., Mekonnen, D. W., Ludewig, F., and Mithöfer A. 2015. Insect herbivory-elicited GABA accumulation in plants is a wound-induced, direct, systemic, and jasmonate-independent defense response. *Front Plant Sci* 6:1–11.

Schwarzenbacher, R. E., Luna, E., and Ton, J. 2014. The discovery of the BABA receptor: Scientific implications and application potential. Animal and plant sciences department. The University of Sheffield, Sheffield, UK.

Shakirova, F. M., Sakhabutdinova, A. R., Bezrukova, M. V., Fatkhutdinova, R. A., and Fatkhutdinova, D. R. 2003. Changes in the hormonal status of wheat seedlings induced by salicylic acid and salinity. *Plant Sci* 164:317–322.

Shi, W., Wang, L., Rousseau, D. P. L., and Lens, P. N. L. 2010. Removal of estrone, 17α-ethinylestradiol, and 17β-estradiol in algae and duckweed-based wastewater treatment systems. *Environ Sci Pollut Res* 17:824–833.

Slaughter, A., Daniel, X., Flors, V., Luna, E., Hohn, B., and Mauch-Mani, B. 2012. Descendants of primed *Arabidopsis* plants exhibit resistance to biotic stress. *Plant Physiol* 158:835–843.

Slaughter, A. R., Hamiduzzaman, M. M., Gindro, K., Neuhaus, J. M., and Mauch-Mani, B. 2008. Beta-aminobutyric acid-induced resistance in grapevine against downy mildew: Involvement of pterostilbene. *Eur J Plant Pathol* 122:185–195.

Srivastava, A. K. 2012. Identification of the thiol induced transcripts, their functional characterization and role in crop improvement. PhD thesis, Bhabha atomic research centre, Homi Bhabha National Institute, Mumbai, India.

Takaishi, M., Kudo, F., and Eguchi, T. 2012. A unique pathway for the 3-aminobutyrate starter unit from L-glutamate through b-glutamate during biosynthesis of the 24 membered macrolactam antibiotic, incednine. *Org Lett* 14:4591–4593.

Thevenet, D., Pastor, V., Baccelli, I. et al. 2017. The priming molecule b-aminobutyric acid is naturally present in plants and is induced by stress. *New Phytol* 213:552–559.

Tian, X. C., Xu, J., and Yang, X. 2005. Normal telomere lengths found in cloned cattle. *Nat Genet* 26:272–273.

Ton, J., and Mauch-Mani, B. 2004. β-amino-butyric acid-induced resistance against necrotrophic pathogens is based on ABA-dependent priming for callose. *Plant J Cell Mol Biol* 38:119–130.

Ton, J., Jakab, G., Toquin, V. et al. 2005. Dissecting the b-aminobutyric acid-induced priming phenomenon in *Arabidopsis*. *Plant Cell* 17:987–999.

Tworkoski, T., Wisniewski, M., and Artlip, T. 2011. Application of BABA and s-ABA for drought resistance in apple. *J Appl Hortic* 13:85–90.

Van Hulten, M., Pelser, M., van Loon, L. C., Pieterse, C. M. J., and Ton, J. 2006. Costs and benefits of priming for defense in *Arabidopsis*. *Proc Natl Acad Sci USA* 103:5602–5607.

Vijayakumari, K., Puthur, J. T. 2014. Drought stress responses in tolerant and sensitive varieties of black pepper (*Piper nigrum* Linn.). *J Plant Crops* 42:78–85.

Vijayakumari, K., and Puthur, J. T. 2016a. γ-Aminobutyric acid (GABA) priming enhances the osmotic stress tolerance in *Piper nigrum* Linn: plants subjected to PEG-induced stress. *J Plant Growth Regul* 78:57–67.

Vijayakumari, K., Jisha, K. C., and Puthur J. T. 2016b. GABA/BABA priming: A means for enhancing abiotic stress tolerance potential of plants with less energy investments on defence cache. *Acta Physiol Plant* 38:230.

Wang, W. X., Vinocur, B., Shoseyov, O., and Altman, A. 2001. Biotechnology of plant osmotic stress tolerance: Physiological and molecular considerations. *Acta Hort* 560:285–292.

Wang, W., Vinocur, B., and Altman, A. 2003. Plant responses to drought, salinity and extreme temperatures: Towards genetic engineering for stress tolerance. *Planta* 218:1–14.

Wang, Y., Gu, W., Meng, Y. et al. 2017. γ-Aminobutyric acid imparts partial protection from salt stress injury to maize seedlings by improving photosynthesis and upregulating osmoprotectants and antioxidants. *Sci Rep* 7:43609.

Wek, R. C., Jiang, H. Y., and Anthony, T. G. 2006. Coping with stress: eIF2 kinases and translational control. *Biochem Soc Trans* 34:7–11.

Wu, C. C., Singh, P., Chen, M. C., and Zimmerli, L. 2010. L-Glutamine inhibits beta aminobutyric acid-induced stress resistance and priming in *Arabidopsis*. *J Exp Bot* 61:995–1002.

Zhong, Y. P., Wang, B., Yan, J. H. et al. 2014. DL-β-amino butyric acid induced resistance in soybean against *Aphis glycines* Matsumura (Hemiptera: Aphididae). *PLoS One* 9:1–11.

Zimmerli, L., Hou, B. H., Tsai, C. H., Jakab, G., Mauch-Mani, B., and Somerville, S. 2008. The xenobiotic beta-aminobutyric acid enhances *Arabidopsis* thermotolerance. *Plant J* 53:144–156.

Zimmerli, L., Jakab, C., Metraux, J. P., and Mauch-Mani, B. 2000. Potentiation of pathogen-specific defense mechanisms in *Arabidopsis* by β-aminobutyric acid. *Proceedings of the National Academy of Sciences of the United States of America* 97:12920–12925.

Zimmerli, L., Metraux, J. P., and Mauch-Mani, B. 2001. β-aminobutyric acid-induced protection of *Arabidopsis* against the necrotrophic fungus *Botrytis cinerea*. *Plant Physiol* 126:517–523.

# 14 Glutamate Sensing in Plants

*Jesús Salvador López-Bucio, Homero Reyes de la Cruz, and A. Arturo Guevara-García*

## CONTENTS



## 14.1 INTRODUCTION

After water, nitrogen (N) is the second most important input factor controlling plant growth and consequently crop production. The N is available to the plants in different forms being nitrate ($NO_3^-$) and ammonium ($NH_4^+$) the most important. N soil concentration has significant temporal and spatial fluctuations, and plants must to adapt its growth to that variable levels. Among the main adaptations to optimize the absorption of this element whose deficiency limits its development, the modification of the architecture of the radical system stands out. Thus, in such conditions of limitations of N, it is not surprising that plants have developed mechanisms that allow them to access and absorb this element from alternative sources such as pools of organic matter resulting from the biological activity of ecosystems (Bloom, 2015; O'Brien et al. 2016; Xuan et al., 2017).

The important signaling role of glutamate (Glu) in animal systems has been known for more than 50 years, but in plants the interest about its role as a signaling molecule just started in 1998 when it was revealed that the *Arabidopsis thaliana* genome contains glutamate receptor-like (GRL) genes closely related to the ionotropic glutamate receptors (iGluR), which mediate the Glu signaling function in animals (Featherstone, 2010; Lam et al. 1998, Lacombe et al. 2001). The similarity of the effects that N inorganic and Glu have on the root architecture is indirect evidence that supports both the role of this amino acid as a source of N efficiently absorbed and metabolized by plants, as well as the existence of a plant Glu signaling pathway similar to that which operates in animals (Forde, 2013, 2014; Forde et al., 2013). In any case, despite all the efforts focused on elucidating the composition and regulation of the signaling network that controls the responses of plants to Glu, the knowledge is yet incipient in many ways and it is far from being completely understood.

Here, we summarize the information available about the signaling pathway of glutamate in plants to finish commenting some of the aspects on which this field of study should focus in the coming years.

## 14.2 GLUTAMATE: AN AMINO ACID WITH ESSENTIAL SIGNALING FUNCTIONS

Glutamic acid is one of the 20 protein-forming amino acids. It is commonly found as anion in neutral solutions, such as physiological fluids, which is usually referred to as glutamate. It is not one of the essential amino acids in humans, meaning the body can synthesize it. In mammals, Glu is a signal molecule essential for the normal development and function of the central nervous system (CNS), where it plays a major role in important activities, including cognition, learning, memory, and many other fundamental brain processes. In presynaptic cells, Glu is released from vesicles and then binds to ionotropic (ligand-gated cation channels, iGluRs) and/or metabotropic receptors (G-protein coupled, mGluR) on the plasma membrane, change the receptor conformation, and allows cations to flow through the plasma membrane, which usually leads to postsynaptic cell depolarization (Kew et al., 2005; Featherstone, 2010). The binding of Glu to some of its multiple ionotropic and metabotropic receptors initiates a cascade of signaling that culminates in cellular responses which include the normal neurotransmission of nerve impulses and the activation of proteases and caspases that lead to cell death, to mention some. These signaling pathways involve classic proteins and signaling molecules as G proteins, phospholipase C, Inositol 1,4,5-triphosphate, diacylglycerol, AMP cyclic, reactive oxygen species, calcium, and protein kinases of several types (MAPK, PKB, PKC, PDK1, TOR, IKK, CaMK), which control the expression of several genes implicated in the regulation of cell cycle, cell growth, and cell proliferation in different ways (Willard and Koochekpour, 2013).

Considering the ubiquitous presence of Glu and its receptors, it seems likely that glutamatergic systems (metabolism, release mechanisms, receptors, and transporters) are somehow involved in almost all aspects of normal brain development, function, and aging, as well as, in most neurological diseases (Danbolt, 2001). Mammalian iGluRs are subdivided into three types: R-amino-3-hydroxyl-5-methyl-4-isoxazole-propionate (AMPA) receptors, N-methyl-D-aspartic acid (NMDA) receptors, and kainate receptors (Kew et al., 2005; Featherstone, 2010). Fast information flow in the nervous system is thought to occur mostly via AMPA receptors. Kainate receptors, though found postsynaptically in some synapses, are (like mGluRs) also present on presynaptic terminal membranes, where they serve to autoregulate synaptic vesicle function and Glu secretion. NMDA receptors tend to remain open longer than AMPA or kainate receptors, allowing significant calcium influx. Thus, NMDA receptor activation can initiate intracellular calcium-dependent signaling cascades that lead to changes in gene expression and synaptic strength, which manifest for example in learning and memory (Kew et al., 2005; Featherstone, 2010). There is little or no biochemical conversion of Glu in the extracellular space, meaning that it is capable of continuously interacting with receptors until it diffuses away or is removed from the extracellular fluid by excitatory amino acid transporters (EAATs) (Danbolt, 2001; Featherstone, 2010).

Unlike the vast amount of information about the role and signaling pathway of Glu in animal systems, very little is known about the mechanism of action of this amino acid in plants. However, plants contain Glu receptors working on a variety of purposes, including light signal transduction and development (Chiu et al., 2002; Li et al., 2006). These observations strongly suggest that Glu signaling function is not exclusive for the animal kingdom, but their also indicate that a primitive signaling mechanism involving amino acids or related small molecules emerged before the divergence of plants and animals. (Chiu et al., 2002; Li et al., 2006) actually should appear that have similar downstream pathways such as Mitogen-Activated Protein Kinase (MAPK) module (Mao et al., 2004; Forde et al., 2013). Recently, evidence has emerged about the role of Glu as a signaling molecule regulating different plant development programs; however, players who mediate plant glutamate responses are unknown, and current information about it is fragmentary yet, so it is to be expected that important contributions about will be made in the coming years.

## 14.3 GLUTAMATE SOURCES AND SIGNALING IN PLANTS

### 14.3.1 Sources

Plant nutrition has important implications for plant development. The adequate acquisition of nutrients determines plant fitness and productivity; thus, the competence of plants to respond to nutrient availability is critical for their adaptation to the environment (López-Bucio et al., 2003). The soil organic nitrogen sources are very important to the plant nutrition so that several researches have been focused to study the role of recycling organic compounds, specifically amino acids, during the growth of plants (Abuarghub and Read, 1988; Dinkeloo et al., 2017; Jones and Darrah, 1994; Kielland, 1994; Näsholm et al., 2009). It has been found that Glu is one of the most represented amino acids in soil with a concentration ranging between 1 and 10 μg per gram of dry soil, although, under certain conditions these concentrations tend to increase, for example, after decomposition of dead organisms (Abuarghub and Read, 1988). Additionally, some soil microorganisms produce biofilms composed of amino acid polymers, such as poly-glutamic acid, which enriches Glu content of soil (Kubota et al., 1996; Richard and Margaritis, 2006; Zhang et al., 2017).

### 14.3.2 Plant GLRs

In *Arabidopsis* cDNAs, encoding proteins with analogies to the animal iGluRs were isolated two decades ago (Lam et al., 1998). Later, based on the annotation of the *Arabidopsis* genome, it was established that it encodes to 20 glutamate-like receptors, which were divided in three clades phylogenetically distinct: Clade I, containing 4 members (GLRs 1.1–1.4); Clade II, composed by 9 GLRs (2.1–2.9); and Clade III consisting of 7 components (GLRs 3.1–3.7) (Lacombe et al., 2001; Chiu et al., 2002). Through retro transcription coupled to polymerase chain reaction (RT–PCR) analysis, it was determined that despite the clade to which they correspond, they share common expression sites in roots, leaves, flowers, and siliques (Chiu et al., 2002). Interestingly, sixteen GLRs are highly expressed in roots, and five of the Clade III are root specific (Chiu et al., 2002). The enhanced and, in some cases, specific expression of GLRs in the root system, suggests that they perform an important role in this organ. For instance, it can be hypothesized that GLRs operate for the perception of ligands present in the rhizosphere to start a signaling pathway modulating root development. Of course, future physiological, biochemical, and molecular experiments must be conducted to verify this suggestion. More recent studies reinforce the existence of at least 20 GLRs in the *Arabidopsis* genome showing detailed expression patterns in specific tissues (Chandra Roy and Mukherjee 2016; Weiland et al., 2016). Several animals' ionotropic glutamate receptors function as ion channels with varying permeability to $Na^+$, $K^+$, and $Ca^{++}$, then considering the homologies between the GLR of plants and the iGluR of animals, the hypothesis that the functioning of these proteins could be similar in both types of organisms was well supported (Price et al., 2012). Results from experiments with chimeric proteins composed of the pore region of a GRL in the context of an iGluR made it clear that GLR can function as ionic channels for $Na^+$, $K^+$, and $Ca^{++}$ (Tapken and Hollmann, 2008). However, more recent studies have pointed out that a GLR of *Arabidopsis* (AtGLR3.4) works as an amino acids-gated ion channel that predominantly transports $Ca^{++}$ (Vincill et al., 2012). Undoubtedly, the functional and biochemical properties of the GLRs of *Arabidopsis* and other plants are only beginning to be elucidated, but the involvement of these proteins in processes such as light signaling (Brenner et al., 2000), calcium homeostasis (Kim et al., 2001), pollen tube growth (Michard et al., 2011), and root gravitropism (Miller et al., 2010) have already been reported and is generally accepted that plant GLRs work like their mammalian counterparts.

It was determined that Glu mediates $Ca^{++}$ increases in vegetal cells (Dennison and Spalding, 2000; Demidchik et al., 2004) supporting the notion that Glu could mediate $Ca^{++}$ increases through GLRs as do the GluRs in mammals. This idea was demonstrated using the GluR antagonist DNQX to abolish the $Ca^{++}$ increments associated with Glu treatments (Dubos et al., 2003). However, several works have been reported that plants' GLRs bind preferably to other amino acids instead of Glu. Comparing the predictions of protein–ligand interaction through a multiscale docking algorithm of the GLR2.9 from *A. thaliana* with the GLR from *Rattus norvegicus*, Dubos et al. (2003) established that Glu binds to the GLR binding–site from *R. norvegicus*, but not to the GLR of *Arabidopsis*. A detailed analysis of the binding site of AtGLR2.9 demonstrates that it differs in a very conserved amino acid (Threonine655/Thr655) among the mammalian GluRs. In AtGLR2.9, Thr655 residue is replaced by phenylalanine, which blocks the ligand-binding pocket for Glu. Additionally, was found that Thr655 is replaced by phenylalanine, leucine, or isoleucine in 18 of 20 *Arabidopsis* GLRs (Dubos et al., 2003). Despite these data, a report shows the induction of the $Ca^{++}$ currents and membrane depolarization by Glu through the participation the GLR3.3 (Qi et al., 2006). But alanine, asparagine, cysteine, and serine also induce membrane depolarization dependent of GLR3.3 (Qi et al., 2006; Stephens et al., 2008). All together, these data indicate that despite the predictions that plant GLRs do not bind Glu, some of them are activated and could mediate responses to Glu. Further studies should be performed to clarify which plant GLRs do respond to and which ones are unresponsive to glutamate.

Not only in *Arabidopsis* but also in other plants, different roles for GLRs during environmental and developmental responses have been reported. Plant GLRs has been involved in defense (Vatsa et al., 2011; Manzoor et al., 2013), wound (Mousavi et al., 2013), touch and cold (Meyerhoff et al., 2005), salt stress (Cheng et al., 2016), light signal transduction (Lam et al., 1998), nutrient and metals signaling (Kim et al., 2001; Kang and Turano, 2003; Sivaguru et al., 2003), and hormonal responses (Kang et al., 2004; Vatsa et al., 2011). Plant GLRs also have been found to regulate root development (Li et al., 2006; Vincill et al., 2013; Singh et al., 2015), hypocotyl elongation (Dubos et al., 2003), and stomatal movement (Kong D et al., 2016). The specific role that each plant GLR plays in each of those processes and the downstream pathway that this induces is pending to be shown in the coming years.

### 14.3.3 Signaling

Some findings indicate that Glu induce ionic gradients in plants as it occurs in mammals through activating GLRs (Dennison and Spalding, 2000; Dubos et al., 2003; Demidchik et al., 2004). Also, a report points that MAP Kinases are activated depending of calcium gradients possibly triggered by GLRs during Microbe Associated Molecular Patterns (MAMPs) and Glu treatments (Kwaaitaal et al., 2011). These evidences suggest that some Glu signaling in plants occurs in a similar fashion as it occurs in mammals through MAPK modules (Mao et al., 2004; Forde et al., 2013); however, the question if MAPK signaling pathways mediating Glu responses depends on the GLRs activity remains to be answered.

Recently, it has been revealed that in rice, Glu induces transcription factors, kinases, phosphatases, and elicitor responsive genes to amplify its signal, as well as to interact with other pathways to regulate growth, metabolism, and defense responses (Kan et al., 2017), just as it was previously reported to happen in *Arabidopsis* (Kwaaitaal et al., 2011). Experimental evidence suggests a broad role of Glu to modulate plant development; however, being that in natural conditions it is the root system that is exposed to the Glu gradients occurring in the rhizosphere, the investigations are focused on root development. In *Arabidopsis* it was reported that in response to aluminum, Glu binds a ligand-gated calcium channel and triggers calcium influx that induces depolimerization of microtubules and depolarization of the plasma membrane to inhibit root elongation (Sivaguru et al., 2003). Besides, Glu treatments alter root architecture, inhibiting primary roots through reducing meristem mitotic activity and inducing lateral root proliferation (Walch-Liu et al., 2006). In mammals, MAPKs are important players of the responses to Glu (Willard and Koochekpour, 2013), and in *Arabidopsis* the Glu effects also are mediated by MAPKs (Forde et al., 2013). A few years ago it was evidenced that Glu's effects on root development are mediated by the MEKK1 (a MAP3K), but surprisingly, it was reported that the kinase activity of MEKK1 is not required for glutamate signaling, it being proposed that this protein could have a scaffolding activity for another MAPKs or, given its known skill to bind to DNA, even to function directly as transcriptional modulator (Forde et al. 2013, Miao et al., 2007; Suarez-Rodriguez et al. 2007). The exact function that MEKK1 plays in the glutamate signaling pathway is awaiting to be determined, but it is a fact that the positioning of this type of enzyme in this signaling pathway attracted the attention of others to explore the possible participation of MAPKs previously associated with development, such as is the case of the root architecture (Forde 2014; López-Bucio et al., 2014). In this regard, MPK6 and the dual specificity serine-threonine-tyrosine phosphatase (MKP1) of *Arabidopsis* have already been involved as active members of the Glu signaling pathway modulating root responses to Glu in plants through a mechanism in which the auxin seems to be involved (Figure 14.1) (López-Bucio et al., 2018). Protein kinase-mediated phosphorylation is considered the most important post-translational modification to impact protein activity, however, is really the reversible protein phosphorylation catalyzed by kinases and phosphatases, which provides the regulatory network controlling most biological process. In that scenario, the identification of the phosphatase MKP1 as a Glu signaling pathway component in plants, results particularly interesting (Bartels et al. 2010). On the other hand, as the MAPK kinases act through a module composed of three kinases (MAP3K, MAP2K and MAPK), which are activated sequentially (Suarez-Rodriguez et al., 2010), the guiding logic of this line of research should focus on the identification of MAP2K and MAP3K acting upstream of MPK6. The characterization of the responses evoked by Glu in mutants for MAP2K and MAP3K coded in the *Arabidopsis* genome would be particularly useful for such purposes. Simultaneously, efforts should be focused to identify the possible targets of the MAPK cascade operating in the plant Glu signaling pathway. In this sense, the characterization of mutants showing similar phenotypes to those of the effects of Glu could be considered among the first candidates to be analyzed.

Cytosolic $Ca^{++}$ serves as a universal second messenger in a myriad of cellular activities. Particularly in plant cells, it plays a key role in signal transduction networks eliciting responses to biotic and abiotic signals (Ranty et al. 2016). As already mentioned, plant GLRs have been implicated in the regulation of several $Ca^{++}$-mediated developmental and physiological responses, and these observations place $Ca^{++}$ and Reactive Oxygen Species (ROS) as components of the Glu signaling pathway in plants.

A graphic summary of different elements hitherto involved in the Glu signaling pathway in relation to their effects on the architecture of the root is shown in Figure 14.2. The specific contribution of each of them, their functional interrelations, and the addition of new elements is something that should be deciphered in the coming years.

## 14.4 FUTURE PROSPECTS

The signaling function that glutamate has in plants is out of any discussion; however, it is far from being completed or even reaching the levels that have been achieved in animals, particularly in mammals. However, perhaps motivated by the clear effects that the Glu has on the architecture of the root that could well be reflected in the plants productivity (Forde, 2013), together with the fact that it has been proposed that this (and other) amino acids could represent an alternative source of N in low-fertile soils of temperate regions (Rothstein, 2009; Tegeder and Masclaux-Daubressel, 2018), several plant research groups have begun to take interest in this topic, and the results of current investigations are promising. Trying to identify other components of the Glu signaling pathway in plants, as well as its crosstalk with other factors affecting the root development, such as phosphorous, auxin, ABA (Abscisic Acid), even the nitrate itself, seem the guidelines of this interesting field of research. The phenotypic, molecular, and functional characterization of single and multiple mutants for the components of the Glu signaling pathway, both of those already identified and those that are being identified in ongoing investigations, complemented with the determination and comparison of the transcriptomes, proteomes, even ionomes of wild-type plants and mutant exposed to glutamate treatments would also be very useful for the identification of the targets of

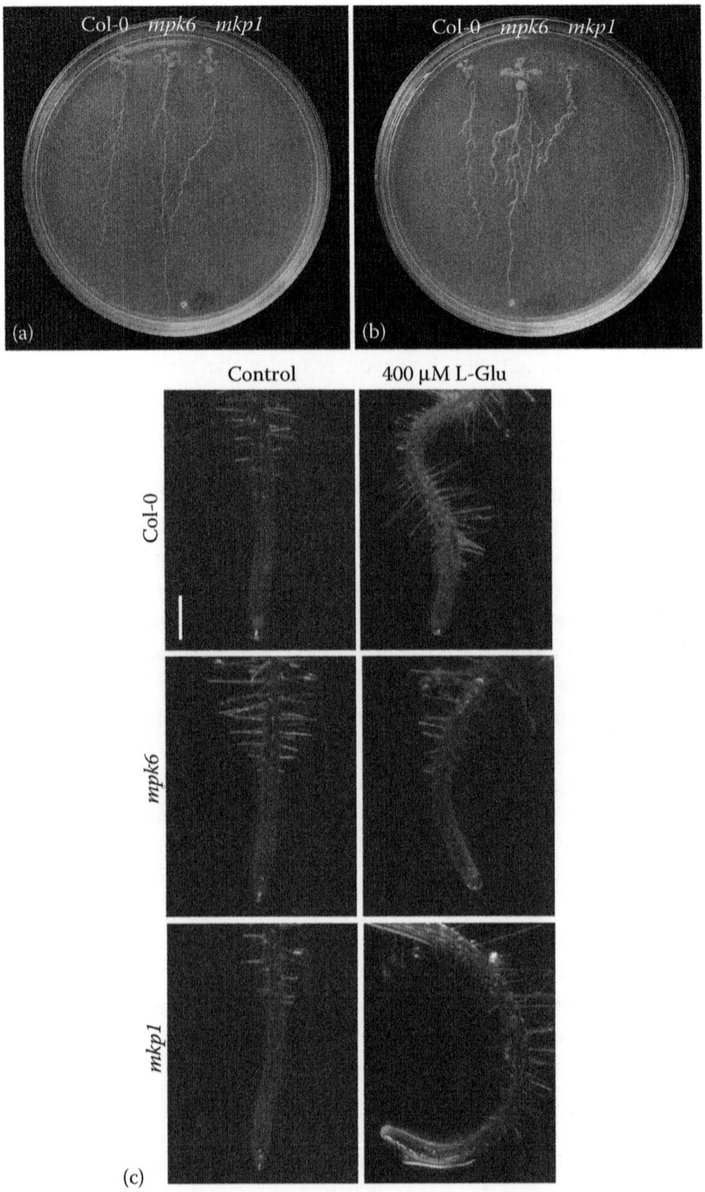


**FIGURE 14.1** *mkp6* and *mkp1* mutants responses to glutamate. One of the main effects of glutamate is the inhibition of the primary root growth. Interestingly, mutants of *Arabidopsis* for MPK6 kinase (*mpk6*) and MKP1 phosphatase (*mkp1*) manifest insensitivity and hypersensitivity phenotypes to glutamate treatment, respectively. The images show wild-type (Col-0) and mutant *mpk6* and *mkp1* seedlings grown side by side under control conditions (a) and in the presence of 400 μM glutamate (b), where the phenotypes of insensitivity (*mpk6*) and hypersensitivity (*mkp1*) to treatment with glutamate are evident. Additionally, a zoom of the root tips from seedlings grown under control conditions and in glutamate treatment are shown (c). Note that the curvature of the root tip caused by glutamate in wild-type (Col-0) plants (upper right panel) is barely perceptible in the *mpk6* mutant (middle right panel), while it is exacerbated in the *mkp1* mutant (lower right panel). The scale in c is 200 μm. Many questions remain to be answered in this regard, but these observations strongly suggest the participation of both enzymes in the glutamate signaling pathway.

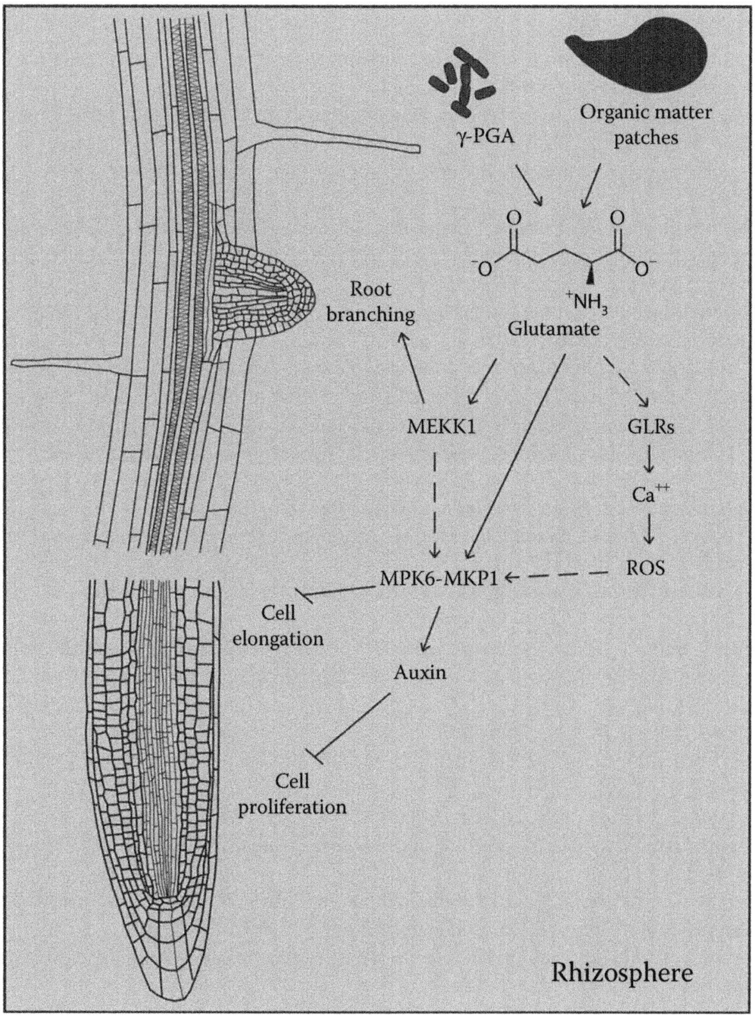


**FIGURE 14.2** Signaling of glutamate in root development. Specific environmental conditions, such as decomposition of organic matter or production of biofilms by microorganisms, increase the concentration and availability of glutamate in the rhizosphere. Glutamate regulates several root system processes including inhibition of primary root growth through a signaling pathway involving receptors (GLR) analogous to animal inotropic glutamate receptors, which function as ligand-gated channels that control the flow of $Ca^{++}$ that impacts the production of reactive oxygen species (ROS). The signal triggered by glutamate that affects elongation and cell proliferation in the primary root involves the participation of the MPK6 kinase and the MKP1 phosphatase, which apparently modulate the transport and/or response to auxins. Additionally, glutamate regulates the growth of the primary root and the formation of lateral roots through a MAP3K (MEKK1) in a process independent of the kinase activity of this enzyme. The cross talk, if any, between these glutamate signaling components and still others to be discovered, remains to be revealed.

these signaling pathways, and they should generate significant advances in this field of study in the next few years. As in any other field of research at present, the analysis and integration of the expected huge amount of information to be generated by these omics studies will be an arduous task that should be approached in an interdisciplinary manner that favors its correct interpretation and eventual application for plant breeding purposes.

## REFERENCES

Abuarghub SM, Read DJ. Quantitative analysis of individual "free" amino acids in relation to time and depth in the soil profile. *New Phytol.* 1988; 108: 433–441.

Bartels S, González Besteiro MA, Lang D, Ulm R. Emerging functions for plant MAP kinase phosphatases. *Trends Plant Sci.* 2010; 15: 322–329.

Bloom A. The increasing importance of distinguish among plant nitrogen sources. *Curr. Opin. Plant Biol.* 2015; 25: 10–16.

Brenner ED, Martinez-Barboza N, Clark AP, Liang QS, Stevenson DW, and Coruzzi GM. Arabidopsis mutants resistant to S(+)-β-methyl-α,β-diaminopropionic acid, a cycad-derived glutamate receptor agonist. *Plant Physiol.* 2000; 124: 1615–1624.

Chandra Roy B, Mukherjee A. Computational analysis of the glutamate receptor gene family of *Arabidopsis thaliana. J. Biomol. Struct. Dyn.* 2016; 35: 2454–2474.

Cheng Y, Tian Q, Zhang WH. Glutamate receptors are involved in mitigating affects of amino acids on seed germination of *Arabidopsis thaliana* under salt stress. *Environ. Exp. Bot.* 2016; 130: 68–78.

Chiu JC, Brenner ED, DeSalle R, Nitabach MN, Holmes TC, Coruzzi GM. Phylogenetic and expression analysis of the Glutamate-Receptor-Like gene family in *Arabidopsis thaliana. Mol. Biol. Evol.* 2002; 19: 1066–1082.

Danbolt NC. Glutamate uptake. *Prog. Neurobiol.* 2001; 65: 1–105.

Demidchik V, Essah PA, Tester M. Glutamate activates cation currents in the plasma membrane of *Arabidopsis* root cells. *Planta* 2004; 219: 167–175.

Dennison KL, Spalding EP. Glutamate-gated calcium fluxes in *Arabidopsis. Plant Physiol.* 2000; 124: 1511–1514.

Dinkeloo K, Boyd S, Pilot G. Update on amino acid transporter functions and on possible amino acid sensing mechanisms in plants. *Semin. Cell. Dev. Biol.* 2017. doi:10.1016/j.semcdb.2017.07.010.

Dubos C, Huggins D, Grant GH, Knight MR, Campbell MM. Role for glycine in the gating of plant NMDA-like receptors. *Plant J.* 2003; 35: 800–810.

Featherstone DE. Intercellular glutamate signaling in the nervous system and beyond. *ACS Chem. Neurosci.* 2010; 1: 4–12.

Forde BG. Glutamate signaling in roots. *J. Exp. Bot.* 2013; 65: 779–787.

Forde BG. Nitrogen signalling pathways shaping root system architecture: An update. *Curr. Opin. Plant Biol.* 2014; 21: 30–36.

Forde BG, Cutler SR, Zaman N, Krysan PJ. Glutamate signalling via a MEKK1 kinase-dependent pathway induces changes in *Arabidopsis* root architecture. *Plant J.* 2013; 75: 1–10.

Jones DL, Darrah PR. Amino-acid influx at the soil-root interface of *Zea mays* L. and its implications in the rhizosphere. *Plant Soil* 1994; 163: 1–12.

Kan CC, Chung TY, Wu HY, Juo YA, Hsieh MH. Exogenus glutamate rapidly induce the expression of genes involved in metabolism and defence responses in rice roots. *BMC Genomics* 2017; 18: 186.

Kang J, Metha S, Turano FJ. The putative glutamate receptor 1.1 (AtGLR1.1) in *Arabidopsis thaliana* regulates abscisic acid biosynthesis and signalling to control development and water loss. *Plant Cell Physiol.* 2004; 10: 1380–1390.

Kang J, Turano FJ. The putative glutamate receptor 1.1 (AtGLR1.1) functions as a regulator of carbon and nitrogen metabolism in *Arabidopsis thaliana. Proceedings of the National Academy of Sciences of the United States of America* 2003; 100: 6872–6877.

Kew JNC, Kemp JA. Ionotropic and metabotropic glutamate receptor structure and pharmacology. *Psychopharmacology* 2005; 179: 4–29.

Kielland K. Amino acid absorption by artic plants: Implications for plant nutrition and nitrogen cycling. *Ecology* 1994; 75: 2373–2383.

Kim SA, Kwak JM, Jae SK, Wang MH, Nam HG. Overexpression of the AtGluR2 gene encoding an Arabidopsis homolog of mammalian glutamate receptors impairs calcium utilization and sensitivity to ionic stress in transgenic plants. *Plant Cell Physiol.* 2001; 42: 74–84.

Kong D, Hu, HC, Okuma E, Lee Y, Lee HS, Munemasa S, Cho D et al. L-Met activates *Arabidopsis* GLR $Ca^{2+}$ channels upstream of ROS production and regulates stomatal movement. *Cell Rep.* 2016; 17: 2553–2561.

Kubota H, Nambu Y, Endo T. Alkaline hydrolysis of poly (γ-glutamic acid) produced by microorganism. *J. Polym. Sci. Polym. Chem.* 1996; 34: 1347–1351.

Kwaaitaal M, Huisman R, Maintz J, Reinstädler A, Panstruga R. Ionotropic glutamate receptor (iGluR)-like channels mediates MAMP-induced calcium influx in *Arabidopsis thaliana. Biochem. J.* 2011; 440: 355–365.

Lacombe B, Becker D, Hedrich R, DeSalle R, Hollmann M, Kwak JM, Schroeder JI et al. The identity of plant glutamate receptors. *Science* 2001; 292: 1486–1487.
Lam HM, Chiu J, Hsieh MH, Meisel L, Oliveira IC, Shin M, Coruzzi G. Glutamate-receptor genes in plants. *Nature* 1998; 396: 125–126.
Li J, Zhu S, Song X, Shen Y, Chen H, Yu J, Yi K et al. A rice glutamate receptor-like gene is critical for the division and survival of individual cells in the root apical meristem. *Plant Cell* 2006; 18: 340–349.
López-Bucio J, Cruz-Ramírez A, Herrera-Estrella L. The role of nutrient availability in regulating root architecture. *Curr. Opin. Plant Biol.* 2003; 6: 280–287.
López-Bucio JS, Dubrovsky JG, Raya-González J, Ugartechea-Chirino Y, López-Bucio J, de Luna-Valdés LA, Ramos-Vega M, León P, Guevara-García AA. *Arabidopsis thaliana* mitogen-activated protein kinase 6 is involved in seed formation and modulation of primary and lateral root development. *J. Exp. Bot.* 2014; 65: 169–183.
López-Bucio JS, Raya-González J, Ravelo-Ortega G, Ruiz-Herrera LF, Ramos-Vega M, León P, López-Bucio J, Guevara-García, AA. Mitogen activated protein kinase 6 and MAP kinase phosphatase 1 are involved in the response of Arabidopsis roots to L-glutamate. *Plant Mol. Biol.* 2018; 96: 339–351.
Manzoor H, Kelloniemi J, Chiltz A, Wendehene D, Pugin A, Poinssot B, García-Brugger A. Involvement of the glutamate receptor atGLR3.3 in plant defence signalling and resistance to *Hyaloperonspora arabidopsisdis. Plant J.* 2013; 76: 466–480.
Mao L, Tang Q, Samsani S, Liu Z, Wang JQ. Regulation of MAPK/ERK phosphorylation via ionotropic glutamate receptors in cultured rat striatal neurons. *Eur. J. Neurosci.* 2004; 19: 1207–1216.
Meyerhoff O, Müller K, Roelfsema MRG, Latz A, Lacombe B, Hedrich R, Dietrich P, Becker D. AtGLR3.4, a glutamate receptor channel-like gene is sensitive to touch and cold. *Planta* 2005; 222: 418–427.
Miao Y, Laun TM, Smykowski A, Zentgraf U. Arabidopsis MEKK1 can take a short cut: It can directly interact with senescencerelated WRKY53 transcription factor on the protein level and can bind to its promoter. *Plant Mol. Biol.* 2007; 65: 63–76.
Michard E, Lima PT, Borges F, Silva AC, Portes MT, Carvalho JE, Gilliham M, Liu LH, Obermeyer G, and Feijó JA. Glutamate receptor-like genes form $Ca^{2+}$ channels in pollen tubes and are regulated by pistil D-serine. *Science* 2011; 332: 434–437.
Miller ND, Durham Brooks TL, Assadi AH, Spalding EP. Detection of a gravitropism phenotype in glutamate receptor-like 3.3 mutants of Arabidopsis thaliana using machine vision and computation. *Genetics* 2010; 186: 585–593.
Mousavi SAR, Chauvin A, Pascuad F, Kellenberger S, Farmer EE. Glutamate receptor-like genes mediate leaf-to-leaf wound signalling. *Nature* 2013; 500: 422–426.
Näsholm T, Kiellan K, Ganeteg U. Uptake of organic nitrogen by plants. *New Phytol.* 2009; 182: 31–48.
O'Brien JA, Vega A, Bouuyon E, Krous G, Gojon A, Coruzzi G, Gutierrez RA. Nitrate transport, sensing, and responses in plants. *Mol. Plant* 2016; 9: 837–856.
Price MB, Jelesko J, Okumoto S. Glutamate receptor homologs in plants: Functions and evolutionary origins. *Front. Plant Sci.* 2012; 3: 235.
Qi Z, Stephens NR, Spalding EP. Calcium entry mediated by GLR3.3, an *Arabidopsis* glutamate receptor with a broad agonist profile. *Plant Physiol.* 2006; 142: 963–971.
Ranty B, Aldon D, Cotelle V, Galaud JP, Thuleau P, Mazars C. Calcium sensors as key hubs in plant responses to biotic and abiotic stresses. *Front. Plant Sci.* 2016; 7: 327.
Richard A, Margaritis A. Kinetics of molecular weight reduction of poly (glutamic acid) by in situ depolymerisation in cell-free broth of *Bacillus subtilis. Biochem. Eng. J.* 2006; 30: 303–307.
Rothstein DE. Soil amino-acid availability across a temperate-forest fertility gradient. *Biogeochemistry* 2009; 92: 201–215.
Singh SK, Chien CT, Chang IF. The *Arabidopsis* glutamate receptor-like gene GLR3.6 controls root development by repressing the Kip-related protein gene KRP4. *J. Exp. Bot.* 2015; 67: 1853–1869.
Sivaguru M, Pike S, Gassmann W, Baskin TI. Aluminum rapidly depolymerizes cortical microtubules and depolarizes plasma membrane: Evidence that these responses are mediated by a glutamate receptor. *Plant Cell Physiol.* 2003; 44: 667–675.
Stephens NR, Qi Z, Spalding EP. Glutamate receptor subtypes evidenced by differences in desensitization and dependence on the *GLR3.3* and *GLR3.4* genes. *Plant Physiol.* 2008; 146: 529–538.
Suarez-Rodriguez MC, Adams-Phillips L, Liu YD, Wang HC, Su SH, Jester PJ, Zhang SQ, Bent AF, Krysan PJ. MEKK1 is required for flg22-induced MPK4 activation in *Arabidopsis* plants. *Plant Physiol.* 2007; 143: 661–669.
Suarez-Rodriguez MC, Petersen M, Mundy J. Mitogen-activated protein kinase signaling in plants. *Annu. Rev. Plant Biol.* 2010; 61: 621–649.

Tapken D, Hollmann M. *Arabidopsis* thaliana glutamate receptor ion channel function demonstrated by ion pore transplantation. *J. Mol. Biol.* 2008; 383: 36–48.

Tegeder M, Masclaux-Daubresse C. Source and sink mechanism of nitrogen transport and use. *New Phytol.* 2018; 217:35–53.

Vatsa P, Chiltz A, Bourque S, Wendehenne D, García-Brugger A, Pugin A. Involvement of putative glutamate receptor in plant defence signalling and NO production. *Biochimie* 2011; 93: 2095–2101.

Vincill ED, Bieck AM, Spalding EP. $Ca^{2+}$ conduction by an amino acid-gated ion channel related to glutamate receptors. *Plant Physiol.* 2012; 159: 40–46.

Vincil ED, Clarin AE, Molenda JN, Spalding EP. Interacting glutamate Receptor-Like proteins regulate lateral root initiation in *Arabidopsis*. *Plant Cell* 2013; 25: 1304–1313.

Walch-Liu P, Liu LH, Remans T, Tester M, Forde BG. Evidence that L-glutamate can act as an exogenous signal to modulate root growth and branching in *Arabidopsis thaliana.Plant Cell Physiol.* 2006; 47:1045–1057.

Weiland M, Mancuso S, Baluska F. Signaling via glutamate and GLRs in *Arabidopsis*. *Funct. Plant Biol.* 2016; 43: 1–25.

Willard SS, Koochekpour S. Glutamate, glutamate receptors, and downstream signaling pathways. *Int. J. Biol. Sci.* 2013; 2009: 948–959.

Xuan W, Beeckman T, Xu G. Plant Nitrogen nutrition: Sensing and signaling. *Curr. Opin. Plant Biol.* 2017; 39: 57–65.

Zhang L, Yang X, Gao D, Wang L, Li J, Wei Z, Shi Y. Affect of Poly-γ-glutamic acid (γ-PGA) on plant growth and its distribution in a controlled plant soil system. *Sci Rep.* 2017; 7: 6090.

# *Section IV*

## *Role of Neurotransmitters in Relationships of Organisms in Biocenosis*

# 15 Role of Acetylcholine System in Allelopathy of Plants

*Rashmi Sharma and Rajendra Gupta*

## CONTENTS



## 15.1 INTRODUCTION

Acetylcholine (ACh) is a main neurotransmitter in the brain, but its non-neuronal role has long been established in animals as well as plants (Wesseler et al., 2001; Wesseler and Kirkpatrick, 2008; Roshchina, 2010; Takahashi, 2015). Many components of cholinergic system like ACh and acetylcholinesterase (AChE), Cholineacetyl transferase, have long been reported in plants (Smallman et al., 1981; Tretyn et al., 1986; Hartmann and Gupta, 1989; Gupta and Gupta, 1997; Gupta et al., 2001; Roshchina, 2001; Horiuchi et al., 2003). Plants are also rich in anticholinesterases (antiChEs), chemicals that inhibit AChE, thus leading to an accumulation of ACh in the system (Mukherjee et al., 2007; Calderon, 2013; Yusoff et al., 2014; Suganthy and Devi, 2016; Tundis et al., 2016, Penumala et al. 2017 ). Anticholinesterase compounds isolated from plants are summarized here (Table 15.2). These antiChEs are the plants' defense against herbivory (Sharma and Gupta, 2007; Mithöfer and Boland, 2012). They are also part of the plants' repertoire of toxic compounds (allelochemicals) that they release to influence the growth, survival, development, and reproduction of other organisms in their vicinity, a phenomenon known as allelopathy (Rizvi et al., 1992; Rice, 1983; Cheng and Cheng, 2015). The allelochemicals present in the plants help them in becoming invasive weeds outside their place of origin (Dhima et al., 2016). The biochemical mechanism of action of many of these allelochemicals has been studied (Einhellig, 1994; Wink et al., 1998; Golovko, 1999; Hejl and Koster, 2004; Rollinger et al., 2004). The mode of action of many allelochemicals is well-known (Chen et al., 2016; Liu et al., 2016; Yan et al., 2016). Some allelochemicals interfere with the cholinergic signal transmission in the plants (Table 15.1). Sharma (2007) was first to suggest disruption of the cholinergic pathway as mode of action of allelochemicals.

**TABLE 15.1**
**Biochemical Targets of Allelochemicals and Their Cholinergic Response**

| Category | Compound | Mode of Allelopathic Action in Plants | References | Remarks on Acetylcholine System in Animals |
|---|---|---|---|---|
| **ALKALOIDS** | | | | |
| Indole Alkaloids | Ergometrine | DNA, RT | Wink et al. (1998) | Inhibits BuChE; agonist of mAChR, nAChR |
| | Gramine | Damages cell wall | Einhellig (1994); Wink et al. (1998) | Inhibits BuChE; agonist of mAChR, nAChR |
| | Harmaline | PrtSyn, DNA, Pol I, RT | Wink et al. (1998) | Inhibits BuChE, AChE; agonist of mAChR |
| Isoquinoline Alkaloids | Berberine | PrtSyn, DNA, Pol I, RT | Wink et al. (1998) | Inhibits BuChE, AChE; agonist of mAChR, nAChR |
| | Sanguinarine | PrtSyn, DNA, Pol I, RT membrane leakage | Wink et al. (1998) | Inhibits AChE, BChE, ChAT; agonist of mAChR, nAChR |
| Purine Alkaloid | Caffeine | Unknown | Wink et al. (1998) | Agonist of mAChR |
| Quinoline Alkaloids | Cinchonine | PrtSyn, DNA, Pol I, RT | Wink et al. (1998) | Inhibits BuChE; agonist of mAChR |
| | Cinchonidine | PrtSyn, DNA, Pol I, RT | Wink et al. (1998) | Agonist of mAChR |
| | Quinidine | PrtSyn, DNA, Pol I, RT | Wink et al. (1998 | Agonist of mAChR |
| | Quinine | PrtSyn, DNA, Pol I, RT | Wink et al. (1998) | Inhibits BuChE; agonist of mAChR |
| Quinolizidine Alkaloids | Cystisine | Unknown | Wink et al. (1998) | Agonist of mAChR, nAChR |
| | Sparteine | Unknown | Wink et al. (1998) | Inhibits BuChE; agonist of mAChR, nAChR |
| Terpene Alkaloids | Aconitine | Unknown | Wink et al. (1998) | Agonist of mAChR |
| Tropane Alkaloids | Hyoscyamine | Unknown | Wink et al. (1998) | Agonist of mAChR |
| | Scopolamine | Unknown | Golovko (1999) | Agonist of mAChR, nAChR |
| Tropolone Alkaloids | Colchicine | Inhibits reverse transcriptase | Wink et al. (1998) | Agonist of mAChR, nAChR |
| **COUMARINS** | | | | |
| | Scopoletin | Decreases mitosis | Einhellig (1994); Golovko (1999); Rollinger et al. (2004) | Inhibits AChE |
| **FLAVANOIDS** | | | | |
| Flavanoid (Flavan-3-ols) | Naringenin | Affects mitochondrial respiration | Einhellig (1994); Heo et al. (2004) | Inhibits AChE |
| Quinone | Juglone | Inhibits redox reaction; free radical damage, affects mitochondrial respiration | | Inhibits ChAT |
| Monoterpene | Cineole | Reduces cell division | Einhellig (1994); Perry et al. (2000) | Inhibits AChE |

## 15.2 NON-NEURONAL ROLE OF ACETYLCHOLINE

Acetylcholine, the molecule primarily identified as a neurotransmitter in mammals, is playing a diverse role in mammals as well as other organisms. In mammals, all the components of the cholinergic system have been found in the epithelial, mesothelial, and immune cells (Sastry and Sadavongvivad, 1978; Wesseler and Kirkpartick, 2008). Wesseler and Kirkpartick (2017) showed that non-neuronal ACh is involved in reproduction in both mammals and non-mammals. Kawashima and Fujii (2003) observed involvement of non-neuronal ACh in the regulation of immune function in lymphocytes. In lungs, ACh participates in bronchoconstriction, activates immune cells, and stimulates mucous secretion and cytokine release (Kummer and Lips, 2006). Yoshida et al. (2006) found ACh in bladder epithelium and implicated it in bladder physiology. Yamamoto et al. (2015) have implicated AChE as a positive regulator of shoot gravitropic response in plants. Bamel et al. (2015) suggested ACh as a natural regulator of morphogenesis in tomatoes. Roshchina (2001) has given a detailed account of the role of acetylcholine in plants.

## 15.3 WEEDS ARE RICH IN ANTICHOLINESTERASES

Weeds are rich in secondary metabolites that help them proliferate vigorously. We undertook an extensive study to look for antiChEs in weeds. Leaves and roots/tubers of the weeds were tested for the presence of antiChEs. Most weeds tested were found to inhibit AChE from both animal (eel AChE) and plants (tomato and wheat AChE). Data is presented in Figure 15.1. Of the methanolic extract of roots of 45 weeds tested, out of which 12 showed 100% and 27 showed 50% inhibition of the enzyme from animal source (Figure 15.1a), while methanolic extract of leaves of 30 weeds showed 50% inhibition and 17 weeds showed 100% inhibition of enzyme from the same source (Figure 15.1d). Root extract of 17 weeds tested showed more than 50% inhibition of AChE from tomato leaves (Figure 15.1b). Interestingly, *Croton bonplandianum*, *Desmodium gangeticum*, *Euphorbia geniculata*, *Heliotropium supinum*, *Polygonum plebium*, and *Rumex dentatus* showed promotion of the enzyme. Similarly, leaf extract of 25 weeds caused 50% inhibition of AChE from tomato leaves (Figure 15.1e). Some weeds could completely inhibit the enzyme, and few, like *Ageratum conyzoides*, *Lantana camara*, and *Polygonum fagopyrum*, had a promotable effect on the tomato AChE. Anticholinesterases from roots of 14 weeds showed 50% inhibition of AChE from wheat. Root extract from weeds like *Bidens biternata*, *Coronopus didymus*, *Prosopis juliflora*, *Ricinus communis*, *Verbascum chinense*, *Veronica agrestis* completely inhibited the enzyme (Figure 15.1c). Leaf extract of 75% of these weeds inhibited the wheat AChE by more than 50%, and 15 weeds caused complete inhibition of the enzyme (Figure 15.1f). Sharma and Gupta (2007) showed that *C. rotundus* inhibited AChE from animal and plant sources and also inhibited germination and seedling growth in wheat and tomato.

Table 15.2 summarizes anticholinesterases already known from other plants. Many antiChEs from plants listed in the table, like galanthamine (Turi et al., 2014), juglone (Hejl and Koster 2004), and α-pinene, 1,8, cineole (Savelev et al., 2003), are established allelochemicals.

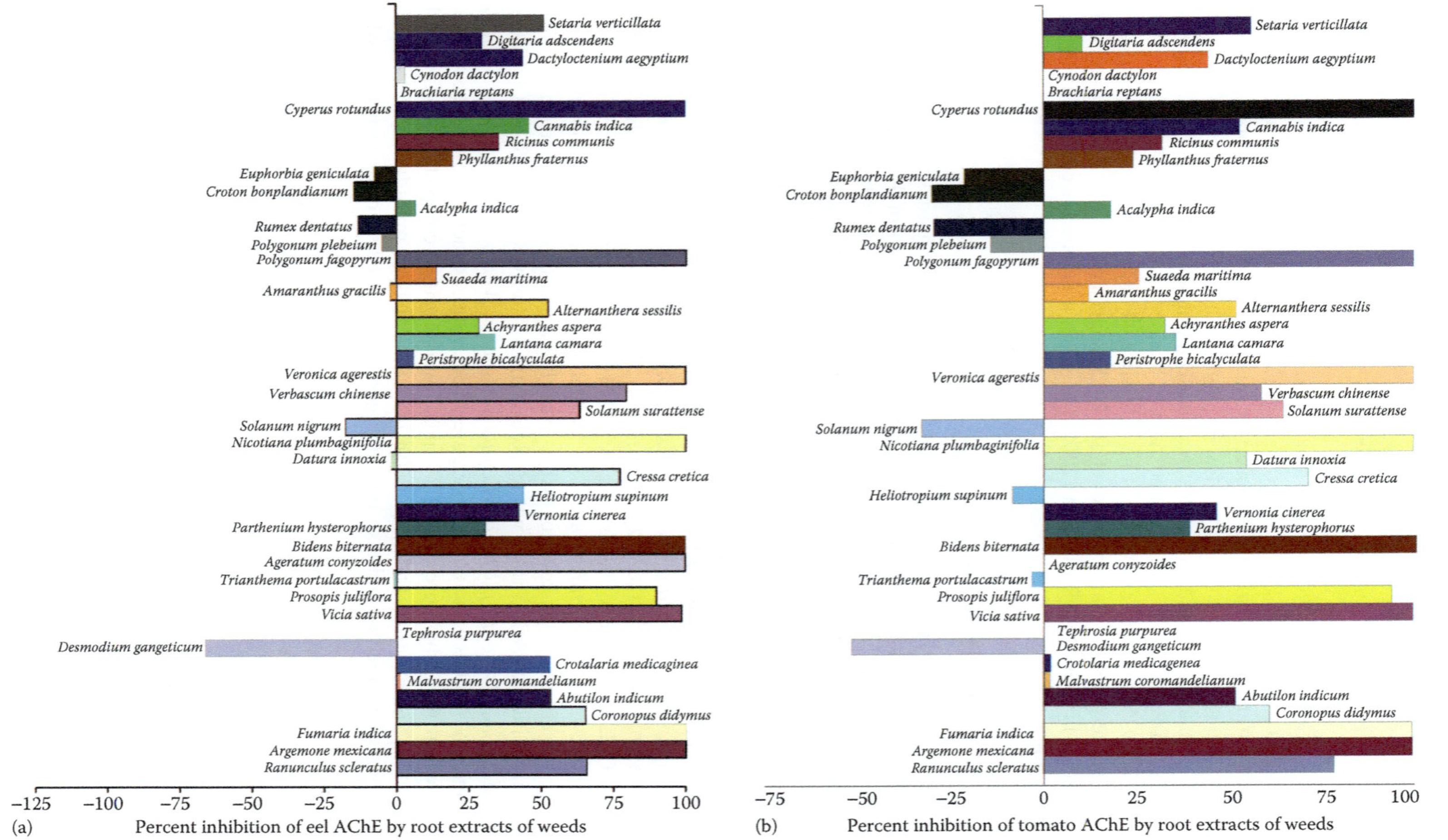


**FIGURE 15.1** Effect of methanolic root and leaf extract of 45 weeds on AChE from eel (a and d), tomato (b and e) and wheat (c and f). The data represents average of three experiments. (*Continued*)

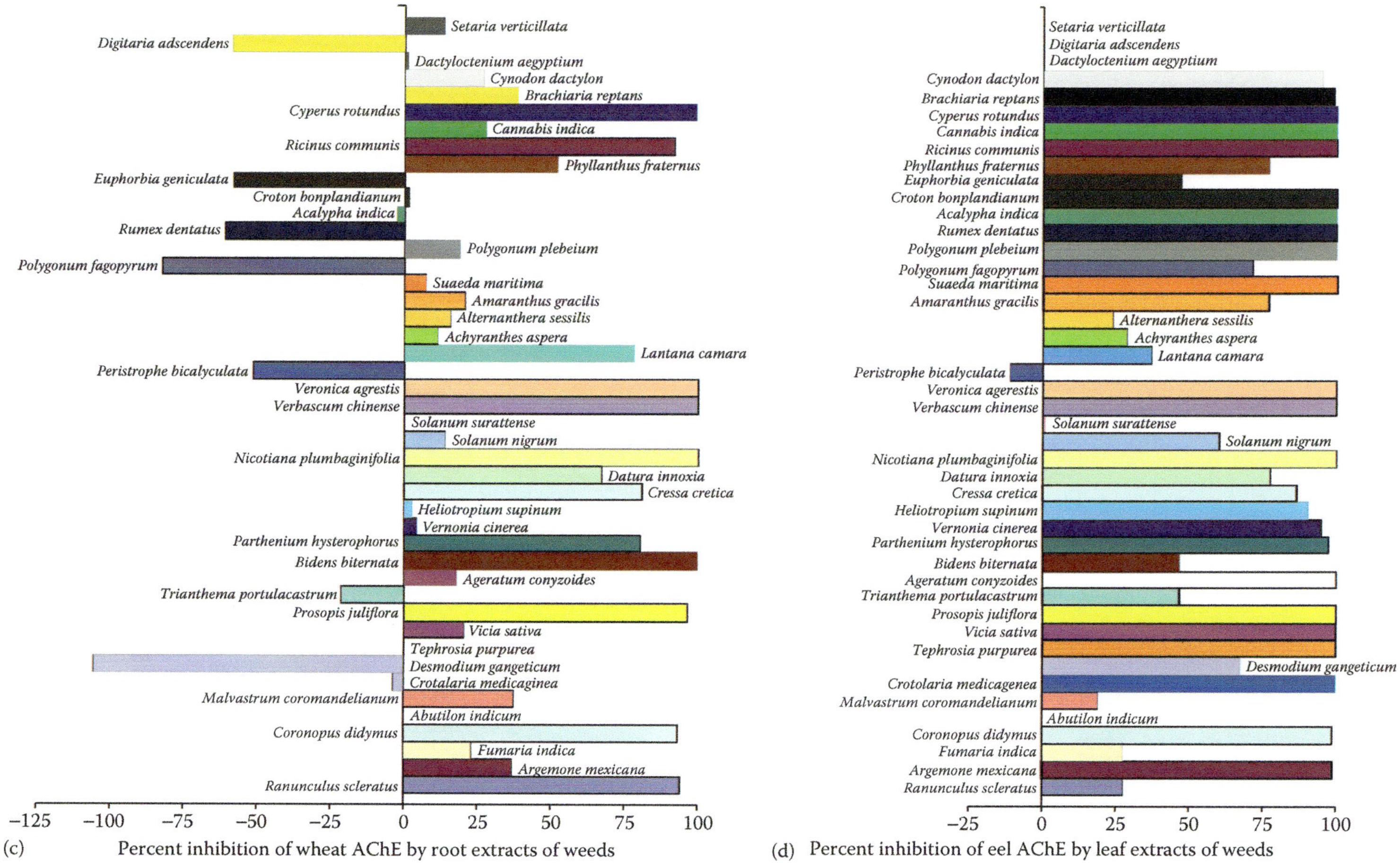


**FIGURE 15.1 (Continued)** Effect of methanolic root and leaf extract of 45 weeds on AChE from eel (a and d), tomato (b and e) and wheat (c and f). The data represents average of three experiments.

*(Continued)*

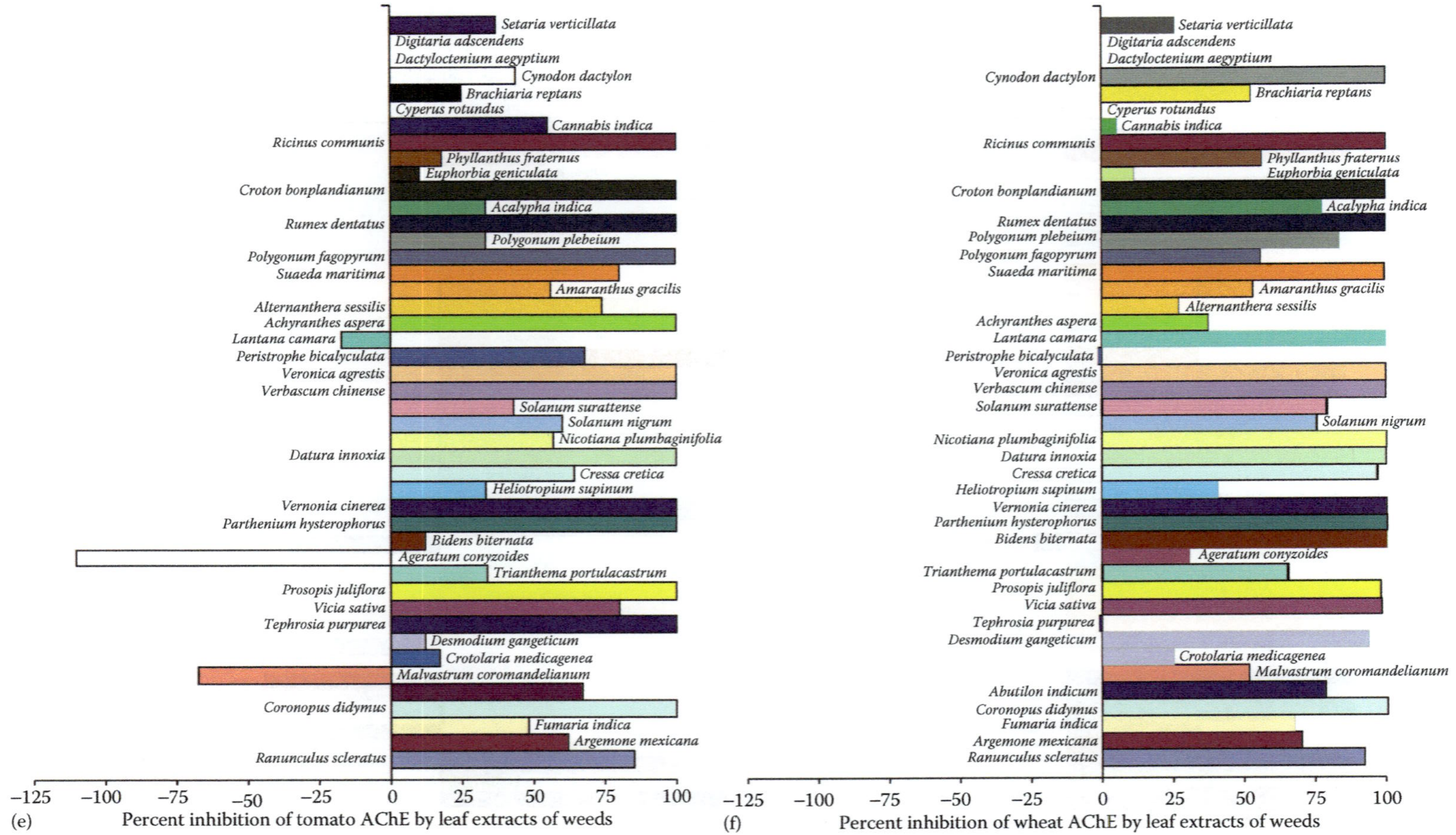


**FIGURE 15.1 (Continued)** Effect of methanolic root and leaf extract of 45 weeds on AChE from eel (a and d), wheat (b and e) and wheat (c and f). The data represents average of three experiments.

**TABLE 15.2**
**Compounds from Plants Showing Anticholinesterase Activity**

| *Plant* | Type of Compound | Compound | $IC_{50}$ (µg/ml) | Kinetics | Type of ChE | Reference |
|---|---|---|---|---|---|---|
| **ALGAE** | | | | | | |
| **Sargassaceae** | | | | | | |
| *Sargassum sagamianum* | Farnesylacetone | (5E,10Z)-6,10,14-trimethylpentadeca-5,10-dien-2,12-dione | | — | ?AChE | Ryu (2003) |
| | | (5E,9E,13E)-6,10,4-trimethyl-pentadeca-5,9,13-trien-2,12-dione | | — | ?AChE | Ryu (2003) |
| | | (5E,10Z)-6,10,14-trimethylpentadeca-5,10-dien-2,12-dione | | — | ?BuChE | Ryu (2003) |
| | | (5E,9E,13E)-6,10,4-trimethyl-pentadeca-5,9,13-trien-2,12-dione | | — | ?BuChE | Ryu (2003) |
| *Sargassum siliquastrum* | Chromene | Sargachromanol G | $82.7^{z}$ | — | ?BuChE | Jang et al. (2005) |
| | | Sargachromanol O | $80.0^{z}$ | — | ?BuChE | Jang et al. (2005) |
| **FUNGI** | | | | | | |
| *Aspergillus flavus* | Polyketide | Aflatoxin $B_1$ | $31.6^{x}$ | NC | mbAChE | Cometa et al. (2005) |
| *Aspergillus terreus* | Lactone | Isoterreulactone A | $2.5^{x}$ | NC | Eel AChE | Yoo et al. (2005) |
| | | Terreulactone A | $0.23^{x}$ | — | ?AChE | Kim et al. (2003) |
| | | Terreulactone B | $0.09^{x}$ | — | ?AChE | Kim et al. (2003) |
| | | Terreulactone C | $0.06^{x}$ | — | ?AChE | Kim et al. (2003) |
| | | Terreulactone D | $0.42^{x}$ | — | ?AChE | Kim et al. (2003) |
| | | Arisugasin A | $1.0^{x}$ | — | ?AChE | Otoguro et al. (1997) |
| | | Arisugasin B | $25.8^{x}$ | — | ?AChE | Otoguro et al. (1997) |
| | | Arisugasin C | $2.5^{x}$ | — | ?AChE | Otoguro et al. (1997) |
| | | Arisugasin D | $3.5^{x}$ | — | ?AChE | Otoguro et al. (1997) |
| *Microsphaeropsis olivacea* | Lactone | Graphislactone A | $27^{x}$ | — | ?AChE | Hormazabal et al. (2005) |
| | | Botrallin | $19^{x}$ | — | ?AChE | Hormazabal et al. (2005) |
| **Hypocreaceae** | | | | | | |
| *Nectria galligena* | Terpenoids | Colletorin B | $286^{x}$ | — | Eel AChE | Gutiérrez et al. (2005) |
| | | Colletochlorin B | $100^{x}$ | — | Eel AChE | Gutiérrez et al. (2005) |

*(Continued)*

**TABLE 15.2 (*Continued*)**
**Compounds from Plants Showing Anticholinesterase Activity**

| *Plant* | Type of Compound | Compound | $IC_{50}$ (μg/ml) | Kinetics | Type of ChE | Reference |
|---|---|---|---|---|---|---|
| | | Ilicicolin E | 73[x] | — | Eel AChE | Gutiérrez et al. (2005) |
| | | Ilicicolin F | >300[x] | — | Eel AChE | Gutiérrez et al. (2005) |
| | | Ilicicolin C | 90[x] | — | Eel AChE | Gutiérrez et al. (2005) |
| | | α,β-Dehydro-curvularin | >300[x] | — | Eel AChE | Gutiérrez et al. (2005) |
| *Penicillium* sp. *FO-4259* | | Arisugasin A | 1.0[x] | — | ?AChE | Otoguro et al. (1997) |
| *Penicillium* sp. *FO-4259* | | Arisugasin B | 25.8[x] | — | ?AChE | Otoguro et al. (1997) |
| **PTERIDOPHYTE** | | | | | | |
| **Lycopodiaceae** | | | | | | |
| *Lycopodium carinatum* | Alkaloid | Carinatumin-A | 4.6 | — | Eel AChE | Choo et al. (2007) |
| *Lycopodium carinatum* | Alkaloid | Carinatumin-B | 7.0 | — | Eel AChE | Choo et al. (2007) |
| *Lycopodium clavatum* | Triterpinoid | α-Onocerin | 5.2 | — | Eel AChE | Orhan et al. (2003) |
| *Lycopodium hamiltonii* | Pentacyclic alkaloid | Lycoperine A | 60.9 | — | bRBC AChE | Hirasawa et al. (2006) |
| *Lycopodium sieboldii* | Tetracyclic alkaloid | Sieboldine A | 2.0 | — | Eel AChE | Hirasawa et al. (2003) |
| **ANGIOSPERM** | | | | | | |
| **Amaryllidaceae** | | | | | | |
| *Crinum bulbispermum* | Alkaloid | 3-O-Acetylhamayne | 594.0 | — | Eel AChE | Elgorashi et al. (2004) |
| | | Crinamine | 697.0 | — | Eel AChE | Elgorashi et al. (2004) |
| | | 6-Hydroxycrinamine | 490.0 | — | Eel AChE | Elgorashi et al. (2004) |
| | | 8α-Ethoxyprecriwelline | 1145.0 | — | Eel AChE | Elgorashi et al. (2004) |
| | | N-Desmethyl-8α-ethoxypretazettine | 234.0 | — | Eel AChE | Elgorashi et al. (2004) |
| | | N-Desmethyl-8β-ethoxypretazettine | 149.0 | — | Eel AChE | Elgorashi et al. (2004) |
| *Crinum glaucum* | Alkaloid | Lycorine | 450 | — | Eel AChE | Houghton et al. (2004) |
| *Crinum jagus* | Alkaloid | Hamayne | 250 | — | Eel AChE | Houghton et al. (2004) |
| *Crinum macowanii* | Alkaloid | Hamayne | 553.0 | — | Eel AchE | Elgorashi et al. (2004) |
| | | Lycorine | 213.0 | — | Eel AChE | Elgorashi et al. (2004) |
| *Crinum moorei* | Alkaloid | Crinine | 461.0 | — | Eel AChE | Elgorashi et al. (2004) |
| | | Epibuphanisine | 547.0 | — | Eel AChE | Elgorashi et al. (2004) |
| | | Crinamidine | 300.0 | — | Eel AChE | Elgorashi et al. (2004) |

(*Continued*)

**TABLE 15.2 (*Continued*)**
**Compounds from Plants Showing Anticholinesterase Activity**

| *Plant* | Type of Compound | Compound | $IC_{50}$ (μg/ml) | Kinetics | Type of ChE | Reference |
|---|---|---|---|---|---|---|
| | | Epivittatine | 239.0 | — | Eel AChE | Elgorashi et al. (2004) |
| | | 1-O-Acetyllycorine | 0.96 | — | Eel AChE | Elgorashi et al. (2004) |
| | | Cherylline | 4.0 | — | Eel AChE | Elgorashi et al. (2004) |
| *Crinum × powellii* | Lipophillic ester | Linoleic acid ethyl ester | ND | — | Eel AChE | Kissling et al. (2005) |
| *Cyrtanthus falcatus* | Alkaloid | Tazettine | 705.0 | — | Eel AChE | Elgorashi et al. (2004) |
| *Eucharis grandiflora* | Alkaloid | Sanguinine | 0.1 | — | Eel AChE | Lopez et al. (2002) |
| | | 11-Hydroxy-galanthamine | 1.6 | — | Eel AChE | Lopez et al. (2002) |
| | | Epinorgalanthamine | 9.6 | — | Eel AChE | Lopez et al. (2002) |
| | | Oxoassoanine | 47.2 | — | Eel AChE | Lopez et al. (2002) |
| | | Assoanine | 3.9 | — | Eel AChE | Lopez et al. (2002) |
| | | Pseudolycorine | 152.3 | — | Eel AchE | Lopez et al. (2002) |
| *Galanthus fosteri* | Alkaloid | Hordenine | 0.39 | | Eel AChE | Emir et al. (2016) |
| *Nerine bowdenii* | | Ungeremine | 0.35 | — | Eel AChE | Rhee et al. (2004) |
| **Apiaceae** | | | | | | |
| *Angelica acutiloba* | Furanocoumarins | Xanthotoxin | 0.58 | — | dmAChE | Miyazawa et al. (2004) |
| | | Isopimpinellin | 0.32 | — | dmAChE | Miyazawa et al. (2004) |
| *Angelica dahurica* | Furanocoumarins | Isoimperatorin | 74.6 | — | mbAChE | Kim et al. (2002) |
| | | Imperatorin | 63.7 | — | mbAChE | Kim et al. (2002) |
| | | Oxypeucedanin | 89.1 | — | mbAChE | Kim et al. (2002) |
| *Angelica gigas* | Coumarine | Decursinol | 28 | — | Eel AChE | Kang et al. (2001) |
| | | Marmesin | 67 | — | Eel AChE | Kang et al. (2001) |
| | | Xanthotoxin | 54 | — | Eel AChE | Kang et al. (2001) |
| | | Isoimperatorin | 69 | — | Eel AChE | Kang et al. (2001) |
| | | Xanthyletin | 150 | — | Eel AChE | Kang et al. (2001) |
| | | 7-Methoxy-5-prenyloxycoumarin | 240 | — | Eel AChE | Kang et al. (2001) |
| | | Decursin | 390 | — | Eel AChE | Kang et al. (2001) |
| | | 7-Hydroxy-6-(2-(R)-hydroxy-3-methylbut-3-enyl)coumarin | 130 | — | Eel AChE | Kang et al. (2001) |
| | | Nodakenin | 68 | — | Eel AChE | Kang et al. (2001) |

(*Continued*)

**TABLE 15.2 (*Continued*)**
**Compounds from Plants Showing Anticholinesterase Activity**

| *Plant* | Type of Compound | Compound | $IC_{50}$ (µg/ml) | Kinetics | Type of ChE | Reference |
|---|---|---|---|---|---|---|
| | | Peucedanone | 180 | — | Eel AChE | Kang et al. (2001) |
| *Cnidium officinale* | Alkylphthalide derivative | (3S)-Butylphthalide | 0.84 | — | dmAChE | Tsukamoto et al. (2005) |
| | | Neocnidilide | ~1.0 | — | dmAChE | Tsukamoto et al. (2005) |
| *Mutellina purpurea* | Dihydropyranocoumarin | Pteryxin | 12.96 | | hsBuChE | Orhan et al. (2017) |
| *Peucedanum ostruthium* | Coumarins | Ostruthin | 2.6* | — | Eel AChE | Urbain et al. (2005) |
| | | Imperatorin | 200* | — | Eel AChE | Urbain et al. (2005) |
| | | Ostruthol | 370* | — | Eel AChE | Urbain et al. (2005) |
| | | Oxypeucedanin hydrate | 980* | — | Eel AChE | Urbain et al. (2005) |
| | Chromone derivative | Peucenin | 3800* | — | Eel AChE | Urbain et al. (2005) |
| *Tabernaemontana australis* | Indole alkaloids | Coronaridine | ND | — | Eel AChE | Andrade et al. (2005) |
| | | Voacangine | ND | — | Eel AChE | Andrade et al. (2005) |
| | | Voacangine hydroxyindolenine | ND | — | Eel AChE | Andrade et al. (2005) |
| | | Rupicoline | ND | — | Eel AChE | Andrade et al. (2005) |
| *Tabernaemontana divaricata* | Bisindole alkaloid | 19,20-Dihydro-dihydro-tabernamine | ND | — | Eel AChE | Ingkaninan et al. (2006a) |
| | | 19,20-Dihydro-ervahanine A | ND | C, S, R | Eel AChE | Ingkaninan et al. (2006a) |
| **Asclepiadaceae** | | | | | | |
| *Cynanchum atratum* | Pregnane glycosides | Cynatroside B | 3.6 | NC, R | ?AChE | Lee et al. (2005) |
| *Leptadenia arborea* | | Syringaresinol | 200 | — | Eel AChE | El-Hassan et al. (2003) |
| **Asteraceae** | | | | | | |
| *Amberboa ramosa* | Triterpenes | (22R)-Cycloart-20, 25-dien-2alpha, 3beta,22alpha-triol 1 | X | — | ?BuChE | Khan et al. (2004) |
| | | (22R)-Cycloart-23-ene-3beta,22alpha,25-triol 2 | X | — | ?BuChE | Khan et al. (2004) |
| *Cichorium intybus* | Sesquiterpene lactones | 8-Deoxylectucin | 308.1 | — | Eel AChE | Rollinger et al. (2005) |
| **Boraginaceae** | | | | | | |
| *Onosma hispida* | Flavanone | Hispidone | 11.6 | — | Eel AChE | Ahmad et al. (2003) |
| | | Hispidone | 15.7 | — | hsBuChE | Ahmad et al. (2003) |

(*Continued*)

**TABLE 15.2 (*Continued*)**
**Compounds from Plants Showing Anticholinesterase Activity**

| *Plant* | Type of Compound | Compound | $IC_{50}$ (µg/ml) | Kinetics | Type of ChE | Reference |
|---|---|---|---|---|---|---|
| | | (2S)-5,2′-Dihydroxy-7,5′-dimethoxy-flavanone | 28.0 | — | Eel AChE | Ahmad et al. (2003) |
| | | (2S)-5,2′-Dihydroxy-7,5′-dimethoxy-flavanone | 7.9 | — | hsBuChE | Ahmad et al. (2003) |
| **Brassicaceae** | | | | | | |
| *Isatis costata* | Oxindole alkaloids | Costinones A | | NC | ?BuChE | Fatima et al. (2006) |
| | | Costinones B | | NC | ?BuChE | Fatima et al. (2006) |
| **Buddlejaceae** | | | | | | |
| *Buddleja crispa* | | | | — | AChE | Ahmad et al. (2005) |
| | | | | — | BuChE | Ahmad et al. (2005) |
| **Buxaceae** | | | | | | |
| *Buxus hyrcana* | Triterpenoid alkaloid | (+)-N-tigloylbuxahyrcanine | 443.6 ± 15.1 | — | Eel AChE | Choudhary et al. (2003) |
| | | (+)-N-tigloylbuxahyrcanine | 31.2 ± 3.0 | — | ?BuChE | Choudhary et al. (2003) |
| | | N-benzoylbuxahyrcanine | 310.6 ± 0.1 | — | ?BuChE | Choudhary et al. (2003) |
| | | N-isobutyroylbuxahyrcanine | 53.7 ± 3.7 | — | ?BuChE | Choudhary et al. (2003) |
| *Buxus papillosa* | Triterpinoid alkaloids | Buxa-kashmiramine | 25.4 | — | Eel AChE | Atta-ur-Rahman et al. (2001) |
| | | Buxa-kashmiramine | 0.74 | — | hsBuChE | Atta-ur-Rahman et al. (2001) |
| | | Buxakarachiamine | 143.0 | — | Eel AChE | Atta-ur-Rahman et al. (2001) |
| | | Buxahejramine | 162.0 | — | Eel AChE | Atta-ur-Rahman et al. (2001) |
| | Triterinoidal bases | Cycloprotobuxine-C | 38.8 | — | Eel AChE | Atta-ur-Rahman et al. (2001) |
| | | Cycloprotobuxine-C | 2.7 | — | hsBuChE | Atta-ur-Rahman et al. (2001) |
| | | Cyclovirobuxiene-A | 105.7 | — | Eel AChE | Atta-ur-Rahman et al. (2001) |
| | | Cyclovirobuxiene-A | 2.0 | — | hsBuChE | Atta-ur-Rahman et al. (2001) |
| | | Cyclomicrophylline-A | 235.0 | — | Eel AChE | Atta-ur-Rahman et al. (2001) |
| | | Cyclomicrophylline-A | 2.4 | — | hsBuChE | Atta-ur-Rahman et al. (2001) |
| *Sarcococca coriacea* | Steroidal alkaloids | Epoxynepapakistamine-A | >200 | — | Eel AChE | Kalauni et al. (2002) |
| | | Epoxynepapakistamine-A | 77.4 | — | hsBuChE | Kalauni et al. (2002) |
| | | Funtumafrine C | 45.7 | — | Eel AChE | Kalauni et al. (2002) |

*(Continued)*

**TABLE 15.2 (*Continued*)**
**Compounds from Plants Showing Anticholinesterase Activity**

| *Plant* | Type of Compound | Compound | $IC_{50}$ (μg/ml) | Kinetics | Type of ChE | Reference |
|---|---|---|---|---|---|---|
| | | Funtumafrine C | 6.5 | — | hsBuChE | Kalauni et al. (2002) |
| | | N-methylfuntumine | 97.6 | — | Eel AChE | Kalauni et al. (2002) |
| | | N-methylfuntumine | 12.7 | — | hsBuChE | Kalauni et al. (2002) |
| *S. coriacea* (Napalese origin) | | Nepapakistamine A | 50.1 | — | Eel AChE | Kalauni et al. (2001) |
| | | Nepapakistamine A | 25.0 | — | hsBuChE | Kalauni et al. (2001) |
| | | Vaganine | 46.8 | — | Eel AChE | Kalauni et al. (2001) |
| | | Vaganine | 10.0 | — | hsBuChE | Kalauni et al. (2001) |
| *Sarcococca hookeriana* | Steroidal alkaloids | Hookerianamide-D | 59.0 | — | Eel AChE | Choudhary et al. (2005a) |
| | | Hookerianamide-D | 100.2 | — | hsBuChE | Choudhary et al. (2005a) |
| | | Hookerianamide-E | 15.9 | — | Eel AChE | Choudhary et al. (2005a) |
| | | Hookerianamide-E | 6.0 | — | hsBuChE | Choudhary et al. (2005a) |
| | | Hookerianamide-F | 1.6 | — | Eel AChE | Choudhary et al. (2005a) |
| | | Hookerianamide-F | 7.2 | — | hsBuChE | Choudhary et al. (2005a) |
| | | Hookerianamide-G | 11.4 | — | Eel AChE | Choudhary et al. (2005a) |
| | | Hookerianamide-G | 1.5 | — | hsBuChE | Choudhary et al. (2005a) |
| | | Na-methylpipachysamine-D | 10.1 | — | Eel AChE | Choudhary et al. (2005a) |
| | | Na-methylpipachysamine-D | 3.2 | — | hsBuChE | Choudhary et al. (2005a) |
| | | Saligenamide-A | 50.6 | — | Eel AChE | Choudhary et al. (2005a) |
| | | Saligenamide-A | 4.6 | — | hsBuChE | Choudhary et al. (2005a) |
| | | Sarcovagine-C | 1.5 | — | Eel AChE | Choudhary et al. (2005a) |
| | | Sarcovagine-C | 0.7 | — | hsBuChE | Choudhary et al. (2005a) |
| | | Sarcovagine-D | 2.2 | — | Eel AChE | Choudhary et al. (2005a) |
| | | Sarcovagine-D | 2.3 | — | hsBuChE | Choudhary et al. (2005a) |
| | | Terminaline | 113.1 | — | Eel AChE | Choudhary et al. (2005a) |
| | | Terminaline | 0.6 | — | hsBuChE | Choudhary et al. (2005a) |
| | | Terminaline (O-alkylated) | 148.2 | — | Eel AChE | Choudhary et al. (2005a) |
| | | Terminaline (O-alkylated) | 34.3 | — | hsBuChE | Choudhary et al. (2005a) |
| | | Terminaline (acetylated) | 13.0 | — | hsBuChE | Choudhary et al. (2005a) |

*(Continued)*

**TABLE 15.2 (*Continued*)**
**Compounds from Plants Showing Anticholinesterase Activity**

| *Plant* | Type of Compound | Compound | $IC_{50}$ (µg/ml) | Kinetics | Type of ChE | Reference |
|---|---|---|---|---|---|---|
| | | Terminaline (oxidized) | 4.1 | — | Eel AChE | Choudhary et al. (2005a) |
| | | Terminaline (oxidized) | 8.2 | — | hsBuChE | Choudhary et al. (2005a) |
| *Sarcococca saligna* | Steroidal alkaloid | 5,14-Dehydro-$N_a$-demethylsaracodine | >200 | — | Eel AChE | Atta-ur-Rehman et al. (2004b) |
| | | 5,14-Dehydro-$N_a$-demethylsaracodine | 25.0 | — | hsBuChE | Atta-ur-Rehman et al. (2004b) |
| | | 14-Dehydro-$N_a$-demethylsaracodine | 183.1 | — | Eel AChE | Atta-ur-Rehman et al. (2004b) |
| | | 14-Dehydro-$N_a$-demethylsaracodine | 10.1 | — | hsBuChE | Atta-ur-Rehman et al. (2004b) |
| | | 16-Dehydrosarcorine | 12.5 | — | Eel AChE | Atta-ur-Rehman et al. (2004b) |
| | | 16-Dehydrosarcorine | 3.95 | — | hsBuChE | Atta-ur-Rehman et al. (2004b) |
| | | 2,3-Dehydro-sarsalignone | 7.0 | — | Eel AChE | Atta-ur-Rehman et al. (2004b) |
| | | 2,3-Dehydro-sarsalignone | 32.2 | — | hsBuChE | Atta-ur-Rehman et al. (2004b) |
| | | Sarcovagenine-C | 187.8 | — | Eel AChE | Atta-ur-Rehman et al. (2004b) |
| | | Sarcovagenine-C | 1.5 | — | hsBuChE | Atta-ur-Rehman et al. (2004b) |
| | | Salignarine-C | 19.7 | — | Eel AChE | Atta-ur-Rehman et al. (2004b) |
| | | Salignarine-C | 1.3 | — | hsBuChE | Atta-ur-Rehman et al. (2004b) |
| **Caesalpiniaceae** | | | | | | |
| *Detarium microcarpum* | Clerodane diterpene | 3,4-Epoxyclerodan-13E-en-15-oic acid | ND | — | Eel AChE | Cavin et al. (2006) |
| | | 5α,8α-(2-Oxokolavenic acid) | ND | — | Eel AChE | Cavin et al. (2006) |
| | | 2-Oxokolavenic acid | ND | — | Eel AChE | Cavin et al. (2006) |
| | | 3,4-Dihydroxyclerodan-13E-en-15oic acid | ND | — | Eel AChE | Cavin et al. (2006) |
| | | 3,4-Dihydroxyclerodan-13Z-en-15oic acid | ND | — | Eel AChE | Cavin et al. (2006) |
| | | Copalic acid | ND | — | Eel AChE | Cavin et al. (2006) |
| **Chenopodiaceae** | | | | | | |
| *Haloxylon recurvum* | Alkylated sterols | Haloxysterols 1 | 8.3 | NC | Eel AChE | Ahmad et al. (2006) |
| | | Haloxysterols 1 | 4.7 | NC | hsBuChE | Ahmad et al. (2006) |
| | | Haloxysterols 2 | 0.89 | NC | Eel AChE | Ahmad et al. (2006) |
| | | Haloxysterols 2 | 2.3 | NC | hsBuChE | Ahmad et al. (2006) |
| | | Haloxysterols 3 | 1.0 | NC | Eel AChE | Ahmad et al. (2006) |
| | | Haloxysterols 3 | 17.8 | NC | hsBuChE | Ahmad et al. (2006) |
| | | Haloxysterols 4 | 17.2 | NC | Eel AChE | Ahmad et al. (2006) |

*(Continued)*

**TABLE 15.2 (*Continued*)**
**Compounds from Plants Showing Anticholinesterase Activity**

| *Plant* | Type of Compound | Compound | $IC_{50}$ (μg/ml) | Kinetics | Type of ChE | Reference |
|---|---|---|---|---|---|---|
| | | Haloxysterols 4 | 2.5 | NC | hsBuChE | Ahmad et al. (2006) |
| | | Haloxysterols 5 | 26.4 | NC | Eel AChE | Ahmad et al. (2006) |
| | | Haloxysterols 5 | 6.9 | NC | hsBuChE | Ahmad et al. (2006) |
| | | Haloxysterols 6 | 19.2 | NC | Eel AChE | Ahmad et al. (2006) |
| | | Haloxysterols 6 | 4.5 | NC | hsBuChE | Ahmad et al. (2006) |
| | | Haloxysterols 7 | 15.2 | NC | Eel AChE | Ahmad et al. (2006) |
| | | Haloxysterols 7 | 3.9 | NC | hsBuChE | Ahmad et al. (2006) |
| | | Haloxysterols 8 | 13.7 | NC | Eel AChE | Ahmad et al. (2006) |
| | | Haloxysterols 8 | 2.0 | NC | hsBuChE | Ahmad et al. (2006) |
| | | Haloxysterols 9 | 3.5 | NC | Eel AChE | Ahmad et al. (2006) |
| | | Haloxysterols 9 | 3.5 | NC | hsBuChE | Ahmad et al. (2006) |
| *Haloxylon salicornicum* | Piperidine alkaloids | Haloxyline A | 25.3 | — | Eel AChE | Ferheen et al. (2005) |
| | | Haloxyline A | 19.0 | — | BuChE | Ferheen et al. (2005) |
| | | Haloxyline B | 20.2 | — | Eel AChE | Ferheen et al. (2005) |
| | | Haloxyline B | 14.7 | — | BuChE | Ferheen et al. (2005) |
| **Crassulaceae** | | | | | | |
| *Rhodiola rosea* | | Gossypetin-7-O-L-rhamnopyranoside | 58 ± 15[a] | — | Eel AChE | Hillhouse et al. (2004) |
| | | Rhodioflavonoside | 38 ± 4[a] | — | Eel AChE | Hillhouse et al. (2004) |
| **Ericaceae** | | | | | | |
| *Vaccinium oldhami* | | Taraxerol | 79 | NC | mbAChE | Lee et al. (2004) |
| | | Scopoletin | 52 | — | mbAChE | Lee et al. (2004) |
| **Fabaceae** | | | | | | |
| *Caragana chamlague* | Stilbene oligomer | (+)-α-Viniferin | 2.0 | NC, R, S | Eel AChE | Sung et al. (2002) |
| | | (+)-α-Viniferin | 16.9* | — | hsBuChE | Sung et al. (2002) |
| | | Kobophenol A | 115.8 | — | Eel AChE | Sung et al. (2002) |
| | | Kobophenol A | 70* | — | hsBuChE | Sung et al. (2002) |
| **Fumariaceae** | | | | | | |
| *Corydalis incisa* | Alkaloid | Corynoline | 30.6 | NC | mbAChE | Kim (2002) |
| *Corydalis speciosa* | | ? | ? | — | ?AChE | Kim et al. (2004) |
| *Corydalis ternata* | Alkaloid | Prototerpine | 50 | C | Eel AChE | Kim et al. (1999) |

(*Continued*)

**TABLE 15.2 (*Continued*)**
**Compounds from Plants Showing Anticholinesterase Activity**

| *Plant* | Type of Compound | Compound | $IC_{50}$ (μg/ml) | Kinetics | Type of ChE | Reference |
|---|---|---|---|---|---|---|
| **Gentianaceae** | | | | | | |
| *Gentiana campestris* | Xanthones | | | — | | Urbain et al. (2004) |
| **Lamiaceae** | | | | | | |
| *Ajuga bracteosa* | Steroids | Withanoloid 1 | 25.2 | — | Eel AChE | Choudhary et al. (2005b) |
| | | Withanoloid 1 | 38.4 | — | hsBuChE | Choudhary et al. (2005b) |
| | | Withanoloid 2 | 22.1 | — | Eel AChE | Choudhary et al. (2005b) |
| | | Withanoloid 2 | 29.0 | — | hsBuChE | Choudhary et al. (2005b) |
| | | Withanoloid 3 | 49.2 | — | Eel AChE | Choudhary et al. (2005b) |
| | | Withanoloid 3 | 40.0 | — | hsBuChE | Choudhary et al. (2005b) |
| *Origanum majorana* | Pentacyclic triterpene acid | Ursolic acid | 7.5[b] | Mixed | | Chung et al. (2001) |
| *Otostegia limbata* | Clerodane diterpenoids | Limbatolide A | 38.5 | — | Eel AChE | Ahmad et al. (2005) |
| | | Limbatolide A | 22.3 | — | hsBuChE | Ahmad et al. (2005) |
| | | Limbatolide B | 47.2 | — | Eel AChE | Ahmad et al. (2005) |
| | | Limbatolide B | 17.5 | — | hsBuChE | Ahmad et al. (2005) |
| | | Limbatolide C | 103.7 | — | Eel AChE | Ahmad et al. (2005) |
| | | Limbatolide C | 14.2 | — | hsBuChE | Ahmad et al. (2005) |
| *Perovskia atriplicifolia* | Isorinic acid derivative | Perovskoate | 42.0 ± 0.02 | — | Eel AChE | Perveen et al. (2006) |
| *P. atriplicifolia* | Isorinic acid derivative | Perovskoate | 25.0 ± 0.001 | — | hsBuChE | Perveen et al. (2006) |
| *Salvia lavandulaefolia* | Essential oils | 1,8-Cineole | 0.06[c] | — | bRBC AChE | Savelev et al. (2003) |
| | | α-Pinene | 0.09[c] | — | bRBC AChE | Savelev et al. (2003) |
| | | β-Pinene | 0.2[c] | — | bRBC AChE | Savelev et al. (2003) |
| | | Camphor | 39 ± 4.0[d] | — | bRBC AChE | Savelev et al. (2003) |
| | | Linalool | 18 ± 2.3[d] | — | bRBC AChE | Savelev et al. (2003) |
| | | Bornyl acetate | 23 ± 4.0[e] | — | bRBC AChE | Savelev et al. (2003) |
| | | Caryophyllene oxide | 35 ± 4.7[e] | — | bRBC AChE | Savelev et al. (2003) |
| | | Borneol | 19 ± 2.6[e] | — | bRBC AChE | Savelev et al. (2003) |
| *Salvia miltiorhiza* | Abietane Diterpenoid | Dihydrotanshinone | 1.0 | — | Eel AChE | Ren et al. (2004) |
| | | Cryptotanshinone | 7.0 | — | Eel AChE | Ren et al. (2004) |
| | | Tanshinone I | >50 | — | Eel AChE | Ren et al. (2004) |

(*Continued*)

**TABLE 15.2 (*Continued*)**
**Compounds from Plants Showing Anticholinesterase Activity**

| *Plant* | Type of Compound | Compound | $IC_{50}$ (µg/ml) | Kinetics | Type of ChE | Reference |
|---|---|---|---|---|---|---|
| | | Tanshinone II A | >140 | — | Eel AChE | Ren et al. (2004) |
| *Salvia santolinifolia* | Triterpenes | Slavin A | 12.5 | — | hsBuChE | Mehmood et al. (2006) |
| | | Slavin B | 65.5 | — | hsBuChE | Mehmood et al. (2006) |
| *Salvia syriaca* | | Ursolic acid | 46.3 | | | Bahadori et al. (2016) |
| | | Corosolic acid | 55.2 | | | Bahadori et al. (2016) |
| | | B-sitosterol | 24.1 | | | Bahadori et al. (2016) |
| | | urs-12-en-2α,3β-diol | 63.0 | | | Bahadori et al. (2016) |
| | | Daucosterol | 34.3 | | | Bahadori et al. (2016) |
| *Vitex negundo* | Lignans | (+)-Diasyringaresinol | 300 | — | hsBuChE | Azhar-ul-Haq et al. (2004) |
| | | (+)-Lynoresinol | 263.3 | — | Eel AChE | Azhar-ul-Haq et al. (2004) |
| | | Nigundins A | >300 | — | Eel AChE | Azhar-ul-Haq et al. (2004) |
| | | Nigundins A | 85.0 | — | hsBuChE | Azhar-ul-Haq et al. (2004) |
| | | Nigundins B | 254.0 | — | Eel AChE | Azhar-ul-Haq et al. (2004) |
| | | Nigundins B | 194.0 | — | hsBuChE | Azhar-ul-Haq et al. (2004) |
| | | Vitrofolal E | 106.5 | — | Eel AChE | Azhar-ul-Haq et al. (2004) |
| | | Vitrofolal E | 35.0 | — | hsBuChE | Azhar-ul-Haq et al., (2004) |
| **Liliaceae** | | | | | | |
| *Colchicum speciosum* | | 12-n-Butoxy-demethoxy-kesselringine | — | — | mRBC AChE | Basova et al. (2006) |
| | | 12-n-Butoxy-demethoxy-kesselringine | — | — | mbsBuChE | Basova et al. (2006) |
| | | Colchamine | — | — | mRBC AChE | Basova et al. (2006) |
| | | Colchamine | — | — | mbsBuChE | Basova et al. (2006) |
| | | Colchicine | — | — | mRBC AChE | Basova et al. (2006) |
| | | Colchicine | — | — | mbsBuChE | Basova et al. (2006) |
| | | Dec-N-dimethylregeline | — | — | mRBC AChE | Basova et al. (2006) |
| | | Dec-N-dimethylregeline | — | — | mbsBuChE | Basova et al. (2006) |
| | | Iolatamine idomethylate | — | — | mRBC AChE | Basova et al. (2006) |
| | | Iolatamine idomethylate | — | — | mbsBuChE | Basova et al. (2006) |
| | | Kesselridine | — | — | mRBC AChE | Basova et al. (2006) |
| | | Kesselridine | — | — | mbsBuChE | Basova et al. (2006) |

(*Continued*)

**TABLE 15.2 (*Continued*)**
**Compounds from Plants Showing Anticholinesterase Activity**

| *Plant* | Type of Compound | Compound | $IC_{50}$ (µg/ml) | Kinetics | Type of ChE | Reference |
|---|---|---|---|---|---|---|
| | | Kesselridine idomethylate | — | — | mRBC AChE | Basova et al. (2006) |
| | | Kesselridine idomethylate | — | — | mbsBuChE | Basova et al. (2006) |
| | | Kesselringine | — | — | mRBC AChE | Basova et al. (2006) |
| | | Kesselringine | — | — | mbsBuChE | Basova et al. (2006) |
| | | Kesselringine idomethylate | — | — | mRBC AChE | Basova et al. (2006) |
| | | Kesselringine idomethylate | — | — | mbsBuChE | Basova et al. (2006) |
| | | Luteidine | — | — | mRBC AChE | Basova et al. (2006) |
| | | Luteidine | — | — | mbsBuChE | Basova et al. (2006) |
| | | Luteidine idomethylate | — | — | mRBC AChE | Basova et al. (2006) |
| | | Luteidine idomethylate | — | — | mbsBuChE | Basova et al. (2006) |
| | | Luteinone | — | — | mRBC AChE | Basova et al. (2006) |
| | | Luteinone | — | — | mbsBuChE | Basova et al. (2006) |
| | | Merenderine | — | — | mRBC AChE | Basova et al. (2006) |
| | | Merenderine | — | — | mbsBuChE | Basova et al. (2006) |
| | | Regelamine | — | — | mRBC AChE | Basova et al. (2006) |
| | | Regelamine | — | — | mbsBuChE | Basova et al. (2006) |
| | | Regeline | — | — | mRBC AChE | Basova et al. (2006) |
| | | Regeline | — | — | mbsBuChE | Basova et al. (2006) |
| | | Regeline idomethylate | — | — | mRBC AChE | Basova et al. (2006) |
| | | Regeline idomethylate | — | — | mbsBuChE | Basova et al. (2006) |
| | | Regeline quaternary base | — | — | mRBC AChE | Basova et al. (2006) |
| | | Regeline quaternary base | — | — | mbsBuChE | Basova et al. (2006) |
| | | Salsolidine idomethylate | — | — | mRBC AChE | Basova et al. (2006) |
| | | Salsolidine idomethylate | — | — | mbsBuChE | Basova et al. (2006) |
| | | Salsoline idomethylate | — | — | mRBC AChE | Basova et al. (2006) |
| | | Salsoline idomethylate | — | — | mbsBuChE | Basova et al. (2006) |
| *Fritillaria imperialis* | Steroidal alkaloid | Delavine | 105.5 ± 1.5 | — | Eel AChE | Atta-ur-Rahman et al. (2002) |
| | | Delavine | 1.7 ± 0.1 | — | hsBuChE | Atta-ur-Rahman et al. (2002) |
| | | Forticine | 100.5 ± 0.4 | — | hsBuChE | Atta-ur-Rahman et al. (2002) |

(*Continued*)

**TABLE 15.2 (*Continued*)**
**Compounds from Plants Showing Anticholinesterase Activity**

| *Plant* | Type of Compound | Compound | $IC_{50}$ (μg/ml) | Kinetics | Type of ChE | Reference |
|---|---|---|---|---|---|---|
| | | Forticine | >500 | — | Eel AChE | Atta-ur-Rahman et al. (2002) |
| | | Imperialine | >500 | — | Eel AChE | Atta-ur-Rahman et al. (2002) |
| | | Imperialine | 121.5 ± 6.6 | — | hsBuChE | Atta-ur-Rahman et al. (2002) |
| | | Impericine | 67.97 ± 2.46 | — | Eel AChE | Atta-ur-Rahman et al. (2002) |
| | | Impericine | 1.6 | — | hsBuChE | Atta-ur-Rahman et al. (2002) |
| | | Persicanimide A | 352.2 ± 4.0 | — | Eel AChE | Atta-ur-Rahman et al. (2002) |
| | | Persicanimide A | 4.2 | — | hsBuChE | Atta-ur-Rahman et al. (2002) |
| **Magnoliaceae** | | | | | | |
| *Magnolia × soulangiana* | | Taspine | 0.33 ± 0.07 | | Eel AChE | Rollinger et al. (2006) |
| | | | 0.54 ± 0.1 | | hAChE | Rollinger et al. (2006) |
| | | | >100 | | BuChE | Rollinger et al. (2006) |
| **Menispermaceae** | | | | | | |
| *Cocculus pendulus* | Bisbenzylisoquinoline alkaloids | Cocsoline | 47.6 | — | Eel AChE | Atta-ur-Rahman et al. (2004a) |
| | | Cocsoline | 6.1 | — | hsBuChE | Atta-ur-Rahman et al. (2004a) |
| | | Cocsuline | 100.0 | — | Eel AChE | Atta-ur-Rahman et al. (2004a) |
| | | Cocsuline | 12.0 | — | hsBuChE | Atta-ur-Rahman et al. (2004a) |
| | | 1,2-Dehydroapateline | 116.5 | — | Eel AChE | Atta-ur-Rahman et al. (2004) |
| | | 1,2-Dehydroapateline | 183.0 | — | hsBuChE | Atta-ur-Rahman et al. (2004a) |
| | | Kurramine-α-N-oxide | 150.0 | — | Eel AChE | Atta-ur-Rahman et al. (2004) |
| | | Kurramine-2′-β-N-oxide | 10.0 | — | Eel AChE | Atta-ur-Rahman et al. (2004a) |
| *Stephania venosa* | | Cyclanoline | 9.23 | — | Eel AChE | Ingkaninan et al. (2006b) |
| | | N-Methyl stepholicline | 31.30 | — | Eel AChE | Ingkaninan et al. (2006b) |
| | | Stepharanine | 14.10 | — | Eel AChE | Ingkaninan et al. (2006b) |
| **Myristicaceae** | | | | | | |
| *Iryanthera megistophylla* | | Iryantherin K | 67.6 ± 3.4[f] | — | Eel AChE | Ming et al. (2002) |
| | | Megislignan | 77.4 ± 6.1[f] | — | Eel AChE | Ming et al. (2002) |
| **Orchidaceae** | | | | | | |
| *Fiatoua villosa* | Purine | Zeatin | 109 | — | Eel AChE | Heo et al. (2002) |

(*Continued*)

**TABLE 15.2 (*Continued*)**
**Compounds from Plants Showing Anticholinesterase Activity**

| *Plant* | Type of Compound | Compound | $IC_{50}$ (μg/ml) | Kinetics | Type of ChE | Reference |
|---|---|---|---|---|---|---|
| **Papaveraceae** | | | | | | |
| *Chelidonium majus* | | 8-Hydroxydihydro chelerythrine | 0.61 | C, S | Eel AChE | Cho et al. (2006) |
| | | 8-Hydroxydihydro chelerythrine | 34.6 | — | hsBuChE | Cho et al. (2006) |
| | | 8-Hydroxydihydro sanguinarine | 1.37 | — | Eel AChE | Cho et al. (2006) |
| | | 8-Hydroxydihydro sanguinarine | 12.8 | — | hsBuChE | Cho et al. (2006) |
| | | Berberine | 1.85 | — | Eel AChE | Cho et al. (2006) |
| | | Berberine | 78.9 | — | hsBuChE | Cho et al. (2006) |
| *Aconitum falconeri* | Norditerpenoid | Faleoconitine | 293.0 | — | Eel AChE | Atta-ur-Rahman et al. (2000) |
| | | Pseudaconitine | 278.0 | — | Eel AChE | Atta-ur-Rahman et al. (2000) |
| *Aconitum heterophyllum* | Diterpinoid | Atidine | 14.6 | C | Eel AChE | Ahmad et al. (2017) |
| | | | 18.1 | C | hsBuChE | Ahmad et al. (2017) |
| | | Hetisinone | 10.1 | NC | Eel AChE | Ahmad et al. (2017) |
| | | | 15.6 | NC | hsBuChE | Ahmad et al. (2017) |
| | | 19-Epiisoatisine | 10.3 | NC | | Ahmad et al. (2017) |
| | | | 14.7 | NC | | Ahmad et al. (2017) |
| **Rubiaceae** | | | | | | |
| *Chimarrhis turbinata* | Indole glucoalkaloids | Turbinatine | 1.86 | — | rbAChE | Cardoso et al. (2004) |
| **Rutaceae** | | | | | | |
| *Citrus junos* | | Naringenin | 66.0%[g] | | PC AChE | Heo et al. (2004) |
| *Citrus paradisi* | Essential oils | Nootkatone | 17%–24%[h] | — | ?AChE | Miyazawa et al. (2001) |
| | | Auraptene | | — | ?AChE | Miyazawa et al. (2001) |
| *Murraya paniculata* | Coumarin | Murranganone | 79.1 | — | Eel AChE | Choudhary et al. (2002) |
| | | Paniculatin | 31.6 | — | Eel AChE | Choudhary et al. (2002) |
| | | Murranganone | 74.3 | — | hsBuChE | Choudhary et al. (2002) |
| | | Paniculatin | >100 | — | hsBuChE | Choudhary et al. (2002) |
| *Zanthoxylum zanthoxyloides* | Peroxide | | | — | | Queiroz et al. (2006) |
| **Scrophulariaceae** | | | | | | |
| *Scrophularia buergeriana* | Phenylpropanoid | E-p-methoxy-cinnamic acid | 12.7 ± 1.9[y] | — | Eel AChE | Kim et al. (2003) |

(*Continued*)

**TABLE 15.2 (*Continued*)**
**Compounds from Plants Showing Anticholinesterase Activity**

| *Plant* | Type of Compound | Compound | $IC_{50}$ (μg/ml) | Kinetics | Type of ChE | Reference |
|---|---|---|---|---|---|---|
| **Solanaceae** | | | | | | |
| *Withania somnifera* | Steroids | Withanoloids 4 | 45.2 | — | Eel AChE | Choudhary et al. (2005a) |
| | | Withanoloids 5 | 20.5 | — | Eel AChE | Choudhary et al. (2005a) |
| | | Withanoloids 4 | 45.1 | — | hsBuChE | Choudhary et al. (2005a) |
| | | Withanoloids 5 | 95.2 | — | hsBuChE | Choudhary et al. (2005a) |
| **Ulmaceae** | | | | | | |
| *Celtis chinensis* | Tyramine | Trans-N-p-Coumaroyl Tyramine | 122.0 | Mixed, R | abAChE | Kim and Lee (2003) |
| **Zygophyllaceae** | | | | | | |
| *Zygophyllum eurypterum* | Pterocarpans | Atricarpan A | 12.5 ± 0.08 | | hsBuChE | Ahmad et al. (2006) |
| | | Atricarpan B | 65.5 ± 0.01 | | hsBuChE | Ahmad et al. (2006) |
| | | Atricarpan C | 19.5 ± 0.05 | | hsBuChE | Ahmad et al. (2006) |
| | | Atricarpan D | 20.5 ± 0.01 | | Eel AChE | Ahmad et al. (2006) |
| | | Atricarpan D | 30.5 ± 0.02 | | hsBuChE | Ahmad et al. (2006) |

*Note:* Values presented in the table are as reported by the investigators. However, due to differing conditions and sensitivity of the assays employed, different sources and purity of enzymes used, and most importantly differences in preparation of plant extracts, the data of different investigators should not be compared.

?, source of AChE not known; -, not determined; NC, non competitive; *, minimum quantity required to produce a white inhibition spot on TLC; a, % inhibition of AChE at 5mg/ml; b, concentration in nM; c, concentration in mg/ml; d, % inhibition of AChE at 0.5 mg/ml; e, % inhibition of AChE at 0.25 mg/ml; f, % inhibition of AChE at 0.8 mg/ml; g, % inhibition of AChE at 210 μg/ml; h, % inhibition of AChE at 1.62 μg/ml; y, % inhibition of AChE at 100 μM; z, % inhibition of AChE at 100 μg/ml; abAChE, animal brain AChE; bRBC AChE, bovine RBC AChE; dm AChE, *Drosophila melanogaster* AChE; mbAChE, mouse brain AChE; PC AChE, PC cell Culture AChE; rbAChE, rat brain AChE; hs BuChE, horse serum butyrylcholinesterase.

## 15.4 ACETYLCHOLINE SYSTEM PLAYS A ROLE IN ALLELOPATHY

Weeds contain strong inhibitors of AChE, which possibly are deterrents against herbivory. The mechanism of action of most of the allelochemicals is not yet known. The biochemical mechanism of some plant allelochemicals as a factor in invasiveness of weeds has been studied (Bais et al., 2003; Weston and Duke, 2003; Li et al., 2010). The methanolic extract of *C. rotundus*, which inhibits germination, root and shoot growth in wheat and tomato, also inhibits AChE from wheat and tomato (Figures 15.2 through 15.7). Of the antiChEs found in *C. rotundus*, **α**-pinene is an allelochemical as well. It inhibits growth (radical elongation) in *Triticum aestivum*, along with *Cicer arietinum*, *Cassia occidentalis*, *Pisum sativum*, and *Amaranthus viridis* (Singh et al., 2006). The effect is reported to be mediated via oxidative stress. Since ACh is also known to cause increased $O_2$ uptake/respiratory quotient, it seems plausible that **α**-pinene acts as allelochemical via cholinergic mechanism. Recent evidences (Zhiqun et al., 2017; Turi et al., 2014)

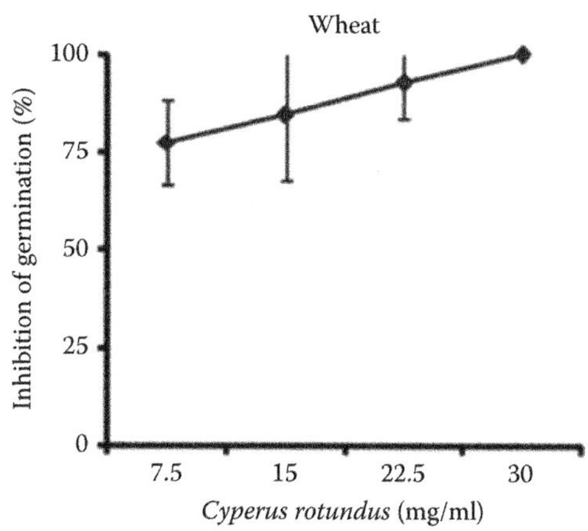


**FIGURE 15.2** Effect of methanolic extract of *Cyperus rotundus* tubers on germination of wheat seeds. 100 seeds were sown per Petri-plate (25 × 150 mm). Observations were taken after 24 hr of imbibition. Bars represent ± S.E.M.

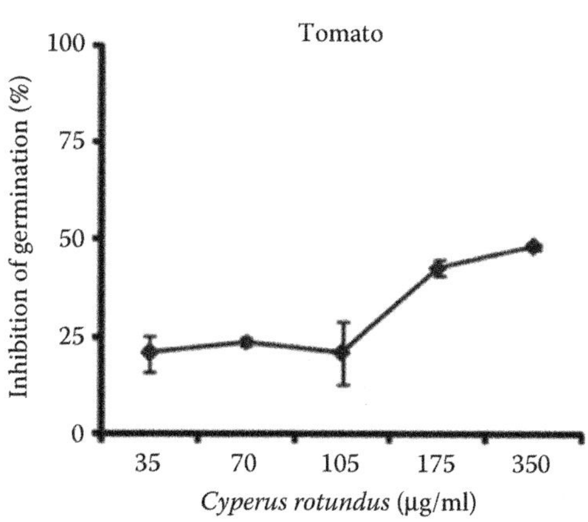


**FIGURE 15.3** Effect of methanolic extract of *Cyperus rotundus* tubers on germination of wheat seeds. 100 seeds were sown per Petri-plate (25 × 150 mm). Observations were taken after 72 hr of imbibition. Bars represent ± S.E.M.

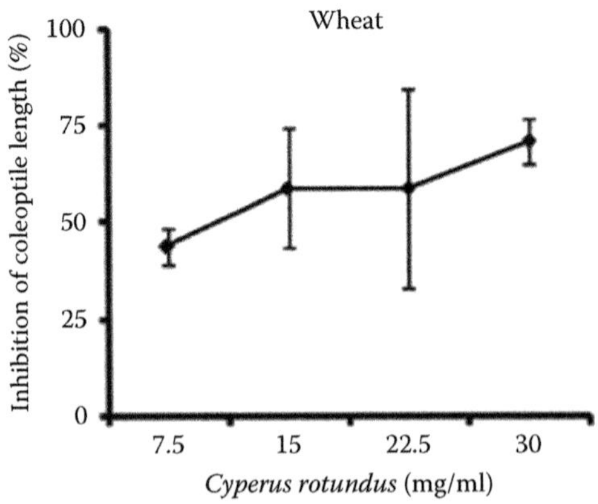


**FIGURE 15.4** Effect of methanolic extract of *Cyperus rotundus* tubers on coleoptile length in wheat seedlings. Observations were taken after 72 hr of imbibition. Bars represent ± S.E.M.

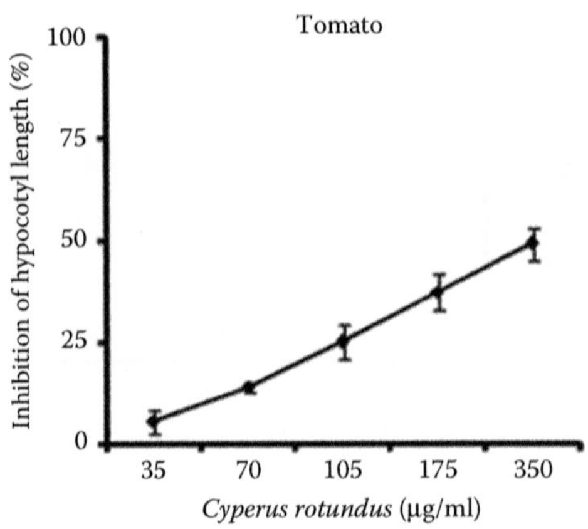


**FIGURE 15.5** Effect of methanolic extract of *Cyperus rotundus* tubers on hypocotyl length in tomato seedlings. Observations were taken after 120 hr of imbibition. Bars represent ± S.E.M.

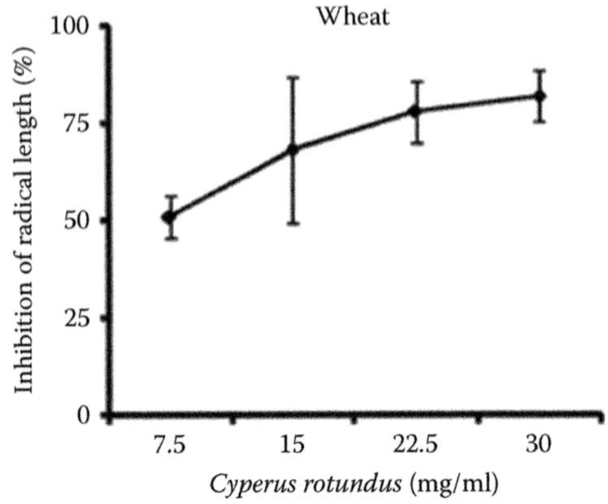


**FIGURE 15.6** Effect of methanolic extract of *Cyperus rotundus* tubers on radical length in wheat seedlings. Observations were taken after 72 hr of imbibition. Bars represent ± S.E.M.

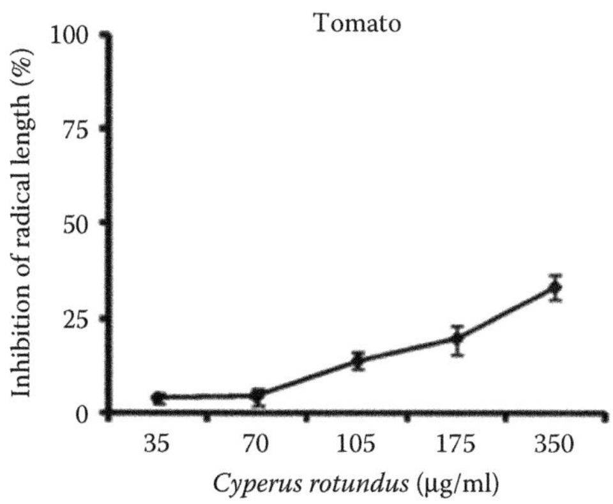


**FIGURE 15.7** Effect of methanolic extract of *Cyperus rotundus* tubers on radical length in tomato seedlings. Observations were taken after 120 hr of imbibition. Bars represent ± S.E.M.

also confirm our long-standing suggestion that these antiChEs could well be acting as allelochemicals functioning via the cholinergic system of the plants aiding in establishing supremacy over other plants trying to grow in the same habitat.

## REFERENCES

Ahmad, H., Ahmad, S., Shah, S.A.A., Latif, A., and Ahmad, M. 2017. Antioxidant and anticholinesterase potential of diterpenoid alkaloids from *Aconitum heterophyllum*. *Bioorganic Med Chem* 25: 3368–3376.

Ahmad, I., Malik, A., Afza, N., Anis, I., Fatima, I., Nawaz, S.A., Tareen, R.B., and Choudhary, M.I. 2003. Cholinesterase inhibitory constituents from *Onosma hispida*. *Chem Pharm Bull* 51: 412–414.

Ahmad, I., Malik, A., Afza, N., Anis, I., Fatima, I., Nawaz, S.A., Tareen, R.B., and Choudhary, MI. 2005. Enzymes inhibitory constituents from *Buddleja crispa*. *Z Naturforsch B* 60: 341–346.

Ahmad, V.U., Iqbal, S., Nawaz, S.A., Choudhary, M.I., Farooq, U., Ali, S.T., Ahmad, A. et al. 2006. Isolation of four new pterocarpans from Zygophyllum eurypterum (Syn. Z. atriplicoides) with enzyme-inhibition properties. *Chem Biodivers* 3: 996–1003.

Ahmad, V.U., Khan, A., Farooq, U., Kousar, F., Khan, S.S., Nawaz, S.A., Abbasi, M.A., and Choudhary, M.I. 2005. Three new cholinesterase-inhibiting cis-clerodane diterpenoids from *Otostegia limbata*. *Chem Pharm Bull* 53: 378–381.

Ahmed, E., Nawaz, S.A., Malik, A., and Choudhary, M.I. 2006. Isolation and cholinesterase-inhibition studies of sterols from *Haloxylon recurvum*. *Bioorg Med Chem Lett* 16: 573–580.

Andrade, M.T., Lima, J.A., Pinto, A.C., Rezende, C.M., Carvalho, M.P., and Epifanio, R.A. 2005. Indole alkaloids from *Tabernaemontana australis* (Muell. Arg) Miers that inhibit acetylcholinesterase enzyme. *Bioorg Med Chem Lett* 13: 4092–4095.

Atta-Ur-Rahman, Akhtar, M.N., Choudhary, M.I., Tsuda, Y., Sener, B., Khalid, A., and Parvez, M. 2002. New steroidal alkaloids from *Fritillaria imperialis* and their cholinesterase inhibiting activities. *Chem Pharm Bull* 50: 1013–1016.

Atta-Ur-Rahman, Atia-Tul-Wahab, Nawaz, S.A., and Choudhary I.M. 2004a. New cholinesterase inhibiting bisbenzylisoquinoline alkaloids from *Cocculus pendulus*. *Chem Pharm Bull* 52: 802–806.

Atta-Ur-Rahman, Fatima, N., Akhtar, F., Choudhary, M.I., and Khalid, A. 2000. New norditerpenoid alkaloids from *Aconitum falconeri*. *J Nat Prod* 63: 1393–1395.

Atta-ur-Rahman, Feroz, F., Naeem, I., Zaheer-ul-Haq, Nawaz, S.A., Khan, N., Khan, M.R., and Choudhary, M.I. 2004b. New pregnane-type their steroidal alkaloids from *Sarcococca saligna* and cholinesterase inhibitory activity. *Steroids* 69: 735–741.

Atta-Ur-Rahman, Parveen, S., Khalid, A., Farooq, A., and Choudhary, M.I. 2001. Acetyl and butyrylcholinesterase-inhibiting triterpenoid alkaloids from *Buxus papillosa*. *Phytochemistry* 58: 963–968.

Azhar-Ul-Haq, Malik, A., Anis, I., Khan, S.B., Ahmed, E., Ahmed, Z., Nawaz, S.A., and Choudhary, M.I. 2004. Enzyme inhibiting lignans from *Vitex negundo*. *Chem Pharm Bull* 52: 1269–1272.

Bahadori, M.B., Dinparast, L., Valizadeh, H., Farimani, M.M., and Ebrahimi, S.N. 2016. Bioactive constituents from roots of *Salvia syriaca* L.: Acetylcholinesterase inhibitory activity and molecular docking studies. *S Afr J Bot* 106: 1–4.

Bais, H.P., Vepachedu, R., Gilroy, S., Callaway, R.M., and Vivanco, J.M. 2003. Allelopathy and exotic plant invasion: From molecules and genes to species interactions. *Science* 301: 1377–1380.

Bamel, K., Gupta, R., and Gupta, S.C. 2015. Nicotine promotes rooting in leaf explants of in vitro raised seedlings of tomato, Lycopersicon esculentum Miller var. Pusa Ruby. *Int Immunopharmacol* 29: 231–234.

Basova, N.E., Rozengart, E.V., and Suvorov, A.A. 2006. Cholinergic activity of isoquinoline alkaloids from the showy autumn crocus (*Colchicum speciosum* Stev.). *Dokl Biochem Biophys* 406: 27–31.

Calderon, A.I., Simithy-Williums, J., Sanchez, R., Espinosa, A., Valdespino, I., and Gupta, M.P. 2013. Lycopodiaceae from Panama: A new source of acetylcholinesterase inhibitors. *Nat Prod Res* 27: 500–505.

Cardoso, C.L., Castro-Gamboa, I., Silva, D.H., Furlan, M., Epifanio-Rde, A., Pinto Ada, C., Moraes de Rezende, C., Lima, J.A., and Bolzani Vda, S. 2004. Indole glucoalkaloids from *Chimarrhis turbinata* and their evaluation as antioxidant agents and acetylcholinesterase inhibitors. *J Nat Prod* 67: 1882–1885.

Cavin, A.L., Hay, A.E., Marston, A., Stoeckli-Evans, H., Scopelliti, R., Diallo, D., and Hostettmann, K. 2006. Bioactive diterpenes from the fruits of *Detarium microcarpum*. *J Nat Prod* 69: 768–773.

Chen, B., Zhou, J., Gou, X., Ma, D.W., Wang, Y.N., Hu, Z.L., and He, Y.Q. 2016. Volatiles from *Chenopodium ambrosioides* L. induce the oxidative damage in maize (*Zea mays* L.) radicles. *Allelopathy J* 38: 171–180.

Cheng, F., and Cheng, Z. 2015. Research progress on the use of plant allelopathy in agriculture and the physiological and ecological mechanisms of allelopathy. *Front Plant Sci* 6: 1020.

Cho, K.M., Yoo, I.D., and Kim, W.G. 2006. 8-hydroxydihydrochelerythrine and 8-hydroxydihydrosanguinarine with a potent acetylcholinesterase inhibitory activity from *Chelidonium majus* L. *Biol Pharm Bull* 29: 2317–2320.

Choo, C.Y., Hirasawa, Y., Karimata, C., Koyama, K., Sekiguchi, M., Kobayashi, J., and Morita, H. 2007. Carinatumins A-C, new alkaloids from *Lycopodium carinatum* inhibiting acetylcholinesterase. *Bioorg Med Chem* 15: 1703–1707.

Choudhary, M.I., Azizuddin, Khalid, A., Sultani S.Z., and Atta-ur-Rahman, S.Z. 2002. A new coumarin from *Murraya paniculata*. *Planta Med* 68: 81–83.

Choudhary, M.I., Devkota, K.P., Nawaz, S.A., Ranjit, R., and Atta-ur-Rahman. 2005a. Cholinesterase inhibitory pregnane-type steroidal alkaloids from *Sarcococca hookeriana*. *Steroids* 70: 295–303.

Choudhary, M.I., Nawaz, S.A., ul-Haq, Z., Lodhi, M.A., Ghayur, M.N., Jalil, S., Riaz, N. et al. 2005b. Withanolides, a new class of natural cholinesterase inhibitors with calcium antagonistic properties. *Biochem Biophys Res Commun* 334: 276–287.

Choudhary, M.I., Shahnaz, S., Parveen, S., Khalid, A., Ayatollahi, S.A.M., and Atta-ur-Rahman, Parvez M. 2003. New triterpenoid alkaloid cholinesterase inhibitors from *Buxus hyrcana*. *J Nat Prod* 66: 739–742.

Chung, Y.K., Heo, H.J., Kim, E.K., Kim, H.K., Huh, T.L., Lim, Y., Kim, S.K., and Shin, D.H. 2001. Inhibitory effect of ursolic acid purified from *Origanum majorana* L. on the acetylcholinesterase. *Mol Cells* 11: 137–143.

Cometa, M.F., Lorenzini, P., Fortuna, S., Volpe, M.T., Meneguz, A., and Palmery, M. 2005. In vitro inhibitory effect of aflatoxin $B_1$ on acetylcholinesterase activity in mouse brain. *Toxicology* 206: 125–135.

Dhima, K., Vasolakoglou, I., Stefanou, S., Gatsis, T., Paschalidis, K., Agellopoulos, S., and Eleftherohorinos, I. 2016. Differential competitive and allelopathic ability of *Cyperus rotundus* on *Solanum lycopersicum*, *Solanum melongena*, and *Capsicum annuum*. *Archives of Agronomy Soil Sci* 62: 1250–1263.

Einhellig, F.A. 1994. Mechanism of action of allelochemicals in allelopathy. In Inderjit D.K.M.M. and Einhellig, F.A. (Eds.). *Allelopathy, Organisms, Processes and Applications*, pp. 96–116. Washington, DC: American Chemical Society.

Elgorashi, E.E., Stafford, G.I., Van Staden, J. 2004. Acetylcholinesterase enzyme inhibitory effects of amaryllidaceae alkaloids. *Planta Med* 70: 260–262.

El-Hassan, A., El-Sayed, M., Hamed, A.I., Rhee, I.K., Ahmed, A.A., Zeller, K.P., and Verpoorte, R. 2003. Bioactive constituents of *Leptadenia arborea*. *Fitoterapia* 74: 184–187.

Emir, A., Emir, C., Bozkurt, B., Onur, M.A., Bastida, J., and Somer, N.U. 2016. Alkaloids from *Galanthus fosteri*. *Phytochem Lett* 17: 167–172.

Fatima, I., Ahmad, I., Nawaz, S.A., Malik, A., Afza, N., Luttfullah, G., and Choudhary, M.I. 2006. Enzyme inhibition studies of oxindole alkaloids from *Isatis costata*. *Heterocycles* 68: 1421–1428.

Ferheen, S., Ahmed, E., Afza, N., Malik, A., Shah, M.R., Nawaz, S.A., and Choudhary, M.I. 2005. Haloxylines A and B, antifungal and cholinesterase inhibiting piperidine alkaloids from *Haloxylon salicornicum*. *Chem Pharm Bull* 53: 570–572.

Golovko, E. 1999. Allelopathy- a history, past, present and future. In Macais, F.A., Galindo, J.C.G., Molinillo, J.M.G., Cutler, H.G. (Eds.). *Recent Advances in Allelopathy* Vol I A Science for the Future, pp. 485–490. Spain: International Allelopathy Society.

Gupta, A., and Gupta, R. 1997. A survey of plants for presence of cholinesterase activity. *Phytochemistry* 46: 827–831.

Gupta, A., Thakur, S.S., Uniyal, P.L., and Gupta, R. 2001. A survey of Bryophytes for presence of cholinesterase activity. *Am J Bot* 88: 2133–2135.

Gutiérrez, M., Theoduloz, C., Rodriguez, J., Lolas, M., and Schmeda-Hirschmann, G. 2005. Bioactive metabolites from the fungus *Nectria galligena*, the main apple canker agent in Chile. *J Agr Food Chem* 53: 7701–7708.

Hartmann, E., and Gupta, R. 1989. Acetylcholine as a signalling system in plants. In: Boss, W.E., and Morre, J.D. (Eds.). *Second Messengers in Plant Growth and Development*, pp. 257–287. New York: A. R. Liss.

Hejl, A.A.M., and Koster, K.L. 2004. Juglone disrupts root plasma membrane $H^+$-ATPase activity and impairs water uptake, root respiration and growth in Soybean (*Glycine max*) and Corn (*Zea mays*). *J Chem Eco* 30: 453–471.

Heo, H.J., Hong, S.C., Cho, H.Y., Hong, B., Kim, H.K., Kim, E.K., and Shin, D.H. 2002. Inhibitory effect of zeatin, isolated from *Fiatoua villosa*, on acetylcholinesterase activity from PC12 cells. *Mol Cells* 13: 113–117.

Heo, H.J., Kim, M.J., Lee, J.M., Choi, S.J., Cho, H.Y., Hong, B., Kim, H.K., Kim, E., and Shin, D.H. 2004. Naringenin from *Citrus junos* has an inhibitory effect on acetylcholinesterase and a mitigating effect on amnesia. *Dement Geriatr Cogn Disord* 17: 151–157.

Hillhouse, B.J., Ming, D.S., French, C.J., and Towers, G.H.N. 2004. Acetylcholine esterase inhibitors in *Rhodiola rosea*. *Pharm Biol* 42: 68–72.

Hirasawa, Y., Kobayashi, J., and Morita, H. 2006. Lycoperine A, A novel C27N3-type pentacyclic alkaloid from *Lycopodium hamiltonii*, inhibiting acetylcholinesterase. *Org Lett* 8:123–126.

Hirasawa, Y., Morita, H., Shiro, M., and Kobayashi J. 2003. Sieboldine A, a novel tetracyclic alkaloid from *Lycopodium sieboldii*, inhibiting acetylcholinesterase. *Org Lett* 5: 3991–3993.

Horiuchi, Y., Kimura, R., Kato, N., Fujii, T., Seki, M., Endo, T., Kato, T., and Kawashima, K. 2003. Evolutionary Study on acetylcholine expression. *Life Sci* 72: 1745–1756.

Hormazabal, E., Schmeda-Hirschmann, G., Astudillo, L., Rodriguez, J., and Theoduloz, C. 2005. Metabolites from *Microsphaeropsis olivacea*, an endophytic fungus of Pilgerodendron uviferum. *Z Naturforsch C* 60: 11–21.

Houghton, P.J., Agbedahunsi, J.M., and Adegbulugbe, A. 2004. Choline esterase inhibitory properties of alkaloids from two Nigerian *Crinum* species. *Phytochemistry* 65: 2893–2896.

Ingkaninan, K., Changwijit, K., and Suwanborirux, K. 2006a. Vobasinyl-iboga bisindole alkaloids, potent acetylcholinesterase inhibitors from *Tabernaemontana divaricata* root. *J Pharm Pharmacol* 58: 847–852.

Ingkaninan, K., Phengpa, P., Yuenyongsawad, S., and Khorana, N. 2006b. Acetylcholinesterase inhibitors from *Stephania venosa* tuber. *J Pharm Pharmacol* 58: 695–700.

Jang, K.H., Lee, B.H., Choi, B.W., Lee, H.S., and Shin, J. 2005. Chromenes from the brown alga *Sargassum siliquastrum*. *J Nat Prod* 68: 716–723.

Kalauni, S.K., Choudhary, M.I., and Khalid, A. 2002. New cholinesterase inhibiting steroidal alkaloids from the leaves of *Sarcococca coriacea* of Nepalese origin. *Chem Pharm Bull* 50: 1423–1426.

Kalauni, S.K., Choudhary, M.I., Shaheen, F., Manandhar, M.D., Atta-ur-Rahman, Gewali, M.B., and Khalid, A. 2001. Steroidal alkaloids from the leaves of *Sarcococca coriacea* of Nepalese origin. *J Nat Prod* 64: 842–844.

Kang, S.Y., Lee, K.Y., and Sung,.S.H. 2001. Coumarins isolated from *Angelica gigas* inhibit acetylcholinesterase: Structure-activity relationships. *J Nat Prod* 64: 683–685.

Kawashima, K., and Fujii, T. 2003. The lymphocytic cholinergic system and its contribution to the regulation of immune activity. *Life Sci* 74: 675–696.

Khan, S.B., Malik, A., Afza, N., Jahan, N., Haq, A.U., Ahmed, Z., Nawaz, S.A., and Choudhary, M.I. 2004. Enzyme inhibiting terpenoids from *Amberboa ramosa*. *Z Naturforsch B* 59: 579–583.

Kim, D.K., and Lee, K. 2003. Inhibitory effect of trans-N-p-coumaroyl tryamine from the twigs of *Celtis chinensis* on the acetylcholinesterase. *Arch Pharm Res* 26: 735–738.

Kim, D.K., Lee, K.T., Baek, N.I., Kim, S.H., Park, H.W., Lim, J.P., Shin, T.Y., Eom, D.O., Yang, J.H., and Eun, J.S. 2004. Acetylcholinesterase inhibitors from the aerial parts of *Corydalis speciosa*. *Arch Pharm Res* 27: 1127–1131.
Kim, D.K., Lim, J.P., Yang, J.H., Eom, D.K., Eun, J.S., and Leem, K.H. 2002. Acetylcholinesterase inhibitors from the roots of *Angelica dahurica*. *Arch Pharm Res* 25: 856–859.
Kim, D.K. 2002. Inhibitory effect of corynoline isolated from the aerial parts of *Corydalis incisa* on the acetylcholinesterase. *Arch Pharm Res* 25: 817–819.
Kim, S.R., Hwang, S.Y., Jang, Y.P., Park, M.J., Markelonis, G.J., Oh, T.H., and Kim, YC. 1999. Protopine from *Corydalis ternata* has anticholinesterase and antiamnesic activities. *Planta Med* 65: 218–221.
Kissling, J., Ioset, J.R., Marston, A., and Hostettmann, K. 2005. Bio-guided isolation of cholinesterase inhibitors from the bulbs of *Crinum × powellii*. *Phytother Res* 19: 984–987.
Kummer, W., and Lips, K.S. 2006. Non-neuronal acetylcholine release and its contribution to COPD pathology. *Drug Discov Today Dis Mech* 3: 47–52.
Lee, J.H., Lee, K.T., Yang, J.H., Baek, N.I., and Kim, D.K. 2004. Acetylcholinesterase inhibitors from the twigs of *Vaccinium oldhami* Miquel. *Arch Pharm Res* 27: 53–56.
Lee, K.Y., Yoon, J.S., Kim, E.S., Kang, S.Y., and Kim, Y.C. 2005. Anti-acetylcholinesterase and anti-amnesic activities of a pregnane glycoside, cynatroside B, from *Cynanchum atratum*. *Planta Med* 71: 7–11.
Li, H.Z., Wang, Q., Ruan, X., Pan, C.D., and Jiang, D.N. 2010. Phenolics and plant allelopathy. *Molecules* 15:8933–8952.
Liu, X.G., Tian, F.J., Tian, Y.Y., Wu, Y.B., Dong, F.S., Xu, J., and Zheng, Y.Q. 2016. Isolation and identification of potential allelochemicals from aerial parts of *Avena fatua* L. and their allelopathic effect on wheat. *J Agric Food Chem* 64: 3492–3500.
Lopez, S., Bastida, J., Viladomat, F., and Codina, C. 2002. Acetylcholinesterase inhibitory activity of some Amaryllidaceae alkaloids and *Narcissus* extracts. *Life Sci* 71: 2521–2529.
Mehmood, S., Riaz, N., Nawaz, S.A., Afza, N., Malik, A., and Choudhary, M.I. 2006. New butyrylcholinesterase inhibitory triterpenes from *Salvia santolinifolia*. *Arch Pharm Res* 29: 195–198.
Ming, D.S., Lopez, A., Hillhouse, B.J., French, C.J., Hudson, J.B., and Towers, G.H.N. 2002. *Bioactive constituents from Iryanthera megistophylla*. *J Nat Prod* 65: 1412–1416.
Mithöfer, A., and Boland, W. 2012. Plant defense against herbivores: Chemical aspects. *Annu Rev Plant Biol* 63: 431–450.
Miyazawa, M., Tougo, H., and Ishihara, M. 2001. Inhibition of acetylcholinesterase activity by essential oil from *Citrus paradisi*. *Nat Prod Lett* 15: 205–210.
Miyazawa, M., Tsukamoto, T., Anzai, J., and Ishikawa, Y. 2004. Insecticidal effect of phthalides and furanocoumarins from *Angelica acutiloba* against Drosophila melanogaster. *J Agr Food Chem* 52: 4401–4405.
Mukherjee, P.K., Kumar, V., and Houghton, P.J. 2007. Screening of Indian medicinal plants for acetylcholinesterase inhibitory activity. *Phytother Res* 21: 1142–1145.
Orhan, E.I., Senol, F.S., Shekfeh, S., Skalicka-Wozniak, K., and Banoglu, E. 2017. Pteryxin-A promising butyrylcholinesterase-inhibiting coumarin derivative from *Mutellina purpurea*. *Food Chem Toxicol* 109: 970–974.
Orhan, I., Terzioglu, S., and Sener, B. 2003. Alpha-onocerin: An acetylcholinesterase inhibitor from *Lycopodium clavatum*. *Planta Med* 69: 265–267.
Otoguro, K., Kuno, F., and Omura, S. 1997. Arisugacins, selective acetylcholinesterase inhibitors of microbial origin. *Pharmacol Ther* 76: 45–54.
Penumala, M., Zinka, R.B., Shaik, J.B., and Gangaiah, DA. 2017. In vitro screening of three Indian medicinal plants for their phytochemicals, anticholinesterase, antiglucosidase, antioxidant, and neuroprotective effects. *BioMed Res Int* Volume 2017, Article ID 5140506, 12 pages. doi:10.1155/2017/5140506.
Perry, N.S., Houghton, P.J., Theobald, A., Jenner, P., and Perry, E.K. 2000. In-vitro inhibition of human erythrocyte acetylcholinesterase by *Salvia lavandulaefolia* essential oil and constituent terpenes. *J Pharm Pharmacol* 52: 203, 895–902.
Perveen, S., Khan, S.B., Malik, A., Tareen, R.B., Nawaz, S.A., and Choudhary, M.I. 2006. Phenolic constituents from *Perovskia atriplicifolia*. *Nat Prod Res* 20: 347–353.
Queiroz, E.F., Hay, A.E., Chaaib, F., van Diemen, D., Diallo, D., and Hostettmann, K. 2006. New and bioactive aromatic compounds from *Zanthoxylum zanthoxyloides*. *Planta Med* 72: 746–750.
Ren, Y., Houghton, P.J., Hider, R.C., and Howes, M.J. 2004. Novel diterpenoid acetylcholinesterase inhibitors from *Salvia miltiorhiza*. *Planta Med* 70: 201–204.
Rhee, I.K., Appels, N., Hofte, B., Karabatak, B., Erkelens, C., Stark, L.M., Flippin, L.A., and Verpoorte, R. 2004. Isolation of the acetylcholinesterase inhibitor ungeremine from *Nerine bowdenii* by preparative HPLC coupled on-line to a flow assay system. *Biol Pharm Bull* 27: 1804–1809.

Rice, E.L. 1983. *Allelpoathy: Physiological Ecology*. 2nd Edition. Academic Press Inc., Orlando, Florida.
Rizvi, S.J.H., Haque, H., Singh, V.K., and Rizvi, V. 1992. A discipline called allelopathy. In: Rizvi, S.J.H. and Rizvi, V. (Eds.). *Allelopathy*. Dordrecht, the Netherlands: Speinger.
Rollinger, J.M., Hornick, A., Langer, T., Stuppner, H., and Prast, H. 2004. Acetylcholinesterase inhibitory activity of scopolin and scopolrtin discovered by virtual screening of natural products. *J Med Chem* 47: 6248–6254.
Rollinger, J.M., Mocka, P., Zidorn, C., Ellmerer, E.P., Langer, T., and Stuppner, H. 2005. Application of the in combo screening approach for the discovery of non-alkaloid acetylcholinesterase inhibitors from *Cichorium intybus*. *Curr Drug Discov Technol* 2: 185–193.
Rollinger, J.M., Schuster, D., Baier, E., Ellmerer, E.P., Langer, T., and Stuppner, H. 2006. Taspine: Bioactivity-guided isolation and molecular ligand-target insight of a potent acetylcholinesterase inhibitor from Magnolia × soulangiana. *J Nat Prod* 69: 1341–1346.
Roshchina, V.V. 2010. Evolutionary considerations of neurotransmitters in microbial, plant and animal cells. In *Microbial Endocrinology: Interkingdom Signaling in Infectious Disease and Health*, pp. 17–52. Cham, Switzerland: Springer Science + Business Media.
Roshchina, V.V. 2001. *Neurotransmitters in Plant Life*. Enfield, NH: Science Publishers.
Ryu, G., Park, S.H., Kim, E.S., Choi, B.W., Ryu, S.Y., and Lee, B.H. 2003. Cholinesterase inhibitory activity of two farnesylacetone derivatives from the brown alga *Sargassum sagamianum*. *Arch Pharm Res* 26: 796–799.
Sastry, B.V., and Sadavongvivad, C. 1978. Cholinergic systems in non-nervous tissues. *Pharmacol Rev* 30: 65–132.
Savelev, S., Okello, E., Perry, N.S.L., Wilkins, R.M., and Perry, E.K. 2003. Synergistic and antagonistic interactions of anticholinesterase terpenoids in *Salvia lavandulaefolia* essential oil. *Pharmacol Biochem Behav* 75: 661–668.
Sharma, R., and Gupta, R. 2007. *Cyperus rotundus* extract inhibits acetylcholinesterase activity from animal and plants as well as inhibits germination and seedling growth in wheat and tomato. *Life Sci* 80: 2389–2392.
Sharma, R. 2007. Studies on anticholinesterases in weeds and the allelopathy acetylcholine nexus. PhD thesis, University of Delhi, New Delhi, India.
Singh, H.P., Batish, D.R., Kaur, S., Arora, K., and Kohli R.K. 2006. α-Pinene inhibits growth and induces oxidative stress in roots. *Annals Bot* 98: 1261–1269.
Smallman, B.N., and Maneckjee, A. 1981. The synthesis of acetylcholine by plants. *Biochem J* 194: 361–364.
Suganthy, N., and Devi, K.P. 2016. In vitro antioxidant and anti-cholinesterase activities of *Rhizophoramuronata*. *Pharm Biol* 15: 118–129.
Sung, S.H., Kang, S.Y., Lee, K.Y., Park, M.J., Kim, J.H., Park, J.H., Kim, Y.C., Kim, J., and Kim, Y.C. 2002. (+)-alpha-viniferin, a stilbene Trimer from *Caragana chamlague* inhibits acetylcholinesterase. *Biol Pharm Bull* 25: 125–127.
Takahashi, T. 2015. Non-neuronal functions of acetylcholine via muscarinic receptor subtypes: Cellular proliferation and differentiation in mammals. *Neurotransmitter* 2: 499–501.
Tretyn, A., Slesak, E., and Kwiatkowska, K. 1986. Cytochemical localization of AChE in plant cells in LM/TEM/SEM. *Folia Histochem Cyto* 24: 328–329.
Tsukamoto, T., Ishikawa, Y., and Miyazawa, M. 2005. Larvicidal and adulticidal activity of alkylphthalide derivatives from rhizome of *Cnidium officinale* against *Drosophila melanogaster*. *J Agr Food Chem* 53: 5549–5553.
Tundis, R., Bonesi, M., Menichini, F., and Loizzo, M.R. 2016. Recent knowledge on medicinal plants as source of cholinesterase inhibitors for the treatment of dementia. *Mini Rev Med Chem* 16: 605–618.
Turi, C.E., Axwik, K.E., Smith, A., Jones, A.M.P., Saxena, P.K., and Murch, S.J. 2014. Galanthamine, an anticholinesterase drug, effects plant growth and development in *Artimisia tridentata* Nutt. via modulation of auxin and neurotransmitter signalling. *Plant Signal Behav* 9: e28645. doi:10.4161/psb.28645.
Urbain, A., Marston, A., and Hostettmann, K. 2005. Coumarins from *Peucedanum ostruthium* as inhibitors of acetylcholinesterase. *Pharm Biol* 43: 647–650.
Urbain, A., Marston, A., Queiroz, E.F., Ndjoko, K., and Hostettmann, K. 2004. Xanthones from *Gentiana campestris* as new acetylcholinesterase inhibitors. *Planta Med* 70: 1011–1014.
Wesseler, I., Kilbinger, H., Bittinger, F., and Kirkpatrick, C.J. 2001. The biological role of non-neuronal acetylcholine in plants and humans. *Jpn J Pharmacol* 85: 2–10.
Wesseler, I., and Kirkpatrick, C.J. 2008. Acetylcholine beyond neurons: The non-neuronal cholinergic system in humans. *Br J Pharmacol* 154: 1558–1571.

Wesseler, I.K., and Kirkpatrick, C.J. 2017. Non-neuronal acetylcholine involved in reproduction in mammals and honeybees. *J Neurochem* 142: 144–150. doi:10.1111/jnc.13953.

Weston, L.A., and Duke, S.O. 2003. Weed and crop allelopathy. *Crit Rev Plant Sci* 22: 367–389.

Wink, M., Schmeller, T., and Latz-Bruning, B. 1998. Modes of action of allelochemical alkaloids: Interaction with neuroreceptors, DNA and other molecular targets. *J Chem Eco* 24: 1881–1937.

Yamamoto, K., Shida, S., Honda, Y., Shono, M., Miyake, H., Oguri, S., Sakamoto, H., and Momonoki, Y.S. 2015. Overexpression of acetylcholinesterase gene in rice results in enhancement of shoot gravitropism. *Biochem Biophys Res Commun* 465: 488–493.

Yan, Z.Q., Wang, D.D., Cui, H.Y., Zhang, D.H., Sun, Y.H., Jin, H., Li, X.Z. et al. 2016. Phytotoxicity mechanisms of two coumarin allelochemicals from *Stellera chamaejasme* in lettuce seedlings. *Acta Physiol Plant* 38: 248. doi:10.1007/s11738-016-2270-z.

Yoo, I.D., Cho, K.M., Lee, C.K., and Kim, W.G. 2005. Isoterreulactone A, a novel meroterpenoid with anti-acetylcholinesterase activity produced by *Aspergillus terreus*. *Bioorg Med Chem Lett* 15: 353–356.

Yoshida, M., Inadome, A., Maeda, Y., Satoji, Y., Masunaga, K., Sugiyama, Y., and Murakami, S. 2006. Non-neuronal cholinergic system in human bladder urothelium. *Urology* 67: 425–430.

Yusoff, M., Hamid, H., and Houghton, P. 2014. Anticholinesterase inhibitory activity of quaternary alkaloids from *Tinosporacrispa*. *Molecules* 19: 1201–1211.

Zhiqun, T., Jian, Z., Junli, Y., Chunzi, W., and Danju, Z. 2017. Allelopathic effects of volatile organic compounds from *Eucalyptus grandis* rhizosphere soil on *Eisenia fetida* assessed using avoidance bioassays, enzyme activity, and comet assays. *Chemosphere* 173: 307–317.

# 16 Production of Neurochemicals by Microorganisms

## *Implications for Microbiota–Plant Interactivity*

*Alexander V. Oleskin and Boris A. Shenderov*

**CONTENTS**



## 16.1 INTRODUCTION

Presently, neuroactive chemical compounds are the focus of attention of several different fields of science, including microbiology, neurology, endocrinology, psychology, and sociology. These important biologically active substances are subdivided into neurotransmitters that are directly responsible for transferring impulses between nervous cells (neurons) across the synaptic cleft; neuromodulators that promote or impede impulse transmission between neurons; and neurohormones that are produced by nervous cells and exert a systemic influence on the whole organism. In the literature, the aforementioned partly overlapping groups of compounds are referred to as *neurochemicals* (Lyte, 2016) or, alternatively, as *biomediators* (Roshchina, 1991, 2010a, b).

This review work concentrates on currently available data on the role of neurochemicals in terms of interaction between plants and their microbial inhabitants, such as prokaryotes and unicellular eukaryotes including various fungi. Diverse species of angiosperms, gymnosperms, pteridophytes, and mosses provide the microbiota with a wide variety of ecological niches, including the surface of the plant organism and the interior of plant organs (colonized by endophytes) that are exemplified by root nodules inhabited, by rhizobia. The relationships between microbiota representatives and the host plant vary between mutualism, commensalism, and parasitism; the same microbiota can play several different roles, depending on the host plant species, the location of the microorganisms involved on/in the plant organism, and other circumstances.

In an analogy to the relatively well-understood microbiota–animal/human organism system (Shenderov, 2008, 2013, 2016; Oleskin et al., 2010, 2016, 2017a, b; Lyte, 2011, 2013, 2016; Oleskin and Shenderov, 2016; Averina and Danilenko, 2017; Oleskin and Rogovsky, 2017), the complex gamut of interactions in the microbiota–plant system is apparently based upon sophisticated chemical communication facilities. Importantly, these facilities are exemplified by *quorum-sensing* (QS) systems that enable prokaryotes to modify their metabolic, genetic, and behavioral activities depending on the density of their populations that is estimated from the concentrations of pheromones (autoinducers) produced by microbial cells. Quorum sensing-dependent processes include secondary metabolism, bioluminescence, protein secretion, motility, virulence factor production, plasmid transfer, and biofilm maturation in diverse bacteria (Fuqua and Greenberg, 2002; Williams, 2007). QS systems are involved in microbiota–host interactivity. For instance, the soil bacterium *Rhodopseudomonas palustris* incorporates plant host-produced *p*-coumarate in its QS pheromone (reviewed, Cooley et al., 2008).

In this work, emphasis is placed upon the functional roles of the subgroup of neurochemicals referred to as *biogenic amines* (including acetylcholine) with respect to the bidirectional neurochemical *dialogue* between the microbiota and the host plant organism. Some of them function as bacterial QS pheromones. For example, the important neurotransmitter serotonin also behaves as the pheromone of the *lasIR* QS system in *Pseudomonas aeruginosa.* It enhances virulence and biofilm formation both *in vitro* and *in vivo*, that is, in the organism of an infected mouse (Knecht et al., 2016).

In the example of acetylcholine and biogenic amines, the following text deals with two interrelated aspects of the functional role of neurochemicals in the microbiota–plant system:

- The microbiota produces neurochemicals that exert an important influence on the host plant organism.
- The host plant releases neurochemicals that impact microbial physiological, morphogenetic, and biochemical processes.

## 16.2 PLANT MICROBIOTA PRODUCES NEUROCHEMICALS THAT EXERT AN INFLUENCE ON THE HOST ORGANISM

Recently, it has been established that various prokaryotic and eukaryotic microorganisms produce a large number of neurochemicals including biogenic amines, such as catecholamines, serotonin, histamine, acetylcholine, and agmatine (Tsavkelova et al., 2000; Özogul, 2004; Shishov et al., 2009; Oleskin et al., 2010, 2014a, b, 2016, 2017a; Özogul et al., 2012; Zhilenkova et al., 2013; Oleskin and Shenderov, 2016). For instance, *Escherichia coli* K-12-synthesized dopamine, norepinephrine, and serotonin were present in the culture liquid during the later culture growth stages at concentrations of ~10–20 nM (Shishov et al., 2009). Some strains of probiotic lactic-acid bacteria tested in the work (Oleskin et al., 2014a, b) also produced these amines; for example, the strain *Lactobacillus helveticus* 100ash released serotonin into the culture liquid (0.4 μM). Neurochemicals-producing prokaryotic and eukaryotic microorganisms inhabit various plant-associated ecological niches. They are exemplified by the species included in the Table 16.1. It is known that virtually every plant has microbial tenants: “All plants are inhabited internally by diverse microbial communities comprising bacterial, archaeal, fungal, and protistic taxa” (Hardoim et al., 2015, p. 293). Therefore, the incomplete list given in the table was limited by the data on microbial species and strains that both (a) produce neurochemicals and (b) represent widespread inhabitants of plant organisms. In an analogy to the human/animal microbiota that exerts a strong multifaceted influence on the host organism with its nervous and immune systems (Oleskin et al., 2016, 2017a, b; Oleskin and Rogovsky, 2017),

**TABLE 16.1**
**Neurochemicals-Producing Microorganisms That Occupy Ecological Niches on/in Plant Organisms**

| Microorganisms | Neurochemicals | Typical Host Plants |
|---|---|---|
| *Escherichia coli* | Dopamine, norepinephrine, serotonin, histamine (Tsavkelova et al., 2000, Shishov et al., 2009) | Lettuce (Rosenblueth and Martinez-Romero, 2006) |
| *Klebsiella pneumonia* | Serotonin (Özogul, 2004) | Giant cactus (International Collection of Microorganisms from Plants, 2018) |
| *Lactobacillus plantarum* **NCFB 2392** | Serotonin (Özogul et al., 2012) | Common onion (International Collection of Microorganisms from Plants, 2018) |
| *Lactococcus lactis* **subsp.** *Lactis* | Dopamine, norepinephrine, serotonin (Stoyanova et al., 2017) | Maize, peas (Alemayehu et al., 2014) |
| *Serratia marcescens* | Dopamine (Tsavkelova et al., 2000) | Common onion, alfalfa, rice (International Collection of Microorganisms from Plants, 2018) |
| *Pseudomonas aeruginosa* | Dopamine (Tsavkelova et al., 2000) | Common onion, maize, chrysanthemum (International Collection of Microorganisms from Plants, 2018) |

*Note:* Two kinds of literature are used in combination: (1) publications that deal with neurochemicals-producing microorganisms and (2) reports on microbial species/strains isolated from plants; the table only contains representative microbial species that are mentioned in both types of publications.

the plant microbiota is expected to produce a considerable effect on the host organism. The following text addresses this subject in some detail.

It should be noted that, apart from prokaryotes, neurochemicals are produced by eukaryotic and, particularly, fungal inhabitants of the plant host organism, including those forming a part of mycorrhiza systems.

### 16.2.1 Microbial Neurochemicals Exert a Regulatory Influence on the Plant Organism with Respect to Growth-Related, Biochemical, and Morphogenetic Activities

Microbial neurochemicals exert a regulatory influence on the plant organism with respect to growth-related, biochemical, and morphogenetic activities. Serotonin (5-hydroxytryptamine) that is formed by some of the prokaryotic inhabitants of plants represents "one of the most active modulators of growth and development both in animals and plants. According to the experiments in which serotonin induced the formation of the growth curvature in oat coleoptiles, the suggestion was put forward that, due to its similarity to indoleacetic acid, this compound performs an analogous growth-stimulating function. An auxin-like effect was later revealed in the hypocotyls of the lupine *Lupinus albus*. It was also established that serotonin stimulates root formation in the tissue culture obtained from *Populus tremuloides* × *P. tremula* leaves–even to a larger extent than indoleacetic acid" (Roshchina, 2010a, electronic version). Hence, serotonin-producing plant microbiota representatives are functionally analogous to the well-studied microorganisms that form plant hormones such as auxins. "The ability to produce auxins…is a typical trait for root-associated endophytes" (Hardoim et al., 2015, p. 302). Interestingly, apart from serotonin per se, a large number of high-performance liquid chromatography (HPLC)-tested microorganisms (including the widespread plant inhabitants *P. aeruginosa* and *Serratia marcescens*) also synthesize 5-hydroxyindoleacetic acid, the product of oxidative deamination of serotonin (Tsavkelova et al., 2000) that represents a substituted auxin. The development of the vegetative microspores of the common horsetail *Equisetum arvense* is stimulated by serotonin, dopamine, and norepinephrine (Roshchina, 2010a) that are formed and released into the medium by a sufficiently large number of microbial species (Tsavkelova et al., 2000). Dopamine and serotonin also promote the development of the pollen of *Hippeastrum hybridum* (Roshchina, 2010a).

Multifarious physiological effects that are partly associated with membrane organelles are caused by acetylcholine, which is produced by diverse prokaryotes including bacilli and lactobacilli (reviewed, Oleskin et al., 2017a). Among the representatives of these taxonomic groups, there are a large number of plant inhabitants (Rosenblueth and Martinez-Romero, 2006; Hardoim et al., 2015; International Collection of Microorganisms from Plants, 2018). Of special interest are the effects of acetylcholine on the motility of intracellular structures, plant cells, and whole plants, including protoplast contraction, cytoplasm cyclosis, stomata opening, nastic movements, and processes in membranes such as ion channel opening. For instance, "In *Nitella* cells, a decrease in cytoplasm movement is observed at an acetylcholine content of $10^{-7}$–$10^{-6}$ M in the medium, and the movement is completely arrested at still higher acetylcholine concentrations... Epinephrine and norepinephrine slow down cyclosis in the cells of this charophyte at approximately the same concentrations" (Roshchina, 2010a, electronic version).

Although emphasis is placed on biogenic amines in this work, it should be noted that microorganisms release a wide variety of other neuroactive compounds, such as amino acids, peptides, steroids, nucleotides, and gaseous substances (reviewed, Lenard, 1992; Oleskin et al., 1998a, 2016, 2017a; Oleskin and Shenderov, 2016). Adenine ribosides that are known to function as neuromediators (binding to the A receptors in the brain) are also formed by bacterial and fungal endophytes. Endophytes growing in meristematic bud tissues of Scots pine (*Pinus sylvestric* L.) such as the bacterium *Methylobacterium extorquens* and the yeast *Rhodotorula minuta* produce adenine ribosides that stimulate the growth and mitigate the browning of the host's tissues (Pirttilä et al., 2004).

### 16.2.2 Microbial Neurochemicals Exert Regulatory and Toxic Effects within the Microbial Consortium That Occupies Ecological Niches on the Surface and in the Tissues of the Plant Host Organism

Microbial neurochemicals exert regulatory and toxic effects within the microbial consortium that occupies ecological niches on the surface and in the tissues of the plant host organism. It is pertinent that the aforementioned work (Hardoim et al., 2015, p. 302) emphasizes that the serotonin-similar compound "indole-3-acetic acid (IAA), a member of the auxin class, increases colonization efficiency..., possibly via interference with the host defense system...and production of this compound or related compounds may be an important property for plant colonization by endophytes." Importantly, various microbial neurochemicals specifically influence not only the plant host but also the microbiota per se. Such neurochemicals can be responsible for intermicrobial communication within the plant microbiota consortium.

Mark Lyte (who coined the term *microbial endocrinology*) and other researchers revealed that addition of catecholamines, especially of norepinephrine, drastically stimulates the growth of a wide variety of bacteria (Lyte, 1993, 2011, 2013, 2016; Lyte and Ernst, 1992; Lyte et al., 1996; Freestone and Lyte, 2008). They are exemplified by the enterohemorrhagic strain *E. coli* O7:H157 (EHEC) in which norepinephrine, apart from promoting growth, also stimulates production of Shiga-like toxins and adhesins (Lyte et al., 1996). Interestingly, norepinephrine, as well as serotonin and histamine, stimulate cell aggregation, the initial stage of biofilm formation, in a non-pathogenic strain of *E. coli* (Shishov et al., 2009). Such facts give grounds for the suggestion that catecholamine- and serotonin-producing microorganisms should stimulate the growth of the plant commensal and/or pathogen *P. aeruginosa* whose development is promoted in the presence of these neurochemicals in the medium. The specific effects of commensals are due to the fact that they "operate as autoinducer AI-3 analogs. Like AI-3, they bind to histidine kinases QseC and QseE" (reviewed, Oleskin et al., 2016, p. 5). To reemphasize, in *P. aeruginosa*, serotonin behaves as the pheromone of the QS system *lasIR* (Knecht et al., 2016). Interestingly, acetylcholine that is produced by a large number of microorganisms affects the chemotaxis of root nodule bacteria (reviewed, Roshchina, 2010a).

If microbiota-produced neurochemicals are present at high concentrations, they exert not only regulatory but also toxic effects on microorganisms, that is, function as antimicrobial agents. At high concentrations (>10 μM), serotonin and histamine suppress the growth of *E. coli* K-12 (Oleskin et al., 1998a, b; Anuchin et al., 2008). Therefore, microbiota-produced neurochemicals are expected to contribute to the antagonistic activity of the microbial inhabitants of plants.

The important role of the antagonistic activity of the plant microbiota is highlighted by the following data: "The frequent isolation of *Curtobacterium flaccumfaciens* as endophytes from asymptomatic citrus plants infected with the pathogen *Xylella fastidiosa* suggested that the endophytic bacteria may help citrus plants to better resist the pathogenic infection... Endophytes from potato plants showed antagonistic activity against fungi...and also inhibited bacterial pathogens belonging to the genera *Erwinia* and *Xanthomonas*... Some of the endophytic isolates produced antibiotics and siderophores in vitro" (Rosenblueth and Martinez-Romero, 2006, p. 830).

This phenomenon and its relationship with neurochemicals still await further research, which also concerns the subject discussed below.

### 16.2.3 Non-Specific Effects of Microbial Neurochemicals: Involvement in Iron Homeostasis

Microbial neurochemicals are also involved in iron homeostasis. Among other potentially important functions of the plant microbiota including endophytes, emphasis should be placed upon production of "vivid siderophores...essential compounds for iron acquisition by soil microorganisms" that "also play important roles in pathogen-host interactions." In particular, "siderophore production was shown to play an important role in the symbiosis of *Epichloë festucae* with ryegrass... It is possible that siderophores modulate iron homeostasis in *E. festucae*-infected ryegrass plants" (Hardoim et al., 2015, p. 302).

Accordingly, one of the potentially important effects of microbial neurochemicals is based on their capacity to help the plant organism maintain Fe homeostasis. Catecholamines can chelate ferric iron, removing it from catecholamine-binding proteins that, in the animal organism, are exemplified by lactoferrin and transferrin of the blood serum and other biological fluids. Catecholamine-bound iron becomes available to bacterial cells that use specific carriers—siderophores (such as enterobactin in *E. coli*) to transfer it into the cell (see review Oleskin et al., 2016, p. 7). For instance, catecholamines stimulate the growth of some strains of lactobacilli and bifidobacteria on the serum-containing SAPI medium (Yunes, 2017) and not on the serum-free BS medium. It seems likely, therefore, that the catecholamines serve as iron chelators in this system (reviewed, Oleskin et al., 2017b).

In all likelihood, the Fe-chelating capacity of catecholamines, including those of microbial origin, should be involved, apart from intermicrobial interaction, in the chemical dialogue between the microbiota and the host plant organism. By influencing the incorporation of iron in cytochromes and other important components, neurochemicals should impact electron transport in the mitochondrial, chloroplast, and plasma membranes of plant cells. This subject should be addressed in future studies.

## 16.3 IMPACT OF HOST NEUROCHEMICALS ON THE MICROBIOTA (IN THE EXAMPLE OF BIOGENIC AMINES)

This short review is chiefly focused upon microbial production of neurochemicals and their influence on the host organism and the microbial consortium associated with it. Nonetheless, there is bidirectional interaction in the host–microbiota system. In the human/animal organism, "microbial cells specifically respond to host-produced neuromediators/neurohormones because they have adapted to them during the course of many millions of years of microbiota–host coevolution" (Oleskin et al., 2017a).

It is known that numerous plant species of various taxonomic groups synthesize neurochemicals, including catecholamines (dopamine, norepinephrine, and epinephrine), serotonin, histamine, and acetylcholine; they contain enzymes that are involved in their biosynthesis, transport, and degradation (Roshchina, 1991, 2010a, b). Microorganisms specifically respond to these neuroactive compounds (Lyte, 1993, 2010, 2011, 2013, 2016; Oleskin et al., 1998a, b, 2010, 2016, 2017a, b; Shenderov, 2008, 2013; Oleskin and Shenderov, 2016). As mentioned above, catecholamines stimulate the growth of a large number of enterobacteria including, unfortunately, pathogens. This results in the development of a vicious circle in an infected animal/human organism: the organism produces catecholamines in response to the infectious stress. These substances stimulate the infection by accelerating the development and increasing the virulence of the pathogen.

There is less data on an analogous effect of host-produced neurochemicals in the microbiota–plant system. It is emphasized in Viktoria Roshchina's work that "catecholamine accumulation may be due to a stress situation, to oxidative processes during fruit maturation, and to leaf and petiole movement in response to mechanical stimuli. The latter is particularly prominent in such plants as *Albizia julibrissin* and *Mimosa pudica*" (Roshchina, 2010a). Could the increased production of neurochemicals in a stressed plant result in stimulating the growth and modifying the properties of the microbiota, including its commensal, mutualistic, and parasitic components? Of relevance might be the fact that plant stem and leaf surfaces produce exudates that attract microorganisms (Compant et al., 2010). Presumably, the microbiota-attracting effect of plant exudates is partly due to the neurochemicals they contain, along with other chemical factors.

Since the data available in the literature testify to the production of sufficiently high concentrations of neurochemicals by some plants, it seems likely that the neurochemicals can produce not only stimulatory and regulatory, but also toxic, that is, antimicrobial, effects, similar to those produced by microbial neurochemicals (see preceding section). An extreme example of neurochemicals' toxic effects concern the alga *Ulvaria obscura* that contains up to 4% of dopamine per raw weight. This seems to be a protective mechanism that prevents marine animals from eating the alga (Roshchina, 2010a) and presumably should also stop most microorganisms from growing on its surface.

It has recently been established that the yeast *Saccharomyces cerevisiae* accumulates serotonin, dopamine, and norepinephrine intracellularly without releasing them into the culture liquid. Therefore, these neurochemicals are unlikely to function as intercellular regulatory signals. Nevertheless, cell proliferation in *S. cerevisiae* is drastically stimulated by dopamine, its agonist apomorphine and, to a lesser extent, by serotonin, but not by norepinephrine. The stimulatory effect of dopamine and serotonin on the yeast is presumably due to the fact that its biofilms, in nature, grow on grape and plum plants and might be evolutionarily adapted to the biogenic amines released by them. Presumably, the neurochemicals are unidirectional signals received by the yeast and produced by other ecosystem components (Malikina et al., 2010; Oleskin et al., 2010, 2015, 2017a, b). In contrast to these effects on the yeast, our recent data seem to indicate that neurochemicals do not significantly influence the growth of the mycelial fungi *Aspergillus oryzae* and *Cunninghamella japonica* (Oleskin et al., 2015). Therefore, it is unlikely that their interaction with plants—if it takes place—involves plant-produced neuromediators that specifically influence the mycobiota.

## 16.4 CONCLUSION

The facts discussed in this review highlight the potential importance of the novel incipient interdisciplinary area of research that focuses on the contributions of neurochemicals such as biogenic amines to the interaction between plants and their microbiota (including the mycobiota). Recent findings demonstrate that neurochemicals are actively produced by plant microbiota, including the inhabitants of the surface of plant organs and of the interior of their tissues. Therefore, it is likely that microbial neurochemicals exert a significant regulatory influence on the morphogenetic, metabolic, and bioenergetic processes in the host plant organism. In addition, it is to be expected that neurochemicals are implicated in communication within the consortium of microorganisms that occupy ecological niches on the surface of plants and inside their organism. Probably, interaction in the microbiota-plant system is bidirectional, and host-produced neurochemicals also exert a specific influence on the microbiota including the mycobiota.

It should be emphasized that the aforementioned data on plant–microbiota interactivity are consistent with Roshchina's idea that universal principles underlie signaling and information transmission in the form of electric and chemical signals in all living organisms (Roshchina, 2010a, b). In terms of biomedical research and biotechnology, the above facts concerning the involvement of neurochemicals, such as biogenic amines and acetylcholine, in plant–microbiota interactivity enable us to develop new strategies of producing drugs including psychopharmacological preparations.

It should be emphasized that *microbial endocrinology*, the field the present work is predominantly concerned with, holds special promise in terms of future research on the mechanisms of adaptation of diverse plant species to their environment. The novel feature of this conceptual approach is that the plant organism is envisaged as an integrated superorganismic system, a consortium that also includes a wide variety of microbial (and fungal) species. An analogous approach has recently received much attention with respect to the interactivity in the human/animal organism–microbiota consortium (reviewed, Lyte, 2011, 2013, 2016; Oleskin et al., 2016; 2017a, b; Shenderov, 2016); to reiterate, this approach seems to be extremely useful both in theoretical and practical terms.

## REFERENCES


Alemayehu, D., J. A. Hannon, O. McAuliffe, and R. P. Ross. 2014. Characterization of plant-derived lactococci on the basis of their volatile compounds profile when grown in milk. *Int J Food Microbiol* 172:57–61.

Anuchin, A. M., D. I. Chuvelev, T. A. Kirovskaya, and A. V. Oleskin. 2008. Effect of neuromediator monoamines on the growth characteristics of *Escherichia coli* K-12. *Microbiology* 77(6):758–765.

Averina, O. V., and V. N. Danilenko. 2017. Human intestinal microbiota and its role in the development and functioning of the nervous system. *Microbiology* 86(1):1–19.

Compant, S., C. Clement, and A. Sessitsch. 2010. Plant growth-promoting bacteria in the rhizo- and endosphere of plants: Their role, colonization, mechanisms involved and prospects for utilization. *Soil Biol Biochem* 42:669–678.

Cooley, M., S. R. Chhabra, and P. Williams. 2008. N-Acyl-homoserine lactone-mediated quorum sensing: A twist in the tail and a blow for host immunity. *Chem Biol* 15:1141–1147.

Freestone, P. P., and M. Lyte. 2008. Microbial endocrinology: Experimental design issues in the study of inter-kingdom signaling in infectious disease. *Adv Appl Microbiol* 64:75–108.

Fuqua, C., and E. P. Greenberg. 2002. Listening in on bacteria: Acyl-homoserine lactone signaling. *Nat Rev Microbiol* 3:685–695.

Hardoim, P. R., L. S. van Overbeek, G. Berg et al. 2015. The hidden world within plants: Ecological and evolutionary considerations for defining functioning of microbial endophytes. *Microbiol Mol Biol Rev* 79:293–320. doi:10.1128/MMBR.00050-14.

International Collection of Microorganisms from Plants (ICMP). 2018. Landcare Research. https://www.landcareresearch.co.nz/resources/collections/icmp.

Knecht, L. D., G. O. O'Connor, R. Mittal et al. 2016. Serotonin activates bacterial quorum sensing and enhances the virulence of *Pseudomonas aeruginosa* in the host. *EBioMedicine* 9:161–169.

Lenard, J. 1992. Mammalian hormones in microbial cells. *Trends Biochem Sci* 17:47–150.

Lyte, M. 1993. The role of microbial endocrinology in infectious disease. *J Endocrinol* 137:343–345.

Lyte, M. 2011. Probiotics function mechanistically as delivery vehicles for neuroactive compounds: Microbial endocrinology in the design and use of probiotics. *Bioessays* 33:574–581.

Lyte, M. 2013. Microbial endocrinology and nutrition: A perspective on new mechanisms by which diet can influence gut-to brain-communication. *Pharma Nutrition* 1:35–39.

Lyte, M. 2016. Microbial endocrinology in the pathogenesis of infectious disease. *Microbiol Spectrum* 4(2). doi:10.1128/microbiolspec.

Lyte, M., and Ernst S. 1992. Catecholamine induced growth of gram negative bacteria. *Life Sci* 50:203–212.

Lyte, M., C. D. Frank, and B. T. Green. 1996. Production of an autoinducer of growth by norepinephrine-cultured *Escherichia coli* O157:H7. *FEMS Microbiol Lett* 139(2–3):155–159.

Malikina, K. D., V. A. Shishov, D. I. Chuvelev, V. S. Kudrin, and A. V. Oleskin. 2010. Regulatory role of neuromediator amines in *Saccharomyces cerevisiae* cells. *Appl Biochem Micro* 46(6):672–677.

Oleskin, A. V., and V. S. Rogovsky. 2017. Role of biogenic amines in the interaction between the microbiota and the nervous and the immune system of the host organism. *J Restorative Med Rehabil* No. 1:41–51.

Oleskin, A. V., I. V. Botvinko, and T. A. Kirovskaya. 1998a. Microbial endocrinology and biopolitics. *Moscow State University Herald* (*Russian*) 4:3–10.

Oleskin, A. V., T. A. Kirovskaya, I. V. Botvinko, and L. V. Lysak. 1998b. Effect of serotonin (5-hydroxytryptamine) on the growth and differentiation of microorganisms. *Microbiology* 67(3):306–311.

Oleskin, A. V., V. I. Shishov, and K. D. Malikina. 2010. *Symbiotic Biofilms and Brain Neurochemistry*, Hauppauge, New York: Nova Science.

Oleskin, A. V., I. R. Vodolazov, K. D. Malikina, and A. V. Kurakov. 2015. Effects of neuroactive amines on the growth characteristics of yeast and mycelial fungi. *Mycoses* 58(Suppl 3):50–51.

Oleskin, A. V., O. G. Zhilenkova, B. A. Shenderov, A. M. Amerhanova, V. S. Kudrin, and P. M. Klodt. 2014a. Lactic-acid bacteria supplement fermented dairy products with human behavior-modifying neuroactive compounds. *J Pharm Nutrit Sci* 4:199–206.

Oleskin, A. V., O. G. Zhilenkova, B. A. Shenderov, A. M. Amerhanova, V. S. Kudrin, P. M. Klodt. 2014b. Starter cultures of lactobacilli as producers of neuromediators such as biogenic amines and amino acids. *Mol Prom* (*Russian*) 9:42–43.

Oleskin, A. V., G. I. El'-Registan, and B. A. Shenderov. 2016. Role of neuromediators in the functioning of the human microbiota: "Business talks" among microorganisms and the microbiota-host dialogue. *Microbiology* 85(1):1–22.

Oleskin, A. V., and B. A. Shenderov. 2016. Neuromodulatory effects and targets of the SCFAs and gasotransmitters produced by the human symbiotic microbiota. *Microb Ecol Health*. doi:10.3402/mehd.v27.30971.

Oleskin, A. V., B. A. Shenderov, and V. S. Rogovsky. 2017a. Role of neurochemicals in the interaction between the microbiota and the immune and the nervous system of the host organism. *Probiotics Antimicro Prot*. doi:10.1007/s12602-017-9262-1.

Oleskin, A. V., O. G. Zhilenkova, and I. R. Vodolazov. 2017b. Biotechnology and microbial neurochemistry: biotechnological implications of the role of neuromediators in microbial systems. *Adv Biotech Micro* 2(4). doi:10.19080/AIBM.2017.02.555594.

Özogul, F. 2004. Production of biogenic amines by *Morganella morganii, Klebsiella pneumonia* and *Hafnia alvii* using a rapid HPLC method. *Eur Food Res Technol* 219:465–469.

Özogul, F., E. Kuley, Y. Özogul, and I. Özogul. 2012. The function of lactic acid bacteria on bio-genic amines production by food-borne pathogens in arginine decarboxylase broth. *Food Sci Technol Res* 18:795–804.

Pirttilä, A. M., P. Joensuu, H. Pospiech, J. Jalonen, and A. Hohtola. 2004. Bud endophytes of Scots pine produce adenine derivatives and other compounds that affect morphology and mitigate browning of callus cultures. *Physiol Plantarum* 121:305–312.

Rosenblueth M., and E. Martinez-Romero. 2006. Bacterial endophytes and their interactions with hosts. *Mol Plant Microbe Interact* 19(8):827–837.

Roshchina, V. V. 1991. *Biomediatory v rasteniyakh. Atsetilkholin i biogennye aminy* (Biomediators in Plants. Acetylcholine and Biogenic Amines), Pushchino: NTS.

Roshchina, V. V. 2010a. *Neirotransmittery–biomediatory i regulyatory rasteniy* (Neurotransmitters: plant Biomediators and Regulators). Pushchino-na-Oke: Institute of Cell Biophysics, Russian Academy of Sciences.

Roshchina, V. V. 2010b. Evolutionary considerations of neurotransmitters in microbial, plant, and animal cells, in *Microbial Endocrinology: Interkingdom Signaling in Infectious Disease and Health*, M. Lyte, and P. P. E. Freestone (Eds.), pp. 17–52. New York: Springer.

Shenderov, B. A. 2008. *Funktsional'noe pitanie i ego rol' v profilaktike metabolicheskogo sindroma* (Functional nutrition and its role in prevention of the metabolic syndrome). Moscow: DeLi Print. 318 pages.

Shenderov, B. A. 2013. Targets and effects of short-chain fatty acids. *Sovrem Med Nauka* (*Russian*) No. 1–2:21–50.

Shenderov, B. A. 2016. The microbiota as an epigenetic control mechanism. Chapter 11. In *The Human Microbiota and Chronic Disease: Dysbiosis as a Cause of Human Pathology*, L. Nibali, and B. Henderson (Eds.), pp. 179–197. Hoboken, NJ: John Wiley & Sons.

Shishov, V. A., T. A. Kirovskaya, V. S. Kudrin, and A. V. Oleskin. 2009. Amine neuromediators, their precursors, and oxidation products in the culture of *Escherichia coli* K-12. *Appl Biochem Micro* 45(5):494–497.

Stoyanova, L. G., I. R. Vodolazov, and A. V. Oleskin. 2017. Probiotic strains of *Lactococcus lactis* subsp. *lactis* produce neuroactive substances. *Food Quality and Safety, Health & Nutrition Conference*. Skopje, Macedonia: NUTRICON.

Tsavkelova, E. A., I. V. Botvinko, V. S. Kudrin, and A. V. Oleskin. 2000. Detection of neuromediator amines in microorganisms by high-performance liquid chromatography. *Dokl Akad Nauk* 372:840–842.

Williams, P. 2007. Quorum sensing, communication and cross-kingdom signalling in the bacterial world. *Microbiology* 153:3923–3938.

Yunes, R. A. 2017. Adaptive significance of the bacteria of the genera *Lactobacillus* and *Bifidobacterium* for humans. Summary of the Cand. Sci. (Biol.) Dissertation. Moscow: RUDN University [In Russian].

Zhilenkova, O. G., B. A. Shenderov, P. M. Klodt, V. S. Kudrin, and A. V. Oleskin. 2013. Dairy products as a potential source of compounds affecting consumer behavior. *Mol Prom* (*Russian*) 10:16–19.

# 17 Possible Role of Biogenic Amines in Plant–Animal Relations

*Victoria V. Roshchina*

## CONTENTS



## 17.1 INTRODUCTION

The sensitivity of every living cell—microbial, plant, animal—to the neurotransmitters is new global problem, especially in aspect of the behavior, growth, and development of organisms in biocenosis (Roshchina 2010, 2016a). The role of neurotransmitters excreted in the plant–animal relations is *terra incognita* as yet, although in a phenomenon named allelopathy (includes all types of chemical relations between organisms in community), the occurrence of biogenic amines—catecholamines and histamine in plant cells—has become of interest. The knowledge is valuable for fundamental biochemistry and ecology as well as for medicine, pharmacology, and food production.

A possible way to study the interactions of the plants and animals include the determination of the neurotransmitter presence in plant species analyzed, and then choosing a model animal object, which is the natural sensor. Biogenic amines—catecholamines and histamine—actively influence the growth and morphogenesis of both of plants (Kamo and Mahlberg 1984; Protacio et al. 1992; Roshchina 1992; Kuklin and Conger 1995; Kulma and Szopa 2007) and animals (Buznikov 1990). Moreover, their presence in plants directly connects with stress (Świędrych et al. 2004).

The first experiments to determine biogenic amines were done by biochemical methods, and then chromatography, mass spectrometry, and fluorescent analysis were used (Roshchina 2016a). For express diagnostics, neurotransmitters such as biogenic amines in cells and cellular secretion fluorescent methods (glyoxylic acid as reagent for catecholamines, such as dopamine, noradrenaline, and adrenaline or *ortho*-phthalic aldehyde as a reagent for histamine) may be used (Roshchina et al. 2016b) via special procedures of the sample drying at 80°C in histochemical assays with induced fluorescence (Eninger et al. 1968; Cross et al. 1971; Lindvall et al. 1974; Markova et al. 1985). Molecular probes form fluorescent products—derivatives of isoquinolines.

After the establishment of the presence of biogenic amines in the object, a researcher needs to model some systems for the study possible interaction between plants evacuated neurotransmitters out cell and animals, which percepts the compounds as chemosignals and regulators of metabolism. One of the valuable objects is pollen contacting with animals and humans. Among water-lived species, it needs to choose appropriate models too. As for animal models, sensitive samples related to the influence of exogenous serotonin and dopamine on salivary glands of cockroach *Periplaneta* species were known (Just and Walz 1996; Walz et al. 2006). In concentrations of $10^{-8}$ $10^{-7}$ M, the compounds stimulated saliva secretion (Just and Walz 1996). Walz and co-workers (2006) summarized current knowledge concerning the aminergic control of cockroach salivary glands and discussed efforts to characterize *Periplaneta* biogenic amine receptors. These monoamines caused increases in the intracellular concentrations of cAMP and $Ca^{2+}$. Stimulation of the glands by serotonin results in the production of a protein-rich saliva, whereas stimulation by dopamine results in saliva that is protein-free. Thus, two elementary secretory processes, namely electrolyte/water secretion and protein secretion, are triggered by different aminergic transmitters. Because of its simplicity and experimental accessibility, cockroach salivary glands have been used extensively as a model system to study the cellular actions of biogenic amines and to examine the pharmacological properties of biogenic amine receptors. Today, we can only suppose the influence of the insect saliva enriched in biogenic amines on plant reply realized in fast electric irritation or even changes in metabolism, growth, and morphogenesis. It is *terra incognita* yet.

The problem of plant exometabolites acting on insects, for example biogenic amines, may be very important for agrobiology, forestry, and common ecology as well as economy. Plant–animal contact may include many mechanisms, including the inclusion of the plant exometabolites in own animal metabolism as well as coding and synthesis of appropriate enzymes, for example the insect overcoming of plant defense (Zhu-Salzman and Liu 2011). We cannot exclude that in the plant excretion, having biogenic amines, acts on surface of animal via dopamine, histamine, or serotonin.

The purpose of the chapter is to analyze suitable plant systems that could contain catecholamines and histamine and to estimate of their possible influence on animals.

## 17.2 BIOGENIC AMINES IN PLANT EXCRETIONS AND THEIR EFFECTS ON SOME ANIMALS AT CONTACTS

As the sources of biogenic amines for the sensing by animal organisms may be volatile and water excretions of plants (Roshchina and Roshchina 1989, 1993, 2012). The main components of volatile excretions should attract insect-pollinators, while in water secretions there are also nutrients. The floral aroma of some species often demonstrates a characteristic aminoid smell (Smith and Meeuse 1966). Insect-pollinators may be attracted by this aroma, including histamine. Aroid inflorescences can be visited by several types of insects, but only a few—and sometimes only one—are the real pollinators for each species of bees, beetles, and flies; although, vertebrates, for example hummingbirds, have been mentioned (Gibernau 2003). Occurrence of catecholamines and serotonin in volatile excretions is absent, but they were found among water exometabolites (Roshchina 2014, 2016a; Roshchina and Yashin 2014; Roshchina et al. 2015).

Meanwhile, biogenic amines may serve both components of the defense and regulators in order to attract (the insects of pollinators or animals spreading seeds) or frighten away (blights, mainly feeding). Plants can accumulate a wide variety of defense compounds in their tissues that confer resistance to herbivorous insects. In its turn, in order to overcome the tolerant features, some insects have insect herbivoryinducible proteins that may be regulated by multiple signaling pathways (Zhu-Salzman and Liu 2011). In the confirmation of the opinion, there are some data related to tick histamine-binding proteins that have been isolated and cloned, and their three-dimensional structure was studied (Paesen et al. 1999). Possible mechanism of exogenous dopamine action on fly *Drosofilla* consists of the formation of its oxidized product by air oxygen known as dopaminechrome whose isomerization leads to the melanin formation that makes the black/white surface (Li and Christensen, 2011).

Interactions between animals and plants may be observed on analogues of mammalians such as unicellular animal species, for example *Paramecium* species, or on multicellular worms, planarians (Corrado et al. 1999; Oviedo et al. 2008; Roshchina 2014) as model biosensors for components contained in water plant excretions. Today, the planarian objects are analyzed as a universal model having characteristics similar with mammalians (Oviedo et al. 2008) and sensitive to dopamine

(Buttarelli et al. 2000) or serotonin (Franquinet and Martelly 1981), which can act on the regeneration of the worms after trauma through the stimulation of RNA synthesis. Planarian *Polycelis tenuis* reacts with exogenous biogenic amines and their antagonists regulating the regeneration (Franquinet and Martelly 1981). Another planarian *Girardinia tigrina* was capable of binding yohimbine, antagonists of catecholamines (Roshchina 2014).

Among insects contacting with plants as models should be chosen objects sensitive to exogenous biogenic amines. The saliva of insects appears to be sensitive to exogenous dopamine and serotonin from pollen, flowers, and leaves of plants because the compounds commonly found in neurons innervate the salivary glands. Moreover, exogenous neurotransmitters may stimulate different aspects of salivation. For example, in the American cockroach *Periplaneta*, serotonin induces secretion of proteinaceous saliva, whereas dopamine induces secretion of nonproteinaceous saliva (Troppmann et al. 2007). The effects of the antagonists of serotonin or dopamine were also studied and shown a difference in comparison with mammalians. In most cases, components of the insect saliva contain biogenic amines capable of regulating plant growth and development after feeding of the plant species. For example, the tick *Boophilus microplus* contains dopamine at 0.74 μg/g and noradrenaline at 0.45 μg/g (Megaw and Robertson 1974), which is approximately 10 to 100 times more than in plants (Roshchina 1991, 2001). Further studies of the connection between the accumulation of biogenic amines in plants and their effects on animal inhabitants in the same biocenosis will be important for ecology.

### 17.2.1 Contacts of Water-Lived Organisms

The water life of organisms is also dependent on the excretions from both of plant and animal inhabitants of the biocenosis that pointed in monograph devoted to model systems (Roshchina 2014). The first information about histamine related to water plant excretions the reader may also find in the first papers of Barwell (1979, 1989) on red algae. Other data connected with the contact of algae excretions enriched in dopamine that acted on animal inhabitants has been described by van Alstyne and co-workers (2006, 2011, 2013, 2014, 2015). The water life of organisms is also dependent on the excretions from both plant and animal inhabitants of the biocenosis. In the excretions there may be biogenic amines and acetylcholine that mutually influence the living organisms, like has been evidently shown for dopamine (van Alstyne et al. 2006) and histamine (Swanson et al. 2004) observed in oceanic conditions. To understand the mechanisms associated with dopamine release, van Alstyne and co-workers (2013) experimentally determined whether light quantity and quality, desiccation, temperature, exudates from conspecifics, and dissolved dopamine caused dopamine release (the concentration may be more than 500 μM). Although dopamine released into seawater can reduce the survival or growth of potential competitors, its release is associated with significant physiological stress and tissue mortality. As an antiherbivore defense, the temperate green alga *Ulvaria obscura* demonstrates dopamine release and may serve as a sea model to study plant–animal interactions (van Alstyne et al. 2006, 2013, 2014). The red color of the algae population is concerned with the formation of red oxidized products of dopamine, named dopachrome or dopaminechrome. Van Alstyne and co-workers (2014) examined the effects of a range of dopamine concentrations on the growth of the green alga *Ulva lactuca* on the germination of zygotes of the brown alga *Fucus distichus*, and on the survival, time to metamorphosis, and time to first molt of crab (*Metacarcinus magister* and *Cancer oregonensis*) larvae and juveniles. Dopamine began to inhibit Fucus germination at concentrations above 5 μM, *Ulva* growth at concentrations above 50 μM, and the survival of *Metacarcinus zoeae* at concentrations above 168 μM. It did not affect the survival of *Cancer megalopae* or juveniles or the time to metamorphosis of megalopae. It had no effect on the time to first molt of *Cancer* juveniles, except at the highest concentration tested (738 μM), where it delayed molting by an average of a day and a half. These toxic effects could have been due to the dopamine or to its oxidation products. Authors concluded that the large-scale release of dopamine by *U. obscura* following stressful environmental

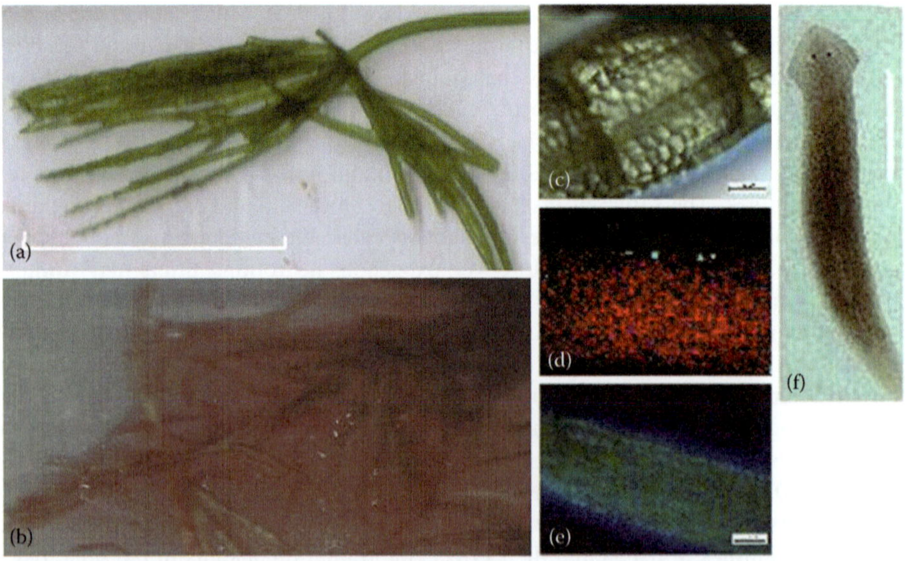


**FIGURE 17.1** Water inhabitants of aquarium: algae *Chara vulgaris* (a–e) and worm planarian *Girardinia tigrina* (f). a and b—images in water and on air, relatively; c—common image of cells; d and e—red and blue-green fluorescence of the cells at excitation by light 360–380 nm before and after the treatment with glyoxylic acid; f—common image of planarian. Bars = 2 cm (a, b), 140 μm (c–e), 300 μm (f).

conditions could significantly affect co-occurring species in intertidal pools as well as intertidal and shallow subtidal marine communities where the alga can form large blooms.

Among other possible models of plant–animal relations in water, larvae of the sea urchin *Holopneustes purpurascens* were suitable objects that undergone by secretions from a host alga containing histamine (Swanson et al. 2004, 2007). This compound induced the settlement of larvae and that's why the object may be a model for the observation relations between plants and animals.

Another example of possible interactions between plants and worms in our experiments were aquarium-lived species: algae *Chara carollina*, *Chara vulgaris* and *Nitella flexilis*, flowering angiosperm plant *Lemna minor*, and worm planarian *G. tigrina*. Figure 17.1 represents the external image of algae *Chara vulgaris* taken out from water without moistening that leads to the fast color change of green to red (a, b) that is perhaps peculiar to the dopamine oxidation in air. After histochemical treatment of green algae (c) with the reagent for dopamine—glyoxylic acid and excitation by UV-light 360–380 nm instead red fluorescence as in control (d) due to chloroplasts blue-green emission (e) arose on the cellular surface. We also represent Table 17.1, where by the histochemical fluorescent method (Roshchina 2016b), the amount of catecholamines (with glyoxylic acid) and histamine (with o-phthalic aldehyde) were analyzed in our model systems. The difference in concentration of the compounds are seen in the objects. The histamine amount was less than $10^{-8}$ M, or traces were seen for all model systems. As for catecholamines (mainly dopamine), *Chara carollina* (in internode cells) and *Nitella* (in nodal cells) or *Lemna* and worm contained it in concentrations of 10–100-fold higher.

**TABLE 17.1**
**Concentration (M) of Dopamine and Histamine in Water-Lived Organisms**

| Biogenic Amine | *Chara carollina* | *Nitella flexilis* | *Lemna minor* | *Girardinia tigrina* |
|---|---|---|---|---|
| Dopamine | Internode $10^{-8}$<br>Nodal—no, traces | Internode—no, traces<br>Nodal $10^{-7}$ | Leaf $10^{-6}$ | Body $10^{-6}$ |
| Histamine | All objects < $10^{-8}$ | | | |

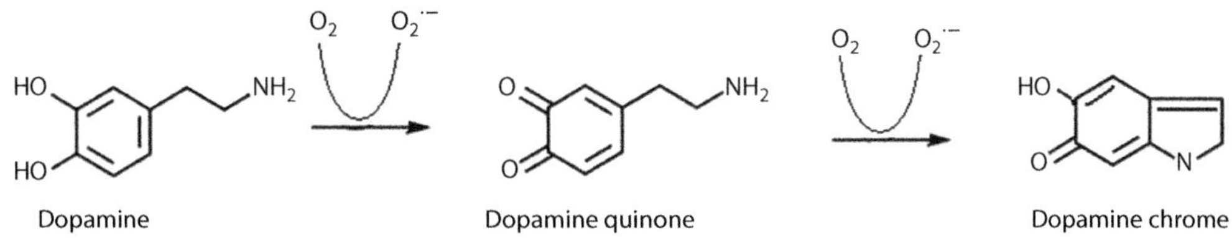


**FIGURE 17.2** Formation of dopaminechrome by air oxygen via the forming superoxide anionradicals.

The experiments on the influence of the biogenic amines on the behavior of planarian have been done too. In a concentration range of $10^{-8}$–$10^{-2}$ M of pure dopamine or histamine, it showed the sensitivity of planarian as animal objects (in small life water volume 200 mL per one worm). At high amount of the biogenic amines (>$10^{-4}$ M) the worm lacked motility and moved to the bottom. However, some worms lived and were restored after the change of the solution without reagent. Analogues of the experiments were repeated with aquarium-lived mollusks *Bulimus tentaculatus*, and all effects were a repeated lack of motility.

One mechanism of the effect on the motility may be explained from data described in experiments of Moshkov et al. (2010) and Shubina et al. (2011), where the interaction of dopamine with model membranes, isolated G-actin, and living cells, such as Mauthner neurons and fibroblast-like BHK-21 cells have been studied. It was found that *in vitro* dopamine passes through the phospholipid membrane and directly polymerizes G-actin due to incorporation into threads as their integral part. In *in vivo* conditions, it penetrates inside the cell and induces the appearance of a network of actin filaments in loci rich in globular actin. The data suggest that there exists a mechanism of dopamine interaction with living cells, which is based on direct polymerization of cytosolic G-actin as its cellular target. At conditions of chronic exposure, dopamine can penetrate into the cytoplasm, inducing actin polymerization and becoming bound to the newly formed actin cytoskeleton. Structurally, this can be apparent as hypertrophy of the cytoskeleton and its derivatives, with significant influences on the overall structure of the cell.

According to van Alstyne and co-workers (2006), the dopamine effects on algae may be related to oxidized derivatives of the compounds. During the first hour, dopamine is oxidized to free superoxide anionradical and semiquinone, then the formation of red pigment (absorbance maxima 480 nm) aminochrome is observed. The following oxidation led to the appearance of black pigment melanin, which is polymerized (quinones as monomers) products formed with participation of polyphenoloxidases. Similar red product dopaminechrome (Figure 17.2) we saw just after 20 min of the *Chara vulgaris* exposure on air without water—algae changed color from green to red color (Figure 17.1).

### 17.2.2 Contacts of Terrestrial Inhabitants

The first information about biogenic amines in plants appeared in the middle of the twentieth century. At the beginning, Werle and Raubb (1948) published a list of plants containing histamine. Later, similar information was revised for catecholamines (Roshchina 1991, 2001). Even Stanley and Linskens (1974) attracted attention to histamine being found in pollen, mainly of wind-pollinated species *Agrostis alba, Alopecurus pratensis, Bromus erectus, Corylus avellana, Cynosurus cristatus, Dactylis glomerata, Gelsemium sempervirens, Lolium perenne, Phleum pratense, Poa pratensis, Secale cereale, Zea mays,* as well as of some insect-pollinated species *Syringa vulgaris, Tilia platyphyllos,* and *T. cordata* (Marquardt and Vogg 1952). Later, pollen *Acer platanoides* L., *Anemone ranunculoides* L., *Anthriscus sylvestris* L., *Betula verrucosa* Ehrh., *Corylus avellana* L., *Larix decidua* Mill., *Hippeastrum hybridum* L., *Narcissus pseudonarcissus* L., *Picea excelsior (excelsa)* L., *Pinus sylvestris* L., *Philadelphus grandiflorus* Willd., *Phleum pratense* L., *Plantago major* L., *Populus balsamifera* L., *Quercus robur* L., *Rubus odoratus* L., *Syringa vulgaris* L., *Taraxacum officinale* L.,

**TABLE 17.2**
**Content of Catecholamines (Cat) and Histamine (Hist) in Some Pollens**

| Pollen from Wind-Pollinated Species | Concentration of Neurotransmitter, M | Pollen from Insect-Pollinated Species | Concentration of Neurotransmitter, M |
|---|---|---|---|
| *Corylus avelliana* L. | Cat $10^{-6}$ | *Acer platanoides* L. | Cat $10^{-6}$ |
| *Larix decidua* Mill. | Cat $5 \times 10^{-7}$ | | Hist $5 \times 10^{-7}$ |
| *Populus balsamifera* L. | Cat $10^{-6}$ | *Anemone ranunculoides* L. | Cat $10^{-6}$ |
| | Hist > $10^{-3}$ | | Hist >$10^{-3}$ |
| *Quercus robur* L. | Cat $10^{-8}$–$10^{-7}$ | *Anthriscus sylvestris* L. | Hist >$10^{-3}$ |
| | | *Syringa vulgaris* L. | Hist >$10^{-2}$ |

*Source:* Roshchina, V.V., *Model Systems to Study Excretory Function of Higher Plants*, Springer, Dordrecht, the Netherlands, 198, 2014; Roshchina, V.V. and Yashin, V.A., *Allelopathy J.*, 34, 1–16, 2014.

*Tilia cordata* Mill., *Tussilago farfara* L., *Tulipa* sp., and *Urtica dioica* L. were collected (from different ecosystems: grasslands, meadow, parks, groves) during June–August, and greenhouse plants *Aloe vera* L., *Epiphyllum hybridum* L., *Hibiscus rosa-sinensis* L., *Hippeastrum hybridum* were studied (Roshchina 2014; Roshchina and Yashin 2014). Dopamine and histamine play roles in signaling at the interaction of pollen and pistil during fertilization (Roshchina and Melnikova 1998).

Modern express methods for the neurotransmitters in pollen are determination-based ELISA kits, using absorbance of solutions from 100 µg of pollen grains/ml at 450 nm (Heidinger et al. 2017), or a fluorescent analysis of single-pollen cells and their excretions, where the emission excited with ultra-violet light is measured at 450-470 nm (Roshchina 2014, 2016b; Roshchina and Yashin 2014; Roshchina et al. 2016). In the following we show examples received by fluorescent methods. Table 17.2 demonstrates the amounts of catecholamines and histamine found (Roshchina 2014; Roshchina and Yashin 2014) in some pollen grains, both from wind-pollinated and insect-pollinated. Among the species, a high concentration (about $10^{-3}$ M) of histamine was shown for pollen of *Populus balsamifera, Taraxacum officinale*, and *Tussilago farfara*. Most wind-pollinated species studied have no, or smaller, amounts of catecholamines ($10^{-7}$–$10^{-6}$ M). Completely catecholamines nor histamine were for pollen from conifer *Larix decidua* and *Betula verrucosa*. The analysis of pollen from species represented in Table 17.2 shows the connection between metabolism of insect-pollinated plants that release an aminoid aroma (*Acer, Anemone, Anthriscus* and *Syringa*) in spring and the content of catecholamines and histamine in the generative microspores.

Knowledge of the accumulation of catecholamines and histamine in pollen is an important problem for medicine because the neurotransmitters play roles as growth regulators, demencion agents, and components of allergic reactions. Histamine found in pollen may be the agent of allergic reaction of humans (Stanley and Linskens 1974). Under stress reactions, a sharp increase of histamine is observed in plants, like in animals. In plants, the rise in histamine is found at drought, for instance for sunflower seeds (Korobova et al. 1988). Recently, stress conditions such as ozone (Roshchina et al. 2015) and ultraviolet irradiation (Heidinger et al. 2017) have been demonstrated to increase the content of histamine in some pollens. Mechanisms of spring allergy of the human ocular apparatus to pollen of black alder *Alnus glutinosa* L. as well as hazel *Corylus avellana* L. may be connected with the action of histamine, allergic agent (Heidinger et al. 2017). Environmental stresses, ozone, and UV light in spring may sharply enhance (Roshchina and Roshchina 2003) the concentration of biogenic amines that is dangerous for mammalians and humans. Allergic rhinitis induced by pollen or other microspores could be also related to histamine-containing objects due to vascular engorgement associated with vasodilation and increased capillary permeability caused by this neurotransmitter. Besides allergenicity of histamine, dopamine produces reactive oxygen species influencing as oxidants demonstrated toxicity for different animal cells (Roshchina 2001).

Interactions of air plant particles with human gelatinous nasal mucus or saliva is not well-studied in the field of their mutual influence at the contact. Nasal mucus serves as a system of defense from intervention into human organisms. In our preliminary experiments with vegetative microspores of horsetail *Equisetum arvense* and pollen of hazel *Coryllus avellana* (contained both histamine and dopamine), spreading by wind in air showed some effects: (1) human mucus and saliva quenched the autofluorescence of the microspores and pollen relatively in a comparison with a control; (2) *in vitro* moistening of plant microspores with these human excretions decreased the germination of the objects during the first 24 hours in a comparison with a control, although many ugly pollen tubes or horsetail rhizoids were formed. In any case, it is an interesting field for study.

## 17.3 CONCLUSION

Interactions of plants enriched in neurotransmitters with animals and human are known as a new perspective direction for the future studies of allelopathic chemical relations between species via these signal compounds that simultaneously serve as regulators of growth and development Today, there is yet little information connected with plant–animal model systems contacted through biogenic amines. New data has been received for water-lived systems, released catecholamines, in particular dopamine, both for marine algae and aquarium-inhabitants, related to effects of dopamine in plant excretions on animal living, motility, and toxicity. On recent stage of the studies donor of histamine is pollen, allergic reactions take place for humans, for example related to ophthalmology or other medicine field problems in a spring care.

## REFERENCES


Barwell, C.J. 1979. The occurrence of histamine in the red alga of *Furcellaria lumbricalis* Lamour. *Botanica Marina* 22: 399–401.

Barwell, C.J. 1989. Distribution of histamine in the thallus *Furcellaria lumbricalis*. *Journal of Applied Phycology* 1: 341–344.

Buttarelli, F.R., Pontieri, F.E., Margotta, V., Palladini, G. 2000. Acetylcholine/dopamine interaction in planaria. *Comparative Biochemistry and Physiology C: Toxicology & Pharmacology* 125(2): 225–231.

Buznikov, G.A. 1990. *Neurotransmitters in Embryogenesis*. Chur, Switzerland: Harwood Academic Press. 526 pp.

Corrado, D.M.U., Politi, H., Trielli, F. et al. 1999. Evidence for the presence of a mammalian-like cholinesterase in *Paramecium primaurelia* (Protista Ciliophora) developmental cycle. *Journal of Experimental Zoology* 283: 102–105.

Cross, S.A.M., Even, S.W.B., Rost, F.W.D. 1971. A study of the methods available for the cytochemical localization of histamine by fluorescence induced with o-phthalaldehyde or acetaldehyde. *Histochemical Journal* 3(6): 471–476.

Eninger, B., Håkanson, R., Owman, C., Sporrong, B. 1968. Histochemical demonstration of histamine in paraffin sections by a fluorescence method. *Biochemical Pharmacology* 17(9): 1997–1998.

Franquinet, R. and Martelly, I. 1981. Effects of serotonin and catecholamines on RNA synthesis in planarians; in vitro and in vivo studies. *Cell Differentiation* 10(4): 201–209.

Gibernau, M. 2003. Pollinators and visitors of aroid inflorescences. *Join the International Aroid Society* 26: 66–83.

Heidinger, A., Rabensteiner, D.F., Rabensteiner, J., Kieslinger, P., Horwath-Winter, J., Stabentheiner E., Riedl, R., Wedrich, A., Scmut, O. 2017. Decreased viability and proliferation of Chang conjunctival epithelial cells after contact with ultraviolet-irradiated pollen. *Cutaneous and Ocular Toxicology* 26. doi:10.1080/15569527.2017.1414226.

Just, F. and Walz, B. 1996. The effects of serotonin and dopamine on salivary secretion by isolated cockroach salivary glands. *Journal of Experimental Biology* 199(2): 407–413.

Kamo, K.K. and Mahlberg, P.G. 1984. Dopamine biosynthesis at different stages of plant development in *Papaver somniferum*. *Journal of Natural Products* 47: 682–686.

Korobova, L.N., Beletskii, Y.D., Karnaukhova, T.B. 1988. The experiment of the selection of salt-tolerant forms of sunflower among the selection material based on the content of histamine in seeds. *Fiziologiya and Biokhimiya Kulturnikh Rastenii* (*USSR*) 20: 403–406.

Kuklin, A.I. and Conger, B.V. 1995. Catecholamines in plants. *Journal of Plant Growth Regulation* 14: 91–97.
Kulma, A. and Szopa, J. 2007. Catecholamines are active compounds in plant. *Plant Science* 172: 433–440.
Lindvall, O., Björklund, A., Svensson, L.A. 1974. Fluorophore formation from catecholamines and related compounds in the glyoxylic acid fluorescence histochemical method. *Histochemistry and Cell Biology* 39(4): 197–227.
Li, J. and Christensen, B.M. 2011. Biological function of insect yellow gene family. In*: Recent Advances in Entomological Research.* (*From Molecular biology to pest management*), Liu, T. and Kang, L. (Eds.). Higher Education Press: Beijing, China, pp. 121–131.
Markova, L.N., Buznikov, G.A., Kovačević, N., Rakić, L., Salimova, N.B., Volina, E.V. 1985. Histochemical study of biogenic monoamines in early ("Prenervous") and late embryos of sea urchins. *International Journal of Developmental Neuroscience* 3(5): 493–495, 497–499.
Marquardt, P. and Vogg, G. 1952. Pharmakologische und Chemische Untersuchungen uber Wirkstoffe in Bienenpollen. *Arzneimittel Forschung* (Germany) 21(353): 267–271.
Megaw, M.W.J. and Robertson, H.A. 1974. Dopamine and noradrenaline in the salivary glands and brain of the tick, *Boophilus microplus*: Effect of reserpine. *Experientia* 30: 1261–1262.
Moshkov, D.A., Pavlik, L.L., Shubina, V.S., Parnyshkova, E.L., Mikheeva, I.B. 2010. Cytoskeletal regulation of the cellular function by dopamine. *Biofizika* 55(5): 850–856.
Oviedo, N.J., Nicolas, C., Adams, D.S., Levin, M. 2008. Planarians: A versatile and powerful model system for molecular studies of regeneration, adult stem cell regulation, aging and behavior. *Cold Spring Harbor Protocols*. doi:10.1101/pdb.emo101.
Paesen, G.C., Adams, P.L., Harlos, K. et al. 1999. Tick histamine-binding proteins: Isolation, cloning, and three-dimensional structure. *Molecular Cell* 3: 661–671.
Protacio, C.M., Dai, Y.R., Lewis, E.F., Flores, H.E. 1992. Growth stimulation by catecholamines in plant tissue/organ cultures. *Plant Physiology* 98(1): 89–96.
Roshchina, V.D. and Roshchina, V.V. 1989. *The Excretory Function of Higher Plants*. Moscow: Nauka, 214 p.
Roshchina, V.V. 1991. *Biomediators in Plants. Acetylcholine and Biogenic Amines*. Pushchino, Russia: Biological Center AN USSR, 192 p.
Roshchina, V.V. 1992. The action of neurotransmitters on the seed germination. *Biologicheskie Nauki* (Biological Sciences, USSR, Moscow) 9: 124–129.
Roshchina, V.V. 2001. *Neurotransmitters in Plant Life*. Enfield, NH: Science Publisher, 283 p.
Roshchina, V.V. 2010. Evolutionary considerations of neurotransmitters in microbial, plant and animal cells. In: *Microbial Endocrinology. Interkingdom Signaling in Infectious Disease and Health*, Lyte, M. and Freeston, R. (Eds.). New York: Springer, pp. 17–52.
Roshchina, V.V. 2014. *Model Systems to Study Excretory Function of Higher Plants*. Dordrecht, the Netherlands: Springer, 198 p.
Roshchina, V.V. 2016a. New trends and perspectives in the evolution of neurotransmitters in microbial, plant, and animal cells. *Advances in Experimental Medicine and Biology* 874: 25–77. doi:10.1007/978-3-319-20215-0_2.
Roshchina, V.V. 2016b. The fluorescence methods to study neurotransmitters (Biomediators) in plant cells. *Journal of Fluorescence* 26(3): 1029–1043. doi:10.1007/s10895-016-1791-6.
Roshchina, V.V. and Melnikova, E.V. 1998. Allelopathy and plant generative cells. Participation of acetylcholine and histamine in a signalling at the interactions of pollen and pistil. *Allelopathy Journal* 5: 171–182.
Roshchina, V.V. and Roshchina, V.D. 1993. *The Excretory Function of Higher Plants*. Springer, Berlin, Germany, 314 p.
Roshchina, V.V. and Roshchina, V.D. 2003. *Ozone and Plant Cell*. Dordrecht, the Netherlands: Kluwer Academic Publishers, 240 p.
Roshchina, V.V. and Roshchina, V.D. 2012. *The Excretory Function of Higher Plants*. Saarbrúcken: Lambert Academic Publishing, 466 p.
Roshchina, V.V. and Yashin, V.A. 2014. Neurotransmitters catecholamines and histamine in allelopathy: Plant cells as models in fluorescence microscopy. *Allelopathy Journal* 34(1): 1–16.
Roshchina, V.V., Yashin, V.A., Kuchin, A.V. 2015. Fluorescent analysis for bioindication of ozone on unicellular models. *Journal of Fluorescence* 25(3): 595–601. doi:10.1007/s10895-015-1540.
Roshchina, V.V., Yashin, V.A., Kuchin, A.V. 2016. Fluorescence of neurotransmitters and their reception in plant cell. *Biochemistry (Moscow) Supplement Series A: Membrane and Cell Biology* 10(3): 233–239.
Shubina, V.S., Lavrovskaya, V.P., Bezgina, E.N., Pavlik, L.L., Moshkov, D.A. 2011. Cytochemical and Ultrastructural characteristics of BHK-21 cells exposed to dopamine. *Neuroscience and Behavioral Physiology* 41(1): 1–5.

Smith, B.N. and Meeuse, B.J.D. 1966. Production of volatile amines and skatole at anthesis in some *Arum* lily species. *Plant Physiology* 41: 343–347.

Stanley, R.G. and Linskens, H.F. 1974. *Pollen: Biology, Biochemistry and Management.* Berlin, Germany: Springer-Verlag.

Swanson, R.L., Marshall, D.J., Steinberg, P.D. 2007. Larval desperation and histamine: How simple responses can lead to complex changes in larval behaviour. *Journal of Experimental Biology* 210: 3228–3235.

Swanson, R.L., Williamson, J.E., De Nys, R., Kumar, N., Bucknall, M.P., Steinberg, P.D. 2004. Induction of settlement of larvae of the sea urchin *Holopneustes purpurascens* by histamine from a host alga. *Biological Bulletin* 206: 161–172.

Świędrych, A., Kukuła, K.L., Skirycz, A., Szopa, J. 2004. The catecholamine biosynthesis route in potato is affected by stress. *Plant Physiology and Biochemistry* 42: 593–600.

Troppmann, B., Walz, B., Blenau, W. 2007. Pharmacology of serotonin-induced salivary secretion in *Periplaneta americana. Journal of Insect Physiology* 53(8): 774–781.

Van Alstyne, K.L., Anderson, K.J., van Hees, D.H., Gifford, S.-A. 2013. Dopamine release by *Ulvaria obscura* (Chlorophyta): Environmental triggers and impacts on the photosynthesis, growth, and survival of the releaser. *Journal of Phycology* 49: 719–727.

Van Alstyne, K.L., Anderson, K.J., Winans, A.K., Gifford, S.-A. 2011. Dopamine release by the green alga *Ulvaria obscura* after simulated immersion by incoming tides. *Marine Biology* 158: 2087–2094.

Van Alstyne, K.L., Harvey, E.L., Cataldo, M. 2014. Effects of dopamine, a compound released by the green-tide macroalga *Ulvaria obscura* (Chlorophyta), on marine algae and invertebrate larvae and juveniles. *Phycologia* 53: 195–202.

Van Alstyne, K.L., Nelson, A.V., Vyvyan, J.R., Cancilla, D.A. 2006. Dopamine functions as an antiherbivore defense in the temperate green alga *Ulvaria obscura. Oecologia* 148: 304–311.

Van Alstyne, K.L., Nelson, T.A., Ridgway, R.L. 2015. Environmental chemistry and chemical ecology of "green tide" seaweed blooms. *International Society for Computational Biology* 55: 518–532.

Walz, B., Baumann, O., Krach, C., Baumann, A., Blenau, W. 2006. The aminergic control of cockroach salivary glands. *Archives of Insect Biochemistry and Physiology* 62: 141–152. doi:10.1002/arch.20128.

Werle, E. and Raub, A. 1948. Über Vorkommen, Bildung und Abbau biogener Amine bei Pflanzen unter besonderer Beruck-sichtigung des Histamins. *Biochemische Zeischrift* (Germany) 318: 538–553.

Zhu-Salzman, K. and Liu, T. 2011. Insect herbivory-inducible proteins confer post-ingestive plant defenses. In: *Recent Advances in Entomological Research. (From Molecular biology to pest management)*, Liu, T. and Kang, L. (Eds.). Higher Education Press and Springer: Beijing, Germany, Dordrecht, the Netherlands, pp. 34–48.

# Section V

## Methodical and Practical Aspects of Neurotransmitters' Enriched Plants for Agronomy, Food and Medicine

# 18 Current Advancements of Serotonin Analysis in Plants

## *Insights into Qualitative and Quantitative Methodologies*

*Soumya Mukherjee and Akula Ramakrishna*

## CONTENTS



## 18.1 INTRODUCTION

Current methods for the investigation of serotonin include thin layer chromatography (TLC), radioimmunoassay (RIA), enzyme-linked immunosorbent assay (ELIZA), high-performance liquid chromatography with ultraviolet detection (HPLC-UV), and high-performance liquid chromatography with mass spectrometry (HPLC-MS). The major difficulty in using chemical or electrochemical means for assaying 5-hydroxytrptophan involve both aqueous and alcoholic extracts, which contain many substances having similar physical and chemical properties. These molecules are commonly tryptophan and its metabolites such as N-acetyltryptophan, 5-methoxytryptophan, N-acetyl-5-hydroxytryptamine (acetylserotonin), melatonin, tryptophol, 5-hydroxytryptophol, 5-methoxytryptophol, 5-hydroxyindoleacetic acid, and 5-methoxyindole acetic acid (Francis and Smith, 1982). The chemical structure of all these molecules contain one or more of the functional groups of -OH, $-NH_2$, and -COOH. Serotonin shares similar spectral properties with tryptamine, tryptophan, and 5-hydroxytryptamine (Garattini and Valzelli, 1965) and exhibit strong absorption between wavelengths of 230 and 300 nm. Fluorescent method of HPLC detection involves better analysis of samples with convenient method of sample preparation. Serotonin detection by using RIA has been reported by Feldman and Lee (1985). Colorimetric assay of 5-hydroxytryptophan has been developed by Udenfriend et al. (1955), which exhibits limitations imposed by interference from other phenolic compounds or indole compounds. Due to their similarity in biochemical nature, they can undergo a similar kind of derivatization reaction. Fellows and Bell (1971) reported extraction, isolation, and crystallization of serotonin to perform a colorimetric assay, although the yield was low. Similar reports (Quay, 1962) also involve limitations of low amount of extraction resulting due to several steps involved in sample processing. The amount of serotonin in some tissues is exceptionally low in quantity. Optimization of the extraction protocol is therefore necessary to obtain maximum yield and detection of serotonin. Stability of serotonin is often regulated by temperature, pH, and metal ions (Huang and Kissinger, 1996). Thus, various analytical methods have been explored to detect and measure serotonin across wide ranges of plant tissues. Various methods implied for serotonin estimation involve HPLC, TLC, liquid chromatography with mass spectrometry (LC-MS), and spectroscopic analysis (Murch et al., 2000; Badria et al., 2002). These methods differ in their sensitivity and pH range toward detection limits of serotonin (Figure 18.1). Gas chromatography with mass spectrometry (GC-MS) method exhibits high sensitivity and specificity for detection.

However, this method requires derivatization of analytes prior to their separation. Among the commonly used methods implied for serotonin detection, high-performance liquid chromatography (HPLC) with UV/VIS photodiode array (PDA), fluorescence detector (FD) or electrochemical detector (ECD), and HPLC coupled to mass spectrometry (LC-MS) are commonly used. HPLC-ECD method, however, possesses some limitations. This involves co-elution of various analytes possessing similar oxidation states and polarity with that of serotonin. Reported Limit of detection (LOD) of serotonin using HPLC-FD is 0.05 ng/ul (Lavizzari et al., 2006) and for HPLC-UV is 0.1 ng/g (Hosseinian et al., 2008) respectively. HPLC-FD is preferable for detection of serotonin. In HPLC-ECD, high concentrations of organic solvent (mobile phase) often reduce the efficiency of electrode in the detector. This imposes limitations in the concentration of solvents, which normally do not exceed 20% and thus increase retention time for analytes. Additionally, HPLC-ECD method is not suitable for analytical purpose of collection of eluents and changes the property of analytes. Based on the aforementioned limitations, effective and precise methods of LC-MS

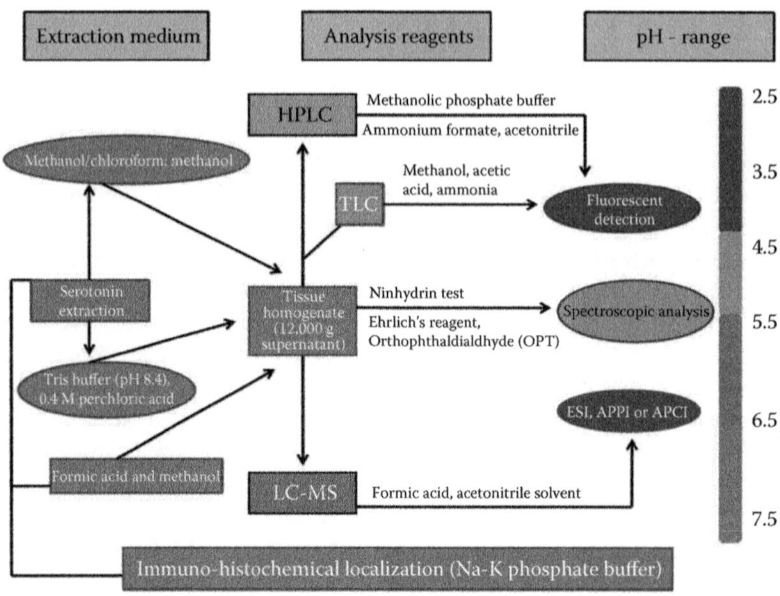


**FIGURE 18.1** Schematic representation of solvent-based serotonin extraction methods in plants and their detection at different pH ranges.

have largely replaced the conventional protocols. The LC-MS methods exhibit high sensitivity and better detection abilities. Furthermore, derivatization of compounds is not required. LCMS methods have been widely used in investigations of both qualitative and quantitative analysis of serotonin in various plant systems. Because of the wide range of LC systems, MS techniques, column types, and solvent mixtures, no single available method is suitable for all applications. The present chapter focuses on the various methodological aspects of serotonin estimation. The limitations and scopes have also been elucidated for the better understanding of the choice of methods for research in biological amines.

## 18.2 BIOSYNTHESIS AND METABOLISM OF SEROTONIN

In animals, serotonin (SER) is synthesized from tryptophan in two successive steps involving tryptophan hydroxylase (TPH; EC 1.14.16.4) and aromatic L-amino acid decarboxylase (AADC; EC 4.1.1.28) in which tryptophan hydroxylase acts as the rate-limiting enzyme (Veenstra-Vander Weele et al., 2000). In plants, SER is synthesized differently in that tryptophan is first catalyzed into tryptamine by tryptophan decarboxylase (TDC; EC 4.1.1.28), followed by the catalysis of tryptamine by tryptamine 5-hydroxylase (T5H) to form SER (Schroder et al., 1999). However, in St. John's wort (*Hypericum perforatum*), SER synthesis has been reported to occur *via* 5-hydroxytryptophan akin to mammals (Murch et al., 2001; Ramakrishna et al., 2009, 2011a, 2011b, 2012a, 2012b, 2012c; Ramakrishna, 2015; Ravishankar and Ramakrishna, 2016). However, another pathway of transformation of tryptophan to SER in plants, which is similar to that found typically in animals, is difficult to rule out. Hydroxylation of tryptophan leads to formation of 5-oxytryptophan in the presence of T5H. Subsequently, 5-oxytryptophan is decorboxylated by decorboxylase to yield SER (Fellows and Bell, 1971). T5H and SER synthesis in rice plants has been reported by Kang et al. (2007). The abundant production of SER in roots (~6.5 mg $g^{-1}$ FW) and (1 mg $g^{-1}$ FW) in stems of rice is reported. Synthesis of SER also occurs in rice seedlings and requires tryptamine as a substrate for the T5H enzyme activity. The tryptamine content is, thus, closely related to SER synthesis (Kang et al., 2007). Serotonin (5-hydroxytryptamine) is a precursor of melatonin in plants with amphiphatic nature. Serotonin is one of the major indoleamines synthesized in plant tissues. Regulation and transport of serotonin in plants is modulated by temporal and spatial separation of organ and developmental age. Serotonin, being an intermediary compound in the biochemical pathway, is metabolized into melatonin. Reports suggest varied condition of extraction and separation medium used for serotonin analysis. The molecule is rightly soluble in polar solvents of methanol or ethanol in aqueous phases. Some reports suggest the extraction of serotonin in aqueous formic acid. The molecule shows stability in low pH of 3.0–4.5 when separated in buffer solvents with 0.01% trifluoroacetic acid. The standard solutions of synthetic serotonin do not show temperature lability and are stable over a wide range of temperatures from 4°C to 30°C. Fluorometric estimations of serotonin have been reported with much higher temperature ranges. However, freshly prepared plant extracts are preferred to be used for crude extracts. A wide range of metabolites (phenols, flanoves, lignin, free radicals, coniferyl alcohol, sinapyl alcohol) are primarily present in the crude extracts, which allow oxidation or derivatization of serotonin. Methylated derivatives of SER and the alkaloids bufotenine, bufotenidine, psilocin, and psilocybin have been reported in plants (Atta-ur-Rahman and Basha 1999). Serotonin also is a precursor of bufotenine reported in seeds Piptadenia sp. (Fellows and Bell, 1971). Bufotenine is also found in the South American and Caribbean tree *Piptadenia peregrine*. In due course of serotonin metabolism there occurs condensation of the isopentyl group along with the indole nucleus of serotonin to produce major alkaloids *viz*. reserpine, yohimbane and yohimbine, vinblastine (Ramakrishna et al., 2011a). The antioxidant capacity of serotonin extends to tryptophan, tryptamine, and its derivatives. Serotonin, due to its structural organization, possesses antioxidant capability of quenching Reactive oxygen species (ROS) similar to that of melatonin. SER derivatives like N-(p-Coumaroyl) serotonin (CS) and N-feruroylserotonin (FS) possess antioxidative activities (Hotta et al., 2002).

## 18.3 IMMUNOLOCALIZATION OF SEROTONIN IN PLANTS

This technique is routinely used in plants and has been used to label diverse plant tissues and organelles including cell walls, vacuoles, chromatin, nuclei, nuclear membranes, and chloroplasts (Kukkola et al., 2004; Park et al., 2004). Immunohistochemistry exploits the binding of antibodies to specific targets to determine the localization of biomolecules within tissues. Immunolocalization using antibodies specific for serotonin has been widely used to localize serotonin in animal tissues (Perrot-Rechenmann and Gadal, 1986). However, localization of serotonin in plants at the cellular and tissue level is also not well understood. Our report suggested that SER was localized in different tissues of *Coffea canephora* including vascular tissues of stems, roots, and somatic embryos, as well as in endocarps (husks) of immature fruits (Ramakrishna et al., 2012c). Moreover, the distribution of SER in various tissues of *Musa paradisica* L, *Allium cepa* L and other plants *viz.* lemon has been reported (Kimura, 1968). Furthermore, localization studies involving the downstream enzyme Serotonin-N-acetyltransferase (SNAT) in susabi-nori (*Pyropia yezoensis*) suggests that serotonin is localized to the choloroplast or is transported there before conversion to N-acetylserotonin and later melatonin (Byeon et al., 2014). In addition, in rice leaves, immunolocalization analysis revealed that serotonin was abundant in the vascular parenchyma cells, including companion cells and xylem parenchyma cells, suggestive of its involvement in maintaining the cellular integrity. Specifically, the serotonin accumulated during pathogenic infection is incorporated into the cell walls leading to strengthening of the wall, whereas serotonin synthesized upon senescence is found in the soluble fraction of senescent tissues, especially in the vascular bundle cells, leading to its role in delaying senescence (Kang et al., 2009). Immunolocalization studies have revealed the presence of serotonin in reserve tissues of leaf blade. The presence of serotonin in the vascular bundle of the leaf blade suggests that tryptophan or 5-hydroxytryptophan synthesized in the leaf or some other organ may be enzymatically decarboxylated for transport to the reserve tissues. Immunohistochemical localization of serotonin and melatonin has been undertaken in the differentiating zone of primary roots of NaCl-stressed sunflower seedlings both in transverse and radial longitudinal sections (Mukherjee et al., 2014). NaCl stress-induced accumulation of serotonin was evident in the pericycle, endodermis, and vascular bundle (protophloem and protoxylem elements) of root sections, increasing with stages of seed germination 2, 4, and 6d. Serotonin was primarily localized in the symplastic region of root cells (Mukherjee et al., 2014).

## 18.4 SOLVENT-BASED EXTRACTION METHODS OF SEROTONIN ANALYSIS IN PLANTS

The process of serotonin extraction from various plant organs have been optimised for various methods of analysis. The success of extraction in qualitative terms does not differ much among the extraction medium. However, the quantity solely depends upon the success of the extraction media. Serotonin extraction in ethanolic medium (80%) supplemented with β-C-5hydroxy tryptamine as an internal standard for HPLC has been reported by Gatineau et al. (1997). Extraction of glucosylated forms of serotonin (serotonin 5-O-b-glucoside) from coffee bean for HPLC-ESI tandem mass analysis has been performed in (0.1%) formic acid at a ratio of 1:50 (w/w); (Servillo et al., 2016). The efficacy of the extraction mediums for estimation of derivative forms of serotonin were tested by comparison with formic acid and a mixture of formic acid and methanol (1:1 and 1:2; v/v). The analysis results have been interpreted with better efficacy of formic acid (0.1%) in water as a better extraction medium for detection and analysis of serotonin derivatives from coffee bean extracts (Servillo et al., 2016). The extraction method, however, consumes a longer time of 3–5 h incubation of tissue homogenate, which are processed to obtain supernatant (18,000 g. Islam et al., 2016) performed extraction and homogenate preparation of plant tissue in 0.2 M perchloric acid for fluorescent HPLC detection of serotonin. The tissue aliquots were mixed with 0.2 M perchloric acid, and 12,000 g supernatant was obtained for HPLC analysis. *Griffonia simplicifolia* seeds of over one year of age have been investigated for serotonin analysis by Lemaire and Adosraku (2002). Optimization of extraction medium has been obtained by use of methanol, ethanol, and their aqueous mixtures. Relative efficacy of each of these solvents was accessed in terms of maximum amount of serotonin detected. Tissue homogenates performed in each of the solvents (1:70; w/v) was filtered before analysis by HPLC separation of serotonin. Buffer-mediated homogenization of plant tissue for serotonin detection has been reported by Murch et al. (2000). The procedure has been further adapted by Ramakrishna et al. (2012a) for serotonin analysis by HPLC and LC-MS ESI in *Coffea canephora* seeds. Samples are extracted in 1 M Tris buffer (pH 8.4) and further homogenized in buffer containing 0.4 M perchloric acid, 0.05% sodium metabisulphite, and 1% EDTA. The supernatant (12,000 rpm) thus obtained was dried and reconstituted in methanol.

## 18.5 METHODOLOGICAL ASPECTS FOR SEROTONIN DETECTION AND ANALYSIS

### 18.5.1 Spectroscopic Methods for Serotonin Analysis

Earlier reports of serotonin estimation involve fluorimetric method of serotonin detection using ninhydrin (Vanable, 1963; Sasse et al., 1982). In incubation of samples with ninhydrin at 75°C for 30 min, the fluorimetric method is based upon an increase in fluorescence intensity upon reaction of serotonin with ninhydrin (excitation 394 nm, emission 505 nm). The reaction medium was kept at pH 4.0, and the lower limit of detection observed was near to 0.01 ug serotonin in 5 mL of reaction mixture (Vanable, 1963). Methods of fluorimetric estimation of serotonin have been reported in various non-plant systems. These protocols bear novel prospects in terms of their adaption in plant systems. Thompson et al. (1970) reported the efficacy of ortho-phthaldialdhyde (OPT) as a suitable reagent for fluorimetric detection of serotonin. This analysis was based upon the changes and enhancement of spectrum obtained after reaction of serotonin with ortho-phthaldialdhyde with emission maxima at 473 nm. This method proved beneficial (LOD 12.3 ng/mL) with respect to the direct method of serotonin estimation at 545 nm, which imposed various spectral interferences. Further extension on the aspects of pH sensitivity, OPT concentrations, and temperature have been reported (De Leiva and Schwratz, 1976; Komesu and Thompson, 1971). Tachiki and Aprison (1975) reported the accuracy of the serotonin-OPT fluorimetric method to measure serotonin content with LOD as low as in 2–5 picomole range in tissues. These methodologies provide contrivances of rapid and accurate measurement by the fluorimetric method of serotonin estimation in plants. However, sensitivity, cross reactivity, and elimination of noise are to be overcome by optimization of various plant extracts. The limit of detection and sensitivity to OPT shall depend upon the extraction medium, crude or pure form of serotonin, and its derivatization in plant extracts. Reports of the spectrophotometric method of serotonin estimation have been reported by Subbaraju et al. (2005). The detection procedure involved formation of a blue-colored complex with Folin-Ciocalteu reagent in presence of sodium carbonate with emission maxima at 736 nm. Jin et al. (2008) reported spectrphotometric estimation for the determination of total serotonin derivatives in the safflower (*Carthamus tinctorius* L.) seeds. The chromogenic reaction involved reaction of *p*-dimethylaminobenzaldehyde (Ehrlich's reagent) with absorption maxima of the complex at 625 nm.

### 18.5.2 High-Performance Liquid Chromatographic Analysis

Separation and analysis of serotonin derivates (methylated and glycosylated forms) have been accomplished by HPLC-ESI tandem mass analysis (Servillo et al., 2016). The HPLC system coupled to LC-MSD SL quadrupole ion trap has been used for separation and analysis. Chromatographic separation of the derivatives has been separated by using C8 column (250 × 3.0 mm, particle size 5 um) following isocratic elution in mobile phase of 0.1% formic acid in water and at a constant flow rate of 0.1 mL/min. Detection of analytes was performed by the electrospray ionization (ESI) method in a positive ion mode, with nitrogen as the nebulizing and drying gas (Servillo et al., 2016). The levels of serotonin derivatives were thus compared among different species of coffee bean used for the analysis. Fluorescent detection of serotonin has been performed by Islam et al. (2016) using variable parameters of mobile phase composition and their percentage mixtures obtained by gradient elution. The process of elution was proceeded with 10 mM ammonium formate, pH 3.4 for 0–5 min followed by linear gradient of 0%–25% acetonitrile in 10 mM ammonium formate, pH 3.4. Atlantis C 18 column (4.6 × 50 mm, 5 µm) was used with elution time of serotonin obtained at 9.2 min as average for various samples. Fluorescence corresponding to the serotonin and other allied metabolites were detected at excitation and emission wavelengths of 300 and 355 nm, respectively. Internal standards were also used to ensure the elution peaks of the chromatogram corresponding to each of the analytes thus obtained. The minimum levels of serotonin detected were 0.5 ug/gm of fresh weight tissue. The validation of this method was performed by limits of serotonin detection, interday variation, and recovery percentages. Fluorescent detection of HPLC thus appears to be a sensitive method for serotonin detection by the HPLC method. Methanolic (3%–10%) phosphate buffer (5 mM; pH 4.8) has been successfully used as a mobile phase for isocratic elution of serotonin from homogenates of *Griffonia* seeds (Lemaire and Adosraku, 2002). Separation has been performed using column dimension of 250 X (4.6 mm; 5 um particle size) followed by detection using 440 diode array detector at a wavelength of 275 nm. Interestingly, better suitability of methanol in water (1:1) as a solvent was attributed to its higher polarity in comparison with ethanol. However, the aqueous phase of both the solvents was found suitable for serotonin separation. The dilution in aqueous phase yielded better resolution and separation of other associated metabolites along with serotonin. The optimization of peak properties in the chromatogram was obtained as a function of flow rate (1.5 mL/min) and 3% methanolic phosphate buffer as a mobile phase. An inverse relation was obtained among elution time and methanol concentration where a decrease in methanol percentage in the phosphate buffer increased the elution time. Ramakrishna et al. (2012a) reported the use of isocratic acidic elution buffer (0.1 M sodium acetate, 0.1 M citric acid, 0.5 mM sodium octanoylsulfonate, 0.15 M EDTA) and 5% methanol as the potent mobile phase. Detection of serotonin peaks were performed at 260 nm. Serotonin derivates have been reported to be separated through reverse phase HPLC (Jang et al., 2004). The analytes were allowed to be separated by isocratic elution with 35% (v/v) methanol in water containing 0.5% trifluoroacetic acid at a flow of 0.8 ml/min. Eluents were detected at a wavelength of 320 nm. Further confirmation of the presence of coumaroylserotonin and feruloylserotonin were obtained by positive ion spray MS spectra. Tanaka et al. (2003) performed HPLC analysis of methanolic extracts, which were air dried and reconstituted in ethanol. The separation was performed in a mixture of acetonitrile:water (18:88). Molecules were detected at a wavelength of 280 nm using fluorescence detector. HPLC-mediated separation of serotonin using acetonitrile (90%) as a solvent with 0.01 M phosphate buffer has also been reported from walnut microcuttings (Gatineau et al., 1997).

### 18.5.3 Thin-Layer Chromatographic Analysis

Recent investigations by Hayashi et al. (2016) have reported the role of serotonin associated with lesion formation in rice. The infected leaf discs with brown lesions were collected and ground to powder in liquid nitrogen. The powder was dispersed in extraction medium containing methanol: 99.7% acetic acid:10% ammonia (6:1:1 by volume. Supernatant 21,500 g) thus obtained was used for thin layer chromatographic separation of serotonin. Sufficient amount of the extract was resolved in silica gel chromatographic plates through with a solvent system containing methanol:99.7% acetic acid:10% ammonia (6:1:1 by volume). The fluorescence of resolved spots were detected under UV light (312 nm). Serotonin dissolved in methanol was exposed to fluorescent light to undergo auto-oxidation. The brown deposit of auto-oxidized serotonin served as control for the fluorescent detection method.

### 18.5.4 Liquid Chromatographic-Mass Spectrometric Analysis

Higher range of serotonin detection limits and characterization of its derivatives has been better obtained by liquid chromatography coupled to tandem mass spectrometry. Methanolic extracts have been reported to be obtained by Cao et al. (2006). Separation was obtained by gradient of 0.45% formic acid: acetonitrile (95:5% (v/v); 5–6 min), followed by 95:5% to 0:100% (v/v) for 6–16 min and 0:100% (v/v) for 6 min. Elution of serotonin was obtained at 16 min at a flow rate of 0.25 mL/min. Serotonin was detected and quantified through electrospray ionization (ESI), atmospheric pressure chemical ionization (APCI), and atmospheric pressure photoionization (APPI) modes respectively. Ion transition of parent molecule 177 u was monitored to the principle daughter ion 160 u in ESI, as well as 178–161 u in APCI and APPI. Inclusion of 0.45% formic acid in the solvent gradient has been reported to be suitable for separation and analyte ionization. The method of detection of serotonin by Cao et al. (2006) involves limit of detection and limit of quantification for serotonin to be 100 pg/mL and 5 ng/mL, respectively. Direct infusion of HPLC fractions into mass spectrometer and analysis of serotonin by ESI-MS method have been reported by Ramakrishna et al. (2012a).

## 18.6 EXTRACTION METHODS FOR SEROTONIN DERIVATIVES

Jang et al. (2004) reported the investigation for serotonin derivative estimations in transgenic rice leaves. The method was adapted according to Ishihara et al. (2000) and involved methanolic extraction of plant tissue. The (12,000 g) supernatant thus obtained was allowed to filter through Sep pak cartridge and eluted with a ratio of chloroform: methanol (30:1). The aliquot obtained was reconstituted in 500 uL methanol for separation through reverse phase HPLC. Methanolic extraction of serotonin derivative has also been reported to be supplemented with 1% acetic acid according to Tanaka et al. (2003).

## 18.7 PURIFICATION AND CHARACTERIZATION OF SEROTONIN DERIVATIVES

Methanolic crude extracts obtained for analysis of serotonin derivatives have also been resolved by silica gel thin layer chromatography using a ratio of chloroform:methanol (1:9) as a solvent (Tanaka et al., 2003). Detection of resolved compounds was obtained by applying Ehrlich's reagent (4-dimethylaminobenzaldehyde 1 g, methanol 75 mL, 36% HCl 25 mL. The detected molecules of interest were eluted with ethanolic or methanolic extracts and separated further by reverse phase HPLC using acetonitrile:water (23:77) as a solvent. The separated molecules were detected at 280 nm and further infused in LC-MS. The solvent used for separation involved a linear gradient of 15% acetonitrile. The MS spectrum was obtained for confirmation of coumaroylserotonin and feruloylserotonin.

## 18.8 HIGH THROUGHPUT METHODS FOR SEROTONIN DETECTION: PERSPECTIVES FOR NUTRACEUTICAL AND INDUSTRIAL RESEARCH

Among various methodologies used for serotonin detection and quantitation in plants, the limit of detection, sample extraction, and noise interference are various factors that affect output. The success of quantification of serotonin amount in plant tissue, therefore, depends upon various biochemical factors along with the choice of detection method. Liquid chromatography coupled to mass spectrometry has emerged as a new integrated method of global detection of metabolites among various plant systems. The metabolomic approach in recent times have been facilitated by the MS2 spectral tag (MS2T) library from the total scan ESI MS/MS data in various plant systems. Rice plant metabolomic analysis has been performed for such purposes. In this context, serotonin derivatives have been linked up with wide range of metabolic avenues. The significance of serotonin detection in various edible plant sources pertains to the understanding of the nutraceutical values. The high output of serotonin detection is achievable by HPLC coupled to ESI-triple quadrupole-linear ion trap (Q TRAP)–MS or ESI–Qq TOF–MS. The LC-MS method has emerged as a convenient step toward serotonin estimation over other techniques. The LC-MS method commonly does not require any derivitization step for separation of analytes. There are variations in the type of solvent, extraction medium, and nature of column. Thus, a varied number of protocols have been generated for serotonin estimation through LC-MS. Comparison of various techniques suggests LC-MS to be more sensitive than HPLC-ECD, FD, or UV (Huang and Mazza, 2011). The recent advancements in mass spectrometry pertain to the progress in LC-MS interface technique and liquid introduction atmospheric-pressure ionization (API). This method of LC-MS involves infusion of analytes from LC into the MS through the ion source, mass analyzer, and to the detector. Various types of mass analyzers implied to determine and quantify serotonin are Single Quadrupole Mass Analyzer, Triple Quadrupole Mass Analyzer, and Ion-Trap Mass Analyzer. These features have led to the exploration of various types of tissues systems in plant materials. Sweet cherry cultivars (*Prunus avium* L.) have been investigated for the presence of serotonin in eight cultivars using the LC-MS technique (González-Gómez et al., 2009). Wild-type rice (*Oryza sativa* cv. Dongjin) seedlings were investigated for the levels of serotonin in different organs of leaf, root, stem, and seeds (Kang et al., 2007). Roots and 19-day old seed hulls exhibited higher serotonin content. Seeds contained the lowest amount of serotonin among all the organs investigated. Investigations by Hosseinian et al. (2008) revealed no serotonin content detected in purple wheat, which, however, exhibited high melatonin accumulation. This was attributed to the intermediary stage of serotonin in the biochemical pathway, which gets metabolized into melatonin. Dietary value of *Prata* banana (*Musa acuminata* × *Musa balbisiana*) has been investigated with respect to serotonin content associated with fruit ripening (Adão and Glória, 2005). Various other systems like spinach, hazel nut, potato, and capsicum have also been investigated for the presence of serotonin (Engstrom et al., 1992; Adão and Glória, 2005, Kang and Back, 2006; Lavizzari et al., 2006). Evidences, therefore, provide promising steps toward commercialization of serotonin quantification among various food crops through the LC-MS method.

## 18.9 CONCLUSION

Various methodological approaches have been adopted in different plant systems from time to time, which have added valuable results to the critical analysis of serotonin. Current understandings of the methodological advances of serotonin analysis pave the way to future perspectives of identification of serotonin transporters in plants. Such understandings shall provide clarity to the aspects of serotonin mobility in plants. Immunolocalization of SER is important to understand its physiological roles during developmental events. Moreover, SER localization studies could be useful to elucidate cellular signaling mechanism of SER in plants. The mechanism of action of serotonin and its biomolecular cross talk require further advancements. There are requirements to design new methodological aspects for investigating serotonin interaction with several other biomolecules. Spectroscopic techniques are promising rapid methods to obtain faster results. Methods for quantitative analysis have undergone ample refinement in terms of their output and efficacy (LOD). Methods of immunofluorescent localization of serotonin in plant tissues require more exploitation. The advantages of fluorescent imaging for serotonin at the subcellular level will provide new clues to its cytosolic and organellar distribution. The temporal and spatial differences of serotonin distribution in plant tissues can be better analyzed by such methods. The fact that serotonin undergoes frequent derivatization is one of the major reasons for variations in output obtained from analytical methods. The application of antibodies raised against animal-based serotonin has proved effective in plants. The epitope specificity can be further exploited for producing target sites for serotonin derivatives. Industrial applications of wide-scale serotonin estimation can be implied by high throughput methods of LC-MS. The need to standardize protocols for optimum tissue extraction is a dynamic process and depends mostly upon the type of plant material used. Serotonin-mediated signaling events like serotonin–protein interaction, serotonin-induced hormonal regulation, or other biomolecular changes are necessary to be deciphered for plant physiological processes. There are requirements to develop specific inhibitor assays for observing the precise role of serotonin in various processes. The efficacy and application of such inhibitors are yet to be obtained in plant techniques. Thus, new techniques can be adopted for fulfilling the quest for fate of phytoserotonin across various plant systems. Model system databases are required to be constructed for serotonin-specific findings in various plants, which may include its structural forms, derivatives, enzymatic activity, and its turnover.

## REFERENCES

Adão RC and Glória MB. 2005. Bioactive amines and carbohydrate changes during ripening of "Prata" banana (*Musa acuminata* × *M. balbisiana*. *Food Chem* 90(4): 705–711.

Atta-ur-Rahman and Basha A. 1999. *Indole Alkaloids*. Berks, PA: Harwood Academic Publishers, p. 324.

Badria FA. 2002. Melatonin, serotonin, and tryptamine in some Egyptian food and medicinal plants. *J Med Food* 5(3): 153–157.

Byeon Y, Yool Lee H, Choi D-W, Back K. 2014. Chloroplast-encoded serotonin N-acetyltransferase in the red alga Pyropia yezoensis: Gene transition to the nucleus from chloroplasts. *J Exp Bot* 66: 709–717.

Cao J, Murch SJ, O'Brien R, Saxena PK. 2006. Rapid method for accurate analysis of melatonin, serotonin and auxin in plant samples using liquid chromatography–tandem mass spectrometry. *J Chromatog A* 1134: 333–337.

De Leiva A and Schwartz S. 1976. Relation of pH to fluorescence of serotonin, melatonin, and other indole compounds reacted with o-phthaldialdehyde. *Clin Chem* 22: 1999–2005.

Engstrom K, Lundgren L, Samuelsson G. 1992. Bioassay-guided isolation of serotonin from fruits of *Solanum tuberosum* L. *Acta Pharmaceutica Nordica* 4(2): 91–92.

Feldman JM and Lee EM. 1985. Serotonin content of foods: Effect on urinary excretion of 5-hydroxy indoleacetic acid. *Am J Clin Nutr* 42: 639–643.

Fellows LE and Bell EA. 1971. Indole metabolism in Piptadenia peregrine. *Phytochemistry* 10: 2083–2091.

Francis PL and Smith I. 1982. High performance liquid chromatography of tryptophan metabolites: Application I biosynthesis and kinetics. *J Chromatogr* 232: 165–169.

Garrattini S and Valzelli L. 1965. *Serotonin*. Elsevier, London, UK, pp. 1–53.

Gatineau F, Fouche JG, Kevers C, Hausman JF, Gaspar T. 1997. Quantitative variations of indolyl compounds including IAA, IAA aspartate and serotonin in walnut microcuttings during root induction. *Biol Plant*. 39:131–137.

González-Gómez DG, Lozano M, Ferńandez-Léon MF, Ayuso MC, Bernalte MJ, Rodŕiguez, AB. 2009. Detection and quantification of melatonin and serotonin in eight sweet cherry cultivars (*Prunus avium* L. *Eur Food Res Technol* 2292: 223–229.

Hayashi K, Fujita Y, Ashizawa T, Suzuki F, Nagamura Y, Hayano Saito Y. 2016. Serotonin attenuates biotic stress and leads to lesion browning caused by a hypersensitive response to *Magnaporthe oryzae* penetration in rice. *Plant J* 85: 46–56.

Hosseinian FS, Li WD, Beta T. 2008. Measurement of anthocyanins and other phytochemicals in purple wheat. *Food Chem* 109(4): 916–924.

Hotta Y, Nagatsu A, Liu W, Muto T, Narumiya C, Lu X et al. 2002. Protective effects of antioxidative serotonin derivatives isolated from safflower against postischemic myocardial dysfunction. *Mol Cell Biochem* 238: 151–162.

Huang TH and Kissinger PT. 1996. Liquid chromatographic determination of serotonin in homogenized dog intestine and rat brain tissue using a 2 mm i.d. PEEK column. *Curr Sep* 14(3–4): 114–119.

Huang X and Mazza G. 2011. Application of LC and LC-MS to the analysis of melatonin and serotonin in edible plants. *Crit Rev Food Sci Nutr* 51: 269–284.

Ishihara A, Kawata N, Matsukawa T, Iwamura H. 2000. Induction of N-hydroxycinnamoyltyramine synthesis and tyramine N-hydroxycinnamoyltransferase (THT) activity by wounding in maize leaves. *Biosci Biotechnol Biochem* 64: 1025–1031.

Islam J, Shirakawa H, Nguyen TK, Aso H, Komai M. 2016. Simultaneous analysis of serotonin, tryptophan and tryptamine levels in common fresh fruits and vegetables in Japan using fluorescence HPLC. *Food Biosci* 13: 56–59.

Jang SM, Ishihara A, Back K. 2004. Production of coumaroylserotonin and feruloylserotonin in transgenic rice expressing pepper hydroxycinnamoyl-coenzyme A: Serotonin N (hydroxycinnamoyl)transferase. *Plant Physiol* 135: 346–356.

Jin Q, Shan L, Yue J, Wang X. 2008. Spectrophotometric determination of total serotonin derivatives in the safflower seeds with Ehrlich's reagent and the underlying color reaction mechanism. *Food Chem* 108: 779–783.

Kang K, Kim YS, Park S, Back K. 2009. Senescence-induced serotonin biosynthesis and its role in delaying senescence in rice leaves. *Plant Physiol* 150: 1380–1393.

Kang S and Back K. 2006. Enriched production of N-hydroxycinnamic acid amides and biogenic amines in pepper (*Capsicum annuum*) flowers. *Scientia Horticulturae* 108(3): 337–341.

Kang S, Kang K, Lee K, Back K. 2007. Characterization of tryptamine 5-hydroxylase and serotonin synthesis in rice plants. *Plant Cell Rep* 26: 2009–2015.

Kimura M. 1968. Fluorescence histochemical study on serotonin and catecholamine in some plants. *Jpn J Pharmacol* 18: 162–168.

Komesu NN and Thompson JH. 1971. Optimum concentration of OPT for the fluorometric measurement of serotonin. *Eur J Pharmacol* 16: 248–250.

Kukkola EM, Koutaniemi S, Pöllänen E, Gustafsson M, Karhunen P, Lundell TK, Saranpää P, Kilpeläinen I, Teeri TH, Fagerstedt KV. 2004. The di-benzodioxocin lignin substructure is abundant in the inner part of the secondary wall in Norway spruce and silver birch xylem. *Planta* 218: 497–500.

Lavizzari T, Veciana-Nogués MT, Bover-Cid S, Marin é -Font A, Vidal-Carou MC. 2006. Improved method for the determination of biogenic amines and polyamines in vegetable products by ion-pair high-performance liquid chromatography. *J Chromatogr A* 1129(1): 67–72.

Lemaire PA and Adosraku RK. 2002. An HPLC method for the direct assay of the serotonin precursor, 5-hydroxytrophan, in seeds of *Griffonia simplicifolia*. *Phytochem Anal* 13: 333–337.

Mukherjee S, David A, Yadav S, Baluška F, Bhatla SC. 2014. Salt stress-induced seedling growth inhibition coincides with differential distribution of serotonin and melatonin in sunflower seedling roots and cotyledons. *Physiol Plant* 152: 714–728.

Murch SJ, Campbell SSB, Saxena PK. 2001. The role of serotonin and melatonin in plant morphogenesis: Regulation of auxin induced root organogenesis in *in vitro*-cultured explants of St. John's wort (*Hypericum perforatum* L. *In Vitro Cell Dev Biol Plant* 37: 786–793.

Murch SJ, Krishna Raj S, Saxena PK. 2000. Tryptophan is a precursor for melatonin and serotonin biosynthesis in in vitro regenerated St. John's wort (*Hypericum perforatum* L. cv. Anthos) plants. *Plant Cell Rep* 19(7): 698–704.

Park M, Kim SJ, Vitale A, Hwang I. 2004. Identification of the protein storage vacuole and protein targeting to the vacuole in leaf cells of three plant species. *Plant Physiol* 134: 625–639.

Perrot-Rechenmann C and Gadal P. 1986. Enzyme immunochemistry. In: Wang, T.L. (Ed) *Immunology in Plant Science*. Cambridge University press, Cambridge, pp. 59–88.

Quay WB. 1962. Differential extractions for the spectrofluorometric measurement of diverse 5-hydroxy and 5-methoxy indoles. *Anal Biochem* 5: 51–59.

Ramakrishna A. 2015. Indoleamines in edible plants: Role in human health effects. In: *Indoleamines: Sources, Role in Biological Processes and Health Effects*. Biochemistry Research Trends Series. (Ed) Catalá, A., Nova Publishers. Biochemistry Research Trends. p. 279.

Ramakrishna A, Dayananda C, Giridhar P, Rajasekaran T, Ravishankar GA. 2011b. Photoperiod influence endogenous indoleamines in cultured green alga Dunaliella bardawil. *Indian J Exp Biol* 49: 234–240.

Ramakrishna A, Giridhar P, Jobin M, Paulose CS, Ravishankar GA. 2012a. Indoleamines and calcium enhance somatic embryogenesis in cultured tissues of *Coffea canephora* P ex Fr. *Plant Cell Tissue Organ Cult* 108: 267–278.

Ramakrishna A, Giridhar P, Kadimi U, Gokare R. 2012c. Endogenous profiles of indoleamines: Serotonin and melatonin in different tissues of Coffea canephora P ex Fr. as analyzed by HPLC and LC-MS-ESI. *Acta Physiol Plantarum* 34: 393–396.

Ramakrishna A, Giridhar P, Ravishankar GA. 2009. Indoleamines and calcium channels influence morphogenesis in *in vitro* cultures of *Mimosa pudica* L. *Plant signal Behav* 12: 1136–1141.

Ramakrishna A, Giridhar P, Ravishankar GA. 2011a. Phytoserotonin: A review. *Plant Signal Behav* 6: 800–809.

Ramakrishna A, Giridhar P, Udaya Sankar K, Ravishankar GA. 2012b. Melatonin and serotonin profiles in beans of *Coffea* sps. *J Pineal Res* 52: 470–476.

Ravishankar GA and Ramakrishna A (Eds.). 2016. *Serotonin and Melatonin: Their Functional Role in Plants, Food, Phytomedicine, and Human Health*. Taylor & Francis Group, CRC Press, Boca Raton, FL, p. 582.

Sasse F, Heckenberg U, Berlin J. 1982. Accumulation of carboline alkaloids and serotonin by cell cultures of *Peganum harmala* L: I. correlation between plants and cell cultures and influence of medium constituents. *Plant Physiol* 169: 400–404.
Schroder P, Abele C, Gohr P, Stuhlfauth-Roisch U, Grosse W. 1999. Latest on the enzymology of serotonin biosynthesis in walnut seeds. *Adv Exp Med Biol* 467: 637–644.
Servillo L, Giovane A, Casale R et al. 2016. Glucosylated forms of serotonin and tryptophan in green coffee beans. *LWT–Food Sci Technol* 73: 117–122.
Subbaraju GV, Kannababu S, Vijayakumar K, Vanisree PBS, Tsay H. 2005. Spectrophotometric estimation of L-5-hydroxytryptophan in *Griffonia simplicifolia* extracts and dosage forms. *IJASE* 3(2): 111–116.
Tachiki KH and Aprison MH. 1975. Fluorometric assay for 5-hydroxytryptophan with sensitivity in the picomole range. *Anal Chem* 47: 7.
Tanaka E, Tanaka C, Mori N, Kuwahara Y, Tsuda M. 2003. Phenylpropanoid amides of serotonin accumulate in witches' broom diseased bamboo. *Phytochemistry* 64: 965–969.
Thompson JH, Spezia CA, Angulo M. 1970. Fluorometric detection of serotonin and serotonin-*O.P.T. Ir J Med Sci* 3: 19.
Udenfriend S, Weissbach H, Clark CT. 1955. The estimation of 5-hydroxytryptamine (serotonin) in biological tissues. *J Biol Chem* 215: 337–344.
Vanable JW, Jr. 1963. A ninhydrin reaction giving a sensitive quantitative fluorescence assay for 5-hydroxytryptamine. *Anal Biochem* 6: 393–403.
Veenstra-Vander Weele J, Anderson GM, Cook EH. 2000. Pharmacogenetics and the serotonin system: Initial studies and future directions. *Eur J Pharm* 410: 165–181.

# 19 Biogenic Amines in Plant Food

*Kamil Ekici and Abdullah Khalid Omer*

## CONTENTS



## 19.1 BIOGENIC AMINES AND HEALTH

Biogenic amines (BAs) are a group of low molecular weight, heat stable, non-volatile, basic nitrogenous compounds with biological activity, formed and degraded as a result of normal metabolic activity in living cells and thus they are omnipresent in humans, animals, plants, and microorganisms, and chiefly created by microbial decarboxylation of amino acid in foodstuffs or by amination and transamination of aldehydes and ketones by amino acid transaminases (Shalaby, 1996; Beneduce et al., 2010; Flasarová et al., 2016). Naturally biogenic amines are present in humans, animals, plants, and microbes. They are related in the natural physiological processes, such as transmission of a nerve impulse across a synapse, control of blood pressure, allergic response, and cellular development control. However, the compounds may be dangerous to people health if their amounts reach a critical threshold (Russo et al., 2010; EFSA, 2011). Dietary polyamines spermidine and spermine participate in an array of physiological roles with both favorable and injurious effects on human health (Kalač, 2014). Aromatic monoamines, such as phenylethylamine, tyramine, and histamine occur in plants as a result of phenylalanine, tyrosine, and histidine enzymatic decarboxylation, respectively (El-Yazal and Rady, 2013). Amines are basic nitrogenous compounds in which 1, 2, and 3 atoms of

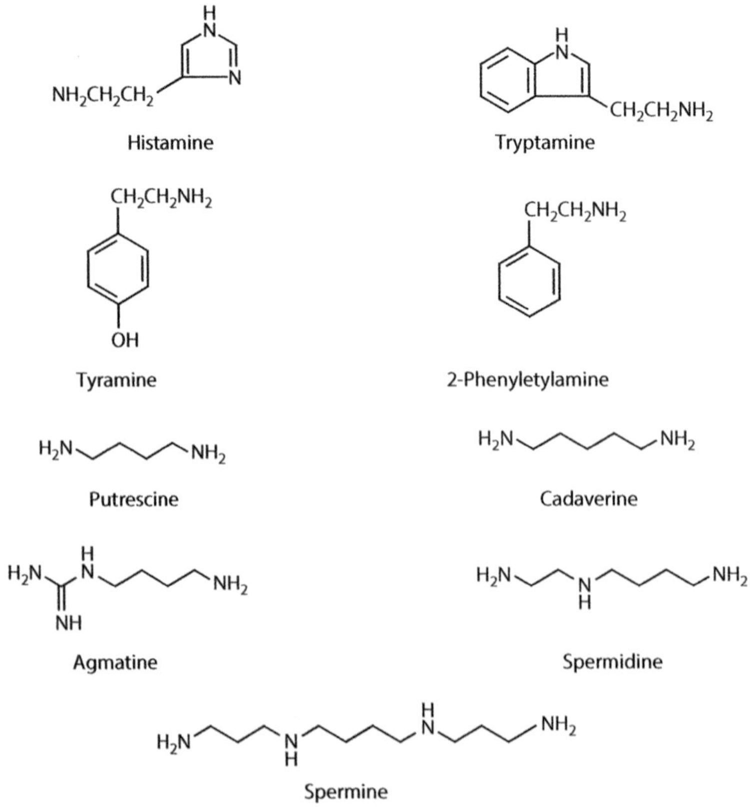


**FIGURE 19.1** Chemical structures of some common biogenic amines.

hydrogen in alkyl groups are replaced by ammonia. They can be classified either by the number of hydrogen atoms substituted in ammonia as primary, secondary, and tertiary amines, or by the number of basic groups (NH or $NH_2$) as mono-, di-, and polyamines (Figure 19.1). The term *biogenic amines* defines decarboxylation products, such as histamine, serotonin, tyramine, phenylethylamine, tryptamine, and also aliphatic polyamines. Aliphatic polyamines such as spermidine and spermine, but also the diamines putrescine, cadaverine, and agmatine, are essential for normal and neoplastic cell growth and proliferation and can be ubiquitously synthesized from their aminoacidic precursors (Righetti et al., 2008). Biogenic amines play various urgent parts in the physiology and advancement of eukaryotic cells; they are wellsprings of nitrogen and precursors for the synthesis of hormones, alkaloids, nucleic acids, and proteins (Santos, 1996; Ladero et al., 2010; Spano et al., 2010).

Biogenic amines are among the most vital modulators of vertebrate and invertebrate nervous system function and behavior that act as significant messengers in the central nervous system (CNS) and peripheral nervous system (PNS). Biogenic amines regulate many functions in the brain, including endocrine secretion, cognitive function, aggression, sleep and waking, emotional states, motivation, reward circuitry, decision making, and learning and memory (Roeder, 1999; Blenau and Baumann, 2003; Farooqui, 2007). The physiological functions of biogenic amines such as catecholamines, serotonin, and histamine known as neurotransmitters found in plants and animals play important roles in human brain activity. However, an excessive oral intake of biogenic amines can induce adverse reactions such as nausea, headaches, rashes and change in blood pressure. Biogenic amines play an essential role in the human body regulating of body temperature, stomach pH, and gastric acid secretions; likewise, they advance development, metabolic action, and immunological arrangement of the gastrointestinal tract and are active in the nervous system and in blood pressure control (Komprda et al., 2001; Ladero et al., 2010; Ercan et al., 2013). The biogenic amine histamine has been recognized as a neurotransmitter in the central nervous system. Dopamine, epinephrine (also called adrenaline), and norepinephrine (also called noradrenaline) are regarded as catecholamines, and the role of catecholamines within the human body was focused on by many researchers because they are involved in many important biological functions and are commonly associated with several physical and mental disorders, for example, Alzheimer's, Parkinson's, schizophrenia, glaucoma, Huntington's, epilepsy, arrhythmias, and so on. Also, acetylcholine, serotonin, nitric oxide, glutamate, and tryptamine considered neurotransmitters (Ribeiro et al., 2016; Hasanzadeh et al., 2017).

Biogenic amines can be naturally present in plant food since they are required in cellular metabolism and in growing tissues (Santos, 1996; Martinez-Villaluenga et al., 2008). Plants can perceive abiotic stresses and elicit appropriate responses with altered metabolism, growth, and development. The regulatory circuits include stress sensors, signaling pathways comprising a network of protein-protein reactions, transcription factors and promoters, and finally the output proteins or metabolites. It is proposed that our understanding of plant stress tolerance can be greatly refined by thorough characterization of individual genes and assessing their contribution to stress tolerance (Bartels and Sunkar, 2005). Polyamine levels have been determined in foods and drinks coming from animal and plant sources, even though for plant-derived food these analyses have not generally taken into consideration the possibility of utilising specific plant-derived foods for patients affected by diseases that need special feeding (Righetti et al., 2008). Autofluorescence of intact secretory cells of plants is considered a possible biosensor for the cellular biomonitoring. The fluorescence of secretory cells was due to the chemical composition of their secretions (Roshchina, 2003).

The substances produced in normal metabolism in any cells of living organisms participate in neural transmission of mammalians (catecholamines, serotonin, melatonin) and as mediators of inflammation (histamine and tyramine) (Önal, 2007; Spano et al., 2010; Nishikawa et al., 2012). The aliphatic polyamines, putrescine, spermidine and spermine, are normal cell constituents that play important roles in cell proliferation and differentiation. The equilibrium between cellular uptake and release and the balanced activities of biosynthetic and catabolic enzymes of polyamines are essential for normal homeostasis in the proliferation and functions of cells and tissues. The main precursor of polyamines is the amino acid ornithine, which is synthesized mainly in mitochondria from glutamate through the acetylation of the amino group, phosphorylation, and reduction of the acetylated derivative to the N-acetylglutamic-ϒ-semialdehyde (Teti et al., 2002).

In plants, diamine putrescine, triamine spermidine, and tetramine spermine are frequently present in amounts varying from micromolar to more than millimolar. Thus, the levels of polyamines in plant cells are significantly higher than those of phytohormones. Polyamines may mediate the action of hormones as part of their signal response and are thus suggested as hormonal second-messengers. Recently, polyamines have also been shown to affect physiological processes by modulating the transduction of the pectic signal. In addition, polyamines are considered as one of the carbon and nitrogen reserves at least in cultured tissues (Kakkar and Sawhney, 2002). When the plant initiates flowering, the conjugates accumulate in the floral organs and disappear from the leaves. Polyamines can also be found as eel wall-bound forms (Martin-Tanguy, 1997).

There is a great amount of data demonstrating that under many types of abiotic stresses, an accumulation of the three main polyamines—putrescine, spermidine, spermine—does occur. These amines, involved with the control of numerous cellular functions, including free radical scavenger and antioxidant activity, have been found to confer protection from abiotic stresses (Groppa and Benavides, 2008). Biogenic amines are stress markers, especially catecholamines which are easily oxidized to various pigments catecholaminechromes. Catecholamines dopamine, noradrenaline and adrenaline may cause dosedependent stimulation or inhibition in germination of seeds, pollen and vegetative microspores. The histamine increases quickly in plants under drought condition (Roshchina and Yashin, 2014). In animal and plant systems, the cellular levels of certain polyamines are directly associated with the rate of cell division (Fienberg et al., 1984). Beside the role of neurotransmitter, histamine is a main participator of allergic reactions, serving as a primary mediator of the immediate symptoms noted in allergic responses, it is a powerful vasodilator in animal tissue, and it also stimulates acid secretion in the stomach (Santos, 1996; Rai et al., 2013). It may also act as a local hormone, induce gastric acid secretion, and regulate cell growth and differentiation, circadian rhythm, body temperature, food intake, learning and memory, and immune response. In mammalians, histamine regulates an assortment of capacities by interfacing with particular receptors on target cells, to be specific $H_1$, $H_2$, and $H_3$ receptors of the G-protein coupled receptor family (Maintz and Novak, 2007; Ladero et al., 2010; EFSA, 2011). In addition, histamine synthesis and release are controlled by $H_3$ receptors and described as presynaptic auto-receptors on neurons (Ladero et al., 2010).

## 19.2 PLANT FOOD

There is a long history of studies on plant intelligence starting with Aristotle in about 280 BC, who was convinced that plants have a soul and feelings. Recent advances in plant cell biology allowed identification of plant synapses transporting the plant-specific neurotransmitter-like molecule, auxin. This suggests that synaptic communication is not limited to animals and humans but seems to be widespread throughout plant tissues (Baluška et al., 2004). Moreover, higher plants generate and transmit rapid electrical signals known as action potentials (Baluška et al., 2005). The world population is increasing at an alarming rate and is expected to reach about six billion by the end of year 2050. On the other hand, food productivity is decreasing due to the effect of various abiotic stresses; therefore, minimizing these losses is a major area of concern for all nations to cope with the increasing food requirements. Abiotic stresses cause losses worth hundreds of million dollars each year due to reduction in crop productivity and crop failure. In fact, these stresses threaten the sustainability of agricultural industry. Cold, salinity, and drought are among the major stresses, which adversely affect plant growth and productivity; hence, it is important to develop stress-tolerant crops (Mahajan and Tuteja, 2005). Biogenic amines are involved in plant development and formative procedures in light of abiotic and biotic anxieties (Bais and Ravishankar, 2002; Walters, 2003; Alcázar et al., 2006). The use of seed sprouts as food has spread in the past few decades from Far Eastern countries to parts of the Western world. Consumers can find on the market an extraordinary variety of different types of sprouts in which the *Cruciferae* family is well represented. Eating the fresh sprouts is the best way of gaining all of the health benefits claimed for cruciferous sprouts because only minor losses in health-promoting components are likely to occur (Martinez-Villaluenga et al., 2008). Sprouting improves the nutritional quality of seeds, by increasing the contents and availability of essential nutrients and decreasing the levels of antinutrients. Although biogenic amines are endogenously produced during the germination process, very little has thus far been discovered about their importance in sprouts and the microbial contribution to their overall load in this product group. They have recently attracted considerable attention from a public health standpoint. Although, the general health risk is limited, unless ingested in large quantities or when their natural degradation is inhibited (Simon-Sarkadi and Holzapfel, 1995).

Plant polyamines are also responsible for characteristics of agro-economical importance, including phytonutrient content, fruit quality, and vine life. In higher plants, earlier studies suggest that polyamine levels are critical for a number of developmental processes, including cell division, somatic embryogenesis, root growth, floral initiation, and flower and fruit development. Recent molecular genetic analyses have shown that altered polyamine levels have profound effects on plant growth and development. Osmotic shock, drought, and salt stress also increase polyamine levels. Salt stress increases polyamine levels in several crop plants, including rice, mung bean, maize, and sorghum. Air pollution has been correlated with elevated polyamine levels in plants as well (Liu et al., 2000). These procedures incorporate incitement of cell division, reaction to ecological anxieties, and control of embryogenesis, decrepitude, floral improvement, and natural product aging (Kakkar and Sawhney, 2002; Kusano et al., 2008). Currently, there is a general agreement that higher plants are not only able to receive diverse signals from the environment but that they also possess mechanisms for rapid signal transmission (Baluška et al., 2004). Plants undergo continuous exposure to various biotic and abiotic stresses in their natural environment. To survive under such conditions, plants have evolved intricate mechanisms to perceive external signals, allowing optimal response to environmental conditions.

Plants possess efficient defence mechanisms to cope with a plethora of environmental stresses, which includes drought, UV, high-salinity, cold stresses, and pathogen attack (Fujita et al., 2006).

In plants, mutant and transgenic plants with altered activity pointed to their involvement with different abiotic and biotic stresses. Furthermore, microarray, transcriptomic, and proteomic approaches have elucidated key functions of different polyamines in signaling networks in plants subjected to abiotic and biotic stresses (Hussain et al., 2011). Changes in polyamines have been related to environmental stresses and various physiologic processes including rhizogenesis and development of flowers (Bais et al., 2000).

## 19.3 FUNCTION OF POLYAMINES IN PLANTS

The biogenic amines in raw vegetables are usually present both as free bases and conjugated with other molecules, like phenolic acid and proteins, with levels depending on variety, ripening stage, and storage conditions (Casal et al., 2004). Since their discovery, the cellular functions of the physiological polyamines—putrescine, spermidine, spermine—have been the focus of much study. Due to their positive charge, these compounds can bind various cellular macromolecules, including DNA, RNA, chromatin, and proteins by electrostatic linkages, which can cause stabilization or destabilisation. Furthermore, covalent linkages can lead to cross-link formation of proteins forming cytotoxic derivatives. Thus, they have been implicated in myriad fundamental cellular processes, including regulation of gene expression, translation, cell proliferation, modulation of cell signaling, and membrane stabilization. Polyamines can also regulate cell death, particularly apoptosis. The three major polyamines are putrescine, spermidine, and spermine. Cadaverine is also present in legumes. Plant polyamines have been suggested to play important roles in morphogenesis, growth, embryogenesis, organ development, leaf senescence, and abiotic and biotic stress response (Liu et al., 2000; Kusano et al., 2008). The genes encoding the polyamines' biosynthetic enzymes have been isolated from a variety of plant species. Polyamines, widely present in living organisms, are now regarded as a new class of growth substances, which includes spermidine (a triamine), spermine (a tetramine), and their obligate precursor putrescine (a diamine), which play a pivotal role in the regulation of plant developmental and physiological processes. In plants, putrescine, spermidine, and spermine are major components that are not only involved in fundamental cellular processes, for example cell proliferation, differentiation, and programmed cell death, but also in adaptive responses to environmental stress (Kusano et al., 2007).

Serotonin is a well-known pineal hormone that in mammals plays a key role in mood. In plants, serotonin is implicated in several physiological roles, such as flowering, morphogenesis, and adaptation to environmental changes (Kang et al., 2011). Fruits and seeds are the major tissues in which serotonin occurs abundantly. As both the melatonin and serotonin are structurally related to the plant hormone, indole-3-acetic acid (IAA), possibly they influence physiology and development in plants. Consumption of edible plant tissues that contain the serotonin and melatonin would be useful, as their antioxidant activity is linked to their medicinal value for treatment of human diseases (Ramakrishna et al., 2012). The antisenescence action of exogenously applied polyamines (PAs) was first noted in freshly prepared mesophyll protoplasts of oat and other cereals (Galston and Sawhrey, 1990). The anti-oxidant function of spermine and spermidine involves direct scavenging of reactive oxygen species (ROS), especially singlet oxygen and hydroxyl radicals. Polyamines readily form chelates with copper and iron ions. These can also contribute to polyamine antioxidant activity. Spermine and spermidine reduce $Fe^{3+}$ to $Fe^{2+}$ and enhance ferric reducing activity of human plasma *in vitro* (Mozdzan et al., 2006).

Occurrence of biogenic amines known as neurotransmitters—catecholamines and histamine—in plant cells and their influence on growth play a major role in allelopathy. Natural compounds such as biogenic amines known as neurotransmitters in animals are also present in all living organisms (animal, plant, and microbial) as ancient components of irritability. They may occur in secretions of plants, animals, and microorganisms and influence the sensitive acceptor-cell viz., in cell–cell contacts regulating various processes, such as physiological responses as cellular growth and development (Roshchina and Yashin, 2014).

It was reviewed by Bouchereaua et al. (2000) and Teti et al. (2002) that amines in plants have been associated with many cell processes, including cell division and differentiation, synthesis of nucleic acids and proteins, membrane stability, pH and thermic or osmotic stress responses, and delay in senescence.

The diamine putrescine, the triamine spermidine, and the tetramine spermine are ubiquitous in plant cells, while other polyamines are of more limited occurrence. Their titer is very responsive to external conditions, such as light, temperature, and various chemical and physical stress agents, and the application of exogenous polyamines to plants or plant parts can produce visible effects such as the prevention of senescence in excised leaves and the formation of embryoids or floral primordia in certain otherwise vegetative tissue cultures. These facts compel us to examine the possible role of polyamines as regulators of physiological processes in plants. The accumulation of putrescine begins very rapidly after the application of stress. Polyamines in their free forms have been reported as anti-senescence agents, from both endogenous and exogenous application, the main effects in fruits being retarded color changes, increased fruit firmness, delayed ethylene and respiration rate emissions, induced mechanical resistance, and reduced chilling symptoms. Climacteric fruit, such as apple, apricot, avocado, banana, peach, plum, and tomato are characterized by their increased respiration and ethylene biosynthesis rates during ripening. By contrast, in non-climacteric fruit, such as citrus, eggplant, grape, pepper, and strawberry, ethylene is not required for the coordination and completion of ripening of these fruits (Valero et al., 2002).

Polyamines are essential compounds for growth and development in plants. Since only polyamines spermidine and spermine share a common precursor S-adenosylmethionine (SAM) with ethylene, they demonstrated competitive effects on functions in fruit development and ripening in many plants. Polyamines are essential compounds for growth and development in plants. Since polyamines only spermidine and spermine share a common precursor S-adenosylmethionine (SAM) with ethylene, they demonstrated competitive effects on functions in fruit development and ripening in many plants. In higher plants generally, the polyamines are found in both free and conjugated forms its endogenous levels depend on the external conditions of light and temperature. Especially in reproductive organs and seeds they appear as conjugated forms. There is no clear role envisaged for these conjugated polyamines in plants, but scanty reports suggest their possible involvement in plant defence (Sridevi et al., 2009). Increased polyamine levels were observed during somatic embryogenesis of carrot cell cultures (Fienberg et al., 1984) and in developing tomato fruits (Heimer et al., 1979). L-Tyrosine decarboxylase (TYDC) belongs to the aromatic L-amino acid decarboxylase enzyme family and catalyses the conversion of tyrosine to tyramine in plants (Park et al., 2012).

## 19.4 BIOGENIC AMINES IN DIFFERENT FOOD PLANTS

### 19.4.1 TEA

Tea is one of the most popular beverages worldwide, which is of great interest due to its beneficial medicinal properties (Zhao et al., 2006). Tea is the most frequently consumed beverage worldwide, besides water. All the three most popular types of tea—green (unfermented), black (fully fermented), and oolong (semifermented)—are manufactured from the leaves of the plant *Camellia sinensis*. Tea possesses significant antioxidative, anti-inflammatory, antimicrobial, anticarcinogenic, antihypertensive, neuroprotective, cholesterol-lowering, and thermogenic properties (Hayat et al., 2015). Its consumption has reached a point where it has become the second most commonly consumed beverage worldwide. This popularity was due to its characteristic aroma, flavor, and most influencing, its health benefits (Butt and Sultan, 2009).

Spizzirri et al. (2016) reported that in tea leaves total biogenic amines concentration ranged from 2.23 $\mu g\ g^{-1}$ to 11.24 $\mu g\ g^{-1}$ and putrescine (1.05–2.25 $\mu g\ g^{-1}$) and spermidine (1.01–1.95 $\mu g\ g^{-1}$) were always present, while serotonin (nd 1.56 $\mu g\ g^{-1}$), histamine (nd 2.44 $\mu g\ g^{-1}$), and spermine (nd–1.64 $\mu g\ g^{-1}$)

were detected more rarely and cadavarine and phenylethylamine determined in a few samples at much concentrations while none of the samples contained tyramine and also tea infusions showed the same trend with total biogenic amines concentrations never exceeding 80.7 μg $L^{-1}$ and black teas showed higher amounts of biogenic amines than green teas and organic and decaffeinated samples always contained much lower biogenic amines levels than their conventional counterparts. Cadavarine and phenylethylamine were determined in a few samples at much lower concentrations while none of the samples contained tyramine. Tea infusions showed the same trend with total biogenic amine concentrations never exceeding 80.7 μg$L^{-1}$. Black teas showed higher amounts of biogenic amines than green teas, and organic and decaffeinated samples always contained much lower biogenic amines levels than their conventional counterparts. In a study by Brückner et al. (2012), fourteen black teas, five green teas, one oolong tea, and one instant tea, harvested in different regions and processed and fermented under varying conditions, were analysed for biogenic amines content. Okamoto et al. (1997), using high-performance liquid chromatography (HPLC) and o-phthaldialdehyde (OPA) for quantification, analyzed hot-water extracts of teas for biogenic amines (ratio dry tea leaf to water 1:50, g/v; 85°C for 4 min). Single black, green, and oolong teas were analysed for putrescine, spermidine, spermine, and cadaverine. Per gram, dry leaves contained: in black tea 1.1 μg putrescine, 1.7 μg spermidine, and <2 μg spermine; in green tea 1.5 μg putrescine, 4.6 μg spermidine, and <1 μg spermine; in oolong tea 2.1 μg putrescine, and 1.5 μg spermidine. Nishimura et al. (2006) analysed 227 foods and drinks for the presence of polyamine s using HPLC and derivatization with OPA. The authors determined polyamines in hot-water extracts (1 g tea leaf and 5 mL water) of a single black and green tea. Quantities of biogenic amines per gram of black tea were 0.5 μg putrescine, 1.2 μg spermidine, 2.2 μg spermine, 2.7 μg cadaverine, and per gram of green tea were 1.6 μg putrescine, 3.3 μg spermidine, 3.8 μg spermine, 3.9 μg cadaverine.

### 19.4.2 Cacao

Cacao, designated *Theobroma cacao* by the eighteenth-century botanist, Carolus Linnaeus, is an important Neotropical, perennial crop, on which the thriving global chocolate industry is based. Cacao's putative center of genetic diversity is at the headwaters of the Amazon River, South America, and it is indigenous to the Amazon and Orinoco rainforests. *Cacao* was the Aztec word for chocolate and Theobroma means *Food of the Gods*, in keeping with the Aztecs' regard for the drink they made from cacao seeds (Badrie et al., 2015).

Commercial cocoa beans are seeds of the tree *T. cacao*, which are harvested, fermented, and dried. Cocoa is appealing in that a food commonly consumed for pure pleasure might also bring tangible benefits for human health and nutrition. Cocoa contains many compounds such as biogenic amines known to influence consumer health. Spermidine, putrescine, histamine, tyramine, β-phenylethylamine, cadaverine, and serotonin have been found in several cocoa-based products (Restuccia et al., 2015b). Biogenic amine content depends mainly on the plant variety and region of cultivation, as well as degree of maturity and the post-harvest manufacturing processes and storage conditions. The main biogenic amines found in cocoa and chocolate are 2-phenylethylamine, tyramine, tryptamine, serotonin, dopamine, and histamine (Oracz and Nebesny, 2014).

Previous researches reported, for cocoa beans, concentrations never exceeding 33 mg $kg^{-1}$, and for chocolate only a few biogenic amines at levels varying from a few mg $kg^{-1}$ to about 35 mg $kg^{-1}$ (Lavizzari et al., 2006; Mayr and Schieberle, 2012; Oracz and Nebesny, 2014). The most divergent data are related to cocoa powder samples showing concentrations of biogenic amines in the range 5.7–72.3 μg $g^{-1}$; other studies, for the same kind of samples, never contained levels of biogenic amines higher than 0.5 mg $kg^{-1}$ (Oracz and Nebesny, 2014).

### 19.4.3 Coffee

Coffee has a long-lasting history and was introduced as an economic crop during the fifteenth century; nowadays it is the most important food commodity worldwide, ranking second among all commodities after crude oil (Restuccia et al., 2015a). Coffee, consumed for its refreshing and stimulating effect, belongs to the tribe Coffee of the subfamily Cinchonoidea of Rubiaceae family. Coffee is a complex chemical mixture composed of several chemicals. It is responsible for a number of bioactivities and a number of compounds accounting for these effects (George et al., 2008).

It would be of interest to know if polyamines exist in green coffee, their behavior during the roasting process, and if coffees with different qualities of beverage within the same variety have the same polyamine content (Amorim et al., 1977). Biogenic amines derived from decarboxylation of amino acids represent undesirable coffee components (Restuccia et al., 2015a).

Amorim et al. (1977) reported free polyamine contents of arabica coffee for the first time, aiming to find correlations between their levels and beverage quality. Özdestan (2014) reported that the total amine levels in the ground coffee varied from 126.0 mg/kg to 352.2 mg/kg and the total amine levels in the brewed coffee varied from 5.679 mg/L to 48.88 mg/L and serotonin was the prevailing bioactive amine in both ground and brewed coffees.

### 19.4.4 Legume

Legumes are an important source of proteins and other nutrients and are commonly used as food and fodder. Particularly in developing countries, legumes represent the major component of daily dietary food stuffs along with bread (Shalaby, 2000). The major sources of putrescine were fruit, cheese, and non-green vegetables. All foods contributed similar amounts of spermidine to the diet, although levels were generally higher in green vegetables, and meat was the richest source of spermine (Valero et al., 2002). It has been reported that the profile and levels of amines in plants can be affected by different cultivars (Cirilo et al., 2003; Gloria et al., 2005). The increase in total bioactive amines observed during germination was previously reported in different seeds. According to Shalaby (2000), amines are endogenously produced during the germination process. The significantly higher levels of spermidine, spermine, and putrescine observed in 48–72 h suggest that this is the period with the greatest cellular multiplication and growth. Many biogenic amines in different levels are available in various juices, nectars, and lemonades, which are made from oranges, raspberries, lemons, grapefruit, mandarins, strawberries, currants, and grapes, and putrescine was the predominant and being the most significant amine in most of them, while histamine was low and tyramine concentrations were high (Santos, 1996; Shalaby, 1996). Putrescine increased during ripening in long-keeping tomato (Yahia et al., 2001).

Halász et al. (1994) found large concentrations of biogenic amines noradrenaline and tryptamine in orange juice; tyramine, tryptamine, and histamine in tomato; tyramine, noradrenaline, tryptamine, and serotonin in banana; tyramine and noradrenaline in plum; and histamine in spinach leaves. Chiacchierini et al. (2006) reported that the most abundant amines were putrescine, tyramine, spermidine, and tryptamine found at highest levels in ketchup, followed by concentrated tomato paste, biological mashed tomato, and conventional mashed tomato.

Płonka and Michalski (2017) reported that gelling sugar contained 1.05 mg/g norepinephrine, 1.70 mg/g normetanephrine, 0.72 mg/g dopamine, 0.68 mg/g tyramine, and 1.68 mg/g serotonin and the contents of normetanephrine, dopamine, tyramine, and serotonin in oranges in the range of 3.41 μg, 1.90 μg, 0.76 μg and 1.53 μg, respectively and cooking processes decreased the amine content by 18% (tyramine) to over 27% (serotonin), juice and nectar pasteurization reduced catecholamine contents up to 60% and freezing had the least impact on amine content and decreased the serotonin content in orange sorbet by only 5% (Płonka and Michalski, 2017).

The presence of biogenic amines has been described in different legume and radish seeds (Simon-Sarkadi and Holzapfel, 1995; Shalaby, 2000; Gloria et al., 2005). Shalaby (2000) and Simon-Sarkadi and Holzapfel (1995) observed an increase in cadaverine levels during the germination of some legume and radish seeds. The variations of biogenic amines during germination of mung bean, radish, and lentil were investigated by Simon-Sarkadi and Holzapfel (1995) in mung bean; the spermine content decreased while the spermidine concentration fluctuated. However, little changes in the spermidine and spermine contents were found during the germination period of radish. Simon-Sarkadi and Holzapfel (1995) found that during sprouting of legumes, Enterobacteriaceae and *Pseudomonas* spp. represented the dominant groups and comprised up to 95% of total microbial population of mung bean and lentil sprouts and more than 99% for radish sprouts and could support the development of biogenic amines in the sprouts. Shalaby (2000) found different biogenic amines in legume seeds during germination. In spite of the fact that the ungerminated legume seeds may contain high measures of biogenic amines, the germination can lead to an increase in the biogenic amine contents. Ungerminated legume seeds (broad bean, chick pea, and lupine) contained tryptamine, putrescine, cadaverine, spermidine, histamine, spermine, tyramine, and phenylethylamine as detected by Shalaby (2000). Tryptamine was the main biogenic amine detected, and its concentration considerably increased during the germination. Chick pea contained the highest amount of tryptamine (30.2 mg/kg) followed by broad bean (24.3 mg/kg) and lupine (11.7 mg/kg). β-phenylethylamine was detected in small amounts, and its concentration slowly increased during germination to reach its maximum amounts at the end of germination period being 10.3, 4.8, and 6.4 mg/kg of broad bean, chick pea, and lupine, respectively (Shalaby, 2000). Kalač et al. (2002) reported that putrescine and spermidine were observed in frozen spinach puree, ketchup, concentrated tomato paste, and frozen green peas, and in most samples at detectable levels, while histamine, spermine, and cadaverine concentrations were frequently below detection limits. Maximum mean levels (33.6 and 52.5 mg $kg^{-1}$ of tyramine and putrescine, respectively) were found in ketchup and 46.6 mg $kg^{-1}$ of spermidine in pea.

### 19.4.5 Broccoli and Radish Sprouts

The utilization of seed sprouts as sustenance has spread in the previous couple of decades from Far Eastern nations to parts of the Western world. Broccoli and radish grows contained quantities of mesophilic, psychrotrophic, total and faecal coliform bacteria, which are the usual counts for minimally processed germinated seeds. Putrescine, cadaverine, histamine, tyramine, spermidine, and spermine increased during sprout production (Martínez-Villaluenga et al., 2008). The contents of putrescine, spermidine, spermine, cadaverine, and agmatine were found in the range of (95–233, 259–565, 42, 27–106, and 12 nmol/g, 501, 188, not detected, 98 and 307 nmol/g, 3, 86, 28, 62 and not detected nmol/g, 38, 50–497, 11, 516 and not detected nmol/g) in broccoli, broccoli sprout, garlic (grown in Japan), and garlic (grown in China), respectively (Nishimura et al., 2006).

### 19.4.6 Fermented Soya Beans and Fermented Bean Curd

Fermented soya beans (or douchi) and fermented bean curd (or sufu) are the most typical traditional fermented soy products. Both are made by microbial fermentation beginning with soybean as raw material (Yang et al., 2014). In various biogenic amines detected in fermented soya beans by Yang et al. (2014), the total biogenic amines content was more than 900 mg/kg in 19 white sufu samples in fermented bean curd; these levels can cause hazard crisis to human consumptions. The concentration of histamine, tyramine, and β-phenylethylamine (PEA) was high enough in some fermented soya bean samples to cause a possible safety threat. The content of biogenic amines in fermented soya bean products should be studied and appropriate limits determined to ensure the safety of eating these foods as determined by Yang et al. (2014). Glória et al. (2005) detected some biogenic amines in soybeans (*Glycine max* L. Merril). They found spermidine, spermine, putrescine, agmatine, and cadaverine, while spermidine was the predominant amine followed by spermine. High concentrations of these amines confirmed soybeans as a rich source of biogenic amines, whereas cadaverine was confirmed to be inherent to soybean. Shukla et al. (2010) found numerous biogenic amines in Korean traditional fermented soybean paste (Doenjang) with different ranges. The contents of biogenic amines (tryptamine, 2-phenylethylamine, putrescine, cadaverine, agmatine, histamine, tyramine, spermidine, and spermine) in 23 Doenjang samples ranged in the mean of 18.37, 82.03, 70.84, 34.24, 47.32, 26.79, 126.66, 74.41, and 244.36 mg%, respectively.

Soybeans, tea leaves, and mushrooms obviously contained a large amount of spermidine, whereas oranges were rich in putrescine. Amid the fermented foods, soy sauces contained large amounts of putrescine and histamine, while Japanese sake contained a sufficient amount of agmatine (Okamoto et al., 1997).

Beatriz et al. (2005) investigated the total levels of amines in the different soybean cultivars harvested in 2003 and 2004, and the study demonstrated that during the first 96 h of germination, there was a significant change in bioactive amines in soybeans, reported on a dry weight basis. The levels of spermidine, spermine, and putrescine were significantly higher in 48 h of germination, whereas cadaverine levels increased significantly up to 96 h. The distribution of amines in soybean seedlings is heterogeneous. Significantly higher spermidine and spermine levels are present in the cotyledon, whereas putrescine and cadaverine accumulated in the radicle. Therefore, germination for 48 h can be used to increase polyamine levels in soybean. Researchers detected five bioactive amines in soybeans: spermidine, spermine, putrescine, agmatine, cadaverine. Spermidine was the predominant amine, followed by spermine, in every cultivar analysed. Cadaverine was observed to be inherent to soybeans, a specific characteristic of plants from the Leguminoseae family. Torrigiani et al. (2004) reported that applied to peach trees (*Prunus persica* L. Batsch cv Stark Red Gold) under open field conditions putrescine (5, 10, and 20 mM), spermidine (0.5, 1, and 2 mM), and aminoethoxyvinylglycine (AVG; 0.32, 0.64, and 1.28 mM) and both polyamines and AVG reduced ethylene production of fruit, delayed loss of firmness, retained titratable acidity, and prevented the increase in dry matter (DM) and soluble solids concentration (SSC).

### 19.4.7 Flaxseed Sourdough

Several biogenic amines were found in flaxseed in different amounts; the highest level (141.9 mg/kg) of phenylethylamine was found in non-fermented flaxseed and (73.2 mg/kg) fermented with *Pediococcus pentosaceus* flaxseed (Bartkiene et al., 2014). Food-fermenting lactic acid bacteria are usually measured as to be non-lethal and non-pathogenic (Spano et al., 2010). Nevertheless, biogenic amines in fermented food items are considered as significant food care elements, particularly when raw materials rich in proteins are used for fermentations. Phenylethylamine was found in the range of 11.9 mg/100 g and 20.9 mg/100 g in flaxseed fermented with *Lactobacillus sakei* and *Pedicoccus acidilactici*, respectively (Bartkiene et al., 2014). Other biogenic amines, such as putrescine, histamine, tyramine, spermidine, and spermine were found in both fermented and non-fermented flaxseeds with different amounts and contain no health hazards for human consumptions in the term of biogenic amines as reported by Bartkiene et al. (2014). Dietary fiber, iron, B-group vitamins, selenium, magnesium, and also carbohydrates can be present in bread and may contain biogenic amines. Tyramine was found in the range of (41.8 mg/kg) in the experimental sourdough bread (Diana et al., 2014). Nishimura et al. (2006) found putrescine, spermidine, spermine, cadaverine, and agmatine in the range of (706, 2.440, 721, 300 and 190 nmol/g) respectively, in wheat germ. The contents of putrescine, spermidine, spermine, and cadaverine were found in the range of (15, 21, 21 and 23 nmol/g, 198–842, 298, 6 and 447–20 nmol/g, 41, 133, 148 and 100 nmol/g, 30, 623, 65 and 158 nmol/g) in bread, corn, buckwheat flour, and grain amaranth (*Amarantus hypochondriacus* L.), respectively (Nishimura et al., 2006).

In this sense, it should be considered that, as for other foods containing biogenic amines, direct overlapping of the data arising from different studies are generally difficult to accomplish as biogenic amines levels and distributions are influenced by many parameters regarding either the hygienic conditions of the raw materials or the production process, as well as the preservation techniques and the packaging of the product (EFSA, 2011).

### 19.4.8 Fermented Beverages

Fermented beverages such as boza (a traditional cereal-based fermented Turkish beverage), beer, cider (an alcoholic beverage made from apple juice), and wine contain histamine and other biogenic amines (Stratton et al., 1991; Ercan et al., 2013). Fermented beverages represent an important category of foodstuff that can supply significant quantities of biogenic amines. Wine is the most frequently fermented alcoholic beverage. Since alcohol is an inhibitor of monoamine oxidases (MAOs), the control of biogenic amines in fermented beverages is of considerable importance for consumer's health (Russo et al., 2010; Ercan et al., 2013). Agmatine, cadaverine, ethanolamine, histamine, putrescine, and tyramine are formed at larger concentrations during fermentation of alcohol (Santos, 1996). In wine biogenic amine may have two various sources, raw materials and fermentation processes (Spano et al., 2010).

The biogenic amine amount in wine can differ over an extensive range and hinges on various oenological factors (including maceration or prolonged contact with yeast dregs enhance the level of free amino acid, resulting in a worse quality of wine, also pH, manufacturing practice, levels of $SO_2$, time and storage conditions, also a presence of precursor amino acids). Histamine, tyramine, and putrescine are the most plentiful and main biogenic amine frequently found in higher levels (Marques et al., 2008). Cadaverine, phenylethylamine, and isoamylamine are also available but at lower levels (Landete et al., 2007; Marques et al., 2008). Similar biogenic amine concentrations have also been identified in ciders (Garai et al., 2006, 2007).

Numerous sorts of biogenic amines have been identified in both white and red wine: tyramine, histamine, tryptamine, mono-methylamine, 2-phenethylamine, putrescine, cadaverine, spermidine, iso- and n-amylamine, pyrrolidine, iso- and n-butylamine, iso- and n-propylamine, and ethylamine (Santos, 1996). The most representative biogenic amines in beer are putrescine, cadaverine, and tyramine (Glória et al., 1999; Loret et al., 2005), and beer has mainly been reported to be a health problem for some consumers, like patients under monoamine oxidase inhibitor (MAOI) drugs (such as painkillers, antidepressant drugs, and so on and also in case of Parkinson's disease) because they can inhibit the natural MAO detoxification enzymes and can cause hypertensive crises (Shulman et al., 1997). Tyramine considered being the cause of this effect. Beers with more than 10 mg/l have been considered hazardous for this type of patients (Tailor et al., 1994).

## 19.5 METHODS TO DETECT BIOGENIC AMINES IN FOOD

Analysis of biogenic amines is significant due to their potential toxicity and their usage as indicators of the degree of freshness or spoilage of food (Önal, 2007). Numerous methods for the quantitation of biogenic amines have already been published. Because amines are very polar, commonly a derivatization of the amino group is performed to reduce their polarity and, also, to provide a chromophore for UV or fluorescence detection. The most common derivatization agents for LC analysis are benzoyl chloride and dansyl chloride, and the derivatives are usually monitored by either fluorescence or diode array detection. However, in more recent studies, liquid chromatography–mass spectrometry (LC-MS) was applied in particular, to increase the reliability of amine identification (Mayr and Schieberle, 2012). Several procedures for the determination of biogenic amines have been published mainly based on chromatographic methods, which include, thin layer chromatography (TLC) (Naguib et al., 1995; Shalaby, 1999; Lapa-Guimarães and Pickova, 2004), liquid chromatography (LC) (Marks and Anderson, 2005), liquid chromatography and mass spectrometry (LC-MS/MS) (Sagratini et al., 2012), high-performance liquid chromatography (HPLC) (Moret and Conte, 1996; Lange et al., 2002; Tassoni et al., 2004; Moret et al., 2005; Oguri et al., 2007; Karmi, 2014), fluorometric method (Tzanavaras et al., 2013), and capillary electrophoresis (CE) (Paproski et al., 2002; Kvasnička and Voldřich, 2006). Chang et al. (1985) portrayed a technique for deciding some naturally dynamic amines incorporating histamine in cheeses by HPLC. The HPLC technique was discovered helpful for screening to recognize cheese tests containing dangerous measures of histamine. Voight et al. (1974) used thin layer chromatography (TLC) and described the use of 7-chloro-4-nitrobenzofurazan (NBO-CI) for the quantitation of amines, and compared with other commonly used detection reagents. The extract obtained from cheese was spotted directly on TLC plates, chromatographed, and NBO-CI derivatives formed. Chambers and Staruszkiewicz (1978) decided histamine substance of cheddar by an authority Association of Official Analytical Chemists (now AOAC International) fluorometric technique, which is the strategy for deciding histamine in fish. Use of the technique to cheddar had no unique troubles. Quick techniques were looked into by Stratton et al. (1991). Ekici and Coskun (2002), using fluorometric technique and o-phthaldialdehyde (OPA), have determined the histamine content of some commercial vegetable pickles at the range of 16.54 and 74.91 mg/kg (average 30.73 mg/kg). The maximum value (74.91 mg/kg) was obtained from a sample of hot pepper pickles.

Numerous enzymatic techniques, including catalyst connected immunosorbent examine Enzyme Linked Immuno-Sorbent Assay (ELISA), have been created to recognize histamine in blood and tissues. Enzymatic strategies use histamine N-methyltransferase and radioactive S-adenosylmethionine. In any case, these strategies and the ELISA procedure have not been connected to sustenances.

## 19.6 MICROORGANISMS PRODUCING BIOGENIC AMINES

Many bacteria, and some of yeast, may show decarboxylase activity in forming biogenic amines. The following bacteria: *Bacillus, Clostridium, Enterobacter, Escherichia, Lactobacillus, Pediococcus, Proteus, Pseudomonas,* and *Salmonella* have decarboxylase activity (Bodmer et al., 1999; Pinho et al., 2001). A crucial component in the development of biogenic amines in sustenances is the existence of microorganisms with the ability to decarboxylate amino acids. This capacity has been portrayed in various genera, species, and strains of microscopic organisms, both Gram positive and Gram negative. The creation of biogenic amines in microscopic organisms might be connected with the supply of vitality and can be utilized as a protective system of bacteria against an acidic situation (Suzzi and Gardini, 2003; Karovičová and Kohajdová, 2005; Ladero et al., 2010; Stadnik and Dolatowski, 2010).

Amino acid decarboxylases are chemicals that exist in numerous microorganisms, which might be either normally recommended in nourishment items or might be prescribed by defilement recently, amid, or after sustenance preparing. Many bacteria and some of yeast may show decarboxylase activity in forming biogenic amines. For example, the following bacteria have decarboxylase activity: *Pseudomonas*, *Clostridium*, *Bacillus*, *Photobacterium*, Enterobacteriaceae, *Escherichia*, *Klebsiella*, *Citrobacter*, *Proteus*, *Shigella*, *Salmonella*, Micrococcaceae, *Staphylococcus*, and *Micrococcus*. Additionally, a considerable measure of lactic acid bacteria (LAB) having a place with the genera *Enterococcus*, *Lactobacillus*, *Carnobacterium*, *Pediococcus*, *Lactococcus*, and *Leuconostoc* can decarboxylate amino acids (Santos, 1996; Stadnik and Dolatowski, 2010).

Naila et al. (2011) found that species of *Bacillus subtilis, Bacillus megaterium, Bacillus thuringiensis, Brevibacterium casei, Enterococcus casseliflavus, Enterococcus faecali* and *Staphylococcus haminis* are identified to have the latent properties to form histamine. Both lactic acid bacteria and coagulase-negative staphylococci are two main groups of bacteria that are considered important in meat fermentation. Generally, *Lactobacillus* species are dominant in sausage. Nevertheless, in some slightly acidified sausages from southern Europe, *Enterococcus* and *Lactobacillus* species cohabit. Among lactobacilli, *L. sakei, L. curvatus,* and *L. plantarum* generally constitute the predominant microbiota during traditional sausage ripening (Mokhtar et al., 2012). *Leuconostoc mesenteroides* has a high latent to form tyramine or histamine in wine (Moreno-Arribas et al., 2003; Landete et al., 2007).

Mokhtar et al. (2012) reported that accumulation of biogenic amines during sausage fermentation inoculated with either, commercial starter, *L. plantarum* plus *B. lactis* Bb12, or *L. plantarum* plus *B. bifidum* alone or *L. plantarum* with both strains was lower compared with that of naturally fermented sausage.

Many strains of enterococci (for example, *E. faecalis* and *E. faecium*), *Lactobacillus* (For example, *L. curvatus,* and *L. brevis*) are able to produce tyramine (Bover-Cid et al., 2001; Suzzi and Gardini, 2003). Strains of *Pediococcus parvalus*, *Pediococcus damnosus*, *Oenococcus oeni*, *Tetragenococcus* species, *Leuconostoc* species, *Lactobacillus hilgardii*, *Lactobacillus buchnerii*, and *Lactobacillus curvatus* are recognized to form histamine in fermented nourishment (Kimura et al., 2001; Lucas et al., 2005; Spano et al., 2010).

In cheese and fermented sausages, the main tyramine makers are Gram-positive microbes inside the genera *Enterococcus* (such as *E. faecalis* and *E. faecium*), *Lactobacillus* (such as *L. curvatus* and *L. brevis*) (Bover-Cid et al., 2000), *Leuconostoc* and *Lactococcus* (Fernández et al., 2004, 2007), and *Carnobacterium* spp. (Masson et al., 1999). *Staphylococcus* may likewise have a part in the creation of tyramine (Ansorena et al., 2002; Martin et al., 2005; De las Rivas et al., 2008; Latorre-Moratalla et al., 2010).

## 19.7 FORMATION AND CONTROL OF BIOGENIC AMINE IN FOODS

Biogenic amines are formed when the alpha carboxyl group breaks away to from free amino acids. They are the chemicals in which one two or three hydrogen atoms in ammonia are displaced by alkyl or aryl groups (Shalaby, 1996; Chong et al., 2011). For example, histamine from histidine, tyramine from tyrosine, tryptamine from tryptophane. During food spoilage, bacteria can form large levels of biogenic amines by decarboxylation the corresponding amino acids precursors (Karvoičová and Kohajdová, 2005; Chong et al., 2011; Ercan et al., 2013). Biogenic amines are generated by chemicals in crude substances or by microbial decarboxylation of substrate amino acids during aging and storage in food and beverages (Santos, 1996; Stadnik and Dolatowski, 2010).

Biogenic amines are present in almost every food in the daily diet, but their concentrations change extensively between and even inside sustenance sorts (Santos, 1996; Ercan et al., 2013). Biogenic amines can aggregate as the after effect of uncontrolled microbial enzymatic action (Halász et al., 1994; Ladero et al., 2010; Ercan et al., 2013). Biogenic amines are available in an extensive variety of sustenance items, including fish and fish items, cheese, meat and meat items, dairy items, wine, lager, vegetables, organic products, nuts, and chocolate. Many types of research showed that various levels of histamine in fermented nourishments had been involved as the causative specialist in a few episodes of sustenance intoxications. Canned fish (Merson et al., 1974), fish burgers (Becker, 2001), Swiss cheese (Taylor et al., 1982), and gouda cheese (Chambers and Staruszkiewicz, 1978) contain excessive measures of histamine. The most well-known biogenic amines in nourishment are histamine, tyramine, tryptamine, phenylethylamine, putrescine, cadaverine, spermine, and spermidine. Biogenic amines have been reported by Sanceda et al. (1999), Valsamaki et al. (2000), Cinquina et al. (2004), Pinho et al. (2001), Marks and Anderson (2005), Moret et al. (2005), Bover-Cid et al. (2006), Chiacchierini et al. (2006), Marcobal et al. (2006), Custódio et al. (2007), Hassanien et al. (2011), Self and Wu (2012), Garai-Ibabe et al. (2013), and Renes et al. (2014) in fish sauce, feta cheese, fish tissue, terrincho cheese, seafood (finfish and shellfish), fresh and preserved vegetables, red and white wine, tomato products, wine, grated Parmesan cheese, fast food (fried and grilled meat, chicken, and fish), seafood products, natural ciders and ewe's milk cheese, respectively with a different contents.

The creation and undesired accretion of biogenic amines in food needs the presence of amino acid precursors, bacteria with amino acid decarboxylases, either derived from environmental pollution or from an added starter culture and favorable situations that permit bacterial development, decarboxylase synthesis, and decarboxylase activity (Suzzi and Gardini, 2003; Karvoičová and Kohajdová, 2005; Ladero et al., 2010). There are many factors influencing biogenic amine formation, such as storage condition (temperature and humidity), pH of the product, nature of the food, additives, manufacturing practices, and presence of amino acids and bacteria with favorable environmental conditions (Maijala and Nurmi, 1995; Stadnik and Dolatowski, 2010).

## 19.8 FACTORS AFFECTING BIOGENIC AMINES FORMATION

Several extrinsic and intrinsic characteristics (for example, redox potential, pH, temperature, NaCl, water activity, and so on) and the extent of the sausage sanitary conditions of manufacturing practices play a significant part in the creation of biogenic amines in foodstuffs (Maijala and Nurmi, 1995; Suzzi and Gardini, 2003; Naila et al., 2010; Stadnik and Dolatowski, 2010).

High temperatures, high pH qualities, and little salt substance can support the collection of free amino acids and in this manner animate the development of biogenic amines (Lorenzo et al., 2007). Also, some sustenance mechanical administrations, for example, salting, aging, and maturation can build the generation of biogenic amines in fish (Visciano et al., 2012). Storage temperature is the most significant reason influencing biogenic amine formations (Chong et al., 2011; Visciano et al., 2012). Amine formation is minimized at reducing temperatures by the way of prevention of bacterial development and the diminishment of enzyme activity (Naila et al., 2010; EFSA, 2011). The best temperature for the production of biogenic amines by microorganism has mainly been recommended to be between 20°C and 37°C, while formation of biogenic amine reduce beneath 5°C or over 40°C (Karvoičová and Kohajdová, 2005; EFSA, 2011).

Besides temperature control, other factors (for example pH, water activity, redox potential, oxygen supply, nature of the food, manufacturing practices, preservatives, additives, and so on) may have a significant effect on the creation of biogenic amines in fermented sausages (Suzzi and Gardini, 2003). pH is the main factor affecting amino acid decarboxylase activity (Santos, 1996; Stadnik and Dolatowski, 2010). In a reduced pH climate, microorganisms are more induced to generate decarboxylase as a part of their protective mechanism upon the acidity (Stadnik and Dolatowski, 2010; EFSA, 2011).

Oxygen supply seems to importantly affect the biosynthesis of biogenic amines. *Enterobacter cloacae* produces about a large portion of the amount of putrescine under anaerobic as contrasted to oxygen consuming, and *Klebsiella pneumoniae* incorporates fundamentally less cadaverine, however, gains the capacity to deliver putrescine under anaerobic conditions (Halász et al., 1994). The redox potential of the medium additionally impacts biogenic amine generations. Conditions bringing about a diminished redox potential fortify histamine creation, and histidine decarboxylase movement is by all accounts inactivated or crushed within the sight of oxygen (Karovičová and Kohajdová, 2005).

Salting appears to be a successful method in reducing biogenic amine levels (Chong et al., 2011). The existence of sodium chloride enacts tyrosine decarboxylase movement and hinders histidine decarboxylase action (Shalaby, 1996).

Additives and preservatives can minimize the production of biogenic amines (Naila et al., 2010). Presence of potassium sorbate and ascorbic acid in sausage seemed to be an important role in abatement or deduction in biogenic amine accretion (Bozkurt and Erkmen, 2004). The additive of sugar can also minimize biogenic amine production in a little form (Bover-Cid et al., 2001). Bozkurt and Erkmen (2004) found that sodium nitrate and sodium nitrite inhibit biogenic amine production. Bover-Cid et al. (2001) reported that sodium sulphite ordinarily acts to minimize the growth of bacteria during sausage ripening and also showed the annexation of sugar may slightly decrease the biogenic amine production.

## 19.9 LEGAL LIMITS OF BIOGENIC AMINES IN FOODS

The toxicological level of biogenic amines is exceptionally hard to set up on the grounds that it relies on upon the capacity of detoxification arrangement of the intestinal tract of individuals and on the existence of different amines (Santos, 1996; Komprda et al., 2001). Despite the fact that the harmfulness of histamine to man is a questionable subject, ingestion of from 70 to 1000 mg histamine will, as a rule, cause clinical manifestations of inebriation (Henry, 1960). Some laboratories have made more general recommendations; for example, the Netherlands Institute of Dairy Research (Ten Brink et al., 1990) sets a limit of 100 mg/kg on histamine in foods. Lawful furthest cut-off points of 100 mg histamine/kg nourishment and 2 mg/L alcoholic beverages have been recommended (Santos, 1996).

For tyramine, legitimate breaking points of 100–800 mg/kg are said for foodstuffs (Schneller et al., 1997; Gardini et al., 2001), and 30 mg/kg for phenylethylamine have been accounted for as the harmful dosage in foods (Santos, 1996; Gardini et al., 2001). An admission of 5–10 mg of histamine can be considered as surrendering to some delicate individuals, 10 mg is considered as a fair farthest point, 100 mg affect a medium poisonous quality, and 1000 mg is exceptionally lethal and considered hazardous for well-being (Santos, 1996; Karovičová and Kohajdová, 2005). Nout (1994) proposed for fermented foods 50–100 ppm, 100–800 ppm, and 30 ppm for histamine, tyramine, and β-phenylethylamine, respectively, or a total of 100–200 ppm. Such levels could be regarded as acceptable also for nonfermented foods. Data obtained show that none of the analysed samples represent a possible risk for consumer health, although additional risk factors such as amine oxidase-inhibiting drugs, alcohol, and gastrointestinal diseases may play an important role in determining the threshold for bioactive amines toxicity.

The European Union has built up directions for histamine levels, as indicated by which histamine level ought to be beneath 100 mg/kg in crude fish and underneath 200 mg/kg in salted fish for species having a place with the Scombridae and Clupeidae families (Karovičová and Kohajdová, 2005). However, the U.S. Food and Drug Administration (FDA) built up 50 mg/kg as the limit for a well-being hazard connected with histamine in fish (FDA, 2011). The nutritional codex of the Slovak Republic had decided the maximal bearable farthest point for the histamine was (20 mg/kg in lager and 200 mg/kg in fish and fish products) and for tyramine (200 mg/kg in cheeses) (Karovičová and Kohajdová, 2005). Taylor et al. (1978) reported that the ingestion of 70–1000 mg of histamine in a single meal is necessary to elicit any symptoms of toxicity depending on age and health of a person. In the Turkish Food Codex (2002), 200 mg/kg of histamine is accepted as an indicator of defect in fishes, and 10 mg/kg for wines. In Germany, the regulation allows for a higher level of histamine in mackerel and herring fishes at approximately of 200 mg/kg. In the US, the limit at which histamine is considered hazardous in tuna 50 mg histamine/100 g. The histamine level at which tuna is considered defective is 10 mg/100. However, more attention is necessary to the regulation of histamine in foods (Stratton et al., 1991). Also, it must be noticed that smaller measures of histamine may bring about harm, especially if the individual is powerless, in view of the hindrance of the histamine detoxification system in the body because of reasons, for example, individual inclination, gastrointestinal maladies, the utilization of certain solutions and liquor consumption, and the presence of different amines (Joosten, 1988; Bodmer et al., 1999).

## 19.10 TOXICOLOGY OF BIOGENIC AMINES

Biogenic amine intoxication is always related with consumption of large contents of biogenic amines in nourishment, which stimulate toxicological risks and health troubles, and they have psychoactive, vasoactive, and hypertensive effects (Bozkurt and Erkmen, 2002; Coïsson et al., 2004; Karvoičová and Kohajdová, 2005). In a detailed study by Til et al. (1997), the acute and subacute toxicity of tyramine, spermidine, spermine, putrescine, and cadaverine were examined in experimental rats following oral and intravenous administration. A relatively low oral toxicity of ≥2 g/kg body weight was found for tyramine, cadaverine, and putrescine, and of 0.6 g/kg body weight (BW) for spermidine and spermine (SPM). Intravenous administration of the biogenic amines caused a decrease of blood pressure with the exception of tyramine where an increase was found. Naila et al. (2011) reported oral toxicity levels for putrescine (2,000 mg/kg), cadaverine (2,000 mg/kg), tryptamine (2,000 mg/kg), spermidine (600 mg/kg), and spermine (600 mg/kg).

Tyramine can cause migraines when more than 100 mg is ingested orally and can induce severe poisoning symptoms when more than 1080 mg/kg is ingested. The maximum allowable concentration range for tyramine is 100–800 mg/kg in food (Nout, 1994).

In terms of histamine poisoning, some families of fishes such as the type of describe Scombridae and Scomberesocidae come to mind in the first order. Therefore, the term *scombroid fish poisoning* has been used to describe a type of fish poisoning (Taylor, 1986).

Histamine exerts its toxicity by binding or interacting with receptors on cellular membranes in the respiratory, cardiovascular, gastrointestinal, and immunological system and skin (Santos, 1996; Lehane and Olley, 2000; Karovičová and Kohajdová, 2005).

There are three sorts of receptors, $H_1$, $H_2$, and $H_3$. The most common symptoms result from the activity on the cardiovascular framework (Karovičová and Kohajdová, 2005). Histamine causes dilatation of peripheral blood vessels, capillaries, and arteries, thus resulting in hypotension, flushing, and headache. Histamine affecting constriction of the intestinal smooth muscle, intervened by $H_1$ receptors, may represent stomach issues, looseness of the bowels, and regurgitating (EFSA, 2011). Gastric acid secretion is controlled by histamine through $H_2$ receptors situated on the parietal cells (Stratton et al., 1991; Ladero et al., 2010). Torment and itching associated with the urticarial lesions might be because of sensory and motor neuron incitement through $H_1$ receptors (EFSA, 2011). Thus, histamine poisoning manifests a wide variety of symptoms, including hypotension, headaches, hot flushes, and a burning sensation in the mouth may occur. The clearest manifestations of utilization of large dosages of biogenic amines are nausea, vomiting, respiratory crisis (dyspnoea), hot flushes, oral burning, and withdrawal of intestinal smooth muscles bringing on stomach spasms and hyper- or hypotension (Santos, 1996; Komprda et al., 2001). Psychoactive amines (e.g., histamine, putrescine, and cadaverine) can bring about some neurotransmission issues because of their activity as false neurotransmitters. Some aromatic amines (such as tyramine, tryptamine, and β-phenylethylamine) demonstrate a vasoconstrictor activity while others (histamine and serotonin) introduce a vasodilator in veins, capillaries, and arteries, causing headaches, hypertension, flushing, gastrointestinal misery, and oedema (Santos, 1996; Bozkurt and Erkmen, 2002; Önal, 2007; Stadnik and Dolatowski, 2010). In the human body, there is a detoxification system that breaks down the biogenic amines to physiologically less active amines. The enzymes diamine oxidase (DAO) and monoamine oxidase (MAO) play an essential role in system detoxification. Nevertheless, upon intake of a high content of biogenic amines in sustenance, this detoxification framework is deactivated and not able to eliminate biogenic amines adequately (Karovičová and Kohajdová, 2005; Ercan et al., 2013). Also, utilization of alcohol and certain medicines (for example, antihistamines, antimalarial agents, psychopharmaceuticals and antidepressants, antihypertensives, antiarrhythmics, and anti-hypotonics) reduce or inhibit the activity of monoamine and diamine oxidase. In that case, this enzyme cannot act leading to absorption of biogenic amines by the human body and show toxic effects (Karovičová and Kohajdová, 2005; Ercan et al., 2013). Also, disease of intestinal mucosa can decrease the activity of biogenic amine detoxification enzyme (Stratton et al., 1991).

### 19.10.1 Histamine

In healthy individuals, histamine is quickly detoxified by amine oxidase enzyme; however, they may build up a serious side effect of histamine inebriation as a consequence of its high sums consumed with nourishment, for example, scombroid fish poisoning or matured cheese (Taylor, 1986; Lehane and Olley, 2000). Diamine oxidase (DAO) is the main enzyme for the metabolism of ingested histamine. It has been proposed that DAO, when functioning as a secretory protein, may be responsible for scavenging extracellular histamine after mediator release. Dysfunction of diamine oxidase (DAO) activity either because of hereditary inclination, gastrointestinal illnesses, or because of medicine with diamine oxidase inhibitors results in histamine intolerance numerous symptoms mimicking an allergic reaction even after the ingestion of little measures of histamine tolerated by healthy peoples (Maintz and Novak, 2007). Normally, during the food intake process in the human gut, low amounts of biogenic amines are metabolized to physiologically less active degradation products. Many inhibitors of DAO have been identified, such as aminoguanidine and some antihistaminic drugs. In case of insufficient and inhibition of diamine oxidase-activity or monoamine oxidases, due to secondary effects of medicines or alcohol, low levels of amines cannot be metabolised efficiently, and then they easily lead toxic effect (Bodmer et al., 1999). A low amount of histamine is not a potential risk. However, if histamine intake is large amount, or the patients use mono amino oxidase inhibitor drugs, or catabolism of amines is inhibited, then various symptoms may result. Histamine intoxication can result from the ingestion of food containing unusually large levels of histamine. Fish of the scombridae and scomberesicideae families, such as tuna, mackerel, saury, seerfish, butterfly kingfish, and bonito, also certain non-scombroid fish including mahi-mahi, anchovies, herring, marline, bluefish, and sardines are commonly implicated in biogenic amine intoxication, especially histamine poisoning (Taylor, 1986; Shalaby, 1996; Chong et al., 2011; Karmi, 2014). Thus, histamine poisoning manifests a wide variety of symptoms including hypotension, headaches, hot flushes, and a burning sensation in the mouth may occur. The side effects of histamine intoxication affecting the integumentary system and smooth muscles include rash, oedema (eyelids), urticaria, pruritus, and localized inflammation. The gastrointestinal association is distinguished by queasiness, vomiting, looseness of the bowels, and stomach issues. Neurological include headache, palpitation, flushing, tingling, burning, and itching. Other symptoms related to the effect on blood vessels include headache, nasal secretion, bronchospasm, tachycardia, hyper- and hypotension, and in extreme cases bronchospasm, suffocation, and serious respiratory misery has been reported (Taylor, 1986; Shalaby, 1996; Lehane and Olley, 2000; Maintz and Novak, 2007; Ladero et al., 2010; Chong et al., 2011; FDA, 2011). Histamine induces a bronchospasm when it is used i.v. at a dose of 0.4 mg/kg in rabbits (Al-Jawad and Al-Jumaily, 2005). In case of insufficient or inhibited of diamine oxidase-activity or monoamine oxidases, or due to side effects of drugs, or alcohol, low levels of amines cannot be metabolized efficiently, and then they become toxic. Furthermore, the levels of biogenic amines in cheese could be useful as indicators of the freshness and hygienic quality of raw materials and manufacturing conditions of food (Pinho et al., 2001).

The results of the tests on cheese showed that some bacteria that grows on the surface of Munster cheeses degrades histamine. During a 4-week maturing process, strains of *Brevibacterium linens* reduced the histamine and tyramine contents in the course of deamination by 55%–70% (Leuschner and Hammes, 1998).

## 19.11 TOXICOLOGY OF BIOGENIC AMINES AND CARCINOGENICITY

Biogenic amines are investigated as a possible mutagenic precursor, since some amines (i.e., putrescine, cadaverine, spermine, and spermidine) may be nitrosated by reaction with nitrite and generate volatile nitrosamines, which are carcinogenic or act as precursors for other substances capable of producing nitrosamines, which are carcinogenic to different kinds of animals and pose a potential health threat to humans (Shalaby, 1996; Önal et al., 2013). The accumulation of biogenic amines in fish is involved in nitrosamine formation (known as carcinogens) (Prester, 2011). The production of nitrosamines in foods is of concern to researchers, consumers, food companies, and health authorities due to their possible association with cancer (Bulushi et al., 2009).

Generally, nitrosamines are produced through reactions between nitrites and secondary amines (such as putrescine, cadaverine, and agmatine, and so on) and/or polyamines (such as spermine and spermidine) (Santos, 1996; Shalaby, 1996; Prester, 2011; Karmi, 2014) in the following manner: (1) nitrite and its subordinates may respond with some biogenic amines to create unpredictable N-nitrosamine; (2) nitrite can be changed over into nitrosating operators that may effectively respond with auxiliary amines to form cancer-causing N-nitrosamines; (3) the existence of some amines, for example, dimethylamine (DMA) and tertiary amines, for example, trimethylamine (TMA) have been observed to be involved in nitrosamine development; (4) essential amines, for example, putrescine and cadaverine have been proposed to cyclize amid warming to secondary amines, for example, pyrrolidine and piperidine, which respond with nitrite to frame cancer-causing nitrosamines (Bulushi et al., 2009; Ercan et al., 2013).

Putrescine and cadaverine can react with nitrite to form heterocyclic carcinogenic nitrosamine, nitrosopyrrolidine, and nitrosopiperidine (Figure 19.2) and may also cause haemoglobinemia (Santos, 1996; Kurt and Zorba, 2009; Chong et al., 2011). Generally, nitrosamines are produced through reactions between nitrites and secondary amines (such as putrescine, cadaverine and agmatine) and/or polyamines (such as spermine and spermidine) (Santos, 1996; Shalaby, 1996; Prester, 2011; Karmi, 2014).

In addition to the reactions between secondary amines present in the fish and nitrites, which is commonly used for coloring, flavoring, and preservation of fish, impure salt and heating may enhance nitrosamine formation in fish (Santos, 1996; Bulushi et al., 2009; Prester, 2011). Putrescine and cadaverine on heating are converted to pyrrolidine and piperidine, respectively, from which *N*-nitrosopyrrolidine and *N*-nitrosopiperidine are formed by heating (Shalaby, 1996; Bulushi et al., 2009; Prester, 2011).

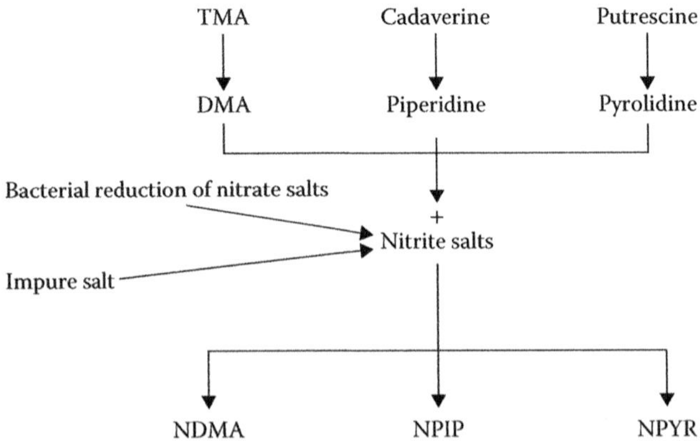


NDMA: nitrosodimethylamine, NPIP: nitrosopiperidine, NPYR: nitrosopyrrolidine

$$\text{Tyramine} + \text{Nitrites} \xrightarrow{\text{Heating}} \text{3-diazotyramine (3-DT)}$$

**FIGURE 19.2** Nitrosation of fish amines and conceivable origins of nitrite salts (Bulushi et al., 2009).

## 19.12 CONCLUSION

Dietary polyamines, spermidine and spermine, participate in an array of physiological roles with both favorable and injurious effects on human health (Kalač, 2014). Polyamines are essential compounds for growth and development in plants. Since only polyamines spermidine and spermine share a common precursor S-adenosylmethionine with ethylene, they demonstrated competitive effects on functions in fruit development and ripening in many plants (Sridevi et al., 2009). Polyamines may also function as stress messengers in plant responses to different stress signals. Polyamines are small ubiquitous polycations involved in many processes of plant growth and development and are well-known for their anti-senescence and anti-stress effects due to their acid neutralizing and antioxidant properties, as well as for their membrane and cell wall stabilizing abilities (Gill and Tuteja, 2010). Biogenic amines may occur in secretions of plants, animals, and microorganisms and influence the sensitive acceptor-cell viz., in cell–cell contacts regulating various processes, such as physiological responses as cellular growth and development (Roshchina and Yashin, 2014). Plant polyamines are also responsible for characteristics of agro-economical importance, including phytonutrient content, fruit quality, and vine life (Liu et al., 2000). Their titer is very responsive to external conditions, such as light, temperature, and various chemical and physical stress agents, and the application of exogenous polyamines to plants or plant parts can produce visible effects, such as the prevention of senescence in excised leaves and the formation of embryoids or floral primordia in certain otherwise vegetative tissue cultures. These facts compel us to examine the possible role of polyamines as regulators of physiological processes in plants. (Galston and Sawhrey, 1990).

Biogenic amines are steadily present in plant foods, and their determination is more important for consumers due to their physiological and toxicological effects also related to freshness, food spoilage, and safety. They are used as a quality indicator that shows the degree of spoilage, use of non-hygienic raw material, and poor manufacturing practice. Under ordinary circumstances in people, biogenic amines ingested from any sustenance are quickly detoxified by the activity of amine oxidases; yet, in those people groups who are under MAOI drug inhibitors or in the case of allergic individuals, the detoxification procedure is bothered and biogenic amines gather in the body.

The biogenic amines dopamine, norepinephrine, epinephrine, serotonin, tyramine, octopamin, and histamine are neurotransmitters or neuromodulators that play vital roles in the regulation of a wide variety of neurophysiological, behavioral, and cognitive processes, such as motor control, affect, motivation, learning, and memory. The consumption of food containing large amounts of these amines can have toxicological consequences, and it is generally assumed that they should not be allowed to accumulate. In fact, although several biogenic amines can play important roles in many human physiological functions, their presence in foods is always undesirable, because if adsorbed at too high concentration, they may induce headaches, respiratory distress, heart palpitations, hypo- or hypertension, and several allergenic disorders (Restuccia et al., 2015b).

Various measures have been taken with the aim of inhibiting or reducing production of biogenic amines during the manufacture of raw-cured materials, such as improved sanitary conditions in production plants, the use of starter cultures formed by lactic acid bacteria with acidifying capacity, and the use of certain preservatives (Bermúdez et al., 2012). A better understanding of the mechanisms by which biogenic amines are being produced is necessary to prevent their formation. In fermented foods, the use of short fermentations with carefully selected active starter cultures instead of wild fermentations will help to prevent the formation of toxic amines (Shukla et al., 2010).

For the future, it can be recommended or proposed and suggested that: (1) shopper ought to be educated about what sustenance contains what levels of biogenic amines; (2) nourishment makers ought to name their item regarding biogenic amines content; (3) doctors and dieticians ought to have enough data on biogenic amines substance of sustenances; (4) records are required for the suspect patients coming to healing facilities; (5) controls must be resolved for all foods in all nations (Bodmer et al., 1999).

## 19.13 SUMMARY

Biogenic amines are a group of low molecular weight, heat-stable, non-volatile, basic nitrogenous compounds with biological activity, formed and degraded as a result of normal metabolic activity in living cells and thus they are omnipresent in humans, animals, plants, and microorganisms and chiefly created by microbial decarboxylation of amino acid in foodstuffs or by amination and transamination of aldehydes and ketones by amino acid transaminases.

Polyamines are essential compounds for growth and development in plants, and their levels are critical for a number of developmental processes, including cell division and differentiation, somatic embryogenesis, root growth, floral initiation, flower and fruit development, synthesis of nucleic acids and proteins, membrane stability, pH and thermic or osmotic stress responses, and delay in senescence. Biogenic amines regulate many functions in the brain, including endocrine secretion, cognitive function, aggression, sleep and waking, emotional states, motivation, reward circuitry, decision making, and learning and memory. The world population is increasing at an alarming rate and is expected to reach about six billion by the end of year 2050. Abiotic stresses cause losses worth hundreds of million dollars each year due to reduction in crop productivity and crop failure. In fact, these stresses threaten the sustainability of agricultural industry.

Biogenic amines are toxic substances that can cause disease in humans; although the toxic effects of biogenic amines differ and depend on many variables. The manifestation of biogenic amines not only indicates the extent of microbial contamination but also makes food a potential toxic hazard for humans. Generally, biogenic amines in all kinds of food can be controlled by strict use of good hygiene in both raw material and manufacturing environments with corresponding inhibition of spoiling microorganisms. In fermented foods, the use of short fermentations with carefully selected active starter cultures instead of wild fermentations will help to prevent the formation of toxic amines.

## ACKNOWLEDGMENTS

The authors acknowledge English support from the Yüzüncü Yıl University, Distance Education Center Directorate (YÜSEM) and Prof. Dr. Suat Ekin assistance for the illustration of Figure 19.1.

## REFERENCES


Alcázar R, Marco F, Cuevas JC. Involvement of polyamines in plant response to abiotic stress. *Biotech Lett.* 2006;28:1867–1876.

Al-Jawad FH, and Al-Jumaily W. The relaxant effect of some drugs and aqueous extract of medicinal plants on bronchial smooth muscle in rabbits. *Tikrit J Pharm Sci.* 2005;1:58–63.

Amorim HV, Basso LC, Crocomo OJ, Teixeira AA. Polyamines in green and roasted coffee. *J Agric Food Chem.* 1977;25:957–958.

Ansorena D, Montel M, Rokka M, Talon R, Eerola S, Rizzo A. Analysis of biogenic amines in northern and southern European sausages and role of flora in amine production. *Meat Sci.* 2002;61:141–147.

Badrie N, Bekele F, Sikora E, Sikora M. Cocoa agronomy, quality, nutritional, and health aspects. *Cri Rev Food Sci Nut.* 2015;55:620–659.

Bais HP, and Ravishankar GA. Role of polyamines in the ontogeny of plants and their biotechnological applications. *Plant Cell Tissue Organ Cult.* 2002;69:1–34.

Bais HP, Sudha GS, Ravishankar GA. Putrescine and $AgNO_3$ influences shoot multiplication, in vitro flowering and endogenous titres of polyamines in Cichorium intybus L. cv. Lucknow local. *J Plant Growth Regul.* 2000;19:238–248.

Baluška F, Mancuso S, Volkmann D, Barlow P. Root apices as plant command centres: The unique 'brain-like' status of the root apex transition zone. *Biol Bratislava.* 2004;59:1–13.

Baluška F, Volkmann D, Menzel D. Plant synapses: Actin-based domains for cell-to-cell communication. *Trends Plant Sci.* 2005;10:106–111.

Bartels D, and Sunkar R. Drought and salt tolerance in plants. *Crit Rev Plant Sci.* 2005;24:23–58.

Bartkiene E, Schleining G, Juodeikiene G, Vidmantiene D, Krungleviciute V, Rekstyte T, Basinskiene L et al. The influence of lactic acid fermentation on biogenic amines and volatile compounds formation in flaxseed and the effect of flaxseed sourdough on the quality of wheat bread. *LWT-Food Sci Technol.* 2014;56:445–450.

Beatriz M, Gloria A, Tavares-Neto J, Labanca RA, Carvalho MS. Influence of cultivar and germination on bioactive amines in soybeans (*Glycine max* L. Merril). *J Agric Food Chem.* 2005;53:7480–7485.

Becker K. Histamine poisoning associated with eating tuna burgers. *J Am Med Assoc.* 2001;285:1327–1330.

Beneduce L, Romano A, Capozzi V, Lucas P, Barnavon L, Bach B, Vuchot P, Grieco F, Spano G. Biogenic amine in wines. *Ann Microbiol.* 2010;60:573–578.

Bermúdez R, Lorenzo J, Fonseca S, Franco I, Carballo J. Strains of staphylococcus and bacillus isolated from traditional sausages as producers of biogenic amines. *Front Microbiol.* 2012;3;1–6.

Blenau W, and Baumann A. Aminergic signal transduction in invertebrates: Focus on tyramine and octopamine receptors. *Recent Res Devel Neurochem.* 2003;6:225–240.

Bodmer S, Imark C, Kneubühl M. Biogenic amines in foods: Histamine and food processing. *Inflamm Res.* 1999;48:296–300.

Bouchereau A, Guénot P, Larher F. Analysis of amines in plant materials. *J Chromatogr B.* 2000;747:49–67.

Bover-Cid S, Izquierdo-Pulido M, Carmen Vidal-Carou M. Changes in biogenic amine and polyamine contents in slightly fermented sausages manufactured with and without sugar. *Meat Sci.* 2001;57:215–221.

Bover-Cid S, Izquierdo-Pulido M, Marine-Font A, Vidal-Carou M. Biogenic mono-, di- and polyamine contents in Spanish wines and influence of a limited irrigation. *Food Chem.* 2006;96:43–47.

Bover-Cid S, Izquierdo-Pulido M, Vidal-Carou M. Influence of hygienic quality of raw materials on biogenic amine production during ripening and storage of dry fermented sausages. *J Food Prot.* 2000;63:1544–1550.

Bozkurt H, and Erkmen O. Effects of starter cultures and additives on the quality of Turkish style sausage (sucuk). *Meat Sci.* 2002;61:149–156.

Bozkurt H, and Erkmen O. Effects of temperature, humidity and additives on the formation of biogenic amines in sucuk during ripening and storage periods. *Food Sci Technol Int.* 2004;10:21–28.

Brückner H, Flassig S, Kirschbaum J. Determination of biogenic amines in infusions of tea (*Camellia sinensis*) by HPLC after derivatization with 9-fluorenylmethoxycarbonyl chloride (FMOC-Cl). *Amino Acids.* 2012;42:877–885.

Bulushi I, Poole S, Deeth H, Dykes G. Biogenic amines in fish: Roles in intoxication, spoilage, and nitrosamine formation—A review. *Crit Rev Food Sci Nutr.* 2009;49:369–377.

Butt S, and Sultan M. Green tea: Nature's defense against malignancies. *Cri Rev Food Sci Nutr.* 2009;49:463–473.

Casal S, Mendes E, Alves RM, Alves RC, Oliveira MBPP, Ferreira MA. Free and conjugated biogenic amines in green and roasted coffee beans. *J Agric Food Chem.* 2004;52:6188–6192.

Casella I, Gatta M, Desimoni E. Determination of histamine by high-pH anion-exchange chromatography with electrochemical detection. *Food Chem.* 2001;73:367–372.

Chambers TL, and Staruszkiewicz WF. Fluorometric determination of histamine in cheese. *J AOAC.* 1978;61:1092–1097.
Chang S, Ayres J, Sandine W. Analysis of cheese for histamine, tyramine, tryptamine, histidine, tyrosine, and tryptophane. *J Dairy Sci.* 1985;68:2840–2846.
Chiacchierini E, Restuccıa D, Vinci G. Evaluation of two different extraction methods for chromatographic determination of bioactive amines in tomato products. *Talanta.* 2006;69:548–555.
Chong CY, Abu Bakar F, Russly AR, Jamilah B, Mahyudin NA. Mini review: The effects of food processing on biogenic amines formation. *Int Food Res J.* 2011;18:867–876.
Cinquina A, Cañ A, Longo F, Santis L, Severoni A, Abballe F. Determination of biogenic amines in fish tissues by ion-exchange chromatography with conductivity detection. *J Chromatogr A.* 2004;1032:73–77.
Cirilo MPG, Coelho AFS, Araújo CM, Gonçalves FRB, Nogueira FD, Glória MA. Profile and levels of bioactive amines in green and roasted coffee. *Food Chem.* 2003;82:397–402.
Coïsson J, Cerutti C, Travaglia F, Arlorio M. Production of biogenic amines in "Salamini italiani alla cacciatora PDO". *Meat Sci.* 2004;67:343–349.
Custódio F, Tavares É, Glória M. Extraction of bioactive amines from grated Parmesan cheese using acid, alkaline and organic solvents. *J Food Comp Anal.* 2007;20:280–288.
De las Rivas B, Ruiz-Capillas C, Carrascosa A, Curiel J, Jiménez-Colmenero F, Muñoz R. Biogenic amine production by Gram-positive bacteria isolated from Spanish dry-cured "chorizo" sausage treated with high pressure and kept in chilled storage. *Meat Sci.* 2008;80:272–277.
Diana M, Rafecas M, Quílez J. Free amino acids, acrylamide and biogenic amines in gamma-aminobutyric acid enriched sourdough and commercial breads. *J Cereal Sci.* 2014;60:639–644.
Ekici K, and Coskun H. 2002. Histamine content of some commercial vegetable pickles. *Proceedings of ICNP-2002 – Trabzon*, Turkiye. pp. 162–164.
El-Yazal M, Rady M. Foliar-applied Dormex™ or thiourea-enhanced proline and biogenic amine contents and hastened breaking bud dormancy in "Ain Shemer" apple trees. *Trees.* 2013;27:161–169.
Ercan SŞ, Bozkurt H, Soysal Ç. Significance of biogenic amines in foods and their reduction methods. *J Food Sci Eng.* 2013;3:395–410.
European Food Safety Authority (EFSA). Scientific Opinion on risk based control of biogenic amine formation in fermented foods. *EFSA J.* 2011;9:2393–2487.
Farooqui T. Octopamine-mediated neuronal plasticity in honeybees: Implications for olfactory dysfunction in humans. *Neuroscientist.* 2007;13:304–322.
Fernández M, Lınares D, Alvarez M. Sequencing of the tyrosine decarboxylase cluster of *Lactococcus lactis* IPLA 655 and the development of a PCR method for detecting tyrosine decarboxylating lactic acid bacteria. *J Food Protect.* 2004;67:2521–2529.
Fernández M, Linares D, Rodríguez A, Alvarez M. Factors affecting tyramine production in *Enterococcus durans* IPLA 655. *Appl Microbiol Biotechnol.* 2007;73:1400–1406.
Fienberg AA, Choi JH, Lubich WP, Sung ZR. Developmental regulation of polyamine metabolism in growth and differentiation of carrot culture. *Planta.* 1984;162:532–539.
Flasarová R, Pachlová V, Buňková L, Menšíková A, Georgová N, Dráb V, Buňka F. Biogenic amine production by Lactococcus lactis subsp. cremoris strains in the model system of Dutch-type cheese. *Food Chem.* 2016;194:68–75.
Food and Drug Administration (FDA). *Fish and Fishery Products Hazards and Controls Guidance*, 4th ed, Chapter 7, Department of Health and Human Services, Food and Drug Administration, Center for Food Safety and Applied Nutrition, 2011, Washington, DC.
Fujita M, Fujita Y, Noutoshi Y, Takahashi F, Narusaka Y, Yamaguchi-Shinozaki K, Shinozaki K. Crosstalk between abiotic and biotic stress responses: A current view from the points of convergence in the stress signaling networks. *Curr Opin Plant Biol.* 2006;9:436–442.
Galston AW, and Sawhrey RK. Polyamines in plant physiology. *Plant Physiol.* 1990;94:406–410.
Garai G, Dueñas M, Irastorza A, Martín-Álvarez P, Moreno-Arribas M. Biogenic amines in natural ciders. *J Food Protect.* 2006;69:3006–3012.
Garai G, Dueñas M, Irastorza A, Moreno-Arribas M. Biogenic amine production by lactic acid bacteria isolated from cider. *Lett Appl Microbiol.* 2007;45:473–478.
Garai-Ibabe G, Irastorza A, Dueñas M, Martín-Álvarez P, Moreno-Arribas V. Evolution of amino acids and biogenic amines in natural ciders as a function of the year and the manufacture steps. *Int J Food Sci Technol.* 2013;48:375–381.
Gardini F, Martuscelli M, Caruso M, Galgano F, Crudele M, Favati F, Guerzoni M, Suzzi G. Effects of pH, temperature and NaCl concentration on the growth kinetics, proteolytic activity and biogenic amine production of *Enterococcus faecalis. Int J Food Microbiol.* 2001;64:105–117.

George SE, Ramalakshmı K, Rao LJM. A perception on health benefits of coffee. *Cri Rev Food Sci Nut.* 2008;48:464–486.

Gill S, and Tuteja N. Polyamines and abiotic stress tolerance in plants. *Plant Signal Behav.* 2010;5:26–33.

Glória M, and Izquierdo-Pulido M. Levels and significance of biogenic amines in Brazilian beers. *J Food Comp Anal.* 1999;12:129–136.

Gloria MBA, Tavares-Neto J, Labanca RA, Carvalho MS. Influence of cultivar and germination on bioactive amines in soybeans (*Glycine max* L var. Merrit). *J Agric Food Chem.* 2005;53:7480–7485.

Groppa MD, and Benavides MP. Polyamines and abiotic stress: Recent advances. *Amino Acids.* 2008;34:35–45.

Halász A, Baráth A, Simon-Sarkadi L, Holzapfel W. Biogenic amines and their production by micro-organisms in food. *Trends Food Sci Technol.* 1994;5:42–49.

Hasanzadeh M, Shadjou N, Guardia M. Current advancement in electrochemical analysis of neurotransmitters in biological fluids. *TrAC Trends Anal Chem.* 2017;86:107–121.

Hassanien FS, Hassan MA, Salem AM, El-Wakeel ME. Demonstration of biogenic amines in fast foods. *BVMJ.* 2011;22:230–237.

Hayat K, Iqbal H, Malik U, Bilal U, Mushtaq S. Tea and its consumption: Benefits and risks. *Cri Rev Food Sci Nutr.* 2015;55:939–954.

Heimer YM, Mizrahi Y, Bachrach U. Ornithine decarboxylase activity in rapidly proliferating plant cells. *FEBS Lett.* 1979;104:146–148.

Henry M. Dosage biologique de l'histamine dans les aliments. *Ann Fals Eexp Chim.* 1960;53:24–33.

Hussain SS, Muhammad A, Maqbool A, Siddique KHM. Polyamines: Natural and engineered abiotic and biotic stress tolerance in plants. *Biotech Advan.* 2011;29:300–311.

Joosten HMLJ. The biogenic amine contents of Dutch cheese and their toxicological significance. *Netherland Milk Dairy J.* 1988;42:25–42.

Kakkar RK, and Sawhney VK. Polyamine research in plants: A changing perspective. *Physiol Plant.* 2002;116:281–292.

Kalač P. Health effects and occurrence of dietary polyamines: A review for the period 2005-mid 2013. *Food Chem.* 2014;161:27–39.

Kalač P, Švecová S, Pelikánová T. Levels of biogenic amines in typical vegetable products. *Food Chem.* 2002;77:349–351.

Kang K, Park S, Natsagdorj U, Kim YS, Back K. Methanol is an endogenous elicitor molecule for the synthesis of tryptophan and tryptophan-derived secondary metabolites upon senescence of detached rice leaves. *Plant J.* 2011;66:247–257.

Karmi M. Determination of histamine and tyramine levels in canned salted fish by using HPLC. *Global Veterinaria.* 2014;12:264–269.

Karovičová J, and Kohajdová Z. Review: Biogenic amines in food. *Chem Pap.* 2005;59:70–79.

Kimura B, Konagaya Y, Fujii T. Histamine formation by *Tetragenococcus muriaticus*, a halophilic lactic acid bacterium isolated from fish sausage. *Int J Food Microbiol.* 2001;70:71–77.

Komprda T, Neznalovà J, Standara S, Bover-Cid S. Effect of starter culture and storage temperature on the content of biogenic amines in dry fermented sausage poličan. *Meat Sci.* 2001;59:267–276.

Kurt Ş, and Zorba Ö. The effects of ripening period, nitrite level and heat treatment on biogenic amine formation of "sucuk"—A Turkish dry fermented sausage. *Meat Sci.* 2009;82:179–184.

Kusano T, Berberich T, Tateda Y, Takahashi Y. Polyamines: Essential factors for growth and survival. *Planta.* 2008;228:367–381.

Kusano T, Yamaguchi K, Berberich T, Takahashi Y. Advances in polyamine research in 2007. *J Plant Res.* 2007;120:345–350.

Kvasnička F, and Voldřich M. Determination of biogenic amines by capillary zone electrophoresis with conductimetric detection. *J Chromatogr A.* 2006;1103:145–149.

Ladero V, Calles-Enríquez M, Fernández M, Alvarez MA. Toxicological effects of dietary biogenic amines. *Curr Nut Food Sci.* 2010;6:145–156.

Landete JM, Pardo I, Ferrer S. Tyramine and phenylethylamine production among lactic acid bacteria isolated from wine. *Int J Food Microbiol.* 2007;115:364–368.

Lange J, Thomas K, Wittmann C. Comparison of a capillary electrophoresis method with high-performance liquid chromatography for the determination of biogenic amines in various food samples. *J Chromatogr B.* 2002;779:229–239.

Lapa-Guimarães J, and Pickova J. New solvent systems for thin-layer chromatographic determination of nine biogenic amines in fish and squid. *J Chromatogr A.* 2004;1045:223–232.

Latorre-Moratalla ML, Bover-Cid S, Talon R, Garrig M, Zanardi E, Ianieri A, Fraqueza MJ, Elias M, Drosinos EH, Vidal-Carou MC. Strategies to reduce biogenic amine accumulation in traditional sausage manufacturing. *Food Sci Technol.* 2010;43:20–25.

Lavizzari T, Veciana-Nogues MT, Bover-Cid S, Marine-Font A, Vidal-Carou MC. Improved method for the determination of biogenic amines and polyamines in vegetable products by ion-pair high-performance liquid chromatography. *J Chromatogr A.* 2006;1129:67–72.

Lehane L, and Olley J. Histamine fish poisoning revisited. *Int J Food Microbiol.* 2000;58:1–37.

Leuschner RGK, and Hammes W. Degradation of histamine and tyramine by *Brevibacterium linens* during surface ripening on Munster cheese. *J Food Production.* 1998;61:874–878.

Liu K, Fu H, Bei Q, Luan S. Inward potassium channel in guard cells as a target for polyamine regulation of stomatal movements. *Plant Physiol.* 2000;124:1315–1326.

Lorenzo J, Martínez S, Franco I, Carballo J. Biogenic amine content during the manufacture of dry-cured lacón, a Spanish traditional meat product: Effect of some additives. *Meat Sci.* 2007;77:287–293.

Loret S, Deloyer P, Dandrifosse G. Levels of biogenic amines as a measure of the quality of the beer fermentation process: Data from Belgian samples. *Food Chem.* 2005;89:519–525.

Lucas PM, Wolken WAM, Claisse O, Lolkema JS, Lonvaud-Funel A. Histamine-producing pathway encoded on an unstable plasmid in *Lactobacillus hilgardii* 0006. *Appl Environ Microbiol.* 2005;71:1417–1424.

Mahajan S, and Tuteja N. Cold, salinity and drought stresses: An overview. *Arch Biochem Biophys.* 2005;444:139–158.

Maijala R, and Nurmi E. Influence of processing temperature on the formation of biogenic amines in dry sausages. *Meat Sci.* 1995;39:9–22.

Maintz L, and Novak N. Histamine and histamine intolerance. *Am J Clin Nutr.* 2007;85:1185–1196.

Marcobal A, de las Rivas B, Moreno-Arribas MV, Muñoz R. Evidence for horizontal gene transfer as origin of putrescine production in *Oenococcus oeni* RM83. *Appl Environ Microbiol.* 2006;72:7954–7958.

Marks HS, and Anderson CR. Determination of putrescine and cadaverine in seafood (finfish and shellfish) by liquid chromatography using pyrene excimer fluorescence. *J Chromatogr A.* 2005;1094:60–69.

Marques AP, Leitão MC, San Romão MV. Biogenic amines in wines: Influence of oenological factors. *Food Chem.* 2008;107:853–860.

Martín MC, Fernández M, Linares DM, Alvarez MA. Sequencing, characterization and transcriptional analysis of the histidine decarboxylase operon of *Lactobacillus buchneri. Microbiol.* 2005;151:1219–1228.

Martin-Tanguy, J. Conjugated polyamines and reproductive development: Biochemical molecular and physiological approaches. *Physiol Plant.* 1997;100:675–688.

Martinez-Villaluenga C, Frias J, Gulewicz P, Gulewicz K, Vidal-Valverde C. Food safety evaluation of broccoli and radish sprouts. *Food Chem Toxicol.* 2008;46:1635–1644.

Massona F, Johanssonb G, Montela MC. Tyramine production by a strain of *Carnobacterium divergens* inoculated in meat–fat mixture. *Meat Sci.* 1999;52:65–69.

Mayr CM, and Schieberle P. Development of stable isotope dilution assays for the simultaneous quantitation of biogenic amines and polyamines in foods by LC-MS/MS. *J Agr Food Chem.* 2012;60:3026–3032.

Merson MH, Baine WB, Gangarosa EJ, Swanson RC. Scombroid fish poisoning. Outbreak traced to commercially canned tuna fish. *J Am Med Assoc.* 1974;228:1268–1269.

Mokhtar S, Mostafa G, Taha R, Eldeep GSS. Effect of different starter cultures on the biogenic amines production as a critical control point in fresh fermented sausages. *Eur Food Res Technol.* 2012;235:527–535.

Moreno-Arribas V, Polo MC, Jorganes F, Muñoz R. Screening of biogenic amine production by lactic acid bacteria isolated from grape must and wine. *Int J Food Microbiol.* 2003;84:117–123.

Moret S, and Conte LS. High-performance liquid chromatographic evaluation of biogenic amines in foods: An analysis of different methods of sample preparation in relation to food characteristics. *J Chromatogr A.* 1996;729:363–369.

Moret S, Smela D, Populin T, Conte LS. A survey on free biogenic amine content of fresh and preserved vegetables. *Food Chem.* 2005;89:355–361.

Mozdzan M, Szemraj J, Rysz J, Stolarek RA, Nowak D. Antioxidant activity of spermine and spermidine re-evaluated with oxidizing systems involving iron and copper ions. *Int J Biochem Cell Biol.* 2006;38:69–81.

Naguib K, Ayesh AM, Shalaby AR. Studies on the determination of biogenic amines in foods. 1. Development of a TLC method for the determination of eight biogenic amines in fish. *J Agric Food Chem.* 1995;43:134–139.

Naila A, Flint S, Fletcher GC, Bremer PJ, Meerdink G. Biogenic amines and potential histamine forming bacteria in Rihaakuru (a cooked fish paste), *Food Chem.* 2011;128:479–484.

Naila A, Flint S, Fletcher G, Bremer P, Meerdink G. Control of biogenic amines in food–existing and emerging approaches. *J Food Sci.* 2010;75:139–150.

Nishikawa H, Tabata T, Kitani S. Simple detection method of biogenic amines in decomposed fish by intramolecular excimer fluorescence. *Food Nutr Sci.* 2012;3:1020–1026.

Nishimura K, Shiina R, Kashiwagi K, Igarashi K. Decrease in polyamines with aging and their ingestion from food and drink. *J Biochem.* 2006;139:81–90.

Nout MJR. Fermented foods and food safety. *Food Res Int.* 1994;27:291–298.

Oguri S, Enami M, Soga M. Selective analysis of histamine in food by means of solid-phase extraction cleanup and chromatographic separation. *J Chromatogr A.* 2007;1139:70–74.

Okamoto A, Sugi E, Koizumi Y, Yanagida F, Udaka S. Polyamine content of ordinary foodstuffs and various fermented foods. *Biosci Biotech Biochem.* 1997;61:1582–1584.

Oracz J, and Nebesny E. Influence of roasting conditions on the biogenic amine content in cocoa beans of different Theobroma cacao cultivars. *Food Res Int.* 2014;55:1–10.

Önal A. A review: Current analytical methods for the determination of biogenic amines in foods. *Food Chem.* 2007;103:75–86.

Önal A, Tekkeli S, Önal C. A review of the liquid chromatographic methods for the determination of biogenic amines in foods. *Food Chem.* 2013;138:509–515.

Öner Z, Osman Sağdiç O, Bedia Şimşek B. Lactic acid bacteria profiles and tyramine and tryptamine contents of Turkish tulum cheeses. *Eur Food Res Technol.* 2004;219:455–459.

Özdestan O. Evaluation of bioactive amine and mineral levels in Turkish coffee. *Food Res Int.* 2014;61:167–175.

Paproski RE, Roy KI, Lucy CA. Selective fluorometric detection of polyamines using micellar electrokinetic chromatography with laser-induced fluorescence detection. *J Chromatogr A.* 2002;946:265–273.

Park S, Lee K, Kim YS, Chi YT, Shin JS, Back K. Induced tyramine overproduction in transgenic rice plants expressing a rice tyrosine decarboxylase under the control of methanol-inducible rice tryptophan decarboxylase promoter. *Bioprocess Biosyst Eng.* 2012;35:205–210.

Pinho O, Ferreira I, Mendes E, Oliveira B, Ferreira M. Effect of temperature on evolution of free amino acid and biogenic amine contents during storage of Azeitão cheese. *Food Chem.* 2001;75:287–291.

Płonka J, and Michalski A. The influence of processing technique on the catecholamine and indolamine contents of fruits. *J Food Comp Anal.* 2017;57:102–108.

Prester L. Biogenic amines in fish, fish products and shellfish: A review. *Food Addit Contam Part A.* 2011;28:1547–1560.

Rai KP, Pradhan HR, Sharma BK, Rijal SK. Review: Histamine in foods: Its safety and human health implications. *J Food Sci Technol Nepal.* 2013;8:1–11.

Ramakrishna A, Giridhar P, Sankar KU, Ravishankar GA. Endogenous profiles of indoleamines: Serotonin and melatonin in different tissues of Coffea canephora P ex Fr. as analyzed by HPLC and LC-MS-ESI. *Acta Physiol Plant.* 2012;34:393–396.

Renes E, Diezhandino I, Fernández D, Ferrazza R, Tornadijo M, Fresno J. Effect of autochthonous starter cultures on the biogenic amine content of ewe's milk cheese throughout ripening. *Food Microbiol.* 2014;44:271–277.

Restuccia D, Spizzirri UG, Parisi OI, Cirillo G, Picci N. Brewing effect on levels of biogenic amines in different coffee samples as determined by LC-UV. *Food Chem.* 2015a;175:143–150.

Restuccia D, Spizzirri G, Puoci F, Picci N. Determination of biogenic amine profiles in conventional and organic cocoa-based products. *Food Add Contam Part A.* 2015b;32:1156–1163.

Ribeiro J, Fernandes P, Pereira C, Silva F. Electrochemical sensors and biosensors for determination of catecholamine neurotransmitters: A review. *Talanta.* 2016;160:653–679.

Righetti L, Tassoni A, Bagni N. Polyamines content in plant derived food: A comparison between soybean and Jerusalem artichoke. *Food Chem.* 2008;111:852–856.

Roeder T. Octopamine in invertebrates. *Progress Neurobiol.* 1999;59:533–561.

Roshchina VV. Autofluorescence of plant secreting cells as a biosensor and bioindicator reaction. *J Fluoresc.* 2003;13:403–420.

Roshchina VV, and Yashin A. Neurotransmitters catecholamines and histamine in allelopathy: Plant cells as models in fluorescence microscopy. *Allelopathy J.* 2014;34:1–16.

Russo P, Spano G, Arena MP, Capozzi V, Grieco F, Beneduece L. Are consumers aware of the risks related to biogenic amines in food? *Curr Res Technol Edu Top Appl Microbiol Microb Biotechnol.* 2010;2:1087–1095.

Sagratini G, Fernandez-Franzon M, Berardinis F, Font G, Vittori S, Manes J. Simultaneous determination of eight underivatised biogenic amines in fish by solid phase extraction and liquid chromatography–tandem mass spectrometry. *Food Chem.* 2012;132:537–543.

Sanceda NG, Suzuki E, Ohashi M, Kurata T. Histamine behavior during the fermentation process in the manufacture of fish sauce. *J Agric Food Chem.* 1999;47:3596–3600.
Santos S. Biogenic amines: Their importance in food. *Int J Food Microbiol.* 1996;29:213–231.
Schneller R, Good P, Jenny M. Influence of pasteurized milk, raw milk and different ripening cultures on biogenic amine concentrations in semi soft cheeses during ripening. *Z Lebensm Unters Forschung A.* 1997;204:265–272.
Self RL, and Wu WH. Determination of eight biogenic amines in selected seafood products by MSPD extraction followed by UHPLC-Orbitrap MS. *J Food Comp Anal.* 2012;27:169–173.
Simon Sarkadi L, and Holzapfel WH. Biogenic amines and microbial quality of sprouts. *Z Lebensm Unters Forsch.* 1995;200:261–265.
Shalaby AR. Changes in biogenic amines in mature and germinating legume seeds and their behaviour during cooking. *Nahrung-Food.* 2000;44:23–27.
Shalaby AR. Significance of biogenic amines in food safety and human health. *Food Res Int.* 1996;29:675–690.
Shalaby AR. Simple, rapid and valid thin layer chromatographic method for determining biogenic amines in foods. *Food Chem.* 1999;65:117–121.
Shukla S, Park H, Kim J, Kim M. Determination of biogenic amines in Korean traditional fermented soybean paste (Doenjang). *Food Chem Toxicol.* 2010;48:1191–1195.
Shulman KI, Tailor SAN, Walker SE, Gardner DM. Tap (draft) beer and monoamine oxidase inhibitor dietary restrictions. *Can J Psychiatry.* 1997;42:310–312.
Spano G, Russo P, Fune A, Lucas P, Alexandre H, Grandvalet C, Coton E et al. Biogenic amines in fermented foods. *Eur J Clin Nutr.* 2010;64:95–100.
Spizzirri U, Picci N, Restuccia D. Extraction efficiency of different solvents and LC-UV determination of biogenic amines in tea leaves and infusions. *J Anal Methods Chem.* 2016;2016:1–10.
Sridevi V, Giridhar P, Ravishankar GA. Endogenous polyamine profiles in different tissues of Coffea sp., and their levels during the ontogeny of fruits. *Acta Physiol Plant.* 2009;31:757–764.
Stadnik J, and Dolatowski ZJ. Biogenic amines in meat and fermented meat products. *Acta Sci Pol Technol Aliment.* 2010;9:251–263.
Stratton JE, Hutkıns RW, Taylor SL. Biogenic amines in cheese and other fermented foods. A review: *J Food Prot.* 1991;54:460–470.
Suzzi G, and Gardini F. Biogenic amines in dry fermented sausages: A review. *Int J Food Microbiol.* 2003;88:41–54.
Tailor S, Shulman K, Walker S, Moss J, Gardner D. Hypertensive episode associated with phenelzine and tap beer: A reanalysis of the role of pressor amines in beer. *J Clin Psychopharmacol.* 1994;14:5–14.
Tassoni A, Germaná MA, Bagni N. Free and conjugated polyamine content in *Citrus sinensis* Osbeck, cultivar Brasilino N. L. 92, a navel orange, at different maturation stages. *Food Chem.* 2004;87:537–541.
Taylor SL. Histamine food poisoning: Toxicology and clinical aspects. *Crit Rev Toxicol.* 1986;17:91–128.
Taylor SL, Keefe JT, Windham SE, Howell JF. Outbreak of histamine poisoning associated with consumption of Swiss cheese. *J Food Prot.* 1982;45:455–457.
Taylor SL, Leatherwood M, Lieber ER. Histamine in sauerkraut. *J Food Sci.* 1978;43:1030–1032.
Ten Brink B, Damink C, Joosten HMLJ. Occurrence and formation of biologically active amines in foods. *Int J Food Microbiol.* 1990;11:73–84.
Teti D, Visalli M, McNair H. Analysis of polyamines as markers of (patho) physiological conditions. *J Chromatogr B.* 2002;781:107–149.
Til HP, Falke HE, Prinsen MK, Willems MI. Acute and subacute toxicity of tyramine, spermidine, spermine, putrescine and cadaverine in rats. *Food Chem Toxicol.* 1997;35:337–348.
Torrigiani P, Bregoli AM, Ziozi V, Scaramagli S, Ciriaci T, Rasori A, Biondi S, Costa G. Pre-harvest polyamine and aminoethoxyvinyglycine (AVG) applications modulate fruit ripening in stark red gold nectarines (*Prunus persica* l. Batsch). *Postharvest Biol Technol.* 2004;33:293–308.
Turkish Food Codex. Türk Gıda Kodeksi, Gıda maddelerinde Belirli Bulaşanların Maksimum Seviyelerinin Belirlenmesi Hakkında Tebliğ, Tebliğ No: 2002/63, Resmi Gazete, Sayı:24885, 23.09.2002.
Tzanavaras P, Deda O, Karakosta T, Themelis D. Selective fluorimetric method for the determination of histamine in seafood samples based on the concept of zone fluidics. *Analytica Chimica Acta.* 2013;778:48–53.
Valero D, Martínez-Romero D, Serrano M. The role of polyamines in the improvement of the shelf life of fruit. *Trends Food Sci Technol.* 2002;13:228–234.
Valsamaki K, Michaelidou A, Polychroniadou A. Biogenic amine production in Feta cheese. *Food Chem.* 2000;71:259–266.
Visciano P, Schirone M, Tofaloand R, Suzzi G. Biogenic amines in raw and processed seafood. *Front Microbiol.* 2012;3:1–10.

Voigt MN, RR Eitenmiller, PH Koehler, MK Hamdy. Tyramine, histamine and tryptamine content of cheese. *J Milk Food Technol.* 1974;37:377–381.

Walters DR. Polyamines in plant disease. *Phytochem.* 2003;64:97–107.

Wang LC. Polyamines in soybeans. *Plant Physiol.* 1972;50:152–156.

Yahia EH, Contreras-Padilla M, Gonzalez-Aguilar G. Ascorbic acid content in relation to ascorbic acid oxidase activity and polyamine content in tomato and bell pepper fruits during development, maturation and senescence. *Lebensm-Wiss u.-Technol.* 2001;34:452–457.

Yang J, Ding X, Qin Y, Zeng Y. Safety Assessment of the biogenic amines in fermented soya beans and fermented bean curd. *J Agric Food Chem.* 2014;62:7947–7954.

Zhao JW, Chen QS, Huang XY, Fang CH. Qualitative identification of tea categories by near infrared spectroscopy and support vector machine. *J Pharmaceut Biomed Anal.* 2006;41:1198–1204.

# 20 Neurotransmitters in Medicinal Plants

*Dimitrii A. Konovalov*

**CONTENTS**



## 20.1 INTRODUCTION

Acetylcholine, dopamine, norepinephrine (noradrenaline), epinephrine (adrenaline), serotonin, histamine, and γ-aminobutyric acid, known as neurotransmitters, have been found not only in animals but also in plants (Roshchina 1991a, 1991b; Roshchina 2001; Murch 2006; Kulma and Szopa 2007; Ramakrishna et al. 2012a; Roshchina 2016a; Erland and Saxena 2017). Today, we have increasing evidence that neurotransmitters are multifunctional substances that play roles of chemosignalizators, trigger the growth and development (seed and pollen germination, embryo formation, etc.); regulate ion permeability, energetics, metabolism, movements (dealt with roots, leaves, stomata cells); and protect against stresses as well as participate in intercellular and intracellular relations. Stimulatory effects of some neurotransmitters on plant growth reactions may be used in agriculture.

The evolutionary look on the role of neurotransmitters at all steps of evolutionary development—from unicellular organisms to multicellular ones—of fungi, plants, and animals shows that non-synaptic functions of the substances have arisen earlier than synaptic functions in animals with developed nervous systems, and so may play a universal role as elementary molecular agents of irritation in any living cell (Roshchina 2001, 2016a). At present, neurotransmitters are considered universal agents of communication that are instrumental in existing relationships between microorganism–microorganism, microorganism–animal, and microorganism–plant.

In this connection, the interaction of human organisms with relative compounds—neurotransmitters contained in plants—becomes actual. There are perspectives to use the medicinal plants enriched in some neuromediators for pharmacology and medicine as an alternative to chemical drugs. The review is devoted to the problem.

## 20.2 PHARMACOLOGICAL AND MEDICINAL PROPERTIES OF NEUROTRANSMITTERS

Interest in the detection of neurotransmitters in plants was associated with the fact that they are contained in significant concentrations in food and medicinal plant raw materials. Up to now, in most degrees this is related to serotonin, which possesses antioxidant, anti-inflammatory, antitumor, antibacterial, and antistress activities. In addition, it has a beneficial effect in the treatment of migraines (Titus et al. 1986). Serotonin plays an important role in the control of appetite, eating behavior, and body weight (Cruzon 1990; Heisler et al. 2002).

Wide-ranging investigations of the compound's distribution in plants were undertaken. Fruits of American cranberry (*Vaccinium macrocarpon*), cranberry (*Vaccinium oxycoccos*), English walnut (*Juglans regia*), banana, plantain, plum (*Prunus domestica*), and tomato (*Lycopersicon esculentum*) contain substantial amounts of serotonin and may have a great number of health benefits if incorporated into a healthy diet. Such data indicates that serotonin in human diets and plant-based medicines can affect human health (Coutts et al. 1986). Fruit food enriched in serotonin may be a beneficial treatment for patients because it can show them how daily exercise, along with a good diet, can help to raise serotonin levels without relying on any type of prescription drugs. This may be useful for a wide range of health problems derived from a lack of brain serotonin, such as insomnia, depression, obesity, eating disorders, panic attacks, alcoholism, anxiety disorders, and bulimia. Serotonin precursor, 5-hydroxytriptophan, has been also used as a natural replacement for prescription antidepressants (Roshchina 2016b). Most interest in serotonin is connected with its use in the treatment of neurodegenerative disorders, such as stroke, Alzheimer's disease, Parkinson's disease, schizophrenia, Down syndrome, and autism, as well as atherosclerosis and atherothrombosis.

The attempts to analyze effects of serotonin-containing plants are also known. Thus, the serotonin derivatives have been isolated from safflower seeds and demonstrated antioxidant and anti-inflammatory activities (Hotta et al. 2002), antitumor (Nagatsu et al. 2000), antibacterial (Kumarasamya et al. 2003), and antistress potentials, as well as being involved in reducing depression and anxiety (Yamamotova et al. 2007; Rinki et al. 2016). Other biogenic amines and acetylcholine of plants in situ were not applied for pharmacy yet. In this connection, plants enriched in neurotransmitters as a perspective for medicine will be considered in the following.

## 20.3 OCCURRENCE OF NEUROTRANSMITTERS IN PLANTS PERSPECTIVE FOR THE HEALTH CARE

### 20.3.1 Acetylcholine

Acetylcholine, a metabolite formed from choline and acetic acid, is identified in more than 80 plant species from 38 families (Roshchina 1991a; 2001; Wessler et al. 2001) and some of them enriched in the compound are seen in Table 20.1. The localization of acetylcholine in plant tissues and cells is usually determined on the basis of the presence of a cholinesterase enzyme catalyzing its degradation. According to this method, cholinesterase was found in the cell wall, exine spores, and plasmalemma, as well as in the nucleus and chloroplasts (Roshchina 2001). In tissues, the enzyme location was observed in different parts. In corn, acetylcholinesterase is predominantly localized in vascular bundles (Yamamoto and Momonoki 2012). Acetylcholine is contained in medicinal species such as the common nettle stinging hairs in a concentration of $10^{-1}$ M or 120–180 nmol/g fresh weight. Kawashima with co-authors (Kawashima et al. 2007) found that at the top of the bamboo shoot the acetylcholine content is approximately 80 times higher (2.9 mmol/g) than in the rat brain. The compound is found in extracts from plant tissues not only in free form (Fluck et al. 2000).

If to mark the concentrations acting in human, then acetylcholine (0.05–0.15 g per day) decreases arterial pressure, acting as a vasodilator, retards the cardiac rhythm, and induces the stenosis of pupils, enhances the contraction of smooth musculature in internal organs, as well as increases

## TABLE 20.1
## Plant Species Enriched in Acetylcholine

| Families and Species | Common Name | Part of Plant and Amount of Acetylcholine in Brackets (nmoles $g^{-1}$ of Fresh [b]Dry Mass) or out Brackets ($\mu g\ g^{-1}$ of Fresh Mass) | References |
|---|---|---|---|
| **Altingiaceae** | | | |
| *Liquidambar styraciflua* L. | | Leaves (0.36) 0.058 | Miura and Shih (1984) |
| **Amaranthaceae** | | | |
| [a]*Spinacia oleracea* | Spinacia | Leaves (~3–7) | Smallman and Maneckjee (1981); Hartmann and Kilbinger (1974a) |
| **Anacardiaceae** | | | |
| *Rhus copallina* L | | Leaves (0.78) 0.13 | Miura and Shih (1984) |
| **Araceae** | | | |
| *Arum specificum* | | (~1.8) | Wessler et al. (2001) |
| *Arum maculatum* | | (~1.8) | Wessler et al. (2001) |
| **Asteraceae (Compositae)** | | | |
| [a]*Helianthus annuus* L. | Helianthus | Stems, shoots (7.9) 1.29, roots (3.5) 0.57 | Hartmann and Kilbinger (1974b) |
| *Porophyllum lanceolatum* DC. | | Leaves (100) 16.32 | Horton and Felippe (1973) |
| | | Leaves (1.31–19.0) 0.21–3.1, stem (1.02–2.4) 0.17–0.39 roots (0.91–4.5) 0.15–0.73 | Ladeira et al. (1982b) |
| *Senecio vulgaris* | Senecio | (~6.5) | Smallman and Maneckjee (1981); Hartmann and Kilbinger (1974a) |
| *Xanthium strumarium* L. | | Roots (1.3–10.2) 0.21–1.66 Leaves (1.32–6.0) 0.22–0.98 Stems (4.5–12.0) 0.73–2.0 | Ladeira et al. (1982b) |
| **Convolvulaceae** | | | |
| *Ipomaea abutiloides* (Carnea) | | Leaves (+), Stems (14) 2.28 Flowers (+), Seeds (85) 13.87 | Villalobos et al. (1974) |
| **Cruciferae (Brassicaceae)** | | | |
| *Capsella bursa pastoris* L. | Capsella | (~4.8) | Wessler et al. (2001) |
| [a]*Raphanus sativus* L. | | Leaves ([b]112–329) 18.28–53.68 Petioles ([b]237–545) 38.67–88.92 Roots ([b]327–510) 53.36–83.22 | Momonoki and Momonoki (1991) |
| [a]*Sinapis alba* L. | | Stems (1.8) 0.29 | Hartmann and Kilbinger (1974b) |
| | | Leaves (3.1–5.0) 0.51–0.82 | Ladeira et al. (1982b) |
| | Sinapis | (~4.8) | Wessler et al. (2001) |
| **Cucurbitaceae** | | | |
| [a]*Cucumis sativus* L. | | Leaves ([b]31–332) 5.06–54.17 Stems ([b]476–2991) 77.67–488.04 Roots ([b]62–247) 10.12–40.3 Nodes ([b]12–284) 1.96–46.34 | Momonoki and Momonoki (1991) |
| [a]*Cucurbita pepo* L. | Cucurbita | Stems (10.5) 1.71, roots (3.3) 0.54 | Hartmann and Kilbinger (1974b) |

*(Continued)*

**TABLE 20.1 (*Continued*)**
**Plant Species Enriched in Acetylcholine**

| Families and Species | Common Name | Part of Plant and Amount of Acetylcholine in Brackets (nmoles $g^{-1}$ of Fresh [b]Dry Mass) or out Brackets ($\mu g\ g^{-1}$ of Fresh Mass) | References |
|---|---|---|---|
| **Equisetaceae** | | | |
| *Equisetum arvense* | Horsetail | (0.039) | Horiuchi et al. (2003); Wessler et al. (2001) |
| **Fabaceae (Leguminosae)** | | | |
| *Phaseolus aureus* Roxb. | | Leaves (0.015–50) 0.002–8.16 | Roshchina (1991a) |
| [a]*Pisum sativum* L. | | Leaves (2.2) 0.36 | Roshchina and Mukhin (1985) |
| | | Stems (8.2) 1.34 | Hartmann and Kilbinger (1974b) |
| | | Roots (1.4) 0.23 | |
| | | Roots (17) 2.77 | Kasturi and Vasantharajan (1976) |
| *Robinia pseudoacacia* L. | | Leaves (traces) | Roshchina (1991a) |
| *Vigna unguiculata* (L.) Walp | | Leaves ([b]43–103) 7.02–16.81 | Momonoki and Momonoki (1991) |
| | | Stems ([b]67–132) 10.93–21.54 | Momonoki and Momonoki (1991) |
| | | Roots ([b]16–44) 2.61–7.18 | |
| | | Primary pulvini ([b]11–390) 1.80–63.64 | |
| | | Secondary pulvini ([b]63–125) 10.28–20.40 | |
| **Gramineae** | | | |
| [a]*Avena sativa* L. | | Green seedlings (10) 1.63 | Tretyn and Tretyn (1988) |
| [a]*Zea mays* L. | | Leaves (6.0) 0.98 | Roshchina (1991a) |
| **Mimosaceae** | | | |
| *Macroptilium atropurpureum* (DC) Urban | | Leaves ([b]21–75) 3.43–12.24 | Momonoki and Momonoki (1991) |
| | | Petioles ([b]23–89) 3.75–14.52 | |
| | | Primary pulvini ([b]41–70) 6.69–11.42 | |
| | | Secondary pulvini ([b]35–164) 5.71–26.76 | |
| | | Stems ([b]30–92) 4.90–15.01 | |
| | | Roots ([b]15–20) 2.45–3.26 | |
| **Pinaceae** | | | |
| *Pinus thunbergii* | Pine | (0.120–0.343) | Hartmann and Kilbinger (1974a) |
| | Pine | Sprout (0.276) | Horiuchi et al. (2003) |
| **Poaceae** | | | |
| *Phyllostachys bambusoides* Siebold & Zucc. | Bamboo | Shoot (top) (2 940) | Kawashima et al. (2007) |
| *Phyllostachys pubescens* | | (600–1 700) | Horiuchi et al. (2003) |
| **Rosaceae** | | | |
| *Urtica dioica* L., *U. parviflora* Roxb., *U. urens* L. | Urtica | Leaves, hairs, stems, roots (~500) | Wessler et al. (2001) |

[a] edible plant.
[b] nmoles $g^{-1}$ of Dry Mass.
(+) -amount was not determined.

the secretion of sudoriferous, lacrimal, and bronchial glands. Physicians (ophthalmologists) recommend the use of anticholinesterase compounds in order to increase the concentration of acetylcholine in the eye. The signaling and restoration of the work of the neuro-muscular system by the neurotransmitter is not estimated yet for the extracts from fresh plant pharmaceutical material enriched in the compound.

As seen from Table 20.1, only genus *Urtica* is in pharmaceutical nomenclature; other species are known as food or used in folk medicine. It should mark that another direction in the application in pharmacy may be a search of native species contained the anticholinesterase compounds (Murray et al. 2013; Deka et al. 2017). Especially since this problem is actually due to the attention to the care in Alzheimer's disease, efforts are being made in search of new molecules with anticholinesterase activity. The fact that naturally occurring compounds from plants are considered to be a potential sources of new inhibitors has led to the discovery of an important number of secondary metabolites and plant extracts with the ability of inhibiting the enzyme that increases the levels of the neurotransmitter acetylcholine in the brain, thus improving cholinergic functions in patients with Alzheimer's disease and alleviating the symptoms of this neurological disorder (Kaufmann et al. 2016; Ranjan and Kumari 2017; Saeedi et al. 2017). Special attention is paid to essential oils, which anticholinesterase activity is related to sesquiterpenes (Owokotomo et al. 2015). For example, essential oil extracts from leaf, seed, stem, and rhizome of four medicinal plants such as *Aframomum melegueta* K. Schum, *Crassocephalum crepidioides* (Benth S. More), *Monodora myristica* (Gaertn.), and *Ocimum gratissimum* (Linn) were tested for acetylcholinesterase inhibitory activity using Ellman's colorimetric method and compared to a reference known acetylcholinesterase inhibitor galantamine. Essential oils from cultivated medicinal plants such as *Salvia officinalis, Mentha piperita, Melissa officinalis,* and others demonstrated high inhibitory activity (over 80%) in many probes (Orhan et al. 2008). To our view, besides the importance of extracts from acetylcholine-enriched species, this is also a new field to study for the application of plant extracts in medicine.

### 20.3.2 Biogenic Amines

Monoamines dopamine and noradrenaline are used at shock states and at acute cardiac insufficiency, while adrenaline stops bronchial asthma and other acute allergic reactions, against glaucoma, and as vasoconstrictive and anti-inflammatory agents in otorhinolaryngologic and ophthalmic practices. Indoleamine serotonin (5–10 mg per day) is useful against hemorrhaging syndromes and increases the capillary stability and decreases the hemorrhage at anemia. It acts on blood vessels and as a hormone. Histamine is known as a component of drugs against polyartria and rheumatism, and in small doses (~$10^{-7}$ M) in order to prevent acute allergic reactions. It excites a secretion of gastric glands and spastic contraction of enteric musculature.

In plants, the presence of neurotransmitters in plant tissues was studied after extraction by various biological, chemical, and physical methods, including modern chromatography and mass spectrometry. At the cellular level, the experiments were started in 1968 by Kimura who discovered catecholamines and serotonin in some vegetables and fruits using histochemical methods. Parenchymal epidermal cells of banana fruit and epidermal cells of onion showed characteristic yellow fluorescence of serotonin. This was observed in the tissues of the leaves and stems of the plant. However, peel cells of lemon did not have similar properties (Kimura 1968).

Presence and localization of neurotransmitters catecholamines and histamine within single plant cells have been studied V. Roshchina and co-authors by fluorescence microscopy and its modification as microspectrofluorimetry using special molecular probes—glyoxylic acid (reagent on catecholamines) and o-phthalic aldehyde (reagent on histamine) (Roshchina et al. 2014). The results of the fluorescence measurement from pollen of 25 species showed that among pollen grains from wind-pollinated plants, such as *Betula verrucosa, Corylus avellana, Larix decidua, Phleum pratense, Populus balsamifera,* and *Quercus robur* fluorescence related to catecholamines were found, mainly, in *C. avellana*, while the emission dealt with catecholamines or histamine were only in microspores of *P. balsamifera*.

By this method, the localization of serotonin in different tissues of *Coffea canephora* was established. It turned out that the largest amount was contained in the conducting tissues of roots, stems, somatic embryos, and endocarps (husks) of immature fruits (Ramakrishna et al. 2012a, 2012b).

#### 20.3.2.1 Biosynthetic Pathways

The biosynthetic pathway of biogenic amines includes the decarboxylation and hydroxylation of corresponding amino acids (Lawrence 2004), and for plants, similar data has been summarized in some publications (Roshchina 1991a, 1991b, 2001; Kulma and Szopa 2007; Ramakrishna et al. 2011; Erlanden and Saxena 2017).

Phenylalanine is a precursor of dopamine, noradrenaline, and epinephrine. Subsequent hydroxylation converts it to tyrosine, and then to dihydroxyphenylalanine (DOPA). These processes are catalyzed by phenylalanine hydroxylase or phenylalanine monoxidase and tyrosine hydroxylase or tyrosine-3-monoxidase. Dopamine, a precursor of norepinephrine and epinephrine, is formed from DOPA by decarboxylation with the decarboxylase dioxyphenylalanine and the decarboxylase of aromatic amino acids. Another way is decarboxylation of tyrosine into tyramine, and then hydroxylation to dopamine and its oxidation to norepinephrine. These transformations are possible with the participation of tyramine hydroxylase and β-hydroxylase 3,4-dioxyphenylethylamine. In conclusion, the transmethylase of phenylethanolamines catalyzes the process of adrenaline formation.

A major factor determining serotonin biosynthesis is availability of its precursor tryptophan (Kang et al. 2008). The metabolism of L-tryptophan, which is unable to accumulate in the cell, is supposed to exist as a mode of detoxication of ammonium. Serotonin is synthesized in plants from tryptophan formed by the shikimate pathway and localized in plastids (Oleskin et al. 1998; Oleskin 2007). This process proceeds by two pathways: either via 5-hydroxytryptophan or tryptamine formation, or the first step of serotonin biosynthesis via decarboxylation of tryptophan, which then transforms in plants to tryptamine by action of the enzyme tryptophan decarboxylase, or by the decarboxylation of aromatic amino acids (Kang et al. 2009).

The limiting factor in the biosynthetic pathway of serotonin in plants is tryptophan decarboxylase, showing strong differences in expression patterns with developmental stage and age (Kang et al. 2009). Tryptamine is transformed to serotonin by hydroxylation with participation of the enzymes tryptamine 5-hydroxylase or 1-tryptophan-5-hydroxylase. Hydroxylation of tryptophan leads to the formation of 5-oxytryptophan in the presence of tryptophan-5-hydroxylase. At the next stage, 5-oxytryptophan is decarboxylated to yield serotonin. Tryptamine 5-hydroxylase was found as a soluble enzyme in rice roots (Kang et al. 2007).

Synthesis of serotonin is also found in the roots (6.5 mg/g fresh weight) and stems (1 mg/g fresh weight) of rice seedlings. This process of the formation of serotonin from tryptamine occurs with the participation of tryptamine 5-hydroxylase (Kang et al. 2007).

#### 20.3.2.2 Dopamine

Dopamine is a monoamine neurotransmitter whose functions are not limited to the central nervous system. It is known that dopamine is a coregulator of the immune system (Franco et al. 2007; Cosentino and Marino 2013; Atkinson et al. 2015), tissues, and organs (Zhang et al. 2012). Disturbances in the dopaminergic system cause many health problems, including high blood pressure (Cuevas et al. 2013), mental disorders (e.g., schizophrenia), and neurodegenerative diseases (e.g., Parkinson's disease) (Arreola et al. 2016).

Dopamine has been found so far in 40 species of 23 families (Roshchina 1991a, 1991b, 2001; Kulma and Szopa 2007). At the cellular level, it was found in vacuoles of medicinal species poppy *Papaver bracteatum* and *P. somniferum* (Kutchan et al. 1986). Table 20.2 presents the data on the level of dopamine in some plants. The greatest amount of dopamine is contained in the peel of banana fruit; in the pulp it is 87 times less. Dopamine is found in all organs of *Aconitum napellus*

**TABLE 20.2**
**The Occurrence of Dopamine in Plants**

| Families and Species | Common Name | Part of Plant and Amount of Dopamine (μg/g of Fresh Mass) | References |
|---|---|---|---|
| **Amaranthaceae** | | | |
| [a]*Spinacia oleracea* L. | Spinach | <1 | Kulma and Szopa (2007) |
| **Asteraceae (Compositae)** | | | |
| [a]*Lactuca sativa* L. | | Seedlings (+) | Kamisaka (1979); Kamisaka and Shibata (1982) |
| **Lauraceae** | | | |
| [a]*Persea americana* Mill. | | Fruit pulp 4 | Kulma and Szopa (2007) |
| **Leguminosae** | | | |
| [a]*Phaseolus vulgaris* | | Leaves 9.67 ng/g<br>Roots 3.86 ng/g | Khozaei et al. (2014) |
| [a]*Pisum sativum* | | Leaves 19.21 ng/g<br>Roots 4.131 ng/g | Khozaei et al. (2014) |
| [a]*Vicia faba* | | Leaves 15.21 ng/g<br>Roots 3.63 ng/g | Khozaei et al. (2014) |
| **Malvaceae** | | | |
| [a]*Theobroma cacao* L. | | Bean powder 1 | Kulma and Szopa (2007) |
| **Musaceae** | | | |
| [a]*Musa acuminata* × *Musa balbisiana* cv. *Prata* | Banana | Fruits 17.6 ± 0.8 mg/kg | Lavizzari et al. (2006) |
| | Banana | Fruits 6.38 ± 0.01 | Płonka and Michalski (2017) |
| | Yellow banana | Fruit pulp 42 | Kulma and Szopa (2007) |
| | Cavendish banana | Fruit pulp 2.5–10, fruit peel 100 | Kulma and Szopa (2007) |
| **Portulacaceae** | | | |
| [a]*Portulaca oleracea* L. | | Leaves 0.69%, stems 0.18%, seeds 0.59% | Chen et al. (2003) |
| **Rutaceae** | | | |
| [a]*Citrus sinensis* | | <1 | Kulma and Szopa (2007) |
| | Orange | fruits 1.96 ± 0.05 | Płonka and Michalski (2017) |
| **Solanaceae** | | | |
| [a]*Lycopersicon esculentum* | | <1 | Kulma and Szopa (2007) |
| [a]*Solanum melanogena* | Aubergine | <1 | Kulma and Szopa (2007) |
| | | Whole cuttings 0.17, aerial parts 0.24, tubers 0.10, roots 0.19 | Hourmant et al. (1998) |

[a] edible plant.
(+) -amount was not determined.

(family Solanaceae), but especially much in flowers and young fruits, whereas in roots its concentration is lowest (Faugeras et al. 1967). In latex of poppy *P. bracteatum*, the content of dopamine is 0.1% (1 mg/mL), and it easily transforms into morphine and other alkaloids (Kutchan et al. 1986).

The amount of dopamine found varies during plant development (Kamo and Mahlberg 1984) and sharply increases during stress (Swiedrych et al. 2004). Of particular note is the finding that

**TABLE 20.3**
**Occurrence of Epinephrine (Adrenaline) in Plants**

| Families and Species | Common Name | Part of Plant and Amount of Epinephrine (mg/g of Fresh Mass) | References |
|---|---|---|---|
| **Amaranthaceae** | | | |
| | [a]Spinach | <1 | Kulma and Szopa (2007) |
| **Fabaceae (Leguminosae)** | | | |
| [a]*Pisum sativum* L. | | Leaves 0.0–34.2 | Roshchina (1991a) |
| [a]*Phaseolus vulgaris* | Beans | <1 | Kulma and Szopa (2007) |
| **Gramineae** | | | |
| [a]*Zea mays* L. | | Leaves 3833 | Roshchina (1991a) |
| **Rutaceae** | | | |
| [a]*Citrus sinensis* | | <1 | Kulma and Szopa (2007) |
| **Solanaceae** | | | |
| [a]*Lycopersicon esculentum* | | <1 | Kulma and Szopa (2007) |
| [a]*Solanum melanogena* | Aubergine | <1 | Kulma and Szopa (2007) |

[a] edible plant.

increased amounts of dopamine (1–4 mg/g fresh mass) are found in flowers and fruits, in particular in Araceae species (Ponchet et al. 1982).

The quantity of dopamine arises sharply at wounding and stresses. At the wounding of cactus *Carnegiea gigantean*, the defensive callus arises and its cortical tissue (pulpy cortex) contains dopamine as a main phenolic component (Steelink et al. 1967). General concentration of dopamine at wounding can achieve 1% of common pulp of callus.

Derivatives of dopamine are also known in plants, for example dopamine betaxanthin in *Portulaca oleracea* (Gandía-Herrero et al. 2009). This demonstrates the important role of the catecholamines as neurotransmitters in fertilization as well as in fruit and seed development.

As follows from the data in the table, in *Ulvaria obscura* (green alga) dopamine concentrations are approximately 0.5%–1% of the fresh mass, or 3%–5% of its dry mass (Van Alstyne et al. 2006, 2011); that is several orders of magnitude higher than concentrations that typically occur in terrestrial plants (Table 20.3).

### 20.3.2.3 Noradrenaline

Noradrenaline was first identified in banana fruits and was considered to be responsible for the therapeutic effect (Waalkes et al. 1958; Wagner 1988). It was found in 30 species of 17 families (Roshchina 1991a, 1991b). Noradrenaline and adrenaline occur in vacuoles but have also been identified in isolated chloroplasts of some plants of families Fabaceae and Urticaceae (Roshchina 1991a, 1991b, 2001).

Some of them enriched in the compound are seen in Table 20.4. In leaves of some leguminous Fabaceae and Urticaceae, its amount is approx. 4000 nmole $g^{-1}$ fresh mass.

The accumulation of noradrenaline in leaves of some plants may be related to stress. Besides, the increase in noradrenaline level in banana fruit peel is from 5.1 to 6.2 in. unmatured and yellow matured to 15.2 μg $g^{-1}$ of fresh mass in yellow-black super-matured ones as well as in pulp from 2.5 in. yellow matured to 10.1 μg $g^{-1}$ of fresh mass in super matured is associated with oxidative processes during ripening (Foy and Parratt 1960). On going to the ripening or overripening state, the concentration of noradrenaline increases two- to fourfold compared to unripe fruits. In plants that are able to respond to mechanical stimuli with motor activity, such as *Albizzia julibrissin* and *Mimosa pudica*, noradrenaline accumulates in motor organs and axes of leaves (Applewhite 1973).

**TABLE 20.4**
**The Occurrence of Norepinephrine (Noradrenaline) in Plants**

| Families and Species/Cultivar | Common Name | Norepinephrine (Noradrenaline) (μg/g of Fresh Mass) | References |
|---|---|---|---|
| **Amaranthaceae** | | | |
| [a]*Spinacia oleracea* L. | Spinach | <1 | Kulma and Szopa (2007) |
| **Brassicaceae** | | | |
| *Brassica olereacea* var. *italica* | | <3.5 | Kulma and Szopa (2007) |
| **Bromeliaceae** | | | |
| | [a]Pineapple | Fruits 0.86 ± 0.02 | Płonka and Michalski (2017) |
| **Lauraceae** | | | |
| [a]*Persea americana* Mill. | Fuerte avocado | Fruit pulp <3.5 | Kulma and Szopa (2007) |
| *Phaseolus aureus* Roxb. | | Leaves 0.49–1234 | Roshchina (1991a) |
| [a]*Phaseolus vulgaris* | Beans | <1 | Kulma and Szopa (2007) |
| | [a]Green pea | | Kalac et al. (2002) |
| [a]*Pisum sativum* L. | | Leaves 0.027–42.7 | Roshchina and Mukhin (1985) |
| *Robinia pseudoacacia* L. | | Leaves 0–6760 | Roshchina (1991a) |
| **Malvaceae** | | | |
| [a]*Theobroma cacao* L. | Cocoa | Bean powder <3.5 | Kulma and Szopa (2007) |
| **Musaceae** | | | |
| | [a]Banana | Fruits 6.07 ± 0.01 | Płonka and Michalski (2017) |
| [a]*Musa acuminata* × *Musa balbisiana* | Yellow banana | Fruit pulp <3.5 | Kulma and Szopa (2007) |
| cv. *Prata* | [a]Cavendish banana | Fruit pulp, fruit peel <3.5 | Kulma and Szopa (2007) |
| **Portulacaceae** | | | |
| [a]*Portulaca oleracea* L. | | Leaves 0.074%, stems 0.029%, seeds 0.054% | Chen et al. (2003) |
| **Rutaceae** | | | |
| [a]*Citrus sinensis* | | <1 | Kulma and Szopa (2007) |
| **Solanaceae** | | | |
| [a]*Lycopersicon esculentum* | | <1 | Kulma and Szopa (2007) |
| [a]*Solanum melanogena* | Aubergine | <1 | Kulma and Szopa (2007) |
| **Urticaceae** | | | |
| *Urtica dioica* L. | | Leaves 591 | Roshchina and Mukhin (1985) |

[a] edible plant.

### 20.3.2.4 Serotonin

Serotonin (5-hydroxytryptamine) is an indoleamine neurotransmitter that has been identified in more, than 130 species of 48 families (Roshchina 1991a, 1991b, 2001; Ramakrishna et al. 2011; Erland et al. 2016; Erland and Saxena 2017). In the past, serotonin in plants was usually determined by paper and thin-layer chromatography (Regula, 1970, 1981), and ten years also by methods of GC, GC-MC, and HPLC in different modifications (Grosse and Artigas 1983; Lembeck and Skofitsch 1984). Although, other analytical techniques, such as capillary electrophoresis, fluorescence, and electrochemical methods have emerged as alternatives to develop faster and cheaper methods. The development of more effective and practical methods is necessary for a reevaluation of many previously studied plant species using modern analytical approaches that take into account such phenomena as derivatization, matrix effects, and potential degradation when using different extraction methods (Erland et al. 2016).

As shown in Table 20.5, the amount of serotonin varies in a dependence on the taxonomic position of the plant (family, species, cultivar), his analyzed part, or edible tissue. In addition, it was found

**TABLE 20.5**
**The Occurrence of Serotonin in Plants**

| Families and Species/ Cultivar | Common Name | Part of Plant and Amount of Serotonin (μg/g of Fresh [b]Dry Mass) | References |
|---|---|---|---|
| **Actinidiaceae** | | | |
| [a]*Actinidia deliciosa* (A. Chev.) C.F. Liang & A.R. Ferguson | Kiwi fruit | Edible tissues 9.52 ± 0.62 | Islam et al. (2016) |
| | | Pulp (edge 6.8, center 3.0), peel 0.3 | Feldman and Lee (1985) |
| **Amaranthaceae** | | | |
| [a]*Spinacia oleracea* L. | Spinach | Edible tissues 0.19 ± 0.07 | Islam et al. (2016) |
| | | Edible tissues 34.4 ± 2.4[b] | Ly et al. (2008) |
| **Amaryllidaceae** | | | |
| [a]*Allium cepa* L. | Onion | Edible tissues 12.2 | Badria (2002) |
| [a]*Allium fistulosum* L. | Green onion | Edible tissues 8 ± 0.8[b] | Ly et al. (2008) |
| [a]*Allium sativum* L. | Garlic | Edible tissues 29.3 | Badria (2002) |
| **Apiaceae** | | | |
| *Daucus carota* L. | Carrot | Roots 31.1 | Badria (2002) |
| **Betulaceae** | | | |
| [a]*Corylus* sp. | Hazelnut | Edible tissues 3.4 ± 0.1 mg/kg | Lavizzari et al. (2006) |
| | [a]Filbert | Edible tissues 2.1 | Feldman and Lee (1985) |
| **Brassicaceae** | | | |
| [a]*Brassica oleracea* var. *capitata* L. | Cabbage | Edible tissues 18.1 | Badria (2002) |
| | | Edible tissues 0.15 ± 0.03 | Islam et al. (2016) |
| [a]*Brassica oleracea* var. *botrytis* L. | Cauliflower | Edible tissues 33.3 | Badria (2002) |
| [a]*Brassica rapa* L. | Chinese cabbage | Edible tissues 110.9 ± 22.5[b] | Ly et al. (2008) |
| | | Edible tissues 32.7 | Badria (2002) |
| [a]*Brassica rapa* ssp. *pekinensis* | Chinese cabbage | Edible tissues 0.07 ± 0.02 | Islam et al. (2016) |
| [a]*Raphanus sativus* L. | Radish | Edible tissues 35.8 | Badria (2002) |
| **Bromeliaceae** | | | |
| [a]*Ananas comosus* | Pineapple | Edible tissues 11.7 | Badria (2002) |
| | | Edible tissues 9.11 ± 0.13 | Islam et al. (2016) |
| | Pineapple | Pulp (edge 31.5, core - center 8.7), peel 2.7 | Feldman and Lee (1985) |
| | Pineapple | Fruits 24.27 ± 0.50 | Płonka and Michalski (2017) |
| **Compositae (Asteraceae)** | | | |
| [a]*Cichorium intybus* L. | Chicory | Edible tissues 8.5 ± 3.2[b] | Ly et al. (2008) |
| *Echinacea purpurea* (L.) Moench | Echinacea | Leaf explants 0.100 | Jones et al. (2007) |
| [a]*Helianthus annuus* L. | Sunflower | Vegetative tissues 9.1–27.8 | Mukherjee et al. (2014) |
| [a]*Lactuca sativa* L. | Lettuce | Edible tissues 0.12 ± 0.03 | Islam et al. (2016) |
| [a]*Lactuca serriola* L. | Lettuce | Edible tissues 3.3 ± 0.06[b] | Ly et al. (2008) |
| **Cucurbitaceae** | | | |
| [a]*Citrullus lanatus* (Thunb.) Matsum. & Naka | Watermelon | Edible tissues 0.06 ± 0.02 | Islam et al. (2016) |
| | [a]Cantaloupe | Edible tissues 0.9 | Feldman and Lee (1985) |

*(Continued)*

**TABLE 20.5 (*Continued*)**
**The Occurrence of Serotonin in Plants**

| Families and Species/ Cultivar | Common Name | Part of Plant and Amount of Serotonin (μg/g of Fresh [b]Dry Mass) | References |
|---|---|---|---|
| | [a]Honeydew melon | Edible tissues 0.9 | Feldman and Lee (1985) |
| [a]*Cucumis sativus* L. | Cucumber | Edible tissues 23.7 | Badria (2002) |
| **Ebenaceae** | | | |
| | [a]Persimmon | Edible tissues <0.1 | Feldman and Lee (1985) |
| [a]Diospyros kaki L.f. | | Edible tissues 0.11 ± 0.05 | Islam et al. (2016) |
| **Ericaceae** | | | |
| *Vaccinium macrocarpon* Ait. | | 106.3 ± 35.9 | Brown et al. (2012) |
| *Vaccinium oxycoccos* L. | | 104.8 ± 14.1[c] | Brown et al. (2012) |
| *Vaccinium vitis-idaea* L. | | 122.0 ± 37.6[c] | Brown et al. (2012) |
| **Hypericaceae** | | | |
| *Hypericum perforatum* L. | St. John's wort | Vegetative tissues 2. | Murch et al. (2001) |
| | | Reproductive tissues 0.0002–0.002 | Murch and Saxena (2002); Murch et al. (2001) |
| **Juglandaceae** | | | |
| [a]*Carya illinoinensis* (Wangenh.) K·Koch | Pecans | Edible tissues 29 ± 4 | Feldman and Lee (1985) |
| [a]*Carya ovalis* (Wangenh.) Sarg. | Sweet pignuts | Edible tissues 25 ± 8 | Feldman and Lee (1985) |
| [a]*Carya ovata* (Mill.) K·Koch | Shagbark | Edible tissues 143 ± 23 | Feldman and Lee (1985) |
| [a]*Carya tomentosa* (Lam.) Nutt. | Mockernut | Edible tissues 67 ± 13 | Feldman and Lee (1985) |
| [a]*Juglans cinerea* L. | Butternuts | Edible tissues 398 ± 90 | Feldman and Lee (1985) |
| [a]*Juglans nigra* L. | Black walnut | Edible tissues 304 ± 46 | Feldman and Lee (1985) |
| [a]*Juglans regia* L. | | Fruits 178–337 | Regula (1985, 1986) |
| | [a]English walnuts | Edible tissues 87 ± 20 | Feldman and Lee (1985) |
| **Lamiaceae (Labiatae)** | | | |
| *Scutellaria baicalensis* Georgi | Skullcap | Vegetative tissues 0.100 | Cole et al. (2008) |
| **Lauraceae** | | | |
| [a]*Persea americana* Mill. | Avocado | Edible tissues 5.37 ± 0.41 | Islam et al. (2016) |
| [a]*Persea americana* cv. *Haas* | Avocado | Edible tissues 1.6 ± 0.40, pulp 1.10, seed 0.06 | Feldman and Lee (1985) |
| [a]*Persea americana* cv. *Fuerte* | | Edible tissues 1.5 ± 0.21 | Feldman and Lee (1985) |
| [a]*Persea americana* cv. *Booth* | | Edible tissues 0.2 ± 0.04 | Feldman and Lee (1985) |
| **Leguminosae (Fabaceae)** | | | |
| *Griffonia simplicifolia* (M. Vahl ex DC.) Baill. | Griffonia | Leaves 0.0017–0.007[b] | Fellows and Bell (1970) |
| | | Seeds 200 000[b] | Fellows and Bell (1970) |
| *Mimosa pudica* L. | *Mimosa* | *in vitro* 8.3 | Ramakrishna et al. (2009) |
| | | Vegetative tissues 17.3 | |
| | | Reproductive tissues 80.4 | |
| [a]*Pisum sativum* | Garden pea | Leaves, stems 0.9–1.0 | Applewhite (1973) |
| | Soy beans | Edible tissues <0.1 | Feldman and Lee (1985) |
| **Lygophyllaceae** | | | |
| *Peganum harmala* L. | Harmal | Culture of leaf tissues 18 200 | Nettleship and Slaytor (1974); Sasse et al. (1982) |
| | | Hairy root cultures | Berlin et al. (1993) |

(*Continued*)

**TABLE 20.5 (*Continued*)**
**The Occurrence of Serotonin in Plants**

| Families and Species/ Cultivar | Common Name | Part of Plant and Amount of Serotonin (μg/g of Fresh [b]Dry Mass) | References |
|---|---|---|---|
| **Lythraceae** | | | |
| [a]*Punica granatum* L. | Pomegranate | Homogenate of edible tissues 11.6 | Badria (2002) |
| **Moraceae** | | | |
| | [a]Figs | Edible tissues 0.2 | Feldman and Lee (1985) |
| **Musaceae** | | | |
| [a]*Ensete ventricosum* (Welw.) Cheesman | Banana | Homogenate of edible tissues 31.4 | Badria (2002) |
| [a]*Musa acuminata* Colla | Banana | Edible tissues 9.48 ± 0.09 | Islam et al. (2016) |
| | Banana | Peel 31.8, pulp (slice 14.0, center 26.6, edge 6.3) | Feldman and Lee (1985) |
| | Banana | Fruits 14.93 ± 0.50 | Płonka and Michalski (2017) |
| [a]*Musa acuminata* × *Musa balbisiana* cv. *Prata* | | Fruits 7.1–21.0 | Adao and Gloria (2005) |
| | Banana | Fruits 1.15 ± 0.04 | Lavizzari et al. (2006) |
| | [a]Plantain | Edible tissues 39.2–51.4 | Huang and Mazza (2011) |
| | [a]Plantain | Edible tissues 30.0 ± 7.5 | Feldman and Lee (1985) |
| **Myristicaceae** | | | |
| | [a]Nutmeg | Edible tissues <0.1 | Feldman and Lee (1985) |
| **Oleaceae** | | | |
| | [a]Olives | Edible tissues 0.2 | Feldman and Lee (1985) |
| **Pinaceae** | | | |
| | [a]Pine nuts | Edible tissues <0.1 | Feldman and Lee (1985) |
| **Poaceae (Gramineae)** | | | |
| [a]*Hordeum vulgare* L. | Barley | Edible tissues 44.9 | Badria (2002) |
| [a]*Oryza sativa* L. | Rice | Edible tissues 77.3 | Badria (2002) |
| [a]*Oryza sativa* cv. *Dongjin* | Rice | Leaves 6.1 ± 1.4, stems 4.3 ± 1.1, roots 19.5 ± 4.8, hulls 3.6 ± 0.8, seeds 0.2 ± 0.1 | Kang et al. (2007) |
| [a]*Zea mays* L. | Corn | Edible tissues 108.2 | Badria (2002) |
| **Rosaceae** | | | |
| [a]*Fragaria* × *ananassa* (Duchesne ex Weston) Duchesne ex Rozier | Strawberry | Edible tissues 0.05 ± 0.03 | Islam et al. (2016) |
| | | Edible tissues 3.77 ± 0.66[b] | Ly et al. (2008) |
| [a]*Fragaria magna* | Strawberry | Edible tissues 8.5 | Badria (2002) |
| [a]*Malus domestica* Borkh. | Apple | Edible tissues 8.7 | Badria (2002) |
| [a]*Malus pumila* Mill. | | Edible tissues 0.15 ± 0.03 | Islam et al. (2016) |
| [a]*Prunus persica* (L.) Batsch | Peach | Edible tissues 0.22 ± 0.17 | Islam et al. (2016) |
| [a]*Pyrus nivalis* Jacq. | Pears | Edible tissues 0.07 ± 0.03 | Islam et al. (2016) |
| **Rutaceae** | | | |
| [a]*Citrus reticulate* Blanco. | Mikan | Edible tissues 2.14 ± 0.08 | Islam et al. (2016) |
| [a]*Citrus paradisi* Macfad | Grapefruit | Edible tissues 0.97 ± 0.42 | Islam et al. (2016) |
| | [a]Orange | Fruits 2.21 ± 0.06 | Płonka and Michalski (2017) |
| | [a]Tangerine | Edible tissues <0.1 | Feldman and Lee (1985) |

(*Continued*)

**TABLE 20.5 (*Continued*)**
**The Occurrence of Serotonin in Plants**

| Families and Species/ Cultivar | Common Name | Part of Plant and Amount of Serotonin (μg/g of Fresh [b]Dry Mass) | References |
|---|---|---|---|
| | [a]Lemon | Edible tissues <0.1 | Feldman and Lee (1985) |
| | [a]Lime | Edible tissues <0.1 | Feldman and Lee (1985) |
| **Sappindaceae** | | | |
| | [a]Buckeye nuts | Edible tissues <0.1 | Feldman and Lee (1985) |
| | [a]Horse chestnuts | Edible tissues <0.1 | Feldman and Lee (1985) |
| **Solanaceae** | | | |
| [a]*Capsicum annuum* L. | Hot pepper | Edible tissues 17.9 ± 4.5[b] | Ly et al. (2008) |
| | Paprika | Edible tissues 1.8 ± 0.3[b] | Ly et al. (2008) |
| *C. annuum* cv. *Subicho* | | Leaves 0.23 ± 0.06, stems 0.28 ± 0.05, fruits 0.31 ± 0.02, flowers 0.86 ± 0.07, roots 0.15 ± 0.04 | Kang and Back (2006) |
| | [a]Eggplant | Edible tissues 0.2 | Feldman and Lee (1985) |
| [a]*Lycopersicon esculentum* Mill. | | Fruits 221.9 ± 3.8[b] | Ly et al. (2008) |
| | Cherry tomato | Fruits 156.1 ± 22.3[b] | Ly et al. (2008) |
| [a]*Solanum pimpinellifolium* L. | Tomato | Edible tissues 14.7 | Badria (2002) |
| [a]*Solanum tuberosum* L. | Potato | Leaves 2.0, fruits 7.5, tuber 0 | Engstrom et al. (1992) |
| **Urticaceae** | | | |
| *U. dioica* | | Stinging trichomes 3.5 μg per 1000 of stinging hairs | Regula and Devide (1980) |
| **Vitaceae** | | | |
| [a]*Vitis vinifera* L. | Grapes | Edible tissues 0.18 ± 0.03 | Islam et al. (2016) |
| | | Fruits 9–10 | Murch et al. (2010) |
| **Zingiberaceae** | | | |
| [a]*Zingiber officinale* Rosc. | Ginger | Homogenate of edible tissues 63.3 | Badria (2002) |

[a] edible plant.
[b] μg/g of Dry Mass.
[c] the value is determined according to the graphs figures.

that serotonin levels in plants are caused by the environment, including light levels, stresses (biotic and biotic), growth conditions, vegetative growth, reproductive development, seasonal rhythms, location, and pathogens (Erland and Saxena 2017). Immunolocalization analysis performed by S. Kang and co-authors showed that serotonin in a significant amount is contained in the xylem cells of conductive bundles of aging leaves of rice plants (Kang et al. 2007; Ramakrishna et al. 2011).

Besides free serotonin, his conjugated derivatives such as N-feruloylserotonin, N-(p-coumaroyl), serotonin, and N-(p-coumaroyl) serotonin mono-β-D-glucopyranoside were isolated from plants. Serotonin interacts with the phenylpropanoid pathway by serving as an amine substrate for the formation of hydroxycinnamic acid amides by enzyme-mediated condensation of a phenolic substrate (e.g., coumaric acid and ferulic acid) to an amine substrate (serotonin) (Macoy et al. 2015; Erland et al. 2016).

Serotonin is not equally spread along the plant (Table 20.5). Serotonin content is highest in the reproductive parts, such as seeds, fruits, and nuts (Lavizzari et al. 2006; Ly et al. 2008).

**TABLE 20.6**
**The Content of Serotonin in Fruits at Maturing (μg/g of Fresh Mass)**

| Plant Species | Part of Fruit | Nonmatured | Matured | Supermatured | References |
|---|---|---|---|---|---|
| *Ananas comosus* | Pulp | 50–60 | 19 | 0 | Foy and Parrat (1960) |
| *Banana* | Outer peel | 74 | 96 | 161 | Udenfriend et al. (1959) |
| | Inner peel | 13 | 38 | 170 | |
| | Pulp | 24 | 36 | 35 | |
| *Lycopersicon esculentum* | Pulp | 0.18 | 3.75 | 2.9 | West (1958) |
| *Musa sapientum* | Pulp | 24 | 36 | 35 | Udenfriend et al. (1959) |
| | Outer peel | 74 | 96 | 161 | -//- |
| | Inner peel | 13 | 38 | 170 | -//- |
| *Prunus avium* L. cv. Sweetheart | Edible tissue | 20.5[a] | 22.7[a] | N/A | González-Gómez et al. (2009) |
| *Prunus avium* L. cv. Pico Negro | | 12.4[a] | 8.5[a] | N/A | |
| *Prunus avium* L. cv. Ambrunés | | 18.7[a] | 19.1[a] | N/A | |
| *Prunus avium* L. cv. Pico Colorado | | 24.2[a] | 17.6[a] | N/A | |
| *Coffea canephora* Pierreex A. Froehner | Zygotic embryo | 28.45 ± 2 | 96.54 ± 5 | N/A | Ramakrishna et al. (2012a) |
| *Coffea canephora* Pierreex A. Froehner | Endosperm | 35.50 ± 3 | 51.08 ± 4 | N/A | |

[a] ng/100 g fresh mass.
N/A – not available.

Analysis of Table 20.5 shows that fruits and seeds of many plants contain 100 or more times bigger amount of serotonin than vegetative organs, and it is comparable with his concentration in animal tissues. The fruits of some plants show a high level of serotonin. Its distribution in tissues is nonuniform. The greatest amount of this substance in the banana is found in the peel (Table 20.6). In the pulp, the amount of serotonin is considerably less. In some fruits during ripening, the serotonin level decreases (e.g., in ananas, Prunus avium L. cv. Pico Negro, Prunus avium L. cv. Pico Colorado), in others – rises sharply (in the banana, tomato, coffee).

### 20.3.2.5 Histamine

E. Werle and A. Raub already in 1948 described 49 plant species from 28 families containing histamine (Roshchina 1991a, 2001). However, in addition to histamine in free form, its derivatives are also found in plants. Histamine was determined by the fluorescence method in marine algae, in particular in thallus of *Furcellaria lumbricalis* where mean values varied from 60 to 500 μg $g^{-1}$ fresh weight (Barwell 1979, 1989). Emmelin and Feldberg (1947) also found this neurotransmitter in stinging emergences of *Urtica dioica* and *Urtica urens.* Some species of plants from the families Urticaceae and Euphorbiaceae have a high histamine content in the stinging hairs (up to 1250 mg histamine per 1000 hairs), which is a protective mechanism from herbivores (Roshchina 2016a). Histamine is also found in pollen of wind-pollinated and insect-pollinated species (Marquardt and Vogg, 1952; Roshchina and Yashin 2014; Roshchina et al. 2014).

Under stress conditions, a sharp increase of histamine is observed in plants, as well as in animals. K. Ekici and H. Coskun (Ekici and Coskun 2002) have determined the histamine content of some commercial vegetable pickles at the range of 16.54 and 74.91 mg/kg (average 30.73 mg/kg). The maximum value (74.91 mg/kg) was obtained from a sample of hot pepper pickles. The amount of histamine varies according to the phase of plant development. For example, in the marine red algae *Furcellaria lumbricalis* (Huds.) Lamour, the content of histamine was from 60 to 500 mg/g fresh mass in both non-fertile and sexual expressed parts (Barwell 1979, 1989). The content of histamine (in mg/g fresh mass) in the male plant was 90–490 (sometimes up to 1100), in the female plant 60–120, and in asexual tetrasporophyte 100–500. The neurotransmitter cells of male plants contained approximately five times more of histamine than female and asexual plants. Additionally, the concentration of histamine at high salt concentration can also be increased (Roshchina 1991; Roshchina and Yashin 2014) (Table 20.7).

**TABLE 20.7**
**Plant Species Enriched in Histamine (Scombrotoxin)**

| Families and Species | Common Name | Part of Plant and Amount of Histamine (μg/g of Fresh Mass) | References |
|---|---|---|---|
| **Actinidiaceae** | | | |
| [a]*Actinidia deliciosa* (A. Chev.) C.F. Liang & A.R. Ferguson | | Leaves 27.2 ± 0.3 mg/kg | Lavizzari et al. (2006) |
| **Amaranthaceae** | | | |
| [a]*Spinacia oleracea* L. | Spinach | Edible tissues 2 mg/100 g | Moret et al. (2005) |
| **Apiaceae** | | | |
| [a]*Daucus carota* subsp. *sativus* | Carrot | Edible tissues ≤ 0.05 mg/100 g | Moret et al. (2005) |
| [a]*Petroselinum crispum* (Mill.) Fuss | Parsley | Edible tissues 0.2 mg/100 g | Moret et al. (2005) |
| **Brassicaceae (Criciferae)** | | | |
| [a]*Brassica oleracea* | Cauliflower | Edible tissues ≤ 0.05 mg/100 g | Moret et al. (2005) |
| [a]*Brassica oleracea* var. *sabauda* L. | Savoy cabbage | Edible tissues ≤ 0.05 mg/100 g | Moret et al. (2005) |
| [a]*Eruca sativa* Mill. | Arugola | Edible tissues ≤ 0.05 mg/100 g | Moret et al. (2005) |
| **Cucurbitaceae** | | | |
| [a]*Cucumis sativus* L. | Cucumber | Edible tissues ≤ 0.05 mg/100 g | Moret et al. (2005) |
| **Fabaceae (Leguminosae)** | | | |
| [a]*Vicia faba* L. | Broad bean | Edible tissues 0.2 mg/100 g | Moret et al. (2005) |
| **Lamiaceae (Labiatae)** | | | |
| *Sphenodesme pentandra* Jack | | 85 μg/100 g[b] | Nakthonga and Muangthai (2015) |
| **Leguminosae** | | | |
| *Caesalpinia sappan* L. | | 181 μg/100 g[b] | Nakthonga and Muangthai (2015) |
| [a]*Pisum sativum* | Green pea | Frozen edible tissues 2.3–3.8 mg/kg | Kalac et al. (2002) |
| **Menispermaceae** | | | |
| *Coscinium fenestratum* (Gaertn.) Colebr. | | 360 μg/100 g[b] | Nakthonga and Muangthai (2015) |
| **Piperaceae** | | | |
| | [a]Pepper | Edible tissues 0.1 mg/100 g | Moret et al. (2005) |
| **Rutaceae** | | | |
| *Toddalia asiatica* L. Lam. | | 863 μg/100 g[b] | Nakthonga and Muangthai (2015) |
| **Smilacaceae** | | | |
| *Smilax corbularia* var. *corbularia* | | 192 μg/100 g[b] | Nakthonga and Muangthai (2015) |
| **Solanaceae** | | | |
| *Solanum lycopersicum* | [a]Tomato | Edible tissues 0.7 mg/100 g | Moret et al. (2005) |

[a] edible plant.
[b] the value is determined according to the graphs figures.

## 20.4 GAMMA-AMINOBUTYRIC ACID

γ-Aminobutyric acid (GABA) is a four-carbon nonprotein amino acid that acts as a major inhibitory neurotransmitter in the central nervous system and is a multifunctional molecule that has different situational functions in the central nervous system, the peripheral nervous system, and some non-neuronal tissues (Watanabe et al. 2002). GABA has various physiological functions in animals and humans, such as neurotransmission and induction of hypotensive, diuretic, and tranquilizing effects (Jiang et al. 2010). In addition, the concentration of GABA in the brain may be related to various neurological disorders including epilepsy, seizures, convulsions, Huntington's disease, and Parkinsonism (Wong et al. 2003).

Stressful conditions (hypoxia, salt stress, heat or cold shock, drought, mechanical injury, etc.) can increase GABA formation manyfold. In plant cells, GABA is synthesized via the α-decarboxylation of glutamate in reaction, which is catalyzed by glutamate decarboxylase. This metabolic pathway is known as GABA shunt. GABA can also be formed via g-aminobutyraldehyde intermediate from polyamine degradation reaction where diamine oxidase is the key enzyme. Polyamine degradation pathway supplied at least 30% of GABA formation in germinating fava bean under hypoxia stress (Yang et al. 2013).

Anoxia caused a ~20-fold increment on GABA concentration, relative to fresh tea leaves. This increment was due to the increase of glutamate decarboxylase and diamine oxidase activities. J. Liao with colleagues infer that about one-fourth of GABA formed in tea leaves under anoxia comes from the polyamine degradation pathway, opening the possibility of producing GABA tea based through the regulation of metabolism (Liao et al. 2017).

G. Zhao with colleagues reported for the first time the molecular mechanisms underlying GABA accumulation in developing giant embryo rice grains. This study suggested that GABA accumulation in developing normal embryo rice grains mainly originated from glutamate-derived and polyamines-derived pathways, while the higher GABA content in developing giant embryo rice grains was mainly from the up-regulated activity of polyamines-derived pathway and down-regulated activity on GABA catabolism but not the glutamate pathway (Zhao et al. 2017).

Due to the physiological functions of GABA, the development of functional foods containing GABA in high concentrations has been actively pursued. Several GABA-enriched foods have been characterized: GABA tea (Wang et al. 2006), rice germs (Saikusa et al. 1994), brown rice (Miwako et al. 1999), black raspberry juice (Kim et al. 2009), etc. GABA-enriched food has been proved to be beneficial in cases of sleeplessness, depression, and autonomic disorders (Okada et al. 2000) and is effective in relieving chronic alcohol-related symptoms (Oh et al. 2003) in addition to possessing sedative and antihypertension properties (Hayakawa et al. 2004). The intake of GABA-enriched food intake also stimulates the immune cells (Oh et al. 2003), has inhibitory action on cancer cells (Oh and Oh 2004), and may prevent diabetic conditions (Hagiwara et al. 2004).

γ-Aminobutyric acid has been identified in more than 50 species of 29 families (Table 20.8).

## 20.5 THE INFLUENCE OF PROCESSING TECHNIQUE ON THE BIOGENIC AMINES CONTENT OF FRUITS RECOMMENDED IN MEDICINE

The study of J. Płonka and A. Michalski determined the relationship between various thermal processes used in fruit processing (drying, boiling, pasteurization, freezing) and the corresponding changes in the contents of selected biogenic amines (dopamine, serotonin) and their derivatives that are produced within the plant during its growth. None of the biogenic amines were shown to possess thermostable character. Processing of food raw materials decreased serotonin content over 27%. Juice and nectar pasteurization reduced catecholamine contents up to 60%. Freezing had the least impact on the serotonin content. The process with the greatest influence on the biogenic amine content was boiling, where the temperature used was the highest and the process time was the longest (Płonka and Michalski 2017).

## TABLE 20.8
## The Occurrence of γ-Aminobutyric Acid (GABA) in Plants

| Families and Species/Cultivar | Common Name | Gamma-Aminobutyric Acid (μg/g of Fresh Mass) | References |
|---|---|---|---|
| **Amaranthaceae** | | | |
| [a]*Spinacia oleracea* L. | | Edible tissues 267 ± 29.3 mg/100 g DW | Yoon et al. (2017) |
| **Araliaceae** | | | |
| *Panax notoginseng* (Burkill) F.H. Chen | | Stem, leaves 0.49%, flower 0.53%[b] | Yang et al. (2014) |
| **Brassicaceae** | | | |
| [a]*Brassica juncea* L. cv. purple-leaf | | Seeds 6.63 ± 0.06 mg/100 g | Li et al. (2013) |
| [a]*Brassica juncea* L. cv. green-leaf | | Seeds 1.46 ± 0.03 mg/100 g | |
| **Compositae** | | | |
| *Chrysanthemum indicum* L. | | Flowers 0.07 ± 0.06 mg/g DW | Bi et al. (2016) |
| *Coreopsis tinctoria* Nutt. | | Leaves 0.32 ± 0.05 mg/g DW | Bi et al. (2016) |
| **Lamiaceae (Labiatae)** | | | |
| *Scutellaria baicalensis* Georgi | | Leaves 0.55 ± 0.19 mg/g DW | Bi et al. (2016) |
| **Leguminosae** | | | |
| [a]*Glycine max* cv. YH-NJ | *Soybean* | Seeds 0.23 ± 0.08 mg/g DW | Yang et al. (2016) |
| [a]*Glycine max* cv. Suxie-1 | | Roots 0.61 ± 0.03 mmol/g FW | Xing et al. (2007) |
| [a]*Vicia faba* L. | Fava bean | Seeds, cotyledons 33, shoots 18, radicles 19 mg/g [c]DW[b] | Yang et al. (2015) |
| [a]*Vicia faba* L. cv. Qidou-2 | | Seeds 2.41 g/kg DW | Li et al. (2010) |
| **Malvaceae** | | | |
| *Hibiscus sabdariffa* L. | | Flowers 1.02 ± 0.21 mg/g DW | Bi et al. (2016) |
| **Oleaceae** | | | |
| *Forsythia suspensa* (Thunb.) Vahl | | Leaves 0.30 ± 0.16 mg/g DW | Bi et al. (2016) |
| **Poaceae (Gramineae)** | | | |
| [a]*Oryza sativa* L. var. *Khao Dok Mali 105* | | Seeds 1.60 mg/100 g DW[b] | Komatsuzaki et al. (2007) |
| [a]*Oryza sativa* L. cv. Tainung 71 (nonpigmented) | | Seeds 25.6 ± 2.9 mg/kg DW | Ng et al. (2013) |
| [a]*Oryza sativa* L. cv. black glutinous (pigmented) | | Seeds 17.5 ± 2.0 mg/kg DW | |
| [a]*Oryza sativa* L. *subsp. indica* cv. *Heinuo* | | Seeds 14.91 ± 0.12 mg/100 g DW | Ding et al. (2016) |
| [a]*Oryza sativa* L. subsp. *indica* cv. Xianhui 207 | | Seeds 14.47 ± 0.15 mg/100 g DW | |
| [a]*Oryza sativa* L. *japonica* cv. 'Shangshida No.5' | | Seeds 25.56 ± 1.90 mg/100 g DW | Zhao et al. (2017) |
| [a]*Oryza sativa* L. *japonica* cv. 'Chao2-10' | | Seeds 12.86 ± 0.37 mg/100 g DW | |
| *Phyllostachys praecox* C.D.Chu & C.S.Chao *f. prevernalis* | | Shoots 18 mg/g[b] FW[c] | Wang et al. (2017) |
| *Setaria italica* (L.) P.Beauv. | | Seeds 25.5 mg/100 g FW[b] | Bai et al. (2009) |
| **Theaceae** | | | |
| *Camellia sinensis* (L.) Kuntze | Green tea | Leaves 3 ± 5, stem 9 ± 3 mg/100 g FW | Sawai et al. (2001) |
| | Green tea | Leaves 16.94 ± 8.46 mg/100 g DW | Wang et al. (2006) |

*(Continued)*

**TABLE 20.8 (*Continued*)**
**The Occurrence of γ-Aminobutyric Acid (GABA) in Plants**

| Families and Species/Cultivar | Common Name | Gamma-Aminobutyric Acid (μg/g of Fresh Mass) | References |
|---|---|---|---|
| | Green tea | Leaves 127 ± 1 mg/100 g FW | Jeng et al. (2007) |
| | | Taiwanese green tea, leaves 24.7 ± 2.6 mg/100 g; Japanese green tea, leaves 53.6 ± 2.6 mg/100 g DW | Hsieh and Chen (2007) |
| | Green tea | Leaves 0–105.4 ± 9.9 mg/100 g DW | Syu et al. (2008) |
| | Green tea | Leaves 3.9–40.1 mg/100 g DW | Zhao et al. (2011) |
| | | Leaves 0.28 ± 0.06 mg/g DW | Bi et al. (2016) |
| **Vitaceae** | | | |
| *Ampelopsis grossedentata* (Hand.- Mazz.) W.T.Wang | | Leaves 2.18 ± 0.38 mg/g DW | Bi et al. (2016) |

[a] edible plant.
[b] mean value.
[c] the value is determined according to the graphs figures.
FW fresh weight.
DW dry weight.

Temperature processes of fruit processing undoubtedly have a major impact on the content of compounds such as biogenic amines. Loss of content depends primarily on the type of processing (temperature), duration, and the type of fruit.

The contents of all determined amines decreased in orange jam compared with fresh fruit. In samples of bananas after the cooking process (e.g., jam), most of the analyzed biogenic amines increased in concentration. However, the concentrations of norepinephrine and dopamine did not change. The largest concentration increase was observed for serotonin (approximately 66%). In pineapple, the concentrations of all detected amines decreased after a 15-minute cooking process.

The concentrations of norepinephrine, normetanephrine, and serotonin after boiling carried the highest downward trend in the product as relative to the fresh fruit.

## 20.6 TOXIC PROPERTIES OF BIOGENIC AMINES

The content of neurotransmitters in plants that are used for food and medical purposes raises concerns about the toxicity of these substances. Since it is known that some microbial contaminants can significantly increase the content of catecholamines and histamine during processing and storage of plant material (Rodriguez et al. 2014).

Biogenic amines are present in low concentrations or are not detected in fresh food. In food of animal origin, such as fish, meat, eggs, cheese, and fermented foods, they can be present in high concentration, above 50 μg/g, capable of inducing a chemical intoxication (Flick and Granata 2005).

Identification and quantification of amines is also an important indicator of the sanitary condition of both the raw material and the final product.

Histamine poisoning is the most common food-borne problem caused by biogenic amines. This intoxication, also termed *Scombroid poisoning*, is an important food-borne disease over the world.

Histamine poisoning causes an allergic reaction characterized by difficult breathing, vomiting, rash, fever diarrhea, hypotension, headache, and pruritus (Fernández et al. 2006).

The symptoms can appear within minutes or up to an hour after ingestion, and they include strange taste, headache, dizziness, nausea, facial swelling and flushing, abdominal pain, and rapid and weak pulse, besides diarrhea (Fernández et al. 2006). It is worthy to note that once histamine is formed, it is not destroyed by cooking (Gerald 2009). The concentration of histamine capable of producing poisoning varies in accordance with the susceptibility of each individual. In susceptible individuals, values between 5 and 10 mg/100 g will cause symptoms (Kalač et al. 2009).

Histamine exerts its effects by binding to cell membrane receptors of skin and respiratory, cardiovascular, gastrointestinal, and immunologic systems (Shalaby 1996). Clinical signs are more severe in people taking drugs that inhibit histamine detoxifying enzymes in the intestine, in immunosuppressed individuals, and individuals who use drugs and/or alcohol (Food and Drug Administration 1996).

The toxic dose of histamine is 10 mg/100 g of food; however, susceptible individuals—asthma and ulcer patients—are more susceptible to the toxic effects of this amine.

High concentrations of catecholamines or histamine released by algae may be hazardous for the development of surrounding species, including sea animals (Van Alstyne et al. 2006, 2011). Catecholamines and histamine release are also of concern in eliciting allergic reaction in humans following contact with pollen (Roshchina et al. 2014).

Oxidation of tryptophan in foods by oxidizing agents or by photooxidation generates toxic compounds. Also, extensive loss of tryptophan occurs when proteins react with oxidized lipids. The carboline compounds are formed when tryptophan-containing food products are heated at high temperatures, such as during deep fat frying. Among these compounds, alpha-carboline and gamma-carboline have mutagenic activities. Carboline compounds also cause hepatogenecity.

These compounds are also formed due to Maillard reactions in foods. The browning reactions in foods produce heterocyclic amines, which are genotoxic and carcinogenic. The heterocyclic amines derived from tryptophan induce liver cancer (Friedman and Cuq 1988). Therefore, food-processing operations should be carried out with caution to avoid the development of these toxic compounds in tryptophan-rich food products. Serotonin syndrome, also known as hyperserotonemia or serotonergic syndrome, is a potentially life-threatening condition where there is an excess of serotonin in the CNS (Frank 2008). In such cases, serotonin agonists and even foods containing tryptophan must be avoided.

Increasing evidence suggests that the production of toxic or deterrent natural products by ulvoid green macroalgae (phylum Chlorophyta, order Ulvales) can be responsible for inhibiting invertebrate settlement (Magre 1974) and herbivore feeding (Van Alstyne and Houser 2003), causing death of larvae (Johnson and Welsh 1985), and slowing phytoplankton growth (Jin and Dong 2003).

K. Van Alstyne and co-authors indicated that dopamine, which constituted an average of 4.4% dry mass or 0.94% ± 0.18% free mass of the algae, was responsible for decreased feeding by sea urchins (*Strongylocentrotus droebachiensis*). Subsequent experiments demonstrated that dopamine also reduced the feeding rates of snails (*Littorina sitkana*) and isopods (*Idotea wosnesenskii*) (Van Alstyne et al. 2006). This is the first experimental demonstration of a plant (algal) catecholamine functioning as a feeding deterrent.

## 20.7 CONCLUSION

Information about medicinal plants containing neurotransmitters, which demonstrated strong biological activity, shows perspectives in the use of plant neurotransmitters for pharmacy and biotechnology. In the future, it should concern with the analysis of (1) medicinal effects of extracts enriched in the compounds and the date of their storage, (2) the preparation of similar plant species in dry and fresh forms from juice to ampoules and tablets or pills, and (3) sterilization and storage of plant extracts where the compounds are found. Besides, the knowledge about useful plants containing neurotransmitters is needed to represent to a wider audience of physicians and pharmacologists.

## REFERENCES

Adao, R. C., and M. B. Gloria. 2005. Bioactive amines and carbohydrate changes during ripening of "Prata" banana (Musa acuminata × M. balbisiana). *Food Chemistry* 90:705–711.

Applewhite, P. B. 1973. Serotonin and norepinephrine in plant tissues. *Phytochemistry* 12:191–192.

Arreola, R., S. Alvarez-Herrera, G. Pérez-Sánchez et al. 2016. Immunomodulatory effects mediated by dopamine. *Journal of Immunology Research* 2016, Article ID 3160486, 31 p. https://doi.org/10.1155/2016/3160486.

Atkinson, K. F., S. H. Kathem, X. Jin et al. 2015. Dopaminergic signaling within the primary cilia in the renovascular system. *Frontiers in Physiology* 6:103.

Badria, F. A. 2002. Melatonin, serotonin, and tryptamine in some Egyptian food and medicinal plants. *Journal of Medicinal Food* 5:153–157.

Bai, Q., M. Chai, Z. Gu, X. Cao, Y. Li, and K. Liu. 2009. Effects of components in culture medium on glutamate decarboxylase activity and γ-aminobutyric acid accumulation in foxtail millet (Setaria italica L.) during germination. *Food Chemistry* 116:152–157.

Barwell, C. J. 1979. The occurrence of histamine in the red alga of Furcellaria lumbricalis Lamour. *Botanica marina* 22:399–401.

Barwell, C. J. 1989. Distribution of histamine in the thallus of Furcellaria lumbricalis. *Journal of Applied Phycology*. 1:341–344.

Berlin, J., C. Rügenhagen, N. Greidziak, I. N. Kuzovkina, L. Witte, and V. Wray. 1993. Biosynthesis of serotonin and β-carboline alkaloids in hairy root cultures of Peganum harmala. *Phytochemistry* 33:593–599.

Bi, W., C. He, Y. Ma et al. 2016. Investigation of free amino acid, total phenolics, antioxidant activity and purine alkaloids to assess the health properties of non-Camellia tea. *Acta Pharmaceutica Sinica B* 6:170–181.

Brown, P. N., C. E. Turi, P. R. Shipley, and S. J. Murch. 2012. Comparisons of large (Vaccinium macrocarpon Ait.) and small (Vaccinium oxycoccos L., Vaccinium vitis-idaea L.) cranberry in British Columbia by phytochemical determination, antioxidant potential, and metabolomic profiling with chemometric analysis. *Planta Medica* 78:630–640.

Chen, J., Y.-P. Shi, and J.-Y. Liu. 2003. Determination of noradrenaline and dopamine in Chinese herbal extracts from Portulaca oleracea L. by high-performance liquid chromatography. *Journal of Chromatography A*, 1003:127–132.

Cole, I. B., J. Cao, A. R. Alan, P. K. Saxena, and S. J. Murch. 2008. Comparisons of Scutellaria baicalensis, Scutellaria lateriflora and Scutellaria racemosa: Genome size, antioxidant potential and phytochemistry. *Planta medica* 74:474–481.

Cosentino, M., and F. Marino. 2013. Adrenergic and dopaminergic modulation of immunity in multiple sclerosis: Teaching old drugs new tricks? *Journal of Neuroimmune Pharmacology* 8:163–179.

Coutts, R. T., G. B. Baker, and F. M. Pasutto. 1986. Foodstuffs as sources of psychoactive amines and their precursors: Content, significance and identification. *Advances in Drug Research* 15:169–232.

Cruzon, G. 1990. Serotonin and appetite. *Annals of the New York Academy of Sciences* 600:521–530.

Cuevas, S., V. A. Villar, P. A. Jose, and I. Armando. 2013. Renal dopamine receptors, oxidative stress, and hypertension. *International Journal of Molecular Sciences* 14:17553–17572.

Deka, P., A. Kumar, B. K. Nayak, and N. Eloziia. 2017. Some plants as a source of acetyl cholinesterase inhibitors: A review. *International Research Journal of Pharmacy* 8:5–13.

Ding, J., T. Yang, H. Feng et al. 2016. Enhancing contents of γ-aminobutyric acid (GABA) and other micronutrients in dehulled rice during germination under normoxic and hypoxic conditions. *Journal of Agricultural and Food Chemistry* 64:1094–1102.

Ekici, K., and H. Coskun. 2002. Histamine content of some commercial vegetable pickles. In: *Proceedings of ICNP*, pp. 162–164. Trabzon, Turkey.

Emmelin, N., and W. Feldberg. 1947. The mechanism of the sting of the common nettle (Urtica urens). *Journal of Physiology* 106: 440–455.

Engstrom, K., L. Lundgren, and G. Samuelsson. 1992. Bioassay-guided isolation of serotonin from fruits of Solanum tuberosum L. *Acta Pharmaceutica Nordica* 4:91–92.

Erland, L. A. E., and P. K. Saxena. 2017. Beyond a neurotransmitter: The role of serotonin in plants. *Neurotransmitter* 4(May): 1–12. http://www.smartscitech.com/index.php/NT/article/view/1538/pdf.
Erland, L. A. E., C. E. Turi, and P. K. Saxena. 2016. Serotonin: An ancient molecule and an important regulator of plant processes. *Biotechnology Advances* 34:1347–1361.
Faugeras, G. M., J. Debelmas, and R. R. M. Paris. 1967. Sur la presence et la repartition d'aminophenols (tyramine, 3-hydroxy-tyramine ou dopamine, et nor-adrenaline), chez l'Aconit Napel: Aconitum napellus L. *Comptes Rendus Des Seances De'l Academie Des Sciences, Paris Serie D* 264:1864–1867.
Feldman, J. M., and E. M. Lee. 1985. Serotonin content of foods: Effect on urinary excretion of 5-hydroxyindoleacetic acid. *American Journal of Clinical Nutrition* 42:639–643.
Fellows, L. E., and E. A. Bell. 1970. 5-Hydroxy-L-tryptophan, 5-hydroxytryptamine and L-tryptophan-5-hydroxylase in Griffonia simplicifolia. *Phytochemistry* 9:2389–2396.
Fernández, M., B. Del Río, D. M. Linares, M. C. Martín, and M. A. Alvarez. 2006. Real-time polymerase chain reaction for quantitative detection of histamine-producing bacteria: Use in cheese production. *Journal of Dairy Science* 89:3763–3769.
Flick, G. J., and L. A. Granata. 2005. Biogenic amines in foods. In: *Toxins in Food (Chemical and Functional Properties of Food Components)*, W. M. Dabrowski, and Z. E. Sikorski (Eds.), pp. 121–154. Boca Raton, FL: CRC Press.
Fluck, R. A., P. A. Leber, J. D. Lieser et al. 2000. Choline conjugates of auxins. I. Direct evidence for the hydrolysis of choline-auxin conjugates by pea cholinesterase. *Plant Physiology and Biochemistry* 38:301–308.
Food and Drug Administration (FDA). 1996. *Fish and Fisheries Products Hazards and Controls Guide.* Office of Seafood, Washington, DC.
Foy, J. M., and J. R. Parratt. 1960. A note on the presence of noradrenaline and 5-hydroxytryptamine in plantain (Musa sapientum, var. paradisiaca). *Journal of Pharmacy and Pharmacology* 12:360–364.
Franco, R., R. Pacheco, C. Lluis, G. P. Ahern, and P. J. O'Connell. 2007. The emergence of neurotransmitters as immune modulators. *Trends in Immunology* 28:400–407.
Frank, C. 2008. Recognition and treatment of serotonin syndrome. *Canadian Family Physician* 54:988–992.
Friedman, M., and J. Cuq. 1988. Chemistry, analysis, nutritional value, and toxicology of tryptophan in food. A review. *Journal of Agricultural and Food Chemistry* 36:1080–1093.
Gandía-Herrero, F., M. Jiménez-Atiénzar, J. Cabanes et al. 2009. Fluorescence detection of tyrosinase activity on dopamine-betaxanthin purified from Portulaca oleracea (common purslane) flowers. *Journal of Agricultural Food and Chemistry* 57:2523–2528.
Gerald, G. M. 2009. Regulatory toxicology. In: *Food Science and Technology*, G. Campbell-Platt (Ed.), pp. 399–410. Ames, IA: Wiley-Blackwell.
González-Gómez, D., M. Lozano, M. F. Fernández-León, M. C. Ayuso, M. J. Bernalte, and A. B. Rodríguez. 2009. Detection and quantification of melatonin and serotonin in eight sweet cherry cultivars (Prunus avium L.). *European Food Research and Technology* 229:223–229.
Grosse, W., and F. Artigas. 1983. Incorporation of N-ammonia into serotonin in cotyledons of maturing walnuts. *Zeischrift Naturforschung C* 38:1057–1058.
Hagiwara, H., T. Seki, and T. Ariga. 2004. The effect of pre-germinated brown rice intake on blood glucose and PAI-1 levels in streptozotocin-induced diabetic rats. *Bioscience, Biotechnology, and Biochemistry* 68:444–447.
Hartmann, E., and H. Kilbinger. 1974a. Gas-liquid chromatographic determination of light-dependent acetylcholine concentrations in moss callus. *Biochemical Journal* 137:249–252.
Hartmann, E., and H. Kilbinger. 1974b. Occurrence of light-dependent acetylcholine concentrations in higher plants. *Experientia* 30:1387–1388.
Hayakawa, K., M. Kimura, K. Kasaha, K. Matsumoto, H. Sansawa, and Y. Yamori. 2004. Effect of a γ-aminobutyric acid-enriched dairy product on the blood pressure of spontaneously hypertensive and normotensive Wistar-Kyoto rats. *British Journal of Nutrition* 92:411–417.
Heisler, L. K., M. A. Cowley, L. H. Tecott et al. 2002. Activation of central melanocortin pathways by fenfluramine. *Science* 297:609–611.
Horiuchi, Y., R. Kimura, N. Kato et al. 2003. Evolutional study on acetylcholine expression. *Life Sciences* 72:1745–1756.
Horton, E. W., and G. M. Felippe. 1973. An acetylcholine-like substance in Porophyllum lanceolatum. *Biologia Plantarum* 15:150–151.
Hotta, Y., A. Nagatsu, W. Liu et al. 2002. Protective effects of antioxidative serotonin derivatives isolated from safflower against postischemic myocardial dysfunction. *Molecular and Cellular Biochemistry* 238:151–162.

Hourmant, A., F. Rapt, J. M. Morzadec, A. Feray, and J. Caroff. 1998. Involvement of catecholic compounds in morphogenesis of in vitro potato plants. Effect of methylglyoxal-bis (guanylhydrazone). *Journal of Plant Physiology* 152:64–69.

Hsieh, M. M., and S. M. Chen. 2007. Determination of amino acids in tea leaves and beverages using capillary electrophoresis with light-emitting diode-induced fluorescence detection. *Talanta* 73:326–331.

Huang, X., and G. Mazza. 2011. Application of LC and LC-MS to the analysis of melatonin and serotonin in edible plants. *Critical Reviews in Food Science and Nutrition* 51:269–284.

Islam, J., H. Shirakawa, T. K. Nguyen, H. Aso, and M. Komai. 2016. Simultaneous analysis of serotonin, tryptophan and tryptamine levels in common fresh fruits and vegetables in Japan using fluorescence HPLC. *Food Bioscience* 13:56–59.

Jeng, K. C., C. S. Chen, Y. P. Fang, R. C. Hou, and Y. S. Chen. 2007. Effect of microbial fermentation on content of statin, GABA, and polyphenols in Pu-Erh tea. *Journal of Agricultural and Food Chemistry* 55:8787–8792.

Jiang, B., Y. Fu., T. Zhang, 2010. γ-Aminobutyric acid. In: *Bioactive Proteins and Peptides as Functional Foods and Nutraceuticals*, Y. Mine, E. Li-Chan, B. Jiang (Eds.), pp. 121–130. Ames, IA: Wiley-Blackwell.

Jin, Q., and S. Dong. 2003. Comparative studies on the allelopathic effects of two different strains of Ulva pertusa on Heterosigma akashiwo and Alexandrium tamarense. *Journal of Experimental Marine Biology and Ecology* 293:41–55.

Johnson, D., and B. Welsh. 1985. Detrimental effects of Ulva lactuca (L.) exudates and low oxygen on estuarine crab larvae. *Journal of Experimental Marine Biology and Ecology* 86:73–83.

Jones, M. P. A., J. Cao, R. O'Brien, S. J. Murch, and P. K. Saxena. 2007. The mode of action of thidiazuron: Auxins, indoleamines, and ion channels in the regeneration of Echinacea purpurea L. *Plant Cell Reports* 26:1481–1490.

Kalač P., E. Dadáková, and T. Pelikánová. 2009. Content of biogenic amines and polyamines in some species of European wild-growing edible mushrooms. *European Food Research Technology* 230:163–171.

Kalač, P., S. Svecova, and T. Pelikánová. 2002. Level of biogenic amines in typical vegetable products. *Food Chemistry* 77:349–351.

Kamisaka, S. 1979. Catecholamine stimulation of the giberellin action that induced lettuce hypocotyl elongation. *Plant and Cell Physiology* 20:1199–1207.

Kamisaka, S., and K. Shibata. 1982. Identification in lettuce seedlings of catecholamine active synergistically enchancing the gibberellin effect on lettuce hypocotyl elongation. *Plant Growth Regulation* 1:3–10.

Kamo K. K., and P. G. Mahlberg. 1984. Dopamine biosynthesis at different stages of plant development in Papaver somniferum. *Journal of Natural Products* 47:682–686.

Kang, K., S. Kang, K. Lee, M. Park, and K. Back. 2008. Enzymatic features of serotonin biosynthetic enzymes and serotonin biosynthesis in plants. *Plant Signaling and Behavior* 3:389–390.

Kang, K., Y. S. Kim, S. Park, and K. Back. 2009. Senescence-induced serotonin biosynthesis and its role in delaying senescence in rice leaves. *Plant Physiology* 150:1380–1393.

Kang, S., and K. Back. 2006. Enriched production of N-hydroxycinnamic acid amides and biogenic amines in pepper (Capsicum annuum) flowers. *Scientia Horticulturae* 108:337–341.

Kang, S., K. Kang, K. Lee, and K. Back. 2007. Characterization of tryptamine 5-hydroxylase and serotonin synthesis in rice plants. *Plant Cell Reports* 26:2009–2015.

Kasturi, R., and V. N. Vasantharajan. 1976. Properties of acetylcholine esterase from Pisum sativum. *Phytochemistry* 15:1345–1347.

Kaufmann, D., A. K. Dogra, A. Tahrani, F. Herrmann, and M. Wink. 2016. Extracts from traditional Chinese medicinal plants inhibit acetylcholinesterase, a known Alzheimer's disease target. *Molecules* 21:1161–1176.

Kawashima, K., H. Misawa, Y. Moriwaki et al. 2007. Ubiquitous expression of acetylcholine and its biological functions in life forms without nervous systems. *Life Sciences* 80:2206–2209.

Khozaei, M., F. Ghorbani, G. Mardani, and R. Emamzadeh. 2014. Catecholamines are active plant-based drug compounds in Pisum sativum, Phaseolus vulgaris and Vicia faba Species. *Journal of HerbMed Pharmacology* 3:61–65.

Kim, J. Y., M. Y. Lee, G. E. Ji, Y. S. Lee, and K. T. Hwang. 2009. Production of γ-aminobutyric acid in black raspberry juice during fermentation by Lactobacillus brevis GABA100. *International Journal of Food Microbiology* 130:12–16.

Kimura, M. 1968. Fluorescence histochemical study on serotonin and catecholamine in some plants. *Japanese Journal of Pharmacology* 18:162–168.

Komatsuzaki, N., K. Tsukahara, H. Toyoshima, T. Suzuki, N. Shimizu, and T. Kimura. 2007. Effect of soaking and gaseous treatment on GABA content in germinated brown rice. *Journal of Food Engineering* 78:556–560.

Kulma, A., and J. Szopa. 2007. Catecholamines are active compounds in plants. *Plant Science* 172:433–440.
Kumarasamya, Y., M. Middletona, R. G. Reida, L. Naharb, and S. D. Sarkera. 2003. Biological activity of serotonin conjugates from the seeds of Centaurea nigra. *Fitoterapia* 74:609–612.
Kutchan, T. M., M. Rush, and C. J. Coscia. 1986. Subcellular localization of alkaloids and dopamine in defferent vacuolar compartment of Papaver bracteatum. *Plant Physiology* 81:161–166.
Ladeira, A. M., S. M. C. Dietrich, and G. M. Felippe. 1982b. Acetylcholine and flowering of photoperiodic plants. *Revta Brasilian Botanica* 5:21–24.
Lavizzari, T., M. T. Veciana-Nogues, S. Bover-Cid, A. Marine-Font, and M. C. Vidal-Carou. 2006. Improved method for the determination of biogenic amines and polyamines in vegetable products by ion-pair high-performance liquid chromatography. *Journal of Chromatography A* 1129:67–72.
Lawrence, S. A. 2004. *Amines: Synthesis Properties and Applications.* Cambridge: Cambridge University Press.
Lembeck, F., and G. Skofitsch. 1984. Distribution of serotonin in Juglans regia seeds during ontogenetic development and germination. *Zeischrift der Pflanzenphysiologie* 114:349–353.
Li, X., Y. B. Kim, M. R. Uddin, S. Lee, S. J. Kim, and S. U. Park. 2013. Influence of light on the free amino acid content and γ-aminobutyric acid synthesis in Brassica juncea seedlings. *Journal of Agricultural and Food Chemistry* 61:8624–8631.
Li, Y., Q. Bai, X. Jin, H. Wen, and Z. Gu. 2010. Effects of cultivar and culture conditions on γ-aminobutyric acid accumulation in germinated fava beans (Vicia faba L.). *Journal of the Science of Food and Agriculture* 90:52–57.
Liao, J., X. Wu, Z. Xing et al. 2017. γ-Aminobutyric acid (GABA) accumulation in tea (Camellia sinensis L.) through the GABA shunt and polyamine degradation pathways under anoxia. *Journal of Agricultural and Food Chemistry* 65:3013–3018.
Ly, D., K. Kang, J.-Y. Choi, A. Ishihara, K. Back, and S.-G. Lee. 2008. HPLC analysis of serotonin, tryptamine, tyramine, and the hydroxycinnamic acid amides of serotonin and tyramine in food vegetables. *Journal of Medicinal Food* 11:385–389.
Macoy, D. M., W. Y. Kim, S. Y. Lee, and M. G. Kim. 2015. Biosynthesis, physiology, and functions of hydroxycinnamic acid amides in plants. *Plant Biotechnology Reports* 9:269–278.
Magre, E. 1974. Ulva lactuca L. negatively affects Balanus balanoides (L.) (Cirripedia Thoracica) in tidepools. *Crustaceana* 27:231–234.
Marquardt, P., and G. Vogg. 1952. Pharmakologische und chemische Untersuchungen uber Wirkstoffe in Bienenpollen. *Arzneimittel forschung* 21 (353): 267–271.
Miura, G. A., and T. M. Shin. 1984. Cholinergic constituents in plants: Characterization and distribution of acetylcholine and choline. *Physiologia Plantarum* 61:417–421.
Miwako, K., S. Miyuki, Y. Akira, and Y. Koji. 1999. Accumulation of GABA in brown rice by high pressure treatment. *Journal of the Japanese Society of Food Science* 46:329–333.
Momonoki, Y. S., and T. Momonoki. 1991. Changes in acetylcholine levels following leaf wilting and leaf recovery by heat stress in plant cultivars. Nippon Sakumotsu Gakkai Kiji. *Japanese Journal of Crop Science* 60:283–290.
Moret, S., D. Smela, T. Populin, and L. S. Conte. 2005. A survey on free biogenic amine content of fresh and preserved vegetables. *Food Chemistry* 89:355–361.
Mukherjee, S., A. David, S. Yadav, F. Baluška, and S. C. Bhatla. 2014. Salt stress-induced seedling growth inhibition coincides with differential distribution of serotonin and melatonin in sunflower seedling roots and cotyledons. *Physiologia Plantarum* 152:714–728.
Murch, S. J. 2006. Neurotransmitters, neuroregulators and neurotoxins in plants. In: *Communication in Plants–Neuronal Aspects of Plant Life*, Baluska, S. Mancuso, and D. Volkmann (Eds.), pp. 137–151. Berlin, Germany: Springer.
Murch, S. J., S. S. B. Campbell, and P. K. Saxena. 2001. The role of serotonin and melatonin in plant morphogenesis: Regulation of auxin-induced root organogenesis in in vitro-cultured explants of St. John's wort (Hypericum perforatum L.). *In Vitro Cellular and Developmental Biology-Plant* 37:786–793.
Murch, S. J., B. A. Hall, C. H. Le, and P. K. Saxena. 2010. Changes in the levels of indoleamine phytochemicals during vėraison and ripening of wine grapes. *Journal of Pineal Research* 49:95–100.
Murch, S. J., and P. K. Saxena. 2002. Mammalian neurohormones: Potential significance in reproductive physiology of St. John's wort (Hypericum perforatum L.)? *Naturwissenschaften* 89:555–560.
Murray, A. P., M. B. Faraoni, M. J. Castro, N. P. Alza, and V. Cavallaro. 2013. Natural AChE inhibitors from plants and their contribution to Alzheimer's disease therapy. *Current Neuropharmacology* 11:388–413.

Nagatsu, A., H. L. Zhang, and H. Mizukami. 2000. Tyrosinase inhibitory and antitumor promoting activities of compounds isolated from safflower (Carthamus tinctorius L.) and cotton (Gossypium hirsutum L.) oil cakes. *Natural Product Letters* 4:153–158.

Nakthonga, P., and Muangthai, P. 2015. Tyramine and Histamine content in some Thai medicinal plants. *Proceedings of the Burapha University International Conference*, pp. 370–376, July 10–12, 2015, Bangsaen, Chonburi, Thailand.

Nettleship, L., and M. Slaytor. 1974. Limitation of feeding experiments in studying alkaloid biosynthesis in Peganum harmala callus cultures. *Phytochemistry* 13:735–742.

Ng, L. T., S. H. Huang, Y. T. Chen, and C. H. Su. 2013. Changes of tocopherols, tocotrienols, γ-oryzanol, and γ-aminobutyric acid levels in the germinated brown rice of pigmented and nonpigmented cultivars. *Journal of Agricultural and Food Chemistry* 61:12604–12611.

Oh, C. H., and S. H. Oh. 2004. Effect of germinated brown rice extracts with enhanced levels of GABA on cancer cell proliferation and apoptosis. *Journal of Medicinal Food* 7:19–23.

Oh, S. H., J. R. Soh, and Y. S. Cha. 2003. Germinated brown rice extract shows a nutraceutical effect in the recovery of chronic alcohol-related symptoms. *Journal of Medicinal Food* 7:115–121.

Okada, T., T. Sugishita, T. Murakami et al. 2000. Effect of the defatted rice germ enriched with GABA for sleeplessness, depression, autonomic disorder by oral administration. *Nippon Shokuhin Kagaku Kogaku Kaishi= Journal of the Japanese Society for Food Science and Technology* 47:596–603.

Oleskin, A. V. 2007. *Biopolitics*. Moscow: Nauchnii Mir.

Oleskin, A. V., T. A. Kirovskaya, I. V. Botvinko, and L. V. Lysak. 1998. Effects of serotonin (5-hydroxytryptamine) on the growth and differentiation of microorganisms. *Microbiology (Russia)* 67:305–312.

Orhan, I., M. Kartal, Y. Kan, and B. Şener. 2008. Activity of essential oils and individual components against acetyl and butyrylcholinesterase. *Zeitschrift fuer Naturforschung C* 63:547–553.

Owokotomo, I. A., O. Ekundayo, T. G. Abayomi, and A. V. Chukwuka. 2015. In-vitro anti-cholinesterase activity of essential oil from four tropical medicinal plants. *Toxicology Reports* 2:850–857.

Płonka, J., and A. Michalski. 2017. The influence of processing technique on the catecholamine and indolamine contents of fruits. *Journal of Food Composition and Analysis* 57:102–108.

Ponchet, M., J. Martin-Tanguy, A. Marais, and C. Martin. 1982. Hydroxycinnamoyl acid amides and aromatic amines in the inflorescences of some Araceae species. *Phytochemistry* 21:2865–2869.

Ramakrishna, A., P. Giridhar, M. Jobin, C. S. Paulose, and G. A. Ravishankar. 2012b. Indoleamines and calcium enhance somatic embryogenesis in Coffea canephora P. ex Fr. *Plant Cell, Tissue and Organ Culture* 108:267–278.

Ramakrishna, A., P. Giridhar, and G. A. Ravishankar. 2009. Indoleamines and calcium channels influence morphogenesis in in vitro cultures of Mimosa pudica L. *Plant Signaling and Behavior* 4:1136–1141.

Ramakrishna, A., P. Giridhar, and G. A. Ravishankar. 2011. Phytoserotonin: A review. *Plant Signaling and Behavior* 6:800–809.

Ramakrishna, A., P. Giridhar, K. U. Sankar, and G. A. Ravishankar. 2012a. Endogenous profiles of indoleamines: Serotonin and melatonin in different tissues of Coffea canephora P. ex Fr. as analyzed by HPLC and LC-MS-ESI. *Acta Physiologiae Plantarum* 34:393–396.

Ranjan, N., and M. Kumari. 2017. Acetylcholinesterase inhibition by medicinal plants: A review. *Annals of Plant Sciences* 6:1640–1644.

Regula, I. 1970. 5-Hidroksitriptamin u ljutoj Koprivi (Urtica pilulifera L.). *Acta Botanica Croatica* 29:69–74.

Regula, I. 1981. Serotonin in the tissues of Loasa vulcanica ed. Andre. *Acta Botanica Croatica* 40:91–94.

Regula, I. 1985. The presence of serotonin in the ambryo of Juglans Mandshurica Maxim. *Acta Botanica Croatica* 44:19–22.

Regula, I. 1986. The presence of serotonin in embryo of black walnut (Juglans nigra). *Acta Botanica Croatica* 45:91–95.

Regula, I., and Z. Devide. 1980. The presence of serotonin in some species of genus Urtica. *Acta Botanica Croatica* 39:47–50.

Rinki, V., A. Agrawal, G. P. Dubey, P. K. Singh, and G. P. I. Singh. 2016. Regulation of serotonin in depression: Efficacy of ayurvedic plants. In: *Serotonin and Melatonin: Their Functional Role in Plants, Food, Phytomedicine, and Human Health*, G. A. Ravishankar, A. Ramakrishna (Eds.), pp. 398–419. Boca Raton, FL: CRC Press/Taylor & Francis Group.

Rodriguez, M., C. Carneiro, M. Feijó, C. Júnior, and S. Mano. 2014. Bioactive amines: Aspects of quality and safety in food. *Food and Nutrition Sciences* 5:138–146.

Roshchina, V. V. 1991a. *Biomediators in Plants. Acetylcholine and Biogenic Amines*. Pushchino: Biological Center of USSR Academy of Sciences.

Roshchina, V. V. 1991b. Neurotransmitters catecholamines and serotonin in plants. *Uspekhi Sovremennoi Biologii (Advances in Modern Biology, Russia)* 111:622–636.
Roshchina, V. V. 2001. *Neurotransmitters in Plant Life*. Enfield, NH: Science Publishers.
Roshchina, V. V. 2016a. New trends and perspectives in the evolution of neurotransmitters in microbial, plant, and animal cells. In *Microbial Endocrinology: Interkingdom Signaling in Infectious Disease and Health*, pp. 25–77. New York: Springer.
Roshchina, V. V. 2016b. Serotonin in living cells: Fluorescent tools and models to study. In *Serotonin and Melatonin: Their Functional Role in Plants, Food, Phytomedicine, and Human Health*, pp. 47–60. Boca Raton, FL: CRC Press/Taylor & Francis Group.
Roshchina, V. V., and E. N. Mukhin. 1985. Acetylcholinesterase activity in chloroplasts and acetylcholine effects on photochemical reactions. *Photosynthetica* 19:164–171.
Roshchina, V. V., and V. A. Yashin. 2014. Neurotransmitters catecholamines and histamine in allelopathy: Plant cells as models in fluorescence microscopy. *Allelopathy Journal* 34:1–16.
Roshchina, V. V., V. A. Yashin, A. V., Kuchin, and V. I. Kulakov. 2014. Fluorescent analysis of catecholamines and histamine in plant single cells. *International Journal of Biochemistry, Photon* 195:344–351.
Saeedi, M., K. Babaie, E. Karimpour-Razkenari et al. 2017. In vitro cholinesterase inhibitory activity of some plants used in Iranian traditional medicine. *Natural Product Research* 31:2690–2694.
Saikusa, T., T. Horino, and Y. Mori. 1994. Accumulation of γ-aminobutyric acid (Gaba) in the rice germ during water soaking. *Bioscience, Biotechnology, and Biochemistry* 58:2291–2292.
Sasse, F., U. Heckenberg, and J. Berlin. 1982. Accumulation of β-carboline alkaloids and serotonin by cell cultures of Peganum harmala. II. Interrelationship between accumulation of serotonin and activities of related enzymes. *Zeischrift der Pflanzenphysiologie* 105:315–322.
Sawai, Y., Y. Yamaguchi, and D. Miyana. 2001. Cycling treatment of anaerobic and aerobic incubation increases the content of γ-aminobutyric acid in tea shoots. *Amino Acids* 20: 331–334.
Shalaby, A. R. 1996. Significance of biogenic amines to food safety and human health. *Food Research International* 29:675–690.
Smallman, B. N., and A. Maneckjee. 1981. The synthesis of acetylcholine by plants. *Biochemical Journal* 194:361–364.
Steelink, C., M. Yeung, and R. L. Caldwell. 1967. Phenolic constituents of healthy and wound tissues in the giant cactus (Carnegiea gigantea). *Phytochemistry* 6:1435–1440.
Swiedrych, A., K. L. Kukuła, A. Skirycz, and J. Szopa. 2004. The catecholamine biosynthesis route in potato is affected by stress. *Plant Physiology and Biochemistry* 42:593–600.
Syu, K. Y., C. L. Lin, H. C. Huang, and J. K. Lin. 2008. Determination of theanine, GABA, and other amino acids in green, oolong, black, and puerh teas with dabsylation and high-performance liquid chromatography. *Journal of Agricultural and Food Chemistry* 56:7637–7643.
Titus, F., A. Dávalos, J. Alom, and A. Codina. 1986. 5-hydroxytryptophan versus methysergide in the prophylaxis of migraine. *European Neurology* 25:327–329.
Tretyn, A., and M. Tretyn. 1988. Diurnal acetylcholine oscillation in green oat seedlings. *Acta Physiologia Plantarum* 10:243–246.
Udenfriend, S., W. Lovenberg, and A. Sjoerdsma. 1959. Physiologically active amines in common fruits and vegetables. *Archives in Biochemistry and Biophysics* 85:487–490.
Van Alstyne, K., K. Anderson, A. Winans, and S.-A. Gifford. 2011. Dopamine release by the green alga Ulvaria obscura after simulated immersion by incoming tides. *Marine Biology* 158:2087–2094.
Van Alstyne, K. L., and L. T. Houser. 2003. Dimethylsulfide release during macroinvertebrate grazing and its role as an activated chemical defense. *Marine Ecology Progress Series* 250:175–181.
Van Alstyne, K. L., A. V. Nelson, J. R. Vyvyan, and D. A. Cancilla. 2006. Dopamine functions as an antiherbivore defense in the temperate green alga Ulvaria obscura. *Oecologia*. 148:304–311.
Villalobos, J., F. Ramirez, and H. Moussatche. 1974. Some observations on the presence of acetylcholine and histamine in plants. *Ciência e Cultura* 26:690–693.
Waalkes, T. P., A. Sjoerdama, C. R. Greveling, H. Weissbach, and S. Udenfriend. 1958. Serotonin, norepinephrine, and related compounds in bananas. *Science* 127:648–650.
Wagner, H. 1988. Non-steroid, cardio active plant constituents. In: *Economic and Medicinal Plant Research*, H. Wagner, H. Hikino, and N. R. Farnsworth (Eds.), pp. 17–38. London, UK: Academic Press.
Wang, D., L. Li, Y. Xu et al. 2017. Effect of exogenous nitro oxide on chilling tolerance, polyamine, proline and gamma-aminobutyric acid and in bamboo shoots (Phyllostachys praecox f. prevernalis). *Journal of Agricultural and Food Chemistry* 65:5607–5613.

Wang, H. F., Y. S. Tsai, M. L. Lin, and A. S. Ou. 2006. Comparison of bioactive components in GABA tea and green tea produced in Taiwan. *Food Chemistry* 96:648–653.

Watanabe, M., K. Maemura, K. Kanbara, T. Tamayama, and H. Hayasaki. 2002. GABA and GABA receptors in the central nervous system and other organs. *International Review of Cytology* 213:1–47.

Wessler, I., H. Kilbinger, F. Bittinger, and C. J. Kirkpatrick. 2001. The biological role of non-neuronal acetylcholine in plants and humans. *Japanese Journal of Pharmacology* 85:2–10.

West, G. B. 1958. Tryptamine in edible fruits. *Journal of Pharmacy and Pharmacology* 10:589–590.

Wong, C. T., T. Bottiglieri, and O. C. Snead. 2003. GABA, γ-hydroxybutyric acid, and neurological disease. *Annals of Neurology* 54:S3–S12.

Xing, S. G., Y. B. Jun, Z. W. Hau, and L. Y. Liang. 2007. Higher accumulation of γ-aminobutyric acid induced by salt stress through stimulating the activity of diamine oxidases in Glycine max (L.) Merr. Roots. *Plant Physiology and Biochemistry* 45:560–566.

Yamamoto, K., and Y. S. Momonoki. 2012. Tissue localization of maize acetylcholinesterase associated with heat tolerance in plants. *Plant Signaling and Behavior* 7(3):301–305.

Yamamotova, A., M. Pometlova, J. Harmatha, H. Raskova, and R. Rokyta. 2007. The selective effect of N-feruloylserotonins isolated from Leuzea carthamoides on nociception and anxiety in rats. *Journal of Ethnopharmacology* 112:368–374.

Yang, J. J., Y. Liu, X. M. Cui, Y. Qu, and L. Q. Huang. 2014. Determination of gamma-aminobutyric acid in aerial part of Panax notoginseng by HPLC. *Zhongguo Zhong yao za zhi= Zhongguo zhongyao zazhi= China Journal of Chinese Materia Medica* 39:606–609.

Yang, R., L. Feng, S. Wang, N. Yu, and Z. Gu. 2016. Accumulation of γ-aminobutyric acid in soybean by hypoxia germination and freeze–thawing incubation. *Journal of the Science of Food and Agriculture* 96:2090–2096.

Yang, R., Q. Guo, and Z. Gu. 2013. GABA shunt and polyamine degradation pathway on γ-aminobutyric acid accumulation in germinating fava bean (Vicia faba L.) under hypoxia. *Food Chemistry* 136:152–159.

Yang, R., Q. Hui, and Zh. Gu. 2015. Effects of ABA and $CaCl_2$ on GABA accumulation in fava bean germinating under hypoxia-NaCl stress. *Bioscience, Biotechnology, and Biochemistry* 80:540–546.

Yoon, Y. E., S. Kuppusamy, K. M. Cho, P. J. Kim, Y. B. Kwack, and Y. B. Lee. 2017. Influence of cold stress on contents of soluble sugars, vitamin C and free amino acids including gamma-aminobutyric acid (GABA) in spinach (Spinacia oleracea). *Food Chemistry* 215:185–192.

Zhang, Y., S. Cuevas, L. D. Asico et al. 2012. Deficient dopamine $D_2$ receptor function causes renal inflammation independently of high blood pressure. *PLoS One* 7(6): 1–11.

Zhao, G. C., M. X. Xie, Y. C. Wang, and J. Y. Li. 2017. Molecular mechanisms underlying γ-aminobutyric acid (GABA) accumulation in giant embryo rice seeds. *Journal of Agricultural and Food Chemistry* 65:4883–4889.

Zhao, M., Y. Ma, and Zh.-zh. Wei. 2011. Determination and comparison of γ-aminobutyric acid (GABA) content in pu-erh and other types of Chinese tea. *Journal of Agricultural and Food Chemistry* 59:3641–3648.

# 21 Phytoextracts and Their Derivatives Affecting Neurotransmission Relevant to Alzheimer's Disease

## *An Ethno-Medicinal Perspective*

*Atanu Bhattacharjee*

**CONTENTS**



## 21.1 INTRODUCTION

Alzheimer's disease (AD) is a progressive age-related neurodegenerative disorder that clinically impairs learning and memory ability and significantly alters regular physiological activities of the patients (George, 2016). AD is the cause of dementia in the elderly, with duration of around nine years between onset of clinical symptoms and death and has been emerged as one of the deadliest disorder in developed nations. By 2040, World Health Organization (WHO) estimated around 71% of the dementia cases will occur in developing countries (Ferri et al., 2006; Puchchakayala et al., 2012; Mahesh et al., 2014).

The assertive role of neurotransmitter acetylcholine (ACh) with cognitive function is clinically well established now (Bush, 2003). One of the pathological hallmarks of AD is loss of cholinergic cells in the basalis of Meynert and hippocampus resulting in a diminished level of ACh in the brain (Rinne et al., 2003). Hence, repletion of ACh levels has been exploited therapeutically during last few decades to achieve symptomatic relief in AD (Orhan et al., 2009). Based on these etiologies, several pharmacological strategies with multiple possible targets are under investigation (Brinton et al., 1998), and the first line of therapeutic approaches involves reestablishment of acetylcholine levels, with the inhibition of cholinesterases viz. acetylcholinesterase (AChE) and butyrylcholinesterase (BChE) and also monoamine oxidase (MAO) enzymes (Kim et al., 2002; Senol et al., 2010). AChE principally acts by termination of nerve impulses at the cholinergic synapses by rapid hydrolysis of ACh to choline and acetic acid. This inhibition alleviates the half-life of ACh and thereby increasing in concentration at the synapse (Adewusi et al., 2011). In fact, tacrine was the first clinically established synthetic AChE inhibitor introduced to improve ACh levels at the nerve ending. Later, second-generation inhibitors were introduced, but the resulting side effects and cost factor made their utilization limited (Mankil et al., 2007). Although AD, in medical dictionary, has existed for around 100 years, age-related mental disturbances along with cognitive decline have been well documented in classical literature for thousands of years (Harvey, 2008). The Indian subcontinent has rich ancient heritage of traditional medicine (Ramila et al., 2011). Medicinal plants and their extracts are playing a vital role in the current therapy of cognitive disorders, either as main-stream or complementary medicine with particular reference to their ethnopharmacological aspects (Newman et al., 2003; Brahmachari, 2010). Essentially, in Ayurveda (traditional Indian system of medicine), these medicinal plants were classified as *medharasayanas* (Sanskrit: *medha* means intellect/cognition and *rasayana* means rejuvenation (Reena, 2012). Medharasayanas include a group of four medicinal plants with multifold benefits, specifically to improve memory and intellect by Prabhava (specific action) viz. Mandukaparni (*Centella asiatica*; Family: Umbelliferae), Yastimadhu (*Glycyrrhiza glabra*; Family: Leguminosae), Guduchi (*Tinospora cordifolia*; Family: Menispermaceae) and Shankhapushpi (*Convolvulus pleuricaulis*; Family: Convolvulaceae) (Sharma et al., 2005). Traditional Indian medicinal plants, like Ashwagandha (*Withania somnifera*; Family: Solanaceae), Brahmi (*Bacopa monnieri*; Family: Scophulariaceae), Jyothishmati (*Celastrus panniculata*; Family: Celastraceae), Kushmanda (*Benincasa hispida*; Family: Cucurbitaceae), Vacha (*Acorus calamus*; Family: Araceae), and Jatamamsi (*Nardostachys jatamamsi*; Family: Valerianaceae) are well documented in Ayurveda and other classical texts as brain tonics and memory enhancers (The Ayurvedic Pharmacopoeia of India, 2004). Indian turmeric (*Curcuma longa*; Family: Zingiberaceae) contains curcumin, a proven antioxidant and anticholinesterase drug, and is found to be very effective in delaying the progression of AD (Jagdeep et al., 2009; Patel et al., 2014).

## 21.2 MECHANISM OF CHOLINERGIC NEUROTRANSMISSION

In the central nervous system (CNS), ACh is believed to be involved in the regulation of learning, memory, and mood. ACh is synthesized in the soma from acetyl-CoA and choline catalyzed by choline acetyltransferase (ChAT) and stored in the pre-synaptic terminal. Pre-synaptic ACh is released into the synaptic cleft upon voltage sensitive membrane-bound $Ca^{2+}$ channel stimulation. Once released, ACh diffuses through the synaptic cleft and binds to the receptor, then it is metabolized into acetate and choline by AChE and BChE (Grazyna et al., 2009; Thomas and Robert, 2012). The resulting choline is transported back into the pre-synapse by choline transporters and recycled (Figure 21.1).

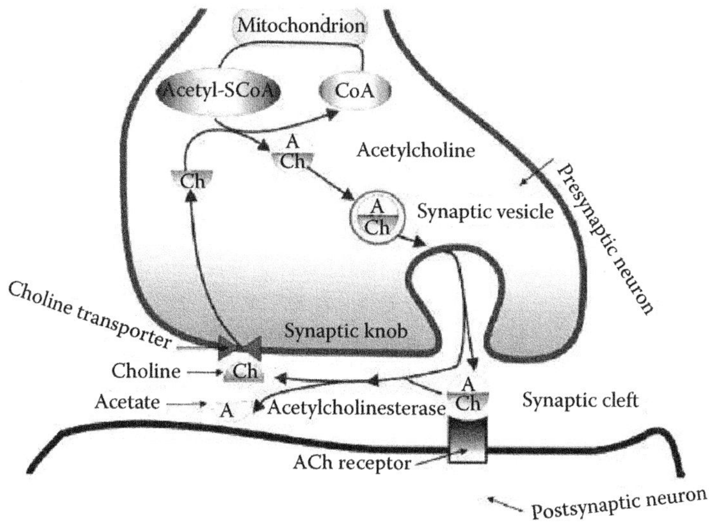


**FIGURE 21.1** Acetylcholine transmission at the synaptic cleft. (Adapted from Grazyna, N. et al., *Prog. NeuroPsychopharmacology. Biol. Psychiatry*, 33, 791–805, 2009. With permission.)

## 21.3 CHOLINERGIC HYPOTHESIS: POSSIBLE THERAPEUTIC APPROACHES AGAINST ALZHEIMER'S DISEASE

Although recent intensive efforts have been made to discover new potential drugs for AD, only a few drugs are available to address the symptoms of cognitive loss, without delaying or modifying the disease progression. Based on the pathological hallmarks, several lines of pharmacological treatment have been investigated. Some of them are discussed in the following.

### 21.3.1 Cholinergic Hypothesis

The cholinergic hypothesis claims that decrease in cognitive function in dementia is predominantly related to a decrease in cholinergic neurotransmission. AD leads to selective degeneration of cholinergic neurons of the basal forebrain and results in the decline of cholinergic markers viz. ChAT, ACh level, and increase in AChE (Jemima et al., 2011; Talic et al., 2014) (Figure 21.2).

Moreover, it has been postulated that ACh acts as a protein kinase inhibitor and thereby reduces tau phosphorylation. This inhibits intracellular neurofibrillary tangle formation and delays the progression of AD (Prerna et al., 2010). Therefore, restoration of the central cholinergic function plays a significant role in improving cognitive impairment in patients with AD. There are three principal approaches by which the cholinergic deficit can be addressed: (1) nicotinic receptor stimulation, (2) muscarinic receptor stimulation, and (3) cholinesterase (AChE and BChE) inhibition.

### 21.3.2 Nicotinic Receptor Stimulation

Nicotine and its agonists act by stimulating nicotinic receptors at the neuromuscular junction and thereby may improve ACh levels at cholinergic nerve endings. This may in turn improve cognitive functions in AD patients as well as healthy elderly people (Jian-zhi and Ze-fen, 2006).

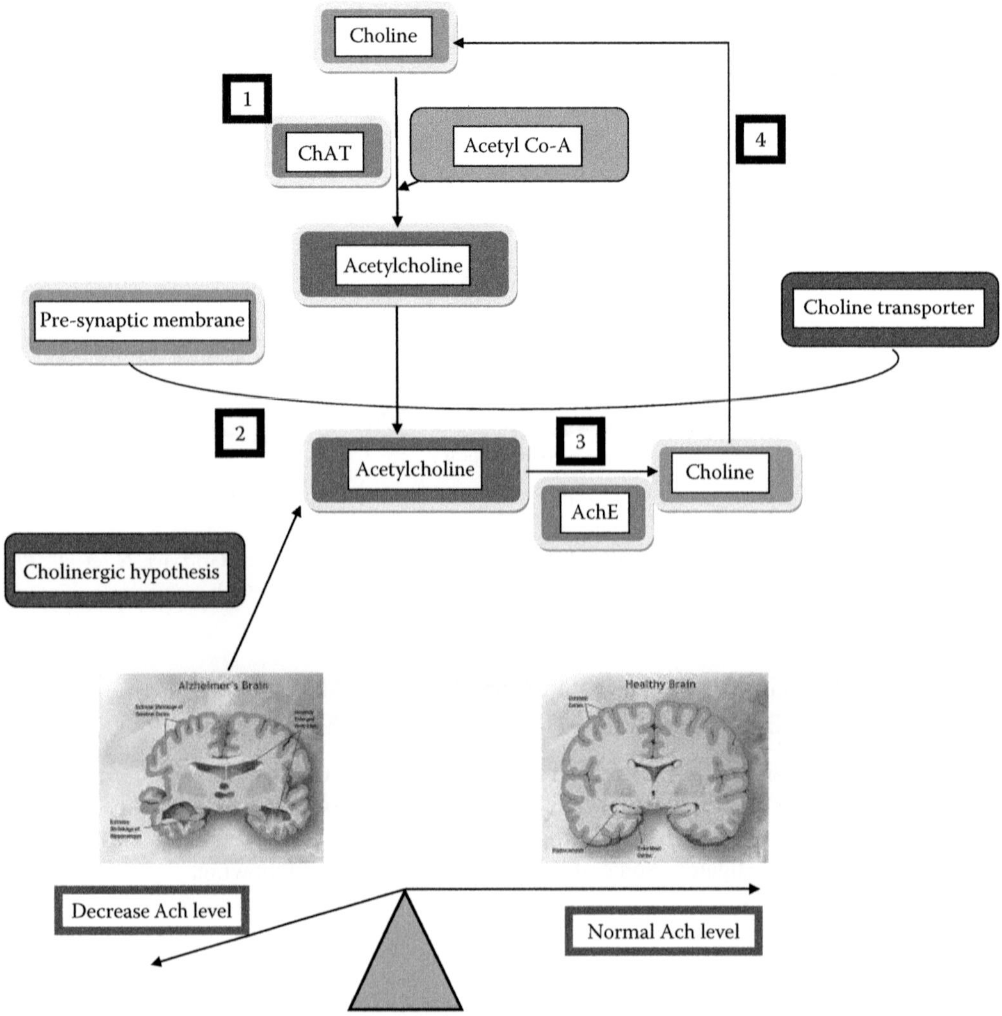


**FIGURE 21.2** Possible therapeutic approaches in modulation of cholinergic transmission. (1) Increase in ChAT activity. (2) Inhibition of ACh release. (3) Inhibition of AChE. (4) Promotes choline transport. (Adapted from Jian-zhi, W. and Ze-fen, W., *Acta Pharmacol Sinica*, 27, 41–49, 2006. With permission.)

### 21.3.3 Muscarinic Receptor Stimulation

Muscarinic agonists act by stimulating muscarinic receptors (M1 and M2) at the neuromuscular junction and thereby may improve ACh levels at cholinergic nerve endings, improving the cognitive ability of Alzheimer's patients. In addition, these agents also inhibit the neurofibrillary tangle formation and amyloid beta (Aβ) production in the brain and thus may help to delay the progression of AD (Jian-zhi and Ze-fen, 2006).

### 21.3.4 Cholinesterase Inhibitors

AChE hydrolyzes ACh to choline and acetate, thereby decreasing ACh levels at cholinergic synapses. Cholinesterase inhibitors inhibit AChE and thereby increase the availability of acetylcholine in the brain, specifically in the cholinergic synapse. However, this approach is limited to patients who have intact and functionally active pre-synaptic neurons capable of synthesizing and releasing ACh. Therefore, AChE inhibitors are useful in the early and moderate stages of AD (Vinutha et al., 2007).

Recent studies indicated that cholinesterase inhibitors might play a modulatory role on Aβ plaque deposition via suppressing amyloid precursor protein (APP) expression (Hyun et al., 2010).

In the field of AD therapy, to date, the greatest successes have resulted from acetyl cholinesterase (AChE) inhibitors (Ahirwar et al., 2012). But the resulting side effects associated with available synthetic drugs have made their utility limited. Therefore, it is worthwhile to explore the utility of traditional medicines in the treatment of AD, as they are considered to be safe and economical (Philomena, 2011).

## 21.4 MEDICINAL PLANTS AFFECTING CHOLINERGIC NEUROTRANSMISSION: A POTENTIAL SOURCE OF ANTI-ALZHEIMER'S DRUG

Ayurvedic medicinal plants have been the single-most productive source of leads for the development of drugs, and over 100 new products are already in clinical development (Rammohan et al., 2012). Indeed, several pharmacological screenings showed the efficacy of these medicinal plant extracts alone or in combination along with their active constituents to encounter AD (Manyam, 1999). The findings of experimental outcomes on various medicinal plants affecting cholinergic neurotransmission relevant to the treatment of AD are given in Table 21.1.

Standardized extracts of *B. monniera* (Family: Scrophulariaceae) showed dose-dependent AChE inhibitory activity (Ahirwar et al., 2012). *In vivo* studies also depicted improvement of rat brain glutamic acid and gamma-aminobutyric acid (GABA) levels (Vohora et al., 2000). A significant increase in the intelligent quotient (IQ) score was observed after the treatment of 110 males aged 10–13 years for 9 months with *B. monniera* containing herbal formulation suksma (Singh et al., 1997; Carlo et al., 2008).

*C. asiatica* (Family: Umbelliferae) aqueous extract showed improved learning and memory potential *in vivo.* Its extract significantly alleviated oxidative stress by decreasing malonaldehyde (MDA) levels and increasing glutathione levels. Further, it also showed AChE inhibitory activity, which combined may be associated with cognition-enhancing efficacy (Anil et al., 2009). In a recent randomized, placebo-controlled, double-blind study, 28 participants (>61 years of age) received either *C. asiatica* extracts (250, 500, and 750 mg daily) or a placebo in order to determine the effect of the extract on cognitive function and mood. After 2 months, cognitive function (as assessed by event-related potential and the computerized assessment battery test) and mood (using BondLader visual analogue) were significantly improved in those receiving the 750 mg dose extract (Kashmira et al., 2010).

*C. pluricaulis* (Family: Convulvolaceae) significantly increased ACh, catecholamine, and 5-hydroxytryptamine (5-HT) levels in brain (Parul et al., 2014). An *in vitro* assay showed its anticholinesterase activity of *C. pluricaulis* extract at 50% inhibitory concentration ($IC_{50}$) 234 ± 38 μg/mL (Maya and Sarada, 2014). Methanolic extracts of *Eugenia caryophyllus* (Family: Myrtaceae), *Glycyrrhiza glabra* (Family: Leguminosae), and *Hydrocotyle asiatica* (Family: Umbelliferae) were found to have significant AChE inhibitory activity in rodent brain homogenates (Akinrimisi and Ainwannde, 1975; Ahmad et al., 2003; Khare, 2007; Muralidharan et al., 2009).

Methanolic extracts of *Magnifera indica* (Family: Anacardiaceae), *Semecarpus anacardium* (Family: Anacardiaceae), *Carum carvi* (Family: Apiaceae), *Pimpinella anisum* (Family: Leguminosae), *A. calamus* (Family: Leguminosae), *Carthamus tinctorius* (Family: Apiaceae), *Terminalia bellirica* (Family: Combretaceae), *Terminalia chebula* (Family: Combretaceae), *Arnica Montana* (Family: Compositae), and *Lycopodium clavatum* (Family: Lycopodiaceae) have been tested positively for cholinesterase inhibitory properties (Orhan et al., 2004; Vinutha et al., 2007).

The extracts of *Corydalis solida* (Family: Papaveraceae) and *Buxus sempervirens* (Family: Buxaceae) showed remarkable AChE inhibitory activity above 50% inhibition rate at 1 mg/mL. *Genera salvia* of Lamiaceae family also exhibited anticholinesterase as well as nicotinic activity, and many of them are successfully screened for clinical trials now as anti-dementia drugs (Perry et al., 2000, 2001).

**TABLE 21.1**
**Medicinal Plants Affecting Cholinergic Neurotransmission**

| Family | Common Name | Source | Parts Used | Type of Extract | Mechanism (% Inhibition) (Concentration of Phytoextract) | | References |
|---|---|---|---|---|---|---|---|
| | | | | | AChE | BChE | |
| Acanthaceae | Sea Holly | *Acanthus ebracteatus* | Aerial parts | Methanol | TLC and 96 well plate; 36.19 ± 8.00 (0.1 mg/mL) | NA | Ingkaninan et al. (2003); Mukherjee et al. (2007a) |
| | Kalmegh | *Andrographis paniculata* | Aerial parts | Methanol | 96 well plate; 50% (222.41 μg/mL)[a] | NA | Mukherjee et al. (2007b) |
| Anacardiaceae | Mango | *Magnifera indica* | Bark | Methanol | TLC and 96 well plate; 8.15 ± 0.77 (100 μg/mL) | NA | Vinutha et al. (2007) |
| | | | Bark | Water | TLC and 96 well plate; 6.29 ± 0.37 (100 μg/mL) | NA | Vinutha et al. (2007) |
| | Marking Nut | *Semecarpus anacardium* | Bark | Methanol | TLC and 96 well plate; 69.94 ± 0.75 (100 μg/mL) | NA | Vinutha et al. (2007) |
| | | | Bark | Water | TLC and 96 well plate; 1.09 ± 0.37 (100 μg/mL) | NA | Vinutha et al. (2007) |
| Apiaceae | Caraway | *Carum carvi* | Radix | Methanol | TLC and 96 well plate; 11.00 ± 0.00 (0.1 mg/mL) | NA | Adsersen et al. (2006) |
| | Parsley | *Petroselinum crispum* | Radix | Methanol | TLC and 96 well plate; 21.00 ± 0.00 (0.1 mg/mL) | NA | Adsersen et al. (2006) |
| | Anise | *Pimpinella anisum* | Fruit | Methanol | TLC and 96 well plate; 3.00 ± 0.00 (0.1 mg/mL) | NA | Adsersen et al. (2006) |
| Apocynaceae | Crape jasmine | *Tabernaemontana divaricata* | Roots | Ethanol | 96 well plate; 50% (2.56 mg/L)[a] | 96 well plate; 50% (76.95 mg/L)[a] | Chattipakorn et al. (2007) |
| | | *Rauvolfia reflexa* | Leaf | Dichloromethane | 96 well plate; 14.65 ± 0.32 (50 μg/mL) | 96 well plate; 8.49 ± 0.92 (50 μg/mL) | Mehran et al. (2013) |
| | | *Himatanthus lancifolius* | Stem bark | Ethyl acetate | 96-well plates; 74.20 ± 2.3.0 (5 mg/mL) | NA | Claudia et al. (2010) |
| Araceae | Sweet Flag | *Acorus calamus* | Rhizome | Methanol | 96 well plate; 50% (791.35 μg/mL)[a] | NA | Ahmed et al. (2009) |

*(Continued)*

**TABLE 21.1 (*Continued*)**
**Medicinal Plants Affecting Cholinergic Neurotransmission**

| Family | Common Name | Source | Parts Used | Type of Extract | Mechanism (% Inhibition) (Concentration of Phytoextract) AChE | BChE | References |
|---|---|---|---|---|---|---|---|
| Asteraceae | Safflower | *Carthamus tinctorius* | Flower | Methanol | TLC and 96 well plate; 30.33 ± 9.22 (0.1 mg/mL) | NA | Ingkaninan et al. (2003) |
| Brassicaceae | White mustard | *Brassica alba* | *Seed* | Ethanol | UV spectroscopy; 50% (84.3 ± 1.36 μg/mL)[a] | NA | Seyed et al. (2014) |
| | Mustard | *Brassica nigra* | *Seed* | Ethanol | UV spectroscopy; 50% (135.0 ± 5.91 μg/mL)[a] | NA | Seyed et al. (2014) |
| Buxaceae | Boxwood | *Buxus sempervirens* | Whole plant | Chloroform: methanol (1:1) | 96 well plate; 61.76 ± 0.76 (1 mg/mL) | NA | Mukherjee et al. (2007a) |
| Caesalpinaceae | False acacia | *Robinia pseudoacacia* | Whole plant | Chloroform: methanol (1:1) | 96 well plate; 26.32 ± 0.82 (1 mg/mL) | 96 well plate; 31.47 ± 0.99 (1 mg/mL) | Orhan et al. (2004) |
| | Silver nail root | *Paronychia argentea* | Aerial parts | Water | UV spectophotometry; 26.10 ± 1.20 (5 mg/mL) | NA | Ferreira et al. (2006) |
| | | | | Essential oil | UV spectophotometry; 49.50 ± 1.00 (1 mg/mL) | NA | Ferreira et al. (2006) |
| | | | | Ethanol | UV spectophotometry; 48.70 ± 6.10 (0.5 mg/mL) | NA | Ferreira et al. (2006) |
| Capparidaceae | Varuna | *Crataeva nurvala* | Stem bark | Chloroform | UV spectophotometry; 2.38 ± 0.160 (μMoles/min/g of tissue) | NA | Atanu et al. (2015) |
| Caryophyllaceae | Clove | *Eugenia caryophyllus* | Flower bud | Essential oil | 96 well plate; 50.45 ± 0.003 (100 μg/mL) | 96 well plate; 49.76 ± 0.005 (100 μg/mL) | Kumar et al. (2012) |
| Celastraceae | Spindle tree | *Euonymus sachalinensis* | Leaf | Methanol | 96 well plate; 10.00 ± 3.00 (5 mg/mL) | 96 well plate; 43.00 ± 1.00 (5 mg/mL) | Sancheti et al. (2010) |

(*Continued*)

**TABLE 21.1 (*Continued*)**
**Medicinal Plants Affecting Cholinergic Neurotransmission**

| Family | Common Name | Source | Parts Used | Type of Extract | Mechanism (% Inhibition) (Concentration of Phytoextract) AChE | BChE | References |
|---|---|---|---|---|---|---|---|
| Combretaceae | Forest bushwillow | *Combretum kraussii* | Leaf | Ethyl acetate | TLC and 96 well plate; 96.00 ± 4.60 (1 mg/mL) | NA | Eldeen et al. (2005) |
| | | | Bark | Ethanol | TLC and 96 well plate; 83.00 ± 4.50 (1 mg/mL) | NA | Eldeen et al. (2005) |
| | | | Root | Ethyl acetate | TLC and 96 well plate; 81.00 ± 4.10 (1 mg/mL) | NA | Eldeen et al. (2005) |
| | Bohera | *Terminalia bellirica* | Fruit | Methanol | TLC and 96 well plate; 39.68 ± 8.15 (0.1 mg/mL) | NA | Ingkaninan et al. (2003) |
| | Haritaki | *Terminalia chebula* | Fruit | Methanol | 96 well plate; 89.00 ± 1.00 (5 mg/mL) | 96 well plate; 95.00 ± 1.00 (5 mg/mL) | Sancheti et al. (2010) |
| Compositae | Arnica | *Arnica montana* | Flower | Methanol | 96 well plate; 67.4 ± 9.3 (400 µg/mL) | 96 well plate; 57.7 ± 0.13 (400 µg/mL) | Talic et al. (2014) |
| Convolvulaceae | Shankhapushpi | *Convolvulus pluricaulis* | Whole plant | Methanol | TLC and 96 well plate; 2.22 ± 1.17 (100 µg/mL) | NA | Vinutha et al. (2007) |
| | Dwarf Morning Glory | *Evalvulus alsinoides* | Whole plant | Hydro alcohol | 96 well plate; 50% (141.76 µg/mL)[a] | NA | Mukherjee et al. (2007b) |
| Crassulaceae | Goethe plant | *Bryophyllum pinnatum* | Leaf | Ethyl acetate | 96 well plate; 50% (0.16 mg/mL)[a] | NA | Feitosa et al. (2011) |
| | Golden root | *Rhodiola rosea* | Root | Methanol | 96 well plate; 42.00 ± 3.20 (10 g/l) | NA | Hillhouse et al. (2004); Mukherjee et al. (2007a) |
| Cupressaceae | Sawara cypress | *Chamaecyparis pisifera* | Whole plant | Methanol | 96 well plate; 59.00 ± 2.00 (5 mg/mL) | 96 well plate; 62.00 ± 2.00 (5 mg/mL) | Sancheti et al. (2010) |
| | Savin | *Juniperus sabina* | Fruit | | 96 well plate; 50% (0.379 ± 9.38 µg/mL) | NA | Seyed et al. (2014) |

(*Continued*)

**TABLE 21.1 (*Continued*)**
**Medicinal Plants Affecting Cholinergic Neurotransmission**

| Family | Common Name | Source | Parts Used | Type of Extract | Mechanism (% Inhibition) (Concentration of Phytoextract): AChE | Mechanism (% Inhibition) (Concentration of Phytoextract): BChE | References |
|---|---|---|---|---|---|---|---|
| Cyperaceae | Common Nut Sedge | *Cyperus rotundus* | Whole plant | Methanol | TLC and 96 well plate; 44.19 ± 2.27 (0.1 mg/mL) | NA | Ingkaninan et al. (2003); Mukherjee et al. (2007a) |
| Dioscoreaceae | Aerial yam | *Dioscorea bulbifera* | Whole plant | Methanol | 96 well plate; 79.00 ± 2.00 (5 mg/mL) | 96 well plate; 82.00 ± 2.00 (5 mg/mL) | Sancheti et al. (2010) |
| Ericaceae | Pontic azalea | *Rhododendron luteum* | Whole plant | Chloroform: methanol (1:1) | 96 well plate; 76.32 ± 0.58 (1 mg/mL) | 96 well plate; 69.14 ± 1.89 (1 mg/mL) | Orhan et al. (2004); Mukherjee et al. (2007a) |
| | Common rhododendron | *Rhododendron ponticum* | Whole plant | Chloroform: methanol (1:1) | 96 well plate; 93.03 ± 1.12 (1 mg/mL) | 96 well plate; 95.23 ± 1.28 (1 mg/mL) | Orhan et al. (2004); Mukherjee et al. (2007a) |
| | Royal azalea | *Rhododendron schlippenbachii* | Whole plant | Methanol | 96 well plate;67.00 ± 1.00 (5 mg/mL) | 96 well plate; 63.00 ± 2.00 (5 mg/mL) | Sancheti et al. (2010) |
| Euphorbiaceae | Antique spurge | *Euphorbia antiquorum* | Stem | Methanol | TLC and 96 well plate; 42.31 ± 9.10 (0.1 mg/mL) | NA | Ingkaninan et al. (2003); Mukherjee et al. (2007a) |
| | Amla | *Emblica officinalis* | Fruit | Methanol | 96 well plate; 73.3 ± 5.9 (0.1 mg/mL) | NA | Dhivya et al. (2014) |
| | Cotton-leaf physicnut | *Jatropha gossypiifolia* | Leaf | Methanol | 96 well plate; 50% (0.05 mg/mL)[a] | NA | Dhivya et al. (2014) |
| Fabaceae | Flat-crown | *Albizia adianthifolia* | Bark | Ethyl acetate | TLC and 96 well plate; 61.00 ± 5.10 (1 mg/mL) | NA | Eldeen IMS et al. (2005) |
| | | | Bark | Ethanol | TLC and 96 well plate; 53.00 ± 2.20 (1 mg/mL) | NA | Eldeen IMS et al. (2005) |
| | | | Root | Ethyl acetate | TLC and 96 well plate; 45.0 ± 2.10 (1 mg/mL) | NA | Eldeen IMS et al. (2005) |
| | | | Root | Ethanol | TLC and 96 well plate; 51.00 ± 3.40 (1 mg/mL) | NA | Eldeen IMS et al. (2005) |
| | Broad bean | *Vicia faba* | Whole plant | Chloroform: methanol (1:1) | 96 well plate; 45.23 ± 1.03 (1 mg/mL) | 96 well plate; 55.85 ± 0.48 (1 mg/mL) | Orhan et al. (2004); Mukherjee et al. (2007a) |

(*Continued*)

**TABLE 21.1 (*Continued*)**
**Medicinal Plants Affecting Cholinergic Neurotransmission**

| Family | Common Name | Source | Parts Used | Type of Extract | Mechanism (% Inhibition) (Concentration of Phytoextract) | | References |
|---|---|---|---|---|---|---|---|
| | | | | | AChE | BChE | |
| Fumariaceae | Indian Fumitory | *Fumaria asepala* | Whole plant | Chloroform: methanol (1:1) | 96 well plate; 91.99 ± 0.70 (1 mg/mL) | 96 well plate; 93.12 ± 0.28 (1 mg/mL) | Orhan et al. (2004); Mukherjee et al. (2007a) |
| | | *Fumaria capreolata* | Whole plant | Chloroform: methanol (1:1) | 96 well plate; 96.89 ± 0.17 (1 mg/mL) | 96 well plate; 89.24 ± 0.83 (1 mg/mL) | Orhan et al. (2004); Mukherjee et al. (2007a) |
| | | *Fumaria cilicica* | Whole plant | Chloroform: methanol (1:1) | 96 well plate; 88.03± 0.65 (1 mg/mL) | 96 well plate; 80.03 ± 0.28 (1 mg/mL) | Orhan et al. (2004); Mukherjee et al. (2007a) |
| Ginkgoaceae | Ginkgo | *Ginkgo biloba* | Whole plant | Ethanol | 96 well plate; 50% (268.33 μg/mL)[a] | NA | Szwajgier et al., 2013 |
| Guttiferae | Surangi | *Mammea harmandii* | Flower | Methanol | TLC and 96 well plate; 33.63 ± 8.00 (0.1 mg/mL) | NA | Ingkaninan et al. (2003); Mukherjee et al. (2007a) |
| Hypericaceae | Sweet Amber | *Hypericum undulatum* | Flower | Essential oil | UV spectophotometry; 30.30 ± 19.70 (1 mg/mL) | NA | Ferreira et al. (2006) |
| | St. John's wort | *Hypericum perforatum* | Aerial part | Methanol | UV spectophotometry; 73.5 ± 2.40 (400 μg/mL) | UV spectophotometry; 50.5 ± 0.70 (400 mg/mL) | Talic et al. (2014) |

(*Continued*)

**TABLE 21.1 (*Continued*)**
**Medicinal Plants Affecting Cholinergic Neurotransmission**

| Family | Common Name | Source | Parts Used | Type of Extract | Mechanism (% Inhibition) (Concentration of Phytoextract) | | References |
|---|---|---|---|---|---|---|---|
| | | | | | AChE | BChE | |
| Lamiaceae | Lavender | *Lavandula angustifolia* | Aerial parts | Ethanol | UV spectophotometry; 64.30 ± 9.00 (1 mg/mL) | NA | Ferreira et al. (2006) |
| | Menthe/Pudina | *Mentha spicata* | Whole plant | Methanol | TLC and 96 well plate; 15.00 ± 0.00 (0.1 mg/mL) | NA | Adsersen et al. (2006) |
| | Mint | *Mentha suaveolens* | Aerial parts | Water | UV spectophotometry; 68.90 ± 2.50 (5 mg/mL) | NA | Ferreira et al. (2006) |
| | Oregano | *Origanum vulgare* | Whole plant | Methanol | TLC and 96 well plate; 3.00 ± 0.00 (0.1 mg/mL) | NA | Adsersen et al. (2006) |
| | Rosemary | *Rosmarinus officinalis* | Whole plant | Methanol | TLC and 96 well plate; 17.00 ± 0.00 (0.1 mg/mL) | NA | Adsersen et al. (2006) |
| | Sage | *Salvia albimaculata* | Whole plant | Petroleum ether | 96 well plate; 89.40 ± 2.07 (1 mg/mL) | 96 well plate; 73.90 ± 0.76 (1 mg/mL) | Orhan et al. (2007) |
| | Germander | *Teucrium polium* | Whole plant | Petroleum ether | 96 well plate; 45.60 ± 4.17 (1 mg/mL) | 96 well plate; 85.00 ± 53.10 (1 mg/mL) | Orhan et al. (2007) |
| | Avishan-e Shirazi | *Zataria multiflora* | Whole plant | Methanol | 96 well plate; 50% (8.2 (50 µg/mL)[a] | NA | Fariba et al. (2012) |
| Lauraceae | Bay Laurel | *Laurus nobilis* | Leaf | Ethanol | UV spectophotometry; 64.30 ± 9.00 (1 mg/mL) | NA | Ferreira et al. (2006) |
| Leguminosae | White siris | *Albizia procera* | Bark | Methanol | TLC and 96 well plate; 40.71 ± 0.46 (0.1 mg/mL) | NA | Ingkaninan et al. (2003); Mukherjee et al. (2007a) |
| | Indian Laburnum | *Cassia fistula* | Root | Methanol | TLC and 96 well plate; 54.13 ± 3.90 (0.1 mg/mL) | NA | Ingkaninan et al. (2003); Mukherjee et al. (2007a) |
| | Touch me not/ Lajjavati | *Mimosa pudica* | Whole plant | Water | TLC and 96 well plate; 1.68 ± 0.22 (100 µg/mL) | NA | Vinutha et al. (2007) |
| | Licorice | *Glycyrrhiza glabra* | Root | Methanol | 96 well plate; 50% (418 ± 30.7 µg/mL)[a] | NA | Maya et al. (2014) |

(*Continued*)

**TABLE 21.1 (*Continued*)**
**Medicinal Plants Affecting Cholinergic Neurotransmission**

| Family | Common Name | Source | Parts Used | Type of Extract | Mechanism (% Inhibition) (Concentration of Phytoextract) AChE | BChE | References |
|---|---|---|---|---|---|---|---|
| Fagaceae | Menthi | *Trigonella foenum graecum* | Seeds | Hydro alcohol | TLC and 96 well plate; 50% (6 ± 0.9 µg/mL) | NA | Dhivya et al. (2014) |
| Lythraceae | Pomegranate | *Punica granatum* | Fruit | Methanol | 96 well plate; 50% (77 ± 6.2 µg/mL) [a] | NA | Maya et al. (2014) |
| Magnoliaceae | Champak | *Michelia champaca* | Leaf | Methanol | TLC and 96 well plate; 34.88 ± 4.56 (0.1 mg/mL) | NA | Ingkaninan et al. (2003); Mukherjee et al. (2007a) |
| Malvaceae | Indian Mallow | *Abutilon indicum* | Whole plant | Methanol | TLC and 96 well plate; 30.66 ± 1.06 (0.1 mg/mL) | NA | Ingkaninan et al. (2003); Mukherjee et al. (2007a) |
| | Asiatic cotton | *Gossypium herbaceum* | Whole plant | Methanol | 96 well plate; 100% (2 mg/mL)[a] | NA | Feitosa et al. (2011) |
| Meliaceae | Neem | *Azadirachta indica* | Bark | Aqueous | TLC and 96 well plate; 5.89 ± 0.33 (100 µg/mL) | NA | Vinutha et al. (2007) |
| | Forest mahogany | *Trichilia dregeana* | Bark | Ethyl acetate | TLC and 96 well plate; 55.00 ± 4.40 (1 mg/mL) | NA | Eldeen IMS et al. (2005) |
| Menispermaceae | Stephania | *Stephania suberosa* | Roots | Methanol | TLC and 96 well plate; 91.93 ± 10.80 (0.1 mg/mL) | NA | Ingkaninan et al. (2003); Mukherjee et al. (2007a) |
| | Tapering-Leaf Tiliacora | *Tiliacora triandra* | Root | Methanol | TLC and 96 well plate; 42.29 ± 2.89 (0.1 mg/mL) | NA | Ingkaninan et al. (2003); Mukherjee et al. (2007a) |
| | Gulbel/ Guduchi | *Tinospora cordifolia* | Stem | Methanol | TLC and 96 well plate; 69.43 ± 0.37 (100 µg/mL) | NA | Vinutha et al. (2007) |
| Mimosaceae | Tomentose Babool | *Acacia nilotica* | Leaf | Ethyl acetate | TLC and 96 well plate; 53.00 ± 3.70 (1 mg/mL) | NA | Eldeen et al. (2005) |

(*Continued*)

**TABLE 21.1 (*Continued*)**
**Medicinal Plants Affecting Cholinergic Neurotransmission**

| Family | Common Name | Source | Parts Used | Type of Extract | Mechanism (% Inhibition) (Concentration of Phytoextract) | | References |
|---|---|---|---|---|---|---|---|
| | | | | | AChE | BChE | |
| Moraceae | Peepal tree | *Ficus religiosa* | Bark | Methanol | TLC and 96 well plate; 54.47 ± 1.28 (100 μg/mL) | NA | Vinutha et al. (2007) |
| Moringacea | Drumstick tree | *Moringa oleifera* | Bark | Methanol | TLC and 96 well plate; 4.99 ± 2.74 (100 μg/mL) | NA | Vinutha et al. (2007) |
| Musaceae | Banana | *Musa sapientum* | Fruit | Methanol | TLC and 96 well plate; 29.14 ± 4.73 (0.1 mg/mL) | NA | Ingkaninan et al. (2003); Mukherjee et al. (2007a) |
| Myristicaceae | Nutmeg | *Myristica fragrans* | Seed | Hydroalcohol | 96 well plate; 50% (133.28 μg/mL)[a] | NA | Mukherjee et al. (2007b) |
| Myrsinaceae | Vidanga | *Embelia ribes* | Fruit | Methanol | TLC and 96 well plate; 15.70 ± 1.19 (100 μg/mL) | NA | Vinutha et al. (2007) |
| | | | Root | Methanol | TLC and 96 well plate; 50.82 ± 0.71 (100 μg/mL) | NA | Vinutha et al. (2007) |
| Nelumbonaceae | Indian lotus | *Nelumbo nucifera* | Stamen | Methanol | TLC and 96 well plate; 23.77 ± 2.83 (0.1 mg/mL) | NA | Ingkaninan et al. (2003); Mukherjee et al. (2007a) |
| Nyctaginaceae | Punarnava | *Boerhavia diffusa* | Whole plant | Methanol | TLC and 96 well plate; 23.78 ± 1.17 (100 μg/mL) | NA | Vinutha et al. (2007) |
| Olacaceae | Muira Puama | *Ptychopetalum olacoides* | Root | Ethanol | Dose dependent activity at doses of 50 and 100 mg/kg, i.p. | NA | Siqueira et al. (2003); Mukherjee et al. (2007a) |
| Papaveraceae | Bhootakeshi | *Corydalis cava* | Whole plant | Water | TLC and 96 well plate; 62.00 ± 0.00 (0.1 mg/mL) | NA | Adsersen et al. (2006) |
| | Fume wort | *Corydalis solida* | Whole plant | Water | TLC and 96 well plate; 48.00 ± 0.00 (0.1 mg/mL) | NA | Adsersen et al. (2006) |
| Piperaceae | Betel pepper | *Piper interruptum* | Stems | Methanol | TLC and 96 well plate; 65.16 ± 8.13 (0.1 mg/mL) | NA | Ingkaninan et al. (2003); Mukherjee et al. (2007a) |
| | Black pepper | *Piper nigrum* | Seeds | Methanol | TLC and 96 well plate; 58.02 ± 3.83 (0.1 mg/mL) | NA | Ingkaninan K et al. (2003); Mukherjee et al. (2007a) |

(*Continued*)

**TABLE 21.1 (*Continued*)**
**Medicinal Plants Affecting Cholinergic Neurotransmission**

| Family | Common Name | Source | Parts Used | Type of Extract | Mechanism (% Inhibition) (Concentration of Phytoextract) | | References |
|---|---|---|---|---|---|---|---|
| | | | | | AChE | BChE | |
| Plumbaginaceae | Rose-colored leadwort | *Plumbago indica* | Root | Methanol | TLC and 96 well plate; 30.14 ± 3.28 (0.1 mg/mL) | NA | Ingkaninan et al. (2003); Mukherjee et al. (2007a) |
| Poaceae | Camel-Hay | *Cymbopogon schoenanthus* | Whole plant | Aqueous | 96 well plate; 50% (0.42 mg/mL)[a] | ND | Khadri et al. (2010) |
| Rosaceae | Garden burnet | *Sanguisorba minor* | Aerial parts | Ethanol | UV spectophotometry; 77.5 ± 2.2 (0.1 mg/mL) | NA | Ferreira et al. (2006) |
| Rubiaceae | Skunk vine | *Paederia linearis* | Whole plant | Methanol | TLC and 96 well plate; 29.31 ± 6.39 (0.1 mg/mL) | NA | Ingkaninan et al. (2003); Mukherjee et al. (2007a) |
| | Indian Madder | *Rubia cordifolia* | Stem | Methanol | TLC and 96 well plate; 22.12 ± 2.22 (100 µg/mL) | NA | Vinutha et al. (2007) |
| Rutaceae | Bael | *Aegle marmelos* | Fruit | Methanol | TLC and 96 well plate; 44.65 ± 3.04 (0.1 mg/mL) | NA | Ingkaninan et al. (2003); Mukherjee et al. (2007a) |
| | Garden Rue | *Ruta graveolens* | Whole plant | Water | TLC and 96 well plate; 22.00 ± 0.00 (0.1 mg/mL) | NA | Adsersen et al.(2006) |
| Saliaceae | Cape silver willow | *Salix mucronata* | Bark | Ethyl acetate | TLC and 96 well plate; 82.00 ± 3.90 (1 mg/mL) | NA | Eldeen IMS et al. (2005) |
| Scrophulariaceae | Brahmi | *Bacopa monniera* | Whole plant | Ethanol | 96 well plate; 42.90 ± 1.20 (0.1 mg/mL) | NA | Ahirwar et al. (2012) |
| Solanaceae | Aswagandha | *Withania somnifera* | Root | Methanol | TLC and 96 well plate; 75.95 ± 0.16 (100 µg/mL) | NA | Vinutha et al. (2007) |
| | | | Root | Aqueous | TLC and 96 well plate; 24.60 ± 0.38 (100 µg/mL) | NA | Vinutha et al. (2007) |
| Symplocaceae | | *Symplocos chinensis* | Whole plant | Methanol | 96 well plate; 74.00 ± 2.00 (5 mg/mL) | 96 well plate; 75.00 ± 2.00 (5 mg/mL) | Sancheti et al. (2009) |

(*Continued*)

**TABLE 21.1 (*Continued*)**
**Medicinal Plants Affecting Cholinergic Neurotransmission**

| Family | Common Name | Source | Parts Used | Type of Extract | Mechanism (% Inhibition) (Concentration of Phytoextract) | | References |
|---|---|---|---|---|---|---|---|
| | | | | | AChE | BChE | |
| Tamariacaceae | Myricaria | *Myricaria elegans* | Aerial parts | Methanol | 96 well plate; 74.80 ± 0.00 (0.2 µg/mL) | NA | Ahmad et al. (2003) |
| Umbelliferae | Mandookparni | *Centella asiatica* | Whole plant | Hydroalcohol | 96 well plate; 50% (106.55 µg/mL)[a] | NA | Mukherjee et al. (2007b) |
| Valerianaceae | Indian Valerian | *Nardostachys jatamansi* | Rhizomes | Methanol | 96 well plate; 50% (562.21 µg/mL)[a] | NA | Ahmed et al. (2009) |
| Verbanaceae | Wild Sage | *Lantana camara* | Aerial parts | Aqueous | TLC and 96 well plate; 3.63 ± 1.20 (100 µg/mL) | NA | Vinutha et al. (2007) |
| Zingiberaceae | Galanga | *Alpinia galanga* | Rhizomes | Methanol | TLC and 96 well plate; 16.98 ± 0.37 (100 µg/mL) | NA | Vinuta et al. (2007) |
| Zygophyllaceae | Land-Caltrops/ Gokhru | *Tribulus terrestris* | Whole plants | Chloroform: methanol (1:1) | 96 well plate; 37.89 ± 0.77 (1 mg/mL) | 96 well plate; 78.32 ± 1.27 (1 mg/mL) | Orhan et al. (2004) |

[a] represents $IC_{50}$
NMDA (N-methyl-D-aspartate receptor)
NA not available

## 21.5 PHYTOCHEMICALS AFFECTING CHOLINERGIC NEUROTRANSMISSION

Bioactive phytoconstituents resulting from isolation and characterization of natural flora and fauna have always been an important reservoir of drugs. The majority of the compounds examined to date affecting cholinergic neurotransmission and aiding protection against AD are primarily derived from plant sources. A list of phytoconstituents having significant AChE inhibitory activity is provided in Table 21.2 and structures of these compounds are shown in (Figure 21.3).

### 21.5.1 Alkaloids

#### 21.5.1.1 Indole Alkaloid

1. Physostigmine:
   Physostigmine [Figure 21.3 (i)] is a reversible AChE inhibitor originally isolated from the seed of *Physostigma venenosum* (Family: Fabaceae). It enhances short-term memory in dementia patients and used to treat mild to moderate AD (Hostettmann et al., 2006).
   a. Rivastigmine:
      Rivastigmine [Figure 21.3 (ii)], an analog of physostigmine, is a potent AChE inhibitor used for the symptomatic treatment of mild to moderately severe AD. Rivastigmine is reported to inhibit AChE in the cortex and hippocampus, brain areas involved in cognition (Thomas, 2007).
   b. Voacangine and voacangine hydroxyindolenine:
      Thin-layer chromatography (TLC) analysis of AChE inhibitory activity (Ellman's method) of voacangine and voacangine hydroxyindolenine [Figure 21.3 (iii and iv)] isolated from *Tabernaemontana australis* (Family: Apocynaceae) showed detectable spot at the minimum concentration of 25 μM (Andrade et al., 2005).
   c. Turbinatine and desoxycordifoline:
      Turbinatine and desoxycordifoline [Figure 21.3 (v and vi)] isolated from *Chimarrhis turbinata* (Family: Rubiaceae) bark and leaves showed moderate inhibitory activity at $IC_{50}$ 0.1 μM and 1.0 μM, respectively (Cardoso et al., 2004).
   d. Legucin A:
      Legucin A isolated from Ayurvedic medicinal plant Shalaparni (*Desmodium gangeticum*; Family: Papilionaceae) showed activity against AChE with $IC_{50}$ 27.0 μM (Houghton et al., 2006).
   e. Geissospermine:
      Bioactive guided fractionation of *Geissospermum vellosii* (Family: Apocynaceae) showed presence of geissospermine [Figure 21.3 (vii)], which exhibited inhibitory activity to rat brain AChE and horse serum BChE in a concentration-dependent manner with mean $IC_{50}$ values of 39.3 and 1.6 μg/mL, respectively. An *in vivo* experiment also supported with AChE inhibitory activity and thereby can revert cognitive deficits in a model of cholinergic hypofunction (Lima et al., 2009).
2. Steriodal alkaloid:
   a. Galanthamine:
      Galanthamine [Figure 21.3 (viii)] and Epinorgalantamine isolated from *Galanthus nivalis* and *Leucojum aestivum* (Family: Amaryllidaceae) respectively are reversible competitive AChE inhibitors and also modulate nicotinic ACh receptors (Vladimir-Knezevic et al., 2014). Both the phytoconstituents are recently introduced in practice in the treatment of AD.
   b. Homomoenjodaramine and moenjodaramine:
      Homomoenjodaramine and moenjodaramine [Figure 21.3 (ix and x)] isolated from *Buxus hyrcana* (Family: Buxaceae) showed non-competitive AChE inhibitory activity with $IC_{50}$ value of 19.2 and 50.8 mM, respectively (Rahman et al., 1998).

**TABLE 21.2**
**Phytoconstituents Affecting Cholinergic Neurotransmission**

| Chemical Nature of Phytoconstituents | | | Alkaloid | | |
|---|---|---|---|---|---|
| | Class | Source | Family | Activity on AChE | References |
| Assoanine | Steroidal alkaloid | *Narcissus assoanus* | Amaryllidaceae | 50% inhibition at 3.87 ± 0.24 pM | Lopez et al. (2002) |
| Buxamine B | Steroidal alkaloid | *Bums hyrcana* | Buxaceae | 50% inhibition at 7.56 ± 0.008 pM | Khalid et al. (2005) |
| Epinorgalantamine | Steroidal alkaloid | *Narcissus confuses* | Amaryllidaceae | 50% inhibition at 9.60 ± 0.65 pM | Lopez et al. (2002) |
| Galanthamine | Steroidal alkaloid | *Galanthus nivalis* | Amaryllidaceae | 50% inhibition at 1.07 ± 0.18 pM | Rhee et al. (2001) |
| Sanguinine | Steroidal alkaloid | *Eucharis grandeora* | Amaryllidaceae | 50% inhibition at 0.10 ± 0.01 pM | Lopez et al. (2002) |
| Oxoassoanine | Steroidal alkaloid | *Narcissus assoanus* | Amaryllidaceae | 50% inhibition at 47.21 ± 1.13 pM | Lopez et al. (2002) |
| N, N-dimethyl buxapapine | Steroidal alkaloid | *Bunts papillosa* | Buxaceae | 50% inhibition at 7.28 ± 0.06 pM | Rahman and Choudhary (2001) |
| Sarsalignone | Steroidal alkaloid | *Sarcococca saligna* | Buxaceae | 44.3% inhibition at 10 pM | Rahman and Choudhary (2001) |
| Homomoenjodaramine | Steroidal alkaloid | *Buxas sempervirens* | Buxaceae | $IC_{50}$ value of 19.2 mM (1 mg/mL) | Rahman A-ur et al. (1998) |
| Moenjodaramine | Steroidal alkaloid | *Buxas sempervirens* | Buxaceae | $IC_{50}$ value of 50.8 mM (1 mg/mL) | Rahman A-ur et al. (1998) |
| a-Solanine | Glycoalkaloid | *Sarcococca saligna* | Buxaceae | 50% inhibition at 8.59 ± 0.155 pM | Rahman and Choudhary (2001) |
| Vaganine | Steroidal alkaloid | *Solanum tuberosum* | Solanaceae | 50% inhibition at 7.028 ± 0.007 pM | Natarajan et al. (2009) |
| Physostigmine | Indole alkaloid | *Physostigma venenosum* | Fabaceae | 50% inhibition at 6 ± 1.0 pM | Peter and Melanie-Jayne (2005) |
| Coronaridine | Indole alkaloid | *Tabernaernontana austral* | Apocynaceae | Minimum concentration of 25 pM to produce detectable spot in TLC | Andrade et al. (2005) |
| Voacangine | Indole alkaloid | *Tabernaentontana austral* | Apocynaceae | Minimum concentration of 25 pM to produce detectable spot in TLC | Andrade et al. (2005) |
| Rupicoline | Indole alkaloid | *Tabernaetnontana australis* | Apocynaceae | Minimum concentration of 25 µM to produce detectable spot in TLC | Andrade et al. (2005) |
| Retamine | Quinolizidine alkaloid | *Genista aucheri* | Fabaceae | AChE and BChE inhibition (%) at 1 mg/mL at 15.0 ± 1.08 and 66.3 ± 0.88 respectively | Orhan et al., 2007 |
| Corynoline | Isoquinoline alkaloid | *Corydalis incisa* | Papaveraceae | 50% inhibition at 30.6 pM | Keyvan et al. (2007) |
| Palmatine | Isoquinoline alkaloid | *Corydath speciosa* | Papaveraceae | 50% inhibition at 5.8 pM | Keyvan et al. (2007) |
| Protopine | Isoquinoline alkaloid | *Corydalis speciosa* | Papaveraceae | 50% inhibition at 16.1 µM | Keyvan et al. (2007) |
| (-)-Huperzine A | Quinolizidine alkaloid | *Huperzia serrate* | Lycopodiaceae | 50% inhibition at 10.1 pM | Wang et al. (2009) |

*(Continued)*

**TABLE 21.2 (*Continued*)**
**Phytoconstituents Affecting Cholinergic Neurotransmission**

| Chemical Nature of Phytoconstituents | Class | Source | Family | Activity on AChE | References |
|---|---|---|---|---|---|
| | | | **Alkaloid** | | |
| Retamine | Quinolizidine alkaloids | *Genista aucheri* | Fabaceae | AChE and BChE [(15.0 ± 1.08) and (66.3 ± 0.88)% respectively | Orhan et al. (2007) |
| Stepharanine | Protoberberine alkaloids | *Stephania venosa* | Menispermaceae | $IC_{50}$ values of 14.10 ± 0.81 mM | Ingkaninan et al. (2006) |
| Cyclanoline | Protoberberine alkaloids | *Stephania venosa* | Menispermaceae | $IC_{50}$ values of 9.23 ± 3.47 μM | Ingkaninan et al. (2006) |
| N- methyl stepholidine | Protoberberine alkaloids | *Stephania venosa* | Menispermaceae | $IC_{50}$ values of 31.30 ± 3.67 μM | Ingkaninan et al. (2006) |
| **Glycoside** | | | | | |
| Cynatroside A | Pregnant glycoside | *Cynanchum frown* | Asclepiadaceae | 50% inhibition at 6.4 pM | Lee et al. (2005) |
| Cynatroside B | Pregnant glycoside | *Cynanchum stratum* | Asclepiadaceae | 50% inhibition at 3.6 pM | Lee et al. (2005) |
| Norswertianolin | Gluco pyranoside | *Gentiana campestris* | Coniferae | Minimum concentration of 1.20 nM to produce detectable spot in TLC | Urbain et al. (2004) |
| Swertianolin | Bellidifolin 8-o-13 glucopyranoside | *Gentians campestris* | Coniferae | Minimum concentration of 0.18 nM to produce detectable spot in TLC | Urbain et al. (2004) |
| **Others** | | | | | |
| Scopoletin | Coumarin | *Vaccinium oldhami* | Ericaceae | $IC_{50}$ value of 79 μM | Lee et al. (2004) |
| Murranganin | Coumarins | *Murraya paniculata* | Rutaceae | $IC_{50}$ value of 79.1 μM | Natarajan et al. (2009) |
| Hainanmurpanin | Coumarins | *Murraya paniculata* | Rutaceae | $IC_{50}$ value of 31.6 μM | Natarajan et al. (2009) |
| Viniferin | Stilbene oligomer | *Caragana chantlague* | Leguminosae | 50% inhibition at 2.0 pM | Natarajan et al. (2009) |
| Bellidin | Xanthone | *Centime campestris* | Coniferae | Minimum concentration of 0.03 nM to produce detectable spot in TLC bioassay | Urbain et al. (2004) |
| Ursolic acid | Triterpenoids | *Origami majorana* | Lamiaceae | 50% inhibition of AChE and BChE at 75.87 ± 0.92 and 32.21 ± 0.88 respectively | Gulacti and Tuba (2014) |
| Capsaicin | Pungent principle | *Capsicum annum* | Solanaceae | 50% inhibition of AChE and BChE at 62.7 ± 0.79 and 75.3 ± 0.98 respectively | Orhan et al. 2007 |
| Viniferin | Polyphenol | *Caragana chamlague* | Leguminosae | $IC_{50}$ values of 2.0 mM | Natarajan et al. (2009) |
| Kobophenol A | Polyphenol | *Caragana chamlague* | Leguminosae | $IC_{50}$ value of 115.8 mM | Natarajan et al. (2009) |
| Curcuminoids | Triterpene | *Curcuma longa* | Ziniberaceae | $IC_{50}$ value of 19.67 mM | Natarajan et al. (2009) |

3. Pregnane-type steroidal alkaloids:
   a. *Sarcococca hookeriana*:
      Bioactive-guided fractionation of *Sarcococca hookeriana* (Family: Buxaceae) resulted in two new pregnane-type steriodal alkaloids hookerianamide H and hookerianamide I [Figure 21.3 (xi and xii)] along with three alkaloids viz. N-α-methyl-epipachysamine D, sarcovagine C and dictyophlebine. All compounds showed significant inhibitory activities against AChE and BChE with $IC_{50}$ value ranging from 2.9 to 34.1 mM and 0.3–3.6 mM, respectively (Khalid et al., 2004).

      Salignenamides C, E, and F, Axillarine-C, Saligcinnamide, Vaganine-A, 5,6-dehydrosarconidine, 2-hydroxysalignarine-E, Salignamine, 2-hydroxysalignamine-E, Epipachysamine-D, Dictyophlebine, iso-N-formylchone-morphine, and Axillaridine-A isolated from *Sarcococca saligna* (Family: Buxaceae) inhibited both AChE and BChE non-competitively with $IC_{50}$ value ranging from 2.65–250.0 μM against AChE and 1.63–30.0 μM against BChE (Jing et al., 1999).
4. Quinolizidine alkaloid:
   a. Huperzine A:
      Huperzine A [Figure 21.3 (xiii)], isolated from the clubmoss *Lycopodium serratum* (Family: Lycopodiaceae), showed AChE inhibitory activity *in vivo*. α-Onocerin [Figure 21.3 (xiv)] a triterpene-type compound, from *L. clavatum* also showed 50% activity against AChE (Orhan et al., 2003). Huperzine A is also a NMDA receptor antagonist, which protects the brain against glutamate-induced damage, and it increases nerve growth factor levels (Eduardo et al., 2012).
5. Isoquinoline alkaloid:
   a. Protopine:
      Bioactivity guided fractionation from tubers of *Corydalis ternate* (Family: Papaveraceae) afforded protopine [Figure 21.3 (xv)], which exhibited reversible competitive inhibition with an $IC_{50}$ value of 30.5 μM. This result was supported by *in vivo* experiments (Keyvan et al., 2007).
   b. Corynoline:
      Corynoline [Figure 21.3(xvi)] isolated from *Corydalis incisa* exhibited reversible non-competitive inhibition with an $IC_{50}$ value of 50 μM (Ma et al., 1999).
   c. Bulbocapnine, Corydaline and Corydine:
      Bulbocapnine, corydaline, and corydine [Figure 21.3 (xvii, xviii, and xix)] were isolated from *Corydalis cava* as active constituents. Bulbocapnine inhibited AChE and BChE in a dose-dependent manner with $IC_{50}$ values of 40 ± 2.0 and 83 ± 3.0 μM, respectively, whereas corydaline inhibited AChE and BChE in a dose-dependent manner $IC_{50}$ value of 15 ± 3.0 μM and 52 ± 4.0 μM, respectively (Adsersen et al., 2007).
   d. Berberine and its derivatives:
      Bioactive-guided fractionation of *Corydalis turtschaninovii* led to the development of five alkaloids viz. 7-Epiber-berine, Pseudocopsitine, Berberine, Pseudoberberine, Pseuduodehydrocorydaline [Figure 21.3 (xx)], which showed AChE inhibitory with $IC_{50}$ values of 6.5 ± 0.5, 8.4 ± 0.5, 4.7 ± 0.2, 4.3 ± 0.3, 4.5 ± 0.2 μM respectively (Hung et al., 2008).
6. Protoberberine alkaloids:
   a. Stepharanine, cyclanoline and N-methyl stepholidine:
      Bioactive-guided fraction of tuber extract of *Stephania venosa* (Family: Menispermaceae) produced stepharanine, cyclanoline, and N-methyl stepholidine [Figure 21.3 (xxi, xxii, and xxiii)], which exhibited inhibitory activity on AChE with $IC_{50}$ values of 14.10 ± 0.81, 9.23 ± 3.47 and 31.30 ± 3.67 μM, respectively (Cousin et al., 1996).

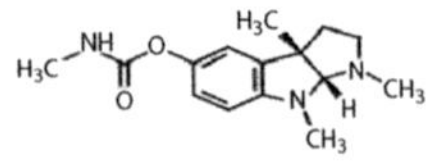
Physostigmine (i)

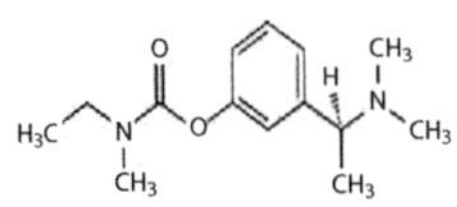
Rivastigmine (ii)

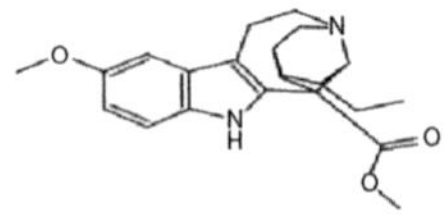
Voacangine (iii)

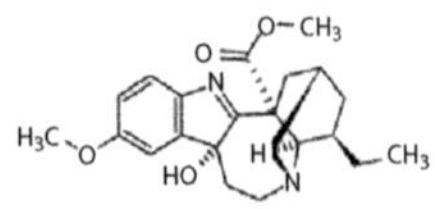
Voacangine hydroxyindolenine (iv)

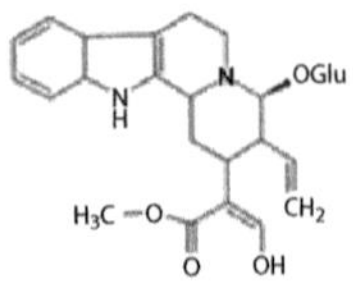
Turbinatine (v)

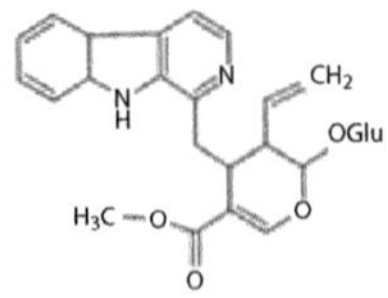
Desoxycordifoline (vi)

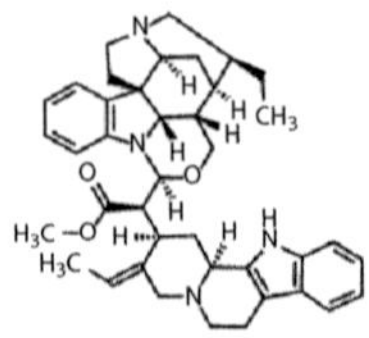
Geissospermine (vii)

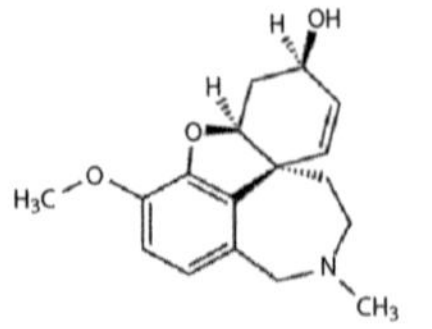
Galanthamine (viii)

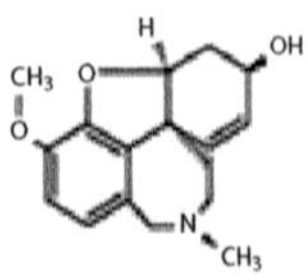
Homomoenjodaramine (ix)

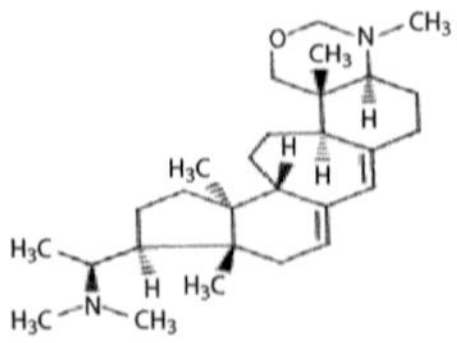
Moenjodaramine (x)

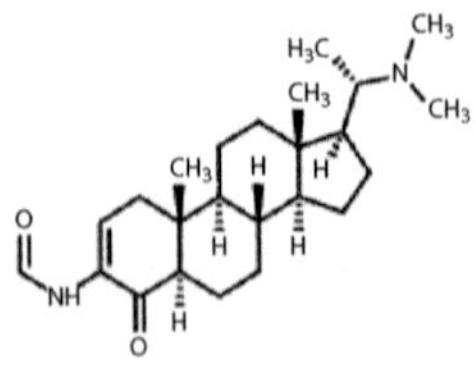
Hookerianamide H (xi)

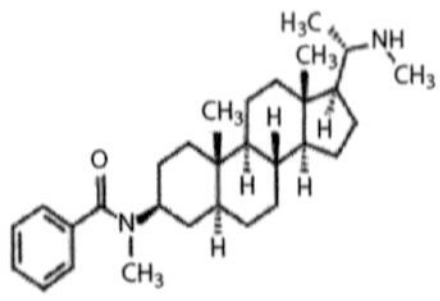
Hookerianamide I (xii)

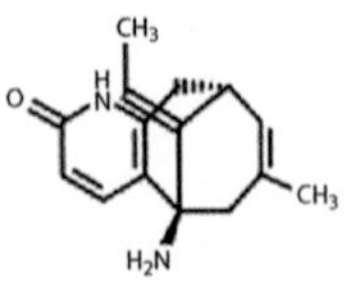
Huperzine A (xiii)

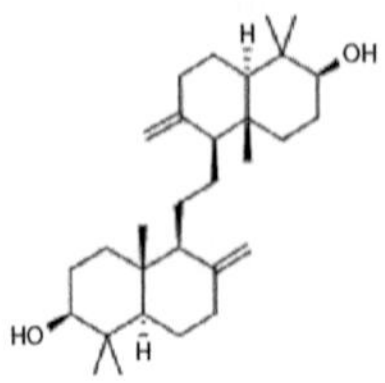
α-Onocerin (xiv)

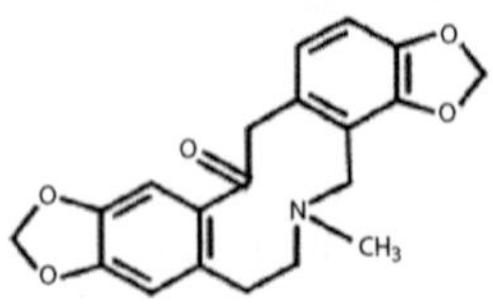
Protopine (xv)

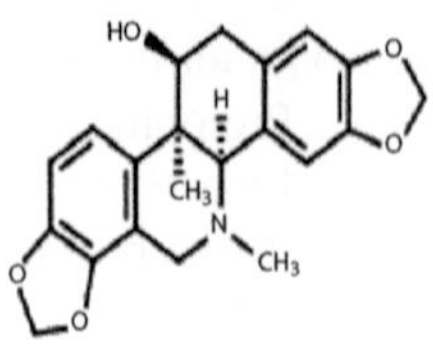
Corynoline (xvi)

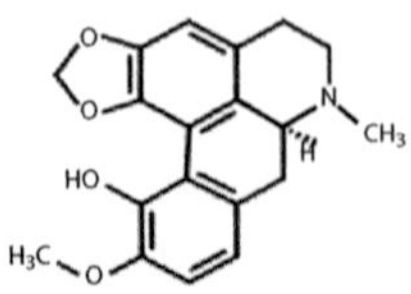
Bulbocapnine (xvii)

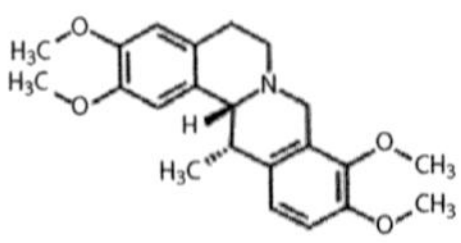
Corydaline (xviii)

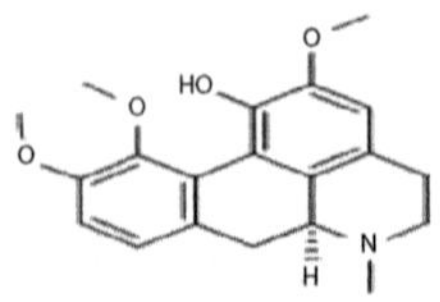
Corydine (xix)

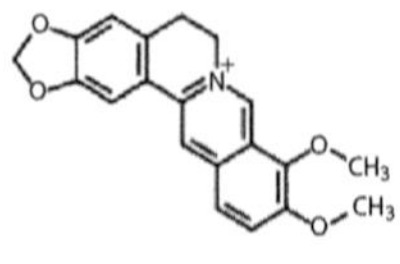
Berebeine (xx)

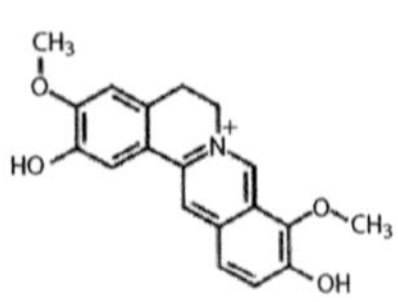
Stepharanine (xxi)

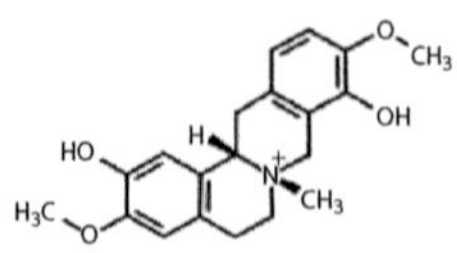
Cyclanoline (xxii)

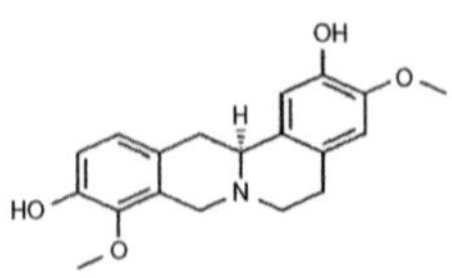
N-methyl stepholidine (xxiii)

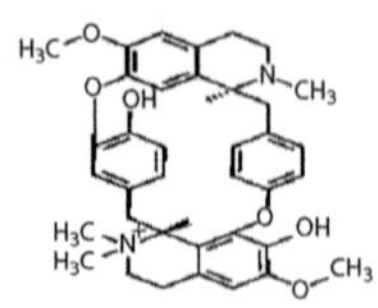
Tubocurarine (xiv)

**FIGURE 21.3** Structure of various phytoconstituents affecting cholinergic neurotransmission.

*(Continued)*

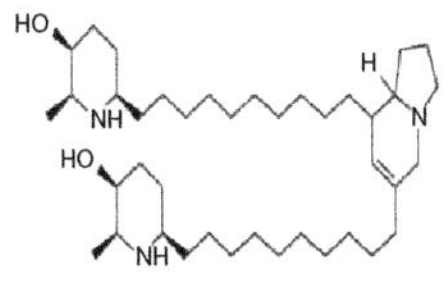
Juliflorine (xv)

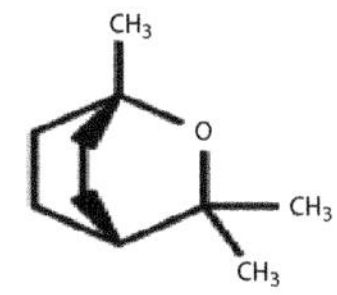
1,8-Cineole (xvi)

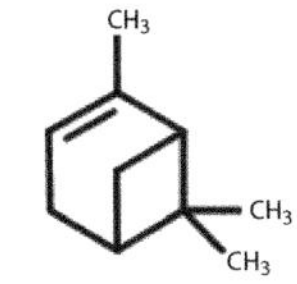
α-Pinene (xvii)

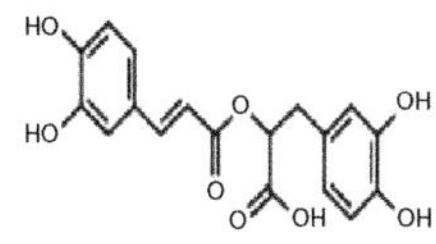
Rosemarinic acid (xviii)

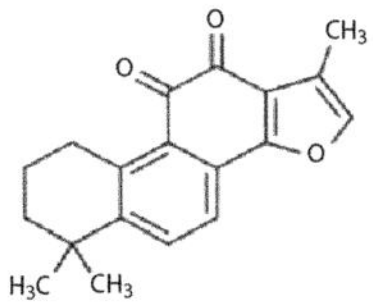
Tanshinone 1A(xix)

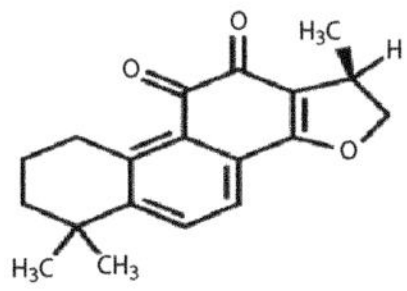
Cryptotanshinone (xxx)

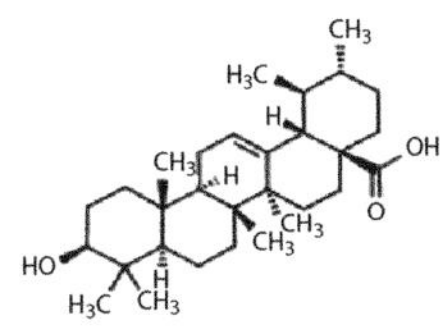
Ursolic acid (xxxi)

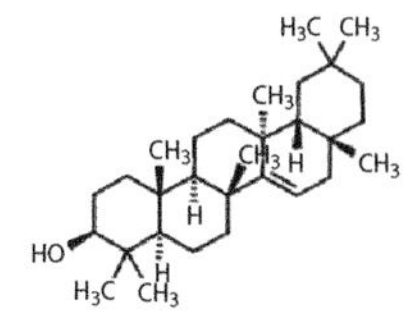
Taraxerol (xxxii)

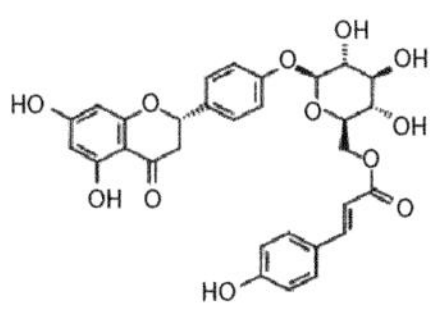
Leufolin A (xxxiii)

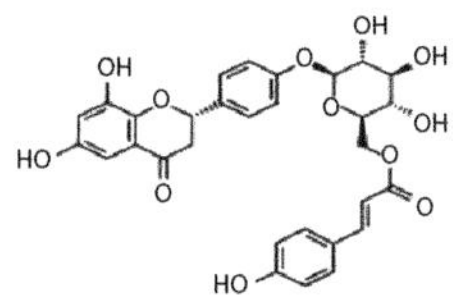
Leufolin B (xxxiv)

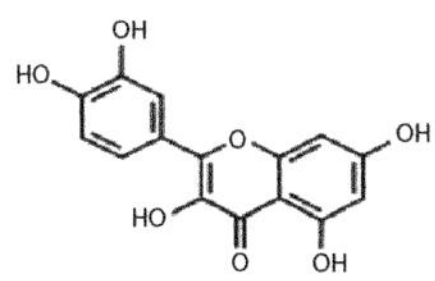
Quercetin (xxxv)

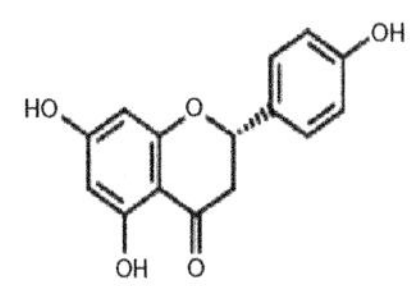
Naringenin (xxxvi)

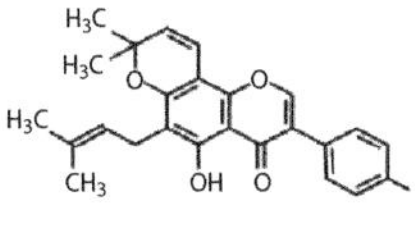
Osajin (xxxvii)

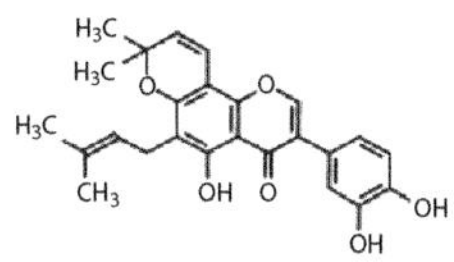
Pomiferin (xxxviii)

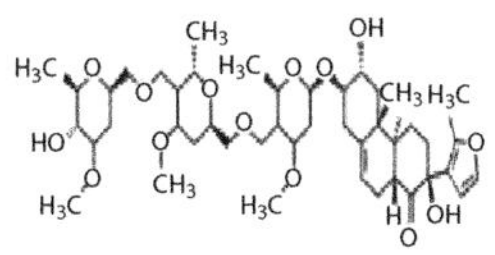
Cynatroside B (xxxix)

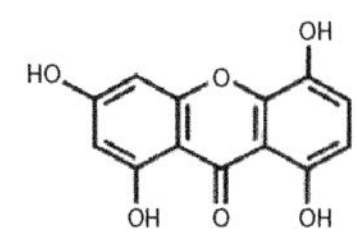
Bellidin (XL)

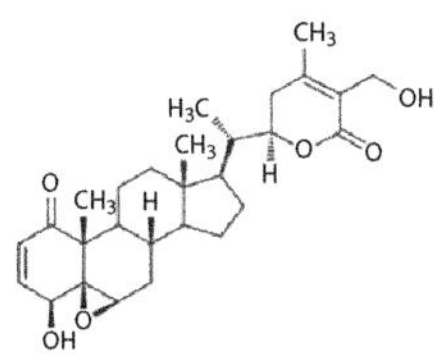
Withaferin A (XLI)

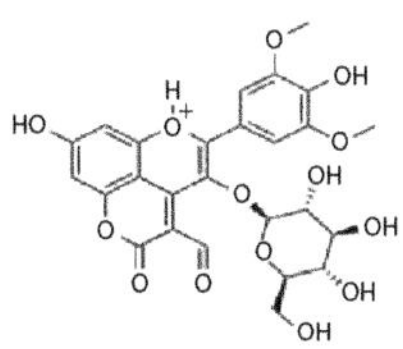
Vitisin A (XLII)

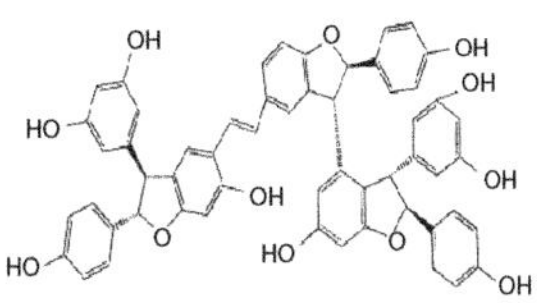
Heyneanol A (XLIII)

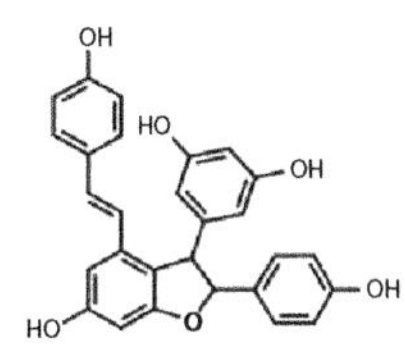
α-Viniferin (XLIV)

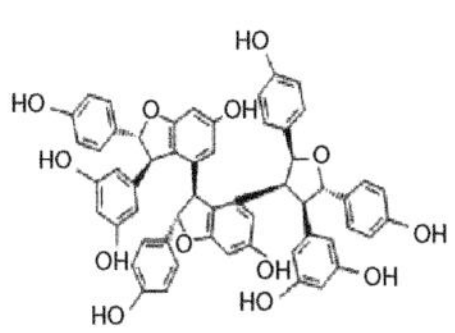
Kobophenol A (XLV)

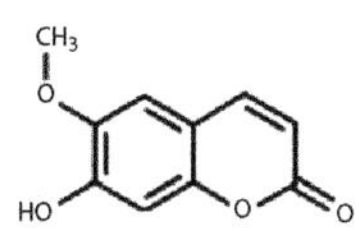
Scopoletin (XLVI)

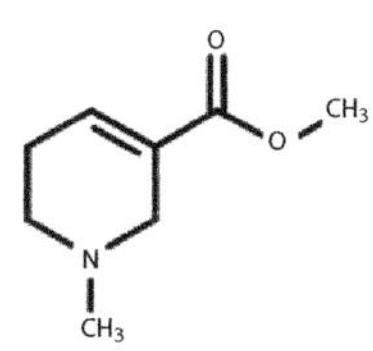
Arecoline (XLVII)

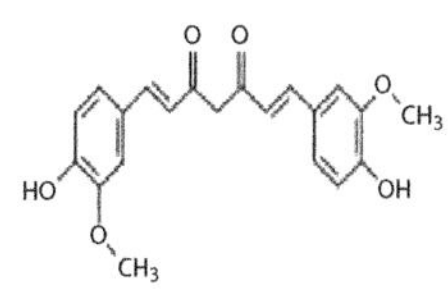
Curcumin (XLVII)

**FIGURE 21.3 (Continued)** Structure of various phytoconstituents affecting cholinergic neurotransmission.

b. Tubocurarine:

Tubocurarine [Figure 21.3 (xiv)] obtained from *Chondodendron tomentosum* (Family: Menispermaceae) showed AChE inhibitory activity *in vivo* (Thandla, 2002).

7. Piperidine alkaloids:

a. Juliflorine:

Juliflorine [Figure 21.3 (xv)] has been isolated from the leaves of *Prosopis juliflora* (Family: Papilionaceae) and it noncompetitively inhibits both AChE ($IC_{50}$ = 0.42 μM) and BChE ($IC_{50}$ = 0.12 μM) (Choudhary et al., 2005).

### 21.5.2 Terpenoids

1. Monoterpenoids:
   *Salvia lavandulaefolia* (Family: Lamiaceae) oil exhibited *in vitro* AChE inhibition due to the presence of cyclic monoterpenes 1,8-cineole and α-pinene [Figure 21.3 (xvi and xvii)] (Kennedy et al., 2011). Oral administration of the essential oil decreased striatal AChE activity in both the striatum and the hippocampus compared to the control rats *in vivo* (Perry et al., 2002). Clinical studies with 11 patients with AD showed oral administration of the essential oil significantly improved cognitive ability (Tildesley et al., 2003). A similar result was obtained with other Lamiceae family plants viz. *Rosmarinus officinalis* and *Melissa officinalis*. Rosmarinic acid [Figure 21.3 (xviii)], a phenolic compound in rosemary, exhibited the highest inhibitory activity of 85.8% toward AChE (Kennedy et al., 2011).
2. Diterpenoids:
   Root extract of *Salvia miltiorrhiza* (Family: Lamiaceae) showed AChE inhibitory effect due to the presence of diterpenes, tanshinones, and cryptotanshinones [Figure 21.3 (xix and xxx)] with $IC_{50}$ = 1.0 and 7.0 μM, respectively (Gulacti and Tuba, 2014).
3. Triterpenoids:
   Ursolic acid [Figure 21.3 (xxxi)] isolated from the *Origanum majorana* (Family: Lamiaceae) extract showed AChE inhibitory activity with $IC_{50}$ value of 7.5 nM (Gulacti and Tuba, 2014).
4. Taraxerol:
   Taraxerol [Figure 21.3 (xxxii)], isolated from twigs of *Vaccinium oldhami* (Family: Ericaceae) showed AChE inhibitory activity with $IC_{50}$ value of 79 μM (Natarajan et al., 2009).

### 21.5.3 Flavonoids and Their Derivatives

1. Flavonoids:
   Ethanolic extract of *Lecuas urticifolia* (Family: Lamiaceae) whole plant exhibited potent BChE inhibitory activity due to the presence of Leufolin A and Leufolin B [Figure 21.3 (xxxiii and xxxiv)] with $IC_{50}$ value of 1.6 ± 0.98 and 3.6 ± 1.7 μM, respectively (Gulacti and Tuba, 2014).
2. Flavones:
   Ginkobilbo EGb 761 standardized extract of *Gingko biloba* (Family: Ginkgoaceae) exhibited potent AChE inhibitory activity equivalent to standard drug tacrine (Wettstein, 2000).
   The ethyl acetate extract of whole plants of *Agrimonia pilosa* (Family: Rosaceae) showed significant AChE inhibitory activity due to presence of quercetin [Figure 21.3 (xxxv)]. Inhibitory activity of quercetin is due to the presence of a catechol moiety on ring B, which facilitates quercetin binding to AChE (Natarajan et al., 2009).
3. Flavanones:
   Naringenin [Figure 21.3 (xxxvi)] isolated from *Citrus junos* (Family: Rutaceae) inhibited AChE *in vitro* in a dose-dependent manner (Heo et al., 2004). Hispidone isolated from *Onosma hispida* (Family: Boraginaceae) was moderately active with $IC_{50}$ value = 11.6 μM (Kumar et al., 2013).
4. Isoflavonoids:
   *Maclura pomifera* (Family: Moraceae) contains two major isoflavonoids osajin and pomiferin [Figure 21.3 (xxxvii and xxxviii)], which showed AChE inhibition at $IC_{50}$ value of 58.02 and 52.87 μM, respectively (Jungand Park, 2007).

5. Pregnane glycosides:
   Roots of *Cynanchum atratum* (Family: Asclepiadaceae) containing cynatroside B [Figure 21.3 (xxxix)] exhibited potent AChE inhibitory activity with $IC_{50}$ value of 3.6 pM (Lee et al., 2005).
6. Xanthones:
   Methanol leaf extract of *Gentiana campestris* (Family: Gentianaceae) exhibited significant inhibition of AChE activity, which may be associated with isolated compound viz. xanthones, bellidin [Figure 21.3 (XL)] bellidifolin, bellidin 8-O-β-glucopyranoside (norswertianolin), and bellidifolin 8-O-β-glucopyranoside (swertianolin) (Urbain et al., 2004).
7. Sitoindosides and Withaferin A:
   *Withania somnifera* (Family: Solanaceae) is a traditional Ayurvedic *Rasayana* plant used as rejuvenative tonic. Standardized root extract containing sitoindosides and withaferin A [Figure 21.3 (XLI)] showed significant AChE inhibitory activity and enhanced muscarinic M1 receptor binding in cortical regions *in vivo*. The extract also reversed the reduction in cholinergic markers viz. ACh, ChAT *in vivo* (Divya et al., 2003).
8. Polyphenols:
   a. Vitisin A and heyneanol A:
      Vitisin A and heyneanol A [Figure 21.3 (XLII and XLIII)], two polymers of reseveratol isolated from the butanolic root extract of *Vitis amurensis* (Family: Vitaceae), inhibited both AChE and BChE in a dose-dependent manner (Jang et al., 2008).
   b. α-Viniferin and Kobophenol A:
      α-Viniferin and kobophenol A [Figure 21.3 (XLIV and XLV)] isolated from methanolic extract of *Caragana chamlague* (Family: Leguminosae) showed reversible noncompetitive AChE inhibition in a dose-dependent manner ($IC_{50}$ values are 2.0 and 115.8 mM), respectively (Natarajan et al., 2009).

### 21.5.4 Shikimate-Derived Compounds

1. Coumarins:
   Scopoletin [Figure 21.3 (XLVI)], a coumarin isolated from *Vaccinium oldhami* (Family: Ericaceae) showed significant anticholinesterase activity with $IC_{50}$ value of 79 μM (Lee et al., 2005).
   Two pyrenlated coumarins murranganin and hainanmurpanin isolated from *Murraya paniculata* (Family: Rutaceae) exhibited weak AChE inhibitory activity (IC50 79.1 and 31.6 μM) (Natarajan et al., 2009).
2. Miscellaneous:
   Standardized extract of *Areca catechu* (Family: Piperaceae) containing arecoline [Figure 21.3 (XLVII)] increase ACh level in the rat brain in a dose-dependent manner (Keyvan et al., 2007).
   Curcuminoids (a mixture of curcumin, bisdemethoxy- curcumin and demethoxycurcumin) [Figure 21.3 (XLVIII)] are the active principles of turmeric (*Curcuma longa*; Family: Zingiberaceae), a well-known Ayurvedic medicinal plant. *In vitro* experiments with curcuminoids showed significant AChE inhibitory potential with $IC_{50}$ value of 19.67 μM, while bisdemethoxycurcumin, demethoxycurcumin, and curcumin inhibited AChE with $IC_{50}$ value of 16.84, 33.14, and 67.69 μM, respectively (Ahmedand Gilani, 2009).
   Methanolic extract of another Ayurvedic plant lemon grass (*Cymbopogon schoenanthus*; Family: Poaceae) exhibited the highest AChE inhibitory activity with $IC_{50}$ 0.23 ± 0.04 mg/mL (Khadri et al., 2010).
   Further, a summary of phytoextracts and their lead molecules, with ChE inhibitory potential progressed in preclinical and clinical development, is given in Table 21.3.

**TABLE 21.3**
**ChE Inhibitors in Preclinical and Clinical Development**

| Drug | Present Status |
|---|---|
| Physostigmine | First drug investigated; however, due to side effects, short half-life, and better treatment options, it is no longer in use |
| Donepezil | Highly selective anticholinesterase. Approved for mild, moderate, and severe AD |
| Rivastigmine | It has dual BChE and AChE inhibitory activities. Approved for mild-to-moderate AD |
| Galanthamine | A lower AChE inhibitory potency with allosteric nicotinic receptor modulation properties. |
| Metrifonate | A highly selective AChE inhibitor but was abandoned after Phase III RCTs because of risk of neuromuscular dysfunction |
| Phenserine | A derivative of physostigmine with anti-Amyloidβ properties as well as AChEI. At present, screened in Phase II trials. |
| Tolserine | A promising physostigmine derivative suggested from clinical trials |
| Esolerine | Another physostigmine derivative in pipeline, with AChE and BChE inhibitory activities |
| Huperzine A | It is selective ChE inhibitor and use as a supplement in the US and China. Phase II trials in the US showed significant effect on cognition in AD |
| Nelumbo nucifera | In pipeline with potent AChE and BChE inhibitory activities |
| Himatanthus lancifolius | In pipeline with potent AChE and BChE inhibitory activities |
| Galangin | In pipeline with potent AChE inhibitory activities. |

*Source:* Mona, M. et al., *Int. J. Alz. Dis.*, 2012, 1–8, 2012.

## 21.6 CONCLUSION

During last decade, research efforts on ethno-medicinal plants affecting cholinergic neurotransmission have generated substantial data and thus provide rationale for them to be used either as main-stream or complementary medicine to provide symptomatic relief in AD. To date, cholinomimetics like tacrine along with other AChE inhibitors has gained substantial progress in AD therapy (Howes and Houghton, 2003; Natalia et al., 2010) but resulting side effects with high toxicity, short duration of action, low bioavailability, and so on have made their utility limited. Hence, from a patient safety point of view, searching for new drugs with low toxic profiles with better potency has become essential. With this background, an array of phytoextracts and bioactive phytoconstituents from those species with AChE and BChE inhibitory activity have been screened to treat AD and other neurodegenerative disorders (Perry et al., 2002; Mukherjee et al., 2007a).

Natural alkaloids with AChE inhibitory activity viz. physostigmine, huperzine-A, and galanthamine, are successfully practiced to encounter progression of AD. Stilbene oligomer viz. α-viniferin and triterpenoid viz. ursolic acid are other highly potential lead molecules with AChE inhibitory activity. The other major classes of phytoconstituents reported to have such activity are the flavonoids, glycosides, and coumarins. Plants belonging to diverse family viz. Acanthaceae, Apocynaceae, Amaryllidaceae, Araceae, Asclepiadaceae, Berberidaceae, Buxaceae, Combretaceae, Compositae, Coniferae, Cyperaceae, Ericaceae, Fumariaceae, Gentianaceae, Lamiaceae, Leguminosae, Lycopodiaceae, Magnoliaceae, Menispermaceae, Moraceae, Papaveraceae, Piperaceae, Rubiaceae, Rutaceae, Solanaceae Tamaricaceae, and Zingiberaceae have been reported to possess anticholinesterase activity (Ingkaninan et al., 2003; Mukherjee et al., 2007b; Kennedy et al., 2011). Although, with enormous efforts, the structural activity relationship (SAR) of a few isolated bioactive compounds have been established, but most of them are yet to be finalized. Moreover, cellular-based mechanism of action, clinical efficacy, and toxicity profile of these phytoextracts and their derivatives require further assessment before proceeding toward clinical trials and final approval to encounter AD.

## ACKNOWLEDGMENTS

The author is grateful to Assam Down Town University, Panihaiti-781026, Assam, India, for providing necessary support for preparation of the manuscript.

## REFERENCES

Adewusi EA, Steenkamp V. In vitro screening for acetylcholinesterase inhibition and antioxidant activity of medicinal plants from southern Africa. *Asian Pac J Trop Med* 2011; 5(2): 829–835.

Adsersen A, Gauguin B, Gudiksen L, Jager AK. Screening of plants used in Danish folk medicine to treat memory dysfunction for acetylcholinesterase inhibitory activity. *J Ethnopharmacol* 2006; 104: 418–422.

Adsersen A, Kjolbye A, Dall O, Jaager AK. Acetylcholinesterase and butyrylcholinesterase inhibitory compounds from Corydalis cava. *J Ethanophramacol* 2007; 113: 179–182.

Ahirwar S, Tembhre M, Gour S, Namdeo A. Anticholinesterase efficacy of Bacopa monnieri against the brain regions of rat: A novel approach to therapy for Alzheimer's disease. *Asian J Exp Sci* 2012; 26(1): 65–70.

Ahmad W, Ahmad B, Ahmad M, Iqbal Z, Nisar M, Ahmad M. In vitro inhibition of acetylcholinesterase, butyrylcholinesterase and lipoxygenase by crude extract of Myriacaria elegans Royle. *J Biol Sci* 2003; 3: 1046–1049.

Ahmed T, Gilani AH. Inhibitory effect of curcuminoids on acetylcholinesterase activity and attenuation of scopolamine-induced amnesia may explain medicinal use of turmeric in Alzheimer's disease. *Pharmacol Biochem Behav* 2009; 91: 554–559.

Akinrimisi ED, Ainwannde AI. Effect of aqueous extract of Eugenia caryolphyllus on brain acetylcholinesterase in rats. *West Afr J Pharmacol Drug Res* 1975; 2: 127–133.

Andrade MT, Lima JA, Pinto AC, Rezende CM, Carvalho MP, Epifanio RA. Indole alkaloids from Tabernaemontana australis (Muell. Arg) Miers that inhibit acetylcholinesterase enzyme. *Bioorg Med Chem* 2005; 13: 4092–4095.

Anil K, Samrita D, Atish P. Neuroprotective effects of Centella asiatica against intracerebroventricular colchicine-induced cognitive impairment and oxidative stress. *Int J Alzheimer's Disease* 2009; 9: 1–8.

Atanu B, Shastry CS, Saha S. Nootropic activity of Crataeva nurvala Buch-ham against scopolamine-induced cognitive impairment. *Excli J* 2015; 14: 335–345.

Ayurvedic Pharmacopoeia of India. Part 1. Vol. 4. Delhi: Controller of Publications; Ministry of Health and Family Welfare, Government of India; 2004.

Brahmachari G. *Handbook of Pharmaceutical Natural Products.* Vol. 1, 1st ed., Wiley-VCH Verlag GmbH & Co. KGaA, Weinheim, Germany, 2010.

Brinton RD, Yamazaki RS. Advances and challenges in the prevention and treatment of Alzheimer's disease. *Pharmaceut Res.* 1998; 15: 386–398.

Bush AI. The metallobiology of Alzheimer's disease. *Trends Neurosci* 2003; 26(4): 207–214.

Cardoso CL, Castro-Gamboa I, Silva DH, Furlan M, Epifanio Rde A, Pinto Ada C et al. Indole glucoalkaloids from Chimarrhis turbinata and their evaluation as antioxidant agents and acetylcholinesterase inhibitors. *J Nat Prod* 2004; 67(11): 1882–1885.

Carlo C, William LG, Michael L, Dale K, Kerry B, Barry O. Effects of a standardized Bacopa monnieri extract on cognitive performance, anxiety, and depression in the elderly: A randomized, double-blind, placebo-controlled trial. *J Altern Complement Med* 2008; 14(6): 707–713.

Chattipakorn S, Pongpanparadorn A, Pratchayasakul W, Pongchaidacha A, Ingkaninan K, Chattipakorn N. Tabernaemontana divaricata extract inhibits neuronal acetylcholinesterase activity in rats. *J Ethnopharmacol* 2007; 110: 61–68.

Choudhary MI, Nawaz SA, Zaheer-ul-Haq, Azim MK, Ghayur MN et al. Juliflorine: A potent natural peripheral anionic-site-binding inhibitor of acetylcholinesterase with calcium-channel blocking potential, a leading candidate for Alzheimer's disease therapy. *Biochem Biophys Res Commun* 2005; 332(4): 1171–1177.

Claudia S, Correia BL, Stinghen AEM, Santos CAM. Acetylcholinesterase inhibitory activity of uleine from Himatanthus lancifolius. *Zeitschrift fur Naturforschung C* 2010; 65(7–8): 440–444.

Cousin SB, Duval N, Bassoulie J, Bon C. Cloning and expression of acetylcholinesterase from Bungarus fasciatus venom: A new type of COOH-terminal domain; involvement of a positively charged residue in the peripheral site. *J Biol Chem* 1996; 271: 15099–15108.

Dhivya PS, Sobiya M, Selvamani P, Latha S. An approach to Alzheimer's disease treatment with cholinesterase inhibitory activity from various plant species. *Int J Pharm Tech Res* 2014; 6(5): 1450–1467.

Divya S, Lakshmi CM. Alzheimer's disease. *Scientific Basis for Ayurvedic Therapies*. 2003. CRC Press, New York, pp. 411–426.

Eduardo LK, Bruna MN, Paula SL, Carolina dos S. Investigation of the in vitro and ex vivo acetylcholinesterase and antioxidant activities of traditionally used Lycopodium species from South America on alkaloid extracts. *J Ethnopharmacol* 2012; 139(1): 58–67.

Eldeen IMS, Elgorashi EE, Van Staden J. Antibacterial, anti-inflammatory, anticholinesterase and mutagenic effects of extracts obtained from some trees used in South African traditional medicine. *J Ethnopharmacol* 2005; 102: 457–464.

Fariba S, Mansour M, Mohammad JA, Ehsan Z. Essential oil and methanolic extract of Zataria multiflora Boiss with anticholinesterase effect. *Pak J Bio Sci* 2012; 15(1): 49–53.

Feitosa CM, Freitas RM, Luz NNN, Bezerra MZB, Trevisan MTS. Acetylcholinesterase inhibition by some promising Brazilian medicinal plants. *Braz J Biol* 2011; 71(3): 783–789.

Ferreira A, Proenca C, Serralheiro MLM, Araujo MEM. The in vitro screening for acetylcholinesterase inhibition and antioxidant activity of medicinal plants from Portugal. *J Ethnopharmacol* 2006; 108: 31–37.

Ferri CP, Prince M, Brayne C, Brodaty H, Fratiglioni L, Ganguli M et al. Global prevalence of dementia: A Delphi consensus study. *Lancet* 2006; 366: 2112–2117.

George A. A role for the regulation of the melatonergic pathways in Alzheimer's disease and other neurodegenerative and psychiatric conditions. In: Gokhare AR, Akula R (Eds.), *Serotonin and Melatonin: Their Functional Role in Plants, Food, Phytomedicine, and Human Health*. New York: Taylor & Francis Group, 2016.

Grazyna N, Marta B, Gernot R. The septo-hippocampal system, learning and recovery of function. *Prog NeuroPsychopharmacology Biol Psychiatry* 2009; 33: 791–805.

Gulacti T, Tuba K. Lamiaceae family plants as a potential anticholinesterase source in the treatment of Alzheimer's disease. *Bezmialem Science* 2014; 1: 1–25.

Harvey AL. Natural products in drug discovery. *Drug Discov Today* 2008; 13: 894–901.

Heo HJ, Kim MJ, Lee JM, Choi SJ, Cho HY, Hong B et al. Naringenin from Citrus junos has an inhibitory effect on acetylcholinesterase and a mitigating effect on amnesia. *Dement Geriatr Cogn Disord* 2004; 17(3): 151–157.

Hillhouse BJ, Ming DS, French CJ, Neil Towers GH. Acetylcholinesterase inhibitors in Rhodiola rosea. *Pharm Biol* 2004; 42: 68–72.

Hostettmann K, Borloz AU, Marston A. Natural product inhibitors of acetylcholinesterase. *Curr Org Chem* 2006; 10: 825–847.

Houghton PJ, Ren Y, Howes MJ. Acetylcholinesterase inhibitors from plants and fungi. *Nat Prod Rep* 2006; 23: 181–199.

Howes MJ, Houghton PJ. Plants used in Chinese and Indian traditional medicine for improvement of memory and cognitive function. *Pharmacol Biochem Behav* 2003; 75: 513–527.

Hung TM, Na MK, Dat NT, Ngoc TM, Youn UJ, Kim HJ et al. Cholinesterase inhibitory and anti-amnesic activity of alkaloids from Corydaliss turtschaninovii. *J Ethanophramacol* 2008; 119: 74–80.

Hyun Ah J, Seong Eun J, Ran Joo C, Dong Hyun K, Yeong Shik K, Jong Hoon R et al. Anti-amnesic activity of neferine with antioxidant and anti-inflammatory capacities, as well as inhibition of ChEs and BACE1. *Life Sci* 2010; 87: 420–430.

Ingkaninan K, Phengpa P, Yuenyongsawad S, Khorana N. Acetylcholinesterase inhibitors from Stephania venosa tuber. *J Pharm Pharmacol* 2006; 58: 695–700.

Ingkaninan K, Temkitthawon P, Chuenchom K, Yuyaem T, Thongnoi W. Screening for acetylcholinesterase inhibitory activity in plants used in Thai traditional rejuvenating and neurotonic remedies. *J Ethnopharmacol* 2003; 89: 261–264.

Jagdeep SD, Prasad DN, Avinash CT, Rajiv G. Role of traditional medicine in neuropsychopharmacology. *Asian J Pharm Clin Res* 2009; 2(2): 72–76.

Jang MH, Piao XL, Kim JM, Kwon SW, Park JH. Inhibition of cholinesterase and amyloid-β aggregation by resveratrol oligomers from Vitis amurensis. *Phytother Res* 2008; 22: 544–549.

Jemima J, Bhattacharjee P, Singhal RS. Melatonin: A review on the lesser known potential nutraceutical. *Int J Pharm Sci Res* 2011; 2(8): 1975–1987.

Jian-zhi W, Ze-fen W. Role of melatonin in Alzheimer-like neurodegeneration. *Acta Pharmacol Sinica* 2006; 27(1): 41–49.

Jing L, Zhang HY, Wang LM, Tang XC. Inhibitory effects of huperzine B on cholinesterase activity in mice. *Acta Pharmacol Sin* 1999; 20: 141–145.

Jung M, Park M. Acetylcholinesterase inhibition by flavonoids from Agrimonia pilosa. *Molecules* 2007; 12: 2130–2139.

Kashmira JG, Jagruti P, Anuradha KG. Pharmacological review on centella asiatica: A potential herbal cure-all. *Ind J Pharm Sci* 2010; 72(5): 546–556.

Kennedy DO, Dodd FL, Robertson BC, Okello EJ, Reay JL, Scholey AB et al. Monoterpenoid extract of sage (Salvia lavandulaefolia) with cholinesterase inhibiting properties improves cognitive performance and mood in healthy adults. *J Psychopharmacol* 2011; 25(8): 1088–1100.

Keyvan D, Damien Dorman HJ, Heikki V, Raimo H. Plants as potential sources for drug development against Alzheimer's disease. *Int J Biomed Pharm Sci* 2007; 1(2): 83–104.

Khadri A, Neffati M, Smiti S, Fale P, Lino ARL, Serralheiro MLM et al. Antioxidant, antiacetylcholinesterase and antimicrobial activities of Cymbopogon schoenanthus L. Spreng (lemon grass) from Tunisia. *LWT-Food Sci Tech* 2010; 43(2): 331–336.

Khalid A, Azim MK, Parveen S, Atta-ur-Rahman, Choudhary MI. Structural basis of acetylcholinesterase inhibition by triterpenoidal alkaloids. *Biochem Biophys Res Commun* 2005; 331(4): 1528–1532.

Khalid A, Zaheer-ul-Haq, Anjum S, Khan MR, Atta-ur-Rahman, Choudhary MI. Kinetics and structure-activity relationship studies on steroidal alkaloids that inhibit cholinesterases. *Bioorg Med Chem* 2004; 12: 1995–2003.

Khare CP. *Indian Medicinal Plants: An Illustrated Dictionary*, 1st ed. New Delhi, India: Springer; 2007.

Kim WG, Cho KM, Lee CK, Yoo ID. Terreulactone A: A novel meroterpenoid with antiacetylcholinesterase activity from Aspergillus terreus. *Tetrahedron Lett* 2002; 43: 3197–3198.

Kumar S, Brijeshlata, Dixit S. Screening of traditional Indian spices for inhibitory activity of acetylcholinesterase and butyrylcholinesterase enzymes. *Int J Pharma Bio Sci* 2012; 3(1): 59–70.

Kumar N, Kumar R, Kishore K. Onosma L. A review of phytochemistry and ethnopharmacology. *Pharmacogn Rev* 2013; 7(14): 140–151.

Lee JH, Lee KT, Yang JH, Baek NL, Kim DK. Acetylcholinesterase inhibitors from the twigs of Vaccinium oldhami Miquel. *Arc Pharmacal Res* 2004; 27: 53–56.

Lee KY, Yoon JS, Kim ES, Kang SY, Kim YC. Anti-acetylcholinesterase and anti-amnesic activities of a pregnane glycoside, cynatroside B, from Cynanchum atratum. *Planta Medica* 2005; 71: 7–11.

Lima J, Costa RS, Epifanio RA, Castro NG, Rocha MS, Pinto AC. Geissospermum vellosii stem bark: Anticholinesterase activity and improvement of scopolamine-induced memory deficits. *Pharmacol Biochem Behav* 2009; 92: 508–513.

Lopez S, Bastida J, Viladomat F, Codina C. Acetylcholinesterase inhibitory activity of some Amaryllidaceae alkaloids and Narcissus extracts. *Life Sci* 2002; 71: 2521–2529.

Ma WG, Fukushi Y, Tahara S. Fungitoxic alkaloids from Hokkaido a Corydalis species. *Fitoter* 1999; 70: 258–265.

Mahesh G, Tasneem S. Biomarker controversies and diagnostic difficulties in Alzheimer's disease. *AJPCT* 2014; 2(4): 463–468.

Mankil J, Moonsoo P. Acetylcholinesterase inhibition by flavonoids from Agrimonia pilosa. *Molecules* 2007; 12: 2130–2139.

Manyam BV. Dementia in ayurveda. *J Altern Complement Med* 1999; 5: 81–88.

Maya M, Sarada S. In vitro screening for anti-cholinesterase and antioxidant activity of methanolic extracts of Ayurvedic medicinal plants used for cognitive disorders. *PLoS One* 2014; 9(1): e86804.

Mehran F, Hamid AH, Yalda K, Alireza B, Vikneswaran M. Cholinesterase enzymes inhibitors from the leaves of Rauvolfia reflexa and their molecular docking study. *Molecules* 2013; 18: 3779–3788.

Mona M, Abdu A, Marwan S. New acetylcholinesterase inhibitors for Alzheimer's disease. *Int J Alz Dis* 2012; 2012; 1–8.

Mukherjee PK, Kumar V, Houghton PJ. Screening of Indian medicinal plants for acetylcholinesterase inhibitory activity. *Phytother Res* 2007b; 21: 1142–1145.

Mukherjee PK, Kumar V, Mal M, Houghton PJ. Acetylcholinesterase inhibitors from plants. *Phytomedicine* 2007a; 14: 289–300.

Muralidharan P, Balamurugan G, Venu B. Cerebroprotective effect of Glycyrrhiza glabra Linn. root extract on hypoxic rats. *Bangladesh J Pharmacol* 2009; 4: 60–64.

Natalia W, Agnieszka K, Anna KK. Screening of traditional European herbal medicines for acetylcholinesterase and butyrylcholinesterase inhibitory activity. *Acta Pharm* 2010; 60: 119–128.

Natarajan S, Shunmugiahthevar KP, Kasi PD. Cholinesterase inhibitors from plants: Possible treatment strategy for neurological disorders—A review. *Int J Biomed Pharm Sci* 2009; 3(1): 87–103.

Newman DJ, Cragg GM, Snader KM. Natural products as sources of new drugs over the period. *J Nat Prod* 2003; 66: 1022–1037.

Orhan I, Aslan M. Appraisal of scopolamine-induced antiamnesic effect in mice and in vitro antiacetylcholinesterase and antioxidant activities of some traditionally used Lamiaceae plants. *J Ethnopharmacol* 2009; 122(2): 327–332.

Orhan I, Kartal M, Naz Q, Ejaz A, Yilmaz G, Kan Y et al. Antioxidant and anticholinesterase evaluation of selected Turkish Salvia species. *Food Chem* 2007; 103: 1247–1254.

Orhan I, Qamar N, Murat K, Fatma T, Bilge S, Iqbal CM. In vitro anticholinesterase activity of various alkaloids. *Z Naturforsch* 2007; 62c: 684–688.

Orhan I, Salih T, Bilge S. An acetylcholinesterase inhibitor from Lycopodium clavatum. *Planta Medica* 2003; 69: 1–3.

Orhan I, Sener B, Choudhary MI, Khalid A. Acetylcholinesterase and butyrylcholinesterase inhibitory activity of some Turkish medicinal plants. *J Ethnopharmacol* 2004; 91: 57–60.

Parul A, Bhawna S, Amreen F, Sanjay KJ. An update on Ayurvedic herb Convolvulus pluricaulis Choisy. *Asian Pac J Trop Biomed* 2014; 4(3): 245–252.

Patel VS, Jivani NP, Patel SB. Medicinal plants with potential nootropic activity: A review. *Res J Pharm Bio Chem Sci* 2014; 5(1): 1–11.

Perry NS, Houghton PJ, Jenner P, Keith A, Perry EK. Salvia lavandulaefolia essential oil inhibits cholinesterase in vivo. *Phytomed* 2002; 9: 48–51.

Perry NSL, Houghton PJ, Sampson J, Theolad AE, Hart S, Lis-Balchin M. In vitro activities of Salvia lavandulaefolia (Spanish Sage) relevant to treatment of Alzheimer's disease. *J Pharm Pharmacol* 2001; 53: 1347–1356.

Perry NSL, Houghton PJ, Theolad AE, Jenner P, Perry EK. In vitro inhibition of human erythrocyte acetylcholinesterase by Salvia lavandulaefolia essential oil and constituent terpenes. *J Pharm Pharmacol* 2000; 52: 895–902.

Peter JH, Melanie-Jayne H. Natural products and derivatives affecting neurotransmission relevant to Alzheimer's and Parkinson's disease. *Neurosignals* 2005; 14: 6–22.

Philomena G. Concerns regarding the safety and toxicity of medicinal plants—An overview. *J Appl Pharm Sci* 2011; 1(6): 40–44.

Prerna U, Vikas S, Mushtaq A. Therapy of Alzheimer's disease: An update. *Afr J Pharm Pharmacol* 2010; 4(6): 408–421.

Puchchakayala G, Akina S, Thati M. Neuroprotective effects of meloxicam and selegiline in scopolamine-induced cognitive impairment and oxidative stress. *Int J Alzheimer's Dis* 2012; 12: 1–8.

Rahman A-ur, Choudhary MI. Bioactive natural products as a potential source of new pharmacophores a theory of memory. *Pure Appl Chem* 2001; 73: 555–560.

Rahman A-ur, Parveen S, Khalid A, Farooq A, Ayattollahi SAM, Choudhary MI. Acetylcholinesterase inhibiting triterpenoidal alkaloids from Buxus hyrcana. *Heterocycle* 1998; 49: 481–488.

Ramila M, Manoharan A. Characteristics of pharmacognostical significance of Erythrina variegate var. and Ficus racemosa Linn. bark. *J Chem Pharm Res* 2011; 3(6): 707–714.

Rammohan VR, Olivier D, Varghese J, Dale EB. Ayurvedic medicinal plants for Alzheimer's disease: A review. *Alzheimer's Res Ther* 2012; 4(22): 1–12.

Reena K, Girish J, Abhimanyu K. Nootropic herbs (Medhya Rasayana) in Ayurveda: An update. *Pharmacogn Rev.* 2012; 6(12): 147–53.

Rhee IK, Meent MV, Ingkaninan K, Verpoorte R. Screening for acetylcholinesterase inhibitors from Amaryllidaceae using silica gel thin-layer chromatography in combination with bioactivity staining. *J Chromatogr* 2001; 915: 217–223.

Rinne JO, Kaasinen V, Jarvenpaa T, Någren K, Roivainen A, Yu M et al. Brain acetylcholinesterase activity in mild cognitive impairment and early Alzheimer's disease. *Neurol Neurosurg Psychiatry* 2003; 74: 113–115.

Sancheti S, Sancheti S, Um BH, Seo SY. 1,2,3,4,6-penta-O-galloyl-β-D-glucose: A cholinesterase inhibitor from Terminalia chebula. *S Afr J Bot* 2010; 76(10): 285–288.

Senol FS, Orhan I, Yilmaz G, Cicek M, Sener B. Acetylcholinesterase, butyrylcholinesterase, and tyrosinase inhibition studies and antioxidant activities of 33 Scutellaria L. taxa from Turkey. *Food Chem Toxicol* 2010; 48: 781–788.

Seyed BJ, Arash A, Naghmeh G, Parvin P, Massoud A. A preliminary investigation of anticholinesterase activity of some Iranian medicinal plants commonly used in traditional medicine. *Daru J Pharm Sci* 2014; 22(17): 1–5.

Sharma PC, Yelne MB, Dennis TJ. *Database on Medicinal Plants Used in Ayurveda and Siddha*, 1st ed. New Delhi, India: CCRAS, Department of AYUSH, Ministry of Health and Family Welfare, Government of India; 2005.

Siqueira IR, Fochesatto C, Lourenco da Silva A, Nunes DS, Battastini AM, Netto CA et al. Ptychopetalum olacoides, a traditional Amazonian nerve tonic possesses anticholinesterase activity. *Pharmacol Biochem Behav* 2003; 75: 645–650.

Szwajgier D, Maria W, Edyta W, Zdzislaw T. Anticholinesterase and antioxidant activities of commercial preparations from Ginkgo biloba leaves. *Acta Sci Pol Hortorum Cultus* 2013; 12(5): 111–125.

Talic S, Dragicevic I, Corajevic L, Martinovic BA. Acetylcholinesterase and butyrylcholinesterase inhibitory activity of extracts from medicinal plants. *Bull Chem Technol Bosnia and Herzegovania* 2014; 43: 11–14.

Thandla R. Neuromuscular blocking drugs: Discovery and development. *J R Soc Med* 2002; 95(7): 363–367.

Thomas M. Rivastigmine in the treatment of patients with Alzheimer's disease. *Neuropsychiatr Dis Treat* 2007; 3(2): 211–218.

Thomas SH, Robert HE. Neurotransmitter corelease: Mechanism and physiological role. *Annu Rev Physiol* 2012; 74: 225–243.

Tildesley NT, Kennedy DO, Perry EK, Ballard CG, Savelev S, Wesnes KA et al. Salvia lavandulaefolia (Spanish sage) enhances memory in healthy young volunteers. *Pharmacol Biochem Behav* 2003; 75: 669–674.

Urbain A, Marston A, Queiroz EF, Ndjoko K, Hostettmann K. Xanthones from Gentiana campestris as new acetylcholinesterase inhibitors. *Planta Medica* 2004; 70: 1011–1014.

Vinutha B, Prashanth D, Salma K, Sreeja SL, Pratiti D, Padmaja R et al. Screening of selected Indian medicinal plants for acetylcholinesterase inhibitory activity. *J Ethnopharmacol* 2007; 109: 359–363.

Vladimir-Knezevic S, Biljana B, Marija K, Jelena V, Agnieszka DL, Adelheid HB. Acetylcholinesterase inhibitory, antioxidant and phytochemical properties of selected medicinal plants of the Lamiaceae family. *Molecules* 2014; 19: 767–782.

Vohora D, Pal SN, Pillai KK. Protection from phenytoin-induced cognitive deficit by Bacopa monniera, a reputed Indian nootropic plant. *J Ethnopharmacol* 2000; 71: 253–259.

Wang BS, Wang H, Wei ZH, Song YY, Zhang L, Chen HZ. Efficacy and safety of natural acetylcholinesterase inhibitor huperzine A in the treatment of Alzheimer's disease: An updated meta-analysis. *J Neural Transm* 2009; 116(4): 457–465.

Wettstein A. Cholinesterase inhibitors and Gingko extracts are they comparable in the treatment of dementia? Comparison of published placebo-controlled efficacy studies of at least six months' duration. *Phytomed* 2000; 6: 393–401.

# 22 Neurotransmitters in Edible Plants

## *Implications in Human Health*

*Paramita Bhattacharjee and Soumi Chakraborty*

## CONTENTS



## 22.1 INTRODUCTION

Plants produce a wide variety of phytochemicals, which govern cell function and respond to environmental factors for their growth and survival, and many of these are human neuroregulatory molecules, such as serotonin, melatonin, gamma-aminobutyric acid (GABA), acetylcholine, and the catecholamines, which include dopamine, epinephrine (adrenaline), and norepinephrine (noradrenaline). Although there are numerous studies on plant neurotransmitters including those present in edible plant parts, their roles *in vivo* in human post consumption of the same is less documented, perhaps less investigated. This chapter reviews the occurrence of neurotransmitters present in edible plants, the levels at which they have been detected in plants. their positive health benefits, and adverse effects, (if any), in humans. It also suggests future scope of research in assessment of bioavailabilities of edible plant neurotransmitters in human when neurotransmitter-enriched (fortified) food products are consumed.

## 22.2 OCCURRENCE OF NEUROTRANSMITTERS IN EDIBLE PLANT PARTS

Table 22.1 lists neurotransmitters that occur in edible plants and their contents in the edible parts.

Serotonin is present in highest amount in seeds (398 $\pm$ 90 μg/g tissue) of butternuts (Feldman and Lee, 1985) followed by seeds of black walnut where it is present at the level of 304 $\pm$ 46 μg/g

**TABLE 22.1**
**Neurotransmitters in Various Edible Plants**

| Serotonin Content | | | | |
|---|---|---|---|---|
| **Name of Plant** | **Common Name** | **Edible Portion of Plants** | **Serotonin Content in Plants** | **References** |
| *Endiandra* sp. | Red plum | Fruit | 10 μg/g | Udenfriend et al. (1959) |
| *Musa* sp. | Banana | Fruit (peel) | 5–150 μg/g | |
| *Musa* sp. | Banana | Fruit (pulp) | 28 μg/g | |
| *Musa* sp. | Plantain | Fruit (pulp) | 45 μg/g | |
| *Persea* sp. | Avocado | Fruit | 10 μg/g | |
| *Prunus* sp. | Blue-red plum | Fruit | 8 μg/g | |
| *Solanum melongena* | Eggplant | Fruit | 2 μg/g | |
| *Actinidia* sp. | Kiwifruit | Fruit | 5.8 μg/g tissue | Feldman and Lee (1985) |
| *Ananas* sp. | Pineapple | Fruit | 17.0 ± 5.1 μg/g tissue | |
| *Brassica* sp. | Broccoli | Flowering head | 0.2 μg/g tissue | |
| *Brassica* sp. | Cauliflower | Flower | 0.1 μg/g tissue | |
| *Caraya ovalis* | Sweet pignut | Seed | 25 ± 8 μg/g tissue | |
| *Caraya ovata* | Shagbark | Seed | 143 ± 23 μg/g tissue | |
| *Caraya tomentosa* | Mocernut | Seed | 67 ± 13 μg/g tissue | |
| *Citrus* sp. | Grapefruit | Fruit | 0.9 μg/g tissue | |
| *Cucumis* sp. | Cantaloupe | Fruit | 0.9 μg/g tissue | |
| *Cucumis* sp. | Honeydew melon | Honeydew melon | 0.6 μg/g tissue | |
| *Ficus* sp. | Figs | Fruit | 0.2 μg/g tissue | |
| *Juglans cinerea* | Butternut | Seed | 398 ± 90 μg/g tissue | |
| *Juglans nigra* | Black walnut | Seed | 304 ± 46 μg/g tissue | |
| *Juglans regia* | English walnut | Seed | 87 ± 20 μg/g tissue | |
| *Musa* sp. | Plantain | Fruit | 30.0 ± 7.5 μg/g tissue | |
| *Musa* sp. | Banana | Fruit | 15.0 ± 2.4 μg/g tissue | |
| *Olea* sp. | Olive (black) | Fruit | 0.2 μg/g tissue | |
| *Persea* sp. | Avocado (var. Haas) | Fruit | 1.6 ± 0.4 μg/g tissue | |
| *Persea* sp. | Avocado (var. Fuerte) | Fruit | 1.5 ± 0.21 μg/g tissue | |
| *Phoenix* sp. | Dates | Fruit | 1.3 μg/g tissue | |
| *Prunus* sp. | Plums | Fruit | 4.7 ± 0.80 μg/g tissue | |
| *Punica* sp. | Pomegranate and Strawberry | Fruit | <0.1 μg/g tissue | |
| *Solanum* sp. | Tomatoes | Fruit | 3.2 ± 0.6 μg/g tissue | |
| *Solanum* sp. | Eggplant | Fruit | 0.2 μg/g tissue | |
| *Spinacia* sp. | Spinach | Leaf | 0.1 μg/g tissue | |
| *Allium cepa* | Onion | Root | 12.2 μg | Badria (2002) |
| *Allium sativum* | Garlic | Flower stalk | 29.3 μg | |
| *Ananas comosus* | Pineapple | Fruit | 11.7 μg | |
| *Brassica oleraceae* | Cabbage (var. capitata) | Leaf | 18.1 μg | |
| *Brassica oleraceae* | Cauliflower (var. botrytis) | Flower | 33.3 μg | |
| *Brassica rapa* | Turnip | Root | 32.7 μg | |
| *Cucumis sativus* | Cucumber | Fruit | 23.7 μg | |

(*Continued*)

**TABLE 22.1 (*Continued*)**
**Neurotransmitters in Various Edible Plants**

| | | | | |
|---|---|---|---|---|
| *Daucus carota* | Carrot | Root | 31.1 μg | Badria (2002) |
| *Fragaria magna* | Strawberry | Fruit | 8.5 μg | |
| *Hordeum vulgare* | Barley | Seed (grain) | 44.9 μg | |
| *Lycopersicon impinellifolium* | Tomatoes | Fruit | 14.7 μg | |
| *Malus domestica* | Apple | Fruit | 8.7 μg | |
| *Musa ensete* | Banana | Fruit | 31.4 μg | |
| *Oryza sativum* | Rice (var. japonica) | Seed (grain) | 77.3 μg | |
| *Punica granatum* | Pomegranate | Fruit | 11.6 μg | |
| *Raphnus sativus* | Radish | Root | 35.8 μg | |
| *Zea mays* | Corn | Seed (grain) | 108.2 μg | |
| *Zingiber officinale* | Ginger | Root (rhizome) | 63.3 μg | |
| *Corylus* sp. | Hazelnut | Seed | 3.4 ± 0.1 mg/kg tissue | Lavizzari et al. (2006) |
| *Musa* sp. | Banana | Fruit | 11.5 ± 0.4 mg/kg tissue | |
| *Prunus avium* L. | Sweet cherry | Fruit | 8.5–37.6 ng/100 g f.w. | González-Gómez et al. (2009) |
| *Ananas comosus* | Pineapple | Fruit | 1.5 μg/g f.w. | Ramakrishna et al. (2011) |
| *Brassica rapa* | Chinese cabbage | Leaf | 110.9 ± 22.5 μg/g f.w. | |
| *Capsicum annuum* | Hot pepper | Fruit | 17.9 μg/g f.w. | |
| *Capsicum annuum* | Paprika | Fruit | 1.8 μg/g f.w. | |
| *Carica papaya* | Papaya | Fruit | 1.1–2.1 μg/g f.w. | |
| *Cichorium intybus* | Chicory | Leaf | 8.5 ± 3.2 μg/g f.w. | |
| *Coffea arabica* | Coffea (Green) | Seed | 2.5 mg/kg d.w. | |
| *Coffea canephora* | Coffea (Green) | Seed | 2.1 mg/kg d.w. | |
| *Fragaria* × *ananassa* | Strawberry | Fruit | 3.77 ± 0.66 μg/g f.w. | |
| *Girardinia heterophylla* | Himalayan nettle | Shoot with leaf | 0.42 μg/g f.w. | |
| *Griffonia simplicifolia* | Griffonia | Leaf | 0.0017–0.007 μg/g | |
| *Griffonia simplicifolia* | Griffonia | Seed | 20,000 μg/g | |
| *Juglans ailanthifolia* | Heartseed walnut | Seed | 95 μg/g f.w. | |
| *Juglans mandshurica* | Japanese walnut | Seed | 251 μg/g f.w. | |
| *Juglans nigra* | Black walnut | Seed | 180 μg/g f.w. | |
| *Lactuca serriola* | Lettuce | Leaf | 3.3 ± 0.6 μg/g f.w. | |
| *Lycopersicon esculentum* | Tomato | Fruit | 221.9 μg/g f.w. | |
| *Lycopersicon esculentum* | Cherry tomato | Fruit | 156.1 μg/g f.w. | |
| *Mimosa tenuiflora, Mimosa pudica* | Mimosa | Seed | 80.4 μg/g f.w. | |
| *Musa sapientum* | French plantain banana | Fruit (peel) | 40–150 μg/g f.w. | |
| *Phaseolus multiflorus*[b] | Scarlet runner bean | Leaf | 0.6 μg/g f.w. | |
| *Pisum sativum*[b] | Garden pea | Leaf | 0.9–1 μg/g f.w. | |
| *Prunus avium* | Sweet cherry (var. Burlat) | Fruit | 12.6 ng/100 g f.w. | |
| *Prunus avium* | Sweet cherry (var. Navalinda) | Fruit | 30.7 ng/100 g f.w. | |

(*Continued*)

**TABLE 22.1 (*Continued*)**
**Neurotransmitters in Various Edible Plants**

| | | | | |
|---|---|---|---|---|
| *Prunus avium* | Sweet cherry (var. Van) | Fruit | 10.6 ng/100 g f.w. | Ramakrishna et al. (2011) |
| *Prunus avium* | Sweet cherry (var. Pico Limon Negro) | Fruit | 27.1 ng/100 g f.w. | |
| *Prunus avium* | Sweet cherry (var. Sweetheart) | Fruit | 19.2 ng/100 g f.w. | |
| *Coffea* sp. | Coffee | Seed (beans) | 1.05–6.15 μg/g | Restuccia et al. (2015) |
| *Prunus avium* | Sweet cherry (var. Pico Negro) | Fruit | 2.8 ng/100 g f.w. | |
| *Prunus avium* | Sweet cherry (var. Ambrunes) | Fruit | 37.6 ng/100 g f.w. | |
| *Prunus avium* | Sweet cherry (var. Pico Colorado) | Fruit | 36.6 ng/100 g f.w. | |
| *Citrus reticulata* | Orange | Fruit | 2.21 ± 0.006 μg/g | Płonka and Michalski (2017) |
| *Musa* sp. | Banana | Fruit | 14.93 ± 0.50 μg/g | |

**Melatonin Content**

| Name of Plant | Common Name | Edible Portion of Plants | Melatonin Content in Plants | References |
|---|---|---|---|---|
| *Basella* sp. | Indian spinach | Leaf | 38.7 ± 5.5 pg/g tissue | Hattori et al. (1995) |
| *Abena sativa* | Oat | Seed (grain) | 1796.1 ± 43.3 pg/g tissue | |
| *Actinidia chinensis* | Kiwi fruit | Fruit | 24.4 ± 1.7 pg/g tissue | |
| *Allium cepa* | Onion | Bulb | 31.5 ± 4.8 pg/g tissue | |
| *Allium fistulosum* | Welsh onion | Bulb | 85.7 ± 8.0 pg/g tissue | |
| *Ananas comosus* | Pineapple | Fruit | 36.2 ± 8 pg/g tissue | |
| *Angelica keiskei* | Carrot | Root | 55.3 ± 11.9 pg/g tissue | |
| *Asparagus officinalis* | Asparagus | Stem | 9.5 ± 3.2 pg/g tissue | |
| *Bassica campestris* | Japanese radish | Root | 657.2 ± 29.0 pg/g tissue | |
| *Brassica oleracea* | Cabbage | Leaf | 107.4 ± 7.3 pg/g tissue | |
| *Cucumis sativus* | Cucumber | Fruit | 24.6 ± 3.5 pg/g tissue | |
| *Fragaria magna* | Strawberry | Fruit | 12.4 ± 3.1 pg/g tissue | |
| *Hordeum vulagare* | Barley | Seed (grain) | 378.1 ± 25.8 ng/100 g | |
| *Lycopersicon esculentum* | Tomato | Fruit | 32.2 ± 2.4 pg/g tissue | |
| *Malus domestica* | Apple | Fruit | 47.6 ± 3.1 pg/g tissue | |
| *Oryza sativa japonica* | Rice | Seed (grain) | 1006.0 ± 58.5 ng/100 g | |
| *Petasites japonica* L. | Japanese butterbur | Overground portion | 49.5 ± 5.6 pg/g tissue | |
| *Raphanus sativus* | Chinese cabbage | Leaf | 112.5 ± 10.3 pg/g tissue | |
| *Zea mays* | Sweet corn | Fruit | 1366.1 ± 465.1 pg/g tissue | |
| *Zingiber officinale* | Ginger | Root (rhizome) | 583.7 ± 50.3 pg/g tissue | |
| *Lycopersicon esculentum* | Tomato | Fruit | 0.5 ng/g | Dubbels et al. (1995) |
| *Hypericum perforatum* | St. John's wort | Leaf | 1750 ng/g d.w. | Murch et al., (2001) |
| *Brassica hirta* | White mustard | Seed | 189 ng/g dry seed | Manchester et al. (2000) |

(*Continued*)

**TABLE 22.1 (*Continued*)**
**Neurotransmitters in Various Edible Plants**

| | | | | |
|---|---|---|---|---|
| *Brassica nigra* | Black mustard | Seed | 129 ng/g dry seed | Manchester et al. (2000) |
| *Helianthus annuus* | Sunflower | Seed | 29 ng/g dry seed | |
| *Popaver somniferum* | Poppy Seed | Seed | 6 ng/g dry seed | |
| *Coriandrum sativum* | Coriander | Seed | 7 ng/g dry seed | |
| *Lycium barbarum* | Wolf berry | Seed | 103 ng/g dry seed | |
| *Elettaria cardamomum* | Green cardamom | Seed | 15 ng/g dry seed | |
| *Medicago sativum* | Alfalfa | Seed | 16 ng/g dry seed | |
| *Prunus amygdalus* | Almonds | Seed | 39 ng/g dry seed | |
| *Pimpinela anisum* | Anise | Seed | 7 ng/g dry seed | |
| *Apium graveolens* | Celery | Seed | 7 ng/g dry seed | |
| *Foeniculum vulgare* | Fennel | Seed | 28 ng/g dry seed | |
| *Trigonella foenum-graecum* | Fenugreek | Seed | 43 ng/g dry seed | |
| *Linum usitatissimum* | Flax Seed | Seed | 12 ng/g dry seed | |
| *Allium cepa* | Onion | Root | 29.9 ng/100 g | Badria (2002) |
| *Allium sativum* | Garlic | Flower stalk | 58.7 ng/100 g | |
| *Ananas comosus* | Pineapple | Fruit | 27.8 ng/100 g | |
| *Brassica oleraceae* | Cabbage (var. capitata) | Leaf | 30.9 ng/100 g | |
| *Brassica oleraceae* | Cauliflower (var. botrytis) | Flower | 82.4 ng/100 g | |
| *Brassica rapa* | Turnip | Root | 50.1 ng/100 g | |
| *Cucumis sativus* | Cucumber | Fruit | 59.2 ng/100 g | |
| *Daucus carota* | Carrot | Root | 49.4 ng/100 g | |
| *Fragaria magna* | Strawberry | Fruit | 13.6 ng/100 g | |
| *Hordeum vulgare* | Barley | Seed (grain) | 87.3 ng/100 g | |
| *Lycopersicon pimpinellifolium* | Tomatoes | Fruit | 30.2 ng/100 g | |
| *Malus domestica* | Apple | Fruit | 16.1 ng/100 g | |
| *Musa ensete* | Banana | Fruit | 65.5 ng/100 g | |
| *Oryza sativum* | Rice (var. japonica) | Seed (grain) | 149.8 ng/100 g | |
| *Punica granatum* | Pomegranate | Fruit | 16.8 ng/100 g | |
| *Raphnus sativus* | Radish | Root | 75.8 ng/100 g | |
| *Zea mays* | Corn | Seed (grain) | 187.8 ng/100 g | |
| *Zingiber officinale* | Ginger | Root (rhizome) | 142.3 ng/100 g | |
| *Prunus* sp. | Sweet cherry | Fruit | $0.6 \pm 0.0 - 22.4 \pm 1.3$ ng/100 g f.w. | González-Gómez et al. (2009) |
| *Brassica oleracea* | Kohlrabi (convar. acephala var. gongylodes) white | Bulb | 0.78 μg/100 g f.w. | |
| *Brassica napus* L. | Rutabaga (var. napobrassica) | Sprouted seed | 0.44–0.59 μg/100 g f.w. | Pasko et al. (2014) |
| *Brassica napus* L. | Rutabaga (var. napobrassica) | Root | 0.56 μg/100 g f.w. | |
| *Brassica oleracea* | Kohlrabi (convar. acephala var. gongylodes) | Sprouted seed | 0.62–0.78 μg/100 g f.w. | |

(*Continued*)

## TABLE 22.1 (*Continued*)
## Neurotransmitters in Various Edible Plants

| | | | | |
|---|---|---|---|---|
| *Brassica oleraceae* | Broccoli | Seed | 0.41 ± 0.04 ng/g d.w. | Aguilera et al. (2015) |
| *Brassica oleraceae* | Red Cabbage | Seed | 0.34 ± 0.04 ng/g d.w. | |
| *Lens sculenta* | Lentil | Seed | 0.07 ± 0.01 ng/g d.w. | |
| *Medicago sativa* | Alfalfa | Seed | 0.05 ± 0.00 ng/g d.w. | |
| *Morus alba* | Mulberry (var. Baiyuwang) black | Fruit | 0.58 ng/g f.w. | |
| *Morus nigra* | Mulberry (var. Hongguo2) white | Fruit | 1.41 ng/g f.w. | |
| *Raphanus sativus* | Radish | Seed | 0.28 ± 0.01ng/g d.w. | |
| *Vigna radiata* | Mung bean | Seed | 0.01 ± 0.0 ng/g d.w. | |
| *Basella* sp. | Spinach | Leaf | 0.04 ng/g f.w. | Meng et al. (2017) |
| *Actinidia chinensis* | Kiwi fruit | Fruit | 0.02 ng/g f.w. | |
| *Ananas comosus* L. | Pineapple | Fruit | 0.04–0.28 ng/g f.w. | |
| *Avena sativa* L. | Oats | Grain | 90.6 ± 7.7 ng/g f.w. | |
| *Avena sativa* L. | Oats | Grain | 1.80 ng/g f.w. | |
| *Babreum coscluea* | Shiya tea-leaf | Flower | 2120 ng/g d.w. | |
| *Beta vulgaris* | Beetroot | Root | 0.002 ng/g | |
| *Bucida* sp. | Black olive | Fruit | 0.01 ng/g d.w. | |
| *Coptis chinensis* Franch | Huanglian | Flower | 1008 ng/g d.w. | |
| *Cucumis sativus* L. | Cucumber (cv. Jingyu-1) | Fruit | 17.3 ng/g f.w. | |
| *Fragaria ananassa* L. | Strawberry (cv. Camarosa) | Fruit | 5.58 ± 0.01 ng/g f.w. | |
| *Fragaria ananassa* L. | Strawberry (cv. Candonga) | Fruit | 5.5 ± 0.6 ng/g f.w. | |
| *Fragaria ananassa* L. | Strawberry (cv. Festival) | Fruit | 11.26 ± 0.13 ng/g f.w. | |
| *Fragaria ananassa* L. | Strawberry (cv. Primoris) | Fruit | 8.5 ± 0.6 ng/g f.w. | |
| *Fragaria magna* L. | Strawberry | Fruit | 12.4 ± 3.1 pg/g tissue | |
| *Fragaria magna* L. | Strawberry | Fruit | 0.14 pg/g tissue | |
| *Fragaria magna* L. | Strawberry | Fruit | 0.01 ng/g f.w. | |
| *Hordeum vulgare* L. | Barley | Seed | 0.87 ng/g f.w. | |
| *Hordeum vulgare* L. | Barley | Seed | 82.3 ± 6.0 ng/g f.w. | |
| *Hordeum vulgare* L. | Barley | Seed | 0.38 ng/g f.w. | |
| *Hordeum vulgare* L. | Barley | Seed | 0.58 ± 0.05 ng/g f.w. | |
| *Juglans regia* | Walnut | Seed | 1.02–1.77 ng/g | |
| *Lens culinaris* L. | Lentils | Seed | 0.5 ng/g d.w. | |
| *Lupinus albus* L. | Lupin (seed-cotyledons) | Seed | 3.83 ± 0.21 ng/g f.w. | |
| *Lupinus albus* L. | Lupin (seed-coat) | Seed | 37.50 ± 2.3 ng/g f.w. | |
| *Lycopersicon pimpinellifolium* | Tomato | Fruit | 0.11 ng/g | |
| *Malus domestica* | Apple | Fruit | N.D. | |
| *Malus domestica* | Apple (Borkh. cv. Red Fuji) | Fruit | 5 ng/g f.w. | |

(*Continued*)

**TABLE 22.1 (*Continued*)**
**Neurotransmitters in Various Edible Plants**

| | | | | |
|---|---|---|---|---|
| *Malus domestica* | Apple | Fruit | 0.16 ng/g f.w. | Meng et al. (2017) |
| *Morus alba* L. | Sangye | Leaf | 1510 ng/g d.w. | |
| *Morus* sp. | Mulberry | Leaves | 40.7–279.6 d.w. | |
| *Oryza sativa* | Rice | Seed (grain) | 1.50–264 ng/100 g | |
| *Capsicum annuum* L. | Pepper (cv. Barranca) | Fruit | 4.48 ng/g f.w.<br>31.01 ng/g d.w. | |
| *Capsicum annuum* L. | Pepper (cv. F26) | Fruit | 11.9 ng/g f.w.<br>93.4 ng/g d.w. | |
| *Periostracum cicadae* | Chantui | Flower | 3771 ng/g d.w. | |
| *Phaseolus vulgaris* L. | Kidney beans | | 1.0 ng/g d.w. | |
| *Phellodendron amurense* Rupr. | Huangbo | Flower | 1235 ng/g d.w. | |
| *Piper nigrum* L. | Black pepper | Flower | 1092.7 ng/g d.w. | |
| *Pistacia vera* | Pistachio | Seed | 226,900–233,000 ng/g d.w. | |
| *Portulaca oleracea* L. | Purslane | Leaf | 19 ng/g f.w. | |
| *Prunus avium* L. | Cherry | Fruit | 0.01–20 ng/g f.w | |
| *Prunus cerasus* L. | Tart cherry (cv. Balaton) | Fruit | 13.46 ± 1.10 ng/g f.w. | |
| *Prunus cerasus* L. | Tart cherry (cv. Montmorency) | Fruit | 2.06 ± 0.17 ng/g f.w. | |
| *Scutellaria biacalensis* | Huang-qin | Flower | 7110 ng/g d.w. | |
| *Solanum lycopersicum* L. | Tomato (cv. Ciliegia) | Fruit | 0.64 ng/g f.w./7.47 ng/g d.w. | |
| *Solanum lycopersicum* L. | Tomato (cv. Optima) | Fruit | 14.77 ng/g f.w./ 249.98 ng/g d.w. | |
| *Solanum lycopersicum* L. | Tomato (cv. Micro-Tom) | Fruit | 1.5–66.6 ng/g f.w. | |
| *Glycine max* | Soybean | Seed | 0.45 ± 0.03 ng/g d.w. | |
| *Triticum aestivum* L. | Wheat | Seed | 124.7 ± 14.9 ng/g f.w. | |
| *Triticum* sp. | Whole grain | Seed | 2–4 ng/g | |
| *Triticum* sp. | Purple wheat | Seed | 4 ng/g d.w. | |
| *Uncaria rhynchophylla* | Gouteng | Flower | 2460 ng/g d.w. | |
| *Vaccinium macrocarpon* Ait. | Cranberry | Whole fruit | 96 ± 26 ug/g d.w. | |
| *Vaccinium oxycoccos* L. | Cranberry | Whole fruit | 40 ± 10 ug/g d.w. | |
| *Vaccinium vitis-idaea* L. | Cranberry | Whole fruit | 25 ± 3 ug/g d.w. | |
| *Viola philipica* Cav. | Diding | Flower | 2368 ng/g d.w. | |
| *Vitis vinifera* L. | Grape (cv. Nebbiolo, Croatina, Barbera, Sangiovese, Marzemino, Cabernet Sauvignon, Merlot, Cabernet Franc; Italy) | Fruit skin | 0.005–0.96 ng/g | |

*(Continued)*

**TABLE 22.1 (*Continued*)**
**Neurotransmitters in Various Edible Plants**

| | | | | |
|---|---|---|---|---|
| *Vitis vinifera* L. | Grape (Malbec Cabernet Sauvignon, Chardonnay) | Fruit | 0.6–1.2 ng/g | Meng et al. (2017) |
| *Vitis vinifera* L. | Grape (cv. Merlot) | Fruit skin | 9.3 ± 0.14 ng/g | |
| *Vitis vinifera* L. | Grape (cv. Malbec) | Fruit skin | 9–158.9 ng/g | |
| *Vitis vinifera* L. | Grape (cv. Merlot) | Seed | 10.04 ± 0.49 ng/g | |
| *Vitis vinifera* L. | Grape (cv. Merlot) | Fruit flesh | 0.2–3.9 ng/g | |
| *Vitis vinifera* L. | Grape (cv. Merlot) | Whole fruit | 100,000–150,000 ng/g | |
| *Vitis vinifera* L. | Grape (cv. Albana, Sangiovese) | Whole Fruit | 1.2–1.5 ng/g | |
| *Epimedium brevicornum Maxim'* | Yinyanghuo | Flower | 1105 ng/g d.w. | |
| *Zea* sp. | Corn (whole, yellow) | Grain | 1.3 ± 0.28 ng/g | |
| *Zea* sp. | Corn (germ meal) | Grain | 1.0 ± 0.10 ng/g | |
| *Zea* sp. | Corn (YM001) | Grain | 10–2034 ng/g d.w. | |
| *Zea* sp. | Corn | Grain | 1.88 ng/g f.w. | |
| *Zea* sp. | Sweet corn | Grain | 1.37 ng/g f.w. | |
| *Brassica campestris* L. | Yellow mustard ($B_9$) | Seed | 660.72 ± 41.05 ng/g d.w. | Chakraborty and Bhattacharjee (2017) |

**GABA Content**

| Name of Plants | Common Name | Edible Portion of Plants | GABA Content in Plants | References |
|---|---|---|---|---|
| *Glycine max* [L.] Merr. | Soybean (cv. Corsoy 79) | Leaf | 0.05 μmol/g f.w. | Wallace et al. (1984) |
| *Camellia sinensis* | Tea (cv. Yabukita) | Leaf | 0.86 μmol/g | Tsushida and Murai (1987) |
| *Vigna unguiculata* | Cowpea | Seed | N.D. | Mayer et al. (1990) |
| *Asparagus sprengeri* | Asparagus | Stem | 3.86 nmol/$10^6$ cells | Crawford et al. (1994) |
| *Phaseolus vulgaris* L. | Bean (cv. Topcrop) | Leaf | 4.44–7.01 μmol/g d.w. | Raggi (1994) |
| *Lycopersicon esculentum* Mill. | Wild tomato | Leaves and roots of seedlings | N.D. | Bolarin et al. (1995) |
| *Oryza sativa* L. cv. Arborio | Rice | Shoot | 16.7 ± 1.3 mg/g tissue f.w. | Aurisano et al. (1995) |
| *Asparagus sprengeri* | Asparagus | Stem | 2.7 nmol/$10^6$ cells | Cholewa et al. (1997) |
| *Glycine max* L. Merr. | Soybean (cv. Corsoy 79) | Leaf | 1.6 μmol/g f.w. | Ramputh and Bown (1996) |
| *Brassica rapa* L. | Turnip (var. Shogoin) | Leaf | 0.31 μmoles/g f.w. | Thompson et al. (1996) |
| *Sesamum indicum* L. | Sesame | Seed | 3.5 ± 0.34 μg/g f.w. | Bor et al. (2009) |
| *Oryza sativa* L. | Rice | Seed (grain) | 3.5 ± 0.34 μmol/g d.w. | Nayyar et al. (2014) |
| *Nephelium lappaceum* L. | Rambutan | Fruit | 0.05–1.0 mg/mL | Meeploy and Deewatthanawong (2016) |
| *Triticum durum* Desf. | Durum wheat (cv. Ofanto) | Seeds | N.D. | Woodrow et al. (2017) |

(*Continued*)

**TABLE 22.1 (*Continued*)**
**Neurotransmitters in Various Edible Plants**

| | | Acetylcholine Content | | |
|---|---|---|---|---|
| **Name of Plants** | **Common Name** | **Edible Portion of Plants** | **Acetylcholine Content in the Plants (nmoles g⁻¹ of fresh [b]dry mass) and out brackets (ng g⁻¹ of fresh mass)** | **References** |
| *Alangium lamarckii* Thw[b] | Sage-leaved alangium | Leaf | N.D. | Roshchina (2001) |
| *Albizzia julibrissin* Durazz.[b] | Persian silk tree | Leaf | N.D. | |
| *Avena fatua* L.[b] | Wild oat | Seed and seedling | N.D. | |
| *Brassica campestris*[b] | Field mustard (var. Gongylodes L.) | Above-ground part | N.D. | |
| *Capsella bursa pastoris* L.[b] | Shepherd's purse | Above-ground part | N.D. | |
| *Carum copticum* Benth.[b] | Ajwain | Overground shoots | N.D. | |
| *Crataegus oxyacantha* L.[b] | Hawthorn | Leaf, fruit | (100–1000) 16.32–163.17 | |
| *Cucumis anguria* L.[b] | Cackrey | Overground part | (2.6) 0.42 | |
| *Cucumis sativus* L.[b] | Cucumber | Leaf | N.D. | |
| *Cucumis sativus* L.[b] | Cucumber | Leaf | (31–332) 5.06–54.17 | |
| *Cucurbita pepo* L.[b] | Pumpkin and squash | Root | (3.3) 0.54 | |
| *Daucus carota*[b] | Wild carrot (var. sativa L.) | Leaf | N.D | |
| *Girardinia heterophylla*[b] | Danskandali | Leaf | (273) 44.55 | |
| *Helianthus annuus* L.[b] | Sunflower | Stem and shoot | (7.9) 1.29 | |
| *Helianthus annuus* L.[b] | Sunflower | Root | (3.5) 0.57 | |
| *Ipomaea abutiloides*[b] | Bush morning glory (Carnea) | Leaf | N.D. | |
| *Phaseolus vulgaris* L.[b] | Common bean and Green bean | Leaf | (2.0) 0.326 | |
| *Pinus sylvestris* L.[b] | Scots pine | Overground shoots | N.D. | |
| *Plantago lanceolata* L.[b] | Ribwort plantain | Seed | N.D. | |
| *Plantago rugelli* Decne.[b] | American plantain | Leaf | (1.65) 0.27 | |
| *Raphanus sativus* L[b] | Radish | Root | (327–510) 53.36–83.22 | |
| *Raphanus sativus* L[b] | Radish | Petioles | (237–545) 38.67–88.92 | |
| *Raphanus sativus* L.[b] | Radish | Leaf | (112–329) 18.28–53.68 | |
| *Rumex obtusifolius* L. | Docks and sorrels | Leaf | (2.8) 0.46 | |
| *Rumex obtusifolius* L.[b] | Docks and sorrels | Seed | N.D. | |
| *Sinapis alba* L.[b] | White mustard | Leaf | (3.1–5.0) 0.51–0.82 | |
| *Smilax hispida* Muhl.[b] | Bristly greenbrier | Leaf | (0.78) 0.13 | |
| *Solanum tuberosum* L.[b] | Potato | Tuber | N.D. | |
| *Triticum vulgare* L.[b] | Wheat | Seed (grain) | N.D. | |

(*Continued*)

**TABLE 22.1 (*Continued*)**
**Neurotransmitters in Various Edible Plants**

| | | | | |
|---|---|---|---|---|
| *Urtica dioica* L.[b] | Stinging nettle | Leaf | N.D. | Roshchina (2001) |
| *Urtica parviflora* Roxb.[b] | Common Nettle | Leaf | N.D. | |
| *Vicia faba* L.[b] | Broad bean | Pollen | N.D. | |
| *Vigna unguiculata* L. Walp[b] | Cowpea | Leaf | (43–103) 7.02–16.81 | |
| *Xanthium strumarium* L.[b] | Cocklebur | Leaf | (1.32–6.0) 0.22–0.98 | |

**Dopamine Content**

| Name of Plants | Common Name | Edible Portion of Plants | Dopamine Content in the Plants | References |
|---|---|---|---|---|
| *Musa* sp. | Banana | Fruit (peel) | 700 μg/g | Udenfriend et al. (1959) |
| *Citrus sinensis*[a] | Oranges | Fruit | <1 μg/g f.w. | Kuklin and Conger (1995) |
| *Lycopersicon esculentum*[a] | Tomatoes | Fruit | <1 μg/g f.w. | |
| *Malus sylvestris*[a] | Apples | Fruit | <1 μg/g f.w. | |
| *Musa acuminata*[a] | Yellow banana | Fruit (pulp) | 42 μg/g f.w. | |
| *Musa sapientum*[a] | Red banana (var. Baracoa) | Fruit | 54 μg/g f.w. | |
| *Persea americana*[a] | Fuerte avocado | Fruit | 4 μg/g f.w. | |
| *Phaseolus vulgaris*[a] | Beans | Fruit | <1 μg/g f.w. | |
| *Pisum sativum*[a] | Peas | Fruit | <1 μg/g f.w. | |
| *Plantago major*[a] | Plantain | Flower | 5.5 μg/g f.w. | |
| *Solanum meiongena*[a] | Eggplants | Fruit | <1 μg/g f.w. | |
| *Spinacia oleracea*[a] | Spinach | Fruit | <1 μg/g f.w. | |
| *Beta vulgaris* L. | Sugar beet | Leaf | N.D. | Roshchina (2001) |
| *Entada pursaetha* DC | African dream herb or snuff box sea bean | Seed | N.D. | |
| *Lactuca sativa* L. | Lettuce | Seedling | N.D. | |
| *Monstera deliciosa* Liebm. | Split leaf philodendron | Fruit | 3900 mg/g of f.w. | |
| *Mucuna pruriens* DC[b] | Velvet bean (Holland) | Leaf | N.D. | |
| *Musa* sp. L. | Bananas and plantain | Fruit (pulp) | 8 mg/g of f.w. | |
| *Musa* sp. L. | Bananas and plantain | Fruit | 1–20 mg/g of f.w. | |
| *Musa* sp. L. | Rough-leaved Pepper | Leaf | N.D. | |
| *Persea* L. | Persea | Fruit | 4–5 mg/g of f.w. | |
| *Piper amalago* L. | Pepper elder | Leaf | 0.10 mg/g of f.w. | |
| *Solanum tuberosum* L. | Potato | Tuber | 0.10 mg/g of f.w. | |
| *Spinacia oleracea* L. | Spinach | Leaf | N.D. | |

(*Continued*)

## TABLE 22.1 (*Continued*)
## Neurotransmitters in Various Edible Plants

**Adrenaline Content**

| Name of Plants | Common Name | Edible Portion of Plants | Adrenaline Content in Plants | References |
|---|---|---|---|---|
| *Pisum sativum* L. | Pea | Leaf | 0.0–34.2 μg/g f.w. | Roshchina (2001) |
| *Brassica olereacea* | Brussel sprouts (var. gemmifera) | Sprouted Seed | <1 mg/g f.w. | Kulma and Szopa (2007) |
| *Citrus sinensis* | Oranges | Fruit | <1 mg/g f.w. | |
| *Lycopersicon esculentum* | Tomatoes | Fruit | <1 mg/g f.w. | |
| *Phaseolus vulgaris* | Beans | Fruit | <1 mg/g f.w. | |
| *Solanum melanogena* | Aubergine | Fruit | <1 mg/g f.w. | |
| *Solanum tuberosum* | Potato | Leaf | N.D. | |
| *Solanum tuberosum* | Potato (var. Desiree-leaves) | Tubers | N.D. | |
| *Spinacia oleracea* | Spinach | Leaf | <1 mg/g f.w. | |

**Noradrenaline Content**

| Name of Plants | Common Name | Edible Portion of Plants | Noradrenaline Content in Plants | References |
|---|---|---|---|---|
| *Musa* sp. | Banana | Fruit (peel) | 122 μg/g f.w. | Udenfriend et al. (1959) |
| *Musa* sp. | Banana | Fruit (pulp) | 2.0 μg/g f.w. | |
| *Solanum* sp. | Potato | Tuber | 0.1–2 μg/g f.w. | |
| *Urtica dioica* L. | Common nettle | Leaf | 591 μg/g f.w. | Roshchina (2001) |
| *Pisum sativum* L. | Pea | Leaf | 0.027–42.7 μg/g f.w. | |
| *Lemna paucicostata* L. | Short-Day Duckweed | Flower | 0.001–0.0002 mM | |
| *Lactuca sativa* L. | Lettuce | Leaf | N.D. | |
| *Albizzia julibrissin* Durazz. | Persian silk tree | Leaf | 2.8 μg/g f.w. | |
| *Mimosa pudica* L. | Shy plant | Petioles | 0.6 μg/g f.w. | |
| *Mucuna pruriens* DC | Velvet bean | Culture of leaf tissue | N.D. | |
| *Pisum sativum* L. | Pea | Leaf | 1.0 μg/g f.w. | |
| *Portulaca oleracea* L. | Common purslane | Overground part of plant | 2500 μg/g f.w. | |

*N.D:* amount not determined; f.w.: fresh weight; d.w.: dry weight.
[a] Norepinephrine was measured ≤3.5 I μg/g f.w.
[b] Parts of plants are commonly used for medicinal purposes.

tissue (Feldman and Lee, 1985). Appreciable amounts of serotonin are also reported in plantain, apricots, cherries, peaches, and Chinese plums (García-Moreno et al., 1983; Garrido et al., 2010; Gónzalez-Flores et al., 2011; Huang and Mazza, 2011; Ramakrishna et al., 2011). Seeds of the plant *Griffonia simplicifolia* cultivated on the west coast of Africa are also reportedly good sources of serotonin (Pathak et al., 2010).

Melatonin was initially thought to be exclusively produced in animals. It was found outside the animal kingdom for the first time in the dinoflagellate algae *Lingulodinium polyedrum*. Since then, melatonin has come to be regarded as a ubiquitous molecule, although its presence is not clearly established in some important taxa, particularly in archaea, mosses, ferns, gymnosperms, sponges, annelids, chelicerates, and echinoderms (Hardeland and Pandi-Perumal, 2005). Melatonin was first identified in edible plants by Hattori et al. (1995). Since then, melatonin has been found in many plants and in different parts of plants, such as the roots of Lupinus albus (Hernandez-Ruiz et al., 2004), sunflower, mustard, and walnut seeds (Manchester et al., 2000), tomato fruit (Van-Tassel and O'Neill, 2001), coffee (Ramakrishna et al., 2012a, 2012b), grape skin (Iriti et al., 2008), and the rind of tart cherries (Burkhardt et al., 2001) at concentrations usually ranging from picograms to nanograms per gram of tissue (Van-Tassel and O'Neill, 2001; Ramakrishna, 2015). Melatonin occurs (227–233 μg/g d.w.) in edible seeds of pistachio (Meng et al., 2017) at the highest level among all edible plants. Rice (Hattori et al., 1995) and corn grains (Meng et al., 2017) are also rich sources of melatonin. Chinese herbs, such as Huang-qin, Chantui, Gouteng, Diding, to name a few, are excellent sources of melatonin and are a part of Chinese traditional herbal medicine (Meng et al., 2017). Rice shoots contain the highest amount of GABA (16.7 mg/g tissue f.w.) in the edible plant kingdom (Aurisano et al., 1995). Roots, leaves, and petioles of radishes contain very small amounts of acetylcholine, where their concentration ranges from 18.28 to 88.92 ng/g of f.w. (Roshchina, 2001). In leaves and fruits of Hawthorn, too, acetylcholine is present in appreciable quantities (16.32–163.17 ng/g of f.w) (Roshchina, 2001). Tomatoes, oranges, aubergine, spinach, and beans have adrenaline contents less than 1 mg/g f.w. (Kulma and Szopa, 2007).

## 22.3 HEALTH BENEFITS OF NEUROTRANSMITTERS POST CONSUMPTION OF EDIBLE PLANT PARTS

### 22.3.1 Serotonin and Melatonin

Serotonin (5-hydroxytryptamine) is an indoleamine monoamine neurotransmitter and is mainly found in the enterochromaffin cells of the gastrointestinal tract (GI tract), in the serotonergic neurons of the central nervous system (CNS), and in blood platelets of animals. In humans, approximately 90% of serotonin is located in the enterochromaffin cells in the GI tract, where it regulates intestinal movements (Berger et al., 2009).

Melatonin (5-methoxy-N-acetyltryptamine), the *hormone of darkness* (Utiger, 1992), is an indole hormone produced in the pineal gland of the brain and controls several functions in our daily lives. Secretion of melatonin commences three months after birth, before which it is primarily supplied through the mother's milk or cow's milk. It is an excellent antioxidant, is beneficial for the immunological system, enhances resistance to infection and diseases, has inhibitory activity on some cancer forms, and induces beneficial effects on neuronal disorders. Several characteristics of melatonin, principally its direct, non-receptor-mediated free-radical scavenging activity, distinguish it from classic hormones (Stege et al., 2010; Rodriguez-Naranjo et al., 2011; Karunanithi et al., 2014). Melatonin is a ubiquitous molecule that plays a decisive role in the development of the brain and body, the regulation of circadian and seasonal rhythms, the sleep-wake process, reproduction, and retinal function (Bhattacharjee et al., 2016).

Walnut is the first common tree nut in which melatonin has been studied from a nutritional perspective (Reiter et al., 2005). Walnuts are reportedly known to increase blood melatonin concentrations, which positively correlate with an increased antioxidative capacity of blood. This increased value of indoleamine in blood could be protective against cardiovascular damage and as well for cancer initiation and growth (Reiter et al., 2005).

Germination of legumes increases the plant levels of melatonin, rendering sprouts too a suitable food source of this hormone. A study on bioavailability of melatonin post legume consumption was conducted by Aguilera et al. (2016). Effect of kidney bean sprout intake on the plasma levels of melatonin and metabolically related compounds (serotonin, 6-sulfatoxymelatonin), total phenolic compounds, and total antioxidant capacity were investigated by the authors. The plasmatic melatonin levels increased after sprout ingestion (16%, $p < 0.05$), and this increment positively correlated with the urinary 6-sulfatoxymelatonin content, the principal biomarker of plasmatic melatonin levels ($p < 0.01$). However, the phenolic compounds and antioxidant capacity levels did not exhibit any significant variation. The researchers concluded that kidney bean sprouts could be a good source of dietary melatonin. Although germinated legume is a suitable natural source of exogenous melatonin, investigations are required to determine the long-term effects of dietary melatonin consumption on antioxidant defense systems and disease prevention.

Effects of 10-day consumption of a Jerte Valley cherry-based nutraceutical product (patent no. ES2342141B1), which contains high levels of tryptophan, serotonin, and melatonin, on the activity/rest rhythms of young and old rats (*Rattus norvegicus*) and ringdoves (*Streptopelia risoria*), as representatives of animals with nocturnal and diurnal habits, respectively, and its possible relationship with their serum levels of melatonin and glucose have been studied. In both young and old rats, the intake of the cherry nutraceutical decreased their diurnal activities, whereas, nocturnal activities increased. The opposite effect was observed for ringdoves. The treatment increased the circulating levels of melatonin in both species and restored the amplitude of the activity rhythm in the old animals to that of the non-treated young groups. The researchers concluded that consumption of a Jerte Valley cherry-based nutraceutical product may help to counteract the impaired activity/rest rhythm found in aged animals and may aid in counteracting the decrease in melatonin and serotonin and can therefore increase in oxidative stress, suggesting potential health benefits of the products especially in aged populations where these parameters have been found to be altered (Delgado et al., 2011; Delgado et al., 2012). However, the author opine that increased plasma melatonin from food sources over prolonged time periods might exert indirect antioxidant actions, as is true for pure synthetic counterparts. The health benefits derived from the dietary intake of melatonin are until now controversial, since it is yet to be recognized that chronic consumption of melatonin through diet has physiological effects (Aguilera et al., 2016).

### 22.3.2 Gamma-Aminobutyric Acid

As a metabolic product of plants and microorganisms produced by the decarboxylation of glutamic acid, GABA functions as the main inhibitory neurotransmitter in the human brain cortex. The GABA receptor is the main target of ethanol in the brain. It plays a crucial role in humans by regulating neuronal excitability throughout the nervous system; it directly affects personality and plays a role in stress management. It is also directly responsible for the regulation of muscle tone and also effective in lowering stress, blood pressure, and hypertension and has positive effects on many metabolic disorders by promoting lean muscle growth, burning fat, and relieving pain. It has an immune-amplifying role in neuroinflammatory diseases such as multiple sclerosis. It reduces symptoms of premenstrual syndrome and is useful in treating attention deficit hyperactivity disorder. When orally administered (the single largest dose of GABA that can be given to a human subject is 3 g), it can increase the quantity of human growth hormone (Braverman et al., 2008; Dhakal et al., 2012; Anju et al., 2014; Rao and Annadana, 2017; Mattila et al., 2018). Generally, the human body can produce its own supply of GABA. However, GABA production is sometimes inhibited by a lack of estrogen, zinc, or vitamins, or by an excess of salicylic acid and food additives. There is a need for GABA-rich or enriched foods, because the GABA content in the typical daily human diet is relatively low (Diana et al., 2014).

A foray was made way back in 1999 by Houghton in assessing efficacy of aqueous extracts of the roots of members of the genus *Valeriana* (Valerianaceae), which contain appreciable amounts of GABA. The extracts had sedative effects on human. GABA when delivered in foods such as in beverages has shown positive results. Participants who received 50 mg of GABA dissolved in a beverage reported less psychological fatigue after completion of arithmetic tasks. It would be more advisable to have intake of GABA from natural sources such as from edible plant parts. GABA is present in ginger (*Zingiber officinale*), which is a commonly used spice consumed regularly. Ginger is highly active therapeutically and its consumption does not have any adverse effects; thus, a minimal rate of GABA uptake is possible through oral absorption. This, however, needs a foolproof experiment to validate the hypothesis. If GABA is to be taken from spinach, which is a relatively rich source one would have to consume 2.34 kg of uncooked spinach to obtain a dose of 100 mg of GABA, equivalent to that in a commercial pharmaceutical tablet (Anju et al., 2014; Boonstra et al., 2015).

There are also positive reports on the intake of green tea, rich in GABA, in decreasing blood pressure in young and old salt-sensitive rats. Extract of GABA-rich Maoyecha tea was shown to have sleep-promoting effect. Mulberry leaf extract rich in GABA, which lowers blood pressure in a dose-dependent manner. Many GABA-enriched food products are cereal based. A study examined the usefulness of defatted rice germ enriched with GABA as a functional food with tranquilizer effects, particularly on sleeplessness, depression, and autonomic disorder (observed during the menopausal or presenile period). There is substantial work on fermentative production of GABA using lactic acid bacteria (LAB) consortium of microorganisms. Numerous studies have reported the potential use of selected LAB on fermented sourdough, resulting in a GABA-enriched sourdough that could increase the GABA content in the final bread making (Okada et al., 2000 Diana et al., 2014; Wu et al., 2014).

Diana et al. (2014) have extensively reviewed usage of a wide variety of cereals, pseudo-cereals, and leguminous flours with well-characterized GABA-producing strains, which have been used to prepare GABA-enriched sourdough and bread (504 mg/kg), with a blend of the most suitable cereal flours, such as chickpea, amaranth, quinoa, and buckwheat. Germinated millets and legumes (particularly beans) are also good sources of GABA and have supported the development of functional cereal-based foods both for the pediatric and geriatric populations.

Besides cereals, legumes, sourdough and breads, there is an enormous variety of other GABA-enhanced food products, such as cheeses, fermented sausages, teas, vegetables, mushrooms, dairy and soy products, alcohol beverages, and traditional Asian fermented foods. Recently, specific processes and fermentation conditions for GABA enrichment in *adzuki* beans have been conducted to design a new, natural functional food product. Black soybeans have also been reportedly used to produce GABA-enhanced fermented milk as an antidepressant candidate. A novel tempeh-like fermented soybean with a high level of GABA has been developed by a specific cultivation procedure. It has been suggested by researchers that GABA production by microorganisms alone can fulfill the demand of GABA through GABA-enriched health beneficial foods (Dhakal et al., 2012; Diana et al., 2014). Table 22.2 lists GABA-enriched food products prepared using edible plant parts and their respective GABA contents. Table 22.3 shows dosages of GABA as a hypotensor agent when delivered from plant parts and when GABA-enriched food products are consumed. GABA-fortified Japonica rice could lower blood pressure in hypertensive rates (Braverman et al., 2008). However, thorough investigations still need to be undertaken to assess whether GABA contents in food matrices and it's bioavailability (*in vivo*) match and in evaluating the effectiveness of GABA (from food matrices) in human metabolic systems.

Food supplements of GABA are widely available. Although many consumers claim that they experience benefits from the use of these products, it is unclear whether these supplements confer benefits beyond a placebo effect. Currently, the mechanism of action behind these products is unknown. It has long been thought that GABA is unable to cross the blood-brain barrier (BBB), but studies that have assessed this issue are often contradictory and range widely in their employed methods. There is some evidence in favor of a calming effect of GABA food supplements, such as one study that reported reduction in the amount of seizures in epileptic patients who were administered GABA (0.8 g/kg daily) at a very high dose. Patients with Huntington's disease also have reduced GABA levels in the brain, but administration of GABA to remedy this deficiency has shown mixed results with regard to the reduction of symptoms. It has been suggested by Boonstra et al. in 2015 that any veridical effects of GABA food supplements on brain and cognition might be exerted through BBB passage or, more indirectly, via an effect on the enteric nervous system. More systematic studies establishing effects of GABA on the human nervous system and its influence on human behavior have to be conducted. Effects, if any, when GABA-enriched (or fortified) foods or supplements are consumed, also needs to be investigated.

Although the administration of GABA does not consistently alter the symptoms in complex and multifaceted disorders, such as epilepsy, Huntington's disease, anxiety disorders, stiff person syndrome, and premenstrual dysphoric disorder, it does not completely rule out the fact that GABA can affect the brain (Wong et al., 2003; Diana et al., 2014).

**TABLE 22.2**
**GABA Contents in Various GABA-Enriched Food Products**

| Food Products | GABA Content in Food Products (mg/kg) |
|---|---|
| Wholemeal wheat sourdough | 258.7 |
| Whole wheat and soya sourdough | 1000 |
| *Bathura* sourdough bread | 22.62 |
| Bread with *Yersinia* GAD supplementation | 115 |
| Cereal bran flakes | 66 |
| Cereal quinoa flakes | 90 |
| Cereal malt flour flakes | 258 |
| Fermented oat by *Aspergillus* sp. | 435.2 |
| Wheat germ | 1630 |
| Barley bran | 948 |
| Foxtail millet | 429 |
| Germinated waxy hull-less barley | 143 |
| Brown rice by proteolytic hydrolysis | 22.6 |
| Brown rice by high pressure treatment | 130 |
| Glutinous brown rice (Laozao) | N.R. |
| Cheese with *L. lactis* spp. *lactis* as starter | 320 |
| Cheddar cheese with probiotic strain | 6773.5 |
| Yogurt | N.R. |
| Fermented soya milk | 424.67 |
| Fermented milk from Tibet | N.R. |
| Fermented milk | 100–970 |
| Fermented goat's milk | 28 |
| Fermented milk by strains isolated on old-style cheese | 5000 |
| Skimmed milk by strains isolated from Italian cheeses | 15–99.9 |
| Fermented skim milk by *L. helveticus* | 113.35 |
| Fermented milk by *L. plantarum* | 77.4 |
| Fermented milk by LAB combination | 144.5 |
| Fermented skim milk by *L. plantarum* | 970 |
| Low fat fermented milk by LAB combination and protease | 806 |
| Black soybean milk | 5420 |
| Meat | 100 |
| Fermented pork sausage | N.R. |
| Fermented pork sausages | 0.124 |
| Japanese lactic-acid fermented fish | 1200 |
| Red mustard leaf | 1780 |
| Cruciferous plants | 300 |
| Vegetables | N.R. |
| *Brassica* product | N.R. |
| *Adzuki* beans | 2012 |
| Soybean nodules | 0.31 |
| Tempeh-like fermented soybean | 3700 |
| White tea | 505 |
| Black raspberry juice | 27600 |
| Fermented grape must | 9 |
| Sugar cane juice-milk | 3200 |
| Fermented-pepper leaves based | 263000 |

*(Continued)*

**TABLE 22.2 (*Continued*)**
**GABA Contents in Various GABA-Enriched Food Products**

| Food Products | GABA Content in Food Products (mg/kg) |
|---|---|
| Fruit juice | 579 |
| Honey-based beverages | N.R. |
| Beverage (not specified) | 200 |
| Rice lees *shochu* distillery liquor | 1560.5 |
| Potato snacks | 1700 |
| Chocolate | 2800 |
| Potatoes | 160-610 |

N.R. signifies not reported
*Source:* Diana, M. et al., *J. Funct. Foods*, 10, 407–420, 2014.

**TABLE 22.3**
**GABA-Enriched Foods and Food Products as Hypotensor Agents**

| Name of Plant Source | GABA Dosage and Duration |
|---|---|
| GABA-rich green tea | 4 mg/day for 4 weeks |
| Mulberry leaf aqueous extract | Single administration of water extract at a dose of 2–20 mg/kg B.W. |
| Dietary supplementation | 80 mg |
| Soy sauce | 0.33 ml/kg B.W. (containing 1% of GABA) |
| Soybean powder | 0.15% |
| Potato snack | 1.7 mg/kg B.W. |
| Tomato | 2 and 10 g tomato/kg B.W. (containing 180% of GABA) |
| Breakfast cereals | 30 g (containing 66 ppm of GABA) |

*Source:* Diana, M. et al., *J. Funct. Foods*, 10, 407–420, 2014.

Thus supplementation of GABA through consumption of edible plant parts in huge quantities may not be a feasible option. Formulation of food supplements using extracts of the said plant parts and their delivery as bioactive components of functional foods or as encapsulated forms such as in liposomal forms (micro and more preferably nano) may be more effective in crossing BBB and in exerting positive health effects in human beings.

### 22.3.3 Acetylcholine

The localization of acetylcholine in plant cells is poorly understood. The concentration of this substance during transmission of electric signals from motor nerves to muscles reaches 1 mM. In cells of human neuroblastoma, it is present at a concentration of 1.9 nmoles/mg protein (Roshchina, 2001).

### 22.3.4 Catecholamines

Catecholamines are a group of amines with a 3,4-dihydroxy-substituted phenyl ring. This class includes dopamine, epinephrine (adrenaline), and norepinephrine (noradrenaline), which are a group of biogenic amines involved in impulse transmission in animal nervous systems, where they are known as neurotransmitters with a glycogen-mobilizing function (Kulma and Szopa, 2007).

Tyrosine hydroxylase is a rate-limiting enzyme involved in the biosynthesis of catecholamine in mammals, which first synthesizes L-dopa, which in turn leads to the synthesis of dopamine and other catecholamines. Their functions in humans after consumption of edible plant parts have not been explored much to date, except for one study that discloses that products rich in L-dopa can effectively replace L-dopa administration in patients with Parkinson's disease. (Quansah, 2009). It is to be investigated whether a few milligrams of dopamine present in banana pulp, where it functions as a strong antioxidant, has favorable or unfavorable effects on human health. If so, it can detoxify the digestive tract and protect the body from a myriad of degenerative diseases, such as atherosclerosis, cancer, diabetes mellitus and Parkinson's disease, and others (Kanazawa and Sakakibara, 2000). The contents of these catecholamine precursors in foods could have major beneficial impact on human health.

## 22.4 ADVERSE EFFECTS OF NEUROTRANSMITTERS *IN VIVO* POST CONSUMPTION OF THE EDIBLE PLANTS

Very little is known of the adverse effects of plant neurotransmitters on human health except for a few studies dated in the early nineteenth century. These effects can be elucidated after consumption of foods fortified with the neurotransmitters singly and in combinations. Synergistic and antagonistic effects among combinations of neurotransmitters also need to be researched upon.

### 22.4.1 Serotonin and Melatonin

There is a hypothesis on endogenous formation of serotonin and its alkaloid derivatives in humans as a possible cause of psychic disorders (Roshchina, 2001).

### 22.4.2 Gamma-Aminobutyric Acid

Adverse effects of GABA when ingested from foods are not documented. Appreciable work needs to be directed to find out bioavailability of GABA in foods *in vivo* and adverse consequences, if any.

### 22.4.3 Acetylcholine

Acetylcholine provokes a pain response and leads to formation of blisters when plants cells secreting the same, such as hairs of the common nettle and the Australian species *Laportea moroides*, come in contact with human skin. In the nineteenth century, burns induced by these plants were reported to be the cause of serious diseases of horses and workers engaged in railway buildings. Some other species of *Laportea* growing in New Guinea reportedly caused serious adverse effects in humans to the extent of being fatal (Roshchina, 2001).

### 22.4.4 Catecholamines

Foy and Parratt in 1960 had reported that ingestion of plantain fruits may lead to the erroneous diagnosis of phaeochromocytoma and of carcinoid tumor by increasing the urinary excretion of noradrenaline and its metabolites. Successive methylation of dopamine in Mexican cactuses *Lophophora williamsii* and *Trichocereus pachanoi* (fam. Cactaceae) results in formation of alkaloids, such as mescaline, anhalamine, pellotine, and lophophorine, which are hallucinating substances for humans. Mescaline binds with adrenoreceptors in the post-synaptic membrane and thereby prevents the action of noradrenaline and adrenaline (Roshchina, 2001).

## 22.5 CONCLUDING REMARKS

Literature on utilization of plant neurotransmitters in development of functional foods is scarce, although extensive work has been carried out by several researchers on the contents of plant neurotransmitters in processed food products, such as in edible oils (olive oil, linseed oil, walnut oil), cocoa powder, and roasted coffee beans, to name a few. However, delivery and bioavailability of these neurotransmitters from frying oils and heterogeneous food matrices need to be elucidated. Extensive research needs to be conducted in development of nutraceutical foods and in designing nanoliposomal formulations using *extracts of edible plant parts* for enhanced selective recovery of plant neurotransmitters therein, and for effective delivery of the same in human metabolic system.

We conclude with the suggestions of Iriti and Faoro (2009) that better comprehension of the bioavailability of dietary phytochemicals is extremely critical in order to correctly evaluate their *in vivo* bioactivity, to interpret the experimental results and to design new approaches. In fact, data supporting their absorption, distribution, metabolism, and excretion in the human body are still fragmentary, despite the enormous amount of indications on their bioactivities.

The earlier indications have been obtained primarily based on studies on pure synthetic plant neurotransmitters and administration of the same as pharmaceutical preparations. However, the effects of these plant neurotransmitters on human health both positive and adverse are yet to be investigated at length.

With increasing advances on green technologies of extraction and food processing, it would be possible to manufacture new designer safe foods based on fortification and/or enrichment with extracts of neurotransmitters from edible plant parts. Studies on bioavailability of plant neurotransmitters from the designer health foods need to be investigated and their possible health benefits and adverse effects in the human system are to be elucidated.

## REFERENCES


Aguilera, Y., Hernanz, M. R., Herrera, T. et al. 2016. Intake of bean sprouts influences melatonin and antioxidant capacity biomarker levels in rats. *Food & Function* 7:1438–1445.

Aguilera, Y., Herrera, T., Benitez, V. et al. 2015. Estimation of scavenging capacity of melatonin and other antioxidants: Contribution and evaluation in germinated seeds. *Food Chemistry* 170: 203–211.

Anju, P., Moothedath, I., and Shree, A. B. R. 2014. Gamma amino butyric acid accumulation in medicinal plants without stress. *Ancient Science of Life* 34:68–72.

Aurisano, N., Bertani, A., and Reggiani, R. 1995. Anaerobic accumulation of 4-aminobutyrate in rice seedlings: Causes and significance. *Phytochemical* 38:1147–1150.

Badria, F. A. 2002. Melatonin, serotonin, and tryptamine in some Egyptian food and medicinal plants. *Journal of Medicinal Food* 5:153–157.

Berger, M., Gray, J. A., and Roth, B. L. 2009. The expanded biology of serotonin. *Annual Review of Medicine* 60:355–366.

Bhattacharjee, P., Ghosh, P. K., Vernekar, M., and Singhal, R. S. 2016. Role of dietary serotonin and melatonin in human nutrition. In *Serotonin and Melatonin: Their Functional Role in Plants, Food, Phytomedicine, and Human Health*, G. A. Ravishankar and A. Ramakrishna (Eds.), pp. 317–333. Boca Raton, FL: CRC Press.

Bolarin, M. C., Santa-Cruza, C., and Perez-Alfocea F. 1995. Short-term solute changes in leaves and roots of cultivated and wild tomato seedlings under salinity. *Plant Physiology* 147:463–468.

Boonstra, E., Kleijn, R. D., Colzato, L. S., Alkemade, A., Forstmann, B. U., and Nieuwenhuis, S. 2015. Neurotransmitters as food supplements: The effects of GABA on brain and behavior. *Frontiers in Psychology* 6:1–6.

Bor, M., Seckin, B., Ozgur, R., Yilmaz, O., Ozdemir, F., and Turkan, I. 2009. Comparative effects of drought, salt, heavy metal and heat stresses on gamma-aminobutryric acid levels of sesame (*Sesamum indicum* L.). *Acta Physiologiae Plantarum* 31:655–659.

Braverman, E. R., Pfeiffer, C. C., Blum, K., and Smayda, R. 2008. Glutamic acid, gamma-aminobutyric acid and glutamine: The brain's three musketeers. In *The Healing Nutrients Within: Facts, Findings, and New Research on Amino Acids*, p. 360. Surry Hills, Australia: Accessible Publishing System.

Burkhardt, S., Tan, D. X., Manchester, L. C., Hardeland, R., and Reiter, R. J. 2001. Detection and quantification of the antioxidant melatonin in Montmorency and Balaton tart cherries (*Prunus cerasus*). *Journal of Agricultural and Food Chemistry* 49:4898–4902.

Chakraborty, S., and Bhattacharjee, P. 2017. Supercritical carbon dioxide extraction of melatonin from *Brassica campestris*: *In vitro* antioxidant, hypocholesterolemic and hypoglycaemic activities of the extracts. *International Journal of Pharmaceutical Sciences and Research* 8:2486–2495.

Cholewa, E., Cholewinski, A. J., Shelp, B. J., Snedden, W. A., and Bown, A. W. 1997. Cold-shock-stimulated γ-aminobutyric acid synthesis is mediated by an increase in cytosolic $Ca^{2+}$, not by an increase in cytosolic $H^{+}$. *Canadian Journal of Botany* 75:375–382.

Crawford, L. A., Bown, A. W., Breitkreuz, K. E., and Cuine, F. C. 1994. The synthesis of γ-aminobutyric acid in response to treatments reducing cytosolic pH. *Plant Physiology* 104:865–871.

Delgado, J., Terron, M. P., Garrido, M. et al. 2011. Diets enriched with a Jerte Valley cherry-based nutraceutical product reinforce nocturnal behaviour in young and old animals of nocturnal (*Rattus norvegicus*) and diurnal (*Streptopelia risoria*) chronotypes. *Journal of Animal Physiology and Animal Nutrition* 97:1–9.

Delgado, J., Terrón, M. P., Garrido, M. et al. 2012. A cherry nutraceutical modulates melatonin, serotonin, corticosterone, and total antioxidant capacity levels: Effect on ageing and chronotype. *Journal of Applied Biomedicine* 10:109–117.

Dhakal, R., Bajpai, V. K., and Baek, K. H. 2012. Production of GABA (γ-aminobutyric acid) by microorganisms: A review. *Brazilian Journal of Microbiology* 43:1230–1241.

Diana, M., Quílez, J., and Rafecas, M. 2014. Gamma-aminobutyric acid as a bioactive compound in foods: A review. *Journal of Functional Foods* 10:407–420.

Dubbels, R., Reiter, R. J., and Klenke, E. 1995. Melatonin in edible plants identified by radioimmunoassay and by high performance liquid chromatography-mass spectrometry. *Journal of Pineal Research* 18:28–31.

Feldman, J. M., and Lee, E. M. 1985. Serotonin content of foods: Effect on urinary excretion of 5-hydroxyindoleacetic acid. *The American Journal of Clinical Nutrition* 42:639–643.

Foy, J. M., and Parratt, J. R. 1960. A note on the presence of noradrenaline and 5-hydroxytryptamine in plantain (*Musa Sapientum,* var. *Paradisiaca*). *Journal of Pharmacy and Pharmacology* 12:360–364.

García-Moreno, C., Rivas-Gonzalo, J. C., Peña-Egido, M. J., and Mariné-Font, A. 1983. Improved method for determination and identification of serotonin in foods. *Journal of Association of Official Analytical Chemists* 66:115–117.

Garrido, M., Paredes, S. D., Cubero, J. et al. 2010. Jerte Valley cherry-enriched diets improve nocturnal rest and increase 6-sulfatoxy melatonin and total antioxidant capacity in the urine of middle-aged and elderly humans. *Journals of Gerontology Series A: Biological Sciences and Medical Sciences* 65:909–914.

Gónzalez-Flores, D., Velardo, B., Garrido, M. et al. 2011. Ingestion of Japanese plums (*Prunus salicina Lindl.* cv. Crimson Globe) increases the urinary 6-sulfatoxy melatonin and total antioxidant capacity levels in young, middle-aged and elderly humans: Nutritional and functional characterization of their content. *Journal of Food and Nutrition Research* 50:229–236.

González-Gómez, D., Lozano, M., Fernández-León, M. F. M. C., Ayuso, M. C., Bernalte, M. J., and Rodríguez, A. B. 2009. Detection and quantification of melatonin and serotonin in eight sweet cherry cultivars (*Prunus avium* L.). *European Food Research and Technology* 229:223–229.

Hardeland, R., and Pandi-Perumal, S. R. 2005. Melatonin, a potent agent in antioxidative defense: Actions as a natural food constituent, gastrointestinal factor, drug and prodrug. *Nutrition and Metabolism* 2:22–37.

Hattori, A., Migitaka, H., Masayake, I., Itoh, M., Yamamoto, K., and Ohtani-kaneko, R. 1995. Identification of melatonin in plant seed and its effects on plasma melatonin levels and binding to melatonin receptors in vertebrates. *International Journal of Biochemistry and Molecular Biology* 35:627–634.

Hernandez-Ruiz, J., Cano, A., and Arnao, M. B. 2004. Melatonin: A growth-stimulating compound present in lupin tissues. *Planta* 220:140–144.

Houghton, P. J. 1999. The scientific basis for the reputed activity of valerian. *Journal of Pharmacy and Pharmacology* 51:505–512.

Huang, X., and Mazza, G. 2011. Simultaneous analysis of serotonin, melatonin, piceid and resveratrol in fruits using liquid chromatography tandem mass spectrometry. *Journal of Chromatography A* 1218:3890–3899.

Iriti, M., and Faoro, F. 2009. Bioactivity of grape chemicals for human health. *Natural Product Communications* 4:611–634.

Iriti, M., Rossoni, M., and Faoro, F. 2008. Melatonin in grape, not just a myth, maybe a panacea. *Journal of the Science of Food and Agriculture* 86:1432–1438.

Kanazawa, K., and Sakakibara. H. 2000. High content of dopamine, a strong antioxidant, in cavendish banana. *Journal of Agricultural and Food Chemistry* 48:844–848.

Karunanithi, D., Radhakrishna, A., Sivaraman, K. P., and Biju, V. M. N. 2014. Quantitative determination of melatonin in milk by LC–MS/MS. *Journal of Food Science and Technology* 51:805–812.

Kuklin, A. I., and Conger, B. V. 1995. Catecholamines in Plants. *Journal of Plant Growth Regulation* 14:91–97.

Kulma, A., and Szopa, J. 2007. Catecholamines are active compounds in plants. *Plant Science* 172:433–440.

Lavizzari, T., Veciana-Nogués, M. T., Bover-Cid, S., Mariné -Font, A., and Vidal-Carou, M. C. 2006. Improved method for the determination of biogenic amines and polyamines in vegetable products by ion-pair high-performance liquid chromatography. *Journal of Chromatography A* 1129:67–72.

Manchester, L. C., Tan, D. X., Reiter, R. J., Park, W., Monis, K., and Qi, W. 2000. High levels of melatonin in the seeds of edible plants: Possible function in germ tissue protection. *Life Sciences* 67:3023–3029.

Mattila, P. H., Marnila, P., and Pihlanto, A. 2018. Wild and cultivated mushrooms. In *Fruit and Vegetable Phytochemicals: Chemistry and Human Health*, E. M. Yahia (Ed.), p. 1291. West Sussex, UK: John Wiley & Sons.

Mayer, R. R., Cherry, J. H., and Rhodes, D. 1990. Effects of heat shock on amino acid metabolism of cowpea cells. *Plant Physiology* 94:796–810.

Meeploy, M., and Deewatthanawong, R. 2016. Determination of γ-Aminobutyric Acid (GABA) in Rambutan Fruit cv. Rongrian by HPLC-ELSD and Separation of GABA from Rambutan Fruit Using Dowex 50W-X8 Column. *Journal of Chromatographic Science* 54:445–452.

Meng, X., Li, Y., Li, S. et al. 2017. Dietary Sources and Bioactivities of Melatonin. *Nutrients* 9:1–64.

Murch, S. J., Campbell, S. S. B., and Saxena, P. K. 2001. The role of serotonin and melatonin in plant morphogenesis: Regulation of auxin-induced root organogenesis in *in vitro*-cultured explants of St. John's wort (*Hypericum perforatum L.*). *In Vitro Cellular & Developmental Biology-Plant* 37:786–793.

Nayyar, H., Kaur, R., Kaur, S., and Singh, R. 2014. γ-Aminobutyric acid (GABA) imparts partial protection from heat stress injury to rice seedlings by improving leaf turgor and upregulating osmoprotectants and antioxidants. *Journal of Plant Growth Regulation* 23:408–419.

Okada, T., Sugishita, T., Murakami, T. et al. 2000. Effect of the defatted rice germ enriched with GABA for sleeplessness, depression, autonomic disorder by oral administration. *Journal of the Japanese Society for Food Science and Technology* 47:596–603.

Pasko, P., Ziaja, K. S., Muszyńska, B., and Zagrodzki, P. 2014. Serotonin, melatonin, and certain indole derivatives profiles in rutabaga and kohlrabi seeds, sprouts, bulbs, and roots. *Food Science and Technology* 59:740–745.

Pathak, S. K., Praveen, T., Jain, N. P., and Banweer, J. 2010. A review on *Griffonia simplicifollia*-an ideal herbal antidepressant. *International Journal of Pharmacy & Life Sciences* 1:174–181.

Płonka, J., and Michalski, A. 2017. The influence of processing technique on the catecholamine and indolamine contents of fruits. *Journal of Food Composition and Analysis* 57:102–108.

Quansah, L. 2009. Molecular basis of catecholamine biosynthesis in banana fruit. PhD thesis, The Hebrew University of Jerusalem, Jerusalem, Israel, http://www.agri.gov.il/download/files/QuansahLydia2009MSc_1.pdf, (accessed January 20, 2018).

Raggi, V. 1994. Changes in free amino acids and osmotic adjustment in leaves of water-stressed bean. *Plant Physiology* 91:427–434.

Ramakrishna, A. 2015. Indoleamines in edible plants: Role in human health effects. In *Indoleamines: Sources, Role in Biological Processes and Health Effects*, A. Catalá (Ed.), p. 279. (Biochemistry Research Trends Series). Hauppage, NY: Nova Publishers.

Ramakrishna, A., Giridhar, P., and Ravishankar, G. A. 2011. Phytoserotonin: A review. *Plant Signaling and Behavior* 6:800–809.

Ramakrishna, A., Giridhar, P., Udaya Sankar, K., and Ravishankar, G. A. 2012a. Endogenous profiles of indoleamines: Serotonin and melatonin in different tissues of Coffea canephora P ex Fr. as analyzed by HPLC and LC-MS-ESI. *Acta Physiologia Plantarum* 34:393–396.

Ramakrishna, A., Giridhar, P., Udaya Sankar, K., and Ravishankar, G. A. 2012b. Melatonin and serotonin profiles in beans of *Coffea* sps. *Journal of Pineal Research* 52:470–476.

Ramputh, A. I., and Bown, A. W. 1996. Rapid γ-aminobutyric acid synthesis and the inhibition of the growth and development of oblique-banded leaf-roller larvae. *Agricultural and Biological Chemistry* 111:1349–1352.

Rao, C. K., and Annadana, S. 2017. Nutrient biofortification of staple food crops: Technologies, products and prospects. In *Phytonutritional Improvement of Crops*, N. Benkeblia (Ed.), p. 135. West Sussex, UK: John Wiley & Sons.

Reiter, R. J., Manchester, L. C., and Tan, D. X. 2005. Melatonin in walnuts: Influence on levels of melatonin and total antioxidant capacity of blood. *Nutrition* 21:920–924.

Restuccia, D., Spizzirri, U. G., Parisi, O. I., Cirillo, G., and Picci, N. 2015. Brewing effect on levels of biogenic amines in different coffee samples as determined by LC-UV. *Food Chemistry* 175:143–150.

Rodriguez-Naranjo, M. I., Gil-Izquierdo, A., Troncoso, A. M., Cantos-Villar, E., and Garcia-Parrilla, M. C. 2011. Melatonin is synthesised by yeast during alcoholic fermentation in wines. *Food Chemistry* 126:1608–1613.

Roshchina, V. V. (Ed.). 2001. Neuromediators, their synthesis and metabolism. In *Neurotransmitters in Plant Life*. Enfield, NH: Science Publishers.

Stege, P. W., Sombra, L. L., Messina, G., Martinez, L. D., and Silva, M. F. 2010. Determination of melatonin in wine and plant extracts by capillary electrochromatography with immobilized carboxylic multi-walled carbon nanotubes as stationary phase. *Electrophoresis* 31:2242–2248.

Thompson, J. F., Stewart, C. R., and Morris, C. J. 1966. Changes in amino acid content of excised leaves during incubation I. The effect of water content of leaves and atmospheric oxygen level. *Plant Physiology* 41:1578–84.

Tsushida, T., and Murai, T. 1987. Conversion of glutamic acid to γ-aminobutyric acid in tea leaves under anaerobic conditions. *Agricultural and Biological Chemistry* 51:2865–2871.

Udenfriend, S., Lovenberg, W., and Sjoerdsma, A. 1959. Physiologically active amines in common fruits and vegetables. *Archives of Biochemistry and Biophysics* 86:487–490.

Utiger, R. D. 1992. Melatonin: The hormone of darkness. *New England Journal of Medicine* 327:1377–1379.

Van-Tassel, D. L., and O'Neill, S. D. 2001. Putative regulatory molecules in plants: Evaluating melatonin. *Journal of Pineal Research* 31:1–7.

Wallace, W., Secor, J., and Schrader, L. E. 1984. Rapid accumulation of γ-aminobutyric acid and alanine in soybean leaves in response to an abrupt transfer to lower temperature, darkness, or mechanical manipulation. *Plant Physiology* 75:170–175.

Wong, C. G., Bottiglieri, T., and Snead, O. C. 2003. GABA, gamma-hydroxybutyric acid, and neurological disease. *Annals of Neurology* 54:3–12.

Woodrow, P., Ciarmielloa, L. F., Annunziatab, M. G. et al. 2017. Durum wheat seedling responses to simultaneous high light and salinity involve a fine reconfiguration of amino acids and carbohydrate metabolism. *Plant Physiology* 159:290–312.

Wu, C., Huang, Y., Lai, X. et al. 2014. Study on quality components and sleep-promoting effect of GABA Maoyecha tea. *Journal of Functional Foods* 7:180–190.

# Index

**Note:** Page numbers followed by f and t refer to figures and tables respectively.

## J



## K



## L



## M

## O



## P

## X



## Y



## Z